T0186556

Protein Misfolding in Neurodegenerative Diseases

Mechanisms and Therapeutic Strategies

CRC Enzyme Inhibitors Series

Series Editors
H. John Smith and Claire Simons
Cardiff University
Cardiff, UK

Carbonic Anhydrase: Its Inhibitors and Activators
Edited by Claudiu T. Supuran, Andrea Scozzafava and Janet Conway

Enzymes and Their Inhibition: Drug Development
Edited by H. John Smith and Claire Simons

Inhibitors of Cyclin-dependent Kinases as Anti-tumor Agents
Edited by Paul J. Smith and Eddy W. Yue

Protein Misfolding in Neurodegenerative Diseases: Mechanisms and Therapeutic Strategies
Edited by H. John Smith, Claire Simons, and Robert D. E. Sewell

CRC Enzyme Inhibitors Series

Protein Misfolding in Neurodegenerative Diseases

Mechanisms and Therapeutic Strategies

Edited by

H. John Smith
Claire Simons
Robert D. E. Sewell

CRC Press
Taylor & Francis Group
Boca Raton London New York

CRC Press is an imprint of the
Taylor & Francis Group, an **informa** business

CRC Press
Taylor & Francis Group
6000 Broken Sound Parkway NW, Suite 300
Boca Raton, FL 33487-2742

First issued in paperback 2019

© 2008 by Taylor & Francis Group, LLC
CRC Press is an imprint of Taylor & Francis Group, an Informa business

No claim to original U.S. Government works

ISBN-13: 978-0-8943-7310-7 (hbk)
ISBN-13: 978-0-367-38812-6 (pbk)

This book contains information obtained from authentic and highly regarded sources. Reasonable efforts have been made to publish reliable data and information, but the author and publisher cannot assume responsibility for the validity of all materials or the consequences of their use. The authors and publishers have attempted to trace the copyright holders of all material reproduced in this publication and apologize to copyright holders if permission to publish in this form has not been obtained. If any copyright material has not been acknowledged please write and let us know so we may rectify in any future reprint.

Except as permitted under U.S. Copyright Law, no part of this book may be reprinted, reproduced, transmitted, or utilized in any form by any electronic, mechanical, or other means, now known or hereafter invented, including photocopying, microfilming, and recording, or in any information storage or retrieval system, without written permission from the publishers.

For permission to photocopy or use material electronically from this work, please access www.copyright.com (http://www.copyright.com/) or contact the Copyright Clearance Center, Inc. (CCC), 222 Rosewood Drive, Danvers, MA 01923, 978-750-8400. CCC is a not-for-profit organization that provides licenses and registration for a variety of users. For organizations that have been granted a photocopy license by the CCC, a separate system of payment has been arranged.

Trademark Notice: Product or corporate names may be trademarks or registered trademarks, and are used only for identification and explanation without intent to infringe.

Library of Congress Cataloging-in-Publication Data

Protein misfolding in neurodegenerative diseases : mechanisms and therapeutic
    strategies / editors, H. John Smith, Claire Simons, and Robert D.E. Sewell.
        p. ; cm. -- (CRC enzyme inhibitors series)
    Includes bibliographical references and index.
    ISBN 978-0-8493-7310-7 (alk. paper)
    1. Nervous system--Degeneration--Etiology. 2.
Nervous system--Degeneration--Treatment. 3. Nervous
system--Degeneration--Pathophysiology, 4. Protein folding. I. Smith, H. J.,
1930- II. Simons, Claire. III. Sewell, Robert D. E. IV. Title. V. Series.
    [DNLM: 1. Neurodegenerative Diseases--etiology. 2. Protein Folding.
3. Neurodegenerative Diseases--physiopathology. 4. Neurodegenerative
Diseases--therapy. WL 359 P9673 2008]

RC365.P78 2008
616.8'0471--dc22   2007020268

Visit the Taylor & Francis Web site at
http://www.taylorandfrancis.com

and the CRC Press Web site at
http://www.crcpress.com

Contents

PART I
ALZHEIMER'S DISEASE

PART II
PARKINSON'S DISEASE

PART III
HUNTINGTON'S DISEASE

PART IV
AMYOTROPHIC LATERAL SCLEROSIS

PART V
TRANSMISSIBLE SPONGIFORM
ENCEPHALOPATHIES

PART VI
OVERVIEW

Series Preface

One approach to the development of drugs as medicines, which has gained considerable success over the past two decades, involves inhibition of the activity of a target enzyme in the body or invading parasite by a small molecule inhibitor, leading to a useful clinical effect.

The CRC *Enzyme Inhibitor* series consists of an expanding series of monographs on this aspect of drug development, providing timely and in-depth accounts of developing and future targets that collectively embrace the contributions of medicinal chemistry (synthesis, design), pharmacology and toxicology, biochemistry, physiology, and biopharmaceutics necessary in the development of novel pharmaceutics.

H. John Smith
Claire Simons

Preface

The chapters in this book provide up-to-date accounts of a number of specific neurodegenerative diseases, namely Alzheimer's disease, Parkinson's disease, amyotrophic lateral sclerosis (motor neuron disease), Huntington's disease, and Creutzfeldt–Jacob disease, and current progress in identification of specific agents as the first step on the way to a cure. A unifying, commonly accepted view is that, in some way, the causal agent is a specific brain protein (for a particular disease), which has been misfolded. Another view emphasizes the importance of oxidative stress as a causal factor. Perhaps future work will clarify whether multicausal agents are involved; this would not be unexpected owing to the complexity and the interrelationship between brain processes, thus, leading to the likelihood of "knock-on" effects between systems. Existing and future strategies to identify the causal factors at the molecular level have been keenly pursued although, so far, usually mixed results have been seen in cellular and animal models. Successful pursuance of these studies may, hopefully, lead to the unequivocal identification of main molecular targets, as their successful manipulation through the rational design of pharmaceuticals could realistically become a way forward to a cure.

H. John Smith
Claire Simons
Robert D.E. Sewell

Editors

H. John Smith is a former reader in medicinal chemistry at the Welsh School of Pharmacy, Cardiff University, United Kingdom. He obtained his Ph.D. in medicinal chemistry at the University of London and received his D.Sc. in 1995. A fellow of the Royal Society of Chemistry and the Royal Pharmaceutical Society of Great Britain, he has spent much of his career studying enzyme inhibitors and their potential use in drugs. John Smith has coauthored several texts on this subject and is editor-in-chief of the *Journal of Enzyme Inhibition and Medicinal Chemistry*.

Claire Simons is a senior lecturer in medicinal chemistry at the Welsh School of Pharmacy, Cardiff University, Unitetd Kingdom. She obtained her Ph.D. in organic chemistry at King's College, University of London, and is a member of the Royal Society of Chemistry. Her main research interests are the design, synthesis, and computational analysis of novel heterocyclic and nucleoside compounds as enzyme inhibitors. Claire Simons has authored a textbook on nucleoside chemistry and its therapeutic application and coedited (with John Smith) a textbook, *Proteinase and Peptidase Inhibition*.

Robert D.E. Sewell is a senior lecturer in pharmacology at the Welsh School of Pharmacy, Cardiff University, United Kingdom and also director of the postgraduate M.Sc. course in clinical research in Cardiff University. He obtained his Ph.D. in pharmacology at the University of Wales Institute of Science and Technology and is a member of the British Pharmacological Society, British Association of Psychopharmacology and the Royal Pharmaceutical Society of Great Britain. His main interests are in central nervous system pharmacology particularly related to neurodegenerative diseases and he is a member of several editorial boards of international biomedical journals.

Contributors

Celine Adessi
CNS Discovery Research
F. Hoffmann-La Roche Ltd
Basel, Switzerland

Luís Almeida
Department of Research and
    Development
Bial Portela & C^a SA
Mamede do Coronado, Portugal

Corinne E. Augelli-Szafran
Laboratory for Experimental Alzheimer
    Drugs
Department of Neurology
Harvard Medical School
Brigham and Women's Hospital
Boston, Massachusetts

Roger A. Barker
Cambridge Centre for Brain Repair
University of Cambridge
Cambridge, United Kingdom

Maria João Bonifácio
Department of Research and
    Development
Bial Portela & C^a SA
Mamede do Coronado, Portugal

Tischa J.M. van der Cammen
Section of Geriatric Medicine
Department of Internal Medicine
Erasmus University Medical Center
Rotterdam, The Netherlands

Rudy J. Castellani
Department of Physiology
Michigan State University
East Lansing, Michigan

Dario Cattaneo
Center for Research on Organ
    Transplantation
Mario Negri Institute for
    Pharmacological Research
Ranica, Italy

Avi Chakabartty
Department of Medical Biophysics
Ontario Cancer Institute
University of Toronto
Toronto, Ontario, Canada

Christian Czech
CNS Discovery Research
F. Hoffmann-La Roche Ltd
Basel, Switzerland

Tom Foltynie
Cambridge Centre for Brain Repair
University of Cambridge
Cambridge, United Kingdom

Pierre Francotte
Drug Research Center, Department
    of Medicinal Chemistry
University of Liège
Liège, Belgium

Michael D. Geschwind
Memory & Aging Center
Department of Neurology
University of California
San Francisco, California

Ann B. Goodman
Department of Psychiatry
Beth Israel Deaconess Medical Center
Harvard Medical School
Boston, Massachusetts

Claire-Anne Gutekunst
Department of Neurosurgery
Emory University School of Medicine
Atlanta, Georgia

Bruno P. Imbimbo
Research and Development
Chiesi Farmaceutici
Parma, Italy

Helmut Jacobsen
CNS Discovery Research
F. Hoffmann-La Roche Ltd
Basel, Switzerland

Jeff Kuret
Center for Molecular Neurobiology
Department of Molecular & Cellular
  Biochemistry
The Ohio State University
College of Medicine
Columbus, Ohio

Hyoung-Gon Lee
Department of Pathology
Case Western Reserve University
Cleveland, Ohio

Giuseppe Legname
International School for Advanced
  Studies
Neurobiology Sector
Trieste, Italy

Harry LeVine III
Center on Aging
Department of Molecular & Cellular
  Biochemistry
University of Kentucky
Lexington, Kentucky

Peter McCaffery
Institute of Medical Sciences
University of Aberdeen
Aberdeen, United Kingdom

Andrew W. Michell
Cambridge Centre for Brain Repair
University of Cambridge
Cambridge, United Kingdom

Fran Norflus
Department of Natural Sciences
Clayton State University
Morrow, Georgia

Akihiko Nunomura
Department of Psychiatry and
  Neurology
Asahikawa Medical College
Asahikawa, Japan

Joana A. Palha
Life and Health Sciences Research
  Institute
School of Health Sciences
University of Minho
Braga, Portugal

P. Nuno Palma
Department of Research and
  Development
Bial Portela & C^a SA
Mamede do Coronado, Portugal

Arthur B. Pardee
Dana Farber Cancer Institute and
  Harvard University
Boston, Massachusetts

George Perry
Department of Pathology
Case Western Reserve University
Cleveland, Ohio

Bernard Pirotte
Drug Research Center
Department of Medicinal Chemistry
University of Liège
Liège, Belgium

Janice Robertson
Department of Laboratory
    Medicine and Pathobiology
Centre for Research in
    Neurodegenerative Diseases
University of Toronto
Toronto, Ontario, Canada

Teresa Sanelli
University of Toronto
Centre for Research in
    Neurodegenerative Diseases
Toronto, Ontario, Canada

Andrea Scozzafava
Department of Chemistry
University of Florence
Florence, Italy

Robert D.E. Sewell
Welsh School of Pharmacy
Cardiff University
Cardiff, United Kingdom

Claire Simons
Welsh School of Pharmacy
Cardiff University
Cardiff, United Kingdom

H. John Smith
Welsh School of Pharmacy
Cardiff University
Cardiff, United Kingdom

Mark A. Smith
Department of Pathology
Case Western Reserve University
Cleveland, Ohio

Patrício Soares-da-Silva
Department of Research and
    Development
Bial Portela & Cª SA
Mamede do Coronado, Portugal

Francesca Speroni
Department of Chemistry
University of Parma
Parma, Italy

Massimo Stefani
Department of Biochemical Sciences
Research Centre on the Molecular Basis
    of Neurodegeneration
University of Florence
Florence, Italy

Jure Stojan
Institute of Biochemistry
Faculty of Medicine
University of Ljubljana
Ljubljana, Slovenia

Michael J. Strong
Department of Clinical Neurological
    Sciences
The University of Western Ontario
London, Ontario, Canada

Claudiu T. Supuran
Department of Chemistry
University of Florence
Florence, Italy

Charlotte E. Teunissen
Department of Molecular Cell Biology
    and Immunology
VU University Medical Center
Amsterdam, The Netherlands

Pascal de Tullio
Drug Research Center, Department of
 Medicinal Chemistry
University of Liège
Liège, Belgium

Xiongwei Zhu
Department of Pathology
Case Western Reserve University
Cleveland, Ohio

Abbreviations

Aβ40(42)	Abeta 40(42)
AADC	aromatic L-amino acid decarboxylase
ACAT	acyl-CoA; cholesterol acyltransferase
ACE	angiotensin-converting enzyme
ACh	acetylcholine
AChE	acetylcholinesterase
AChEI	acetylcholinesterase inhibitor
AD	Alzheimer's disease
ADAM	a disintegrin and metalloproteinase
ADAPT	Alzheimer's Disease Anti-inflammatory Prevention Trial
ADDLs	micro-aggregates of amyloid-β
ADME	adsorption, disposition, metabolism, and excretion
AEI	bicarbonate/chloride anion exchanger
AFM	atomic force microscopy
AGEs	advanced glycation end products
AICD	cytoplasmic domain of APP
ALS	amyotrophic lateral sclerosis
ACT	α-1-antichymotrypsin
APL	acute promyelocytic leukemia
ApoE4	apolipoprotein ε4
APP	amyloid precursor protein
ATP	adenosine triphosphate
ATPase	adenosine triphosphatase
ATRA	all-*trans*-retinoic acid
BACE	beta-site amyloid-cleaving enzyme
BBB	blood–brain barrier
bCA	bovine carbonic anhydrase
BDNF	brain-derived neurotrophic factor
bid	twice daily
BSE	bovine spongiform encephalopathy
BuChE	butyryl cholinesterase
CA	carbonic anhydrase
CAA	carbonic anhydrase activator
CAD	coronary artery disease
CAI	carbonic anhydrase inhibitor
CBP	CREB-binding protein
CD	circular dichroism
CDK	cyclin-dependent kinases
CETP	cholesteryl ester transfer protein
CJD	Creutzfeldt–Jacob disease

CK1	casein kinase-1
C_{max}	maximum plasma concentration
CML	N^ε-carboxymethyl lysine
COMT	catechol-O-methyltransferase
COX	cyclooxygenase
CPEB	cytoplasmic polyadenylation binding protein
CRABP	cellular retinoic acid binding protein
CRBP	cellular retinol binding protein
CSF	cerebrospinal fluid
CYP	cytochrome P450
GtBP	C-terminal binding protein
CVD	cardiovascular disease
CWD	chronic wasting disease
DFP	diisopropyl fluorophosphates
DMPP	dimethyl-allyl-pyrophosphate
DMSO	dimethylsulfoxide
DOPA	3,4-dihydroxyphenylalanine
DOPAC	3,4-dihydroxyphenylacetic acid
DTNB	dithio-bis-nitrobenzoic acid
DZ	dizygotic
E2-25K	His 2, ubiquitin conjugating enzyme
ECG	electrocardiograph
EGF	epidermal growth factor
EI	enzyme–inhibitor complex
EIF2α	eukaryotic initiation factor
EM	electron microscopy
EOAD	early onset familial Alzheimer's disease
EPR	electron paramagnetic resonance
ER	endoplasmic reticulum
ErbB	erythroblastic leukemia viral oncogene
ES	embryonic stem cells
fALS	familial amyotrophic lateral sclerosis
FAP	familial amyloidotic polyneuropathy
FENIB	neuroserpin inclusion bodies
FFI	fatal familial insomnia
FGF	fibroblast growth factor
FPP	farnesyl-pyrophosphate
FTD	frontotemporal dementia
FTIR	Fourier transform infrared spectroscopy
GABA	γ-aminobutyric acid
GAPDH	glyceraldehydes-3-phosphate dehydrogenase
GDNF	glial cell line-derived neurotrophic factor
GEF	guanine nucleotide exchange factor
GPI	glycosylphosphatidylinositol
GPP	geranyl-pyrophosphate
GSK	glycogen synthase kinase

GTPase	guanidine triphosphatase
H	huntingtin
HAP	huntingtin-associated protein
HCIs	hyaline conglomerate inclusions
HD	Huntington's disease
HDAC	histone deacetylase
HDL	high-density lipoprotein
HEK	human embryonic kidney cells
HETE	15-(R)-hydroxyeicosatetraenoic acid
HFIP	hexafluoroethanol
HIP	huntingtin-interacting protein
HMG-CoA	3-hydroxy-3-methylglutaryl coenzyme A reductase
HMW	high molecular weight
Hsf	heat shock protein
HSS	hypotrichosis simplex (scalp)
Htr	serotonin receptor
IC_{50}	concentration of agent to inhibit 50%
IDE	insulin degrading enzyme
IF	intermediate filament
IGF	insulin growth factor
IL	interleukin
InsP3R	inositol 1,4,5-triphosphate receptor
IP	intraperitoneal
IPP	isopenylpyrophosphate
JNK	c-Jun N-terminal kinase
K_a	affinity constant
k_{cat}	catalytic constant
K_i	inhibition constant
KDa	kiloDalton
K_M	Michaelis constant
LCAT	lecithin-cholesterol acyl transferase
L-DOPA	levodopa
LFA	leukocyte function antigen
LID	levodopa-induced dyskinesias
LMW	low molecular weight
LOAD	late onset Alzheimer's disease
LOD	log of the odds ratio
LPL	lipoprotein lipase
LRAT	lecithin:retinol acyl transferase
LRP	lipoprotein receptor-related protein
LRRK	leucine rich repeat kinase
LTD	long-term depression
LTP	long-term potentiation
M	muscarinic receptor
MAP	microtubule-associated protein
MAO	monoamine oxidase

MAPK	mitogen-activated protein kinase
MB-COMT	membrane COMT
MBD	membrane binding domain
MCI	mild cognitive impairment
MCP	monocyte chemoattractant protein
MDMA	3,4-methylenedioxy-N-methylamphetamine, Ecstasy
MHC	major histocompatability complex
MIP	macrophage inflammatory protein
MPP$^+$	1-methyl-4-phenylpyridinium ion
MPTP	1-methyl-4-phenyl-1,2,3,6-tetrahydropyridine
MSF	methane sulfonyl fluoride
3-MT	3-methoxytyramine
mtSOD	mutant SOD
MZ	monozygotic
nAChR	nicotinic acetylcholine receptor
NCAM	neuronal cellular adhesion molecule
N-CoR	nuclear receptor co-repressor
NFTs	neurofibrillary tangles
NICD	Notch intracellular domain
NMDA	N-methyl-D-aspartate
NMR	nuclear magnetic resonance
nNOS	neuronal nitric oxide synthase
NRSE	neuron restrictive silencer element
NSAID	nonsteroidal anti-inflammatory drug
NSE	neuron specific enolase
NUP	naturally unfolded protein
OR	odds ratio
ORF	open reading frame
PAGE	polyethyleneglycol electrophoresis
PAP	poly(A) binding protein
PBS	phosphate buffer system
PD	Parkinson's disease
PDB	Protein Data Bank
PDGF	platelet derived growth factor
PEN	presenilin enhancer
PET	positron emission tomography
PG	prostaglandin
PHFs	paired helical filaments
PK	pharmacokinetics
PKA	protein kinase A
PKC	protein kinase C
PLM	posterior lateral microtubule
PMCA	protein misfolding cyclic amplification
PME1	PP2A methyl transferase
PolyQ	polyglutamine
PP2A	protein phosphatase

PPAR	peroxisome proliferator activated receptor
PPMT	PP2A methyl transferase
PrP^C	prion protein, cellular
PrP^{SC}	prion protein, scrapie
PS1/2	presenilin $\frac{1}{2}$
QSAR	quantitative structure–activity relationship
RA	retinoic acid
RALDH	retinoic acid synthesis enzyme
RAMBA	retinoic acid metabolism blocking agent
RAR	retinoic acid receptor
RARE	retinoic acid response element
RAS	rennin-angiotensin system
RBP	retinol binding protein
RCC	regulator of chromatin condensation
RNAi	RNA inhibitor
RNS	reactive nitrogen species
ROCK	Rho-associated kinase
ROS	reactive oxygen species
RR	relative risk
RXR	retinoid X receptors
SAAp	human serum amyloid A protein
sALS	sporadic amyotrophic lateral sclerosis
SAM	S-adenosyl-L-methionine
SAR	structure–activity relationship
S-COMT	soluble COMT
SDS	sodium dodecyl sulfate
SE	size exclusion
SFI	sporadic fatal insomnia
shRNA	short hairpin RNA
SNc	substantia nigra pars compacta
SNPs	single nucleotide polymorphisms
SP-C	surfactant protein C
SPECT	single photon emission computed tomography
SREBP	sterol response element binding protein
SAP	serum amyloid protein
SOD	copper/zinc superoxide dismutase
SPP	signal peptide peptidase
SUMO	small ubiquitin modifier
$T\frac{1}{2}$	terminal elimination half-life
T_4	thyroxine
TBP	TATA-bindng protein
TEM	transmission electron microscopy
TH	tyrosine hydroxylase
ThT	thioflavin T
TMD	transmembrane domain
TNF	tumor necrosis factor

TSE	transmissible spongiform encephalopathy
TTR	transthyretin
UCH	ubiquitin carboxyhydrolase
UPDRS	Unified Parkinson's Disease Rating Scale
UPP	ubiquitin–proteosome pathway
UTR	untranslated region
vCJD	variant Creutzfeldt–Jacob disease
VEGF	vascular endothelium growth factor
VLDL	very low density lipoprotein
VM	fetal ventral mesencephalon
VPS	vacuolar protein sorting

1 Protein Folding and Misfolding, Relevance to Disease, and Biological Function

Massimo Stefani

CONTENTS

The last decade has provided significant achievements in the field of protein folding thus highlighting key issues in the closely linked areas of protein misfolding and aggregation. Presently, protein folding and protein aggregation are considered in close competition to each other relying on the same physicochemical basis.

Therefore, protein evolution must have selected amino acid sequences favoring protein folding into compact states under the environmental conditions found in cells, avoiding the inherent tendency of the peptide backbone to aggregate into ordered polymers rich in β-sheet stabilized by intermolecular hydrogen bonds. Present knowledge supports the idea that, under suitable conditions, any polypeptide chain may display such an inherent tendency; furthermore, there is increasing awareness that early oligomers in the path of fibrillization are endowed with higher intrinsic cytotoxicity than mature fibrils. Overall, the data appearing in the last few years have prompted the idea that any protein possesses a hidden face leading it to transform into a misfolded rogue able to impair cell viability either by itself or in its oligomeric states. A large wealth of data points also to the existence of shared basic mechanisms of toxicity of early protein aggregates. These data support the suggestion that, as the tendency to aggregate of any polypeptide chain, even aggregate toxicity, at least in most cases, may be considered a generic behavior of the toxic oligomers resulting from a shared basic toxic fold of the polypeptide chain.

These considerations further stress the key importance of the evolution of molecular chaperones not only to improve the efficiency of protein folding but, perhaps more importantly, to avoid the buildup, in the organism, of misfolded species able to nucleate toxic protein aggregates. They also provide a new way to consider the molecular basis of protein misfolding diseases and suggest that protein aggregation into amyloids may be more common in nature than previously believed, as supported by recent reports. Indeed, many degenerative diseases have recently been traced back to the presence, in the affected tissues, of intracellular or extracellular deposits of peptides or proteins previously unknown to be able to aggregate in vivo; other findings indicate that amyloids can also be exploited to perform specific physiological functions in an increasing number of biological systems.

Overall, the body of scientific literature, which has appeared in this field in the last decade, has stressed new chemical, biological, and medical significance of protein aggregation. It has also raised ideas, hypotheses, and theories that represent a significant step forward in the knowledge of the molecular basis of protein aggregation and aggregate toxicity. Finally, it has provided a theoretical framework for the search for molecules and treatments to counteract the most invalidating and prevalent protein deposition diseases.

1.1   INTRODUCTION

The themes of structural biology focussed around protein folding, misfolding, and aggregation are among the most exciting new frontiers in protein chemistry, cell biology, molecular biology, and molecular medicine. The current interest in this topic arises from several issues. The elucidation of a folding code could be of extreme importance in the postgenomic era, where a large number of orphan genes have been identified for which no clear biological function has yet been established. The possibility to confidently predict a protein's three-dimensional structure, not necessarily at the atomic level, from its amino acid sequence and to infer biological function from the predicted fold would be of immense value in biochemistry, structural biology, and cell biology. In addition, it is thought that the knowledge of the molecular basis of protein misfolding and aggregation may help to elucidate the physicochemical features of protein folding and vice versa. It is also expected to shed light on the molecular and biochemical basis of a number of diseases of dramatic social impact including Alzheimer's and Parkinson's diseases, type 2 diabetes, cystic fibrosis, some forms of emphysema, and others recently believed to arise from disturbances of the folding of specific proteins and peptides. A better knowledge of the physicochemical basis of protein folding, misfolding, and aggregation will also be of great value for the biotechnological industry, allowing to design new or modified amino acid sequences with minimal propensity to aggregate. Finally, the growing body of knowledge on these themes will presumably soon lead to the development of molecules and therapeutic strategies to cure protein misfolding diseases or, at least, to delay their appearance and to counteract their clinical signs.

The researches of the last decade have highlighted that protein folding and protein aggregation depend on the same physicochemical features of the polypeptide chains. High mean hydrophobicity, low net charge, and high propensity to α- or β-structure can favor either protein folding and protein aggregation into ordered polymeric assemblies, thus supporting the idea that, in principle, both processes are in close competition with each other. This is best exemplified by natively unfolded proteins (NUPs), a large family of proteins characterized by high content of charged residues and a low content of hydrophobic residues, which are unable to reach any compact, functional folded state after synthesis. Another important achievement has been the recent awareness that protein aggregation is the expression of a general behavior of polypeptide chains that basically does not depend on specific amino acid sequences. However, amino acid sequence is important to favor or disfavor protein aggregation under the physicochemical conditions found in the living systems; they also appear to have been selected by evolution to minimize the inherent tendency to aggregation of polypeptide chains in the environment where each protein folds and performs its biological function.

The general consensus on the idea that protein aggregation is basically the expression of an inherent property of the backbone of every polypeptide chain is also confirmed by two lines of evidence. The first is that an increasing number of degenerative diseases other than the classical protein misfolding diseases such as serpinopathies and amyloidoses can be traced back to the presence, in the affected tissues, of aggregates of proteins and peptides previously not associated with the disease. The other is that in the microbial world (but examples have also been reported in mammals) amyloids may perform specific biological functions, further highlighting the general importance of protein aggregation in biology.

Despite the complexity of the issue, significant data have also been reported on the biochemical and molecular mechanisms underlying toxicity of protein aggregates. For example, the key importance of either the adsorption of protein molecules on surfaces in favoring alternative folds and aggregate nucleation or the interaction of early nonfibrillar aggregates with cell membranes in triggering the cascade of cytotoxic effects is emerging clearly. Similarly, substantial information is presently available on the biochemical perturbations to cells experiencing the presence of protein aggregates eventually leading to cell death; knowledge is also growing on the molecular mechanisms and machineries aimed at reducing the impact of the aggregates on cell functioning or at clearing protein aggregates or their misfolded precursors. In particular, an increasing wealth of information is accumulating on the cell biology of the complex quality control mechanisms of protein folding either in the cytosol or in the endoplasmic reticulum (ER).

The evolving information on protein folding, misfolding, aggregation, and aggregate toxicity to living systems is rapidly improving our knowledge on the molecular basis of protein misfolding diseases, an important field of molecular medicine. Such an increasing body of information is giving rise to the proposal and development of molecular tools and therapeutic strategies to treat a number of such diseases. This chapter is aimed at providing an overview on the nature and origins of protein misfolding and aggregation diseases on the basis of our current knowledge on protein folding.

1.2   ESSENTIALS OF PROTEIN FOLDING

Protein folding, the process leading a disordered polypeptide chain to fold in its unique, stable and biologically active conformation, can be considered one of the most striking examples of biological self-assembly (Pain, 2000). It is increasingly apparent that protein folding is not only fundamental to achieve functional players of biological activity such as proteins, indeed it is also coupled to several other intracellular processes ranging from protein trafficking to the regulation of cell growth and differentiation. The theoretical number of different protein molecules of the same average length as that of cellular proteins (300 amino acid residues) would be 20^{300}, exceeding the total number of atoms in the universe. Accordingly, the proteins found in the living systems are a tiny fraction of the total possible amino acid sequences and represent a very carefully selected group of molecules with special features with respect to those shown by any random amino acid sequence (Dobson, 2004). Such features include the ability to fold rapidly and efficiently into unique structures, thus generating a vast array of different functions characterized by an exceptionally high degree of selectivity and specificity (Dobson, 2003a).

Until recently, the mechanism by which a polypeptide chain folds into its specific compact three-dimensional structure has been shrouded in mystery. Protein native states usually correspond to free energy minima and hence to the structures most energetically stable under physiological conditions (Dobson et al., 1998). However, the number of possible conformations a polypeptide chain can adopt in each of its segments is astonishingly high; therefore, it would take infinite time for an unfolded polypeptide to reach its correct three-dimensional structure simply on the basis of a systematic trial-and-error search of the correct local structure in every peptide segment. In 1968, Levinthal first highlighted such a paradox (Levinthal, 1968), clearly demonstrating that the folding of a polypeptide chain cannot be an ordered sequence of mandatory steps involving specific partially folded states but, rather, must involve a stochastic search of the many conformations that are accessible to a polypeptide chain (Wolynes et al., 1995; Dill and Chan, 1997; Dobson et al., 1998).

In this new view of protein folding, the random fluctuations of the conformation of polypeptide chain segments would allow residues separated in the amino acid sequence to come in close contact with each other. In general, native-like inter-actions are energetically favored and hence much more stable and persistent than the nonnative ones; therefore, they could add to each other greatly limiting the conformational space each element of the polypeptide chain should sample in the search of the overall lowest energy structure. The energetic features of such a behavior can be represented in terms of energy landscapes that funnel the polypep-tide chain to the stable, lowest energy fold through one or more local minima representing partially folded, relatively stable folding intermediates and less stable transition states. Such an energy landscape is substantially specific for each poly-peptide chain and is fully encoded in its amino acid sequence (reviewed in Dobson, 2003b). This means that the information needed for every protein to fold correctly into its specific three-dimensional structure is also totally encoded in its amino acid sequence (Anfinsen, 1973) and consequently in the nucleotide sequence of the gene encoding it. Therefore, biological evolution must have produced nucleotide sequences encoding amino acid stretches able to fold rapidly and efficiently under the working conditions of the corresponding proteins.

This conceptualization has stimulated intense theoretical and experimental research leading to investigation and description of the elementary steps of protein folding. Most simple proteins fold from their unfolded states in a few tenths of μs (Yang and Gruebele, 2003; Mayor et al., 2003), whereas others may require much more time (Plaxco et al., 1998). In addition, computer simulations and biophysical measurements have shown that local structural elements such as α-helices or β-turns can be formed in times of the order of μs (or less) (Eaton et al., 1998; Snow et al., 2002), whereas β-strands can take much more time. Overall, these studies have provided data leading to explain the differing times of folding of different polypep-tide chains on the basis of the structural features of the folded molecules. Possible evolutionary adaptations aimed at matching protein synthesis and protein folding rates have also been suggested by these studies. For example, it is known that in prokaryotes protein synthesis is around one order of magnitude faster than in eukaryotes. Accordingly, it has been shown that, in general, prokaryotic proteins fold much faster than their eukaryotic counterparts (Widmann and Christen, 2000) so as to avoid the persistence in the cell of partially folded molecules for significant lengths of time in the cell before complete folding is attained.

Valuable information on describing the most basic steps of the structural transi-tions occurring during protein folding has been provided by investigating simple proteins less than 100 amino acids in length as folding systems. These molecules usually fold in a very simple way without the complications arising from the accu-mulation of intermediates. The folding behavior of small proteins has been investi-gated by using mutants carrying specific amino acid substitutions specifically designed to probe the role of single residues in the folding process and to analyse and describe transition states (Fersht, 2000). The latter have also been investigated by computer simulation studies, often carried out by incorporating experimentally deter-mined parameters into the theoretical models (Vendruscolo et al., 2001). Protein folding transition states or intermediates exhibit a high degree of disorder; however, their structural characteristics resemble closely those of the final folded states in terms of overall topology and architecture. Taken together, these studies have revealed that the folding behavior of most proteins involves the interaction of a relatively reduced number of residues forming an ordered folding nucleus around which the rest of the structure rapidly condenses by a hydrophobic collapse (Fersht, 2000).

Folding of larger proteins (more than 100 residues in length) usually populates one or more intermediates. These have been considered states assisting further folding into the final structure or, in some cases, traps that stop further advance of the process (Roder and Colon, 1997; Khan et al., 2003; Sanchez and Kiefhaber, 2003). Anyway, the study of the structural features of such intermediates has provided important information on folding and misfolding of large proteins. For example, differing modules or domains in the same polypeptide chain have been shown to fold in a largely independent way (Dinner et al., 2000), native-like interactions among residues being established independently in different regions of the polypeptide. Only in the final cooperative step these regions interact correctly with each other ensuring that all residues have come into close contact in the final compact state and water molecules have been excluded from the interior (Cheung et al., 2002).

Presently, it is not yet clearly understood how the amino acid sequence determines the folding features of a polypeptide chain as well as the final three-dimensional structure it will attain. However, much evidence supports the idea that relatively simple principles underlie protein folding (Baker, 2000), thus increasing our confidence that protein folds can be predicted directly from amino acid sequences. For example, a protein's fold appears somehow related to specific patterns of hydrophobic and hydrophilic residues in the amino acid sequence favoring preferential interactions among specific residues as far as the structure becomes increasingly more compact in the folding path. Another key factor, referred to as contact order correlating fold complexity with experimentally determined folding rates has also been proposed (Plaxco et al., 1998). Contact order is calculated as the average separation, in the amino acid sequence, of the residues that are in contact in the folded state. The rationale of the contact order effect can be explained considering that the stochastic search of native contacts in the folding polypeptide chain will be as much more time consuming as more separated in the amino acid sequence are the residues that will eventually interact and vice versa (Tsai et al., 2002). The final correct structure is reached almost inevitably provided previous contacts and overall topology are the correct ones; otherwise, the polypeptide chain cannot complete its folding into the final stable, compact globular structure thus avoiding gaining a final structure different from the correct one (misfolding).

1.3   KNOWLEDGE OF PROTEIN FOLDING IS FUNDAMENTAL TO UNDERSTAND THE MOLECULAR BASIS OF PROTEIN MISFOLDING DISEASES

The knowledge of protein folding mechanisms and paths as well as of the effects on it of environmental and chemical factors is of great importance to understanding protein folding and misfolding in living organisms, providing clues on the origin of protein misfolding diseases. Polypeptide chains are synthesized in the cell on the ribosomes, which translate the genetic information built in the nucleotide sequences of the corresponding mRNAs. A polypeptide chain starts to fold in the cytosol or in specific compartments (ER, mitochondria) as soon as it emerges from the ribosome (cotranslational fold); in other cases, folding starts only after peptide translation is completed (posttranslational fold); some proteins require specific microenvironments such as the interior of membranes for proper folding, although the basic principles of protein folding are universal. In any case, environmental conditions such as macromolecular crowding and the presence of molecular chaperones play a fundamental role in affecting the ability of proteins to fold rapidly and efficiently while minimizing folding errors. Conversely, the presence of membranes or other charged or uncharged surfaces may disturb correct folding or induce less compact states in folded proteins (see Section 1.12).

Macromolecular crowding is the total concentration of macromolecules within a given biological compartment. In a typical cell, the total concentration of macromolecules averages 300–400 mg/mL. Indeed, it is known that intracellular macromolecular crowding can perturb the folding and binding of proteins (reviewed in Minton, 1994; Ellis, 2001). In this context, it has been estimated that an increase in macromolecular crowding from 30% to 33% (in terms of physical volume occupied by molecules out of the total space) could cause a rise by as much as an order of magnitude in molecular binding affinities (Ellis, 2001). Such a change, which could, for example, be associated with alterations of cellular properties as a result of aging (Nagy et al., 1982) or of the progression through the cell cycle (Conlon et al., 2001), would favor more compact states of macromolecules; it would also increase the population of aggregated species including those that nucleate rapid growth of protein polymers.

The main physiological function of the molecular chaperones both in the cytosol (heat-shock proteins, crystallins, prefoldin, Hsc70) and in the ER (Bip, Grp94, calnexin) is to favor folding of nascent polypeptide chains or refolding of inappropriately folded molecules, to guide later stages of the folding process and to avoid inappropriate interactions of misfolded or incompletely folded polypeptides (Bukau and Horwich, 1998; Hartl and Hayer-Hartl, 2002). When failing to achieve such a task, molecular chaperones act by promoting degradation of polypeptides with nonnative structures. Many proteins require the intervention of molecular chaperones to fold rapidly and efficiently. However, chaperones do not provide any structural information in advance to that built in the amino acid sequence of the polypeptide chain; rather, they provide a microenvironment where a polypeptide can fold escaping competing reactions such as aggregation and avoiding any improper interaction with other molecules or with biological surfaces (Bukau and Horwich, 1998; Hartl and Hayer-Hartl, 2002). Molecular chaperones can also rescue misfolded or even aggregated proteins giving them a second chance to correctly fold before targeting them for degradation (Bukau and Horwich, 1998; Hartl and Hayer-Hartl, 2002). Finally, some chaperones, such as peptidylprolyl isomerases and protein disulfide isomerases, act by accelerating slow folding steps of particular proteins (Schiene and Fischer, 2000).

The importance of molecular chaperones is further stressed by the findings that their amount is substantially increased in cells exposed to stressing conditions and by the effects of surfaces, either biological or artificial, on protein folding, misfolding, and aggregation (Section 1.12). Recent data strengthen our knowledge on this topic. For example, it is assumed that chaperone molecules can assist protein folding even using a hydrophobic-induced stabilization of folding intermediates (Sethuraman and Belfort, 2005).

A similar mechanism could be operating in protein translocation across membranes. In 1988, it was suggested that the molten globule states of proteins are important in enabling proteins to cross cell membranes (Bychkova et al., 1988). This idea is presently supported by a number of findings showing that protein or peptide prefibrillar aggregates, whose exposed hydrophobic surfaces have similarities to those of molten globules (Section 1.8), are able to associate with, and to penetrate into, cell membranes (Bucciantini et al., 2002, 2004; Kayed et al., 2004; Cecchi et al., 2005; Demuro et al., 2005). More generally, it is believed that cytotoxic peptides such as those found in venoms, antimicrobial secretions, the mammalian immune system, together with antibacterial toxins, viral proteins, misfolded proteins or their early aggregates produce malfunctioning and impairment of viability in exposed cells by a shared molecular mechanism involving interaction with cell membranes and membrane damage (Kourie and Shorthouse, 2000; Kourie and Henry, 2002; Demuro et al., 2005).

Such a scenario may explain the toxic effects on tumoral cells of a partially unfolded state of α-lactalbumin at acidic pH (see Section 1.10); it also agrees with the proposed mechanism of toxicity of perfringolysin, a bacterial toxin. The insertion of the latter into the plasma membrane of the target cells requires a conformational change resulting from the conversion of the six α-helices in domain 3 into four amphipathic β-strands (Hotze et al., 2002). It is not known whether such a change may be equivalent to those leading other proteins to assemble into prefibrillar amyloid structures (see Section 1.8); however, it envisages an overall mechanism similar to that proposed for the toxic effects of the prefibrillar aggregates of amyloid type. Similar results have recently been reported with apoptin, a disordered protein encoded by chicken anemia virus. Stable globular aggregates are formed by apoptin specifically in tumoral cells leading to cell death, whereas the intracellular environment of the healthy cells seems to prevent their cytoplasmic accumulation (Zhang et al., 2003).

The key importance of the environment in modulating the tendency of proteins to misfold and aggregate is a principal cause of the well-known difficulty in purifying many proteins expressed in heterologous cells. Examples include the formation of inclusion bodies resulting from the expression of heterologous proteins in bacteria and the precipitation of proteins during laboratory purification procedures. Indeed, protein aggregation is well established as one of the major difficulties associated with the production and handling of proteins in the biotechnology and pharmaceutical industries (Clark, 1998).

Despite the presence of a vast array of molecular chaperones and other quality control tools, protein misfolding does occur in cells. This may happen as a result of several factors. First of all, as outlined earlier, a folded protein is stabilized by a large number of weak interactions that continuously break and reform in the environment where the protein works. This envisions a scenario whereby a protein is present in solution as a population of folded, partially folded, or even unfolded conformational states in dynamic equilibrium where the states closely averaging the native one are largely predominant (Figure 1.1). The very reduced times a protein molecule spends in nonnative conformations usually hinders the occurrence of inappropriate intermolecular interactions resulting in the appearance of aggregation nuclei. However, conditions increasing the number of misfolded molecules in a protein's population such as changes in the environmental conditions, chemical modifications, overexpression of the protein, or the presence of specific mutations may favor aggregate nucleation when the buffering action of chaperones and the other quality control tools is overwhelmed (Calloni et al., 2005) (Figure 1.1). An increasing number of disorders are associated with the failure of specific proteins or peptides to fold correctly and hence with the inability to perform their biological functions (see Section 1.4). In some cases, misfolding or a slower folding of the involved proteins is followed by a rapid clearance of the latter, with lack of function. In other cases, protein misfolding results in the deposition of toxic aggregates in the central nervous system or in peripheral tissues and organs (see Sections 1.5–1.7). Protein misfolding diseases include neurodegenerative diseases (Alzheimer's and Parkinson's diseases) several systemic amyloidoses, cystic fibrosis, familial hypercholesterolemia, serpinopathies and, possibly, many types of cancers.

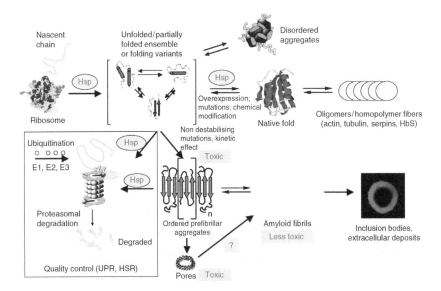

FIGURE 1.1 The conformational states accessible to a protein after synthesis. In the population of protein molecules in solution, usually the equilibrium between compactly folded and partially folded or misfolded molecules, including folding intermediates and conformational variants, is almost completely shifted toward the folded states. The load of partially folded or misfolded molecules can increase as a result of protein overexpression, mutation, chemical destabilization, or any loss of functionality of the quality control of protein folding. Under these conditions, the nucleation of oligomers in the path of fibrillation is favored. The latter is also increased in the presence of mutations that do not shift the above-mentioned equilibrium, but favor it kinetically. UPR, unfolded protein response; HSR, heat-shock response.

1.4 PROTEIN MISFOLDING DISEASES

It is well established that proteins that are misfolded, that is, that are not in their functional conformation, are devoid of their specific biological activity. Many degenerative diseases, often referred to as protein misfolding diseases or conformational diseases, ultimately result from the intracellular or extracellular presence of protein molecules with incorrect structures differing from those displayed by the normally folded counterparts and thus biologically inactive (Thomas et al., 1995). Protein misfolding diseases can therefore be considered as loss of function disorders. However, misfolded proteins often gain the ability to aggregate or to interact inappropriately with other cellular components leading to cell function impairment and eventually to cell death. Accordingly, most protein misfolding diseases have been considered as gain of function conditions. Indeed, in many cases, as for serpinopathies, the final clinical outcome may rely on either a loss of function or a gain of function mechanism (see Section 1.5).

Protein misfolding diseases include conditions where a specific protein, or protein complex, fails to fold correctly (e.g., cystic fibrosis, Marfan syndrome, amyotonic lateral sclerosis) or is not sufficiently stable to perform its normal function (e.g., many forms of cancer). They also include conditions where aberrant folding behavior results in the failure of a protein to be correctly trafficked (e.g., familial hypercholesterolemia, α1-antitrypsin deficiency, and some forms of retinitis pigmentosa) (Thomas et al., 1995). Finally, a number of clinically relevant degenerative pathologies (amyloidoses) are characterized by the aggregation of the misfolded proteins into oligomers and polymers. The latter eventually result in the appearance of ordered fibrils that become deposited intra- or extracellularly, often together with other material, giving rise to what is commonly known as amyloid. Amyloid diseases include conditions of clinical relevance such as Alzheimer's and Parkinson's diseases, type II diabetes mellitus, and a number of systemic amyloidoses (see Section 1.7).

Protein misfolding diseases can be familial or sporadic conditions. The familial forms are early onset, dominantly inherited disorders whose clinical signs appear in childhood or well before late age; they result from the presence of specific mutations in the genes encoding the involved proteins leading to their expression in a less stable or severely destabilized form thus populating conformational states intrinsically prone to aggregation. Such mutations usually consist of single residue substitutions, although insertions or deletions of longer stretches of amino acids are also known. This is the case of triplet repeat diseases including Huntington disease and many ataxias, whose hallmark is the presence of intracellular aggregates of one out of several proteins with poly(Q) or, less frequently, poly(A) or poly(L) extensions often at their N-termini exceeding a threshold number (Ross, 2002). In other cases, the mutation affects one of the proteins involved in the biochemical processing of a precursor giving rise to the aggregated peptide, or the precursor itself (as APP in many forms of familial Alzheimer's disease or BRI in the Danish and British dementias (see Section 1.9). Other familial forms can be traced back to mutant genotypes affecting one of the proteins involved in the quality control of a specific protein. For example, parkin is a ubiquitin ligase involved in α-synuclein degradation whose mutant is responsible for a form of familial Parkinson's disease. Hyperdosage of specific genes may also be involved in some cases. The early onset Alzheimer's disease affecting Down's people results from the presence of the supernumerary chromosome 21 harboring, among others, the APP gene (reviewed in Selkoe, 2001) whose product is the precursor of the Aβ peptides found aggregated in Alzheimer's dementia.

Finally, some protein misfolding diseases including prion diseases such as Creutzfeldt–Jakob disease, spongiform encephalopathies, and fatal familial insomnia are transmissible. In prion diseases, aggregates of a misfolded form (PrP^{sc}) of the prion protein (PrP^c) are present in the affected tissues. PrP^{sc}, which results not necessarily from the presence of a specific destabilizing mutation of PrP^c, is able to recruit the natively folded PrP^c molecules providing them with the structural information to form the aggregating PrP^{sc} structure thus propagating it (Prusiner, 1998). This behavior accounts for the transmissibility of the phenotypes determined by passing among individuals even minute amounts of stable, protease-resistant PrP^{sc} aggregates. Recent data suggest that other proteins or peptides are able to propagate a toxic conformation to the natively folded counterparts (Lundmark et al., 2002; Xing et al., 2002; Giasson et al., 2003) or even to different proteins with low tendency to aggregate (heterogeneous seeding) (Xing et al., 2002).

Conversely, the sporadic forms, which are much more widespread, are generally late onset and apparently not directly associated with specific mutant genotypes. In these, the appearance of protein aggregates can be traced back to a general malfunctioning of cellular tools such as antioxidant defenses and the quality control machineries including the ubiquitin-proteasome pathway (reviewed in Sherman and Goldberg, 2001; Stefani and Dobson, 2003). However, apart from increasing age, which appears the main condition resulting in sporadic protein deposition diseases, a number of other risk factors have been claimed to explain why not all aged people manifest such pathologies. The proposed factors associated with increased risk for sporadic amyloidoses include specific genotypes (such as the ApoE4 variant in the Alzheimer's disease) (Myers and Goate, 2001) or incorrect lifestyles leading to hypercholesterolemia or to excess caloric intake (Austen et al., 2002).

1.5 SERPINOPATHIES

The so-called serpinopathies are pathological conditions arising from altered behavior of a number of serine-protease inhibitors (serpins), a family of proteins including α1-antitrypsin, antithrombin, and plasminogen activator inhibitor type 2. The members of this class of proteins share a common molecular architecture displaying a dominant five-stranded β sheet (A sheet) surrounded by other sheets and by a mobile reactive center loop. Serpins are noncompact, inherently metastable molecules that can readily convert into a more stable form on insertion of the highly mobile loop into a suitable β-sheet, as it normally happens in the complex they form with the target protease (reviewed in Lomas and Carrell, 2002). Such behavior also accounts for the formation of an alternative intermolecular complex formed exclusively among serpin molecules. The complex results from the insertion, as an additional strand, of the reactive center loop of one serpin molecule into the five-stranded A sheet of another molecule; the resulting loop-sheet polymers are insoluble and lack the inhibitory activity of the soluble monomers (Lomas et al., 1992). Native wild-type serpin molecules are metastable structures whose polymerization may be induced only on exposure to mild denaturing conditions. However, some naturally occurring pathological mutants result in A sheet desta-bilization and opening with increasing mobility of the constituting strands allowing insertion of the reactive center loop into the sheet and polymerization even under physiological conditions (Stein and Carrell, 1995). The molecular features of serpin polymerization and the structure of serpin polymers have subsequently been described (Huntington et al., 1999).

Pathological conversions of a loop or, in other cases, an α-helix to a β-strand are also found in other misfolding diseases such as amyloidoses (see Section 1.7). For example, the infectious properties of prions seems to arise from an induced α-helix to β-strand transition (Prusiner, 1998); similarly, β-strand formation appears to be involved in polyglutamine expansion diseases including Huntington disease and spinocerebellar ataxias (Ross, 2002). Recently, domain swapping has also been proposed, at least in some cases, as a possible molecular basis of protein polymer-ization into amyloid assemblies (Newcomer, 2002; Chen and Dokhalyan, 2005).

Serpinopathies include α1-antitrypsin deficiency, thrombosis, angioedema, immune hypersensitivity, familial encephalopathy with neuroserpin inclusion bodies (FENIB), as well as emphysema and liver cirrhosis. They are characterized by the presence inside cells of fibrous serpin polymers arising from such a propagated structural swapping and result in cell impairment by either a loss of function (the lack of active inhibitory protein) and a toxic gain of function (protein aggregate cytotoxicity) (reviewed in Lomas and Carrell, 2002). For example, in α1-antitrypsin deficiency, one out over 70 α1-antitrypsin mutants polymerizes into fibrillar assem-blies found as large inclusion bodies into the ER of hepatocytes, although most of the mutant misfolded protein is eliminated. These features lead to liver disease and to very reduced plasma levels of the serpin thus explaining the appearance of early onset emphysema. The latter results from the reduced protection of the lung paren-chyma against the enzyme neutrophil elastase as a consequence of the lack of active inhibitory protein (reviewed in Lomas and Carrell, 2002). Support for this mechanism of pathogenesis has been provided by in vitro experiments showing that the same α1-antitrypsin mutants associated with disease are also able to form polymers when incubated in vitro under physiological conditions (Dafforn et al., 1999). α1-antitrypsin polymers have also been shown in inclusion bodies from the cirrhotic liver of patients carrying homozygous mutations of the protein (Lomas et al., 1992).

The onset and severity of the symptoms associated with the serpinopathies depend on the type of mutation, its effect on the conformational stability of the protein, and the extent of fibrillar aggregate accumulation; they also demonstrate that intracellular protein aggregation is by itself sufficient to cause the clinical symptoms (reviewed in Lomas and Carrell, 2002). Mutant α1-antitrypsin may be secreted when expressed in model cells as expression systems (Sidhar et al., 1995) and the rate at which α1-antitrypsin polymers are degraded in the ER determines the degree of toxicity (Wu et al., 1994). These findings may explain why subjects carrying the same mutation display different risk of liver disease (Wu et al., 1994).

The serpinopathies bear similarities with the amyloid deposition diseases although they are not classified among the classical amyloidoses and serpin aggregates do not display the typical cross-β structure of the amyloids (see Section 1.8). Therefore, the features of protein polymerization underlying these conditions provide a model for understanding other conformational diseases resulting from aberrant β-linkages. Recently, a new autosomal-dominant dementia known as FENIB has been described (Davis et al., 1999). FENIB is caused by a mutation of neuroserpin, a serpin specifically found in the brain. The subsequent conformational destabilization of neuroserpin leads to the appearance of intracellular protein aggregates that are sequestered into inclusion bodies (Collins' bodies) found in cells of the deeper layer of the cerebral cortex and in the substantia nigra (Davis et al., 2002).

1.6 ER STORAGE DISEASES

The chemical environment in the ER lumen resembles the extracellular space; for example, it is characterized by high free Ca^{2+} concentration, is more oxidizing, and contains a lower macromolecular concentration with respect to the cytosol (reviewed in Sitia and Braakman, 2003). It also plays a key role in subcellular calcium storage and signaling (reviewed in Paschen and Mengesdorf, 2005). Hence, protein folding in the ER occurs under physicochemical conditions substantially different from those found in the cytosol. In addition, the ER contains a specialized set of chaperones, transporters and enzymes involved in the control of folding and maturation of the proteins targeted to membranes or to secretion. Protein folding in the ER appears quite slow and inefficient possibly owing to the complex posttranslational chemical modifications neosynthesized proteins must undergo in this compartment (including cleavage of signal peptides, glycosylation, formation of disulfide bonds, addition of glycosylphosphatidylinositol anchors, membrane insertion). Further complication is provided by the intervention of specific chaperones in most of these steps and by the need to coordinate all these chemical activities. Indeed, the interaction of a nascent chain in the ER with molecular chaperones such as Bip or the lectin chaperones calreticulin and calnexin that bind monoglucosylated glycoproteins, is likely to play a key role in the choice between secretion, retention, and retrotranslocation for degradation (reviewed in Sitia and Braakman, 2003).

Despite such a complexity, the ER protein factory can attain an extraordinarily high efficiency in specialized organs such as exocrine glands and other secretory cells as well as in plasma cells. Such a complexity can also explain why a number of protein misfolding diseases are associated with folding defects of proteins maturating in the ER and with the failure of the quality control mechanisms that ensure proper protein folding in this compartment as well as clearing of misfolded proteins. In particular, most of the polypeptide chains found aggregated in amyloid diseases are proteins maturating into the ER or their degradation fragments. It could be that such proteins are intrinsically more prone to aggregation and are able to escape more easily the cellular mechanisms that protect against misfolding and aggregation. It is also possible that additional steps in which errors can occur or accumulate are represented by processing in the ER before secretion or by the events involved in the retrograde translocation to the cytosol of polypeptides that have failed the quality control tests in the ER (Plemper and Wolf, 1999).

As mentioned earlier (Section 1.5), serpinopathies are ER storage diseases where the aggregated serpin molecules are retained into the ER of the affected cells. This is not the only case where ER is directly involved in protein misfolding diseases. The tools in the ER aimed at performing the quality control of protein folding normally lead to degradation of a significant proportion of polypeptide chains before they have attained their native conformations; ER protein degradation is accomplished through the ubiquitin-proteasome pathway after retrograde translocation to the cytosol (Reits et al., 2000; Schubert et al., 2000). Besides being important in avoiding the appearance of misfolded proteins, the high efficiency of such a quality control may also be beneficial even in other cases. For example, it can improve the promptness of the immune response against viral infections by ensuring exposure of viral antigens on immune cells soon after infection (Reits et al., 2000).

However, in some cases, the high efficiency of such a control may, by itself, be at the origin of disease. Indeed, the most common mutation (ΔF508) of the cystic fibrosis transmembrane conductance regulator (CFTR), a chloride channel associated with cystic fibrosis interferes with the correct folding of the polypeptide chain. Accordingly, much of the mutated protein is not secreted but is retained in the ER, retrotranslocated and rapidly degraded in the cytosol even though, when folded, it could retain some biological activity at the cell surface (Plemper and Wolf, 1999 and references therein). In other cases, diseases can originate from defective degradation, for example as a consequence of an increased rate of synthesis of a specific protein exceeding the combined rates of folding and degradation. Under these conditions, deposits of a fraction of these proteins ubiquitinylated known as aggresomes appear in the pericentriolar region (Kopito and Sitia, 2000); inefficient protein dislocation can also result in the appearance of deposits in dilated ER cysternae (Bence et al., 2001).

It is still debated whether the deposits of aggregated proteins into the ER or in other cellular locations are cytotoxic per se or, rather, they can be considered a form of protection against dangerous misfolded proteins performed by segregating their aggregates into specialized cellular subcompartments. Whatever is the answer to this question, it is important to consider that the sequestration of ubiquitin and chaperones into these aggregates may by itself damage the cell inducing a prolonged state of stress leading to apoptosis (reviewed in Sitia and Braakman, 2003). In some cases, negative selection against proteins with a high tendency to aggregate must occur in a context where folding-defective proteins are produced rapidly as in the case of the production of immunoglobulins by lymphocytes. Indeed, the occurrence of light chain amyloidosis as a consequence of enhanced Ig production is quite rare, even in severe inflammatory conditions. This finding implies the existence, in plasma cells, of particularly efficient quality control mechanisms that enable selection against such mutants (Stevens and Argon, 1999).

Cells are equipped with several responses to the ER stress, a condition where the ER folding machinery cannot cope with its protein load leading to the accumulation of misfolded proteins inside the ER lumen (reviewed in Paschen and Mengesdorf, 2005). The main function of the unfolded protein response (UPR) in the ER is to restore ER function by reducing the load of proteins to be folded and processed in the ER lumen and by enhancing the capacity of protein folding and processing. The UPR proceeds through the coordinate synthesis of various enzymes and transcription factors including the IRE1, PERK ATF4, and ATF6. Under normal conditions, the activities of PERK, ATF6, and IRE1 are suppressed by binding to the chaperone GRP78. However, under ER stress conditions with accumulating unfolded proteins in the ER lumen, GRP78 binds to the unfolded proteins to promote their refolding. Once free from GRP78, IRE1, and PERK become active following oligomerization and autophosphorylation, whereas ATF6 is cut by proteases releasing a transcription factor. Activated PERK gains kinase activity that specifically phosphorylates the α subunit of the eukaryotic initiation factor 2 (eIF2α) thus stopping the initiation step of translation. Phosphorylated IRE1 gains endonuclease activity that induces the splicing of the xbp1 mRNA resulting in the transcriptional upregulation of several genes encoding chaperones and other components of the secretory pathway as well as of factors favoring ER-associated degradation (reviewed in Sitia and Braakman, 2003; Paschen and Mengesdorf, 2005). When such a response fails to clear the ER restoring its normal functionality, several apoptotic pathways are induced (Ma and Hendershot, 2001; Kaufman, 2002; Schröder and Kaufman, 2005).

1.7 AMYLOID DISEASES

Amyloid diseases are, by far, the most clinically relevant types of protein misfolding pathologies due to the high prevalence, in the population, of some of them, including Alzheimer's and Parkinson's diseases and type II diabetes mellitus. The term amyloid was first introduced by Virchow in the nineteenth century to designate a substance apparently amorphous found in the tissue in specific pathological conditions with staining features similar to those displayed by starch (*amylon* in greek). Indeed, amyloid is substantially made of proteinaceous material, although carbohydrates and complex lipids may also be present. A number of degenerative diseases are referred to as amyloidoses, a term first used by Rokitansky in the late nineteenth century. Amyloidoses are characterized by the intracellular or extracellular deposition in specific tissues and organs of proteinaceous deposits of fibrillar aggregates of one out of around 20 peptides or proteins, each found characteristically deposited in a specific disease. Presently, over 20 different amyloid diseases, either familial, sporadic, or transmissible, are known (reviewed in Stefani and Dobson, 2003) although other degenerative diseases with amyloid deposition have recently been described (see Section 1.15).

Amyloid diseases affect different tissues and organs and are characterized by specific clinical signs. Some involve the brain and the central nervous system (e.g., Alzheimer's, Parkinson's, Huntington, and Creutzfeldt–Jakob diseases, amyotrophic lateral sclerosis, many ataxias and other neurodegenerative diseases), whereas others affect peripheral tissues and organs (type-2 diabetes mellitus, systemic amyloidoses). In the latter, proteinaceous deposits are found in several organs such as heart and kidney (primary or secondary systemic amyloidoses) as well as in skeletal tissue and joints (e.g., hemodialysis-related amyloidosis). Other components, including collagen, glycosaminoglycans, and proteins (e.g., SAP) are often present in such deposits, protecting them against degradation (Diaz-Nido et al., 2002; Pepys et al., 2002; van Horssen et al., 2002). Similar proteinaceous deposits are found intracellularly in other diseases, either in the cytoplasm, in the form of specialized aggregates known as aggresomes, Lewy or Russell bodies, or in the nucleus (Kelly, 1998; Dobson, 2001) (Table 1.1).

The presence in tissue of proteinaceous deposits of a specific protein or peptide is a shared hallmark of amyloidoses, suggesting the existence of a causative link between aggregate formation and clinical symptoms (the amyloid hypothesis), as supported by many biochemical and genetic studies (Kelly, 1998; Reilly, 1998; Dobson, 2001; van Horssen et al., 2002). However, the specific nature of the pathogenic species, and the molecular basis of their cytotoxicity are still under intense investigation (Lambert et al., 1998; Hartley et al., 1999; Walsh et al., 1999, 2002; Conway et al., 2000; Reixach et al., 2004). For example, in neurodegenerative

---

TABLE 1.1
A Summary of the Main Amyloidoses

Clinical Syndrome	Fibril Component
Alzheimer's disease	Aβ peptides (1–40, 1–41, 1–42, 1–43)
Spongiform encephalopathies	Prion (whole or fragments)
Parkinson's disease	α-synuclein (wild-type or mutant)
Familial Danish dementia	ADan (34 AA fragment of Bri-277)
Familial British dementia	ABri (34 AA fragment of Bri-277)
Fronto-temporal dementias	Tau (wild-type or mutant)
Hereditary cerebral amyloid angiopathy	Cystatin C (minus a 10-residue fragment)
Amyotrophic lateral sclerosis	Superoxide dismutase
Dentatorubro-pallido-Luysian atrophy	Atrophin 1
Huntington disease	Huntingtin (whole or poly(Q) fragment)
Cerebellar ataxias	Ataxins (whole or poly(Q) fragments)
Kennedy disease	Androgen receptor (whole or poly(Q) fragment)
Spino cerebellar ataxia 17	TATA box-binding protein (whole or poly(Q) fragment)
Myoclonic epilepsy type 1	Cystatin B (mutant or fragment)
Primary systemic amyloidosis	Ig light chains (whole or fragments)
Secondary systemic amyloidosis	Serum amyloid A (whole or 76-residue fragment)
Senile systemic amyloidosis	Transthyretin (whole or fragments)
Familial amyloid polyneuropathy I	Transthyretin (over 45 variants)
Hemodialysis-related amyloidosis	β2-microglobulin
Familial amyloid polyneuropathy III	Apolipoprotein A1 (fragments)
Finnish hereditary systemic amyloidosis	Gelsolin (71/53 amino acid fragments)
Type II diabetes	IAPP (fragment)
Medullary carcinoma of the thyroid	Calcitonin (fragment)
Atrial amyloidosis	ANF
Hereditary nonneuropathic systemic amyloidosis	Lysozyme (whole or fragments)
Injection-localized amyloidosis	Insulin
Hereditary renal amyloidosis	Fibrinogen (fragments)

disorders it is very likely that cell function is directly impaired following the interactions of cellular components with misfolded proteins or their early aggregates (Lorenzo and Yankner, 1994; Thomas et al., 1996). On the other hand, in the systemic nonneurological diseases the accumulation in the affected organs of extensive amyloid deposits can by itself cause at least some of the clinical symptoms (Pepys, 1995), although other more specific effects of aggregates on cell viability might also be present.

The polypeptides involved in the differing amyloid diseases include full length proteins (e.g., α-synuclein, superoxide dismutase, lysozyme, β2-microglobulin, transthyretin, lysozyme), biological peptides (amylin, insulin, atrial natriuretic peptide, calcitonin) or fragments of larger proteins produced as a result of their processing (e.g., the Alzheimer β-peptide, ABri, ADan, gelsolin, apolipoprotein A1) or of a more general degradation (e.g., poly(Q) stretches cleaved from proteins with poly(Q) extensions such as huntingtin, ataxins and the androgen receptor). The peptides and proteins associated with known amyloid diseases are listed in Table 1.1. They can display either wild type sequences, as in the sporadic forms of the amyloidoses, or be variants resulting from genetic mutations associated with early onset, familial forms. Some amyloid diseases are transmissible as is the case with the neurodegenerative conditions associated with aggregates of the prion protein (Creutzfeldt–Jakob disease, fatal familial insomnia). The existence of familial forms of a number of amyloidoses has provided significant clues to the origins of these pathologies. For example, there is significant association between the age of the onset of the familial forms and the effects of the mutations in terms of increase of the propensity of the involved proteins to aggregate in vitro; such an association further supports the existence of a link between protein aggregation and clinical manifestations of a specific disease (Clarke et al., 2000; Perutz and Windle, 2001).

In the past, the formation of amyloid aggregates in vivo was thought to be primarily triggered by proteolysis of the parent protein after it was ascertained that lysosomal enzymes, at acidic pH were able to convert amyloidogenic proteins into amyloid fibrils (Glenner et al., 1971). However, over 10 years ago it was reported that, by itself, acid-induced conformational change was able to generate in vitro amyloid fibrils from transthyretin (Colon and Kelly, 1992). This finding showed for the first time that a structural alteration of a protein was sufficient to significantly populate aggregation-prone conformers. This finding was not immediately accepted, possibly as a consequence of the fact that the peptide found aggregated in Alzheimer's disease resulted from proteolysis of the Alzheimer's precursor protein (APP). However, the behavior of transthyretin was subsequently extended to many other proteins known to aggregate in vivo thus confirming that fibrillar aggregates could be obtained in vitro as a result of induced conformational changes. Overall, these data reinforced the idea that the molecular basis of protein aggregation was an abnormal behavior of the peptides and proteins associated with amyloid diseases resulting from unusual conformational changes specifically related to their structural features and hence to their amino acid sequences.

This idea was first challenged, in 1998, when it was reported independently by two groups that two different proteins unrelated to any amyloid disease were able to aggregate in vitro into ordered polymers indistinguishable from the amyloid fibrils produced by disease-associated peptides and proteins (Gujiarro et al., 1998; Litvinovich et al., 1998). Soon after, it was shown that a similar conversion could be achieved deliberately for other proteins by a rational choice of solution conditions (Chiti et al., 1999, 2001). Since then many studies have confirmed such a behavior for an increasing number of proteins and peptides, also including synthetic peptides, even as short as a few residues, and amino acid homopolymers (Chiti et al., 2001 and references therein; Fändrich and Dobson, 2002) (Table 1.2). In most cases, aggregation was found to require mild denaturing solution conditions (low pH, lack of specific ligands, high temperature, moderate concentrations of salts or cosolvents such as trifluoroethanol) such that the native structure was partially or completely disrupted although the hydrogen-bonded secondary interactions were still maintained.

Presently, a large consensus of opinion favors the idea that amyloid aggregation is a generic property of peptides and proteins and that amyloid fibrils result from physicochemical properties inherent to the covalent peptide backbone (Fändrich and Dobson, 2002); the latter is a common feature of all polypeptide chains, differently from the sequences of their side-chains, whose interactions dictate the large number of different folds characteristic of native proteins (Dobson, 2003a). Indeed, the properties of the side-chains are important in determining amyloid structures although they do not contribute to define the core structure of the latter. Nevertheless, the sequence of a peptide or a protein affects profoundly its propensity to form amyloid structures under given conditions; in addition, the presence of different side-chains influences the details of amyloid fibrils and can give rise to variations of their basic ordered structure (see Section 1.8).

The evolution of the highly cooperative nature of native protein structures appears to be a key factor allowing proteins to escape aggregation for significant lengths of time (Dobson, 1999). However, the functional state of a polypeptide chain is in dynamic equilibrium with other conformational states including partially folded, misfolded, and unfolded states, aggregated states, and their precursors (Figure 1.1). Therefore, the formation of aggregates, and in particular the buildup of the precursor species that nucleate rapid aggregate formation, can be triggered by changes in environmental conditions. In this regard, a crucial factor is simply the concentration of aggregation prone species present under different circumstances. For example, any destabilizing mutation affecting the equilibrium between partially folded or misfolded and natively folded molecules may, by itself, enhance the probability to nucleate fibril formation.The latter may also be favored by mutations neutral with regard to structural stability but able to intrinsically accelerate fibril nucleation from partially folded states. Fibril nucleation can also result from accelerated synthesis of a wild-type protein or peptide leading to an increase in the absolute amount of aggregation-prone species in equilibrium with the natively folded molecules. Aggregation can be affected by other factors such as any perturbation of the environment where aggregation-prone species are present. For example, the change of the total macromolecular concentration within a given biological compartment, can perturb the folding and binding of proteins (Minton, 1994), favoring more compact states of the macromolecules and increasing the population of aggregated species including those that nucleate rapid growth of amyloid fibrils. Synthetic or biological surfaces

TABLE 1.2

Non-Disease-Associated Amyloidogenic Proteins or Peptides

Protein or Peptide	Year
SH3 p85 α PI3 kinase (*Bovine*)	1998
Fibronectin type III module (*Mouse*)	1998
Betabellins 15D, 16D (*Synthetic*)	1998
GAGA factor (*Drosophila*)	1999
Acylphosphatase (*Horse, S. solfataricus*)	1999/2003
Monellin (*Dioscoreophyllum camminsii*)	1999
CspB (1–22 fragment) (*Bacillus subtilis*)	1999
Glycoprotein B (fragment) (*Herpes simplex virus*)	2000
Fiber protein peptide 355–396 (*Adenovirus*)	2000
B1 Ig binding domain (*Streptococcus*)	2000
Phosphoglycerate kinase (*Yeast*)	2000
Apolipoprotein C-II (*Human*)	2000
ADA2H (*Human*)	2000
Met aminopeptidase (*Pyrococcus furiosus*)	2000
Lysozyme (*Hen*)	2000
Apocytochrome c_{552} (*Haemophilus thermophilus*)	2001
HypF N-terminal domain (*E. coli*)	2001
Apomyoglobin (*Horse*)	2001
Amphoterin (*Human*)	2001
Polyamino acids (Synthetic)	2002
Ure2P (*Yeast*)	2002
Complement receptor (18–34 fragment) (*Human*)	2002
Curlin CgsA subunit (*E. coli*)	2002
VI domain (*Mouse*)	2002
Acidic FGF (*Notophthalmus viridescens*)	2002
Stefin B (*Human*)	2002
α-Lactalbumin (*Bovine*)	2002
Prothymosin α (*Human*)	2002
Barstar (*Bacillus amyloliquefaciens*)	2003
Endostatin (*Human*)	2003
Pmel17 (*Human*)	2003
Albumin (*Bovine, glycated*)	2003
Albebetin (*Synthetic*)	2004
Acetylcholinesterase (586–599 fragment) (*Human*)	2004
Core histones (*Bovine*)	2004
α-Chymotrypsin	2004

can also accelerate protein aggregation by either favoring misfolding, increasing the local concentration of misfolded molecules or specifically binding the initial oligomers in the aggregation pathway (see Section 1.12).

Finally, protein aggregation into amyloid is favored under conditions resulting in the impairment or overwhelming of the molecular machineries aimed at performing the quality control of protein folding either in the ER (see Section 1.6) or in the cytosol. The latter comprises the cytosolic molecular chaperones, the ATP-dependent proteolytic complexes in mitochondria and the components of the ubiquitin-proteasome pathway in the cytosol and in the nucleus (Goldberg, 2003). Presently, a large consensus has been reached on the central role performed by these machineries in ensuring that the intracellular steady-state concentration of folding intermediates or unfolded molecules is maintained at negligible levels by binding and refolding them by the molecular chaperones or by allowing them to be degraded by the ubiquitin-proteasome machinery (reviewed in Sherman and Goldberg, 2001). Specific inactivating mutations of any of the components of the quality control or harsh environmental conditions such as heat shock, oxidative stress, or chemical modification may impair the activity of the clearing machinery components or engulf them with excess misfolded molecules. As a consequence, the number of misfolded or unfolded protein molecules further increases, resulting in overwhelming both the molecular chaperones and the proteasome (reviewed in Stefani and Dobson, 2003). The accumulation, in the cytosol, of aberrant proteins triggers the heat-shock response, which results in the de novo synthesis, among others, of heat shock proteins (Hsp) and of proteins involved in the ubiquitin-proteasome pathway (reviewed in Sherman and Goldberg, 2001). Under these conditions, when the response fails to rescue the ubiquitin-proteasome machinery from engulfment by misfolded proteins and their early aggregates, the accumulation of misfolded proteins eventually culminates in the activation of the apoptotic pathway.

1.8   STRUCTURE OF AMYLOID FIBRILS AND THEIR PRECURSORS

Despite the differences in size, amino-acid composition, sequence and structure of the parent peptides and proteins, amyloid fibrils share a basic, common ordered core structure and display marked similarities in their appearance (Figure 1.2). Most often, proteins and peptides display a substantially higher content in β-structure when aggregated into amyloid assemblies than in their soluble monomeric states, as shown by spectroscopic techniques such as CD and FTIR. This is true even when the monomeric peptides or proteins are poor in overall secondary structure or are rich in α-helical structure. The shared core structure of amyloid fibrils can be traced back to the physicochemical properties of the peptide backbone common to all polypeptide chains, as outlined earlier.

The structural features of amyloid fibrils have been investigated mainly by different electron microscopy (EM) techniques and, more recently, by solid-state NMR and by atomic force microscopy (AFM). Typically, amyloid fibrils either produced in vitro or extracted from in vivo amyloid deposits appear long, straight, and unbranched, 6–12 nm in diameter and usually consist of 2–6 elementary units rich in β-structure known as protofilaments. Protofilaments are about 2.0 nm in diameter and often are twisted around each other to form supercoiled rope-like structures (Serpell et al., 2000; Serpell, 2000). Presently, there is a substantial lack of information about the detailed arrangement of the polypeptide chains within amyloid fibrils, either those parts involved in the core β-strands or the regions connecting the various β-strands. In fact, it has not yet been possible to obtain a

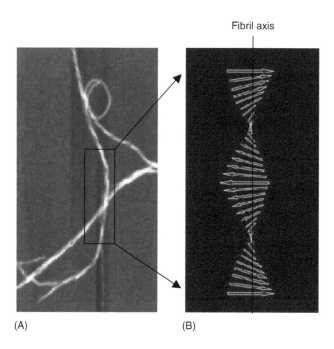

Fibril axis

(A)   (B)

FIGURE 1.2 Atomic force microscopy (AFM) image of amyloid fibrils (A) and outline of the ordered core structure (B) of one of the interwoven protofilaments constituting the fibril. In (B), each *arrow* represents a couple of parallel or antiparallel β-strands. The stacking of β-strand couples generates a double twisted β-sheet that propagates along the fibril axis and whose strands run perpendicular to the latter.

detailed definition of the molecular structure of any amyloid fibril at the atomic level due to the intrinsic difficulties to investigate these monodimensional crystals by the techniques used for proteins in solution or in crystals. Nevertheless, the inner ordered core structure of each protofilament has been investigated by x-ray diffraction, solid-state NMR, molecular dynamics, and molecular modeling studies (Petkova et al., 2002; reviewed in Tycko, 2004; Lührs et al., 2005; Nelson et al., 2005). These studies have shown that amyloid fibrils display highly ordered core structure, with some parts or all of every polypeptide chain showing a characteristic arrangement. In the latter, referred to as cross-β structure, pairs of β-strands running parallel or antiparallel to each other and perpendicular to the protofilament main axis are stacked along the main fibril axis resulting in a double β-sheet that propagates along the direction of the fibril (Figure 1.2).

Recent data suggest that, in the cross β-structure, the sheets are relatively untwisted and may exist, at least in some cases, in quite specific supersecondary structure motifs such as β-helices (Jiménez et al., 1999; Wetzel, 2002) or the proposed μ-helix (Monoi et al., 2000). Significant differences in the pathway of aggregation and in the way the strands are assembled can arise from the conditions of fibrillization (Petkova et al., 2004; Yamaguchi et al., 2004; Chen and Dokholyan, 2005; Pedersen et al., 2006) or from specific characteristics of the polypeptide chain involved (Jiménez et al., 1999; Chamberlain et al., 2000). Self-propagating, molecular level polymorphism in Aβ fibrils associated with different toxicity and arising from subtle differences in the conditions of fibril growth has also been highlighted (Petkova et al., 2005). Factors including length, sequence, and in some cases, the presence of disulfide bonds or posttranslational modifications such as glycosylation, may be important in determining details of the structures.

Several recent papers report structural models for amyloid fibrils made from different polypeptide chains including the Aβ$_{40}$ peptide, insulin, and fragments of the prion protein, based mainly on cryo-electron microscopy and solid-state NMR spectroscopy data (Jiménez et al., 2002; Petkova et al., 2002). These models display many similarities to each other and appear to reflect the fact that the structures of different fibrils are likely to be variations of a common structural theme (Wetzel, 2002). This idea is confirmed by a recent report describing the generation of two monoclonal antibodies specific for a conformational epitope shared among amyloid fibrils made from Aβ$_{40}$ and from other peptides and proteins with unrelated sequences (O'Nuallain and Wetzel, 2002).

A key issue in the investigation of the amyloid structures is the description of either the mechanism by which they are formed from their soluble precursors and the structural features of their intermediates. A number of reports have highlighted the structural modifications of amyloid assemblies made from different disease-related and disease-unrelated peptides and proteins in the path of fibrillization from unfolded monomers and aggregation nuclei to mature fibrils. Several more-or-less well-defined steps are involved in the assembly process of proteins and peptides either associated or not associated with amyloid diseases (Hirakura and Kagan, 2001; Lin et al., 2001; Quintas et al., 2001; Lashuel et al., 2002a; Relini et al., 2004). EM and AFM studies image the initial presence of small, roundish, or tubular particles 2.5–5.0 nm in diameter generally enriched in β-structure often called "amorphous aggregates" or sometimes "micelles" (Quintas et al., 2001; Lashuel et al., 2002a; Poirier et al., 2002; Relini et al., 2004). These species are often associated into bead-like chains or small, doughnut shaped annular rings and appear to be precursors of longer protofilaments and mature fibrils. Other common and highly organized amyloids appear as large closed loops and ribbons (Lin et al., 2001; Lashuel et al., 2002a; Poirier et al., 2002; Relini et al., 2004) (Figure 1.3). These assemblies are not alternative to the mature fibrils but appear as intermediates in the path of fibrillization although some authors consider the small annular oligomers as dead end products of the process.

Presently, a better knowledge of the early aggregates preceding the appearance of mature fibrils is considered very important in understanding the nature and origins of the pathological properties of amyloid structures associated with disease, particularly with neurodegenerative conditions (see Section 1.7). The data provided by O'Nuallain and Wetzel (2002) describing a conformational antibody recognizing a generic amyloid fibril epitope are complemented and substantiated by a recent report. The latter provides evidence that the soluble prefibrillar aggregates of differing peptides and proteins are equally recognized by polyclonal antibodies raised against prefibrillar assemblies made from Aβ$_{1-40}$ or Aβ$_{1-42}$ peptides, whereas the same antibodies are unable to recognize the corresponding fibrillar aggregates or the

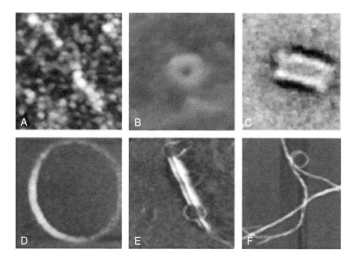

FIGURE 1.3 Morphologies of globular, protofibrillar, and fibrillar amyloid aggregates. Similar images have been reported for aggregates of different peptides and proteins (A) globules and chains of globules; (B) doughnut-shaped oligomers; (C) tubular oligomers; (D) large rings made of globular units; (E) ribbons; (F) mature fibrils. A, B, D, F represent AFM images; C represents electron microscopy (EM) image. A, D–F (From Relini, A., Torrassa, S., Rolandi, R., Ghiozzi, A., Rosano, C., Canale, C. et al., *J. Mol. Biol.*, 338, 943, 2004. With permission.); B (From Ding, T.T., Lee, S.-J., Rochet, J.-C., and Lansbury, P.T., *Biochemistry*, 41, 10209, 2002. With permission.); C (From Lashuel, H.A., Petre, B.M., Wall, J., Simon, M., Nowak, R.J., Walz, T., and Lansbury, P.T., *J. Mol. Biol.*, 322, 1089, 2002a. With permission.).

soluble monomers from which they are formed (Kayed et al., 2003). These findings indicate that prefibrillar aggregates of different proteins and peptides share common structural features that are different from those displayed by the monomeric or fibrillar counterparts; they also provide a structural basis of the toxic potential of these assemblies (see Section 1.10). It has been proposed that the α-pleated sheet structure is the only backbone conformation that may explain the different recognition, by the same antibody, of soluble oligomeric and insoluble fibrillar amyloids (Armen et al., 2004).

1.9 STRUCTURAL FEATURES FAVORING PROTEIN FIBRILLIZATION

The awareness that many peptides and proteins that are not associated with amyloid disease display a generic ability to form amyloid fibrils (see Section 1.7) has important consequences for understanding the fundamental origins of the deposition of protein aggregates in disease. It has also raised a more general discussion on the various states polypeptides can adopt following their synthesis in vivo and on the evolutionary processes set up by the living world to avoid any increase of those states endowed with propensity to aggregation. Finally, the generic aggregation potential of peptides and proteins has very great practical relevance as it increases dramatically the number of proteins and peptides that one can investigate to discover general features underlying the mechanisms of protein misfolding and aggregation.

Besides structural stability, the connections between the physicochemical and structural properties of a protein and its aggregation propensity are far from being known in detail and a number of possible aggregation mechanisms are under investigation (Kelly, 2000). Recent data have shown that the tendency of a unfolded or partially folded polypeptide chain to aggregate is affected by general physico-chemical features such as mean hydrophobicity, net charge, and propensity to α- and β-structure formation (Chiti et al., 2003). This is particularly true when one considers the rate of aggregation from partially folded or unfolded states, the same from which it is thought that in most cases aggregation starts in vivo. A more recent study has provided an empirical correlation also including residue dipole moment, accessible surface area, and aromaticity (Tartaglia et al., 2004, 2005b). Recently, evidence has been provided that the conversion of a soluble nonamyloi-dogenic protein into an amyloidogenic one can be triggered by a nondestabilizing six-residue amyloidogenic insertion in a particular structural environment (Esteras-Chopo et al., 2005). Finally, it has recently been suggested that two-state folders can be less resistant to aggregation whereas cooperative folders are more resistant to aggregation (Clark, 2005). Overall, these data may contribute to explain the molecular basis of a number of familial amyloidoses where the mutant proteins or peptides display amino acid substitutions neutral for their effect on the confor-mational stability.

The same physicochemical features favoring protein aggregation may also account for the molecular basis of the ability of the so-called NUPs (or natively unstructured proteins) to maintain their unfolded state. NUPs are a large group of cellular proteins including many transcription factors as well as proteins involved in the control of the cell cycle and in intracellular signaling, which are in part or completely devoid of ordered structure after their synthesis (reviewed in Uversky, 2002). NUPs become more ordered and compact when they combine with the specific partner, which provides the surfaces needed to reach a stable folded state (Wright and Dyson, 1999; Uversky and Fink, 2004). NUPs can be easily recognized from their amino acid content resulting in low mean hydrophobicity and high net charge in addition with low propensity to gain secondary structure. These character-istics are thought to be the molecular basis of the persistence of these proteins in an unfolded state in the absence of their specific partners; they are also able to reduce the intrinsic tendency of these proteins to aggregate in the highly crowded intracel-lular milieu (Minton, 1994). The unstructured state of these proteins is likely to favor their rapid degradation by the cellular clearance mechanisms, as supported by their rapid intracellular turnover (Wright and Dyson, 1999). Overall, present knowledge on the molecular basis of protein folding and aggregation indicates that both processes depend on the same physicochemical features and therefore can be con-sidered in close competition with each other. From these consideration it results that protein evolution must have carefully selected amino acid sequences with physico-chemical features leading them, under the cellular conditions, to rapidly fold into compact, stable, and biologically active states yet escaping the competition of the alternative amyloid aggregation pathway.

The conformational modifications underlying protein aggregation into amyloid assemblies are favored by the presence of nuclei of the same aggregated protein or, in some cases, of different proteins. The seeding effect relies on the basic nucleation–polymerization mechanism of formation of amyloid polymers and is reflected by the lag time seen in most cases between the onset of aggregation conditions and the appearance of amyloid assemblies. Such an effect is also confirmed by the increase of the rate of aggregation resulting from the presence of preformed oligomers in the solution of aggregating monomers. The propagation of the conformation leading a protein to deposit on preformed nuclei is best exemplified by the prion story. However, this is a more general feature that can be extended to the aggregation process of most proteins or peptides even when preformed nuclei of a protein favor aggregation of a different protein. This is the case of the heterogeneous seeding, a phenomenon described for a number of proteins (Xing et al., 2002) that confirms the common basic structural features of the inner ordered core of amyloid fibrils.

Another important issue underlying protein fibrillization is intramembrane proteolysis. It is known that several peptides found aggregated in tissue from patients affected by a number of amyloid diseases arise from proteolytic processing of membrane proteins. The best known example is provided by Aβ peptides, produced from the processing of a membrane protein, the amyloid precursor protein (APP) along the secretory pathway (reviewed in Selkoe, 2001). APP processing involves several proteins, including large proteolytic complexes known as α-, β-, and gamma-secretases predominantly found, together with APP, in specific membrane micro-domains enriched in cholesterol and gangliosides known as lipid rafts (Marlow et al., 2003; Vetrivel et al., 2004). It is believed that the peptides produced by such a complex pathway are retained into the membrane or can penetrate it from the soluble state thus experiencing a hydrophobic environment favoring their aggregation (Bokvist et al., 2004; Ambroggio et al., 2005).

Other disease-associated peptides are also produced along the secretory pathway of membrane proteins. This is the case of ABri and ADan, whose aggregates are a hallmark of familial amyloid angiopathy conditions known as British dementia and Danish dementia, respectively (Srinivasan et al., 2003; Gibson et al., 2004). Mutations in the precursor membrane-anchored protein, termed BRI precursor protein, favor its aberrant proteolytic cleavage by furin, a proprotein convertase. The resulting anomalous, aggregation-prone peptides are deposited in the affected areas of the central nervous system, notably leptomeninges, blood vessels, and parenchyma. Another example is provided by a mutant form of gelsolin, whose proteolytic fragment is found aggregated in the familial amyloidosis of Finnish type. Slightly different forms of gelsolin are present either intracellularly and extracellularly. Specific mutations of residue D187 appear to be at the origin of the disease allowing aberrant proteolysis of the protein along its secretory pathway with production of a 71-residue fragment with high amyloidogenic potential. Recent studies have shown that gelsolin proteolysis is carried out in the *trans*-Golgi by the furin proprotein convertase during secretion. The susceptibility of the gelsolin variants found in Finnish amyloidosis results from their inability to bind Ca^{2+} ions in a specific domain with a reduction of protein stability and an increase of its susceptibility to general proteolysis (reviewed in Huff et al., 2003). Another example is provided by the 50-residue fragment medin originating from the internal cleavage of lactadherin, a cell surface membrane glycoprotein secreted as a part of the milk fat globule membrane (Häggqvist et al., 1999). In elderly people, amyloid aggregates of medin are deposited in the aortic media of vessels mainly localized in the upper part of the body, a condition with no well established clinical significance (Peng et al., 2005).

Finally, a striking example of proteolysis of a membrane protein leading to amyloid aggregation of the resulting peptides is provided by the formation of Pmel17 fibers in the melanosomal matrix during melanosome biogenesis. Pmel17 is a type I integral membrane protein whose expression correlates closely with melanine production (reviewed in Huff et al., 2003). The maturation of Pmel17 into fibers bears striking resemblance to the mechanism of amyloidogenesis of the gelsolin variant in the Finnish amyloidosis. In fact, it has been shown that Pmel17 is proteolytically processed in the post-Golgi into two fragments by furin or a related proprotein convertase and that premelanosome maturation requires such a cleavage for the appearance of the striations onto which melanin polymerization occurs (Berson et al., 2001). The Pmel17 story is also important as an example of amyloid with a physiological significance; other examples will be given later (see Section 1.16).

1.10 STRUCTURAL RELATIONS BETWEEN MISFOLDED PROTEINS AND THEIR TOXIC EFFECTS

Presently, the amyloid hypothesis is supported by a large number of data on many in vivo and in vitro amyloidogenic proteins indicating a direct cytotoxic effect of amyloid aggregates (Selkoe, 2003). However, it is not generally accepted that the latter are always represented by mature fibrils; rather, an increasing number of reports indicate that, at least in most cases, the prefibrillar assemblies are the main or even the sole cytotoxic species (Hirakura and Kagan, 2001; Lin et al., 2001; Quintas et al., 2001; Bucciantini et al., 2002; Hirakura et al., 2002; Lashuel et al., 2002a; Poirier et al., 2002; Walsh et al., 2002; Wang et al., 2002; Reixach et al., 2004). Indeed, a unifying model for protein fibrillization under physiological conditions has not yet been proposed for all forms of amyloidoses and for many of these no information is currently available on either the identity of the supramolecular aggregates responsible for tissue damage and the molecular mechanism(s) of cell impairment.

Until recently, mature amyloid fibrils were considered cytotoxic proteinaceous aggregates since they were the form of aggregates commonly detected in pathological deposits. Therefore, it appeared likely that the pathogenic features underlying amyloid diseases were a consequence of the interaction with cells of extracellular or intracellular deposits of aggregated material. Besides providing a theoretical frame to understand the molecular basis of these diseases, such a scenario stimulated the exploration of therapeutic approaches to amyloidoses that were mainly focused at searching for molecules able to impair the growth and deposition of fibrillar forms of aggregated proteins. However, increasing evidence indicates that, at least in many cases, the prefibrillar aggregates (sometimes referred to as amorphous aggregates, early aggregates, protein micelles, or protofibrils) preceding the appearance of the mature fibrils are the most highly cytotoxic species, This evidence concerns proteins and peptides either nonassociated (Bucciantini et al., 2002; Yang et al., 2002; Kranenburg et al., 2003; Sirangelo et al., 2004) or associated with disease. The latter include Aβ peptides, α synuclein, amylin, β2-microglobulin, transthyretin, and others (Bhatia et al., 2000; Conway et al., 2000; Conlon et al., 2001; Nilsberth et al., 2001; Sousa et al., 2001; Walsh et al., 2002; Cleary et al., 2005). The presence of such species has also been reported for huntingtin (Poirier et al., 2002), superoxide dismutase (Chung et al., 2003), and possibly for the androgen receptor in diseased transgenic mice (Katsuno et al., 2002). If we accept the idea that prefibrillar aggregates are the most highly cytotoxic species, mature fibrils could be considered as inert, harmless deposits of the latter and hence their growth could be interpreted as a cell defense mechanism. The high cytotoxicity of the early oligomeric aggregates can also explain the lack of direct correlation between density of fibrillar plaques in the brains of victims of Alzheimer's disease and the severity of their clinical symptoms (Dickson, 1995).

Assessing the nature of the species directly involved in the cytotoxic effects underlying the clinical symptoms in amyloid diseases is of the outmost importance. If we consider that the prefibrillar aggregates of most disease-associated proteins and peptides are the main cytotoxic species, any future design of therapeutic interventions against the differing amyloid diseases should be primarily aimed at reducing the load of misfolded molecules populated at equilibrium thus avoiding the appearance of their early aggregates. In this view, hindering the formation of mature fibrils without reducing protein misfolding or increasing the efficiency of the mechanisms aimed at their clearing could be detrimental rather than beneficial. Therefore, deeper information on the structural organization responsible for the cytotoxicity of aggregates of any protein or peptide and on the biological and molecular basis of the latter is needed to design specific interventions aimed at counteracting amyloidogenesis and its toxic effects.

The increasing information supporting the cytotoxicity of early aggregates of peptides and proteins not associated with amyloid disease supports the idea that any amyloid aggregate in its prefibrillar organization can be intrinsically toxic to living cells. It also implies that amyloid cytotoxicity may arise from shared characteristics of the supramolecular structure of the aggregates rather than from any specific features of the amino acid sequences of the parent soluble polypeptides. These features are in direct contrast to the properties of functional proteins, whose native structure and biological function are specifically determined by their amino acid sequences. Indeed, such a remarkable result follows from the conclusion discussed earlier that the structure, and hence the properties of species other than the native state of a protein are not directly determined by the specific interactions among side-chains; that the cytotoxicity of early aggregates made from different peptides and proteins relies on shared structural features of the aggregates themselves is supported by a recent report mentioned earlier (Kayed et al., 2003). The latter has provided evidence that the soluble prefibrillar aggregates, but not mature fibrils, of different peptides and proteins associated with amyloid disease are equally recognized by a polyclonal antibody raised against a molecular mimic of soluble prefibrillar $Aβ_{1-40}$ or $Aβ_{1-42}$ oligomers. In addition, it has demonstrated that the cytotoxicity of these aggregates was substantially reduced in the presence of such antibodies.

Recent reports highlight a specific cytotoxic effect on a large number of differing tumoral cells of a partially unfolded state of the milk protein α-lactalbumin formed at acidic pH, where the protein adopts the apo state by losing its bound calcium ions (reviewed in Svanborg et al., 2003). The apo state contains a high affinity fatty acid binding site harboring an oleic acid molecule that stabilizes and traps the altered protein conformation, possibly displaying molten-globule-like features. It has been proposed that the antitumoral activity of such a folding variant may protect against the appearance of tumoral cells in the breast-fed infants, whose immune system is not yet efficient (Svensson et al., 1999). These findings suggest that protein folding variants, whose formation in peripheral tissues may depend on changes of the folding conditions, can perform specific biological functions; they have also led to propose the use of this folding variant of α-lactalbumin as a possible anticancer tool. Once added to the proposal that the molten globule states can be important in allowing proteins to cross cell membranes (Section 1.3), these data suggest that protein conformational states believed to be precursors of amyloid aggregates may exploit some biological functions. Such an idea also raises the question as to whether altered conformational states of specific proteins with functional significance can be present in other biological systems (see Section 1.16).

Further studies on a wider range of peptides and proteins are required to explore the generality of these observations; however, it may well be that the cytotoxicity of protein folding variants and their early aggregates is generic (Bucciantini et al., 2002). Indeed, such toxicity is likely to arise from the misfolded nature of the aggregated species and their precursors and from the exposure, in them, of regions normally buried in the compact native state that are likely to be aggregation-prone including patches of hydrophobic residues and the polypeptide main-chain. The latter idea is supported by the intrinsic instability of prefibrillar species that enables them to further organize into more ordered structures and to interact with cellular components. Indeed, prefibrillar assemblies have been shown to interact with synthetic phospholipid bilayers (Arispe et al., 1993; Mirzabekov et al., 1996; Lin et al., 1997; Kourie, 1999; Hirakura and Kagan, 2001; Lin et al., 2001; Volles and Lansbury, 2001) and with cell membranes (Lin et al., 2001; Ding et al., 2002; Hirakura et al., 2002) (see Sections 1.12 and 1.13). Moreover, such interactions are believed to result in membrane destabilization with derangement of selective permeability and impairment of specific membrane-bound protein function (Kourie and Shorthouse, 2000; Zhu et al., 2000). The interaction of prefibrillar amyloid aggregates with cell membranes is reminiscent of the action of a number of prokaryotic or eukaryotic peptides or proteins (e.g., several bacterial toxins, the complement membrane attack complex, the apoptosome) that oligomerize into the membranes of the target cells forming pore-like assemblies that destabilize cell membranes and impair ion balance across them leading to cell death (see Section 1.13).

Overall, these considerations might explain the origin of the sporadic, age-related, amyloid diseases in terms of increased presence of aggregation-prone states of peptides or proteins (either native-like or chemically modified). The impairment of the cellular machineries aimed at performing the quality control of protein folding associated with this condition would reinforce the presence of misfolded molecules eventually leading to cell sufferance and death. These ideas also suggest that minute amounts of amyloid aggregates of proteins or peptides presently not associated with any amyloid disease could form in tissues and organs during aging resulting in subtle cell impairments in the absence of a clear amyloid phenotype. The latter suggestion is reinforced by the recent findings indicating that the human genome exons contain around 60,000 single nucleotide polymorphisms (The international SNP map working group 2001) resulting in a higher allelic variance of the human gene pool than previously suspected. Some of these variants could encode natural proteins with reduced structural stability, and hence with a higher propensity to aggregate. Indeed, a structural destabilization of the native fold by as few as 2 kcal/mol (such as one associated with a single residue mutation or with a chemical modification) can enhance the concentration, at equilibrium, of the unfolded molecules by over 30 fold, and hence the probability of aggregate nucleation by over 10^5 fold (Perutz et al., 2002). These figures support the more general possibility that many proteins can nucleate amyloid aggregates under physiological conditions and that a higher number of degenerative diseases than presently known could result from amyloid deposition (see Section 1.15).

1.11   MECHANISMS OF AMYLOID TOXICITY

The presence of toxic aggregates inside or outside cells can impair a number of cell functions eventually leading to cell death by apoptosis or, less frequently, by necrosis (Watt et al., 1994; Lane et al., 1998; Morishima et al., 2001; Velez-Pardo et al., 2001; Ross, 2002) although both cell death features can be carried out simultaneously (Nicotera and Melino, 2004). However, in most cases initial perturbations of fundamental cellular processes appear to underlie the impairment of cell function induced by aggregates of disease-associated polypeptides. Increasing information points to a central role performed by alterations of the intracellular redox status and free Ca^{2+} levels in cells exposed to toxic aggregates (Zhu et al., 2000; Butterfield et al., 2001; Kourie, 2001; Lin et al., 2001; Hyun et al., 2002; Milhavet and Lehmann, 2002; Wyttenbach et al., 2002) although other causes, such as transcriptional derangement in poly(Q) extension diseases, have also been proposed (Ross, 2002). In cells experiencing toxic aggregates any modification of the intracellular redox status is associated with a sharp increase in the levels of reactive oxygen species (ROS). Changes have also been observed in reactive nitrogen species (RNS), with lipid peroxidation, deregulation of NO metabolism (Kourie, 2001), protein nitrosylation (Guentchev et al., 2000) and upregulation of heme oxygenase-1, a specific marker of oxidative stress (Choi et al., 2000).

The key role of oxidative stress following exposure to the early aggregates involved in amyloid diseases is stressed by an increasing number of experimental data. For example, it has been shown that cells can be protected against aggregate toxicity by treatment with antioxidants such as tocopherol, lipoic acid, or reduced glutathione (Zhang et al., 2001). In this regard, an interesting study in prion-infected mice suggests that an increase in free radical production with a reduction of the efficacy of antioxidant defenses in mitochondria is responsible for brain damage, and that this effect could contribute to the development of prion diseases (Lee et al., 1999). Recent data also point to direct effects of aging and oxidative stress on the activity and expression levels of the proteasome and hence on cell viability in nerve tissue (Keller et al., 2002). Inhibition of the proteasome is likely to result in the accumulation of oxidized or otherwise damaged or misfolded proteins thus increasing the detrimental effects of ROS. In this regard, a possible role of Hsp27 in preventing polyglutamine cytotoxicity by suppressing ROS production has recently been proposed in cells transiently transfected with vectors expressing exon 1 to produce varying lengths of glutamine repeats in huntingtin (Wyttenbach et al., 2001, 2002).

It is not yet clear why protein aggregation is followed by increased ROS production either in the cell or in vitro. Specific mechanisms may be involved as in the case of $A\beta_{42}$, where Met35, Gly29, and Gly33 have been suggested to be responsible for this effect (Brunelle and Rauk, 2002). The vulnerability to the oxidative stress of dopaminergic neurons in the substantia nigra, where aggregates of α-synuclein can be present appears also to be increased by the oxidant species (catechol quinones, hydroxyl radicals) produced in the metabolism of dopamine (reviewed in Orth and Schapira, 2002). In other cases, the generation of hydroxide radicals from hydrogen peroxide by metal ions such as Fe, Cu, and Zn has been proposed as a cause of oxidative stress (Turnbull et al., 2001; Tabner et al., 2002). An upregulation of the activity of membrane enzymes producing hydrogen peroxide, such as the plasma membrane NADPH oxidase and the cytochrome P450 reductase in the ER, has also been reported in $A\beta$-induced cytotoxicity to microglia and to cortical neurons (Pappolla et al., 2001; Qin et al., 2002). More generally, intracellular oxidative stress can be related to some form of destabilization of cell membranes by toxic species with loss of regulation of plasma membrane proteins such as receptors and ion pumps (Mattson, 1999) or impairment of mitochondrial function. Mitochondria play a well-recognized role in oxidative stress and apoptosis; in this regard, a key factor in $A\beta$ peptide neurotoxicity could be the opening of permeability transition pores on Ca^{2+} entry in neuronal mitochondria (Moreira et al., 2002) followed by release of strong inducers of apoptosis such as cytochrome c and apoptosis inducing factor.

It has been suggested that intracellular ROS elevation following exposure to amyloid aggregates is a consequence of Ca^{2+} entry into cells following aspecific membrane permeabilization. Increased levels of free Ca^{2+} are able to stimulate the oxidative metabolism, for example by activating the dehydrogenases of the Krebs cycle thus providing the ATP needed to support the increased activity of membrane ion pumps involved in clearing the excess Ca^{2+} (Squier, 2001). In turn, ROS elevation would oxidize membrane ion pumps and their regulatory proteins such as calmodulin, with downregulation of the activity of the Ca^{2+}-ATPase eventually resulting in the increase of intracellular free Ca^{2+} (Squier, 2001). A similar chain of events could be present in old age, where cells are more susceptible to oxidative stress and their energy load is lower.

Such a hypothesis can explain the relationship between ROS and intracellular free Ca^{2+} increase, mitochondrial damage and apoptosis described in cells exposed to toxic amyloid aggregates (Kawahara et al., 2000; Selkoe, 2001; Wyttenbach et al., 2001; Kourie and Henry, 2002). Indeed, many studies support a close relationship between Alzheimer's, Parkinson's, and prion diseases and the dysregulation of calcium homeostasis. As pointed out earlier, the increased intracellular free Ca^{2+} levels can result from the alteration of membrane permeability possibly following the presence of unspecific amyloid pores or other structural modifications provided by the interaction with the aggregates or their misfolded monomers (see Sections 1.12 and 1.13). Membrane lipid peroxidation with production of reactive alkenals and chemical modification of membrane ion pumps could also contribute to the increase of Ca^{2+} levels in cells experiencing toxic aggregates (Varadarajan et al., 2000; Butterfield et al., 2001).

Very recent data have shown that increased ROS and free Ca^{2+} levels are also found in cells exposed to early aggregates of proteins unrelated to disease; even in this case, cell death by apoptosis or, less frequently, by necrosis is the final outcome (Bucciantini et al., 2004, 2005; Sirangelo et al., 2004; Cecchi et al., 2005). It has also been suggested that the choice between apoptosis or necrosis depends on the cell type and can be related with the timing of mitochondria damage (Bucciantini et al., 2005). Finally, the biochemical features involved in the different vulnerability of varying cell types exposed to the same toxic prefibrillar aggregates have been studied. Significant correlations between cell resistance, total antioxidant capacity and Ca^{2+}-ATPase activity were shown in the investigated cell types, further underscoring the importance of early modifications of free Ca^{2+} and redox status in triggering the chain of events culminating with cell death (Cecchi et al., 2005).

Taken together, these findings suggest that, in general, the impairment of cell viability resulting from the presence of toxic aggregates of peptides and proteins not associated with amyloid diseases can be related to the disruption of the same biochemical processes that are affected by similar aggregates of disease-related proteins and peptides. They also support the idea that, at least in most cases, a common mechanism of cytotoxicity is shared among early aggregates of structurally different peptides and proteins. Therefore, amyloid toxicity does not appear to depend on the specific three-dimensional structure or amino acid sequence of the constituting monomers but, rather, on common structural features displayed in their prefibrillar organization (Kayed et al., 2003). Such an idea is strongly supported by the findings indicating that antibodies raised against prefibrillar aggregates of Aβ peptides are able to recognize similar assemblies made by other proteins and peptides, thus reducing their cytotoxicity, as discussed earlier (Section 1.10). Although further studies on a wider range of peptides and proteins are needed to explore the generality of these observations, it can be proposed that the cytotoxicity of these types of aggregates is generic (Bucciantini et al., 2002), arising from the misfolded nature of the aggregated species and their precursors with exposure of regions (e.g., hydrophobic residues and the polypeptide main-chain) normally buried into the compactly folded native state of the protein. Many of these regions are likely to be aggregation-prone (or sticky) and to interact easily with membranes and other cellular components (Bucciantini et al., 2002, 2004) thus triggering the complex cascade of biochemical events eventually leading to cell death.

Recently, a new hypothesis on AD pathogenesis underscores the importance of the association with a pathogen as a trigger for the disease (reviewed in Robinson et al., 2004). Since 1982, it was proposed that reactivated herpesvirus could be involved in AD (Ball, 1982). Indeed, AD has been found in conjunction with herpes infection of the brain (reviewed in Itzhaki et al., 2004). Infectious agents such as *Chlamydia pneumoniae* and spirochaetes have also been proposed to be associated with AD at least in mice (Miklossy, 1993; Little et al., 2004). Finally, an association has been proposed between senile plaques in AD brains and HSV-1 noninfectious virus-derived enveloped tegument structures lacking capsids and viral DNA (L-particles) (Kammerman et al., 2006). The pathogen hypothesis is far from being substantiated by solid experimental data; however, it does not necessarily contradict the well-established amyloid hypothesis. In fact, neuroinflammation has been recognized as a prominent feature in AD associated with aberrant processing and deposition of amyloid (Lue et al., 1996). Therefore, subclinical brain infection might be an additional risk factor for AD following the brain response to the infectious state. For example, it is known that APP is an acute phase reactant and its upregulation in response to inflammation as well as to a number of cellular stresses could increase the amount of Aβ peptides produced during APP processing.

1.12 BIOLOGICAL SURFACES AS PRIMARY SITES OF AMYLOID ASSEMBLY AND TOXICITY

As mentioned earlier (Section 1.2), proteins are synthesized and fold in a very complex environment where they are in close contact with other molecules and with biological surfaces such as membranes and macromolecular assemblies. Therefore, protein interaction with biological surfaces can affect the conformational states of proteins and increase their local concentration; in particular, structural perturbations may be expected to occur in proteins following their interaction with membranes. Indeed, in the adsorbed state, protein residues are involved in interactions with surface-exposed functional groups favoring nonnative structural states. Thus, hydrophobic or charged surfaces may induce local or more extensive unfolding with interaction of normally buried hydrophobic groups with the surface-exposed hydrophobic clusters without paying the energy penalty of exposing to the aqueous environment the same residues (Sethuraman and Belfort, 2005). As discussed earlier, these considerations apply to the behavior of chaperones in assisting protein folding as well as to the trafficking of protein molecules across membranes (Sections 1.2 and 1.3). They also account for the increasing interest in investigating the physicochemical features of protein interaction with natural or artificial surfaces as well as the effect of the latter on protein aggregation. In most cases, the interaction of a misfolded or unfolded species with a membrane is likely to occur via a two-step mechanism involving the electrostatic interaction of the positively charged residues with negatively charged or polar lipid head groups followed by the insertion of hydrophobic regions inside the membrane (Kourie and Henry, 2002).

In general, the hydrophobic interior of the plasma membrane favors structural changes in proteins and peptides leading to their aggregation, as it has been shown for prion protein and Aβ peptides (Yip et al., 2001; Kazlauskaite et al., 2003; Bokvist et al., 2004). It is also believed that surfaces are able to catalyze the formation of amyloid aggregates by a mechanism substantially different from that occurring in bulk solution (Zhu et al., 2002). In fact, in protein molecules adsorbed on surfaces local unfolding populating aggregation-prone conformers can be favored

(Bokvist et al., 2004). The increase of the concentration of misfolded conformers at the bilayer surface is also favored thus accelerating aggregate nucleation, in most cases the rate-limiting step of fibrillogenesis (Zhu et al., 2002; Sethuraman and Belfort, 2005) and enhancong membrane disruption (Yip et al., 2001; Porat et al., 2003).

These considerations, based on studies carried out mainly with synthetic surfaces, have prompted increasing interest on the role performed in protein aggregation by biological surfaces, notably membranes, as well as on the relation to their structure and lipid composition. There is strong evidence that protein polymerization into toxic aggregates is stimulated by membranes or anionic surfaces (Necula et al., 2002; Zhu et al., 2002; Zhao et al., 2004) as well as by the increased content of anionic phospholipids in neuronal membranes of AD patients (see later). These data provide a possible explanation of the variable susceptibility of neuronal populations to the toxic insult given by protein folding variants and their early aggregates. Indeed, prefibrillar assemblies have been shown to interact with synthetic phospholipid bilayers (Arispe et al., 1993; Mirzabekov et al., 1996; Lin et al., 1997; Kourie, 1999; Lin et al., 2001; Volles and Lansbury, 2001) and with cell membranes (Lin et al., 2001; Ding et al., 2002; Hirakura et al., 2002), modifying their structure and permeability and impairing the function of specific membrane-bound proteins and signaling pathways (see Section 1.13). However, not only membrane surfaces can be involved in protein aggregation. For example, it has been proposed that $\beta 2$-microglobulin is able to bind the collagen triple helix, thus providing a possible explanation of the tissue-specificity of dialysis-related amyloidosis (Homma, 1998); more recently it has been suggested that binding affinity fluctuations could influence the concentration in the proximity of collagen fibers of wild-type and N-truncated $\beta 2$-microglobulin and hence their susceptibility to aggregation (Giorgetti et al., 2005).

The importance of the relation between membrane lipid composition and the ability of early aggregates of peptides and proteins to bind to and to disassemble membranes has also been investigated. Several reports highlight the key role of either anionic surfaces and membranes containing anionic phospholipids such as phosphatidylserine (PS) and phosphatidylglycerol (PG) in triggering fibrillization of different proteins and peptides (Necula et al., 2002) and, in the case of cell membranes, as sites of preferential interaction with prefibrillar aggregates. In this regard, it has been proposed that negatively charged membrane surfaces are potent inductors of β-sheet structures by acting as conformational catalysts for amyloids (Jayakumar et al., 2004; Zhao et al., 2004). Recently, it has been reported that PS-containing liposomes are able to induce in vitro amyloid aggregation of a variety of proteins such as lysozyme, glyceraldehyde-3-phosphate dehydrogenase, insulin, myoglobin, transthyretin, cytochrome c, histone H1, α-lactalbumin, and endostatin (Kazlauskaite et al., 2003).

In the past, several candidate cell surface proteins, including RAGE, have been considered as possible receptors of amyloids or their precursors, particularly in the case of Aβ peptides (Yan et al., 1998). However, increasing evidence suggests that additional neuronal binding sites must be involved in the interaction of the plasma membrane with amyloids of different peptides and proteins (Lin and Kagan, 2002; Kazlauskaite et al., 2003; Zhao et al., 2005) and that raising the content of negatively charged lipids results in increased channel formation in synthetic lipid bilayers (Jayakumar et al., 2004). Indeed, anionic membrane phospholipids such as PS and PG have been shown to interact with amyloid aggregates possibly by recognizing a shared fold (Zhao et al., 2004). These data suggest that amyloid aggregates can display different toxicity depending as to whether they are produced into cells or outside cells, considering that PS is almost exclusively found in the cytosolic leaflet of the plasma membrane in all cells other than apoptotic cells, cancer cells and vascular endothelial cells in tumors where it moves to the outer leaflet (Utsugi et al., 1991; Ran and Thorpe, 2002). These ideas are confirmed by recent findings showing that annexin-5 protects against Aβ-peptide cytotoxicity by competing at a common PS-rich site (Lee et al., 2002).

The previously mentioned data could also explain the selective antitumoral specificity of either endostatin (Zhao et al., 2004, 2005) and specific folding variants of some proteins such as α-lactalbumin (see Section 1.10) and apoptin (see Section 1.3). It must also be remembered that amyloid aggregates are found in a wide variety of malignancies (Zhao et al., 2005) suggesting that antitumoral proteins such as endostatin might perform their effects through the formation of toxic aggregates at the surface of cancer cells and vascular endothelial cells in tumors; in such a case, aggregation could be favored by the presence of PS at the outer membrane surface of cancer cells. The recent observation that a number of antimicrobial peptides known to target PG in microbial membranes are also able to kill cancer cells agrees with this idea (Hancock and Scott, 2000). Recent research also reports modifications of the structure of the mitochondrial membranes on interaction with β-amyloids (Aleardi et al., 2005), although there is a substantial lack of information on the effective occurrence and molecular features of the interaction of amyloid aggregates with cell organelles.

The roles of cholesterol and gangliosides in modulating either protein aggregation at the membrane level and aggregate interaction with cell membranes have also been investigated in some detail. A preferential localization of APP and secretases into ganglioside and cholesterol-rich membrane microdomains (lipid rafts) has been reported (Lee et al., 1998; Ehahalt et al., 2003; Kakio et al., 2003) such that aggregation of soluble Aβ peptides has been proposed as a raft-associated process (Ehahalt et al., 2003). It has also been reported that an alteration of cholesterol homeostasis could be a shared primary cause of differing neurodegenerative diseases (Kakio et al., 2003). Aβ aggregates appear to localize at the surface of the plasma membrane and interaction of Aβ with cell surface, particularly with membrane cholesterol, is considered an important requirement for neurotoxicity (Kakio et al., 2002; Koudinov and Koudinova, 2005). On the other hand, reduced levels of cholesterol are present in brains from AD patients (Mason et al., 1992) and a loss of cholesterol in brain leads to neurodegeneration (Arispe and Doh, 2002). Low brain cholesterol could possibly result from downregulation of seladin-1, the neuronal enzyme catalyzing the conversion of 7-dehydrocholesterol to cholesterol (Iivonen et al., 2002). Finally, recent reports indicate that loss of cholesterol in neuronal membranes enhances amyloid peptide generation (Abad-Rodriguez et al., 2005) and that cell interaction with prefibrillar aggregates supplemented to the culture media and aggregate cytotoxicity are reduced on enriching the cell membrane in cholesterol (Cecchi et al., 2005). Although requiring more extensive research, these data support the idea that, in general, a higher membrane rigidity following increased cholesterol content is protective against aggregate interaction with cell membranes and membrane perturbation.

Finally, membranes may also be important in amyloid fibrillogenesis as the primary sources of the aggregating peptide monomers. Membrane environment is of fundamental importance in regulating membrane protein degradation. This is best exemplified by the Aβ peptides resulting from APP processing, the ABri and ADan peptides resulting from BRI precursor protein processing, medin and the gelsolin peptide produced from proteolysis of lactadherin and gelsolin, respectively, as well as other peptides (Section 1.9). The increased presence of APP and secretases in lipid rafts may provide a theoretical basis of the observed increased risk for AD in hypercholesterolemic people, although a mechanistic explanation for such association has not yet been provided and in spite of the previously mentioned data suggesting a protective effect of membrane cholesterol against aggregate cytotoxicity.

Much information is still lacking about a number of key points. For example, further knowledge is required on the relation between chemical composition of membranes and other biological surfaces and the ability of protein or peptide monomers to interact with, misfold on, and polymerize on the latter or the ability of the early aggregates to stick on and to disassemble, destabilize, and permeabilize the membranes. Similarly, more information on the site(s) of a cell where those interactions occur is needed; for example, are there specific membrane receptors for the shared structural features of the prefibrillar aggregates? Are there specific membrane domains where those interactions occur? Is the ER membrane involved in those interactions and which are the biochemical consequences? What is the difference between the inner and outer leaflet of the plasma membrane with regard to those interactions? In which case, are mitochondrial membranes involved and which are the chemical features of those membranes favoring interaction? Nonetheless, in spite of such a substantial lack of information, the idea is steadily gaining support that, at least in most cases biological surfaces, particularly membranes, play a key role in the onset of the chain of events culminating with protein aggregation and cell death.

1.13 ION CHANNEL HYPOTHESIS

In spite of the large number of reports appearing in the last few years, much has still to be learnt on the molecular, biochemical, and biological basis of the effects of the amyloid aggregates on living systems (see Sections 1.11 and 1.12). Much interest has recently been raised by the possibility that a subpopulation of protofibrils, also referred to as amyloid pores (see Section 1.8) may account for the toxicity of the amyloid aggregates, thus envisaging a basically common early biochemical mechanism of aggregate toxicity through the permeabilization of cell membranes (Kourie and Henry, 2002). Indeed, since 1993, a ''channel hypothesis'' of amyloid toxicity was proposed, suggesting that the toxic aggregated species form nonspecific pore-like channels in the membranes of the exposed cells (Arispe et al., 1993) (Figure 1.4).

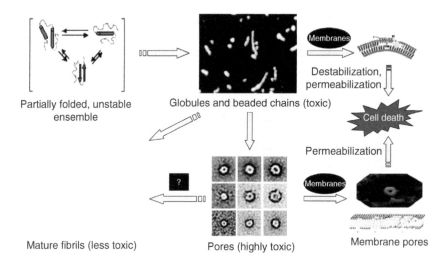

FIGURE 1.4 Prefibrillar amyloid aggregate cytotoxicity involves aggregate interaction with membranes, membrane disassembly, and permeabilization. (Electron microscopy (EM) images are from Harper, J.D., Wong, S.S., Lieber, C.M., and Lansbury, P.T., *Biochemistry*, 38, 8972, 1999. With permission; From Lashuel, H.A., Hartley, D., Petre, B.M., Walz, T., and Lansbury, P.T., *Nature*, 418, 291, 2002b; the AFM image is from Ding, T.T., Lee, S.-J., Rochet, J.-C., and Lansbury, P.T., *Biochemistry*, 41, 10209, 2002. With permission.) The question mark between annular assemblies and mature fibrils indicates that it is not ascertained whether doughnut-shaped assemblies arise in the path of fibril formation.

Early aggregates featuring small annular rings with a central pore have been imaged by either AM and AFM for a large number of proteins and peptides in their fibrillization path. These include Aβ peptides, α-synuclein, ABri peptide, transthyretin, β2-microglobulin, and others (Hirakura and Kagan, 2001; Quintas et al., 2001; Lashuel et al., 2002a, 2002b; Wang et al., 2002; Chung et al., 2003; Srinivasan et al., 2004; Vendrely et al., 2005; ions.s, M.D., Ryu, C. ciated with disease (Reloini et al.). Prefibrillar, doughnut-shaped assemblies of acute-phase isoforms of human and murine serum amyloid A (SAAp and SAA2.2, respectively) have also been shown to permeabilize eukaryotic and bacterial cells (Hirakura et al., 2002) (see also Section 1.16). This behavior is reminiscent of the action of several bacterial pore-forming toxins such as perfringolysin (Hotze et al., 2002) (see Section 1.3), although eukaryotic counterparts of this mechanism have also been described. For example, perforin, the C5b-8 or 9 complement assembly in the membrane attack complex and the BCL-2 family of proapoptotic proteins act by forming aspecific channels in the plasma or mitochondrial membranes of the target cells (Morgan, 1999; Fraser et al., 2000; Reed, 2000; Hotze et al., 2002). Taken together, these similarities suggest the evolution of a general death mechanism performed by protein oligomers, not necessarily of amyloid type, which act as biological bullets forming nonspecific membrane pores resulting in unbalance of the ion content and the redox status in the target cells.

Although the previously mentioned doughnut-shaped prefibrillar aggregates have been imaged during the fibrillization of many peptides or proteins, it is not clear whether they are directly responsible for ion homeostasis impairment in the exposed cells. Indeed, in most cases modifications of membrane permeability have been determined in cells exposed to misfolded monomers or their prefibrillar aggregates with no pore-like appearance. In general, it is believed that the interaction of the aggregates or their precursors with the cell membranes is, by itself, able to destabilize the phospholipid bilayer creating disordered areas allowing reduction or destruction of the ion gradients (Kayed et al., 2003, 2004; Demuro et al., 2005).

A large number of experimental data obtained using either biological membranes or artificial phospholipid bilayers have shown that prefibrillar aggregates of proteins and peptides associated with disease are intrinsically able to permeabilize these membranous systems; and ion currents have also been measured with differing biophysical features depending on the type of membrane or aggregate used (reviewed in Kourie, 2001; Kourie and Henry, 2002). In addition, it has been shown that such a permeabilizing effect is shared with early aggregates of proteins not associated with the disease (Bucciantini et al., 2004; Relini et al., 2004). Taken together, the large wealth of data obtained in recent years depicts a scenario whereby the toxic effects of amyloid aggregates may generally be traced back to a few key features involving the interaction with biological surfaces of misfolded proteins and their early aggregates (see Section 1.12). These features include (possibly among others) the effect of surfaces, notably biological membranes as triggers of amyloid aggregation (see Section 1.12); the interaction of early aggregates or their precursors with membranes (possibly with specific membrane domains); the subsequent membrane permeabilization or the modification of their structural or biochemical features; and the increase of ROS and free Ca^{2+} contents in cells exposed to such toxic species.

1.14 TWO FACES OF PROTEIN BIOLOGY

The potential cytotoxicity of many aggregated proteins or peptides suggests that, in addition to providing cells with mechanisms to clear unfolded and misfolded proteins and to minimize their ability to induce toxicity, biological evolution must also have eliminated protein sequences with a high intrinsic propensity to aggregate (Dobson, 2001). Thus, many potential mutations that are neutral with respect to protein function could have been selected against owing to their enhancement of the tendency of proteins to aggregate under physiological conditions. These considerations are made even more critical in view of the fact that protein folding and protein aggregation are in close competition with each other, being favored basically by the same physicochemical features (see Section 1.9). Actually, proofs of such a pressure and of the constraints operating in the evolution of natural proteins have been highlighted by several studies (see later).

In general, the evolution of the highly cooperative nature of native protein structures appears, by itself, to be a critical step in allowing proteins to avoid aggregation for significant lengths of time (Dobson, 1999). Furthermore, it has been shown that protein evolution disfavors the presence, in natural proteins, of long stretches of either alternating polar–nonpolar residues or hydrophobic residues, both favoring aggregation before a protein is completely folded (Broome and Hecht, 2000; Schwarz et al., 2001). Very recently, a database survey has also shown that the overall tendency to β-aggregation decreases significantly with increasing organism complexity and longevity (Tartaglia et al., 2005a). Finally, it has been highlighted that natural β-sheet proteins carry structural adaptations aimed at hindering β-sheet stacking into an amyloid-like structural organization by avoiding intermolecular association of the edge β-strands in their native states (Richardson and Richardson, 2002). These surveys in the protein databases show clearly that protein evolution must have faced previously unsuspected constraints to avoid the evolution of biologically active but aggregation-prone molecules easily converting into misfolded forms rapidly nucleating cytotoxic oligomers.

These considerations, together with the inherent aggregation potential of most peptides and proteins and the generic toxicity of their aggregates shed a new light on the protein world. Indeed, protein molecules besides being the fundamental molecular tools of the living world, also possess a dark side that can be traced back to their ability to transform into rogues able to kill cells under proper conditions (usually not easily found in biological systems) (reviewed in Stefani, 2004) (Figure 1.5). This new way to consider polypeptide chains has important consequences for an in-depth understanding of the fundamentals of the origin of protein deposition diseases as well as very significant practical outcomes. For example, protein engineers must take into account this potential behavior when designing new or modified peptides and proteins. Moreover, the pharmacological strategies aimed at treating the various amyloidoses must reconsider their targets to develop molecules aimed at reducing

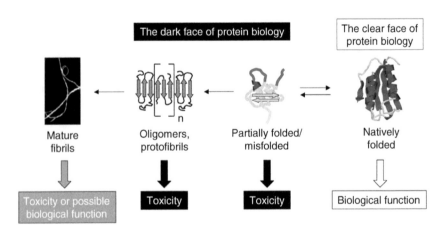

FIGURE 1.5 The dark side of proteins. Under suitable conditions, polypeptide chains can misfold or unfold thus manifesting their inherent tendency to organize into toxic prefibrillar amyloid assemblies eventually leading to the appearance of less toxic mature fibrils. Accordingly, proteins or peptides may be considered potential toxins to cells. In some cases, nature has exploited such a tendency by using amyloid fibrils made from particular polypeptides to perform specific biological functions.

the toxicity of prefibrillar aggregates or their formation, and hence the buildup of misfolded proteins, rather than at hindering the growth of mature fibrils (less toxic).

The toxicity of prefibrillar aggregates discussed earlier further highlights the significance, in biological evolution, of the development of the complex machinery responsible for both the UPR in the ER and the heat shock response in the cytosol as well as the ubiquitin-proteasome pathway of protein degradation. The key biological importance of these cellular pathways is not limited to favoring the presence, inside or outside the cell, of proteins with full biological activity; rather, these molecular machineries are undoubtedly aimed at preventing the intracellular accumulation of potentially toxic materials such as misfolded proteins or their early aggregates by counteracting the inherent tendency of proteins to misfold and aggregate (Muchowski et al., 2000; Bence et al., 2001; Morgan et al., 2001; Tofaris et al., 2001; Waelter et al., 2001; Auluck et al., 2002; Bonini, 2002; David et al., 2002; Fonte et al., 2002; Ma et al., 2002; McNaught et al., 2002; Sakahira et al., 2002; Trinh et al., 2002; Zoghbi and Botas, 2002). Should the molecular chaperone binding capacity be overwhelmed, this "chaperone overload" would allow the accumulation in cells of unchaperoned misfolded proteins enhancing their aggregation. This process would further impair proteasome activity favoring the appearance of intracellular inclusion bodies (see Section 1.6).

The key role of molecular chaperones against protein aggregation is further stressed by recent data describing a number of chaperonepathies, rare inherited diseases involving modifications of the structure or the expression of genes encoding molecular chaperones (Macario and de Macario, 2002). Abnormal levels of molecular chaperones have recently been found in the brains of AD patients (Yoo et al., 2001) suggesting the opportunity to search for the presence of abnormalities in the regulation of the corresponding genes in protein deposition disease (Macario and de Macario, 2002). Finally, the information on the exact role of the ubiquitin-proteasome system in intracellular proteolysis has grown rapidly in the last few years, providing data that point to this pathway as a key tool for cellular homeostasis regulation. In effect, many of the major human degenerative diseases can collectively be described as "ubiquitin protein catabolic disorders," where anomalous protein turnover resulting from defects in the ubiquitin-proteasome system can cause or contribute to the progression of these diseases (Chappie et al., 2001; Layfield et al., 2001 and references therein).

1.15 DEGENERATIVE CONDITIONS WITH PROTEIN AGGREGATES MAY BE MUCH MORE COMMON THAN PREVIOUSLY BELIEVED

As pointed out previously (Section 1.7) our current knowledge supports the idea that most proteins and peptides under suitable medium conditions are able to misfold or unfold thus populating conformers with high propensity to aggregate. It also appears that even single point mutations may destabilize the native state of a protein so as to significantly increase the population of partially unfolded molecules at equilibrium (Section 1.7). In this view, it is well known that a higher number of single nucleotide polymorphisms than previously suspected are present in the exons of the human genome (The international SNP map working group 2001). The human genome also encodes a large number of NUPs that, inspite of their structural adaptations, appear endowed with higher aggregation potential with respect to the compactly folded proteins. Finally, it is apparent that the increasing prevalence of protein deposition diseases in humans is primarily associated with the higher life expectancy, particularly in the developed countries. Indeed, proteins are not necessarily optimized to maintain their correctly folded states under progressively declining environmental conditions and impaired biological safety mechanisms as happens during aging. Therefore, it is likely that we are starting to experience the work of protein evolution in optimizing the resistance of proteins against aggregation to the biological lifetime by which during evolution humans have passed their genes to the offspring.

On the basis of these considerations and of an increasing number of studies on molecular medicine, one can expect that not only the incidence of well-known amyloidoses, but also the number of degenerative diseases one can trace back to the appearance of amyloid aggregates in the affected tissues will rise in the near future (reviewed in Stefani, 2004). Actually, an increasing number of reports suggest that the molecular basis of degenerative diseases of previously unknown origin relies on cell damage by aggregates of known proteins (or of proteins that have not yet been identified) previously unsuspected to be deposited in amyloid assemblies (Table 1.3). For instance, the presence of protein deposits of amyloid type in varying organs and tissues such the basement membrane and lamina propria of the gastro-intestinal tract, in the walls of the gastrointestinal blood vessels and the liver (Demirhan et al., 2002). Lactoferrin amyloid has been found in varying tissues including seminal vesicles, a condition common in elderly males with uncertain clinical significance (Tsutsumi et al., 1996). Deposits of amyloid type in the pituitary of elderly people have been known for several years (Westermark et al., 1997) together with the well known localized amyloid of human endocrine organs arising from aggregation of hormones such as amylin, atrial natriuretic factor, and calcitonin (Johansson and Westermark, 1990; Khurana et al., 2004). Finally, as pointed out earlier (Section 1.9), a localized deposition of medin fibrils in the aortic media in elderly people (Häggqvist et al., 1999) is considered to be involved in the age-related loss of elasticity of the vessels.

Recently, many other degenerative diseases have been proposed as amyloidoses since they are characterized by the presence, in the affected tissues, of amyloid deposits of proteins and peptides previously unknown to aggregate (reviewed in Stefani, 2004). Most often, these are rare inherited diseases following the presence of mutant proteins, although some, such as many cancers and atherosclerosis are much more important and widespread. A detailed survey of all of these diseases is outside the scope of this chapter; however, a glimpse of at least some of these pathologies can be useful to understand the increasing importance of amyloidoses in medicine.

Many degenerative conditions recently reconsidered as amyloid diseases affect eye tissues. These include several forms of retinitis pigmentosa, retinal and corneal dystrophies, as well as inherited and sporadic cataract, pseudoexfoliation syndrome, and heredo-oto-ophthalmo-encephalopathy. These conditions have been shown to be associated with the deposition of amyloid aggregates of proteins as diverse as rhodopsin (Kopito, 2000; Saliba et al., 2002), γ-crystallins, βig-h3 and others not

TABLE 1.3
New Degenerative Diseases with Protein Aggregates

Disease	Aggregated Protein	Tissue or Cell
Cronic lung diseases	Surfactant protein C	Lung
Retinal dystrophies	Rhodopsin	Retina
Macular dystrophy	Peripherin or rds	Retina
Cataract (sporadic or congenital)	γ-crystallins	Lens
Inherited corneal dystrophies	βig-h3	Cornea
Familial subepithelial corneal amyloidosis	Lactoferrin	Cornea
Pseudoexfoliation syndrome	?	Aqueous humor
Heredo-oto-ophthalmo encephalopathy	?	Retinal vessels
Sporadic inclusion-body myositis	APP, Abeta	Skeletal muscle
Hypotrichosis simplex of the scalp	Corneodesmosin	Hair follicle
Atherosclerosis	LDL or ApoB100	Vessel walls
Desmin-related cardiomyopathy	α-B-crystallin	Myocardium
FENIB	Neuroserpin	Cerebral cortex, NS
DFNA9	Cochlin	Inner ear
Hirschprung disease	RET	Enteric nervous system
Cutaneous lichen amyloidosis	RET	Papillary dermis
Charcot-Marie-Tooth-like diseases	Myelin protein 22/0	Peripheral nerve tissue
Short-chain acylCoA DH deficiency	SCAD	Skeletal muscle, brain
Cancers	p53	Tumoral tissues
Huntington disease-like 2	Junctophilin-3	CNS
Focal segmental glomerulosclerosis	α−actinin-4	Kidney
Infantile spastic syndrome or MR	ARX	Brain
Oculopharyngeal muscular dystrophy	PAPB2 or PABPN1	Skeletal muscle

yet identified. Recent findings have shown the presence of electron-dense aggregates of rhodopsin carrying destabilizing mutations both in humans and in animal models of autosomal dominant retinitis pigmentosa, suggesting a molecular mechanism for disease dominance (Saliba et al., 2002). The key role in the molecular pathogenesis of a number of retinal degenerative conditions played by specialized opsin chaperones has also recently emerged (reviewed in Chappie et al., 2001). In humans, various forms of cataract, including juvenile-onset cataract (Pande et al., 2001; Kosinski-Collins and King, 2003), have recently been associated with misfolding and aggregation of specific lens proteins including γ-crystallins (Santhiya et al., 2002). Recently, a role for α-crystallin, a protein with chaperone properties, in preventing the formation of cytoplasmic aggregates of mutant γB-crystallin and other mutated γ-crystallins has been proposed (Sandilands et al., 2002). Several corneal dystrophies for which an amyloid basis has been documented include a group of hereditary conditions caused by specific missense mutations in the *BIGH3* gene (Munier et al., 1997; Clou and Hohenester, 2003) and familial subepithelial corneal amyloidoses, where the cornea contains amyloid deposits of lactoferrin or its fragments (Kintworth et al., 1997; Nilsson and Dobson, 2003). Finally, amyloid aggregates of a yet unidentified protein have been found in the retina of patients with heredo-oto-ophthalmo encephalopathy (Bek, 2000) and in the aqueous humor of patients with pseudoexfoliation syndrome; in the latter, focal fibrillogranular aggregates are scattered throughout the body (Berlau et al., 2001).

It is well known that amyloid fibril formation by lung surfactant protein C (SP-C) is the molecular basis of lung diseases such as pulmonary alveolar proteinosis (Gustaffson et al., 1999). Recently, mutations in the gene encoding SP-C have been shown to be associated with chronic lung disease in children and adults; in the latter, proprotein misfolding and toxicity to epithelial cells are induced, with increased BiP transcription, proprotein trapping in the ER, and its rapid degradation via the ubiquitin-proteasome pathway (Bridges et al., 2003).

Several degenerative conditions affecting the skeletal muscle and the myocardium have recently been reported to involve protein aggregates. These include desmin-related cardiomyopathy and, possibly, other hypertrophic cardiomyopathies, where aggregates containing a mutant form of the small heat shock protein α-B-crystallin have been found (Sanbe et al., 2004). The autosomal dominant form of oculopharyngeal muscular dystrophy is associated with a short (GLG)8–13 expansion in the *PABP2* gene encoding the poly(A) binding protein 2 (PAPB2) (Görlach et al., 1994); this condition is accompained by the appearance of filamentous inclusions in the nucleus of the affected cells (Calado et al., 2000). Sporadic inclusion-body myositis is the most common muscle degenerative disease in aged people (reviewed in Askanas and Engel, 1998). The molecular basis of the disease remains substantially unknown; however, recently a number of reports have shown the presence, within the muscle fibers of affected people, of several abnormally accumulated proteins including the amyloid β-precursor protein and its proteolytic amyloidogenic fragments (Askanas and Engel, 2003). Pathological protein aggregates of amyloid type have also been found in the epithelial tissues of diseased people. Recent findings highlight the possible amyloid basis of hypotrichosis simplex of the scalp (HSS), an autosomal dominant form of isolated alopecia caused by mutations in the *CDSN* gene encoding corneodesmosin (Jonca et al., 2002; Levy-Nissenbaum et al., 2003).

Recently, atherosclerosis has also been proposed as a new protein misfolding disease. The conformational features of apo-B100 in low-density lipoproteins (LDL) are affected by an alteration of the water–lipid interface following chemical modifications such as lipid oxidation possibly leading to protein misfolding as it has been observed in a fraction of oxidatively modified LDL (LDL$^-$) isolated in vivo (reviewed in Lusis, 2000; Ursini et al., 2002). The latter displays apo-B100 misfolding with almost complete loss of α-structure and increased β-structure leading to apo-B100 intracellular accumulation in high molecular weight cytotoxic insoluble aggregates, resistant to proteolysis, which inhibit the ubiquitin-proteasome pathway (reviewed in Ursini et al., 2002).

Many cancers display abnormal accumulation of aggregates of wild-type and mutant p53, a protein characterized by high conformational flexibility resulting from the presence of partially folded and natively unfolded segments (Bell et al., 2002; Dawson et al., 2003). Mutant p53 is found in over 50% of all human cancers (Bullock and Fersht, 2001; Ridiger et al., 2002); however, an inactive form, possibly a conformational variant, of wild-type p53 can also be present in tumors such as neuroblastoma, retinoblastoma, breast cancer, and colon cancer (Moll et al., 1996; Ostermeyer et al., 1996), where p53 nuclear and cytoplasmic aggregates have been described (Bullock and Fersht, 2001; Ridiger et al., 2002). This folding variant displays dominant negative effects since it appears able to drive the active, wild-type protein into an inactive mutant conformation (Ishimaru et al., 2003a). Although the nature of the intracellular aggregates of p53 remains unclear, it has recently been reported that, under mild denaturing conditions, two separated p53 domains can be induced to take a mutant-like conformation in vitro leading them to assemble into β-sheet rich, toxic fibrillar aggregates (Ishimaru et al., 2003b; Lee et al., 2003). It has therefore been suggested that p53 may be involved in tumoral genesis by either a loss of its antitumor function or a gain of function leading it to assemble into aggregates that are able to recruit the functional molecules still present in the cell (Milner and Medcalf, 1991). Presently, a direct association between p53 amyloid and cancer remains speculative. However, the data recently reported have led to hypothesize that tumors could arise in the elderly even in the absence of mutations of p53; this could simply follow a loss of functionality of the ubiquitin-proteasome system allowing p53 folding variants to appear and to nucleate aggregates able to recruit other wild-type molecules similarly to the effect of some tumor-promoting p53 mutations.

During the last decade, retention in the ER of proteins targeted to the plasma membrane carrying missense mutations has emerged as a generic molecular mechanism for several genetic diseases. Hirschprung disease is characterized by abnormal development of the enteric nervous system caused by mutations in RET, a protein with a relatively high propensity to misfold (Carlomagno et al., 1996), affecting its folding and processing in the ER (Cosma et al., 1998). It is not clear whether Hirschprung disease may be considered a true amyloidosis with amyloid aggregates of the mutant RET. However, a recent report describes a cutaneous lichen amyloidosis with amyloid deposition in the papillary dermis in a few families harboring RET proto-oncogene mutations (Verga et al., 2003) suggesting the possibility that mutant RET may indeed be deposited in amyloid assemblies.

The presence of intracellular aggregates as aggresomes of the peripheral myelin protein-22 and myelin protein zero and of short-chain acyl-CoA dehydrogenase have been reported in familial diseases such as Charcot-Marie-Tooth-related disorders (Shames et al., 2003) and short-chain acyl-CoA dehydrogenase deficiency (Pedersen et al., 2003), respectively. In such diseases the involved proteins, are overexpressed or carry destabilizing mutations leading to folding defects or misfolding, although it has not been established whether these aggregates are true amyloids.

1.16 AMYLOIDS MAY PERFORM BIOLOGICAL FUNCTIONS

The possibility that amyloid fibrils may perform, in some cases, specialized biological functions (Table 1.4) was first clearly demonstrated in 2002, when it was shown that the β-sheet rich highly aggregated fibers known as curli produced by *Escherichia coli* cells were indeed amyloid fibrils morphologically identical to those described in amyloid diseases (Chapman et al., 2002). Curli fibrils are deposited at the bacterial surface, where they promote cell adhesion to the substrate, thus favoring the colonization of inert surfaces, and mediate the binding to various host proteins

TABLE 1.4
Physiological Functions Performed by Amyloids

Protein or Peptide	Function
Microcyn E492	Noncytotoxic phenotype
Curli or Tafi	Cell adhesion to substrates
Chaplins or Hydrophobins	Aerial hyphae or fungal coat
Yeast prions (Ure2p, Sup35, Het)	Phenotype changes
Pmel17	Melanin polymerization
Aβ peptides	Small cerebral vessel sealing
Crystallins	Refractive index of the lens
Serum amyloid A	
Bacterial toxins	
Complement MAC	Killing target cells by membrane pores (amyloid?)
Bcl-2 protein family	
Perforin	

(Vidal et al., 1998; Chapman et al., 2002). Curli arise from a protein, curlin, by a nucleation–precipitation process involving the participation of chaperone-like and nucleator proteins encoded by the *csgAB* and *csgDEFG* operons whose proposed function is to prevent curlin polymerization within the cell while accelerating it at the cell surface (Prigent-Combaret et al., 2000). A function similar to that exploited by Curli amyloid is performed in *Salmonella* spp by Tafi amyloid whose formation involves AfsA a protein homologous to CsgA (reviewed in Gebbink et al., 2005).

The Curli story raises the possibility that the proteinaceous filaments found at the cell surfaces of other microorganisms may be amyloid in nature (reviewed in Gebbink et al., 2005). Indeed, recently it has been reported that the formation of aerial hyphae in *Streptomyces coelicolor* requires the assembly into amyloid-like fibrils with possible amyloid features of a chaplin, a protein belonging to a class of secreted hydrophobic highly surface-active proteins (Claessen et al., 2003). The amyloid-like fibrils of these proteins lower the water surface tension of the aqueous environment and provide the aerial structures with a hydrophobic coat enabling hyphae to grow into the air. It has been proposed that in other fungal microorganisms hyphae may also be coated with amyloid-like fibrils of proteins with the same function as chaplins. For example, the hydrophobins of filamentous fungi are a group of proteins functionally similar to chaplins. Hydrophobins are assembled into a mosaic of amyloid-like fibrils giving rise to highly insoluble films that help hyphae to breach the water–air interface by the same mechanism as chaplins (Butko et al., 2001).

Recent work carried out on prion-like proteins in microorganisms has led to the proposal that prions play a much broader role in biology than previously suspected (reviewed in Krishnan and Lindquist, 2005). Among these prions, Ure2p and Sup35 in *Saccharomyces cerevisiae*, HET-s in the filamentous fungus *Podospora anserina* and CPEB in the marine snail *Aplysia* have been investigated in depth. Under physiological conditions, these prions may undergo self-perpetuating structural changes with the appearance of novel physiological functions. For example, in the presence of nitrogen-rich sources, Ure2p undergoes self-perpetuating structural modification associated with phenotypic changes affecting the nitrogen catabolism (the Ure3 phenotype). Such a phenotypic change is accomplished by downregulating the expression of genes whose products are involved in the use of nitrogen sources (Blinder et al., 1996). The self-perpetuating structure of Sup35 is associated with a variety of phenotypic changes (the Psi^+ phenotype) associated with the ability to read-through nonsense codons (Wickner, 1996; Lindquist, 1997; Edskes et al., 1999). When aggregating into amyloid fibrils, HET-s performs some function associated with heterokaryon incompatibility and hence in self recognition (Dos Reis et al., 2002). Finally, a neuronal form of CEPB, implicated in long-term memory in the marine snail *Aplysia*, can switch into a self-perpetuating prion conformation that activates the protein, suggesting that, in this organism, the prion conformation is involved in long-term maintaining of synapses (reviewed in Krishnan and Lindquist, 2005).

As mentioned earlier, some bacterial pore-forming toxins kill the target cells by assembling into pore-like oligomers that penetrate the plasma membrane permeabilizing the cell that eventually dies. In some cases, protein insertion into the plasma membrane of the target cell requires a conformational rearrangement leading to changes in the secondary structure elements (see Section 1.3). In this respect, another recently reported example of a possible functional role of the aggregation of a bacterial protein involves microcin E492 (Mcc) from *Klebsiella pneumoniae*. Mcc is a naturally produced low molecular weight bacteriocin that normally kills bacteria by forming pores in their cytoplasmic membranes (Destoumieux-Garzon et al., 2003). However, the antibacterial activity of Mcc suddenly drops during the stationary phase of cell growth although the levels of Mcc are not changed. The subsequent demonstration that Mcc is able to aggregate in vivo into amyloid-like fibrils devoid of toxicity has lead to propose that the sudden disappearance of Mcc activity can result from a change of the conformational or the oligomerization state of the protein (Bieler et al., 2005). The data on Mcc show that amyloid formation in vivo can provide phenotypic changes even in prokaryotes.

Examples of possible physiological functions of amyloids have also been reported in higher organisms such as insects and mammals. A peptide from the A- and B-family of the silkmoth chorion proteins has been found to account for 30% of the proteinaceous material found as amyloid aggregates in the eggshell. This finding suggests that the silkmoth chorion is a natural amyloid with protective properties necessary for survival and development of the oocyte and embryo (Iconomidou et al., 2000a, 2000b). As reported earlier (Section 1.13), it has recently been shown that acute-phase isoforms of human and murine serum amyloid A (SAAp and SAA2.2, respectively) are present in solution as hexameric assemblies permeabilizing synthetic and cell membranes (Hirakura et al., 2002). The onset of an acute inflammation results in a dramatic increase of SAA concentration in the plasma that becomes persistent during chronic inflammation (Gillmore et al., 2001). Such an increase is believed to contribute to the development of reactive amyloidosis where amyloid deposits of the whole protein or its 1–76 *N*-terminal fragment (AA) are found in the kidney, liver, and spleen (Gillmore et al., 2001). It has been proposed that the toxic effects of the SAA and AA aggregates are a drawback of the proposed physiological function of the hexameric assemblies to protect against infection by forming toxic channels in the cell membranes of the invading bacteria (Hirakura et al., 2002).

A large body of literature supports the idea that Aβ aggregates are key elements in the process of sealing capillaries and arterioles or in maintaining regional integrity following traumatic insults (reviewed in Craig et al., 2003). As mentioned earlier (Section 1.11), APP is an acute phase reactant upregulated in response to inflammation as well as to a number of cellular stresses including oxidative stress, calcium dysregulation and energy shortage; the latter also appears to favor the production of amyloido-genic APP derivatives by modifying APP processing (Mattson et al., 1998). Indeed, the rapid deposition of Aβ following severe head trauma suggests the possible involvement of these aggregates in maintaining vascular integrity thus protecting brain parenchyma against hemorrhage. These data question the opportunity to remove Aβ peptides from the brain as it happens in the immunologic approach against AD (Craig et al., 2003).

Some reports which appeared in recent years have shown that the major proteins in normal mammalian lenses exist predominantly as β-pleated sheets in vivo; more recently, it has been reported that in the normal lens a positivity to the classical amyloidophilic stains such as Congo red and thioflavine T does exist. This finding supports the idea that protein β-sheet arrays are physiologically organized in an amyloid-like supramolecular order in the interior fiber cells of the mammalian lens (Frederikse, 2000). In such a case, the inherent high stability of amyloids could have been exploited by evolution to ensure the long-term structural integrity and transparency of the lens.

The importance of the amyloid deposition of Pmel17 fragments in melanosome maturation and melanine biogenesis has been discussed earlier (Section 1.9). These data have led to the proposal that similar mechanisms could be present in other tissue-specific organelles, thus providing insight on the molecular mechanisms governing their biogenesis as well as the formation of their unique structural features (Berson et al., 2003).

Finally, as mentioned earlier (Section 1.13), a number of mammalian proteins comprising perforin and the membrane attack complex made by the C5b-8 or 9 of the activated complement produce cytotoxic ring-shaped oligomeric assemblies. The formation of these complexes requires partial unfolding of the involved proteins with exposure of hydrophobic surfaces leading to their deep insertion into the lipid bilayer of the membranes of the target cells where they form small leaky pores. However, it is not known whether the resulting inter-subunit association requires structural modifications similar to those found in proteins assembled into amyloid aggregates.

1.17   FINAL CONSIDERATIONS

The increasing knowledge of the mechanistic and physicochemical features of protein folding has provided a theoretical basis to explain protein misfolding and aggregation allowing a better understanding of the molecular basis of protein deposition diseases. The data reported in the last few years strongly suggest that the conversion of normally soluble proteins into amyloid fibrils is a generic feature basically associated with the covalent backbone of all polypeptide chains. The molecular basis of the toxicity of small aggregates appearing during the early stages of fibrillization also appears to display shared characteristics in the different systems that have been studied so far. Indeed, the topics discussed in this chapter provide evidence supporting the idea that an intrinsic potential to misfold and to form structures such as amyloid fibrils is inherent to proteins and, more generally, to polypeptide chains. Recent research has also shown that nature has exploited such a potential to perform specific physiological functions in selected biological systems.

The possibility of many, perhaps all, polypeptides to misfold and polymerize into toxic aggregates suggests that one of the most important driving forces in protein evolution must have been the negative selection against the appearance of amino acid sequences endowed with significant tendency to aggregate. Recent surveys in protein structure databases support such an idea. Nevertheless, as a stochastic process, protein folding can fail sometimes and misfolded proteins or protein folding intermediates in equilibrium with the natively folded molecules are likely to arise continuously within cells. Thus, mechanisms to deal with such species must have coevolved with proteins. Recent work in this field has established that molecular chaperones act by keeping under control protein misfolding and the associated tendency to aggregate so as to make harmless the resulting species by either assisting their refolding or triggering their degradation by the cellular clearance machineries (Muchowski et al., 2000; Bence et al., 2001; Chappie et al., 2001; Dul et al., 2001; Waelter et al., 2001; Auluck et al., 2002; Bonini, 2002; Fonte et al., 2002; McNaught et al., 2002; Zoghbi and Botas, 2002). Therefore, molecular chaperones could be even more important than was thought in the past, being aimed not only at improving the efficiency of protein folding (Hartl and Hayer-Hartl, 2002) but also at controlling the harmful effects of misfolded or aggregated proteins, as suggested by their high levels (about 1% of the total protein content) in normally functioning cells (reviewed in Ellis et al., 1989). Therefore, it is very likely that a "chaperone overload" is associated with failure to control the accumulation of toxic forms of proteins and hence with cell damage (Nardai et al., 2002).

Such considerations have recently led to hypothesize that molecular chaperones can act as capacitors of morphological evolution for their ability to conceal by their binding, and hence to render harmless, mutations that could otherwise damage the cell. Evidence for this idea has come from studies carried out in *Drosophila* (Rutherford and Lindquist, 1998; Roberts and Fader, 1999), yeast (True and Lindquist, 2000), and *Arabidopsis* (Queiltsch et al., 2002) and this function could be important even in humans. Such a function appears of considerable value in reducing the constraints on protein evolution driven by random mutations imposed by the intrinsic potential of many of them to enhance the tendency to fibrillize of the affected proteins. The probability of both the phenotypic manifestation of such mutations and aggregate nucleation should increase during aging (Ross, 2002), when modifications of the intracellular milieu are also likely to occur. Under these conditions, the impairment of chaperone induction, the reduction of the efficiency of either the proteasome and ubiquitinating or deubiquitinating enzymes, and over-whelming of chaperone function would lead to a sudden buildup of protein aggregates within the cell. In turn, such a situation would cause further problems by impairing the function of the ubiquitin-proteasome system in a type of positively reinforcing loop. The latter could explain the dramatic loss of neuronal function that frequently characterizes the progression of many neurodegenerative diseases.

Extension of these ideas has led to the hypothesis that a larger number of degenerative and other diseases than are presently known could result from significant impairment of cell viability following continuous exposure to even minute amounts of protein aggregates (Bucciantini et al., 2002). In such a case, the latter could result from specific mutations or from general effects of aging including those mentioned earlier and from processes such as chemical modification. Indeed, the number of diseases shown to be associated with misfolding and aggregation is increasing steadily (Section 1.15). It is generally accepted that the rising prevalence of amyloid diseases is associated with recent increases in human life expectancy, particularly in highly developed countries. However, it is also interesting to note that several outbreaks of transmissible amyloid diseases associated with prions, notably the bovine spongiform encephalopathy (BSE) have occurred recently as a result of unnatural practices such as the feeding of young cows with the remains of old ones. Such a situation is analogous to the outbreak of kuru associated with ritual cannibalism in Papua New Guinea in the 1950s and of Creutzfeldt–Jakob disease in individuals treated in the past with contaminated growth hormone purified from human cadavers. These cases reflect the violation of practices to which evolutionary pressure has applied (Dobson, 2002). Even modern medical treatment may induce amyloid diseases as a drawback; this is the case of insulin amyloidosis affecting diabetic people at the site of injection (Brange et al., 1997) or, more seriously, of β2-microglobulin amyloidosis associated with prolonged hemodialysis. Taken together, these considerations support the view of most sporadic amyloid conditions as civilization or postevolutionary diseases as they have become prevalent as a result of our recent abilities to prolong our lifespans or of the introduction of new medical and agricultural practices (Csermely, 2001; Dobson, 2002; Dodart et al., 2003).

REFERENCES

Abad-Rodriguez, J., Ledesma, M.D., Craessaerts, K., Perga, S., Medina, M., Delacourte, A. et al. (2005) Neuronal membrane cholesterol loss enhances amyloid peptide generation. *Journal of Cell Science*, 167, 953–960.

Aleardi, A.M., Bernard, G., Augereau, O., Malgat, M., Talbot, J.C., Mazat, J.P. et al. (2005) Gradual alteration of mitochondrial structure and function by beta-amyloids: importance of membrane viscosity changes, energy deprivation, reactive oxygen species production, and cytochrome c release. *Journal of Bioenergetics and Biomembranes*, 37, 07–225.

Ambroggio, E.E., Kim, D.H., Separovic, F., Barrow, C.J., Bamham, K.J., Bagatolli, L.A., and Fidelio, G.D. (2005) Surface behaviour and lipid interaction of Alzheimer β-amyloid peptide 1–42: a membrane-disrupting peptide. *Biophysics Journal*, 88, 2706–2713.

Anfinsen, C.B. (1973) Principles that govern the folding of protein chains. *Science*, 181, 223–230.

Arispe, N. and Doh, M. (2002) Plasma membrane cholesterol controls the cytotoxicity of Alzheimer's disease Aβ(1–40) and (1–42) peptides. *FASEB Journal*, 16, 1526–1536.

Arispe, N., Rojas, E., and Pollard, H.D. (1993) Alzheimer's disease amyloid beta protein forms calcium channels in bilayer membranes: blockade by tromethamine and aluminium. *Proceedings of the National Academy of Sciences of the United States of America*, 89, 10940–10944.

Armen, R.S., DeMarco, M.L., Alonso, D.O., and Daggett, V. (2004) Pauling and Corey's α-pleated sheet structure may define the prefibrillar amyloidogenic intermediate in amyloid disease. *Proceedings of the National Academy of Sciences of the United States of America*, 101, 11622–11627.

Askanas, V. and Engel, W.K. (1998) Sporadic inclusion-body myositis and its similarities to Alzheimer's disease brain. Recent approaches to diagnosis and pathogenesis and relation to aging. *Scandinavian Journal of Rheumatology*, 27, 389–405.

Askanas, V. and Engel, W.K. (2003) Proposed pathogenetic cascade of inclusion-body myositis: importance of amyloid-beta, misfolded proteins, predisposing genes, and ageing. *Current Opinion in Rheumatology*, 15, 737–744.

Auluck, P.K., Chan, H.Y.E., Trojanowski, J.Q., Lee, V.M.-Y., and Bonini, N.M. (2002) Chaperone suppression of α-synuclein toxicity in a *Drosophila* model for Parkinson's disease. *Science*, 295, 865–868.

Austen, B., Christodoulou, G., and Terry, J.E. (2002) Relation between cholesterol levels, statins and Alzheimer's disease in the human population. *Journal of Nutrition, Health and Aging*, 6, 377–382.

Baker, D. (2000) A surprising simplicity for protein folding. *Nature*, 405, 39–42.

Ball, M.J. (1982) Limbic predilection in Alzheimer's dementia: is reactivated herpesvirus involved? *Canadian Journal of Neurology*, 9, 303–306.

Bek, T. (2000) Ocular changes in heredo-oto-ophthalmo-encephalopathy. *British Journal of Ophthalmology*, 84, 1298–1302.

Bell, S., Klein, C., Müller, L., Hansen, S., and Buchner, J. (2002) p53 contains large unstructured regions in first native state. *Journal of Molecular Biology*, 322, 917–927.

Bence, N.F., Sampat, R.M., and Kopito, R. (2001) Impairment of the ubiquitin-proteasome system by protein aggregation. *Science*, 292, 1552–1555.

Berlau, J., Lorenz, P., Beck, R., Makavitzky, J., Schlotzer-Schrehardt, U., Thiesen, H.J., and Guthoff, R. (2001) Analysis of aqueous humor proteins of eyes with and without pseudoexfoliation syndrome. *Graefes Archive for Clinical and Experimental Ophthalmology*, 239, 743–746.

Berson, J.F., Harper, D.C., Tenza, D., Raposo, G., and Marks, M.S. (2001) Pmel17 initiates promelanosome morphogenesis within multivesicular bodies. *Molecular Biology of the Cell*, 12, 3451–3464.

Berson, J.F., Theos, A.C., Harper, D.C., Tenza, D., Raposo, G., and Marks, M.S. (2003) Preprotein convertase cleavage liberates a fibrillogenic fragment of a resident glycoprotein to initiate melanosomes biogenesis. *Journal of Cell Biology*, 16, 521–533.

Bhatia, R., Lin, H., and Lal, R. (2000) Fresh and nonfibrillar amyloid β protein(1–42) induces rapid cellular degeneration in aged human fibroblasts: evidence for AβP-channel-mediated cellular toxicity. *FASEB Journal*, 14, 1233–1243.

Bieler, S., Estrada, L., Lagos, R., Baeza, M., Castilla, J., and Soto, C. (2005) Amyloid formation modulates the biological activity of a bacterial protein. *Journal of Biological Chemistry*, 280, 26880–26885.

Blinder, D., Coschigano, P.W., and Magasanik, B. (1996) Interaction of the GATA factor Gln3p within the nitrogen regulator Ure2p in Saccharomyces cerevisiae. *Journal of Bacteriology*, 178, 4734–4736.

Bokvist, M., Lindström, F., Watts, A., and Gröbner, G. (2004) Two types of Alzheimer's β-amyloid (1–40) peptide membrane interactions: aggregation preventing transmembrane anchoring versus accelerated surface fibril formation. *Journal of Molecular Biology*, 335, 1039–1049.

Bonini, N.M. (2002) Chaperoning brain degeneration. *Proceedings of the National Academy of Sciences of the United States of America*, 99 Suppl. 4, 16407–16411.

Brange, J., Andersen, L., Laursen, E.D., Meyn, G., and Rasmussen, E. (1997) Toward understanding insulin fibrillation. *Journal of Pharmaceutical Science*, 96, 517–525.

Bridges, J.P., Wert, S.E., Nogee, L.M., and Weaver, T.E. (2003) Expression of a human SP-C mutation associated with interstitial lung disease disrupts lung development in transgenic mice. *Journal of Biological Chemistry*, 278, 52739–52746.

Broome, B.M. and Hecht, M.H. (2000) Nature disfavours sequences of alternating polar and non-polar amino acids: implications for amyloidogenesis. *Journal of Molecular Biology*, 296, 961–968.

Brunelle, P. and Rauk, A. (2002) The radical model of Alzheimer's disease: specific recognition of Gly29 and Gly33 by Met35 in a beta-sheet model of Abeta: an ONIOM study. *Journal of Alzheimers Disease*, 4, 283–289.

Bucciantini, M., Giannoni, E., Chiti, F., Baroni, F., Formigli, L., Zurdo, J. et al. (2002) Inherent toxicity of aggregates implies a common mechanism for protein misfolding diseases. *Nature*, 416, 507–511.

Bucciantini, M., Calloni, G., Chiti, F., Formigli, L., Nosi, D., Dobson, C.M., and Stefani, M. (2004) Pre-fibrillar amyloid protein aggregates share common features of cytotoxicity. *Journal of Biological Chemistry*, 279, 31374–31382.

Bucciantini, M., Rigacci, S., Berti, A., Pieri, L., Cecchi, C., Nosi, D. et al. (2005) Patterns of cell death triggered in two different cell lines by HypF-N pre-fibrillar aggregates *FASEB Journal*, 19, 437–439.

Bukau, B. and Horwich, A.L. (1998) The Hsp70 and Hsp60 chaperone machines. *Cell*, 92, 351–366.

Bullock, A.N. and Fersht, A.R. (2001) Rescuing the function of mutant p53. *Nature Reviews Cancer*, 1, 68–76.

Butko, P., Buford, J.P., Goodwin, J.S., Stroud, P.A., McCormick, C.L., and Cannon, G.C. (2001) Spectroscopic evidence for amyloid-like interfacial self-assembly of hydrophobin Sc3. *Biochimica et Biophysica Research Communications*, 280, 212–215.

Butterfield, A.D., Drake, J., Pocernich, C., and Castegna, A. (2001) Evidence of oxidative damage in Alzeimer's disease brain: central role for amyloid β-peptide. *Trends in Molecular Medicine*, 7, 548–554.

Bychkova, V.E., Pain, R.H., and Ptitsyn, O.B. (1988) The molten globule state is involved in the translocation of protein across membranes. *FEBS Letters*, 238, 231–234.

Calado, A., Tomé, F.M.S., Brais, B., Rouleau, G.A., Kühn, U., Wahle, E., and Carmo-Fonseca, M. (2000) Nuclear inclusions in oculopharyngeal muscular dystrophy consist of poly(A) binding protein 2 aggregates which sequester poly(A) RNA. *Human Molecular Genetics*, 9, 2321–2328.

Calloni, G., Zuffoli, S., Stefani, M., Dobson, C.M., and Chiti, F. (2005) Investigating the effects of mutations on protein aggregation in the cell. *Journal of Biological Chemistry*, 280, 10607–10612.

Carlomagno, F., De Vita, G., Berlingieri, M.T., de Franciscis, V., Melillo, R.M., Colantuoni, V. et al. (1996) Molecular heterogeneity of RET loss of function in Hirschsprung's disease. *EMBO Journal*, 15, 2717–2725.

Cecchi, C., Baglioni, S., Fiorillo, C., Pensalfini, A., Liguri, G., Nosi, D. et al. (2005) Insights into the molecular basis of the differing susceptibility of varying cell types to the toxicity of amyloid aggregates. *Journal of Cell Science*, 118, 3459–3470.

Chamberlain, A.K., MacPhee, C.E., Zurdo, J., Morozova-Roche, L.A., Hill, H.A., Dobson, C.M., and Davis, J.J. (2000) Ultrastructural organization of amyloid fibrils by atomic force microscopy. *Biophysics Journal*, 79, 3282–3293.

Chapman, M.R., Robinson, L.S., Pinkner, J.S., Roth, R., Heuser, J., Hammar, M. et al. (2002) Role of *Escherichia coli* curli operons in directing amyloid fiber formation. *Science*, 295, 851–855.

Chappie, J.P., Grayson, C., Hardcastle, A.J., Saliba, R.S., van der Spuy, J., and Cheetham, M.E. (2001) Unfolding retinal dystrophies: a role for molecular chaperones? *Trends in Molecular Medicine*, 7, 414–421.

Chen, Y. and Dokhalyan, N.V. (2005) A single disulfide bond differentiates aggregation pathways of β2-microglobulin. *Journal of Molecular Biology*, 354, 473–482.

Cheung, M.S., Garcia, A.E., and Onuchic, J.N. (2002) Protein folding mediated by solvation: water expulsion and formation of hydrophobic core occur after the structural collapse. *Proceedings of the National Academy of Sciences of the United States of America*, 99, 685–690.

Chiti, F., Webster, P., Taddei, N., Clark, A., Stefani, M., Ramponi, G., and Dobson, C.M. (1999) Designing conditions for in vitro formation of amyloid protofilaments and fibrils. *Proceedings of the National Academy of Sciences of the United States of America*, 96, 3590–3594.

Chiti, F., Bucciantini, M., Capanni, C., Taddei, N., Dobson, C.M., and Stefani, M. (2001) Solution conditions can promote formation of either amyloid protofilaments or mature fibrils from the HypF *N*-terminal domain. *Protein Science*, 10, 2541–2547.

Chiti, F., Stefani, M., Taddei, N., Ramponi, G., and Dobson, C.M. (2003) Rationalization of the effects of mutations on peptide and protein aggregation rates. *Nature*, 424, 805–808.

Choi, Y.G., Kim, J.L., Lee, H.P., Jin, J.K., Choi, E.K., Carp. R.I., and Kim, Y.S. (2000) Induction of heme oxygenase-1 in the brain of scrapie-infected mice. *Neuroscience Letters*, 11, 173–176.

Chung, J., Yang, H., de Beus, M.D., Ryu, C.Y., Cho, K., and Colòn, W. (2003) Cu/Zn superoxide dismutase can form pore-like structures. *Biochemical and Biophysical Research Communications*, 312, 873–876.

Claessen, D., Rink, R., de Jong, W., Siebring, J., de Vreungd, P., Hiddee Borsma, F.G. et al. (2003) A novel class of secreted hydrophobic proteins is involved in aerial hyphae formation in *Streptomyces coelicolor* by forming amyloid-like fibrils. *Genes and Development*, 17, 1714–1726.

Clark, E.D.B. (1998) Refolding of recombinants proteins. *Current Opinion in Biotechnology*, 9, 157–163.

Clark, L.A. (2005) Protein aggregation determinants from a simplified model: cooperative folders resist aggregation. *Protein Science*, 14, 653–662.

Clarke, G., Collins, R.A., Leavitt, B.R., Andrews, D.F., Hayden, M.R., Lumsden, C.J., and McInnes, R.R. (2000) A one-hit model of cell death in inherited neuronal degeneration. *Nature*, 406, 195–199.

Cleary, J.P., Walsh, D.M., Hofmeister, J.J., Shankar, G.M., Kuskowski, M.A., Selkoe, D.J., and Ashe, K.H. (2005) Natural oligomers of the amyloid-β protein specifically disrupt cognitive function. *Nature Neuroscience*, 8, 79–84.

Clou, N.J. and Hohenester, E. (2003) A model of FAS1 domain 4 of the corneal protein βig-h3 gives a clearer view on corneal dystrophies. *Molecular Vision*, 9, 440–448.

Colon, W. and Kelly, J.W. (1992) Partial denaturation of transthyretin is sufficient for amyloid fibril formation *in vitro*. *Biochemistry*, 31, 8654–8660.

Conlon, I.J., Dunn, G.A., Mudge, A.W., and Raff, M.C. (2001) Extracellular control of cell size. *Nature Cellular Biology*, 3, 918–921.

Conway, K.A., Lee, S.-J., Rochet, J.C., Ding, T.T., Williamson, R.E., and Lansbury, P.T. (2000) Acceleration of oligomerization not fibrillization is a shared property of both alpha-synuclein mutations linked to early-onset Parkinson's disease. Implication for pathogenesis and therapy. *Proceedings of the National Academy of Sciences of the United States of America*, 97, 571–576.

Cosma, M.P., Cardone, M., Carlomagno, F., and Colantuoni, V. (1998) Mutations in the extracellular domain cause RET loss of function by a dominant negative mechanism. *Molecular Cell Biology*, 18, 3321–3329.

Craig, S., Bowen, R.L., Smith, M.A., and Perry, G. (2003) Cerebrovascular requirement for sealant, anti-coagulant and remodelling molecules that allow for the maintenance of vascular integrity and blood supply. *Brain Research Reviews*, 41, 164–178.

Csermely, P. (2001) Chaperone overload is a possible contributor to "civilization diseases." *Trends in Genetics*, 17, 701–704.

Dafforn, T.R., Mahadeva, R., Elliott, P.R., Sivasothy, P., and Lomas, D.A. (1999) A kinetic mechanism for the polymerization of alpha1-antitrypsin. *Journal of Biological Chemistry*, 274, 9548–9555.

David, D.C., Layfield, R., Serpell, L., Narain, Y., Groedert, M., and Spillantini, M.G. (2002) Proteasomal degradation of tau protein. *Journal of Neurochemistry*, 83, 176–185.

Davis, R.L., Shrimpton, A.E., Holohan, P.D., Bradshaw, C., Feiglin, D., Collins, G.H. et al. (1999) Familial dementia caused by polymerization of mutant neuroserpin. *Nature*, 401, 376–379.

Davis, R.L., Shrimpton, A.E., Carrell, R.W., Lomas, D.A., Gerhard, L., Baumann, B. et al. (2002) Association between conformational mutations in neuroserpin and onset and severity of dementia. *Lancet*, 359, 2242–2247.

Dawson, R., Müller, L., Dehner, A., Klein, C., Kessler, H., and Buchner, J. (2003) The *N*-terminal domain of p53 is natively unfolded. *Journal of Molecular Biology*, 332, 1131–1141.

Demirhan, B., Bilezikci, B., Kiyici, H., and Boyacioglu, S. (2002) Globular amyloid deposits in the wall of the gastrointestinal tract: report of six cases. *Amyloid*, 9, 42–46.

Demuro, A., Mina, E., Kayed, R., Milton, S.C., Parker, I., and Glabe, C.G. (2005) Calcium dysregulation and membrane disruption as a ubiquitous neurotoxic mechanism of soluble amyloid oligomers. *Journal of Biological Chemistry*, 280, 17294–17300.

Destoumieux-Garzon, D., Thomas, X., Santamaria, M., Goulard, C., Barthelemy, M., Boscher, B. et al. (2003) Microcin E492 antibacterial activity: evidence for a TonB-dependent inner membrane permeabilization on *Escherichia coli*. *Molecular Microbiology*, 49, 1031–1041.

Diaz-Nido, J., Wandosell, F., and Avila, J. (2002) Glycosaminoglycans and beta-amyloid, prion and tau peptides in neurodegenerative diseases. *Peptides*, 23, 1323–1332.

Dickson, D.W., (1995) Correlation of synaptic and pathological markers with cognition of the elderly. *Neurobiology of Aging*, 16, 285–298.

Dill, K.A. and Chan, H.S. (1997) From Levinthal paradox to funnels. *Nature Structural Biology*, 4, 10–19.

Ding, T.T., Lee, S.-J., Rochet, J.-C., and Lansbury, P.T. (2002) Annular α-synuclein protofibrils are produced when spherical protofibrils are incubated in solution or bound to brain-derived membranes. *Biochemistry*, 41, 10209–10217.

Dinner, A.R., Sali, A., Smith, L.J., Dobson, C.M., and Karplus, M. (2000) Understanding protein folding via free energy surfaces from theory and experiment. *Trends in Biochemical Science*, 25, 331–339.

Dobson, C.M. (1999) Protein misfolding, evolution and disease. *Trends in Biochemical Sciences*, 24, 329–332.

Dobson, C.M. (2001) The structural basis of protein folding and its links with human disease. *Philosophical Transactions of the Royal Society B—Biological Sciences*, 356, 133–145.

Dobson, C.M. (2002) Getting out of shape. *Nature*, 418, 729–730.

Dobson, C.M. (2003a) Protein folding and misfolding. *Nature*, 426, 884–890.

Dobson, C.M. (2003b) Protein folding and disease: a view from the first Horizon Symposium. *Nature Reviews Drug Discovery*, 2, 154–160.

Dobson, C.M. (2004) Experimental investigation of protein folding and misfolding. *Methods*, 34, 4–14.

Dobson, C.M., Sali, A., and Kopito, R. (1998) Protein folding: a perspective from theory and experiments. *Angewandte Chemie International Edition*, 37, 868–893.

Dodart, J.C., Bales, K.R., Bales, K.R., and Paul, S.M. (2003) Immunotherapy for Alzheimer's disease: will vaccination work? *Trends in Molecular Medicine*, 9, 85–87.

Dos Reis, S., Coulary-Salin, B., Forgel, V., Lascu, I., Bégueret, J., and Saupe, S.J. (2002) The HET-s prion protein of the filamentous fungus *Podospora anserina* aggregates *in vitro* into amyloid-like fibrils. *Journal of Biological Chemistry*, 277, 5703–5706.

Dul, J.L., Davis, D.P., Williamson, E.K., Satevens, F.J., and Argon, Y, (2001) Hsp70 and antifibrillogenic peptides promote degradation and inhibit intracellular aggregation of amyloidogenic light chains. *Journal of Cell Biology*, 19, 705–715.

Eaton, W.A., Munoz, V., Thompson, P.A., Henry, E.R., and Hofrichter, J. (1998) Kinetics and dynamics of loops, α-helices, β-hairpins and fast-folding proteins. *Accounts of Chemical Research*, 31, 745–753.

Edskes, H.K., Gray, V.T., and Wickner, R.B. (1999) The (URE3) prion is an aggregated form of Ure2p that can be cured by overexpression of Ure2p fragments. *Proceedings of the National Academy of Sciences of the United States of America*, 96, 1498–1503.

Ehahalt, R., Keller, P., Haass, C., Thiele, C., and Simons, K. (2003) Amyloidogenic processing of the Alzheimer beta-amyloid precursor protein depends on lipid rafts. *Journal of Cell Biology*, 160, 113–123.

Ellis, R.J. (2001) Macromolecular crowding: an important but neglected aspect of the intracellular environment. *Current Opinion in Structural Biology*, 11, 114–119.

Ellis, R.J., van der Vies, S.M. and Hemmingsen, S.M. (1989) The molecular chaperone concept. *Biochemistry Society Symposium*, 55, 145–153.

Esteras-Chopo, A., Serrano, L., and Lòpez de la Paz, M. (2005) The amyloid stretch hypothesis: recruiting proteins toward the dark side. *Proceedings of the National Academy of Sciences of the United States of America*, 102, 16672–16677.

Fändrich, M. and Dobson, C.M. (2002) The behaviour of polyamino acids reveals an inverse side chain effect in amyloid structure formation. *EMBO Journal*, 21, 5682–5690.

Fersht, A.R. (2000) Transition-state structure as a unifying basis in protein folding mechanisms: contact order, chain topology, stability and the extended nucleus mechanism. *Proceedings of the National Academy of Sciences of the United States of America*, 97, 1525–1529.

Fonte, V., Kapulkin, V., Taft, A., Fluet, A., Friedman, D., and Link, C.D. (2002) Interaction of intracellular β amyloid peptide with chaperone proteins. *Proceedings of the National Academy of Sciences of the United States of America*, 99, 9439–9444.

Fraser, S.A., Karimi, R., Michalak, M., and Hudig, D. (2000) Perforin lytic activity is controlled by calreticulin. *Journal of Immunology*, 164, 4150–4155.

Frederikse, P.H. (2000) Amyloid-like protein structure in mammalian ocular lenses. *Current Eye Research*, 20, 462–468.

Gebbink, M.F.B.G., Claessen, D., Bouma, B., Dijkhuizen, L., and Wösten, H.A.B. (2005) Amyloids—A functional coat for microorganisms. *Nature Reviews Microbiology*, 3, 333–341.

Giasson, B.I., Forman, M.S., Higuchi, M., Golbe, L.I., Graves, C.L., Kotzbauer, P.T., Trojanowski, J.Q., and Lee, V.M.-Y. (2003) Initiation and synergistic fibrillization of tau and alpha-synuclein. *Science*, 300, 636–640.

Gibson, G., Gunasekera, N., Lee, M., Lelyveld, V., El-Agnaf, O.M.A., Wright, A., and Auste, B. (2004) Oligomerization and neurotoxicity of the amyloid ADan peptide implicated in familial Danish dementia. *Journal of Neurochemistry*, 88, 281–290.

Gillmore, J.D., Lovat, L.B., Persey, M.R., Pepys, M.B., and Hawkins, P.N. (2001) Amyloid load and clinical outcome in AA amyloidoses in relation to circulating concentration of serum amyloid A protein. *Lancet*, 358, 24–29.

Giorgetti, S., Rossi, A., Mangione, P., Raimondi, S., Marini, S., Stoppini, M. et al. (2005) Beta2-microglobulin isoforms display an heterogeneous affinity for type I collagen. *Protein Science*, 14, 696–702.

Glenner, G.G., Ein, D., Eanes, E.D., Bladen, H.A., Terry, W., and Page, D.L. (1971) Creation of amyloid fibrils from Bence Jones proteins *in vitro*. *Science*, 174, 712–714.

Goldberg, A.L. (2003) Protein degradation and protection against misfolded or damaged proteins. *Nature*, 426, 895–899.

Görlach, M., Burd, C.G., and Dreyfuss, G. (1994) The mRNA poly(A)-binding protein: localization, abundance and RNA-binding specificity. *Experimental Cell Research*, 211, 400–407.

Guentchev, M., Voigtlander, T., Haberler, C., Groschup, M.H., and Budka, H. (2000) Evidence for oxidation stress in experimental prion disease. *Neurobiology of Disease*, 7, 270–273.

Gujiarro, J.I., Sunde, M., Jones, J.A., Campbell, I.D., and Dobson, C.M. (1998) Amyloid fibril formation by an SH3 domain. *Proceedings of the National Academy of Sciences of the United States of America*, 95, 4224–4228.

Gustaffson, M., Thyberg, J., Näslund, J., Eliasson, E., and Johansson, J. (1999) Amyloid fibril formation by pulmonary surfactant protein C. *FEBS Letters*, 464, 136–142.

Häggqvist, B., Näslund, J., Sletten, K., Westernmarks, G.T., Mucchiano, G., Tjernberg, L.O. et al. (1999) Medin: an integral fragment of aortic smooth muscle cell-produced lactadherin forms the most common human amyloid. *Proceedings of the National Academy of Sciences of the United States of America*, 96, 8669–8674.

Hancock, R.E.W. and Scott, M.G. (2000) The role of antimicrobial peptides in animal defences. *Proceedings of the National Academy of Sciences of the United States of America*, 97, 8856–8861.

Harper, J.D., Wong, S.S., Lieber, C.M., and Lansbury, P.T. (1999) Assembly of Abeta amyloid protofibrils: an in vitro model for a possible early event in Alzheimer's disease. *Biochemistry*, 38, 8972–8980.

Hartl, F.U. and Hayer-Hartl, M. (2002) Molecular chaperones in the cytosol: from nascent chain to folded protein. *Science*, 295, 1852–1858.

Hartley, D., Walsh, D.M., Ye, C.P., Diehl, T., Vasquez, S., and Vassilev, P.M. (1999) Protofibrillar intermediates of amyloid β-protein induce acute electrophysiological changes and progressive neurotoxicity in cortical neurons. *Journal of Neuroscience*, 19, 8876–8884.

Hirakura, Y. and Kagan, B.L. (2001) Pore formation by beta-2-microglobulin: a mechanism for the pathogenesis of dialysis-associated amyloidosis. *Amyloid*, 8, 94–100.

Hirakura, Y., Carreras, I., Sipe, J.D., and Kagan, B.L. (2002) Channel formation by serum amyloid A: a potential mechanism for amyloid pathogenesis and host defense. *Amyloid*, 9, 13–23.

Homma, N. (1998) Collagen-binding affinity of B2-microglobulin, a preprotein of hemodialysis-associated amyloidosis. *Nephron*, 53, 37–40.

Hotze, E.M., Heuck, A.P., Czajkowsky, M., Shao, Z., Johnson, A.E., and Tweten, R. (2002) Monomer–monomer interactions drive the prepore to pore conversion of a β-barrel-forming cholesterol-dependent cytolysin. *Journal of Biological Chemistry*, 277, 11597–115605.

Huff, M.E., Balch, W.E., and Kelly, J.W. (2003) Pathological and functional amyloid formation orchestrated by the secretory pathway. *Current Opinion in Structural Biology*, 13, 674–682.

Huntington, J.A., Pannu, N.S., Hazes, B., Read, R.J., Lomas, D.A., and Carrell, R.W. (1999) A 2.6? structure of a serpin polymer and implications for conformational disease. *Journal of Molecular Biology*, 293, 449–455.

Hyun, D.-H., Lee, M.H., Hattori, N., Kubo, S.-I., Mizuno, Y., Halliwell, B., and Jenner, P. (2002) Effect of wild-type or mutant parkin on oxidative damage, nitric oxide, antioxidant defenses, and the proteasome. *Journal of Biological Chemistry*, 277, 28572–28577.

Iconomidou, V.A., Vriend, G., and Hamodrakas, S.J. (2000a) Amyloids protect the silkmoth oocyte embryo. *FEBS Letters*, 479, 141–145.

Iconomidou, V.A., Cryssikos, G.D., Gionis, V., Vriend, G., Hoenger, A., and Hamodrakas, S.J. (2000b) Amyloid-like fibrils form a 18-residue analogue of a part of the central domain of the B-family of silkmoth chorion proteins. *FEBS Letters*, 499, 268–273.

Iivonen, S., Hiltunen, M., Alafuzoff, L., Mannermaa, A., Kerokoski, P., Puolivati, J. et al. (2002) Seladin-1 transcription is linked to neuronal degeneration in Alzheimer's disease. *Neuroscience*, 113, 301–310.

Ishimaru, D., Maia, L.F., Maiolino, L.M., Quesado, P.A., Lopez, P.C.M., Almeida, F.C.L. et al. (2003a) Conversion of wild-type p53 core domain into a conformation that mimics a hot-spot mutant. *Journal of Molecular Biology*, 333, 443–451.

Ishimaru, D., Andrade, L.R., Teixeira, L.S.P., Quesado, P.A., Maiolino, L.M., Lopez, P.M. et al. (2003b) Fibrillar aggregates of the tumor suppressor p53 core domain. *Biochemistry*, 42, 9022–9027.

Itzhaki, R.F., Wozniak, M.A., Appelt, D.M., and Balin, B.J. (2004) Infiltration of the brain by pathogen causes Alzheimer's disease. *Neurobiology of Aging*, 25, 619–627.

Jayakumar, R., Jayaraman, M., Koteeswarl, D., and Gomath, K. (2004) Cytotoxic and membrane perturbation effects of a novel amyloid forming model, peptide poly (leucine-glutamic acid). *Journal of Biochemistry (Tokyo)*, 136, 457–462.

Jiménez, J.L., Guijarro, J.I., Orlova, E., Zurdo, J., Dobson, C.M., Sunde, M., and Saibil, H.R. (1999) Cryo-electron microscopy structure of an SH3 amyloid fibril and model of the molecular packing. *EMBO Journal*, 18, 815–821.

Jiménez, J.L., Nettleton, E.J., Bouchard, M., Robinson, C.V., Dobson, C.M., and Saibil, H.R. (2002) The protofilament structure of insulin amyloid fibrils. *Proceedings of the National Academy of Sciences of the United States of America*, 99, 9196–9201.

Johansson, B. and Westermark, P. (1990) The relation of atrial natriuretic factor to isolated atrial amyloid. *Experimental and Molecular Patholgy*, 52, 266–678.

Jonca, N., Guerrin, M., Hadjiolova, K., Caubet, C., Gallinaro, H., Simon, M., and Serre, G. (2002) Corneodesmosin, a component of epidermal corneocyte desmosomes, displays homophilic adhesive properties. *Journal of Biological Chemistry*, 277, 5024–5029.

Kakio, A., Nishimoto, S.-L., Yanagisawa, K., Kozutsumi, Y., and Matsuzaki, K. (2002) interaction of amyloid β-protein with various gangliosides in the raft-like membranes: importance of GM1 ganglioside-bound form as an endogenous seed for Alzheimer amyloid. *Biochemistry*, 41, 7385–7390.

Kakio, A., Nishimoto, S.-L., Kozutsumi, Y., and Matsuzaki, K. (2003) Formation of membrane-active form of amyloid β-protein in raft-like model membranes. *Biochimica et Biophysica Research Communications*, 303, 514–518.

Kammerman, E.M., Neumann, D.M., Ball, M.J., Lukiw, W., and Hill, J.M. (2006) Senile plaques in Alzheimer's diseased brains: possible association of β-amyloid with herpes simplex virus type 1 (HSV-1) L-particles. *Medical Hypotheses*, 66, 294–299.

Katsuno, M., Adachi, H., Kume, A., Li, M., Nakagomi, Y., Niwa, H. et al. (2002) Testosterone reduction prevents phenotypic expression in a transgenic mouse model of spinal and bulbar muscular atrophy. *Neuron*, 35, 843–854.

Kaufman, R.J. (2002) Orchestrating the unfolded protein response in health and disease. *Journal of Clinical Investigations*, 110, 1389–1398.

Kawahara, M., Kuroda, Y., Arispe, N., and Rojas, E. (2000) Alzheimer's β-amyloid, human islet amylin, and prion protein fragment evoke intracellular free calcium elevation by a common mechanism in a hypopthalamic GnRH neuronal cell line. *Journal of Biological Chemistry*, 275, 14077–14083.

Kayed, R., Head, E., Thompson, J.L., McIntire, T.M., Milton, S.C., Cotman, C.W., and Glabe, C.G. (2003) Common structure of soluble amyloid oligomers implies common mechanisms of pathogenesis. *Science*, 300, 486–489.

Kayed, R., Sokolow, Y., Edmonds, B., McIntire, T.M., Milton, S.C., Hall, J.E., and Glabe, C.G. (2004) Permeabilization of lipid bilayers is a common conformation-dependent activity of soluble amyloid oligomers in protein misfolding diseases. *Journal of Biological Chemistry*, 279, 46363–46366.

Kazlauskaite, J., Senghera, N., Sylvester, I., Vénien-Bryan, C., and Pinheiro, T.J.T. (2003) Structural changes of the prion protein in lipid membranes leading to aggregation and fibrillization. *Biochemistry*, 42, 3295–3304.

Keller, J.N., Huang, F.F., and Markesbery, W.R. (2002) Decreased levels of proteasome activity and proteasome expression in aging spinal cord. *Neuroscience*, 98, 149–156.

Kelly, J. (1998) Alternative conformation of amyloidogenic proteins and their multi-step assembly pathways. *Current Opinion in Structural Biology*, 8, 101–106.

Kelly, J.W. (2000) Mechanisms of amyloidogenesis. *Nature Structural Biology*, 7, 824–826.

Khan, F., Chuang, J.L., Gianni, S. and Fesht, A.R. (2003) The kinetic pathway of folding of barnase. *Journal of Molecular Biology*, 333, 169–186.

Khurana, R., Agarwal, A., Bajpai, V.K., Verma, N., Sharma, A.K., Gupta, R.P., and Madhusudan, K.P. (2004) Unraveling the amyloid associated with human medullary thyroid carcinoma. *Endocrinology*, 145, 5465–5470.

Kintworth, G.K., Valnickova, Z., Kielar, R.A., Baratz, K.H., Campbell, R.J., and Enghild, J.J. (1997) Familial subepithelial corneal amyloidoses—α lactoferrin-related amyloidoses. *Investigative Ophthalmology and Vision Science*, 38, 2756–2763.

Kopito, R.R. (2000) Aggresomes, inclusion bodies and protein aggregation. *Trends in Cell Biology*, 10, 524–530.

Kopito, R. and Sitia, R. (2000) Aggresomes and Russell bodies. *EMBO Reports*, 1, 225–231.

Kosinski-Collins, M. and King, J. (2003) *In vitro* unfolding, refolding, and polymerization of human γD crystallin, a protein involved in cataract formation. *Protein Science*, 12, 480–490.

Koudinov, A.L. and Koudinova, N.V. (2005) Cholesterol homeostasis failure as a unifying cause of synaptic degeneration. *Journal of Neurological Science*, 229–230, 233–240.

Kourie, J.I. (1999) Synthetic C-type mammalian natriuretic peptide forms large cation selective channels. *FEBS Letters*, 445, 57–62.

Kourie, J.I. (2001) Mechanisms of amyloid β protein-induced modification in ion transport systems: implications for neurodegenerative diseases. *Cellular and Molecular Neurobiology*, 21, 173–213.

Kourie, J.I. and Henry, C.L. (2002) Ion channel formation and membrane-linked pathologies of misfolded hydrophobic proteins: the role of dangerous unchaperoned molecules. *Clinical and Experimental Pharmacology and Physiology*, 29, 741–753.

Kourie, J.I. and Shorthouse, A.A. (2000) Properties of cytotoxic peptide-induced ion channels. *American Journal of Physiology Cell Physiology*, 278, C1063–C1087.

Kranenburg, O., Kroon-Batenburg, L.M.J., Reijerkerk, A., Wu, Y.-P., Voest, E.E., and Gebbink, M.F.B.G. (2003) Recombinant endostatin forms amyloid fibrils that bind and are cytotoxic to murine neuroblastoma cells *in vitro*. *FEBS Letters*, 539, 149–155.

Krishnan, R. and Lindquist, S.L. (2005) Structural insights into a yeast prion illuminate nucleation and strain diversity. *Nature*, 435, 765–772.

Lambert, M.P., Barlow, A.K., Chromy, B.A., Edwards, C., Freed, R., Liosatos, M. et al. (1998) Diffusible nonfibrillar ligands derived from Aβ-42 are potent central nervous system neurotoxins. *Proceedings of the National Academy of Sciences of the United States of America*, 95, 6448–6453.

Lane, N.J., Balbo, A., Fukuyama, R., Rapoport, S.I., and Galdzick, Z. (1998) The ultrastructural effects of beta-amyloid peptide on cultured PC12 cells: changes in cytoplasmic and intramembranous features. *Journal of Neurocytology*, 27, 707–718.

Lashuel, H.A., Petre, B.M., Wall, J., Simon, M., Nowak, R.J., Walz, T., and Lansbury, P.T. (2002a) α-Synuclein, especially the Parkinson's disease-associated mutants, forms pore-like annular and tubular protofibrils. *Journal of Molecular Biology*, 322, 1089–1102.

Lashuel, H.A., Hartley, D., Petre, B.M., Walz, T., and Lansbury, P.T. (2002b) Neurodegenerative disease: amyloid pores from pathogenic mutations. *Nature*, 418, 291.

Layfield, R., Alban, A., Mayer, R.J., and Lowe, J. (2001) The ubiquitin protein catabolic disorders. *Neuropathology and Applied Neurobiology*, 27, 171–179.

Lee, A.S., Galea, C., DiGiammarino, E.L., Jun, B., Murti, G., Ribeiro, R.C. et al. (2003) Reversible amyloid formation by the p53 tetramerization domain and a cancer-associated mutant. *Journal of Molecular Biology*, 327, 699–709.

Lee, D.W., Sohn, H.O., Lim, H.B., Lee, Y.G., Kim, Y.S., Carp, R.J., and Wisnievski, H.M. (1999) Alteration of free radical metabolism in the brain of mice infected with scrapie agent. *Free Radical Research*, 30, 499–507.

Lee, G., Pollard, H.B., and Arispe, N. (2002) Annexin 5 and apolipoprotein E2 protect against Alzheimer's amyloid-β-peptide cytotoxicity by competitive inhibition at a common phosphatidylserine interaction site. *Peptides*, 23, 1249–1263.

Lee, S.J., Liyanage, U., Bickel, P.E., Xia, W., Lansbury, P.T., and Kosik, K.S. (1998) A detergent-insoluble membrane compartment contains Aβ in vivo. *Nature Medicine*, 4, 730–734.

Levinthal, C. (1968) Are there pathways for protein folding? *Journal of Chemical Physics*, 85, 44–45.

Levy-Nissenbaum, E., Betzm, R.C., Frydmanm, M., Simonm. M., Lahatm, H., Bakhanm, T. et al. (2003) Hypotrichosis simplex of the scalp is associated with nonsense mutations in CDSN encoding corneodesmosin. *Nature Genetics*, 34, 151–153.

Lin, H., Bhatia, R., and Lal, R. (2001) Amyloid β protein forms ion channels: implications for Alzheimer's disease pathophysiology. *FASEB Journal*, 15, 2433–2444.

Lin, M.C. and Kagan, B. (2002) Electrophysiologic properties of channels induced by Aβ25–35 in planar lipid bilayers. *Peptides*, 23, 1215–1228.

Lin, M.C., Mirzabekov, T., and Kagan, B.L. (1997) Channel formation by a neurotoxic prion protein fragment. *Journal of Biological Chemistry*, 272, 44–47.

Lindquist, S. (1997) Mad cows meet psi-chotic yeast: the expansion of the prion hypothesis. *Cell*, 89, 495–498.

Little, C.S., Hammond, C.J., MacIntyre, A., Balin, B.J., and Appelt, D.M. (2004) Chlamydia pneumoniae induces Alzheimer-like amyloid plaques in brains of BALB/c mice. *Neurobiology of Aging*, 25, 419–429.

Litvinovich, S.V., Brew, S.A., Aota, S., Akiyama, S.K., Haudenschild, C., and Ingham, K.C. (1998) Formation of amyloid-like fibrils by self-association of a partially unfolded fibronectin type III module. *Journal of Molecular Biology*, 280, 245–258.

Lomas, D.A. and Carrell, R.W. (2002) Serpinopathies and the conformational dementias. *Nature Reviews Genetics*, 3, 759–768.

Lomas, D.A., Evans, D.L., Finch, J.T., and Carrell, R.W. (1992) The mechanism of Z alpha 1-antitrypsin accumulation in the liver. *Nature*, 357, 605–607.

Lorenzo, A. and Yankner, B.A. (1994) β-Amyloid neurotoxicity requires fibril formation and is inhibited by Congo red. *Proceedings of the National Academy of Sciences of the United States of America*, 91, 12243–12247.

Lue, L.F., Brachova, I., Civin, W.H., and Rogers, J. (1996) A beta deposition and neurofibrillary tangle formation as correlates of Alzheimer's disease neurodegeneration. *Journal of Neuropathology and Experimental Neurology*, 55, 1083–1088.

Lührs, T., Ri, C., Adrian, M., Riek-Loher, D., Bohrmann, B., Döbeli, H. et al. (2005) 3D structure of Alzheimer's amyloid-β(1–42) fibrils. *Proceedings of the National Academy of Sciences of the United States of America*, 102, 17342–17347.

Lundmark, K., Westermark, G.T., Nyström, S., Murphy, C.L., Solomon, A., and Westermark, P. (2002) Transmissibility of systemic amyloidosis by a prion-like mechanism. *Proceedings of the National Academy of Sciences of the United States of America*, 99, 6979–6984.

Lusis, A.J. (2000) Atherosclerosis. *Nature*, 407, 232–241.

Ma, Y. and Hendershot, L.M. (2001) The unfolding tale of the unfolding protein response. *Cell*, 107, 827–830.

Ma, J., Wollmann, R., and Lindquist, S. (2002) Neurotoxicity and neurodegeneration when PrP accumulates in the cytosol. *Science*, 298, 1781–1785.

Macario, A.J.L. and de Macario, E.C. (2002) Sick chaperones and ageing: a perspective. *Ageing Research Reviews*, 1, 295–311.

Marlow, L., Cain, M., Pappolla, M.A., and Sambamurti, K. (2003) Beta secretase processing of the Alzheimer's amyloid protein precursor (APP). *Journal of Molecular Neuroscience*, 20, 233–239.

Mason, R.P., Shoemaker, W.J., Shajenko, L., Chambers, T.E., and Herbette, L.G. (1992) Evidence for changes in the Alzheimer's disease brain cortical membrane structure mediated by cholesterol. *Neurobiology of Aging*, 13, 413–419.

Mattson, M.P. (1999) Impairment of membrane transport and signal transduction systems by amyloidogenic proteins. *Methods in Enzymology*, 309, 733–768.

Mattson, M.P., Guo, Q., Furukawa, K., and Pedersen, W.A. (1998) Presenilins, the endoplasmic reticulum, and neuronal apoptosis in Alzheimer's disease. *Journal of Neurochemistry*, 70, 1–14.

Mayor, U., Guydosh, N.R., Johnson, C.M., Grossmann, J.G., Sato, S., Jas, G.S. et al. (2003) The complete folding pathway of a protein from nanoseconds to microseconds. *Nature*, 421, 863–867.

McNaught, K.S.P., Mytilinieou, C., JnoBaptiste, R., Yabut, J., Shashidharan, P., Jenner, P., and Olanov, C.W. (2002) Impairment of the ubiquitin-proteasome system causes dopaminergic cell death and inclusion body formation in ventral mesencephalic cultures. *Journal of Neurochemistry*, 81, 301–306.

Miklossy, J. (1993) Alzheimer's disease—a spirochetosis? *Neuroreport*, 4, 841–848.

Milhavet, O. and Lehmann, S. (2002) Oxidative stress and the prion protein in transmissible spongiform encephalopathies. *Brain Research Reviews*, 38, 328–339.

Milner, J. and Medcalf, E.A. (1991) Cotranslation of activated mutant p53 with wild-type drives the wild-type p53 protein into the mutant conformation. *Cell*, 65, 765–774.

Minton, A.P. (1994) Influence of macromolecular crowding on intracellular association reactions: possible role in volume regulation. In Strange, K. (Ed.) *Cellular and Molecular Physiology of Cell Volume Regulation*, CRC Press, Boca Raton, Fl, 181–190.

Mirzabekov, T.A., Lin, M.C., and Kagan, B.L. (1996) Pore formation by the cytotoxic islet amyloid peptide amylin. *Journal of Biological Chemistry*, 271, 1988–1992.

Moll, U.M., Ostermeyer, A.G., Haladay, R., Winkfield, B., Frazier, M., and Zambetti, G. (1996) Cytoplasmic sequestration of wild-type p53 protein impairs the G1 checkpoint after DNA damage. *Molecular Cell Biology*, 16, 1126–1137.

Monoi, H., Futaki, S., Kugimyia, S., Minakata, H., and Yoshihara, K. (2000) Poly-L-glutamine forms cation channels: relevance to the pathogenesis of the polyglutamine diseases. *Biophysics Journal*, 78, 2892–2899.

Moreira, P.I., Santos, M.S., Moreno, A., Rego, A.C., and Oliveira, C. (2002) Effect of amyloid beta-peptide on permeability transition pore: a comparative study. *Journal of Neuroscience Research*, 15, 257–267.

Morgan, B.P. (1999) Regulation of the complement membrane attack pathway. *Critical Reviews in Immunology*, 19, 173–198.

Morgan, C.J., Gelfrand, M., Atreya, C., and Miranker, A. (2001) Kidney dialysis-associated amyloidosis: a molecular role for copper in fiber formation. *Journal of Molecular Biology*, 309, 339–345.

Morishima, Y., Gotoh, Y., Zieg, J., Barrett, T., Takano, H., Flavell, R. et al. (2001) Beta-amyloid induces neuronal apoptosis via a mechanism that involves the c-Jun *N*-terminal kinase pathway and the induction of Fas ligand. *Journal of Neuroscience*, 21, 7551–7560.

Muchowski, P.J., Schaffar, G., Sittler, A., Wanker, E.E., Hayer-Hartl, M.K., and Hartl, U. (2000) Hsp70 and Hsp40 chaperones can inhibit self-assembly of polyglutamine proteins into amyloid-like fibrils. *Proceedings of the National Academy of Sciences of the United States of America*, 97, 7841–7846.

Munier, F.L., Korvatska, E., Djenai, A., Le Paslier, D., Zografos, L., Pescia, G., and Schorderet, D.F. (1997) Kerato-epithelin mutations in four 5q31-linked corneal dystrophies. *Nature Genetics*, 15, 247–251.

Myers, A.J. and Goate, A.M. (2001) The genetics of late-onset Alzheimer's disease. *Current Opinion in Neurology*, 14, 433–440.

Nagy, I.Z., Nagy, K., and Lustyik, G. (1982) Protein and water content of aging brain. *Experimental Brain Research*, Suppl. 5, 118–122.

Nardai, G., Csermely, P., and Soti, C. (2002) Chaperone function and chaperone overload in the aged. A preliminary analysis. *Experimental Gerontology*, 37, 1257–1262.

Necula, M., Chirita, C., and Kuret, J. (2002) Rapid anionic micelle-mediated α-synuclein fibrillization in vitro. *Journal of Biological Chemistry*, 278, 46674–46680.

Nelson, R., Sawaya, M., Balbirnie, M., Madsen, A.Ø., Riekel, C. Grothe, R., and Eisenberg, D. (2005) Structure of the cross-β spine of amyloid fibrils. *Nature*, 435, 773–778.

Newcomer, M.E. (2002) Protein folding and three-dimensional domain swapping: a strained relationship? *Current Opinion in Structural Biology*, 12, 48–53.

Nicotera, P. and Melino, G. (2004) Regulation of the apoptosis-necrosis switch. *Oncogene*, 12, 2757–2765.

Nilsberth, C., Westlind-Danielssnon, A., Eckman, C.B., Condron, M.M., Axelman, K., Forsell, C. et al. (2001) The "arctic" APP mutation. (E693G) causes Alzheimer's disease by enhanced Aβ protofibril formation. *Nature Neuroscience*, 4, 887–893.

Nilsson, M.R. and Dobson, C.M. (2003) *In vitro* characterization of lactoferrin aggregation and amyloid formation. *Biochemistry*, 42, 375–382.

O'Nuallain, B. and Wetzel, R. (2002) Conformational Abs recognizing a generic amyloid fibril epitope. *Proceedings of the National Academy of Sciences of the United States of America*, 99, 1485–1490.

Orth, M. and Shapira, A.H.V. (2002) Mitochondrial involvement in Parkinson's disease. *Neurochemistry International*, 40, 533–541.

Ostermeyer, A.G., Runko, E., Winkfield, B., Ahn, B., and Moll, U.M. (1996) Citoplasmically sequestered wild-type p53 protein in neuroblastoma is relocated to the nucleus by a C-terminal peptide. *Proceedings of the National Academy of Sciences of the United States of America*, 93, 15190–15194.

Pain, R.H. (ed.) (2000) Protein folding, 2nd Edition, Oxford University Press, Oxford.

Pande, A., Pande, J., Asherie, N., Lomakin, A., Ogfun, O., King, J., and Benedek, G.B. (2001) Crystal cataracts: human genetic cataract caused by protein crystallization. *Proceedings of the National Academy of Sciences of the United States of America*, 98, 6116–6120.

Pappolla, M.A., Omar, R.A., Chyan, Y.-J., Ghiso, J., Hsiao, K., Bozner, P. et al. (2001) Induction of NADPH cytochrome P450 reductase by the Alzheimer β-protein. Amyloid as a "foreign body." *Journal of Neurochemistry*, 78, 121–128.

Paschen, W. and Mengesdorf, T. (2005) Endoplasmic reticulum stress response and neurodegeneration. *Cell Calcium*, 38, 409–415.

Pedersen, C.B., Bross, P., Winter, V.S., Corydon, T.J., Bolund, L., Bartlett, K. et al. (2003) Misfolding, degradation and aggregation of variant proteins—the molecular pathogenesis of short chain acyl-CoA dehydrogenase (SCAD) deficiency. *Journal of Biological Chemistry*, 278, 47449–47458.

Pedersen, J.S., Dikov, D., Flink, J.L., Hjuler, H.A., Christiansen, G., and Otzen, D.E. (2006) The changing face of glucagon fibrillation: structural polymorphism and conformational imprinting. *Journal of Molecular Biology*, 355, 501–523.

Peng, S., Glennert, J., and Westermark, P. (2005) Medin-amyloid: a recently characterized age-associated arterial amyloid form affects mainly arteries in the upper part of the body. *Amyloid*, 12, 96–102.

Pepys, M.B. (1995) Amyloid familial Mediterranean fever and acute phase response: Amyloidosis. In Wheaterall, D.J., Ledingham, J.G., and Warrel, D.A. (eds), Oxford Textbook of Medicine 3rd Edition, Oxford University Press, Oxford, 1512–1524.

Pepys, M.B., Herbert, J., Hutchinson, W.L., Tennent, G.A., Lachmann, H.J., Gallimore, J.R. et al. (2002) Targeted pharmaceutical depletion of serum amyloid P component for treatment of human amyloidosis. *Nature*, 417, 254–259.

Perutz, M.F. and Windle, A.H. (2001) Cause of neuronal death in neurodegenerative diseases attributable to expansion of glutamine repeats. *Nature*, 412, 143–144.

Perutz, M.F., Pope, B.J., Owen, D., Wanker, E.E., and Scherzinger, E. (2002) Aggregation of proteins with expanded glutamine and alanine repeats of the glutamine-rich and asparagine-rich domains of Sup35 and of the amyloid β-peptide of amyloid plaques. *Proceedings of the National Academy of Sciences of the United States of America*, 99, 5596–5600.

Petkova, A.T., Ishii, Y., Balbach, J.J., Antzutkin, O.N., Leapman, R.D., Delaglio, F., and Tycko, R. (2002) A structural model for Alzheimer's β-amyloid fibrils based on experimental constraints from solid state NMR. *Proceedings of the National Academy of Sciences of the United States of America*, 99, 16742–16747.

Petkova, A.T., Buntkowsky, G., Dyda, F., Leapman, R.D., Yau, W.-M., and Tycko, R. (2004) Solid state NMR reveals a pH-dependent antiparallel β-sheet registry in fibrils formed by a β-amyloid peptide. *Journal of Molecular Biology*, 335, 247–260.

Petkova, A.T., Leapman, R.D., Guo, Z., Yau, W.-M., Mattson, M.P., and Tycko, R. (2005) Self-propagating, molecular level polymorphism in Alzheimer's β-amyloid fibrils. *Science*, 307, 262–265.

Plaxco, K.W., Simons, K.T., and Baker, D. (1998) Contact order, transition state placement and the refolding rates of single domain proteins. *Journal of Molecular Biology*, 277, 985–994.

Plemper, R.K. and Wolf, D.H. (1999) Retrograde protein translocation: eradication of secretory proteins in health and disease. *Trends in Biochemical Sciences*, 24, 266–270.

Porat, Y., Kolusheva, S., Jelinek, R., and Gazit, E. (2003) The human islet amyloid polypeptide forms transient membrane-active prefibrillar assemblies. *Biochemistry*, 42, 10971–10977.

Poirier, M.A., Li, H., Macosko, J., Cail, S., Amzel, M., and Ross, C.A. (2002) Huntingtin spheroids and protofibrils as precursors in polyglutamine fibrillization. *Journal of Biological Chemistry*, 277, 41032–41037.

Prigent-Combaret, C., Prensier, G., Le Thi, T.T., Vidal, O., Lejeune, P., and Dorel, C. (2000) Developmental pathway for biofilm formation in curli-producing *Escherichia coli* strains: role of flagella, curli and colanic acid. *Environmental Microbiology*, 2, 450–464.

Prusiner, S.B. (1998) Prions. *Proceedings of the National Academy of Sciences of the United States of America*, 95, 13363–13383.

Queiltsch, C., Sangster, T.A., and Lindquist, S. (2002) Hsp90 as a capacitor of phenotypic variation. *Nature*, 417, 618–624.

Qin, L., Liu, Y., Cooper, C., Liu, B., Wilson, B., and Hong, J.-S. (2002) Microglia enhance β-amyloid peptide-induced toxicity in cortical and mesencephalic neurons by producing reactive oxygen species. *Journal of Neurochemistry*, 83, 973–983.

Quintas, A., Vaz, D.C., Cardoso, I., Saraiva, M.J.M., and Brito, R.M.M. (2001) Tetramer dissociation and monomer partial unfolding precedes protofibril formation in amyloidogenic transthyretin variants. *Journal of Biological Chemistry*, 276, 27207–27213.

Ran, S. and Thorpe, P.E. (2002) Phosphatidylserine is a marker of tumour vasculature and a potential target for cancer imaging and therapy. *International Journal of Radiation Oncology Biology Physics*, 54, 1479–1484.

Reed, J.C. (2000) Mechanisms of apoptosis. *American Journal of Pathology*, 157, 1415–1430.

Reilly, M.M. (1998) Genetically determined neuropathies. *Journal of Neurology*, 245, 6–13.

Reits, E.A.J., Vos, J.C., Grommé, M., and Neefjes, J. (2000) The major substrates for TAP in vivo are derived from newly synthesized proteins. *Nature*, 404, 774–778.

Reixach, N., Deechingkit, S., Jiang, X., Kelly, J.W., and Buxbaum, J.N. (2004) Tissue damage in the amyloidoses: transthyretin monomers and nonnative oligomers are the major cytotoxic species in tissue culture. *Proceedings of the National Academy of Sciences of the United States of America*, 101, 2817–2822.

Relini, A., Torrassa, S., Rolandi, R., Ghiozzi, A., Rosano, C., Canale, C. et al. (2004) Monitoring the process of HypF fibrillization and liposome permeabilization by protofibrils. *Journal of Molecular Biology*, 338, 943–957.

Richardson, J.S. and Richardson, D.C. (2002) Natural β-sheet proteins use negative design to avoid edge-to-edge aggregation. *Proceedings of the National Academy of Sciences of the United States of America*, 99, 2754–2759.

Ridiger, S., Freund, S.M.V., Veprintsev, D.B., and Fersht, A.R. (2002) CRINEPT-TROSY NMR reveals p53 core domain bound in an unfolded form to the chaperone Hsp90. *Proceedings of the National Academy of Sciences of the United States of America*, 99, 11085–11090.

Roberts, S.P. and Fader, M.E. (1999) Natural hyperthermia and expression of the heat shock protein Hsp70 affect developmental abnormalities in *Drosophila melanogaster*. *Oecologia*, 121, 323–329.

Robinson, S.R., Dobson, C., and Lyons, J. (2004) Challenges and directions for the pathogen hypothesis of Alzheimer's disease. *Neurobiology of Aging*, 25, 629–637.

Roder, H. and Colon, W. (1997) Kinetic role of early intermediates in protein folding. *Current Opinion in Structural Biology*, 7, 15–28.

Ross, C.A. (2002) Polyglutamine pathogenesis: emergence of unifying mechanisms for Huntington's disease and related disorders. *Neuron*, 35, 819–822.

Rutherford, S.L. and Lindquist, S. (1998) Hsp90 as a capacitor for morphological evolution. *Nature*, 396, 336–342.

Sakahira, H., Breuer, P., Hayer-Hartl, M.K., and Hartl, F.U. (2002) Molecular chaperones as modulators of polyglutamine protein aggregation and toxicity. *Proceedings of the National Academy of Sciences of the United States of America*, 99 Suppl 4, 16412–16418.

Saliba, R.S., Munro, P.M.G., Luthert, P.J., and Cheetham, M.E. (2002) The cellular fate of mutant rhodopsin: quality control, degradation and aggresome formation. *Journal of Cell Science*, 115, 2907–2918.

Sanbe, A., Osinska, H., Saffitz, J.E., Glabe, C.G., Kayed, R., Maloyan, A., and Robbins, J. (2004) Desmin-related cardiomyopathy in transgenic mice: a cardiac amyloidosis. *Proceedings of the National Academy of Sciences of the United States of America*, 101, 10132–10136.

Sanchez, I.E. and Kiefhaber, T. (2003) Evidence for sequential barriers and obligatory intermediates in apparent two-state protein folding. *Journal of Molecular Biology*, 325, 367–376.

Sandilands, A., Hutcheson, A.M., Long, H.A., Prescott, A.R., Vrensen, G., Löster, J. et al. (2002) Altered aggregation properties of mutant γ-crystallins cause inherited cataract. *EMBO Journal*, 21, 6005–6014.

Santhiya, S.T., Shyam Manobar, M., Rawley, D., Vijayalaskshmi, P., Namperumalsamy, P., Gopinath, P.M. et al. (2002) Novel mutations in the γ-crystallin genes cause autosomal dominant congenital cataracts. *Journal of Medical Genetics*, 39, 352–358.

Schiene, C. and Fischer, G. (2000) Enzymes that catalyze the restructuring of proteins. *Current Opinion in Structural Biology*, 7, 523–529.

Schröder, M. and Kaufman, R.J. (2005) ER stress and the unfolded protein response. *Mutation Research*, 569, 9–63.

Schubert, U., Antòn, L.C., Gibbs, J., Norbury, C.C., Yewdell, J.W., and Bennink, J.R. (2000) Rapid degradation of a large fraction of newly synthesized proteins by proteasomes. *Nature*, 404, 770–774.

Schwarz, R., Istrail, S., and King, J. (2001) Frequencies of amino acid strings in globular protein sequences indicate suppression of blocks of consecutive hydrophobic residues. *Protein Science*, 10, 1023–1031.

Selkoe, D.J. (2001) Alzheimer's disease: genes, proteins, and therapy. *Physiology Reviews*, 81, 741–766.

Selkoe, D.J. (2003) Folding proteins in fatal ways. *Nature*, 426, 900–904.

Serpell, L.S. (2000) Alzheimer's amyloid fibrils: structure and assembly. *Biochimica et Biophysica Acta*, 1502, 16–30.

Serpell, L.C., Sunde, M., Benson, M.D., Tennent, G.A., Pepys, M.B., and Fraser, P.E. (2000) The protofilament substructure of amyloid fibrils. *Journal of Molecular Biology*, 300, 1033–1039.

Sethuraman, A. and Belfort, G. (2005) Protein structural perturbation and aggregation on homogeneous surfaces. *Biophysics Journal*, 88, 1322–1333.

Shames, I., Fraser, A., Colby, J., Orfali, W., and Snipes, G.J. (2003) Phenotypic differences between peripheral myelin protein-22 (PMP22) and myelin protein zero (P0) mutations associated with Charcot-Marie-Tooth-related diseases. *Neuropathology and Experimental Neurology*, 62, 751–764.

Sherman, M.Y. and Goldberg, A.L. (2001) Cellular defenses against unfolded proteins: a cell biologist thinks about neurodegenerative diseases. *Neuron*, 29, 15–32.

Sidhar, S.K., Lomas, D.A., Carrell, R.W., and Foreman, R.C. (1995) Mutations which impede loop/sheet polymerization enhance the secretion of human alpha 1-antitrypsin deficiency variants. *Journal of Biological Chemistry*, 270, 8393–8396.

Sirangelo, I., Malmo, C., Iannuzzi, C., Mezzogiorno, A., Bianco, M.R., Papa, M., and Irace, G. (2004) Fibrillogenesis and cytotoxic activity of the amyloid-forming apomyoglobin mutant W7FW14F. *Journal of Biological Chemistry*, 279, 13183–13189.

Sitia, R. and Braakman, I. (2003) Quality control in the endoplasmic reticulum protein factory. *Nature*, 426, 891–894.

Snow, C.D., Nguyen, H., Pande, V.S., and Gruebele, M. (2002) Absolute comparison of simulated and experimental protein-folding dynamics. *Nature*, 420, 102–106.

Sousa, M.M., Cardoso, I., Fernandes, R., Guimaraes, A., and Saraiva, M.J. (2001) Deposition of transthyretin in early stages of familial amyloidotic polyneuropathy. *American Journal of Pathology*, 159, 1993–2000.

Squier, T.C. (2001) Oxidative stress and protein aggregation during biological aging. *Experimental Gerontology*, 36, 1539–1550.

Srinivasan, R., Jones, E.M., Liu, K., Ghiso, J., Marchant, R.E., and Zagorski, M.G. (2003) pH-dependent amyloid and protofibril formation by the ABri peptide of familial British dementia. *Journal of Molecular Biology*, 333, 1003–1023.

Srinivasan, R., Marchant, R.E., and Zagorski, M.G. (2004) ABri peptide associated with familial British dementia forms annular and ring-like protofibrillar structures. *Amyloid*, 11, 10–13.

Stefani, M. (2004) Protein misfolding and aggregation: new examples in medicine and biology of the dark side of the protein world. *Biochimica et Biophysica Acta*, 1739, 5–25.

Stefani, M. and Dobson, C.M. (2003) Protein aggregation and aggregate toxicity: new insights into protein folding, misfolding diseases and biological evolution. *Journal of Molecular Medicine*, 81, 678–699.

Stein, P.E. and Carrell, R.W. (1995) What do dysfunctional serpins tell us about molecular mobility and disease? *Nature Structural Biology*, 2, 96–113.

Stevens, F.J. and Argon, Y. (1999) Pathogenic light chains and the B-cell repertoire. *Immunology Today*, 20, 451–457.

Svanborg, C., Agerstam, H., Aronson, A., Bjerkvig, R., Duringer, C., Fischer, W. et al. (2003) HAMLET kills tumor cells by an apoptosis-like mechanism—cellular, molecular, and therapeutic aspects. *Advances in Cancer Research*, 88, 1–29.

Svensson, M., Sabharwal, H., Hakansson, A., Mossberg, A.-K., Lipniunas, P., Leffler, H. et al. (1999) Molecular characterization of α-lactalbumin folding variants that induce apoptosis in tumor cells. *Journal of Biological Chemistry*, 274, 6388–6396.

Tabner, B.J., Turnbull, S., El-Agnaf, O.M.A., and Allsop, D. (2002) Formation of hydrogen peroxide and hydroxyl radicals from Aβ and α-synuclein as a possible mechanism of cell death in Alzheimer's disease and Parkinson's disease. *Free Radical Biology and Medicine*, 32, 1076–1083.

Tartaglia, G.G., Cavalli, A., Pellarin, R., and Caflisch, A. (2004) The role of aromaticity, exposed surface, and dipole moment in determining protein aggregation rates. *Protein Science*, 13, 1939–1941.

Tartaglia, G.G., Pellarin, R., Cavalli, A., and Caflisch, A. (2005a) Organism complexity anti-correlates with proteomic beta-aggregation propensity. *Protein Science*, 14, 2735–2740.

Tartaglia, G.G., Cavalli, A., Pellarin, R., and Catflish, A. (2005b) Prediction of aggregation rate and aggregation-prone segments in polypeptide sequences. *Protein Science*, 14, 2723–2734.

The International SNP map working group (2001) A map of human sequence variation containing 1.4 million single nucleotide polymorphisms. *Nature*, 409, 928–933.

Thomas, P.J., Qu, B.-H., and Pedersen, P.L. (1995) Defective protein folding as a basis of human disease. *Trends in Biochemical Sciences*, 20, 456–459.

Thomas, T., Thomas, G., McLendon, C., Sutton, T., and Mullan, M. (1996) β-Amyloid-mediated vasoactivity and vascular endothelial damage. *Nature*, 380, 168–171.

Tofaris, G.K., Layfield, R., and Spillantini, M.G. (2001) α-Synuclein metabolism and aggregation is linked to ubiquitin-independent degradation by the proteasome. *FEBS Letters*, 509, 22–26.

Trinh, C.H., Smith, D.P., Kalverda, A.P., Phillips, S.E.V., and Radford, S. (2002) Crystal structure of monomeric human β-2-microglobulin reveals clues to its amyloidogenic properties. *Proceedings of the National Academy of Sciences of the United States of America*, 99, 9771–9776.

True, H.L. and Lindquist, S. (2000) A yeast prion provides a mechanism for genetic variation and phenotypic diversity. *Nature*, 407, 477–483.

Tsai, B., Ye, Y., and Rapoport, T.A. (2002) Retrotranslocation of proteins from the endoplasmic reticulum into the cytosol. *Nature Reviews Molecular Cell Biology*, 3, 246–255.

Tsutsumi, Y., Serizawa, A., and Hori, S. (1996) Localized amyloidosis of the seminal vesicle: identification of lactoferrin immunoreactivity in the amyloid material. *Pathology International*, 46, 491–497.

Turnbull, S., Tabner, B.J., El-Agnaf, O.M.A., Moore, S., Davies, Y. and Allsop, D. (2001) α-Synuclein implicated in Parkinson's disease catalyzes the formation of hydrogen peroxide *in vitro*. *Free Radical Biology and Medicine*, 30, 1163–1170.

Tycko, R. (2004) Progress towards a molecular-level structural understanding of amyloid fibrils. *Current Opinion in Structural Biology*, 14, 96–103.

Ursini, F., Davies, K.J.A., Maiorino, M., Parasassi, T., and Sevanian, A. (2002) Atherosclerosis: another protein misfolding disease? *Trends in Molecular Medicine*, 8, 370–374.

Utsugi, T., Schroit, A.J., Connor, J., Bucana, C.D., and Fidler, I.J. (1991) Elevated expression of phosphatidylserine inn the outer membrane leaflet of human tumour cells and recognition by activated human blood monocytes. *Cancer Research*, 51, 3062–3066.

Uversky, V.N. (2002) Natively unfolded proteins: a point where biology waits for physics. *Protein Science*, 11, 739–756.

Uversky, V.N. and Fink, A.L. (2004) Conformational constraints for amyloid fibrillation: the importance of being unfolded. *Biochimica et Biophysica Acta*, 1698, 131–153.

van Horssen, J., Wilhelmus, M.M., Heljasvaara, R., Pihlajaniemi, T., Wesseling, P., de Waal, R.M., and Verbeek, M.M. (2002) Collagen XVIII: a novel heparan sulfate proteoglycan associated with vascular amyloid depositions and senile plaques in Alzheimer's disease brains. *Brain Pathology*, 12, 456–462.

Varadarajan, S., Yatin, S., Aksenova, M., and Butterfield, D.A. (2000) Alzheimer's amyloid β-peptide-associated free radical oxidative stress and neurotoxicity. *Journal of Structural Biology*, 130, 184–208.

Velez-Pardo, C., Arroyave, S.T., Lopera, F., Castano, A.D., and Jimenez Del Rio, M. (2001) Ultrastructure evidence of necrotic neural cell death in familial Alzheimer's disease brains bearing presenilin-1 E280A mutation. *Journal of Alzheimer Disease*, 3, 409–415.

Verga, U., Fugazzola, L., Cambiaghi, S., Tritelli, C., Alessi, E., Cortellazzi, D. et al. (2003) Frequent association between MEN 2A and cutaneous lichen amyloidosis. *Clinical Endocrinology*, 59, 156–161.

Vendrely, C., Valadie, H., Bednarova, L., Cardin, L., Pasdeloup, M., Cappadoro, J. et al. (2005) Assembly of the full-length recombinant mouse prion protein I. Formation of soluble oligomers. *Biochimica et Biophysica Acta*, 1724, 355–366.

Vendruscolo, M., Paci, E., Dobson, C.M., and Karplus, M. (2001) Three key residues form a critical contact network in a transition state for protein folding. *Nature*, 409, 641–646.

Vetrivel, K.S., Chen, H., Lin, W., Sakurai, T., Li, T., Nukina, N. et al. (2004) Association of gamma-secretase with lipid rafts in post-Golgi and endosome membranes. *Journal of Biological Chemistry*, 279, 44945–44954.

Vidal, O., Longin, R., Prigent-Combaret, C., Dorel, C., Hooreman, M., Lejeune, P. et al. (1998) Isolation of an *Escherichia coli* K-12 mutant strain able to form biofilms on inert surfaces: involvement of a new ompR allele that increases curli expression. *Journal of Bacteriology*, 180, 2442–2449.

Volles, M.J. and Lansbury, P.T. (2001) Vesicle permeabilization by protofibrillar α-synuclein: comparison of wild-type with Parkinson's disease linked mutants and insights in the mechanisms. *Biochemistry*, 40, 7812–7819.

Waelter, S., Boeddrich, A., Lurz, R., Scherzinger, E., Lueder, G., Lebrach, H., and Wanker, E.E. (2001) Accumulation of mutant huntingtin fragments in aggresome-like inclusion bodies as a result of insufficient protein degradation. *Molecular Biology of the Cell*, 12, 1393–1407.

Walsh, D.M., Hartley, D.M., Kusumoto, Y., Fezoui, Y., Condron, M.M., Lomakin, A. et al. (1999) Amyloid β-protein fibrillogenesis. Structure and biological activity of protofibrillar intermediates. *Journal of Biological Chemistry*, 274, 25945–25952.

Walsh, D.M., Klyubin, I., Fadeeva, J.V., Cullen, W.K., Anwyl, R., Wolfe, M.S. et al. (2002) Naturally secreted oligomers of amyloid β protein potently inhibit hippocampal long-term potentiation *in vivo*. *Nature*, 416, 535–539.

Wang, L., Lashuel, H.A., Walz, T., and Colòn, W. (2002) Murine apolipoprotein serum amyloid A in solution forms a hexamer containing a central channel. *Proceedings of the National Academy of Sciences of the United States of America*, 99, 15947–15952.

Watt, J.A., Pike, C.J., Walencewicz-Wasserman, A.J., and Cotman, C.W. (1994) Ultrastructural analysis of beta-amyloid-induced apoptosis in cultured hippocampal neurons. *Brain Research*, 661, 147–156.

Westermark, P., Eriksson, L., Engström, U., Eneström, S., and Sletten, K. (1997) Prolactin-derived amyloid in the aging pituitary gland. *American Journal of Pathology*, 150, 67–73.

Wetzel, R. (2002) Ideas of order for amyloid fibril structure. *Structure*, 10, 1031–1036.

Widmann, M. and Christen, P. (2000) Comparison of folding rates of homologous prokaryotic and eukaryotic proteins. *Journal of Biological Chemistry*, 275, 18619–18622.

Wickner, R.B. (1996) Prions and RNA viruses of Saccharomyces cerevisiae. *Annual Reviews in Genetics*, 30, 109–139.

Wolynes, P.G., Onuchic, J.N., and Thirumalai, D. (1995) Navigating the folding routes. *Science*, 267, 1619–1620.

Wright, P.E. and Dyson, H.J. (1999) Intrinsically unstructured proteins: reassessing the protein structure-function paradigm. *Journal of Molecular Biology*, 293, 321–331.

Wu, Y., Whitman, I., Molmenti, E., Moore, K., Hippenmeyer, P., and Perlmutter, D.H. (1994) A lag in intracellular degradation of mutant alpha 1-antitrypsin correlates with the liver disease phenotype in homozygous PiZZ alpha 1-antitrypsin deficiency. *Proceedings of the National Academy of Sciences of the United States of America*, 91, 9014–9018.

Wyttenbach, A., Sauvageot, O., Carmichael, J., Diaz-Latoud, C., Arrigo, A.-P., and Rubinsztein, D.C. (2002) Heat shock protein 27 prevents cellular polyglutamine toxicity and suppresses the increase of reactive oxygen species caused by Huntingtin. *Human Molecular Genetics*, 11, 1137–1151.

Wyttenbach, A., Swartz, J., Kita, H., Thykjaer, T., Carmichael, J., Bradley, J. et al. (2001) Polyglutamine expansions cause decreased CRE-mediated transcription and early gene expression changes prior to cell death in an inducible cell model of Huntington's disease. *Human Molecular Genetics*, 10, 1829–1845.

Xing, Y., Nakamura, A., Korenaga, T., Guo, Z., Yao, J., Fu, X. et al. (2002) Induction of protein conformational change in mouse senile amyloidosis. *Journal of Biological Chemistry*, 277, 33164–33169.

Yamaguchi, K., Katou, H., Hishino, M., Hasegawa, K., Naiki, H., and Goto, Y. (2004) Core and heterogeneity of β_2-microglobulin amyloid fibrils as revealed by H/D exchange. *Journal of Molecular Biology*, 338, 559–571.

Yan, S.D., Stern, D., Kane, M.D., Kuo, Y.M., Lampert, H.C., and Roher, A.E. (1998) RAGE-Abeta interactions in the pathophysiology of Alzheimer's disease. *Restorative Neurology and Neuroscience*, 12, 167–173.

Yang, W., Dunlap, J.R., Andrews, R.B., and Wetzel, R. (2002) Aggregated polyglutamine peptides delivered to nuclei are toxic to mammalian cells. *Human Molecular Genetics*, 11, 2905–2917.

Yang, W.Y. and Gruebele, M. (2003) Folding at the speed limit. *Nature*, 423, 193–197.

Yip, C.M., Elton, E.A., Darabie, A.A., Morrison, M.R., and McLaurin, J. (2001) Cholesterol, a modulator of membrane-associated Aβ-fibrillogenesis and neurotoxicity. *Journal of Molecular Biology*, 311, 723–734.

Yoo, B.C., Kim, S.H., Cairns, N., Fountoulakis, M., and Lubec, G. (2001) Deranged expression of molecular chaperones in brains of patients with Alzheimer's disease. *Biochimica et Biophysica Research Communications*, 280, 249–258.

Zhang, L., Xing, G.Q., Barker, J.L., Chang, Y., Maric, D., Ma, W. et al. (2001) α-Lipoic acid protects rat cortical neurons against cell death induced by amyloid and hydrogen peroxide through the Akt signalling pathway. *Neuroscience Letters*, 312, 125–128.

Zhang, Y.-H., Leliveld, S.R., Kooistra, K., Molenaar, C., Rohn, J.L., Tanke, H.J. et al. (2003) Recombinant apoptin multimers kill tumor cells but are non-toxic and epitope-shielded in a normal-cell-specific fashion. *Experimental Cell Research*, 289, 36–46.

Zhao, H., Tuominen, E.K.J., and Kinnunen, P.K.J. (2004) Formation of amyloid fibers triggered by phosphatidylserine-containing membranes. *Biochemistry*, 43, 10302–10307.

Zhao, H., Jutila, A., Nurminen, T., Wickström, S.A., Keski-Oja, J., and Kinnunen, P.K.J. (2005) Binding of endostatin to phosphatidylserine-containing membranes and formation of amyloid-like fibers. *Biochemistry*, 44, 2857–2863.

Zhu, Y.J., Lin, H., and Lal, R. (2000) Fresh and nonfibrillar amyloid β protein (1–40) induces rapid cellular degeneration in aged human fibroblasts: evidence for AβP-channel-mediated cellular toxicity. *FASEB Journal*, 14, 1244–1254.

Zhu, M., Souillac, P.O., Ionesco-Zanetti, C., Carter, S.A., and Fink, A.L. (2002) Surface-catalyzed amyloid fibril formation. *Journal of Biological Chemistry*, 277, 50914–50922.

Zoghbi, H.Y. and Botas, J. (2002) Mouse and fly models of neurodegeneration. *Trends in Genetics*, 18, 463–471.

Part I

Alzheimer's Disease

2 Alzheimer's Disease

Charlotte E. Teunissen and
Tischa J.M. van der Cammen

CONTENTS

2.1 INTRODUCTION

Age-related neurodegenerative diseases such as Alzheimer's disease (AD) are posing an increasing impact on society, mainly due to increasing life expectancy. AD is characterized by the presence of neurofibrillary tangles and neuritic plaques in

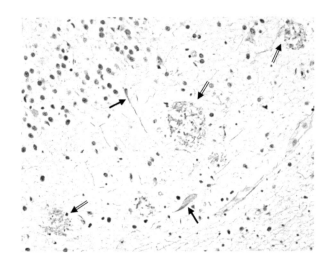

FIGURE 2.1 Characteristic pathological hallmarks in postmortem Alzheimer brain tissue. *Open arrows* represent plaques, *closed arrows* represent fibrillary tangles. 400× magnification.

postmortem brain tissue (Figure 2.1). Plaques consist of extracellular aggregates of amyloid beta (Abeta) peptides derived from amyloid precursor protein (APP), accompanied by inflammation and astrogliosis. Tangles are intraneuronal accumulations of insoluble and hyperphosphorylated tau, a microtubule-binding protein (Hardy, 2003).

2.2   EPIDEMIOLOGY OF AD

AD is the most common form of dementia, accounting for 53%–80% of dementia cases (Rocca et al., 1991; Hebert et al., 2003) (Table 2.1). The recognized risk factors for AD are age and family history. Age is the strongest risk factor for AD, which is age-related rather than age-dependent (Evans, 1996; Gao et al., 1998). With regard to family history, persons who have a first degree relative with AD have a 10%–30% increased risk of developing AD. Similarly, the risk of developing AD is increased for individuals with a first degree relative with Parkinson's disease or Down's syndrome (van Duijn et al., 1991). Several genes are known to be involved in both early and late-onset AD (see Genetic Risk Factors).

There is converging evidence that physical activity can reduce the risk of cognitive decline and demention (Barnes et al., 2007). Recent studies have shown that aerobic fitness reduces brain tissue loss in aging humans (Colcombe et al., 2003; Larson and Wang, 2004). The authors of the latter study (Larson and Wang, 2004) explained the positive effect of walking by an improvement in blood flow to areas of the brain used for memory. These findings strongly reinforce calls to continue exercise throughout life and to maintain healthy lifestyle habits to reduce morbidity (Hubert et al., 2002; Morley and Flaherty, 2002). A high educational level may delay the diagnosis of AD (Ott et al., 1995) and, furthermore, centenarians with high levels of high density lipoprotein (HDL) have better cognitive function and exceptional

TABLE 2.1
Statistics of AD

Prevalence of dementia in the European population over 65 years ($n = 25.810$): 6.4%
Prevalence AD as percentage of dementia: 53.7 (38.5–78)%
Odds ratio women compared to men aged >80 years: 1.2–1.3
Incidence AD people aged 65–69 years: 1.2 (0.6–2.3)%
Incidence AD people aged 70–74 years: 3.3 (2.2–4.3)%
Incidence AD people aged 75–79 years: 9.1 (7.1–10.7)%
Incidence AD people aged 80–84 years: 21.8 (17.6–24.0)%
Incidence AD people aged 85–89 years: 35.3 (25.5–36.3)%
Incidence AD people aged >90 years: 53.5 (36.5–55.8)%
Prevalence mutations as a genetic cause: ~2%
Prevalence ApoE4 allele in general population: 16%
Prevalence ApoE4 allele in late-onset AD patients: 50%
Accuracy clinical based diagnosis: 88%[a]

[a] Percentage of AD patients diagnosed based on clinical criteria that also fulfilled the neuropathological criteria.

Source: Reprinted from Teunissen, C.E., de Vente, J., Steinbusch, H.W.M., and de Bruijn, C., *Neurobiol Aging*, 23, 485, 2002. With permission from Elsevier.

longevity (Atzmon et al., 2002). Cholesterol-lowering, especially in midlife, may result in a decline in atherothrombotic brain infarction, which may, in itself, be a causative factor for AD (Aronow et al., 2002a, 2002b).

The increase in the number of patients with AD will not only lead to an elevation of the emotional burden for more and more patients and their social environment, but also to an increase in the economic costs to society in general. For example, in the Netherlands, the annual cost for an AD patient living at home is estimated to be €5000–6000. The annual cost for admission to a residential home is estimated to be €22,000, whereas the cost for admission to nursing homes is estimated to be €45,000–47,000 (prices in 1996) (van der Roer et al., 2000). In the USA, between 2.4 and 3.1 million spouses, relatives, and friends take care of people with AD (Ernst et al., 1997; Whitehouse, 1997).

2.3 GENETIC RISK FACTORS

The cause of sporadic AD is unknown. Several genetic susceptibility factors for familial AD are known and account for about 5% of AD cases. These factors include mutations in the APP gene on chromosome 21 (a total of 18 missense mutations are known) or in the genes coding for presenilin-1 on chromosome 14 (142 mutations) and presenilin-2 on chromosome 1 (10 mutations) (Lendon et al., 1997). APP mutations all reside within the region coding for Abeta. An extra copy of APP in patients with trisomy 21 leads to earlier onset of AD (age 20). Presenilin-1 mutations lead to the most aggressive forms of AD, sometimes with an age of onset below 30. These mutations are autosomal dominant and highly penetrant. The clinical pheno-type of these familial cases closely resembles that of the sporadic late-onset cases. Generally, presenilin-1 and presenilin-2 mutations lead to selective increase in Abeta1-42 species, whereas APP mutations can lead to increased production of all Abeta species, selective Abeta1-42 increase, or generate highly fibrillogenic Abeta variants.

The Apolipoprotein ε4 (ApoE4) allele is a major risk factor for getting sporadic AD at an early age. This allele is found in only a small part of the general population and in only a proportion of AD patients (Strittmatter et al., 1993). Before the discovery that ApoE has a role in the etiology of AD, its function had been studied extensively in lipid metabolism. The ApoE4 allele is associated with increased serum total cholesterol levels and with increased risk of atherosclerosis and coronary artery disease. In AD, ApoE4 allele has been associated with a decreased age at onset, increased senile plaque density, decreased choline acetyltransferase activity, and cholinergic neuron density. In a study of the association between ApoE4 genotype and cardiovascular disease in AD patients, the diagnosis of probable AD was more frequent in ε4 allele-carrying patients with cardiovascular disease. When comparing homozygotes for ε4 with homozygotes for ε3, a ninefold increase in prevalence of cardiac ischemia on ECG was found in the former. When grouping parameters of left ventricular dysfunction, the prevalence was 7.2 times greater in probable AD patients with ApoE4/4. Thus, in patients with probable AD, ApoE4 is associated with cardiac disease indicative of left ventricular dysfunction (van der Cammen et al., 1998). Further research on the interrelation between ApoE4 allele, and cardiovascular disease is ongoing, as is research on the effect of hypertension, cardiac dysfunction, hyperhomocysteinemia, hypercholesterolemia, diabetes melli-tus, smoking, head trauma, depression, educational level, organic solvent exposure, and gender, on age at onset and clinical course of AD (Polvikoski et al., 1995; Breteler, 2000).

2.4 ASSESSMENT

Dementia is a syndrome, and diagnostic tests to confirm or refute the diagnosis of the various subtypes are lacking. Clinical diagnosis of dementia can be made accord-ing to standardized international criteria such as the DSM-IV-criteria (American Psychiatric Association, 1991). The subdiagnosis of probable AD is based on clinical and neuropsychological examination as described in the NINCDS-ADRDA-criteria (McKhann et al., 1984). Vascular dementia is suspected based on the medical history (e.g., cerebrovascular disease), which is confirmed by magnetic resonance imaging (MRI) or single photon emission computed tomography (SPECT) (Scheltens and Hijdra, 1998; Davis et al., 1999). For a diagnosis of vascular dementia to be made, a time relationship (<3 months) between the cerebrovascular accident and the onset of memory decline is required (NINDS-AIREN-criteria) (Roman et al., 1993).

The prodromal form of AD is mild cognitive impairment (MCI). The specific recommendations of the International Working Group on Mild Cognitive Impairment for the general MCI criteria include the following: (1) the person is neither normal nor demented; (2) there is evidence of cognitive deterioration shown either by objectively measured decline over time or a subjective report of decline by self or an informant in conjunction with objective cognitive deficits; and (3) activities of daily living are preserved and complex instrumental functions are either intact or minimally impaired (Winblad et al., 2004).

Genetic testing is not yet ready for routine use in the assessment of the AD patient (Corder et al., 1993; Henderson et al., 1995). Although knowledge about the etiology of AD has grown after the identification of several genes and the subsequent finding of the proteins they encode, many gaps in the knowledge of AD's genetic background remain. Genetic testing is still lacking as a diagnostic tool in the assessment of AD cases, whether sporadic or familial. This caution does not preclude the relevance of genetic testing in the prediction and counseling of early-onset familial AD in which a causal gene has been identified, and for scientific purposes. Research questions to be answered in the coming years include the pathogenicity and penetrance of all mutations and their possible effects on accelerating the process of AD. In the future, knowledge of a patient's genetic background might have implications for predictions of the progression of the disease and for treatment (van der Cammen et al., 2004).

2.4.1 Diagnostic Route

The initial appointment in a patient with suspected dementia should focus on the history. Preferably, an informant is available to give an adequate history of cognitive and behavioral changes (Table 2.2). A drug history is particularly important; use of drugs that impair cognition (e.g., analgesics, anticholinergics, psychotropic medications, and sedative-hypnotics) should be obtained.

A full dementia evaluation cannot be completed in a routine 30 min. visit; adequate time should be arranged. The initial step is to assess cognitive function by a short mental status questionnaire such as the Mini-Mental State Examination (MMSE) (Folstein et al., 1975). This should be followed by a complete physical

---

TABLE 2.2
Symptoms of AD [http://www.uptodate.com]

AD is characterized by one or more of the following clinical manifestations

—A general decrease in the level of cognition, especially memory (often the first symptom)
—Behavioral disturbances
—Interference with daily function and independence

Patients with dementia may have difficulty with one or more of the following

—Learning and retaining new information (e.g., trouble remembering events)
—Handling complex tasks (e.g., balancing a check book)
—Reasoning (e.g., unable to cope with unexpected events)
—Spatial ability and orientation (e.g., getting lost in familiar places)
—Language (e.g., word finding)
—Behavior

examination, including a neurological examination. The subsequent workup may include laboratory and imaging studies.

Neurological examination: Agreement between the history and the mental status examination is strongly suggestive of the diagnosis of dementia. When the history suggests cognitive impairment but the mental status examination is normal, possible explanations include mild dementia, high intelligence or education, depression, or rarely, misrepresentation on the part of the informants (Knopman, 1998). Conversely, when the mental status examination suggests cognitive impairment but the family and patient deny any problems, possible explanations include an acute confusional state, very low intelligence or education, or inadequate recognition by the family (Knopman, 1998). Psychometric testing is the next step in the diagnostic process.

The physical examination is done to rule out an atypical presentation of a medical illness and should be combined with a neurological examination. This should focus on focal neurological deficits that may be consistent with prior strokes, signs of Parkinson's disease (e.g., cogwheel rigidity and tremors), gait, and eye movements. AD patients generally have no motor deficits at presentation.

Laboratory assessment: The American Academy of Neurology recommends screening for B12 deficiency and hypothyroidism in patients with dementia (Knopman et al., 2001). In the Dutch consensus, hemoglobin, hematocrite, mean corpuscular volume, sedimentation rate, glucose, and creatinine level are recommended as routine tests, and on indication—folate, vitamins B1, B6, B12, sodium, and potassium levels [Guideline on the diagnosis and treatment of dementia, 2005, http://www.cbo.nl]

Imaging: In a change from prior guidelines, the American Academy of Neurology now recommends structural neuroimaging with either a noncontrast head CT or MRI in the routine initial evaluation of all patients with dementia (Knopman et al., 2001). However, it is worth noting that there have been no substantive changes in the evidence base to support this particular change.

In the Dutch consensus, neuroimaging is recommended in all patients below the age of 65 and in patients suspected of a treatable neurosurgical disorder [Guideline on the diagnosis and treatment of dementia, 2005, http://www.cbo.nl]. In addition, neuroimaging is recommended when there is a suspicion of vascular dementia, frontotemporal dementia (FTD), Creutzfeld Jakob Disease (CJD), or another rapidly progressive dementia, and when there is doubt about the diagnosis of AD [Guideline on the diagnosis and treatment of dementia, 2005, http://www.cbo.nl]. In general, most patients do not require head CT scans or MRI to make a diagnosis of AD. It may be possible in the near future, however, to use structural MRI SPECT with specific ligands to detect AD at an early stage before the advanced clinical picture becomes apparent, which may have implications for early therapy (Killiany et al., 2000).

2.4.2 SENSITIVITY AND SPECIFICITY OF ASSESSMENT

Detecting dementia is a problem in routine, day-to-day medical practice (Kawas, 2003). The diagnosis is missed in 21% of demented or delirious patients on a general medical ward, whereas 20% of nondemented patients may be judged to be demented

(Barrett et al., 1997). The clinical diagnosis of AD in specialized centers, such as a Memory Clinic, is reasonably accurate. In an autopsy study, for example, 92 out of 106 cases (87%) of clinically diagnosed AD were confirmed pathologically (Gearing et al., 1995). A conclusive diagnosis of AD may only be made after postmortem examination of brain tissue for the presence of large numbers of plaques and tangles. However, it has to be realized that the mere presence of plaques and tangles in the brain is not sufficient evidence for a conclusive diagnosis of AD as these have also been reported in the cognitively normal elderly (Davis et al., 1999).

The Memory Clinic is a diagnostic facility where accurate workup and diagnosis, as well as tailor-made counseling can be provided (van der Cammen et al., 1987). Most patients with AD do not present themselves with a complaint of memory loss; it is often a spouse or other informant who brings the problem to the physician's attention. The problem with the early recognition of AD is that its onset is insidious and often patchy, so that symptoms are not recognized by the patient's social environment until they are structural. Informant's data have been shown to be a reliable primary source for the detection of dementia in geriatric outpatients (van der Cammen et al., 1992) and in community-dwelling older persons (Jorm, 2004).

Elderly patients with subjective cognitive complaints may suffer from depression rather than dementia, although depressive complaints and other neuropsychiatric symptoms also frequently constitute a part of the dementia syndrome (Aalten et al., 2005). Follow-up of elderly depressed patients with cognitive dysfunction after treatment with antidepressants is imperative, including repeated neuropsychological testing, to establish whether the cognitive dysfunction was part of the depression or part of a dementia onset. Cognitive dysfunction, whatever its cause, is associated with a high rate of medical comorbidity (Doraiswamy et al., 2002) and earlier mortality (Bittles et al., 2002). The effects on mortality are worsened when cognitive dysfunction and depression coexist (Mehta et al., 2003).

2.5 PATHOLOGY

There are two main hypotheses explaining the cause and mechanism of AD, the amyloid cascade hypothesis and the tau hypothesis. The amyloid hypothesis implicates a central role of Abeta formation and aggregation, which leads to plaque formation, microglial activation, gliosis, neuronal dysfunction, and death (Hardy and Allsop, 1991). The tauopathy-hypothesis implies a more prominent role of tau pathology (Iqbal and Grundke-Iqbal, 2005). Other mechanisms involved in the pathology of AD are oxidative stress, inflammation, and altered cholesterol homeostasis.

2.5.1 AMYLOID PATHOLOGY

APP is a transmembrane protein with a large ectodomain containing the N-terminal site, residing in the extracellular compartment or inside organelles. The C-terminal site is very short and lies within the membrane.

APP may be cleaved by alpha, beta, or gamma secretases. Epsilon- or zeta-secretase APP cleaving activity have also been reported to precede the gamma-secretase cleavage, yielding an Abeta 1–46 intermediate (Zhao et al., 2004). Alpha-secretase cleavage of APP leads to nonamyloidogenic products. This enzyme cleaves APP between amino acid 16 and 17 of the Abeta peptide, yielding soluble alpha APP, and a truncated Abeta (17–42/43, called P3) fragment after subsequent gamma-cleavage of the transmembrane region of APP. Alpha-secretase cleavage can be performed by several enzymes, including A Disintegrin and Metalloproteinase (ADAM) family members, ADAM9, ADAM10, and ADAM17, and the aspartyl protease beta-site amyloid-cleaving enzyme (BACE-2). The alpha-secretase pathway accounts for 90% of APP metabolism and may serve a neuroprotective role, in contrast to the toxic beta-secretase pathway (Kobayashi and Chen, 2005).

Beta secretase has aspartyl protease activity and cleaves APP at the N-terminal site of Abeta (amino acid 1). Only one enzyme is known for its beta-secretase activity, namely BACE1. This enzyme nevertheless has only a low affinity toward APP and it is believed that APP is not its only substrate.

After subsequent gamma-secretase activity, Abeta1–40 and 1–42 are formed, which are fibrillogenic and form extracellular aggregates. While Abeta 1–40 is present in higher quantities, Abeta1-42 is more amyloidogenic (Iwatsubo et al., 1994). Gamma secretase is a complex of different proteins, including presenilins 1 and 2, nicastrin, anterior pharynx-defective-1 (APH-1), and presenilin enhancer-2 (PEN-2). The presenilins also play crucial roles in cleaving other membrane proteins, such as Notch1 receptor, p75 neurotrophin receptor, CD44, Delta dn Jagged2, low-density lipoprotein receptor-related protein, cell adhesion molecules N- and E-cadherins, and heparan sulfate proteoglycan syndecan-3. The major Abeta degrading enzymes are neprilysin, endothelin-converting enzyme, and insulin-degrading enzyme (Selkoe, 2001).

The increased presence of the aggregated Abeta1–40 or 1–42 species in AD can be due to increased APP formation, altered APP processing, or decreased clearance of deposited Abeta. Furthermore, the amyloid cascade hypothesis has recently been refined by including the formation of soluble oligomeric isoforms of Abeta, that is, Abeta oligomers or Abeta-derived diffusible ligands (ADDLs). These soluble oligomers consist of 4 to 12 Abeta monomers and are more toxic than Abeta1–40 or 1–42 (Walsh et al., 2002). Oligomers may be part of a completely nonfibrillogenic toxic pathway or an early intermediate in fibrillogenesis (Klein et al., 2001; Hardy and Selkoe, 2002; Barghorn et al., 2005). They may also be a result of dissociation of Abeta from fibrils (Sato et al., 2006). Using specific antibodies, the presence of these oligomers in AD plaques has been shown (Takahashi et al., 2004) and the toxicity could be mediated by membrane permeabilization (Glabe and Kayed, 2006). Immunotherapy with antibodies against oligomers could be very promising. For example, such immunotherapy cleared Abeta and tau pathology in triple transgenic mice (Oddo et al., 2006).

2.5.2 TAU PATHOLOGY

The main component of the tangles is the hyperphosphorylated microtubule associated protein (MAP)-tau. Tau is one of four MAPs, the others are MAP-1 (a and b) and MAP-2. There are six isoforms of tau, derived from a single gene at chromosome 17 and resulting from alternative mRNA splicing. These isoforms vary in the number of microtubule-binding domains containing 31 or 32 amino acids at the C-terminus, that is, 3 or 4 domains or repeats, and the number of inserts of 29 amino acids at the N-terminus (two, one, or none). Tau is normally phosphorylated at a maximum of 3 amino acid residues, but in tangles up to 5–9 residues can be phosphorylated. Since the number of tangles is correlated with the degree of cognitive decline in AD, tau phosphorylation likely is an important mechanism defining the clinical phenotype in AD (Iqbal and Grundke-Iqbal, 2005).

Tau pathology probably acts on neuropathology in two ways: via filament and subsequent tangle formation and via disruption of axonal transport. Since microtubuli are important for fast axonal transport, it can be anticipated that dysfunctional tau will inhibit such transport and axonal integrity. Indeed, knock-out animals for at least two MAPs have shown defects in axonal elongation and neuronal migration. It is further thought that the MAP-proteins act in redundancy, since no phenotype changes have been observed in single knock-out animals (Teng et al., 2001).

Hyperphosphorylation of microtubule-binding repeat domains prohibits binding to microtubuli and the microtubule assembly activity of tau. Hyperphosphorylation of tau precedes the formation of tangles, and seems to be sufficient for assembly of tau into filaments (paired helical and straight filaments). Interestingly, tau is not the only protein in AD that is hyperphosphorylated. Other cytoskeletal proteins that are hyperphosphorylated include neurofilaments, MAP1b and tubulin (Iqbal et al., 2005). The basic charge of certain regions in the tau protein seems to prohibit self-polymerization, and hyperphosphorylation can neutralize this basic inhibitory charge of tau leading to polymerization. The regions flanking the microtubule-binding domains in tau are inhibitory to assembly of tau into filaments (Alonso et al., 2001), and it might be that truncated tau lacking these regions is more prone to filament formation.

There are several factors influencing the degree of tau phosphorylation. Several kinases can phosphorylate tau, including glycogen synthase kinase 3 (GSK-3), cyclin-dependent kinase 5 (cdk5), protein kinase A, calcium or calmodulin-dependent protein kinase II (CaMKII), mitogen-activated protein kinase ERK-1 and -2, and stress-activated protein kinases. In addition, phosphatase activity is decreased in AD, thus leading to decreased dephosphorylation activity. Hyperphosphorylation is also dependent on the presence of three or four repeats, the extent of glycosylation, the residue phosphorylated, the conformation, and free tau is more easily phorphorylated than microtubule-bound tau (Iqbal et al., 2005).

Deregulation of cdk5 can play a major role in phosphorylation of proteins in AD. Increased levels of cdk5, P25, and active calpain are detected in AD brains and are shown to induce tau pathology (Cruz and Tsai, 2004). A major element determining the increased activity of cdk5 is the P25-binding partner. Cdk5 needs to be activated by binding partners, such as P35 and P39. Under certain conditions, for example when calpain is activated, P35 is cleaved into a P25 and P10 fragment. The half-life of the binding partners determines the activity of cdk5, and P25 has a fivefold longer half-life than P35. Furthermore, P25 determines the localization of P25-cdk5 activity, due to the absence of the N-terminal myristoylation signal motif, which is important for membrane association, and it more efficiently and preferentially phosphorylates tau and APP. This complex could provide a link between Abeta and tau pathology, since increased activity of Cdk5 could be induced by Abeta, leading to tau hyperphosphorylation. For example, increased cdk5 activity was observed in APP overexpressing mice Tg2576 (Otth et al., 2002) and Abeta has been shown to induce the cleavage of P35 into P25 in primary neurons (Lee et al., 2000). Cdk5 activity is also related to several other death-inducing mechanisms, such as apoptosis and ischemic damage (Cruz and Tsai, 2004).

2.5.3 OXIDATIVE STRESS

Evidence of oxidative stress in AD is obtained by the presence of increased levels of lipid, protein, and DNA oxidation products and by the inactivation of oxidation-sensitive enzymes, such as glutamine synthase or creatine kinase in AD brain tissue (Hensley et al., 1995; Lovell et al., 1999; Smith et al., 2000). Studies in body fluids have also shown increased concentrations of the DNA oxidation product 8-hydroxy-2′-deoxyguanosine in AD patients (Lovell and Markesbery, 2001), and increased levels of isoprostanes, lipid peroxidation products derived from arachidonic acid, in specific brain regions and in CSF have also been shown in AD (Markesbery et al., 2005; Montine et al., 2005). Oxidative stress could be induced by microglial activation. Moreover, Abeta has also been shown to induce radical formation in vitro and to induce microglia activation. Oxidative stress could also be the result of reduced axonal transport of antioxidants, reduced production of antioxidant enzymes, or neuronal mitochondrial dysfunction (Moreira et al., 2005; Reddy, 2006).

2.5.4 INFLAMMATION

In the late 1980s, the first reports appeared showing that microglia in senile plaques contained a protein normally associated with inflammatory processes, that is, HLA-DR, a protein of the class II major histocompatibility complex (MHC) (McGeer et al., 1987, 1996). Other known mediators of inflammation have also been found in plaques, such as interleukin-1β (IL-1β), interleukin-6 (IL-6), and tumor necrosis factor- (TNF)-α (Griffin et al., 1995; Gonzalez-Scarano and Baltuch, 1999). Additional evidence for the involvement of inflammation in AD has been provided by epidemiological data and retrospective clinical data showing positive effects of long-term use of nonsteroidal anti-inflammatory drugs (NSAIDs) on risk of developing AD and time of onset (Breitner et al., 1995; McGeer et al., 1996; In't Veld et al., 2001; Townsend and Pratico, 2005), though a protective effect of these drugs in clinical trials has not been unequivocally shown to date (Hoozemans and O'Banion, 2005). Classical targets of NSAIDs include cycloxygenase (COX)-1 and -2, nuclear factor kappaB, and peroxisome proliferator-activated receptors (PPARs). For example, certain NSAIDs have been reported to decrease BACE1 gene transcription via activation of PPAR-gamma, which is a nuclear transcriptional regulator (Sastre et al., 2006).

2.5.5 CHOLESTEROL HOMEOSTASIS

Cholesterol is the main lipid constituent of neuronal membranes and myelin. It is known that cholesterol is synthesized in the brain in situ (Spady and Dietschy, 1983) and that extracerebral cholesterol does not contribute significantly to brain cholesterol content (Jurevics and Morell, 1995). Excess brain cholesterol have to be removed into the periphery. The mechanism underlying such transport is unclear, but it may be mediated by ApoE (responsible for 1–2 mg/day) and also by facilitated transport of oxidized products like 24S-hydroxycholesterol (responsible for 5–7 mg/day) (Lütjohann et al., 1996; Björkhem et al., 1999). However, an additional, as yet unknown, transport mechanism may be involved [reviewed by Dietschy and Turley (2001)]. It has been hypothesized that during neurodegenerative processes an increased efflux of cholesterol-metabolites from the brain occurs (Björkhem et al., 1998). Aberration of cholesterol homeostasis may indeed be involved in AD. For example, the unesterified cholesterol to phospholipid ratio was decreased in the temporal gyrus of AD patients, whereas the total cholesterol concentration was unchanged. Further, it was shown that use of cholesterol synthesis lowering drugs was associated with a decreased prevalence of AD (Wolozin et al., 2000b). Research on serum cholesterol levels in relation to AD is ongoing. In some studies, decreased serum HDL-cholesterol concentration in patients with AD was observed compared with controls, while others observed increased total and LDL-cholesterol in AD patients (Teunissen et al., 2002). Cellular cholesterol content seems to influence BACE1 cleavage of APP or amyloidogenesis, likely by modulating the association of these enzymes in lipid rafts (Vetrivel and Thinakaran, 2006).

The risk-modulating effects of ApoE mentioned earlier provide another link for the role of cholesterol homeostasis in AD. ApoE is involved in lipid transport and cholesterol homeostasis. ApoE in the circulation is associated with very low-density lipoprotein (VLDL) and involved in reverse cholesterol transport mediated by HDL from cells to the liver. The mechanism underlying the role of ApoE4 allele in AD is still poorly understood (Huang, 2006). Considerable evidence suggests that ApoE isoforms interact differently with Abeta, resulting in isoform-specific effects on Abeta deposition or clearance (Ye et al., 2005).

2.6 ANIMAL MODELS OF AD

No natural animal models of AD exist, and thus DNA modifying techniques have been very helpful in developing animal models for at least certain aspects of AD. Animal models usually reflect only a limited number of characteristics of a disease and for the models of AD this is not less so. Several transgenic models have nevertheless been useful in investigating the role of gene mutations in AD. These models include mice or invertebrate models as *Drosophila melanogaster* flies or the *Caenorhabditis elegans* worms, models that overexpress key proteins in AD pathology and knock-out animals. Furthermore, these animal models have been used in preclinical drug testing. We summarize the behavioral and pathological characteristics of several animal models in AD research, based on recent reviews (Kobayashi and Chen, 2005; Link, 2005).

2.6.1 MICE

2.6.1.1 PDAPP

This was the first mouse model containing plaque-like pathology (Games et al., 1995; Kobayashi and Chen, 2005). The human APP mutation V717F (Indiana) was introduced into a vector driven by the platelet-derived growth factor promoter

(PDGF), using the C57/Bl6, DBA2J, and Swiss-Webster mouse strains. Abeta1–40 and 1–42 levels were 14 times higher in these mice than nontransgenic controls. Importantly, amyloid plaque-like depositions containing Abeta immunoreactive protein aggregates were present in the cortex and hippocampus after the age of 1 year. Numerous dense core and diffuse amyloid plaques were present. These were found throughout several cortical areas, hippocampus, subiculum, and cerebellum. Plaques were associated with adjacent gliosis and dystrophic neurites around dense amyloid cores. Hyperphosphorylated tau proteins were present in older animals, but no tangles were present. Importantly, there was a significant loss of synapses whereas total numbers of neurons were preserved. A reduction in volume of hippocampus, fornix, and corpus callosum was observed, which positively correlated with impairments in spatial memory (Kobayashi and Chen, 2005). These changes were visible already at young ages and before amyloid deposition, suggesting that amyloid plaque formation may not precede cognitive deficits. Other behavioral changes of these mice included neophobia, disturbed sleep-wake rhythms, and learning deficits in the eye blink conditioning paradigm, which is also disturbed in AD patients (Woodruff-Pak, 2001).

2.6.1.2   Tg2576

The second major mutant human APP transgenic mouse was the Tg2576 mouse, developed by Hsiao (Hsiao et al., 1996). The model overexpresses the Swedish double mutant form of APP695 (K670 and M671L) under the control of the hamster prion protein cosmid vector on a background of C57/Bl6 and SJL mice. The brain levels of hAPP in the mutants were 5.5 times higher than native murine APP levels. Like the PDAPP, after about 1 year plaque-like deposits of Abeta1–40 and 1–42/43 were present in several cortical areas (frontal, temporal, and entorhinal), hippocampus, subiculum, and cerebellum. In contrast to the PDAPP mice, no hippocampal atrophy or synaptic loss was present in this model, whereas neuronal numbers were similarly unaffected. Behaviorally, the Tg2576 animals present a memory deficit progressing during aging. Similarly, brain levels of Abeta1–40 and 1–42/43 increased with age from 5- to 14-fold (Hsiao et al., 1996). Similar to the PDAPP mice, spatial memory impairments were present already before plaque deposition (after 6 months of age). However, soluble Abeta levels were found in brain tissue at these ages, in line with the neurotoxicity hypothesis of soluble Abeta oligomers (Westerman et al., 2002). It has to be noted that some controversial data are present in the literature regarding the memory deficits in these mice, which might be due to variation between Tg2576 colonies used by each group (Kobayashi and Chen, 2005).

2.6.1.3   APP23

The APP23 animals overexpress the Swedish (K670N and M671L) mutations in addition to the London (V717I) mutation, in C57/Bl6 × DBA2J mice, under the control of a Thy-1 expression cassette. It was the first animal model containing multiple human mutations and showing neuronal loss (Sturchler-Pierrat et al., 1997; Bondolfi et al., 2002). The mice display a seven-times increased expression of hAPP, localized to the hippocampus and several cortical regions. After 12 months of age, the cerebrovascular depositions of amyloid are coincident with decreased blood flow, microhemorrhages, and even vessel elimination. Dense core plaques containing Abeta are present at 6 months of age, thus much earlier than in PDAPP and Tg2576 mice, and their numbers heavily increase with age. The plaques are accompanied by gliosis, dystrophic neurites, loss of synapses, cholinergic cell loss, and hyperphosphorylated tau proteins. Progressive memory deficits and anxiety behavior were present as early as 3 months of age, whereas hyperactivity was observed already at 6–8 weeks. Insoluble brain Abeta1–42 levels increased heavily between 3 and 6 months (up to 4 times). The severe cerebrovascular abnormalities are an important discriminative feature of this model, which may be the underlying cause of the behavioral disturbances.

2.6.1.4   TgCRND8

This model contains the Swedish (K670N and M671L) mutations in addition to the Indiana (V717F) mutation of APP, under the control of a Hamster prion protein promoter in C3H/B6 strains (Chishti et al., 2001). The model presents with both cognitive deficits and amyloid deposition at a very early age (3 months). Likewise, spatial memory is affected already at this age. hAPP levels in tissue are increased by about five times, leading to a 200-fold increase in Abeta1–40 and 1–42 levels between 1 and 3 months of age. Dense core plaques and dystrophic neurites are detected as early as 3–5 months, in the hippocampus and cortex, progressively spreading to other brain areas. Plaques are accompanied by gliosis, whereas no hippocampal atrophy is visible. Postnatal survival is very low in these animals.

2.6.1.5   J20

These mice contain the same mutations as the TgCRND8 mice, but are under the control of the PDGF minigene in a C57/Bl6 × DBA2J background (Hsia et al., 1999). Plaque-like depositions are present in the hippocampus and cortex as early as 5–7 months of age and are present in about 90% of the animals. Further characteristics include decreased synaptic density, a decrease in calcium-binding proteins in the dentate gyrus and hippocampus, in addition to spatial memory deficits.

2.6.1.6   Knock-Out Animals

The physiological function of APP has been studied using APP knock-out mice, showing neuronal abnormalities (long-term potentiation deficits and seizures). Similarly, presenilin 1 knock-out mice have a lethal phenotype and gross CNS abnormalities, whereas knock-outs in which the PS1 is conditionally deleted in the postnatal brain have no such severe phenotype and only a modest cognitive impairment is expressed (Kobayashi and Chen, 2005).

2.6.1.7   HPS1

Mice overexpressing several human PS1 mutants have no plaque-like accumulations or any clear impairment of spatial memory, though Abeta1–42 levels are increased (Kobayashi and Chen, 2005). This may be surprising, in view of the severe phenotype of the AD patients carrying a presenilin mutation.

As indicated previously, cleavage of APP at the N-terminal site of Abeta occurs via BACE1, and BACE1 knock-out animals have a lack of beta-site generated amyloid peptides. However, no other biochemical, morphological, or behavioral phenotypes have been observed (Luo et al., 2001; Roberds et al., 2001). Other studies using mice with neuronal-specific overexpression of human BACE1 or BACE1 knock-out mice suggest that there is a role for this protease in anxiety behavior (Harrison et al., 2003). Moreover, mice overexpressing both APP and BACE1 show accelerated plaque formation in a.o. cortical and hippocampal areas (Mohajeri et al., 2004). On the other hand, Tg2576 mice overexpressing APP with knock-out of BACE1 fail to develop plaques and cognitive deficits (Ohno et al., 2004, 2006).

APP overexpressing mice that have a disturbed alpha-secretase cleavage site show increased levels of APP and Abeta peptides, are resistant to glutamate excito-toxicity, and sensitive to kainic acid induced seizures. These changes could be related to the increased levels of Abeta as well as loss of alpha-cleaved peptides. Knock-out animals for ADAMs have differential phenotypes, ranging from no changes to lethal phenotypes. Others reveal that ADAM10 overexpression could rescue the cognitive deficits and plaque formation induced by APP overexpression, whereas deletion of ADAM10 in APP overexpressing animals increases plaque formation (Postina et al., 2004).

2.6.2 WORMS AND FLIES

Worms and flies have been very instrumental in unraveling the functions of APP, presenilins, and tau (reviewed by Link, 2005). Induction of an Abeta1–41 minigene in *Caenorhabditis elegans* leads to the formation of amyloid deposits in vivo and a paralysis phenotype, suggesting that this is a good model to study the role of fibrillogenesis and oxidative stress in Abeta toxicity. Research using the fly homolog of APP, Appl, has provided support for a role of APP in axonal transport (Gunawardena and Goldstein, 2001). RNAi knockdown of the worm APP homolog, apl-1, yields a severe uncoordinated phenotype, and genetic deletion of apl-1 results in early lethality (WormBase release WS123). Mutation of sel-1 in *C. elegans*, a gene with high similarity to presenilins essential for normal egg laying, is used as a model to test novel gamma-secretase inhibitors (Ellerbrock et al., 2004). The fly and worm models have also been used to identify protein partners of presenilins that are required for gamma-secretase activity (Link, 2005). Knock-out of tau homologs in worms or flies does not lead to any phenotype, but overexpression of human tau forms (e.g., three-repeat tau, tau mutations) in flies gives rise to specific neuronal cell pathology and toxicity (Williams et al., 2000).

2.7 BIOMARKERS OF AD

Biomarkers in body fluids could be very helpful in obtaining an objective diagnosis of AD, in discriminating AD from other dementias, or in predicting the conversion from MCI into AD. Since the pathological processes of AD are probably ongoing long before the clinical symptoms of AD become apparent, detection of any of these pathological changes in body fluids could be useful in predicting the incidence of the disease. The more specific a pathological change and the more specific the biomarker for the CNS is, the more informative changes in the body fluid levels of this biomarker are likely to be. In view of the heterogeneity of disease processes and the lack of knowledge of the exact cause of AD, several studies have adopted the idea that the combined analysis of several markers related to different disease mechanisms could yield the most informative biomarker assay (Teunissen and Scheltens, 2004; Wiltfang et al., 2005).

Analysis of biomarkers for the key pathological mechanisms in AD, such as tau and Abeta, has facilitated quite accurate discrimination between AD patients and controls, as well as between AD and other diseases related to dementia. CSF levels of tau are about 200%–300% in AD patients compared with healthy controls, but the difference is less than other diseases associated with massive axonal damage, such as CJD or other forms of dementia. Increased CSF levels of hyperphosphorylated tau isoforms are more specific for AD. The CSF levels of Abeta1–42 are reduced by approximately 50% in AD compared with healthy controls, though this decrease can also be observed in other diseases, including CJD or amyotrophic lateral sclerosis. Furthermore, the combined analysis of increased tau and decreased Abeta has improved the sensitivity (up to 96%) and specificity (up to 100%) of the diagnosis (Blennow, 2004). The results of combined assays have been shown to be modified already in MCI and as such are helpful in identifying MCI patients at high risk of progression to AD (Riemenschneider et al., 2002). The measurement of truncated Abeta isoforms has been considered, but does not increase sensitivity compared with measurement of Abeta alone (Schoonenboom et al., 2005). Interestingly, nanoparticle-based detection of soluble Abeta oligomers identified increased levels of these oligomers in the CSF of AD patients compared with nondemented age-matched controls (Georganopoulou et al., 2005). In addition, a number of laboratories independently described alterations in APP metabolism or concentration in blood platelets of AD patients when compared with control subjects matched for demographic characteristics (Cattabeni et al., 2004). Platelets exhibit concentrations of APP isoforms equivalent to those found in the brain and could therefore serve as a surrogate marker for AD pathology, for example, in evaluating therapeutic effects (Zimmermann et al., 2005).

Other possible markers investigated include a broad range of inflammatory proteins, but so far no consistent differences have been obtained such that they can yield diagnostic or prognostic information as a single marker (Teunissen et al., 2002). For example, the CSF or blood levels of the inflammatory markers interleukin (IL)-6 and TNF-α have been subject to variability. Recently, a polymorphism of the TNF-gene has been associated with an increased risk of AD in ApoE4 carriers and is linked to higher Abeta 1–42 levels in the CSF (Laws et al., 2005). Serum concentration of α-antichymotrypsin (ACT) is a much investigated inflammatory marker (Teunissen et al., 2002) however the analysis of such inflammatory proteins in serum combined with proteins related to other mechanisms has improved the correct classification of AD patients (Teunissen et al., 2003). The relationship between changes in body fluid levels of these markers to CNS pathology is still not clear.

Increased serum or plasma levels of homocysteine have been shown to be a risk factor for AD (Ravaglia et al., 2005) although CSF levels of homocysteine are reported to be similar in AD patients compared with elderly controls (Serot et al., 2005). The value of homocysteine as a diagnostic marker may be limited since it has a relatively unspecific relation with the disease process and a large overlap exists between its levels in patients and their controls. It is not clear therefore what the mechanism is that is responsible for the increases in AD incidence (Teunissen et al., 2005).

Biomarkers of oxidative stress have also been investigated. Studies on the CSF levels of the oxidative stress markers 3-nitrotyrosine and the DNA oxidation product 8-hydroxy-2′-deoxyguanosine have been reported to be increased in AD patients (Teunissen et al., 2002; Ahmed et al., 2005). In addition, CSF levels of specific isoprostanes have been shown to be elevated in MCI (Pratico et al., 2002). In addition, total plasma antioxidant capacity is decreased in both MCI and AD but not in vascular dementia patients (Guidi et al., 2006). Concerning antioxidant vitamins, decreased levels of these markers have been observed, but these markers have low specificity for AD and should therefore be analyzed in combination with other markers for potential use in diagnosis (Teunissen et al., 2002).

Another promising marker could be the brain-specific cholesterol metabolite 24S-hydroxycholesterol. Serum or CSF levels of this marker are almost exclusively related to brain-specific cholesterol metabolism and altered serum and CSF levels have also been described in AD patients. In one study, serum 24S-hydroxycholesterol concentration was increased in mildly affected AD patients compared with controls, though there was considerable overlap (Lütjohann et al., 2000). The 24S-hydroxycholesterol concentrations correlated negatively with severity, that is, the highest levels were observed in the less affected patients while other studies showed decreased 24S-OH-cholesterol levels in the plasma of severely affected AD patients (Papassotiropoulos et al., 2000; Heverin et al., 2004). Thus, increased levels of this oxysterol may be observed during early stages of AD and decreased levels in more advanced states. However, we did not observe any correlation between oxysterol levels and cognitive decline in the healthy elderly (Teunissen et al., 2002). Thus, changes may be typical for early stages of AD or very severe cognitive decline in the elderly.

2.7.1 RECENT DEVELOPMENTS

It has been reported that the CSF levels of the adhesion molecules L1 and neuronal cellular adhesion molecule (NCAM) were increased during AD (Strekalova et al., 2006).

Interestingly, the CSF levels of the Abeta degrading enzyme neprilysin were decreased in MCI and mild AD patients compared with age-matched controls though the levels were normalized in moderate AD patients (Maruyama et al., 2005).

Proteomic analysis of CSF is a current topic in AD and only a few studies have been performed so far (Davidsson and Sjogren, 2006). The effect of preclinical variables, variation within the methods, differences in sampling protocols as well as optimization of the methods for CSF analysis are important aspects in proteomics, and these have been scarcely addressed (Hu et al., 2005). When these issues are resolved, proteomics may provide a powerful tool for biomarker discovery.

2.8 THERAPIES

Therapies affecting any of the key components of the mechanisms involved in AD may be effective in preventing (further) disease development. The currently available therapy is the use of acetylcholinesterase inhibitors (e.g., tacrine hydrochloride, donepezil hydrochloride, rivastigmine tartrate, and galantamine hydrobromide), leading to elevated levels of acetylcholine, a major excitatory neurotransmitter. These drugs marginally ameliorate the cognitive and behavioral symptoms in AD and it is likely that they only act on disease symptoms. For example, donepezil seems to be most effective during the prodromal phases of AD (Rosenberg, 2005). For moderately advanced AD, memantine, an N-methyl-D-aspartate (NMDA) receptor antagonist, is currently in use. This drug has positive effects in moderate to severe patients improving activities of daily living. Hormone replacement therapy has also been proposed to be effective in AD. However, this therapy has been shown in further investigations to have adverse effects on a.o. cognition, and the idea has now been abandoned. Vitamin E supplementation, alone or in combination with vitamin C, does not have convincing positive effects on cognition and even increases mortality and morbidity risks (Boothby and Doering, 2005; Pham and Plakogiannis, 2005).

The use of statins, inhibitors of the rate-limiting enzyme in cholesterol synthesis HMG-CoA, has been linked to a reduced risk of AD in retrospective epidemiological studies (Wolozin et al., 2000a). This effect appears to be dependent on the type of statin used. In a recent open label study, simvastatin increased the CSF levels of nonamyloidogenic metabolites of APP (Hoglund et al., 2005).

Other targets for therapy development include the enzymes involved in APP cleavage, such as alpha, beta, and gamma secretases, Abeta degrading enzymes, or metal chelators such as clioquinol, which are supposed to chelate metals in plaques (Rosenberg, 2005).

Only a few randomized double blind placebo-controlled trials have been performed using anti-inflammatory drugs (COX-2 or nonspecific COX-inhibitors). As indicated earlier, a protective effect of these drugs in clinical trials has not been unequivocally established (McGeer and McGeer, 2007), so the evidence favoring their application is somewhat limited.

Immunization of APP transgenic animals with Abeta yielded such promising results that trials on humans have been started. For example, immunization of PDAPP mice with Abeta1–42 led to a reduction of amyloid pathology (Schenk et al., 1999). Similarly vaccination of CRND8 APP overexpressing mice led to restored cognitive function in the Morris water maze test and reduced amyloid pathology (Janus et al., 2000). Passive immunization may also reduce cognitive deficits, though the effects on plaque deposition are variable (Kotilinek et al., 2002). The effects of immunization between the various APP models probably depend on the severity of plaque deposition and timing.

A randomized clinical trial testing immunization of humans with preaggregated Abeta1–42 has been conducted (Hock et al., 2002). Patients developed specific antibodies against Abeta, which could react with Abeta present in plaques, diffuse Abeta deposits, and Abeta in brain blood vessels. The rate of cognitive decline and decline in daily activities was also slowed. Unfortunately, about 6% of the immunized patients developed autoimmune meningoencephalitis, and the trial was stopped. In deceased patients (due to unrelated causes) almost complete absence of Abeta plaques was observed, whereas neurofibrillary tangles were still present and demyelination was also observed in these brains. Since the vaccination was successful, apart from the side effects, new strategies are being developed, such as DNA vaccination or the use of different epitope or adjuvant combinations (He et al., 2005; Kim et al., 2005).

ACKNOWLEDGMENT

We would like to thank Dr P. van der Valk (VU University Medical Center Amsterdam) for providing AD brain sections for Figure 2.1.

REFERENCES

Aalten, P., de Vugt, M.E., Jaspers, N., Jolles, J., and Verhey, F.R. (2005) The course of neuropsychiatric symptoms in dementia. Part I: findings from the two-year longitudinal Maasbed study. *International Journal of Geriatric Psychiatry*, 20, 523–530.

Ahmed, N., Ahmed, U., Thornalley, P.J., Hager, K., Fleischer, G., and Munch, G. (2005) Protein glycation, oxidation and nitration adduct residues and free adducts of cerebrospinal fluid in Alzheimer's disease and link to cognitive impairment. *Journal of Neurochemistry*, 92, 255–263.

Alonso, A., Zaidi, T., Novak, M., Grundke-Iqbal, I., and Iqbal, K. (2001) Hyperphosphorylation induces self-assembly of tau into tangles of paired helical filaments/straight filaments. *Proceedings of the National Academy of Sciences of the United States of America*, 98, 6923–6928.

American Psychiatric Association. (1991) Diagnostic and Statistical Manual of Mental Disorders. American Psychiatric Association, Washington DC.

Aronow, W.S., Ahn, C., and Gutstein, H. (2002a) Incidence of new atherothrombotic brain infarction in older persons with prior myocardial infarction and serum low-density lipoprotein cholesterol > or = 125 mg/dl treated with statins versus no lipid-lowering drug. *Journal of Gerontology*, 57, M333–M335.

Aronow, W.S., Ahn, C., and Gutstein, H. (2002b) Reduction of new coronary events and new atherothrombotic brain infarction in older persons with diabetes mellitus, prior myocardial infarction, and serum low-density lipoprotein cholesterol >/= 125 mg/dl treated with statins. *Journal of Gerontology*, 57, M747–M750.

Atzmon, G., Gabriely, I., Greiner, W., Davidson, D., Schechter, C., and Barzilai, N. (2002) Plasma HDL levels highly correlate with cognitive function in exceptional longevity. *Journal of Gerontology*, 57, M712–M715.

Barghorn, S., Nimmrich, V., Striebinger, A., Krantz, C., Keller, P., Janson, B. et al. (2005) Globular amyloid beta-peptide oligomer—a homogenous and stable neuropathological protein in Alzheimer's disease. *Journal of Neurochemistry*, 95, 834–847.

Barrett, J.J., Haley, W.E., Harrell, L.E., and Powers, R.E. (1997) Knowledge about Alzheimer disease among primary care physicians, psychologists, nurses, and social workers. *Alzheimer Disease and Associated Disorders*, 11, 99–106.

Bittles, A.H., Petterson, B.A., Sullivan, S.G., Hussain, R., Glasson, E.J., and Montgomery, P.D. (2002) The influence of intellectual disability on life expectancy. *Journal of Gerontology*, 57, M470–M472.

Björkhem, I., Diczfalusy, U., and Lütjohann, D. (1999) Removal of cholesterol from extrahepatic sources by oxidative mechanisms. *Current Opinion in Lipidology*, 10, 161–165.

Björkhem, I., Lütjohann, D., Diczfalusy, U., Stahle, L., Ahlborg, G., and Wahren, J. (1998) Cholesterol homeostasis in human brain: turnover of 24S- hydroxycholesterol and evidence for a cerebral origin of most of this oxysterol in the circulation. *Journal of Lipid Research*, 39, 1594–1600.

Blennow, K. (2004) Cerebrospinal fluid protein biomarkers for Alzheimer's disease. *NeuroRx*, 1, 213–225.

Bondolfi, L., Calhoun, M., Ermini, F., Kuhn, H.G., Wiederhold, K.H., Walker, L. et al. (2002) Amyloid-associated neuron loss and gliogenesis in the neocortex of amyloid precursor protein transgenic mice. *Journal of Neuroscience*, 22, 515–522.

Boothby, L.A. and Doering, P.L. (2005) Vitamin C and vitamin E for Alzheimer's disease. *Annals of Pharmacotherapy*, 39, 2073–2080.

Breitner, J.C., Welsh, K.A., Helms, M.J., Gaskell, P.C., Gau, B.A., Roses, A.D. et al. (1995) Delayed onset of Alzheimer's disease with nonsteroidal anti-inflammatory and histamine H2 blocking drugs. *Neurobiology of Aging*, 16, 523–530.

Breteler, M.M. (2000) Vascular risk factors for Alzheimer's disease: an epidemiologic perspective. *Neurobiology of Aging*, 21, 153–160.

Barnes, D.E., Whitmer, R.A., Yaffe, K. (2007) Physical activity and dementia: the need for prevention trials. *Excercise and Sport Sciences Reviews*, 35, 24–29.

Cattabeni, F., Colciaghi, F., and Di, L.M. (2004) Platelets provide human tissue to unravel pathogenic mechanisms of Alzheimer disease. *Progress in Neuro-Psychopharmacology and Biological Psychiatry*, 28, 763–770.

Chishti, M.A., Yang, D.S., Janus, C., Phinney, A.L., Horne, P., Pearson, J. et al. (2001) Early-onset amyloid deposition and cognitive deficits in transgenic mice expressing a double mutant form of amyloid precursor protein 695. *Journal of Biological Chemistry*, 276, 21562–21570.

Colcombe, S.J., Erickson, K.I., Raz, N., Webb, A.G., Cohen, N.J., McAuley, E. et al. (2003) Aerobic fitness reduces brain tissue loss in aging humans. *Journal of Gerontology*, 58, 176–180.

Corder, E.H., Saunders, A.M., Strittmatter, W.J., Schmechel, D.E., Gaskell, P.C., Small, G.W. et al. (1993) Gene dose of apolipoprotein E type 4 allele and the risk of Alzheimer's disease in late onset families. *Science*, 261, 921–923.

Cruz, J.C. and Tsai, L.H. (2004) Cdk5 deregulation in the pathogenesis of Alzheimer's disease. *Trends in Molecular Medicine*, 10, 452–458.

Davidsson, P. and Sjogren, M. (2006) Proteome studies of CSF in AD patients. *Mechanisms of Ageing and Development*, 127, 133–137.

Davis, D.G., Schmitt, F.A., Wekstein, D.R., and Markesbery, W.R. (1999) Alzheimer neuropathologic alterations in aged cognitively normal subjects.1. *Journal of Neuropathology and Experimental Neurology*, 58, 376–388.

Dietschy, J.M. and Turley, S.D. (2001) Cholesterol metabolism in the brain. *Current Opinion in Lipidology*, 12, 105–112.

Doraiswamy, P.M., Leon, J., Cummings, J.L., Marin, D., and Neumann, P.J. (2002) Prevalence and impact of medical comorbidity in Alzheimer's disease. *Journal of Gerontology*, 57, M173–M177.

Ellerbrock, B.R., Coscarelli, E.M., Gurney, M.E., and Geary, T.G. (2004) Screening for presenilin inhibitors using the free-living nematode, *Caenorhabditis elegans*. *Journal of Biomolecular Screening*, 9, 147–152.

Ernst, R.L., Hay, J.W., Fenn, C., Tinklenberg, J., and Yesavage, J.A. (1997) Cognitive function and the costs of Alzheimer disease. An exploratory study. *Archives of Neurology*, 54, 687–693.

Evans, D.A. (1996) The epidemiology of dementia and Alzheimer's disease: an evolving field. *Journal of the American Geriatrics Society*, 44, 1482–1483.

Folstein, M.F., Folstein, S.E., and McHugh, P.R. (1975) Mini-mental state. A practical method for grading the cognitive state of patients for the clinician 12. *Journal of Psychiatric Research*, 12, 189–198.

Games, D., Adams, D., Alessandrini, R., Barbour, R., Berthelette, P., Blackwell, C. et al. (1995) Alzheimer-type neuropathology in transgenic mice overexpressing V717F beta-amyloid precursor protein. *Nature*, 373, 523–527.

Gao, S., Hendrie, H.C., Hall, K.S., and Hui, S. (1998) The relationships between age, sex, and the incidence of dementia and Alzheimer disease: a meta-analysis. *Archives of General Psychiatry*, 55, 809–815.

Gearing, M., Mirra, S.S., Hedreen, J.C., Sumi, S.M., Hansen, L.A., and Heyman, A. (1995) The consortium to establish a registry for Alzheimer's disease (CERAD). Part X Neuropathology confirmation of the clinical diagnosis of Alzheimer's disease. *Neurology*, 45, 461–466.

Georganopoulou, D.G., Chang, L., Nam, J.M., Thaxton, C.S., Mufson, E.J., Klein, W.L. et al. (2005) Nanoparticle-based detection in cerebral spinal fluid of a soluble pathogenic biomarker for Alzheimer's disease. *Proceedings of the National Academy of Sciences of the United States of America*, 102, 2273–2276.

Glabe, C.G. and Kayed, R. (2006) Common structure and toxic function of amyloid oligomers implies a common mechanism of pathogenesis. *Neurology*, 66, S74–S78.

Gonzalez-Scarano, F. and Baltuch, G. (1999) Microglia as mediators of inflammatory and degenerative diseases. *Annual Review of Neuroscience*, 22, 219–240.

Griffin, W.S., Sheng, J.G., Roberts, G.W., and Mrak, R.E. (1995) Interleukin-1 expression in different plaque types in Alzheimer's disease: significance in plaque evolution. *Journal of Neuropathology and Experimental Neurology*, 54, 276–281.

Guidi, I., Galimberti, D., Lonati, S., Novembrino, C., Bamonti, F., Tiriticco, M. et al. (2006) Oxidative imbalance in patients with mild cognitive impairment and Alzheimer's disease. *Neurobiology of Aging*, 27, 262–269.

Gunawardena, S. and Goldstein, L.S. (2001) Disruption of axonal transport and neuronal viability by amyloid precursor protein mutations in Drosophila. *Neuron*, 32, 389–401.

Hardy, J. and Allsop, D. (1991) Amyloid deposition as the central event in the aetiology of Alzheimer's disease. *Trends in Pharmacological Sciences*, 12, 383–388.

Hardy, J. (2003) The relationship between amyloid and tau. *Journal of Molecular Neuroscience*, 20, 203–206.

Hardy, J. and Selkoe, D.J. (2002) The amyloid hypothesis of Alzheimer's disease: progress and problems on the road to therapeutics. *Science*, 297, 353–356.

Harrison, S.M., Harper, A.J., Hawkins, J., Duddy, G., Grau, E., Pugh, P.L. et al. (2003) BACE1 (beta-secretase) transgenic and knockout mice: identification of neurochemical deficits and behavioral changes. *Molecular and Cellular Neurosciences*, 24, 646–655.

He, Y., Sun, S.H., Chen, R.W., Guo, Y.J., He, X.W., Huang, L. et al. (2005) Effects of epitopes combination and adjuvants on immune responses to anti-Alzheimer disease DNA vaccines in mice. *Alzheimer Disease and Associated Disorders*, 19, 171–177.

Hebert, L.E., Scherr, P.A., Bienias, J.L., Bennett, D.A., and Evans, D.A. (2003) Alzheimer disease in the US population: prevalence estimates using the 2000 census. *Archives of Neurology*, 60, 1119–1122.

Henderson, A.S., Easteal, S., Jorm, A.F., Mackinnon, A.J., Korten, A.E., Christensen, H. et al. (1995) Apolipoprotein E allele epsilon 4, dementia, and cognitive decline in a population sample. *Lancet*, 346, 1387–1390.

Hensley, K., Hall, N., Subramaniam, R., Cole, P., Harris, M., Aksenov, M. et al. (1995) Brain regional correspondence between Alzheimer's disease histopathology and biomarkers of protein oxidation. *Journal of Neurochemistry*, 65, 2146–2156.

Heverin, M., Bogdanovic, N., Lutjohann, D., Bayer, T., Pikuleva, I., Bretillon, L. et al. (2004) Changes in the levels of cerebral and extracerebral sterols in the brain of patients with Alzheimer's disease. *Journal of Lipid Research*, 45, 186–193.

Hock, C., Konietzko, U., Papassotiropoulos, A., Wollmer, A., Streffer, J., von Rotz, R.C. et al. (2002) Generation of antibodies specific for beta-amyloid by vaccination of patients with Alzheimer disease. *Nature Medicine*, 8, 1270–1275.

Hoglund, K., Thelen, K.M., Syversen, S., Sjogren, M., von, B.K., Wallin, A. et al. (2005) The effect of simvastatin treatment on the amyloid precursor protein and brain cholesterol metabolism in patients with Alzheimer's disease. *Dementia and Geriatric Cognitive Disorders*, 19, 256–265.

Hoozemans, J.J. and O'Banion, M.K. (2005) The role of COX-1 and COX-2 in Alzheimer's disease pathology and the therapeutic potentials of non-steroidal anti-inflammatory drugs. *Current Drug Targets CNS and Neurological Disorders*, 4, 307–315.

Hsia, A.Y., Masliah, E., McConlogue, L., Yu, G.Q., Tatsuno, G., Hu, K. et al. (1999) Plaque-independent disruption of neural circuits in Alzheimer's disease mouse models. *Proceedings of the National Academy of Sciences of the United States of America*, 96, 3228–3233.

Hsiao, K., Chapman, P., Nilsen, S., Eckman, C., Harigaya, Y., Younkin, S. et al. (1996) Correlative memory deficits, Abeta elevation, and amyloid plaques in transgenic mice. *Science*, 274, 99–102.

Hu, Y., Malone, J.P., Fagan, A.M., Townsend, R.R., and Holtzman, D.M. (2005) Comparative proteomic analysis of intra- and interindividual variation in human cerebrospinal fluid. *Molecular & Cellular Proteomics*, 4, 2000–2009.

Huang, Y. (2006) Apolipoprotein E and Alzheimer disease. *Neurology*, 66, S79–S85.

Hubert, H.B., Bloch, D.A., Oehlert, J.W., and Fries, J.F. (2002) Lifestyle habits and compression of morbidity. *Journal of Gerontology*, 57, M347–M351.

In't Veld, B.A., Ruitenberg, A., Hofman, A. et al. (2001) Nonsteroidal antiinflammatory drugs and the risk of Alzheimer's disease. *New England Journal of Medicine*, 345, 1515–1521.

Iqbal, K., Alonso, A.C., Chen, S., Chohan, M.O., El-Akkad, E., Gong, C.X. et al. (2005) Tau pathology in Alzheimer disease and other tauopathies. *Biochimica Et Biophysica Acta*, 1739, 198–210.

Iqbal, K. and Grundke-Iqbal, I. (2005) Metabolic/signal transduction hypothesis of Alzheimer's disease and other tauopathies. *Acta Neuropathologica*, 109, 25–31.

Iwatsubo, T., Odaka, A., Suzuki, N., Mizusawa, H., Nukina, N., and Ihara, Y. (1994) Visualization of A beta 42(43) and A beta 40 in senile plaques with end-specific A beta monoclonals: evidence that an initially deposited species is A beta 42(43). *Neuron*, 13, 45–53.

Janus, C., Pearson, J., McLaurin, J., Mathews, P.M., Jiang, Y., Schmidt, S.D. et al. (2000) A beta peptide immunization reduces behavioural impairment and plaques in a model of Alzheimer's disease. *Nature*, 408, 979–982.

Jorm, A.F. (2004) The Informant Questionnaire on cognitive decline in the elderly (IQCODE): a review. *International Psychogeriatrics*, 16, 275–293.

Jurevics, H. and Morell, P. (1995) Cholesterol for synthesis of myelin is made locally, not imported into brain. *Journal of Neurochemistry*, 64, 895–901.

Kawas, C.H. (2003) Clinical practice. Early Alzheimer's disease. *New England Journal of Medicine*, 349, 1056–1063.

Killiany, R.J., Gomez-Isla, T., Moss, M., Kikinis, R., Sandor, T., Jolesz, F. et al. (2000) Use of structural magnetic resonance imaging to predict who will get Alzheimer's disease. *Annals of Neurology*, 47, 430–439.

Kim, H.D., Maxwell, J.A., Kong, F.K., Tang, D.C., and Fukuchi, K. (2005) Induction of anti-inflammatory immune response by an adenovirus vector encoding 11 tandem repeats of Abeta1-6: toward safer and effective vaccines against Alzheimer's disease. *Biochemical and Biophysical Research Communications*, 336, 84–92.

Klein, W.L., Krafft, G.A., and Finch, C.E. (2001) Targeting small Abeta oligomers: the solution to an Alzheimer's disease conundrum? *Trends in Neurosciences*, 24, 219–224.

Knopman, D.S. (1998) The initial recognition and diagnosis of dementia. *American Journal of Medicine*, 104, 2S–12S.

Knopman, D.S., DeKosky, S.T., Cummings, J.L., Chui, H., Corey-Bloom, J., Relkin, N. et al. (2001) Practice parameter: diagnosis of dementia (an evidence-based review). Report of the Quality Standards Subcommittee of the American Academy of Neurology. *Neurology*, 56, 1143–1153.

Kobayashi, D.T. and Chen, K.S. (2005) Behavioral phenotypes of amyloid-based genetically modified mouse models of Alzheimer's disease. *Genes, Brain, and Behavior*, 4, 173–196.

Kotilinek, L.A., Bacskai, B., Westerman, M., Kawarabayashi, T., Younkin, L., Hyman, B.T. et al. (2002) Reversible memory loss in a mouse transgenic model of Alzheimer's disease. *Journal of Neuroscience*, 22, 6331–6335.

Larson, E.B. and Wang, L. (2004) Exercise, aging, and Alzheimer disease. *Alzheimer Disease and Associated Disorders*, 18, 54–56.

Laws, S.M., Perneczky, R., Wagenpfeil, S., Muller, U., Forstl, H., Martins, R.N. et al. (2005) TNF polymorphisms in Alzheimer disease and functional implications on CSF beta-amyloid levels. *Human Mutation*, 26, 29–35.

Lee, M.S., Kwon, Y.T., Li, M., Peng, J., Friedlander, R.M., and Tsai, L.H. (2000) Neurotoxicity induces cleavage of p35 to p25 by calpain. *Nature*, 405, 360–364.

Lendon, C.L., Ashall, F., and Goate, A.M. (1997) Exploring the etiology of Alzheimer disease using molecular genetics. *The Journal of the American Medical Association*, 277, 825–831.

Link, C.D. (2005) Invertebrate models of Alzheimer's disease. *Genes, Brain, and Behavior*, 4, 147–156.

Lovell, M.A., Gabbita, S.P., and Markesbery, W.R. (1999) Increased DNA oxidation and decreased levels of repair products in Alzheimer's disease ventricular CSF. *Journal of Neurochemistry*, 72, 771–776.

Lovell, M.A. and Markesbery, W.R. (2001) Ratio of 8-hydroxyguanine in intact DNA to free 8-hydroxyguanine is increased in Alzheimer disease ventricular cerebrospinal fluid. *Archives of Neurology*, 58, 392–396.

Luo, Y., Bolon, B., Kahn, S., Bennett, B.D., Babu-Khan, S., Denis, P. et al. (2001) Mice deficient in BACE1, the Alzheimer's beta-secretase, have normal phenotype and abolished beta-amyloid generation. *Nature Neuroscience*, 4, 231–232.

Lütjohann, D., Breuer, O., Ahlborg, G., Nennesmo, I., Siden, A., Diczfalusy, U. et al. (1996) Cholesterol homeostasis in human brain: evidence for an age-dependent flux of 24S-hydroxycholesterol from the brain into the circulation. *Proceedings of the National Academy of Sciences of the United States of America*, 93, 9799–9804.

Lütjohann, D., Papassotiropoulos, A., Bjorkhem, I., Locatelli, S., Bagli, M., Oehring, R.D. et al. (2000) Plasma 24S-hydroxycholesterol (cerebrosterol) is increased in Alzheimer and vascular demented patients. *Journal of Lipid Research*, 41, 195–198.

Markesbery, W.R., Kryscio, R.J., Lovell, M.A., and Morrow, J.D. (2005) Lipid peroxidation is an early event in the brain in amnestic mild cognitive impairment. *Annals of Neurology*, 58, 730–735.

Maruyama, M., Higuchi, M., Takaki, Y., Matsuba, Y., Tanji, H., Nemoto, M. et al. (2005) Cerebrospinal fluid neprilysin is reduced in prodromal Alzheimer's disease. *Annals of Neurology*, 57, 832–842.

McGeer, P.L., Itagaki, S., Tago, H., and McGeer, E.G. (1987) Reactive microglia in patients with senile dementia of the Alzheimer type are positive for the histocompatibility glycoprotein HLA-DR. *Neuroscience Letters*, 79, 195–200.

McGeer, P.L. and McGeer, E.G. (2007) NSAIDs and Alzheimer disease: epidemiological, animal model and clinical studies. *Neurobiology of Aging*, 28, 639–647.

McGeer, P.L., Schulzer, M., and McGeer, E.G. (1996) Arthritis and anti-inflammatory agents as possible protective factors for Alzheimer's disease: a review of 17 epidemiologic studies. *Neurology*, 47, 425–432.

McKhann, G., Drachman, D., Folstein, M., Katzman, R., Price, D., and Stadlan, E.M. (1984) Clinical diagnosis of Alzheimer's disease: report of the NINCDS-ADRDA Work Group under the auspices of Department of Health and Human Services Task Force on Alzheimer's Disease. *Neurology*, 34, 939–944.

Mehta, K.M., Yaffe, K., Langa, K.M., Sands, L., Whooley, M.A., and Covinsky, K.E. (2003) Additive effects of cognitive function and depressive symptoms on mortality in elderly community-living adults. *Journal of Gerontology*, 58, M461–M467.

Mohajeri, M.H., Saini, K.D., and Nitsch, R.M. (2004) Transgenic BACE expression in mouse neurons accelerates amyloid plaque pathology. *Journal of Neural Transmission*, 111, 413–425.

Montine, T.J., Montine, K.S., McMahan, W., Markesbery, W.R., Quinn, J.F., and Morrow, J.D. (2005) F2-isoprostanes in Alzheimer and other neurodegenerative diseases. *Antioxidants & Redox Signaling*, 7, 269–275.

Moreira, P.I., Honda, K., Liu, Q., Santos, M.S., Oliveira, C.R., Aliev, G. et al. (2005) Oxidative stress: the old enemy in Alzheimer's disease pathophysiology. *Current Alzheimer Research*, 2, 403–408.

Morley, J.E. and Flaherty, J.H. (2002) It's never too late: health promotion and illness prevention in older persons. *Journal of Gerontology*, 57, M338–M342.

Oddo, S., Caccamo, A., Tran, L., Lambert, M.P., Glabe, C.G., Klein, W.L. et al. (2006) Temporal profile of amyloid-beta (Abeta) oligomerization in an in vivo model of Alzheimer disease. A link between Abeta and tau pathology. *Journal of Biological Chemistry*, 281, 1599–1604.

Ohno, M., Chang, L., Tseng, W., Oakley, H., Citron, M., Klein, W.L. et al. (2006) Temporal memory deficits in Alzheimer's mouse models: rescue by genetic deletion of BACE1. *European Journal of Neuroscience*, 23, 251–260.

Ohno, M., Sametsky, E.A., Younkin, L.H., Oakley, H., Younkin, S.G., Citron, M. et al. (2004) BACE1 deficiency rescues memory deficits and cholinergic dysfunction in a mouse model of Alzheimer's disease. *Neuron*, 41, 27–33.

Ott, A., Breteler, M.M., van, H.F., Claus, J.J., van der Cammen, T.J., Grobbee, D.E. et al. (1995) Prevalence of Alzheimer's disease and vascular dementia: association with education. The Rotterdam study. *British Medical Journal*, 310, 970–973.

Otth, C., Concha, I.I., Arendt, T., Stieler, J., Schliebs, R., Gonzalez-Billault, C. et al. (2002) AbetaPP induces cdk5-dependent tau hyperphosphorylation in transgenic mice Tg2576. *Journal of Alzheimer's Disease*, 4, 417–430.

Papassotiropoulos, A., Lütjohann, D., Bagli, M., Locatelli, S., Jessen, F., Rao, M.L. et al. (2000) Plasma 24S-hydroxycholesterol: a peripheral indicator of neuronal degeneration and potential state marker for Alzheimer's disease. *Neuroreport*, 11, 1959–1962.

Pham, D.Q. and Plakogiannis, R. (2005) Vitamin E supplementation in Alzheimer's disease, Parkinson's disease, tardive dyskinesia, and cataract: Part 2. *Annals of Pharmacotherapy*, 39, 2065–2072.

Polvikoski, T., Sulkava, R., Haltia, M., Kainulainen, K., Vuorio, A., Verkkoniemi, A. et al. (1995) Apolipoprotein E, dementia, and cortical deposition of beta-amyloid protein. *New England Journal of Medicine*, 333, 1242–1247.

Postina, R., Schroeder, A., Dewachter, I., Bohl, J., Schmitt, U., Kojro, E. et al. (2004) A disintegrin-metalloproteinase prevents amyloid plaque formation and hippocampal defects in an Alzheimer disease mouse model. *Journal of Clinical Investigation*, 113, 1456–1464.

Pratico, D., Clark, C.M., Liun, F., Rokach, J., Lee, V.Y., and Trojanowski, J.Q. (2002) Increase of brain oxidative stress in mild cognitive impairment: a possible predictor of Alzheimer disease. *Archives of Neurology*, 59, 972–976.

Ravaglia, G., Forti, P., Maioli, F., Martelli, M., Servadei, L., Brunetti, N. et al. (2005) Homocysteine and folate as risk factors for dementia and Alzheimer disease. *American Journal of Clinical Nutrition*, 82, 636–643.

Reddy, P.H. (2006) Amyloid precursor protein-mediated free radicals and oxidative damage: implications for the development and progression of Alzheimer's disease. *Journal of Neurochemistry*, 96, 1–13.

Riemenschneider, M., Wagenpfeil, S., Diehl, J., Lautenschlager, N., Theml, T., Heldmann, B. et al. (2002) Tau and Abeta42 protein in CSF of patients with frontotemporal degeneration. *Neurology*, 58, 1622–1628.

Roberds, S.L., Anderson, J., Basi, G., Bienkowski, M.J., Branstetter, D.G., Chen, K.S. et al. (2001) BACE knockout mice are healthy despite lacking the primary beta-secretase activity in brain: implications for Alzheimer's disease therapeutics. *Human Molecular Genetics*, 10, 1317–1324.

Rocca, W.A., Hofman, A., Brayne, C., Breteler, M.M., Clarke, M., Copeland, J.R. et al. (1991) Frequency and distribution of Alzheimer's disease in Europe: a collaborative study of 1980–1990 prevalence findings. The EURODEM-Prevalence Research Group. *Annals of Neurology*, 30, 381–390.

Roman, G.C., Tatemichi, T.K., Erkinjuntti, T., Cummings, J.L., Masdeu, J.C., Garcia, J.H. et al. (1993) Vascular dementia: diagnostic criteria for research studies. Report of the NINDS-AIREN International Workshop. *Neurology*, 43, 250–260.

Rosenberg, R.N. (2005) Translational research on the way to effective therapy for Alzheimer disease. *Archives of General Psychiatry*, 62, 1186–1192.

Sastre, M., Dewachter, I., Rossner, S., Bogdanovic, N., Rosen, E., Borghgraef, P. et al. (2006) Nonsteroidal anti-inflammatory drugs repress beta-secretase gene promoter activity by the activation of PPARgamma. *Proceedings of the National Academy of Sciences of the United States of America*, 103, 443–448.

Sato, N., Okochi, M., Taniyama, Y., Kurinami, H., Shimamura, M., Takeuchi, D. et al. (2006) Development of new screening system for Alzheimer disease, in vitro Abeta sink assay, to identify the dissociation of soluble Abeta from fibrils. *Neurobiology of Disease*, 22, 487–495.

Scheltens, P. and Hijdra, A.H. (1998) Diagnostic criteria for vascular dementia 5. *Haemostasis*, 28, 151–157.

Schenk, D., Barbour, R., Dunn, W., Gordon, G., Grajeda, H., Guido, T. et al. (1999) Immunization with amyloid-beta attenuates Alzheimer-disease-like pathology in the PDAPP mouse. *Nature*, 400, 173–177.

Schoonenboom, N.S., Mulder, C., Van Kamp, G.J., Mehta, S.P., Scheltens, P., Blankenstein, M.A. et al. (2005) Amyloid beta 38, 40, and 42 species in cerebrospinal fluid: more of the same? *Annals of Neurology*, 58, 139–142.

Selkoe, D.J. (2001) Alzheimer's disease: genes, proteins, and therapy. *Physiological Reviews*, 81, 741–766.

Serot, J.M., Barbe, F., Arning, E., Bottiglieri, T., Franck, P., Montagne, P. et al. (2005) Homocysteine and methylmalonic acid concentrations in cerebrospinal fluid: relation with age and Alzheimer's disease. *Journal of Neurology, Neurosurgery and Psychiatry*, 76, 1585–1587.

Smith, M.A., Nunomura, A., Zhu, X., Takeda, A., and Perry, G. (2000) Metabolic, metallic, and mitotic sources of oxidative stress in Alzheimer disease. *Antioxidants & Redox Signaling*, 2, 413–420.

Spady, D.K. and Dietschy, J.M. (1983) Sterol synthesis in vivo in 18 tissues of the squirrel monkey, guinea pig, rabbit, hamster, and rat. *Journal of Lipid Research*, 24, 303–315.

Strekalova, H., Buhmann, C., Kleene, R., Eggers, C., Saffell, J., Hemperly, J. et al. (2006) Elevated levels of neural recognition molecule L1 in the cerebrospinal fluid of patients with Alzheimer disease and other dementia syndromes. *Neurobiology of Aging*, 27, 1–9.

Strittmatter, W.J., Saunders, A.M., Schmechel, D., Pericak-Vance, M., Enghild, J., Salvesen, G.S. et al. (1993) Apolipoprotein E: high-avidity binding to beta-amyloid and increased frequency of type 4 allele in late-onset familial Alzheimer disease. *Proceedings of the National Academy of Sciences of the United States of America*, 90, 1977–1981.

Sturchler-Pierrat, C., Abramowski, D., Duke, M., Wiederhold, K.H., Mistl, C., Rothacher, S. et al. (1997) Two amyloid precursor protein transgenic mouse models with Alzheimer disease-like pathology. *Proceedings of the National Academy of Sciences of the United States of America*, 94, 13287–13292.

Takahashi, R.H., Almeida, C.G., Kearney, P.F., Yu, F., Lin, M.T., Milner, T.A. et al. (2004) Oligomerization of Alzheimer's beta-amyloid within processes and synapses of cultured neurons and brain. *Journal of Neuroscience*, 24, 3592–3599.

Teng, J., Takei, Y., Harada, A., Nakata, T., Chen, J., and Hirokawa, N. (2001) Synergistic effects of MAP2 and MAP1B knockout in neuronal migration, dendritic outgrowth, and microtubule organization. *Journal of Cell Biology*, 155, 65–76.

Teunissen, C.E., de Vente, J., Steinbusch, H.W.M., and de Bruijn, C. (2002) Biochemical markers related to Alzheimer's dementia in serum and cerebrospinal fluid. *Neurobiology of Aging*, 23, 485–508.

Teunissen, C.E., Lutjohann, D., von Bergmann, K., Verhey, F., Vreeling, F., Wauters, A. et al. (2003) Combination of serum markers related to several mechanisms in Alzheimer's disease. *Neurobiology of Aging*, 24, 893–902.

Teunissen, C.E. and Scheltens, P. (2004) Biomarkers for Alzheimer's disease. Which way to go? *Neurobiology of Aging*, 25, 695–696.

Teunissen, C.E., van Boxtel, M.P., Jolles, J., de Vente, J., Vreeling, F., Verhey, F. et al. (2005) Homocysteine in relation to cognitive performance in pathological and non-pathological conditions. *Clinical Chemistry and Laboratory Medicine*, 43, 1089–1095.

Townsend, K.P. and Pratico, D. (2005) Novel therapeutic opportunities for Alzheimer's disease: focus on nonsteroidal anti-inflammatory drugs. *FASEB Journal*, 19, 1592–1601.

van der Cammen, T.J., Croes, E.A., Dermaut, B., de Jager, M.C., Cruts, M., Van, B.C. et al. (2004) Genetic testing has no place as a routine diagnostic test in sporadic and familial cases of Alzheimer's disease. *Journal of the American Geriatrics Society*, 52, 2110–2113.

van der Cammen, T.J., Simpson, J.M., Fraser, R.M., Preker, A.S., and Exton-Smith, A.N. (1987) The memory clinic. A new approach to the detection of dementia. *British Journal of Psychiatry*, 150, 359–364.

van der Cammen, T.J., van, H.F., Stronks, D.L., Passchier, J., and Schudel, W.J. (1992) Value of the Mini-Mental State Examination and informants' data for the detection of dementia in geriatric outpatients. *Psychological Reports*, 71, 1003–1009.

van der Cammen, T.J., Verschoor, C.J., van Loon, C.P., van, H.F., de, K.I., Schudel, W.J. et al. (1998) Risk of left ventricular dysfunction in patients with probable Alzheimer's disease with APOε4 allele. *Journal of the American Geriatrics Society*, 46, 962–967.

van der Roer, N., Goes, E.S., Blom, M., and Busschbach, J.J. (2000) Model of costs of care for dementia: community-dwelling vs. institutionalization. *Tijdschr Gerontol Geriatr*, 31, 55–61.

van Duijn, C.M., Clayton, D., Chandra, V., Fratiglioni, L., Graves, A.B., Heyman, A. et al. (1991) Familial aggregation of Alzheimer's disease and related disorders: a collaborative re-analysis of case-control studies. EURODEM Risk Factors Research Group. *International Journal of Epidemiology*, 20 Suppl 2, S13–S20.

Vetrivel, K.S. and Thinakaran, G. (2006) Amyloidogenic processing of beta-amyloid precursor protein in intracellular compartments. *Neurology*, 66, S69–S73.

Walsh, D.M., Klyubin, I., Fadeeva, J.V., Cullen, W.K., Anwyl, R., Wolfe, M.S. et al. (2002) Naturally secreted oligomers of amyloid beta protein potently inhibit hippocampal long-term potentiation in vivo. *Nature*, 416, 535–539.

Westerman, M.A., Cooper-Blacketer, D., Mariash, A., Kotilinek, L., Kawarabayashi, T., Younkin, L.H. et al. (2002) The relationship between Abeta and memory in the Tg2576 mouse model of Alzheimer's disease. *Journal of Neuroscience*, 22, 1858–1867.

Whitehouse, P.J. (1997) Pharmacoeconomics of dementia. *Alzheimer Disease and Associated Disorders*, 11 Suppl 5, S22–S32.

Williams, D.W., Tyrer, M., and Shepherd, D. (2000) Tau and tau reporters disrupt central projections of sensory neurons in Drosophila. *Journal of Comparative Neurology*, 428, 630–640.

Wiltfang, J., Lewczuk, P., Riederer, P., Grunblatt, E., Hock, C., Scheltens, P. et al. (2005) Consensus paper of the WFSBP Task Force on Biological Markers of Dementia: the role of CSF and blood analysis in the early and differential diagnosis of dementia. *The World Journal of Biological Psychiatry*, 6, 69–84.

Winblad, B., Palmer, K., Kivipelto, M., Jelic, V., Fratiglioni, L., Wahlund, L.O. et al. (2004) Mild cognitive impairment–beyond controversies, towards a consensus: report of the International Working Group on Mild Cognitive Impairment. *Journal of Internal Medicine*, 256, 240–246.

Wolozin, B., Kellman, W., Ruosseau, P., Celesia, G.G., and Siegel, G. (2000a) Decreased prevalence of Alzheimer disease associated with 3-hydroxy-3-methyglutaryl coenzyme A reductase inhibitors. *Archives of Neurology*, 57, 1439–1443.

Wolozin, B., Kellman, W., Ruosseau, P., Celesia, G.G., and Siegel, G. (2000b) Decreased prevalence of Alzheimer disease associated with 3-hydroxy-3-methyglutaryl coenzyme A reductase inhibitors 5. *Archives of Neurology*, 57, 1439–1443.

Woodruff-Pak, D.S. (2001) Eyeblink classical conditioning differentiates normal aging from Alzheimer's disease. *Integrative Physiological and Behavioral Science*, 36, 87–108.

Ye, S., Huang, Y., Mullendorff, K., Dong, L., Giedt, G., Meng, E.C. et al. (2005) Apolipoprotein (apo) E4 enhances amyloid beta peptide production in cultured neuronal cells: apoE structure as a potential therapeutic target. *Proceedings of the National Academy of Sciences of the United States of America*, 102, 18700–18705.

Zhao, G., Mao, G., Tan, J., Dong, Y., Cui, M.Z., Kim, S.H. et al. (2004) Identification of a new presenilin-dependent zeta-cleavage site within the transmembrane domain of amyloid precursor protein. *Journal of Biological Chemistry*, 279, 50647–50650.

Zimmermann, M., Borroni, B., Cattabeni, F., Padovani, A., and Di, L.M. (2005) Cholinesterase inhibitors influence APP metabolism in Alzheimer disease patients. *Neurobiology of Disease*, 19, 237–242.

WEBSITES:

www.uptodate.com
www.cbo.nl

3 Improving Cholinergic Transmission

Pierre Francotte, Pascal de Tullio,
Bernard Pirotte, and Jure Stojan

CONTENTS

3.1 CHOLINERGIC TRANSMISSION AND ACETYLCHOLINE RELEASE ENHANCERS

Pierre Francotte, Pascal de Tullio, and Bernard Pirotte

Nearly 30 years ago, the decline in learning and memory ability observed in Alzheimer's disease (AD) was advanced to be derived from a deficiency in cholinergic neurotransmission, linked to the loss of cholinergic neurons in the Nucleus Basalis of Maynert and other nuclei projecting to the hippocampus and mesial temporal regions (Cummings et al., 1998; Geula, 1998). This speculation was logically termed "cholinergic hypothesis." A consequent theory predicted that drugs able to potentiate central cholinergic function should be of therapeutic interest in cognition treatment and even behavioral problems experienced with AD.

It is now commonly accepted that this cholinergic hypothesis may explain the majority of the cognitive deficits in AD, mild cognitive impairment (MCI), or dementia. Hence, many strategies have been investigated to improve or restore the cholinergic transmission in the brain.

The conception of potential targets for the drug design is based on the knowledge of acetylcholine (ACh, **3.1**) pharmacology at the synaptic level. As can be deduced from Figure 3.1, several strategies have been explored to restore the central cholinergic function: the use of ACh-releasing agents, the stimulation of acetylcholine uptake, the activation of cholinergic receptors (postsynaptic M_1 muscarinic) by synthetic agonists, and the lowering of acetylcholine metabolic breakdown by inhibition of acetylcholinesterase (AChE). Acetylcholine precursors such as

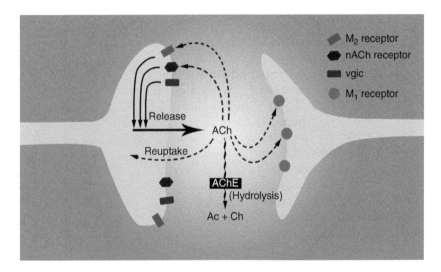

FIGURE 3.1 A simplified view of what happens to ACh in the synapse. After its liberation in the synaptic cleft, ACh may bind with its receptors (muscarinic or nicotinic). Activation of nAChRs and M_2 receptors enhance ACh release from the axon terminal (blocking voltage-gated ion channels—"vgic"—also enhance ACh release). ACh action is ended either owing to a reuptake mechanism or the neurotransmitter's hydrolysis. Hydrolysis is achieved by acetylcholinesterases (AChE), and results in the formation of acetic acid and one choline molecule.

lecithin and choline constituted another possible strategy, but preliminary trials showed their ineffectiveness in AD treatment; however, this approach would probably have to be reconsidered with other molecules (Amenta et al., 2001).

Restoring the transmission by a transmitter-replacement therapy remains an evident strategy. This may be achieved by cholinergic agonists that specifically bind to M_1 subtype receptors, located at the postsynaptic level. One other interesting solution to restore the cholinergic transmission consists in enhancing the release of neurotransmitter at the presynaptic level. This approach has been investigated by designing nicotinic agonists, M_2 muscarinic antagonists, and modulators of voltage-gated ion channels.

3.1.1 ACH RELEASE ENHANCERS: NICOTINIC STIMULATION

Nicotinic acetylcholine receptors (nAChRs) belong to a superfamily of ligand-gated ion channels composed of a pentagonal array of subunit proteins. Neuronal nAChRs may be composed of $\alpha2$–10 and $\beta2$–4 subunits; they are mainly represented by the $\alpha4\beta2$ and the $\alpha7$ subtypes: the first one are homo-oligomers built with $\alpha7$ subunits, whereas the latter are made with 2 $\alpha4$ and 3 $\beta2$ subunits. During AD development, the number of $\alpha4\beta2$ nAChRs is reduced, impairing among others presynaptic ACh release and postsynaptic depolarization (signal transduction). Moreover, recent studies have demonstrated that Aβ interacts directly with the $\alpha7$ nAChR subtype, suggesting that Aβ might have a function as an endogenous ligand for this receptor. The resulting impairment of $\alpha7$ nAChRs could be critical, since they have a high relative calcium permeability and regulate numerous events in nervous cells (Liu et al., 2001). Thus, the approach of nicotinic stimulation could be useful not only for treating symptoms, but also the underlying causes of AD.

Hence, the increase of nicotinic signals appears of therapeutic value in the treatment of cognitive disorders, namely against AD. Nicotinic enhancement may be achieved by two pharmacological classes of compounds: nicotinic agonists or positive allosteric modulators (PAMs) (Figure 3.2).

3.1.1.1 Nicotinic Agonists

After many years of complex work in the quest for selective $\alpha4\beta2$ ligands, medicinal chemists are now increasingly focusing on $\alpha7$ nAChRs ligands (Mazurov et al., 2006). However, both $\alpha4\beta2$ and $\alpha7$ nicotinic receptors appear to be critical for memory function. A recently performed study highlighted the complex actions of nicotine on its receptors, since a selective agonist for $\alpha4\beta2$ receptors was shown to reduce the level of long-term potentiation (LTP) in the dentate gyrus in vivo (Wang et al., 2006) although contradictory results were previously obtained (Matsuyama et al., 2000; Matsuyama and Matsumoto, 2003).

In the first rank of nicotinic agonists stands the prototype ligand, nicotine (**3.2**) itself. This natural alkaloid (initially isolated from *Tobacco* sp.) has been extensively studied and was shown to improve cognitive function in a variety of experiments in both animals and humans (Levin et al., 2006). However, its use in therapy has been precluded because of its abundant side effects. Hence, efforts have been done to develop nicotine-related compounds, with improved pharmacological properties.

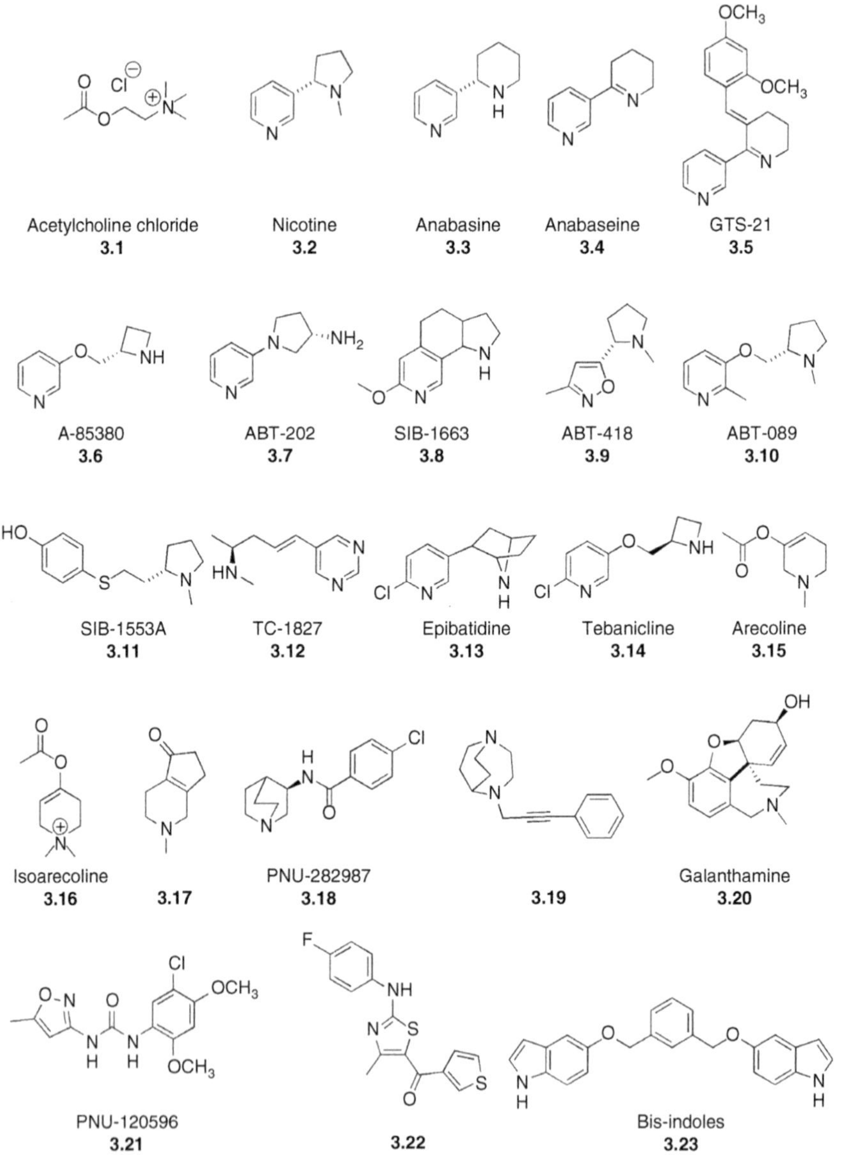

FIGURE 3.2 nAChRs ligands.

Starting from the nicotine structure, several pharmacomodulation strategies were followed. Work was reported focusing attention on the pyrrolidine ring of nicotine. This approach led to the performance of studies with compounds such as anabasine derivatives (anabasine (**3.3**) was previously known as a *Tobacco* alkaloid) and anabaseine derivatives (anabaseine (**3.4**) is a nonselective nicotinic agonist, isolated from a marine worm) (Kem, 2002). Amongst all anabaseine derivatives stands the selective α7 partial agonist GTS-21 (**3.5**), for which intensive evaluations were achieved (de Fiebre et al., 1995; Briggs et al., 1997). It is worth noting that GTS-21 was shown to protect against Aβ-induced neuronal death (Kihara et al., 1997). Other examples are given by compounds A-85380 (**3.6**) or ABT-202 (**3.7**), developed as a pain treatment (Jain, 2004).

Resulting from another research approach, came SIB-1663 (**3.8**), a conformationally restricted analog of nicotine, currently investigated as a pain treatment, since it exhibited in vivo antinociception effects (Vernier et al., 1998).

Although all these compounds bear the pyridine ring of nicotine, other nicotinic agonists were designed conserving the pyrrolidine ring of nicotine. For instance, we may cite here the α4β2 agonists ABT-418 (**3.9**) (Decker et al., 1994), ABT-089 (**3.10**) (Lin et al., 1997), SIB-1553A (**3.11**) (Vernier et al., 1999), currently under study for the treatment of AD.

TC-1827 (**3.12**) is a pyrimidinic derivative (inspired from metanicotine structure), which has been recently described as a selective and potent activator of brain α4β2 receptors with good oral absorption (Bohme et al., 2004a).

Mother Nature also offered to medicinal chemists epibatidine (**3.13**). This compound is another natural alkaloid isolated from an Ecuadorian frog (*Epipedobates tricolor*) and is recognized as one of the most potent, although nonselective agonists; work based on this molecule's structure led to the synthesis of compounds such as tebanicline (also found in the literature as ABT-594, **3.14**). This compound binds with greater specificity to α3β4 receptors and has displayed antinociceptive activity in pain models (Meyer et al., 2000).

Recently, the affinity of novel rigid analogs of arecoline (**3.15**) and isoarecoline (**3.16**) such as (**3.17**) for central nAChRs was evaluated in vitro, providing information on the molecular requirements for developing new active compounds (Guandalini et al., 2002).

On the other hand, quinuclidine derivatives are currently investigated as nicotinic agonists (Bunnelle et al., 2006), leading to the publication of a great number of patents. For instance, PNU-282987 (**3.18**) is a quinuclidine benzamide which was characterized as a very active agonist at α7 nAChRs in cultured rat neurons and was shown to reverse an amphetamine-induced gating deficit in rats (Bodnar et al., 2005); diazabicyclononane derivatives such as (**3.19**) were also recently patented (Ernst et al., 2005).

3.1.1.2 Positive Nicotinic Modulators

One other approach to enhance nicotinic signals consists in allosterically modulating the nAChR. Beside the well-known galanthamine (**3.20** which is also recognized as an AChE inhibitor, see later) (Woodruff-Pak et al., 2002), several new structures were recently described as nAChRs potentiators: the urea PNU-120596 (**3.21**), discovered in a high-throughput screening, could have potential utility to treat AD (Hurst et al., 2005). On the other hand are found compounds such as (**3.22**) (Broad et al., 2006) or bis-indoles (**3.23**) (Balestra et al., 2001).

3.1.2 ACh Release Enhancers: Voltage-Gated Ion Channel Modulators

Blockers of voltage-gated ion channels have been studied for their ability to enhance ACh release. Amongst all, potassium channel blockers have been the most

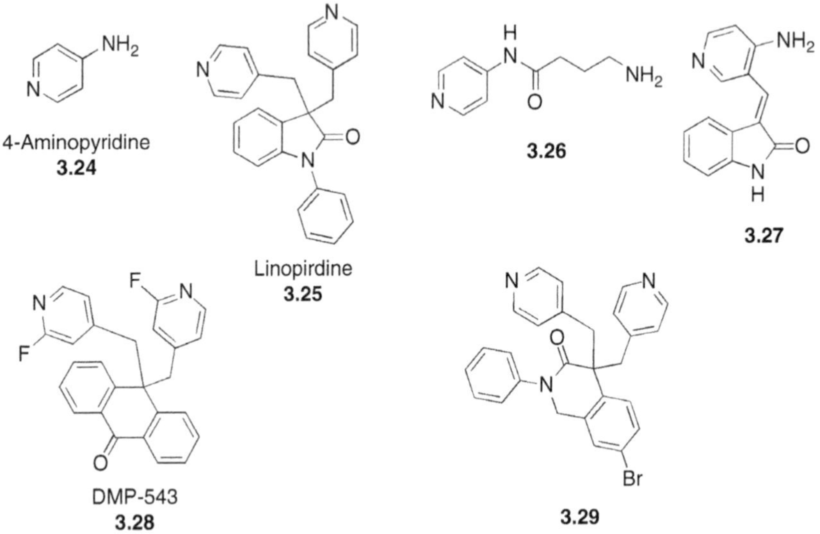

FIGURE 3.3 Voltage-gated ion channel blockers.

studied. Based on the structure of 4-aminopyridine (**3.24**) and linopirdine (**3.25**), the 4-aminopyridine derivatives (**3.26**) and (**3.27**) were reported as potent antiamnesics by Andreani et al. (2000). Other structures were inspired by linopirdine's structure: we can cite here DMP-543 (**3.28**) (Earl et al., 1998), which may represent a potential palliative therapeutic for treatment of AD. Another example is furnished by a series of isoquinolinones from which emerged (**3.29**) (Lin et al., 2002) (Figure 3.3).

3.1.3 ACH RELEASE ENHANCERS: M₂ MUSCARINIC ANTAGONISTS

Another possible mechanism for augmenting central cholinergic activity is to increase acetylcholine release by blockade of inhibitory presynaptic M_2 autoreceptors in the central nervous system (CNS). It has been shown in animals that ACh levels were increased by this means and that cognition was improved. Since M_2 receptors are also found in cardiac tissue, M_2 antagonists may increase heart rate.

The major challenge is designing compounds that have selectivity for M_2 versus the other types of muscarinic receptors, especially the postsynaptic M_1 receptors. Only a few compounds have this property.

The pyridobenzodiazepinone BIBN-99 (**3.30**, Figure 3.4) may be considered as the benchmark selective M_2 antagonist (Doods, 1995); this compound was shown to significantly improve performance in the aged cognitively impaired rat (Quirion et al., 1995); BIBN-99 effects were shown to persist in the absence of the drug (Rowe et al., 2003).

Schering-Plough Research Institute has reported the design of M_2 antagonists (Figure 3.4). The first lead compound of this series was SCH-57790 (**3.31**) (Billard et al., 2000), which showed promising in vivo results, but was abandoned because of the presence of the chemically labile benzylic cyano group (Camps and Munoz-Torrero,

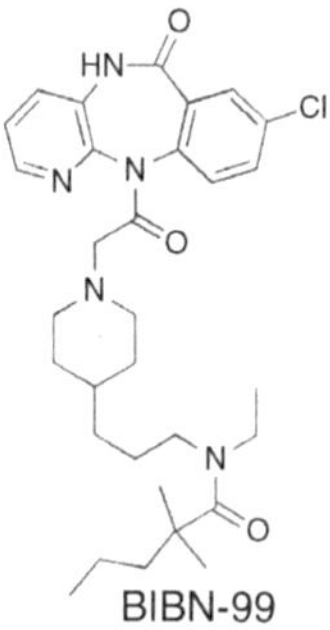

BIBN-99
3.30

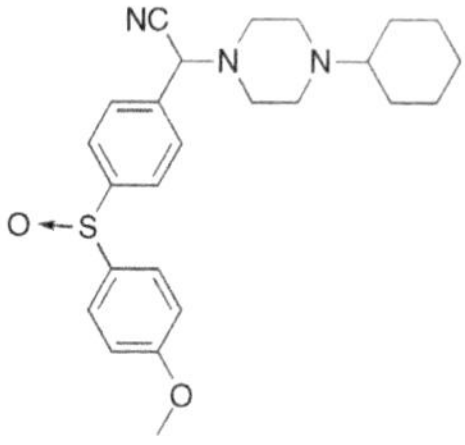

SCH 57790
3.31

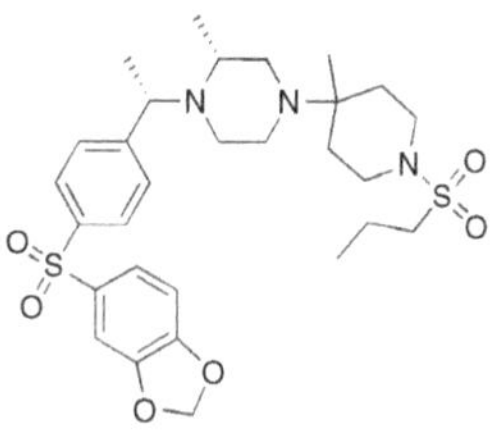

SCH 72788
3.32

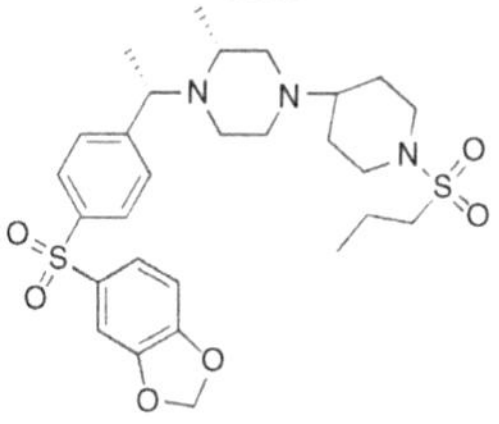

3.33

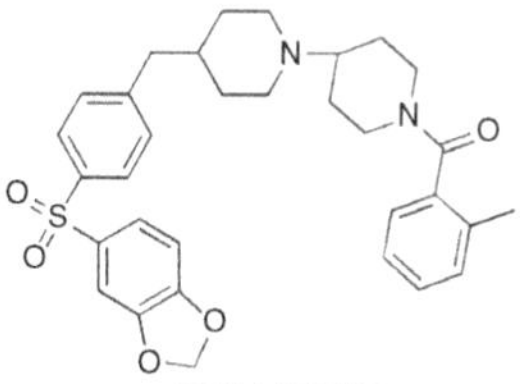

SCH 76050
3.34

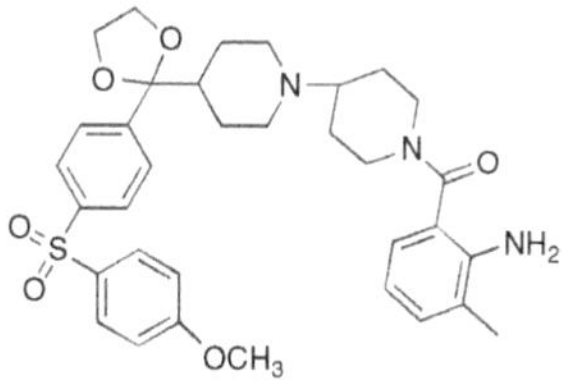

SCH 217443
3.35

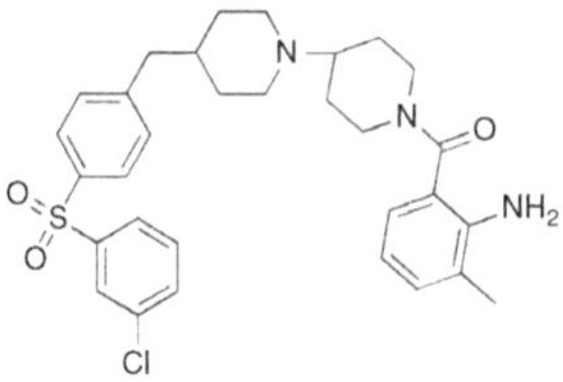

3.36

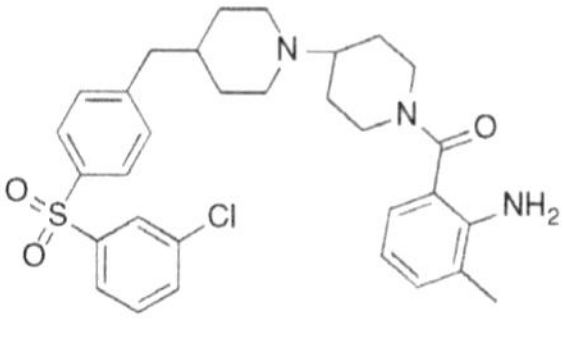

SCH 211803
3.37

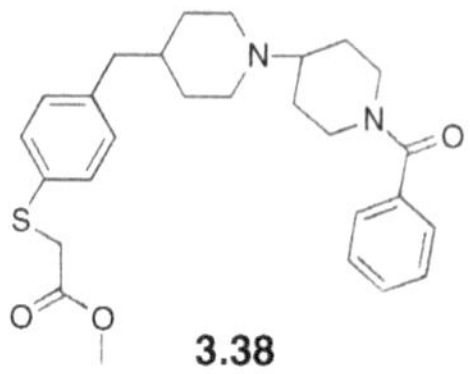

3.38

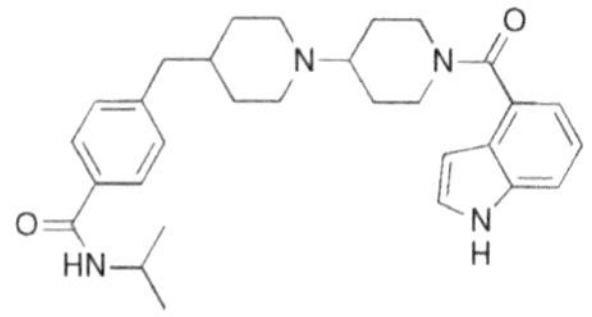

3.39

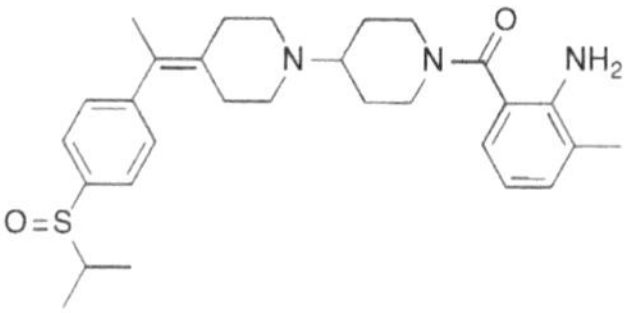

3.40

FIGURE 3.4 M_2 antagonists.

2002). Based on the observation that the replacement of the cyclohexyl ring with a substituted piperidine moiety improved M_2 selectivity, several alternatives were prepared, giving SCH-72788 (**3.32**) with an M_2 K_i of 0.5 nM, an 84-fold M_1 or M_2 selectivity and which had an interesting in vivo activity (Lachowicz et al., 2001).

Another structural modulation starting from SCH-57790 (consisting in the replacement of sulfoxide and cyano moieties were replaced with sulfone and (*S*)-methyl, respectively), led to the synthesis of new compounds such as (**3.33**, Figure 3.4) displaying an M_1 or M_2 selectivity of 109-fold with an M_2 $K_i = 0.7$ nM. Moreover, this compound induced a sustained increase in ACh levels when administered orally in the microdialysis paradigm, in rats (Kozlowski et al., 2002). Additional data about the structural requirements have been reported (McCombie et al., 2002).

Commencing from SCH-57790's (**3.31**) structure, piperidinylpiperidine analogs have been designed (Figure 3.4). Among the novel compounds emerged the potent and selective antagonist SCH-76050 (**3.34**), which was shown to have a poor pharmacokinetic profile (low plasma levels and extensive metabolism). Hence, derivatives with enhanced bioavailability were designed (introduction of a ketal group and the replacement of the (2-methyl)benzamido group by an anthranilamide, resulting in SCH-217443 (**3.35**)). The tachycardiac side effects were observed at doses 30-fold higher than those required for cognition activity in rats, showing how promising is this compound (Greenlee et al., 2001). The most recent development in the "anthranilamide" series was published in 2004, with the discovery of the highly selective M_2 antagonist (**3.36**); this compound also showed good oral bioavailability and in vivo activity (Clader et al., 2004).

SCH-211803 (**3.37**) represented another attractive selective M_2 antagonist and has been shown to be active in animal models of cognition (Asberom et al., 2001). Efforts to simplify the left-hand portion of SCH-211803 (**3.37**) were achieved, obtaining antagonists such as (**3.38**) (Wang et al., 2002) or (**3.39**) (Palani et al., 2004) (Figure 3.4). This type of compounds, with a lower molecular weight, may represent new lead compounds. Compound (**3.40**), a conformationally restricted analog of the previous compounds, was reported to have a superior oral efficacy in animal models (Wang et al., 2002).

On the other hand, based on theoretical structural requirements for M_2 antagonism, compounds containing the benzofulvene core were synthesized; from this series emerged derivatives with a quinuclidine moiety and a pyridine residue such as (**3.41**) (Figure 3.5) (Böhme et al., 2001).

Studying the M_2 antagonistic activity of dimethindene (a well-known H_1 antagonist), a German team discovered (**3.42**), which exhibited a better pharmacological

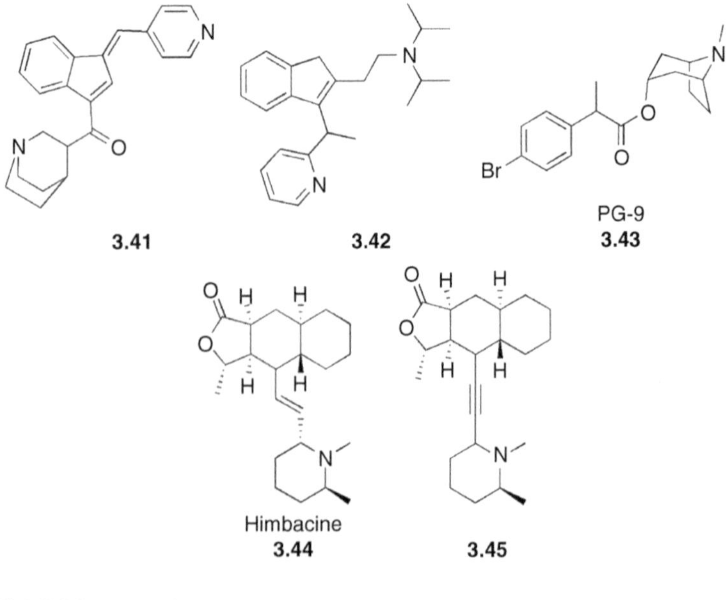

3.41   3.42   PG-9
3.43

Himbacine
3.44   3.45

FIGURE 3.5 M_2 antagonists.

profile than the starting molecule (improved selectivity pattern concerning M_2 receptors and lowered affinity at H_1 receptors) (Böhme et al., 2003).

PG-9 (**3.43**) (Figure 3.5), an M_2 antagonist structurally related to atropine, was reported to have central antinociceptive and antiamnesic effects in mice and rats (Ghelardini et al., 2000).

Another class of M_2 antagonists is constituted by derivatives of himbacine (**3.44**), a piperidine alkaloid originally extracted from *Galbulimima baccata*, which showed potent and selective inhibition of M_2 receptors (Kozikowski et al., 1992). Compound (**3.45**) (Figure 3.5) was reported to be an M_2 antagonist, but exhibited lower affinity and showed a 10-fold selectivity for M_2 or M_1 (Gao et al., 2002). Another study has described the total synthesis of the enantiomeric pair of himbacine and the preparation of himbacine congeners, thus permitting a structure–activity relationship (SAR) study (Takadoi et al., 2002).

3.2 AChE AND ITS INHIBITION

Jure Stojan

Cholinesterases are involved in terminating nerve impulses in the CNS and at the periphery. The high enzyme concentration at the cholinergic synapses is achieved by the oligomerization on an anchoring, collagen-like, peptide. The three-dimensional structure of a monomer shows a buried active site, responsible for the effectiveness and specificity of the enzyme, but also for various pseudocooperative phenomena. Due to its important role, AChE is a target for different inhibitors. Reversible inhibitors prevent substrate accomodation by sterically blocking the active site and the irreversible ones act as acylating agents that modify the active site serine, the major participant in covalent catalysis. When irreversibly inhibited by potent nerve gases, cholinesterases can be reactivated, unless partial dealkylation of the bound inhibitor occured. This particular phenomenon, called aging of cholinesterases together with generalized cytotoxic effect, appears to be the cause of organophosphate-induced delayed neurotoxicity in nerve gas poisoned casualties.

3.2.1 INTRODUCTION

The primary interest in acetylcholine esterase (AChE) is in its involvement in the termination of transmission at the cholinergic synapse. Following the stimulus of a neuron, acetylcholine is released into the synaptic cleft. It then diffuses across the cleft and binds reversibly to its targeted receptor, to trigger depolarization of the postsynaptic membrane. The majority of acetylcholine, however, is rapidly hydrolyzed by AChE, into its inactive products, choline and acetate. However, there is another possible fate for this neurotransmitter; it can be diluted into the neighboring interstitium and into blood, from where two types of choline esterases (ChEs) are able to remove it. These are an erythrocyte-membrane-bound AChE and a less-specific serum ChE, which is also known as a butyrylcholine esterase (BuChE). While the role of AChEs appears well understood, the presence of this serum BuChE is puzzling. In particular, people who either lack or have an inactive BuChE variant display prolonged apnoea after the administration of succinylcholine, while they are otherwise healthy. On the other hand, and against all expectations, AChE nullizygotic mice survive by compensating for the loss of function of AChE by that of BuChE, or of other serine esterases (Xie et al., 2000).

3.2.2 STRUCTURE AND SUBSTRATES OF ACHE

AChE is expressed in different tissues in various molecular forms that differ in their quaternary structure (Dudai et al., 1973; Silman and Futerman, 1987). Its functional localization depends on alternative splicing at the C-terminus and its association with anchoring proteins (Massoulié, 2002). The major variant that is expressed in the brain and muscle tissue of normal adult mammals can generate monomers, dimers and tetramers, as well as collagen-tailed and hydrophobic-tailed (Inestrosa et al., 1987) hetero-oligomers. In the latter cases, up to three tetramers are attached to strands of a collagenous or proline-rich transmembrane anchoring protein, whereby each monomer has its helical tryptophan amphiphilic tetramerization sequence at the C-terminus (Figure 3.6). In each hetero-pentamer, the four parallel tryptophan amphiphilic tetramerization chains form a left-handed superhelix around an antiparallel, left-handed helix of the proline-rich anchoring protein (Dvir et al., 2004, PDB code 1VZJ). Such hetero-pentameric supramolecular structures are enforced by two disulfide bridges between the two tryptophan amphiphilic tetramerization helices, and by another two disulfide bridges between the anchoring protein and the remaining two tetramerization helices, using the same sulfhydryl group. However, this quaternary structure forms even when all of the appropriate cysteines are substituted (Bon et al., 2003).

The kinetic behavior of AChE is peculiar, although it does not depend on the degree of oligomerization. Dimers and tetramers show the same deviations from Michaelis-Menten mechanism as monomers that are devoid of the tryptophan amphiphilic tetramerization domain. Indirect electron paramagnetic resonance (EPR) studies (Šentjurc et al., 1976), which were later supported by the first crystallographic determinations of the three-dimensional structure of AChE (Sussman et al., 1991), suggested that its active site lies at the bottom of a deep and narrow gorge in the center of the protein. Hence, it appears that the pseudocooperative phenomena of all of the ChEs are a consequence of this unusual tertiary structure, which also has another substrate binding site at the entrance to the catalytic site, known as the peripheral anionic site.

AChE belongs to the superfamily of α/β-hydrolases and contain over 530 amino acids. It forms a 12-stranded mixed β-sheet that is surrounded by 14 α-helices and an unusual Ser–His–Glu catalytic triad [PDB ID code: 1EA5]. The catalytic site itself is well adapted for its natural substrate, acetylcholine, so that in contrast to the related BuChE (Nicolet et al., 2003), the turnover number of carboxylic ester substrates is sharply decreased for acyl groups larger than propionyl (Rosenberry, 1975). Soaking of a nonhydrolyzable substrate analog, trimethyl(4-oxo-pentyl)ammonium (Figure 3.7), into the crystals of native AChE shows that the quaternary ammonium ion is bound not to a negatively charged anionic site, but rather that it makes cation–PI interactions with the tryptophan indole ring (Colletier et al., 2006; PDB ID code: 2CF5). At very high, submolar, substrate concentrations, the active site gorge of the ChEs are fully occupied each accepting two substrate molecules. In addition, the arriving substrate molecules block the deacylation step (Stojan et al., 2004; Colletier et al., 2006; PDB ID code: 2C4H).

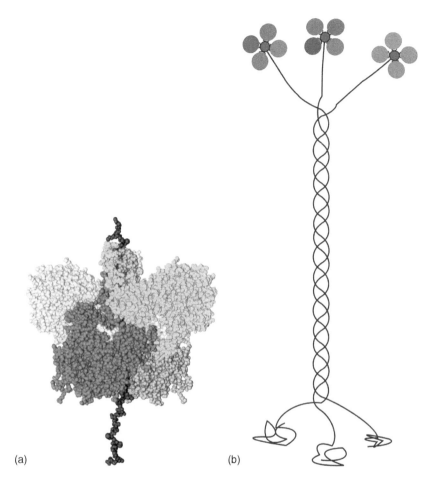

(a)   (b)

FIGURE 3.6   (See color insert following page 296.) (a) The AChE hetero-pentamer (red: part of the proline-rich anchoring protein; white: rim of the active-site gorge). (b) The three tetramers on a proline-rich collagenous anchoring protein.

Apart from all of these kinetic complications that arise from the specificity of the active site architecture, the ChEs are hydrolases that act by combining acid–base and covalent catalytic mechanisms (Krupka, 1966a). In the first step, through

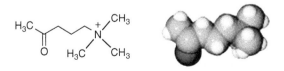

FIGURE 3.7 Trimethyl-(4-oxo-pentyl) a non-hydrolyzable substrate analogue. Two- and three-dimensional representation.

the His-Glu charge-relay system, the partially deprotonated catalytic serine attacks the approaching substrate, to exchange the choline part of the substrate with itself, in a process known as acylation. During the escape of the alcoholic leaving group, the covalent intermediate, the acylated enzyme, is hydrolyzed. In this step, the nearest water molecule is used, and again, it appears to be activated by the same His-Glu charge-relay system. Finally, the small acetate moiety is released very rapidly.

Of course, much of our knowledge about these events during the catalytic turnover of their substrates by ChEs has come from kinetic studies. The information has been summarized in a number of kinetic schemes that represent the reaction mechanism, and as new evidence has become available, this scheme has been enlarged on. Currently, it is generally agreed that the main catalytic pathway in all ChEs has at least three intermediates, one more than the classical double-intermediate scheme proposed by Wilson many years ago (Wilson, 1951). Following the structural evidence of the buried active site, an interpretation was adopted where the first contact between substrate and enzyme occurs at the peripheral site. Here, the substrate is positioned such that productive binding is promoted when it reaches the bottom of the active-site gorge. Unfortunately, this happens only at very low, micromolar, substrate concentrations. At higher, millimolar, concentrations of substrate, a second substrate molecule can bind to the peripheral site and close the active-site gorge while the first substrate molecule is still processed. This situation leads to two opposite effects: the exit of the alcohol is sterically hindered, while the activation of water trapped in the active-site gorge is enhanced. Together with the demonstration that at submolar substrate concentrations the ChEs are completely blocked, these particulars make kinetic studies of these enzymes interesting, but difficult. Nevertheless, a kinetic scheme (Scheme 3.1) that summarizes the current kinetic and structural information would be as follows (Stojan et al., 2004):

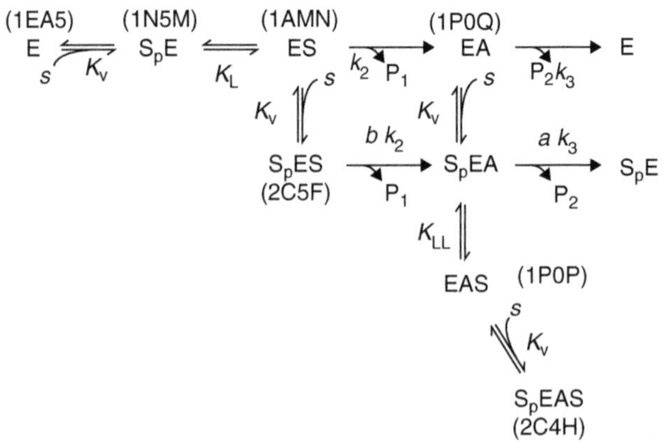

SCHEME 3.1 Substrate hydrolysis by ChEs. E represents the enzyme, S the substrate, and P the products. S_p represents the substrate bound to the peripheral site of the enzyme. There are seven intermediates, and seven kinetic parameters: K_p is the dissociation constant for the binding of the substrate to the peripheral site; K_L and K_{LL} are the partition coefficients; k_2 and k_3 are the overall rate constants for the acylation and deacylation, respectively; and a and b are the proportional factors. The PDB codes of the solved structures of the intermediates are given in the brackets.

Assuming mixed equilibrium and steady-state conditions, this kinetic scheme gives the following rate equation (Stojan et al., 2004):

$$v = \frac{k_{cat}[E]_0[S]}{[S] + K_M} = \frac{\dfrac{k_2 k_3}{k_2 \dfrac{1 + \dfrac{[S]}{K_p} + \dfrac{[S]}{K_p K_{LL}} + \dfrac{[S]^2}{K_p^2 K_{ll}}}{1 + a\dfrac{[S]}{K_p}} + k_3 \dfrac{1 + K_{LL} + \dfrac{[S]}{K_p}}{1 + b\dfrac{[S]}{K_p}}}[E]_0[S]}{S + \dfrac{k_3 K_p K_L \dfrac{1 + b\dfrac{[S]}{K_p}}{}}{k_2 \dfrac{1 + \dfrac{[S]}{K_p} + \dfrac{[S]}{K_p K_{LL}} + \dfrac{[S]^2}{K_p^2 K_{ll}}}{1 + a\dfrac{[S]}{K_p}} + k_3 \dfrac{1 + K_{LL} + \dfrac{[S]}{K_p}}{1 + b\dfrac{[S]}{K_p}}}}. \tag{3.1}$$

Although this equation resembles the Michaelis–Menten equation, it allows for the very special kinetic behavior of the ChEs, as shown in Figure 3.8 for electric eel AChE. To evaluate all of the parameters, Equation 3.1 can be fitted to the experimental data using an appropriate nonlinear fitting program (Stojan, 2005).

There are many reasons why ChEs are so interesting to study. Physiologically irrelevant but of great practical importance is the methodology for activity

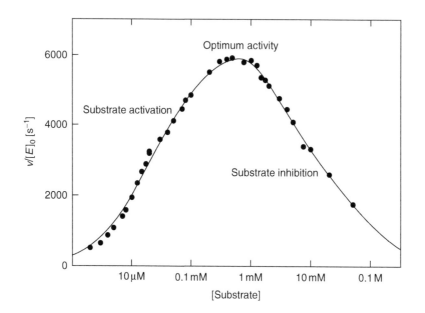

FIGURE 3.8 The dependence of electric eel acetylcholinesterase activity on the concentration of the substrate acetylthiocholine.

determination. The colorimetric method, introduced by Ellman et al. in 1961, simplified kinetic data acquisition and made it user friendly. It is based on a very accurate detection of sulfhydryl groups, stoichiometrically released on hydrolysis of an artificial substrate analog, acetylthiocholine. The method allows for the ChE activity measurements from micromolar to submolar concentrations and dithio-bis-dinitro benzoic acid (DTNB), a specific sulfhydryl reagent, exerts no side effects on ChEs. Additionally, acetylthiocholine is cheap and behaves very similar as the true substrate itself.

The acetylcholine turnover by the vertebrate AChEs is very high, with up to 10,000 per second. It is difficult to imagine how this can occur in an enzyme that has such a deep and narrow active-site gorge. Previous studies have put forward several ideas concerning a putative additional exit of the products through an existing back door (Gilson et al., 1994; Bartolucci et al., 1999). Indeed, some movements of the active-site tryptophan have been observed in the crystal structure of native *Drosophila* AChE (Harel et al., 2000), and remaining activity has been seen for several vertebrate ChEs when completely blocked at the entrance to the active-site gorge by the mamba venom component fasciculin II (Rosenberry et al., 1996; Goličnik and Stojan, 2002). But this "back-door hypothesis" is still not widely accepted.

Acetylcholine   Phenylacetate

Although AChE is a very specific enzyme, it is also able to hydrolyze many other esters, with both different alcohol (phenylacetate) and acid groups (butyrylcholine) (Krupka, 1966b). However, while the hydrolysis of its natural substrate proceeds with similar rates through all of the steps, there is usually a limiting step in the conversion of other substrates. If the substrate is too large, it will not be accommodated properly and will not reach the transition state, as in the case of the anticancer prodrug CPT-11 (Harel et al., 2005; PDB entry: 1U65) and butyrylcholine (Harel et al., 1992). In contrast, some small substrate analogs will bind and acylate the enzyme, but will deacylate very slowly, or even not let the enzyme be regenerated by hydrolysis at all. Both of these groups of potential substrates inhibit AChE, the first reversibly, and the second, irreversibly.

3.2.3 REVERSIBLE AChE INHIBITORS

The gorge-like active site of the ChEs can accept a number of molecules, as long as they are able to fit into it. Of course, the inhibition constant is not only a function of the inhibitor, but also of the type of ChE. However, all ChEs prefer largely hydrophobic ligands that bear a permanent positive charge. A typical representative is tetramethylammonium, a quaternary nitrogen compound that is in fact part of the choline moiety of the substrate acetylcholine. It accommodates between the tryptophan and phenylalanine side chains in a similar way as seen for the substrate, with a binding constant in the millimolar range.

It also became clear very early on that decamethonium, which consists of two tetramethylammoniums connected by 10 methylene repeats, is some 1000 times more potent. Crystallography has revealed that its two charged heads interact in the same manner, one at the peripheral site and the other in the center, thus filling the active-site gorge (Harel et al., 1993; PDB entry: 1ACL). Decamethonium is now a prototype in the design of reversible ChE inhibitors that can span from the entrance to the bottom of the active-site gorge.

Decamethonium   Propidium

Another way to increase the affinity of tetramethylammonium is to mimic choline along its whole length by simultaneously increasing the hydrophobicity. Edrophonium, an aromatic monoquaternary alcohol that is used clinically to diagnose myasthenia gravis, occupies the entire bottom of the aromatic gorge. Its *m*-hydroxyl group is positioned between two of the three members of the catalytic triad, forming hydrogen bonds with each of them (Harel et al., 1993; PDB entry: 2ACK). Again, the affinity increases ~1000 times, thus making it a powerful classical competitive inhibitor.

Sometimes, a very similar molecule, like trimethylammonio-trifluoro-dihydrox-yethylbenzene (Brodbeck et al., 1979), can become a transition state analog and bind to the AChE active site with an affinity that can be 10 orders of magnitude higher [PDB entry: 1AMN]. This is also the most suitable titrating compound for the determination of ChE active-site concentrations.

The third class of reversible inhibitors is the so-called peripheral ligands. These are too large to reach the bottom of the active-site gorge, but the enzyme's strong electrostatic dipole aligned with the gorge (Ripoll et al., 1993) attracts such positively charged compounds that become tightly stacked at the entrance. The most widely used ligands of this type are propidium, gallamine, and D-tubocurarine (Changeux, 1966; Bourne et al., 2003; PDB entries: 1N5M, 1N5R). The last of these inhibits AChEs instantaneously, but acts on horse BuChE in a double, fast–slow, manner (Stojan and Pavlič, 1991). Since reactivation after its dilution is also slow, it seems that D-tubocurarine only fits into the larger BuChE active-site gorge on a slow conformational adaptation.

Finally, a 62-residue-long, neurotoxic, three-finger protein, known as fasciculin II, binds in a stoichiometric ratio to the rim of the active-site gorge, and by covering the entrance in this way it inhibits fish and mammalian AChEs effectively (Marchot et al., 1998; Goličnik and Stojan, 2002). In contrast, the different amino acids around the entrances to the active-site gorges of BuChEs and insect AChEs result in more than a million times lower affinity toward fasciculin II.

The evaluation of an inhibition constant for a reversible inhibitor leads to similar problems as for the characterization of the reaction with the substrate.

For instance, when small active site ligands are tested, they probably interact with the peripheral substrate binding site, too. Taking into account all of the possible enzyme–ligand complexes makes the analysis very complex, although the system is well defined under a variety of conditions (Szegletes et al., 1998; Stojan et al., 2004). However, when only the approximate inhibitory power of a new compound is to be determined, the analysis can be limited to classical kinetic approaches. Substrate concentrations below the inhibitory region sometimes allow an analysis according to Michaelis-Menten principles, especially in the case of small, active-site-directed, instantaneous inhibitors. Usually, when the effect of the combination of inhibitors is studied, similar assumptions are made. When tight binders, like the anti-Alzheimer's candidates are tested, the substrate is only used to follow the remaining activity, and hence the analysis is no longer so complex (Goličnik and Stojan, 2002).

3.2.4 Irreversible ACHE Inhibitors

The so-called acid-transferring inhibitors are the most potent anticholinesterase agents (Kitz et al., 1967). They are esters, halides, or similar unstable compounds of substituted phosphoric, carbamic, and sulfonic acids. Their reaction mechanisms have been well studied and are essentially the same as that of the substrates, although with a long-lasting deacylation step. These acid-transferring inhibitors are effective on all of the ChEs, provided that their size is appropriate. Physiologically, they fall into the class of insecticides and pesticides, or of chemical warfare agents, and in particular, of nerve gases.

Among all of the anti-ChE agents, the nerve gases are considered as the most toxic. The best known of these are diisopropyl fluorophosphate (DFP), isopropyl methyl-phosphonofluoridate (sarin), pinacolyl methylphosphonofluoridate (soman) and ethyl-*N*-dimethyl phosphoramidocyanidate (tabun). These are all phosphoric acid derivatives, and when they react with a ChE, the most acidic, fluoride or nitrile (tabun), group is the leaving group. The two other substituents are basic only relative to the leaving group. Usually they are alkyl (sarin, soman), alkoxy (DFP), or amino (tabun) groups, but they can also be aryl, as in the ethyl-nitro-phenyl phenyl-phosphonothiolate (EPN), or aryloxy (EPN, paraoxon) groups. When the carbonyl oxygen is substituted by sulfur or selenium, these compounds are inactive, and biological activation is required for them to become effective.

Diisopropyl fluorophosphate   Soman   Tabun

Carbamates are widely used as insecticides, but eserine has become important as a diagnostic drug for the ChEs. Engelhard et al. (1967) defined the ChEs as "all hydrolases that cleave choline esters and are inhibited by eserine in a concentration of 10^{-5} M." All of the well-known carbamates have one methyl group on the nitrogen, and the other group is usually methyl or hydrogen. In contrast, the leaving group varies. There has been some controversy with regard to the reversibility of carbamate inhibition, but it is now clear that the carbamoylic acid is covalently bound to the active serine, and that the prolonged decarbamoylation step is a consequence of an H-bond between the carbamoyl nitrogen and N_ε of the histidine in the His-Glu charge-relay system (Bartolucci et al., 1999; PDB code 1OCE).

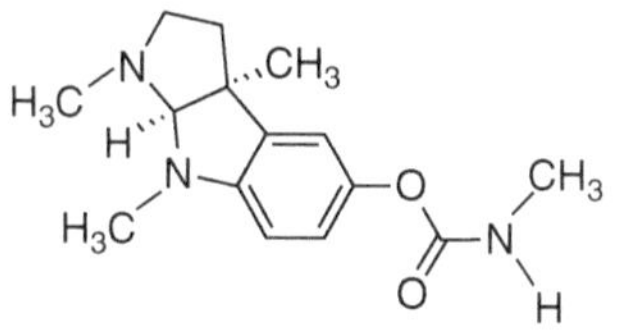

Eserine 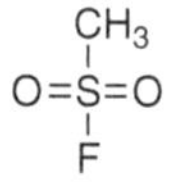　　　　　　　　　　Methanesulfonylfluoride

Analogous to the organophosphates, there are the less-toxic organosulfonates. Many studies have indicated advantages in compounds such as methane sulfonyl fluoride (MSF) because of its small size and the irreversibility of the inhibited enzyme complex (Metzger et al., 1967; Pavlič, 1973). Methane sulfonyl fluoride acts in great excess to the enzyme, but slowly, and it can be used as a probe in studies of acylation (Goličnik et al., 2002).

Kinetic investigations of the action of an irreversible inhibitor usually involve studies of the time course of enzyme inhibition and result in the determination of the corresponding rate constant. In most cases, the concentrations of the irreversible inhibitors of ChEs are much greater than the concentrations of the enzyme. Under such conditions, inhibition occurs in two steps: a reversible step that is followed by an irreversible one (Scheme 3.2).

The idea of a two-step mechanism came from the observation that the bimolecular rate constant decreases with time (nonlinear semilogarithmic plots). The formation of the reversible complex is described by the affinity constant ($K_a = k_{-1}/k_1$) and the formation of the irreversible complex by the acylation constant, k_2. The relationship between these constants and the second-order rate constant of the simplified reaction scheme is $k_i = k_2/K_a$ (Scheme 3.3).

This bimolecular rate constant is generally applicable under usual inhibition conditions, where the concentration of the inhibitor is so high that it does not

$$E + I \xleftrightarrow{\ K_a\ } EI \xrightarrow{\ k_2\ } E - I$$

SCHEME 3.2 Reaction between ChE and an irreversible inhibitor involving initial reversible complex. EI represents the enzyme–inhibitor complex, and E–I the irreversibly inhibited enzyme.

$$E + I \xrightarrow{\ k_i\ } E - I$$

SCHEME 3.3 Single-step bimolecular reaction between ChE and an irreversible inhibitor.

effectively change during the reaction. This can be estimated using the equation derived by Aldridge (1950):

$$k_i = \frac{2.303}{t[I]} \times \log \frac{v_0}{v_t}, \tag{3.2}$$

where $[I]$ is the steady concentration of the inhibitor, v_0 is the initial velocity of the enzymatic reaction, v_t is the velocity after time t.

Irreversible inhibitors interact with the enzyme at equimolar ratios and are thus suitable for determination of unknown enzyme concentrations, as titrating agents. Unfortunately, this is almost always not true for irreversible inhibitors of the ChEs, as the rate of inhibition is so slow that a complete loss of the enzymatic activity is rarely achieved. Despite this, 7-(methylethoxyphosphinyloxy)1-methyl-quinolinium iodide (MEPQ) is sometimes used for these purposes.

3.2.5 Reactivation and Aging of AChE

If the enzyme active site and an irreversible inhibitor are in stoichiometrically equal amounts, the inhibition will increase until the enzymatic activity is completely lost. However, at least some recovery can almost always be seen. This can occur spontaneously as a consequence of an interaction with the solvent or with a compound deliberately added to the solution with the inhibited enzyme. Spontaneous reactivation of phosphorylated ChEs under physiological conditions is negligible. However, Wilson found that nucleophilic agents, like hydroxylamine and choline, can recover the enzymatic activity effectively (Wilson, 1951). Later, he reported that pyridine-2-aldoxime methiodide (Pralidoxime, 2-PAM), an oxime, is a much more potent agent (Wilson and Ginsburg, 1955).

Most of the known effective reactivators of the ChEs resemble choline, the hydrolytic product of the natural substrate (Musilek et al., 2005). They accommodate between the indole moiety of the active side tryptophan and the Ser-His from the catalytic triad. As they incorporate conjugated iminium and oxime moieties that are electron affinic, their hydroxyl groups are good acceptors for the phosphoric acid that originally bound to the serine of the ChE active site.

Choline   Pralidoxime (2-PAM)

In 1955, Hobbinger noted that the longer an organophosphate was in contact with a ChE before it was reactivated, the less enzymatic activity could be recovered when the inhibited enzyme was subjected to the action of a nucleophilic reactivator. This phenomenon is known as aging, and it is general to all of the ChEs; again, it is inhibitor- and enzyme-dependent. The fastest aging of electric eel AChE occurs when it is inhibited by soman, with a half-time of only 8 s (Michel et al., 1967) (Scheme 3.4).

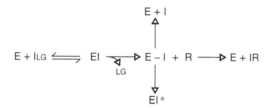

SCHEME 3.4 Summarized events involved in irreversible inhibition, spontaneous reactivation, reactivation by the reactivator and aging of ChEs. E is enzyme, I_{LG} is irreversible inhibitor with a leaving group, EI reversible complex between the enzyme and irreversible inhibitor, E−I is irreversibly inactivated enzyme, EI* aged enzyme and R is reactivator.

Different theories for the aging mechanism have been discussed (Wagner-Jauregg et al., 1953; Davies and Green, 1956). In the case of a DFP-inhibited ChE, Oosterbaan et al., proposed that this occurs by the removal of an isopropyl group from the organophosphate bound at the active serine, which they referred to as dealkylation (Oosterbaan et al., 1958). Recently, it was shown that tabun-inhibited mouse AChE is deaminated (Ekstrom et al., 2006; PDB ID codes: 2C0P, 2C0Q). As in all of the crystallographically documented cases of aging (Millard et al., 1999; PDB ID codes: 1SOM, 1VXO, 1VXR), a hydroxyl group exchanges for one of the organic substituents of the organophosphate-inhibited ChE, thus suggesting that a wrong bond in the covalent complex is attacked and hydrolytically cleaved by the His-Glu charge-relay system. Once aging has occurred, the activity of an organophosphorylated ChE can no longer be recovered. This might be because of a very stable hydrogen bond between the N_ε of the catalytic triad histidine and the newly emerging oxygen from the hydroxyl group on the phosphorus. This hypothesis is supported by the observation that inhibition of sulfonylated ChEs definitely does not involve aging.

When determining the recovery of an irreversibly inhibited ChE by a reactivator, it is important to remember that the reactivator, per se, could also be an inhibitor, could interfere with the substrate or could react with an anti-ChE to form a new potent inhibitor. Therefore, it is necessary to remove excess inhibitor before the addition of a reactivator, and to remove the excess of the reactivator before the activity measurement. Often, it is possible to achieve such conditions only by dilution; in particular, when highly concentrated ChE solutions are available. Similarly, the rate of reactivation and the percentage of nonrecoverable activity depend on the time before the addition of a reactivator, the concentration of the inhibitor and reactivator, the source of the ChE, and the general conditions, like temperature and ionic strength. If all of these effects can be standardized, the time course of reactivation is usually represented by an exponential curve, similar but opposite to that of inhibition. The effectiveness of a nucleophile is represented by the bimolecular rate constant for the hydrolysis of an acid-transferring inhibitor.

The study of aging is technically still more demanding. The rate is determined by removing aliquots from the solution of an inhibited enzyme after various times of incubation, and then by determining, usually, the maximum amount of reactivation that is possible under optimal conditions. Heilbronn (1963) evaluated the rate constant for the aging reaction from the slope of the line, by plotting:

$$\log \frac{(E - E_i) \times 100}{E_0 - E_i} \text{ vs. time,} \qquad (3.3)$$

where E is the activity of the reactivated enzyme, E_i is the activity of the inhibited enzyme, E_0 is the activity of the control incubated with the same amount of reactivator.

The last item corrects for the ChE inhibition as a result of a reaction with the reactivator. In general, BuChEs age more rapidly than AChEs. For instance, the half-times for aging of DFP-inhibited AChE and BuChE are 4.4 h and 28 min, respectively.

Due to this aging phenomenon, the risk of organophosphorus poisoning increases with time. Thus, the treatment of casualties requires not only ventilatory assistance and control, but also acute pharmacological therapy (de Jong, 2003). Conventionally, atropine is administered as an antidote every few hours, and the oximes are used as in vivo reactivators (Kuca et al., 2005). As sarin and tabun age very slowly, pralidoxime (2-PAM) can act even hours after the exposure. In contrast, soman ages within minutes, so that only an instant 2-PAM self-administration will help. However, if nerve gas intoxication is expected, as under war conditions, carbamate prophylaxis is used. The most widely used protector is pyridostigmine bromide, although human recombinant BuChE, which has recently been produced in huge amounts, is becoming the drug of choice in prophylaxis of nerve gas poisoning (Cerasoli et al., 2005). It needs also to be remembered that nerve gas poisoning not only inhibits AChE, thus leading to sudden death, but it can also affect many organs in survivors, which in addition to the stress and the accompanying hazardous cause, can promote permanent disorders, like the dangerous Gulf War syndrome (Loewenstein-Lichtenstein et al., 1995; Moss, 2001). It should also be stressed that natural organophosphate scavenging esterases, like serum paraoxonase, are not effective enough to prevent acute organo-phosphate AChE inhibition (Amitai et al., 2006) nor can protect from delayed neurotoxicity (Damodaran and Abou-Donia, 2000).

3.3 AChE INHIBITORS AND THEIR CLINICAL ASSESSMENT

Pierre Francotte, Pascal de Tullio, and Bernard Pirotte

After its liberation within the synaptic level, acetylcholine (ACh) binds with its (muscarinic or nicotinic) receptors. Its action ends due to the work of enzymes called cholinesterases, whose function consists of hydrolyzing the neurotransmitter, resulting in the formation of one acetate and choline. As these enzymes block cholinergic transmission, they are considered essential to control cholinergic excitation under physiological conditions. Hence, inhibiting ACh turnover will result in an increase in ACh synaptic levels and augmentation of its effects.

Compounds able to inhibit AChE have been found to be of therapeutic interest in the treatment of AD in 1976, since the cholinergic hypothesis was advanced. Since then, AChE inhibitors (AChEIs) have been the most studied class of compounds in the search for a symptomatic treatment of AD.

Interestingly, a growing body of evidence suggests that cholinergic activities might be involved in the processing of amyloid precursor protein, which is the main protein implicated in the development of AD. As AChE may play a role in accelerating Aβ plaques deposition (Inestrosa et al., 1996), it is likely that AChE inhibition protects neurons against a variety of insults and thus positively modulates the disease course of AD (Liu et al., 2005; Verhoeff, 2005; Nordberg, 2006; Racchi et al., 2006).

This section is divided into two parts; the first one presents a wide range of synthetic and natural AChEIs that are studied as potential drugs for AD treatment, whereas the second part is devoted to the three AChEIs currently marketed for the treatment of AD patients.

3.3.1 Synthetic and Natural AChEIs

3.3.1.1 Natural AChEIs

An intensive search has been done in phytochemistry to find new compounds potentially active in AD treatment. To be exhaustive always remains illusive, here more than in other sections: so citing all the natural compounds isolated which have been reported to express an inhibitory activity on AChE is not feasible and here we give the broadest possible overview of currently studied structures, citing for each reference one typical compound. Interested readers are invited to consult excellent specific reviews available (Viegas et al., 2005a; Barbosa Filho et al., 2006; Hostettmann et al., 2006; Houghton et al., 2006). The natural compounds have a wide variety of structures: molecules that have been characterized as AChEIs have been described in a broad spectrum of natural compounds; from alkaloids to coumarins, from flavonoids to terpenoids.

Should the natural products described here not be sufficiently active, or have an inappropriate pharmacological profile (i.e., active on too many pharmacological targets), they can play the role of lead compounds and inspire medicinal chemists in the design of new potent compounds.

An important group of novel natural AChEIs reported are the alkaloids and the first to be mentioned here is galanthamine (**3.46**), isolated from *Galanthus*, that has been marketed in the European Union and in the United States (Raskind et al., 2000).

Knowing the efficacy of galanthamine, several pure Amaryllidaceae alkaloids and extracts from *Narcissus* were tested to discover new AChEIs; among them, only a few compounds showed inhibitory effects, the most effective were alkaloids with the galanthamine structural type. Other active alkaloids were of the lycorine-type, for example, assoanine (**3.47**), which showed a fourfold lower IC_{50} value than galanthamine in a microplate assay for AChE activity; their activity could be explained by a certain planarity of the molecules (López et al., 2002). Other structurally related alkaloids, obtained from two West African *Crinum* species, were recently tested (Houghton et al., 2004).

Huperzine A (**3.48**) is an alkaloid isolated from the Chinese traditional herb *Huperzia serrata*, available in the U.S. market as a dietary supplement. This is a potent selective and long-acting AChEI, and it has high efficacy in improving memory in different animal models and in clinical trials (Tang and Han, 1999; Bai et al., 2000).

Galanthamine
3.46

Assoanine
3.47

Huperzine A
3.48

Salignenamide C
3.49

Salonine C
3.50

Epoxynepapakistamine-A
3.51

(−)-Hookerianamide A
3.52

Faleoconitine
3.53

Turbinatine
3.54

Voacangine
3.55

Berberine
3.56

19,20-Dihydroervahanine A
3.57

Sieboldine A
3.58

Other novel alkaloids are postulated to be potential lead compounds in AD treatment. Indeed, the team directed by Pr. Atta-ur-Rahman has intensively studied Buxaceae plants to discover new alkaloids. In this context, new steroidal alkaloids were found in *Sarcococca* species. The main part of the work consisted in the isolation, structure determination, and pharmacological evaluation of the novel compounds: at least 23 steroidal alkaloids (such as salignenamide C (**3.49**) (Atta-ur-Rahman et al., 2002) or salonine-C (**3.50**) (Atta-ur-Rahman et al., 2004)) were extracted from *Sarcococca saligna*. Other types of steroidal alkaloids such as epoxynepapakistamine-A (**3.51**) were isolated from leaves of Nepalese *Sarcococca coriacea* (Kalauni et al., 2002), whereas work on *Sarcococca hookeriana* permitted the discovery of new compounds such as $(-)$-hookerianamide A (**3.52**) (Choudhary et al., 2004). The new steroidal alkaloids were assayed for AChE and butyrylcholinesterase (BuChE) inhibition, and were shown to inhibit both enzymes. Additional molecular docking studies permitted identification of hydrophobic interactions inside the aromatic gorge area as the major stabilizing factor in enzyme–inhibitor complexes of these alkaloids (Zaheer-ul-haq et al., 2003a). Moreover, significant 3D-QSAR studies by comparative molecular field analysis (CoMFA) were performed with pregnane-type steroidal alkaloid inhibitors (Zaheer-ul-Haq et al., 2003b).

Turning their interest to Iranian *Buxus hyrcana*, the same team also identified three novel triterpenoid alkaloids, which were found to be active against AChE and BuChE (Choudhary et al., 2003). Also at the University of Karachi, research on *Aconitum falconeri* has yielded two new norditerpenoid alkaloids; one of them, faleoconitine (**3.53**) was characterized as a moderate AChEI (Atta-ur-Rahman et al., 2000). Indole glucoalkaloids acting as AChEIs and antioxidants were mentioned after extraction from *Chimarrhis turbinata*; one of the reported compound, turbinatine (**3.54**), was shown to have moderate activity in comparison with galanthamine (Cardoso et al., 2004). On the other hand, 10 indole alkaloids were identified from the chloroform extract of the stalk of *Tabernaemontana australis* (Mueell. Arg) Miers; of these, four compounds (among which was voacangine (**3.55**)) were identified as AChEIs by TLC assay using a modified Ellman's method (Andrade et al., 2005).

Corydalis speciosa was submitted to a bioassay-guided search for AChEIs; this work permitted the identification of four isoquinoline alkaloids, from which emerged berberine (**3.56**) during the in vitro test (Kim et al., 2004).

Tabernaemontana divaricata was investigated for its AChEI content; in this context, 19,20-dihydroervahanine A (**3.57**) isolation was achieved. In addition, a SAR based on other bisindole alkaloids was conducted (Ingkaninan et al., 2006).

Lycopodium sp. is also currently the centre of interest in pharmacognosy studies: a first example is provided by the club moss *Lycopodium sieboldii* from which has been extracted sieboldine A (**3.58**), recognized as a potent AChEI with modest cytotoxicity (Hirasawa et al., 2003). Lycoperine A (**3.59**) was recently discovered by the same team in *Lycopodium hamiltonii*; preliminary biological evaluation highlighted its AChE inhibitory activity (Hirasawa et al., 2006). Finally, we can cite here work achieved on *Lycopodium clavatum*, which led to the identification of the triterpenoid α-onocerin (**3.60**) as the compound responsible for the activity found with extracts (Orhan et al., 2003).

Flavonoids have also been studied as AChEIs. An example is furnished by studies on *Onosma hispida* from which was identified a new flavanone, hispidone (**3.61**) (Ahmad et al., 2003). Xanthones are of interest, too: bellidifolin (**3.62**), extracted from *Gentiana campestris*, was recently reported to show similar inhibitory activity to galanthamine in an enzyme assay (Urbain et al., 2004).

Fatoua villosa emerged during the screening of plant extracts for their inhibiting effects on AChE; the active fraction was purified permitting the isolation of zeatin (**3.63**) (Heo et al., 2002).

Lycoperine A
3.59

α-Onocerin
3.60

Hispidone
3.61

Bellidifolin
3.62

Zeatin
3.63

Cryptotanshinone
3.64

R=H or OCH₃

Limbatolides
3.65

Decursin
3.67

Coumarins
3.68

Knowing that for over 1000 years Chinese *Salvia miltiorhiza* was used in particular for the treatment of cerebrovascular diseases and CNS deterioration in old age, researchers tried to find the compounds responsible for its activity; they isolated from the acetone extract four AChEIs among which was the diterpenoid cryptotanshinone (**3.64**) (Ren et al., 2004). Other diterpenoids (limbatolides (**3.65**)) were recently isolated from the roots of *Otostegia limbata*; these tricyclic *cis*-clerodane type diterpenoids displayed inhibitory potential against both AChE and BuChE (Ahmad et al., 2005).

Seeking novel natural AChEIs, a Korean team isolated (+)-α-viniferin (**3.66**) from *Caragana chamlague*. This stilbene trimer was shown to be a specific, reversible, and noncompetitive AChEI (Sung et al., 2002). The same team identified coumarins from *Angelica gigas* (Kang et al., 2001); among the reported compounds stands decursin (**3.67**), for which an antiamnestic activity was recently confirmed in vivo (Kang et al., 2003). Moreover, based on the finding that hybridation of an AChEI with an irreversible monoamine oxidase (MAO) inhibitor inhibits both enzymes and using coumarinic MAO-B inhibitors, an exploratory study of AChE inhibition identified coumarins (**3.68**) as noncompetitive AChEIs. Although proof of in vivo activity needs to be obtained, optimization of these novel structures could lead to a new potent family of compounds (Brühlmann et al., 2001). Other coumarins were recently isolated from plants and characterized as moderate AChEIs; for instance imperatorin (**3.69**) was found in *Angelica dahurica* (Kim et al., 2002) and has also been reported in *Peucedanum ostruthium* (Urbain et al., 2005).

In another program searching for natural AChEIs, an extract of *Origanum majorana* L. was selected based on its high inhibitory effects. Further investigation led to the isolation of the active component which was ursolic acid (**3.70**) (Chung et al., 2001).

Searching for new natural cognitive enhancers with low toxicity, Park et al., reported a *Polygala tenuifolia* extract to have neuroprotective effects. This activity includes reduction of cell death induced by glutamate and Aβ, a C-terminal fragment of APP but also noncompetitive inhibition of AChE (Park et al., 2002).

Arisugacin A
3.71

Territrem B
3.72

Quinolactacin A1
3.73

Terferol
3.74

3.75

R = CH$_6$: Physostigmine **3.76**
R = C$_7$H$_{15}$: Eptastigmine **3.78**
R = C$_6$H$_5$: Phenserine **3.79**

Eseridine
3.77

Quilostigmine
3.80

3.81

Ganstigmine
3.82

Miotine
3.83

Rivastigmine
3.84

3.85

Xanthostigmine
3.86

3.87

Like higher plants, microorganisms have also been studied for new AChEIs; *Penicillium* sp. gave a series of compounds called arisugacins; among them, four showed substantial inhibition of AChE (Otoguro et al., 2000). A great effort has been made toward the total synthesis of arisugacin A (**3.71**) (Cole et al., 2002; Sunazuka et al., 2002), and the total enantioselective synthesis of (−)-arisugacin A has been reported (Cole and Hsung, 2002). Structurally close to the arisugacins, the myco-toxin territrem B (**3.72**) is a noncovalent irreversible AChEI; its innovative mech-anism could be useful for the design of a new class of AChEIs (Chen et al., 1999). On the other hand, two AChEIs such as quinolactacin A1 (**3.73**) were isolated from a solid state fermentation of *Penicillium citrinum* 90648 (Kim et al., 2001).

During screening for new AChEIs derived from microorganisms, a team from China isolated a potent noncompetitive reversible AChEI, terferol (**3.74**), already known as an inhibitor of cyclic adenosine-3′,5′-monophosphate phosphodiesterase (Dong et al., 2002).

The last example is given by (**3.75**) isolated from *Chrysosporium* sp.; even if not potent nor selective, this compound could be useful as a lead compound (Rao et al., 2001).

3.3.1.2   Synthetic Compounds

3.3.1.2.1   Carbamates

At the "origin" of this chemical class stands physostigmine (**3.76**), a natural deriva-tive isolated form *Physostigma venenosum*, from which another active alkaloid, eseridine (geneserine) (**3.77**), has also been extracted. Carbamate AChEIs have been demonstrated to nonselectively inhibit both AChE and BuChE (Costagli and Galli, 1998; Ogura et al., 2000).

With the goal of developing potential Alzheimer's drugs, many derivatives based on the physostigmine structure have been designed from which emerged the in vivo active eptastigmine (**3.78**) (Imbibo, 2001), phenserine (**3.79**) (Thatte, 2005), quilos-tigmine (**3.80**) (Sramek et al., 1999), which have been reported to be under clinical investigation. In 2002, novel physostigmine and eserine analogs were prepared that permitted a definition of the structural requirements for differential inhibition between AChE and BuChE in this series (Yu et al., 2002); the same team more recently reported the preparation of a new series of carbamates such as (**3.81**) (Luo et al., 2005).

Ganstigmine (CHF-2819) (**3.82**) is an orally active AChEI inspired from the structure of eseridine (Cassano et al., 2002; Trabace et al., 2002). Interestingly, ganstigmine was shown to provide a protection against β-amyloid 25–35 neurotoxicity on chicken cortical neurons, independently from its cholinergic activity (Windisch et al., 2003). The ganstigmine development for AD treatment prompted evaluation of other eseridine derivatives, but none were as potent or selective as their physostigmine analog counterparts (Yu et al., 2002). Recently, the crystal structure of the ganstigmine conjugate with *Torpedo californica* AChE was determined, thus providing a structural framework for the design of novel compounds with improved binding affinity and pharmacological properties (Bartolucci et al., 2006).

Miotine (**3.83**), another natural derivative also served as a structural template for the design of novel AChEIs; leading to the development of rivastigmine (**3.84**), one of the three AChEIs currently marketed for AD treatment. New phenylcarba-mates structurally related to rivastigmine were recently reported; among these derivatives emerged the orally active (**3.85**), which possessed a good pharmacoki-netic profile with low toxicity (Mustazza et al., 2002).

On the other hand, research based on xanthostigmine (**3.86**) as a lead compound permitted the synthesis of new long-lasting carbamate AChEIs such as (**3.87**) and BuChEIs. A SAR study permitted the clarification of the role of the different moieties of this subclass of compounds (Rampa et al., 2001). Continuing this program, research led to the discovery of novel analogs, which were found to be able to simultaneously block both the catalytic and the β-amyloid (Aβ) proaggre-gatory activities of AChE. This may represent a potential new strategy for the treatment of AD (Belluti et al., 2005).

Finally, in a search for less-flexible analogs of caproctamine (**3.88**), Bolognesi et al. designed (**3.89**); in vitro evaluation showed that it was as potent as physos-tigmine (Bolognesi et al., 2001).

Caproctamine
3.88

3.89

Tacrine
3.90

Velnacrine
3.91

Amiridine
3.92

SM-10888
3.93

CI-1002
3.94

3.95

ITH-4012
3.96

Donepezil
3.97

Zanapezil
3.98

3.99

3.100

3.101

3.102

3.103

3.3.1.2.2 Acridines

The prototype of this structural family is tacrine (**3.90**), the first agent approved for treating the cognitive symptoms of AD. However, besides its interesting activity, its therapeutic use was limited since it was shown to induce severe adverse effects such as hepatotoxicity and gastrointestinal upset (Gracon et al., 1998), and hence, tacrine was eventually abandoned.

Since the demonstration of efficacy of tacrine, pharmacomodulations around its structure were attempted to find novel active derivatives. One of the first examples is given by velnacrine (**3.91**), which was rapidly found toxic and devoid of efficacy (Birks and Wilcock, 2004). From other work emerged amiridine (HCl) or ipidacrine (**3.92**), SM-10888 (**3.93**) (Anonymous, 1991), and CI-1002 (**3.94**) (Emmerling et al., 1995), the latter bearing halogen atoms.

More recently, a study was performed with novel tetrahydroacridine isostere derivatives such as (**3.95**) in an attempt to derive a comprehensive SAR picture for tacrine analogs (Recanatini et al., 2000).

Also based on the tacrine structure, a great synthetic and pharmacological effort was made to design AChEIs possessing other properties, such as the modulation of voltage-dependent Ca^{2+} channels (de los Rios et al., 2002; Marco et al., 2004). From these works emerged ITH-4012 (**3.96**), which is an AChEI also acting as a calcium promoter, a property leading to neuroprotection through the induction of antiapoptotic proteins (Orozco et al., 2004).

Based on the pharmacomodulation strategy of dimeric or hybrid compound design, many tacrine derivatives have been reported in the literature; they are discussed later.

3.3.1.2.3 N-Benzylpiperidines

Donepezil (E2020) (**3.97**), the lead compound of the *N*-benzylpiperidines, comes from pharmacomodulation of an *N*-benzylpiperazine compound found during random screening for AChEIs (Sugimoto et al., 2000).

Great interest has been shown in the design of new *N*-benzylpiperidines, principally by changing the indane core with other heterocyclic systems (Martinez et al., 2000). Therefore, zanapezil (TAK-147) (**3.98**) was synthesized, and is under phase II clinical trials. On the other hand, inventors of donepezil recently described halogeno-derivatives of donepezil such as (**3.99**) (Takeuchi et al., 2002) and the benzylpyridinium salt (**3.100**) (Iimura and Kosasa, 2001); both of these new derivatives have enhanced activity in vitro compared with donepezil.

T-82 (**3.101**), a new quinoline structurally related to donepezil was recently shown to ameliorate the impairment of memory in rats (Isomae et al., 2003).

More recently, bioisosteric replacement of the benzyl group by the benzisoxazole heterocycle was achieved and led to the synthesis of novel active AChEIs such as (**3.102**) (Rangappa and Basappa, 2005).

To end with the *N*-benzylpiperidines class, it is worth mentioning here that a French team trained in indanone-chemistry has reported the synthesis of novel donepezil derivatives; among the prepared compounds emerge (**3.103**), which has in vitro activity of the same rank order as donepezil. In vivo studies need to be conducted to confirm the potential interest of selected compounds (Omran et al., 2005).

3.3.1.2.4 Other Chemical Classes

A series of 2,5-piperazinediones were synthesized using combinatorial techniques, to find highly selective AChE or BuChE inhibitors. This approach led to the identification of several lead compounds; among these, DKP80 (**3.104**) was the most active AChEI, with no significant inhibition of BuChE; moreover binding interactions of this compound to the active center of AChE were characterized (Carbonell et al., 2002).

Encouraged by the potential interest of ensaculin (**3.105**) as a new antidementia agent (Hoerr and Noeldner, 2002), a team reported the synthesis of novel coumarins, enabling a SAR study (Shen et al., 2005). While several coumarins were shown to be mixed inhibitors (AChE or BuChE), (**3.106**) was characterized as a noncompetitive AChEI.

Piperidine compounds were recently designed, based on the structure of natural AChEIs found in *Senna spectabilis*; one of the semisynthetic analogs, (**3.107**), was shown to reverse scopolamine-induced amnesia in mice, at nontoxic doses (Viegas et al., 2005b).

DKP80
3.104

Ensaculin
3.105

3.106

3.107

Bis(7)-tacrine
3.108

3.109

3.110

3.111

3.112

3.113

3.114

Huprine X
3.115

TV3326
3.116

Rasagiline
3.117

3.3.1.2.5   Bivalent Compounds (Dimeric and Hybrid Compounds)
These types of compounds are based on the concept of combining two structural units coming from one or two active monomers. Monomers may be spaced by a spacer of suitable length or not.

Hence the dimerization concept has been applied in the search for novel AChEIs. Based on the good results for bis(7)-tacrine (**3.108**) (Wang et al., 1999), new bis-tacrine congeners such as (**3.109**) have been designed (Wu et al., 2002). Further studies with tacrine dimers permitted validation of the hypothesis of extra sites of interaction in the AChE and BuChE active-site gorges (Savini et al., 2003). The enhanced activity of these type of compounds seems to be due to binding to the active and peripheral sites of AChE (Carlier et al., 1999). Recently, (−)-huperzine A analogs dimers like (**3.110**) were studied with *Torpedo californica* AChE and showed increased affinity to that of huperzine A. Nevertheless, when tested with rat AChE, such dimers showed lower inhibitory potency (Wong et al., 2003).

Tacrine–thiadiazolidinone hybrids have been described (Dorronsoro et al., 2005). Several reported compounds, such as (**3.111**), may be considered as new leads in the optimization of AD modifying agents. On the other hand, new tacrine–melatonin hybrids (**3.112**) have been reported (Rodriguez-Franco et al., 2006); interestingly, some compounds displayed higher in vitro properties than the sum of their parts.

In the attempt of designing AChEIs able to bind simultaneously to the peripheral and catalytic sites of the enzyme, donepezil–tacrine hybrids were prepared. The initial goal of the work seems to have been achieved with (**3.113**) (Alonso et al., 2005).

The preparation of heterodimers based on galanthamine structure like (**3.114**) permitted the investigation of SAR (Mary et al., 1998).

Huprines are synthetic hybrids resulting from the combination of the 4-aminoquinoline substructure (for tacrine) with the carbocyclic substructure of huperzine. Among the 30 different compounds synthesized emerged huprine X (**3.115**), which is one of the most potent reversible inhibitors known. It is worth mentioning here that research suggested that huprine X also had agonist effects on both M_1 and nicotinic receptors (Roman et al., 2002). In vitro investigations on newly synthesized regioisomeric (±)-*syn*-huprines showed that they were slightly less active than the corresponding antiderivatives, this difference in effects has been explained on the basis of an inverse solvation effect (Camps et al., 2001).

TV-3326 (**3.116**) results from the combination of the moieties from rasagiline (**3.117**) and rivastigmine, and exhibits interesting neuroprotective effects, including antiapoptotic activity, activation of α-secretase (Youdim and Weinstock, 2002). Based on its administration in rat models of anxiety and depression, TV-3326 seems to be a potentially valuable drug for the treatment of dementia in patients with depression (Weinstock et al., 2002).

3.3.2   THE AChEIs CURRENTLY IN CLINICAL USE

The important work achieved in the field of AChEIs led to the approval of four AChEIs by the U.S. Food and Drug Administration's for the treatment of AD.

As was described earlier, tacrine was the first drug approved by the FDA for the treatment of AD. Even if progress was made with this first drug, the frequent side effects, combined with the poor pharmacokinetic properties (requiring four times daily dosing), forced its abandonment. The main lessons learned from tacrine were that second-generation AChEIs should be less toxic, more selective, and more efficacious.

3.3.2.1 Donepezil

Donepezil, in addition to having a similar efficacy to tacrine in mild-to-moderate AD, appeared to have a major advantage in that its use has been associated with a lower incidence of cholinergic-like side effects and no liver toxicity.

Therapeutic plasmatic concentrations are observed after doses of 5–10 mg/day. The drug is well absorbed, unaffected by food and peak plasma concentrations are obtained 3–4 h after oral administration; side effects described during clinical trials include nausea, vomiting, diarrhea, insomnia, muscle cramps, and sleep disorders.

The fact that treatment and placebo groups became indistinguishable following a 6 week placebo washout suggests that donepezil is not a disease-modifying drug (Rogers et al., 1998b). However, additional pharmacological properties were found for donepezil, that is, a modulation on nicotinic receptors of the *Substantia nigra* dopaminergic neurones (Di Angelantonio et al., 2004).

Moreover, other work recently reported a protective activity against NMDA toxicity in cortical neurons, and this neuroprotection seemed to be partially mediated by inhibition of the increase of $[Ca^{2+}]$ (Akasofu et al., 2006).

A great number of original papers reporting randomized controlled trials with donepezil have been published since 1996 (Rogers et al., 1996, 1998a,b; Burns et al., 1999; Greenberg et al., 2000; Homma et al., 2000; Mohs et al., 2001; Tariot et al., 2001; Winblad et al., 2001; AD2000 Collaborative Group, 2004; Holmes et al., 2004), showing the important benefits, compared with placebo, across functional, cognitive, and behavioral symptoms with good tolerability in the treatment of mild-to-moderate AD.

Additional clinical studies showed the efficacy and safety of donepezil in advanced severe AD stages (Feldman et al., 2001; Winblad et al., 2006).

However, donepezil is also reported in a single-center clinical trial as showing that the drug improved memory in multiple sclerosis patients with initial cognitive impairment (Krupp et al., 2004). Donepezil is also expected to be therapeutically useful and safe in treating Dementia with Lewy bodies (Satoru et al., 2006). Moreover, the combined analysis of two large-scale clinical trials concluded that donepezil should be considered as an important therapeutic element in the overall management of patients with vascular dementia (Erkinjuntti et al., 2004; Roman et al., 2005).

3.3.2.2 Rivastigmine

As stated earlier, rivastigmine is a phenylcarbamate which has been developed from the miotine structure and an initial interesting feature of this drug is that it inhibits both AChE and BuChE, the latter being an enzyme mainly associated with glial cells and which have been shown to be more active in patients with AD (Ballard, 2002). AChE activity decreased progressively whereas BuChE activity remained stable or even increased, probably due to glyal proliferation (Scarpini et al., 2003). This AChEI is brain-region selective and has a long duration of action. It inhibits preferentially the G_1 enzymatic form of AChE, which predominates in the brain of AD patients (Polinsky, 1998). Moreover based on studies with rats, it was advanced that rivastigmine was more potent at inhibiting AChE in the cortex and the hippocampus than in other brain regions, this latter property having potential clinical implications since these two regions are the main brain regions affected during AD.

Therapeutic dosing is 6–12 mg/day given twice daily, with higher doses having the potential for greater benefits. The majority of side effects that have been reported are related to the gastrointestinal system (e.g., nausea, vomiting, and anorexia) and necessitate monitoring of the patients' weight. Side effects occur predominantly during the initial dose-titration phase and may be minimized by slower dose-escalation intervals and administration with a full meal (Desai and Grossberg, 2005).

Based on its dual inhibitory effect on both acetyl- and butyrylcholinesterase, rivastigmine may represent an interesting alternative treatment option for patients not responding to therapy with other AChEIs (Plosker and Keating, 2004; Bartorelli et al., 2005). Switching from donepezil or galanthamine to rivastigmine may improve cognition and behavior may be even more effective if combined with memantine (Dantoine et al., 2006). However, combining two AChEIs is not recommended and more research is needed to enable the development of practice guidelines.

Rivastigmine has beneficial effects on global function, the activities of daily living, and behavioral problems (Plosker and Keating, 2004). A pilot study was recently conducted by Potkin et al. and the data obtained in this work suggested that rivastigmine treatment may have beneficial effects in the treatment of patients with mixed dementia (Potkin et al., 2006). Additional controlled studies have highlighted the potential use of rivastigmine treatment in dementia with Lewy bodies and Parkinson's disease dementia.

3.3.2.3 Galanthamine

Galanthamine is the most recent AChEI marketed for AD treatment. This alkaloid has been characterized as a reversible competitive AChE inhibitor and also as an allosteric modulator of nicotinic ACh receptors.

As for rivastigmine, time is needed to reach a therapeutic dose after the start of titration. The starting dosage is 4 mg (hydrobromide salt) twice daily and the dosage should be increased to 8 mg twice daily and then to 12 mg twice daily at 4-weekly intervals. Galanthamine extended-release capsules (16 or 24 mg/day) have been developed to ameliorate patient compliance (Hing et al., 2005). Galanthamine is about 90% bioavailable and displays linear pharmacokinetics. It has a relatively large volume of distribution and low protein-binding potential (Farlow, 2003). Typical side effects during galanthamine treatment are weight loss, headache, and abdominal pain. Increased mortality was recently reported to be associated with galanthamine treatment, but this finding remains under investigation (Lleo et al., 2006).

Clinical studies have been published with galanthamine (Raskind et al., 2000; Tariot et al., 2000; Wilcock et al., 2000; Rockwood et al., 2001; Wilkinson and Murray, 2001).

Summarizing the clinical data coming from more than 2000 subjects with mild-to-moderate AD, meta-analysis concluded that there was a modest but statistically significant improvement in the behavioral symptoms (Herrmann et al., 2005). Pooling data from four randomized trials, Orgogozo et al. also demonstrated a cognition improvement after 6 months of galanthamine 24 mg/day therapy (Orgogozo et al., 2004.)

Studies were also performed with galanthamine to test its possible use in other disorders; for instance, galanthamine was reported to have no clinical use in the treatment of attention deficit or hyperactivity disorder (Biederman et al., 2006), as well as in the treatment of patients with chronic fatigue syndrome (Blacker et al., 2004); Galanthamine was also proposed to be a possible treatment for vascular dementia (Small et al., 2003). Moreover, a 12-week trial indicated a potential interest for galanthamine in the treatment of patients with dementia with Lewy bodies (Edwards et al., 2004).

3.4 CONCLUSION ON THE CLINICAL USE

As it is still impossible to identify patients subgroup who will respond to treatment before treatment (Birks, 2006), research is currently focusing on the early definition of treatment responders (Bizzarro et al., 2005; Pola et al., 2005).

Controversial meta-analysis have been performed with AChEIs, some remaining cautious and asserting that benefits after AChEIs treatment are minimal (Kaduszkiewicz et al., 2005) or modest (Lanctôt et al., 2003), whereas others demonstrating the effectiveness and the good tolerability of such drugs. A recent review of the clinical data by Takeda concluded that the AChEIs currently marketed were able to delay cognitive impairment for at least 6 months (Takeda et al., 2006).

On the other hand, a study was carried out to compare the three AChEIs: despite the slight variations in the mode of action of the three drugs, no significant differences in efficacy were found between donepezil, galanthamine, and rivastigmine. The three AChEI demonstrated similar efficacy in short-term use (6 months) (Lopez-Pousa et al., 2005).

As long-term data were needed such studies have been performed; from these studies, it appeared that AChEIs treatment could offer continued therapeutic benefit for up to 2 years in patients with moderate AD. (Rogers et al., 2000; Doody et al., 2001; Raskind et al., 2004; Bullock et al., 2005).

However, it was found that rapidly reversible ChEIs appear to increase AChE activity over the longer term but irreversible or very slowly reversible ChEIs do not seem to have this effect. As chronic increases in AChE activity may exacerbate neurodegenerative processes, it is believed that the rationale and expectations of some drugs in the long-term management of AD should be reconsidered (Lane et al., 2004).

REFERENCES

AD2000 Collaborative Group (2004) Long-term donepezil treatment in 565 patients with Alzheimer's disease (AD2000): randomised double-blind trial. *Lancet*, 363, 2105–2115.
Ahmad, I., Anis, I., Malik, A., Nawaz, S., Ahmad, C. and Muhammad, I. (2003) Cholinesterase inhibitory constituents from *Onosma hispida*. *Chemical & Pharmaceutical Bulletin*, 51, 412–414.

Ahmad, V.U., Khan, A., Farooq, U., Kousar, F. et al. (2005) Three new cholinesterase-inhibiting *cis*-clerodane diterpenoids from *Otostegia limbata*. *Chemical & Pharmaceutical Bulletin*, 53, 378–381.

Akasofu, S., Kimura, M., Kosasa, T., Ogura, H. and Sawada, K. (2006) Protective effect of donepezil in primary-cultured rat cortical neurons exposed to *N*-methyl-D-aspartate (NMDA) toxicity. *European Journal of Pharmacology*, 530, 215.

Aldridge, W.N. (1950) Some properties of specific cholinesterase with particular reference to the mechanism of inhibition by diethyl-*p*-nitrophenyl triphosphate (E605) and analogues. *Biochemical Journal*, 46, 451–460.

Alonso, D., Dorronsoro, I., Rubio, L., Munoz, P., Garcia-Palomero, E., Del Monte, M. et al. (2005) Donepezil–tacrine hybrid related derivatives as new dual binding site inhibitors of AChE. *Bioorganic & Medicinal Chemistry*, 13, 6588–6597.

Amenta, F., Parnetti, L., Gallai, V. and Wallin, A. (2001) Treatment of cognitive dysfunction associated with Alzheimer's disease with cholinergic precursors. Ineffective treatments or inappropriate approaches? *Mechanisms of Ageing and Development*, 122, 2025–2040.

Amitai, G., Gaidukov, L., Adani, R., Yishay, S., Yacov, G., Kushnir, M. et al. (2006) Enhanced stereoselective hydrolysis of toxic organophosphates by directly evolved variants of mammalian serum paraoxonase. *FEBS Journal*, 273, 1906–1919.

Andrade, M.T., Lima, J.A., Pinto, A.C., Rezende, C.M., Carvalho, M.P. and Epifanio, R.A. (2005) Indole alkaloids from *Tabernaemontana australis* (Mueell. Arg) Miers that inhibit acetylcholinesterase enzyme. *Bioorganic & Medicinal Chemistry*, 13, 4092–4095.

Andreani, A., Leoni, A., Locatelli, A., Morigi, R., Rambaldi, M., Pietra, C. and Villetti, G. (2000) 4-Aminopyridine derivatives with antiamnesic activity. *European Journal of Medicinal Chemistry*, 35, 77–82.

Anonymous (1991) SM-10888. Agent for cognition disorders, Acetylcholinesterase inhibitor. *Drugs Future*, 16, 33.

Asberom, T., Billard, W., Binch, H., Clader, J.W., Cox, K., Crosby, G. et al. (2001) Discovery of SCH 211803: a potent and highly selective muscarinic M2 antagonist and a promising new approach to the treatment of Alzheimer's disease. Abstracts of Papers, 221st ACS National Meeting, San Diego, MEDI-169.

Atta-ur-Rahman, Fatima, N., Akhtar, F., Choudhary, M.I. and Khalid, A. (2000) New norditerpenoid alkaloids from *Aconitum falconeri*. *Journal of Natural Products*, 63, 1393.

Atta-Ur-Rahman, Zaheer-Ul-Haq, Khalid, A., Anjum, S., Khan, M.R. and Choudhary, M.I. (2002) Pregnane-type steroidal alkaloids of *Sarcococca saligna*: a new class of cholinesterase inhibitors. *Helvetica Chimica Acta*, 85, 678–688.

Atta-ur-Rahman, Zaheer-ul-Haq, Feroz, F., Khalid, A., Nawaz, S., Ahmad, K. et al. (2004) New cholinesterase-inhibiting steroidal alkaloids from *Sarcococca saligna*. *Helvetica Chimica Acta*, 87, 439–448.

Bai, D.L., Tang, X.C. and He, X.C. (2000) Huperzine A, a potential therapeutic agent for treatment of Alzheimer's disease. *Current Medicinal Chemistry*, 7, 355–374.

Balestra, M., Gurley, D. and Rosamond, J. (2001) Preparation of bis-indoles for pharmaceutical use as positive modulators of nicotinic receptor agonists. PCT Int. Appl., WO 2001032620 A1.

Ballard, C.G. (2002) Advances in the treatment of Alzheimer's disease: benefits of dual cholinesterase inhibition. *European Neurology*, 47, 64.

Barbosa Filho, J.M., Medeiros, K.C., Paula, D., Margareth de Fatima, F.M., Batista, L.M., Athayde-Filho, P.E. et al. (2006) Natural products inhibitors of the enzyme acetylcholinesterase. *Revista Brasileira de Farmacognosia*, 16, 258–285.

Bartolucci, C., Perola, E., Cellai, L., Brufani, M. and Lamba, D. (1999) "Back door" opening implied by the crystal structure of a carbamoylated acetylcholinesterase. *Biochemistry*, 38, 5714–5719.

Bartolucci, C., Siotto, M., Ghidini, E., Amari, G., Bolzoni, P.T., Racchi, M. et al. (2006) Structural determinants of *Torpedo californica* acetylcholinesterase inhibition by the novel and orally active carbamate based anti-Alzheimer drug ganstigmine (CHF-2819). *Journal of Medicinal Chemistry*, 49, 5051–5058.

Bartorelli, L., Giraldi, C., Saccardo, M., Cammarata, S., Bottini, G., Fasanaro, A.M. and Trequattrini, A. (2005) Effects of switching from an AChE inhibitor to a dual AChE-BuChE inhibitor in patients with Alzheimer's disease. *Current Medical Research and Opinion*, 21, 1809.

Belluti, F., Rampa, A., Piazzi, L., Bisi, A., Gobbi, S., Bartolini, M. et al. (2005) Cholinesterase inhibitors: xanthostigmine derivatives blocking the acetylcholinesterase-induced b-amyloid aggregation. *Journal of Medicinal Chemistry*, 48, 4444–4456.

Biederman, J., Mick, E., Faraone, S., Hammerness, P., Surman, C., Harpold, T. et al. (2006) A double-blind comparison of galantamine hydrogen bromide and placebo in adults with attention-deficit/hyperactivity disorder: a pilot study. *Journal of Clinical Psychopharmacology*, 26, 163.

Billard, W., Binch, H., Bratzler, K., Chen, L.Y., Crosby, G., Duffy, R.A. et al. (2000) Diphenylsulfone muscarinic antagonists: piperidine derivatives with high m2 selectivity and improved potency. *Bioorganic & Medicinal Chemistry Letters*, 10, 2209–2212.

Birks, J. (2006) Cholinesterase inhibitors for Alzheimer's disease. *Cochrane Database of Systematic Reviews (Online)*, CD005593.

Birks, J. and Wilcock, G.G.W. (2004) Velnacrine for Alzheimer's disease. *Cochrane Database of Systematic Reviews (Online)*, 2, CD004748.

Bizzarro, A., Marra, C., Acciarri, A., Valenza, A., Tiziano, F.D., Brahe, C. and Masullo, C. (2005) Apolipoprotein E e4 allele differentiates the clinical response to donepezil in Alzheimer's disease. *Dementia and Geriatric Cognitive Disorders*, 20, 254–261.

Blacker, C.V.R., Greenwood, D.T., Wesnes, K.A., Wilson, R., Woodward, C., Howe, I. and Ali, T. (2004) Effect of galantamine hydrobromide in chronic fatigue syndrome. A randomized controlled trial. *Journal of the American Medical Association*, 292, 1195.

Bodnar, A.L., Cortes-Burgos, L.A., Cook, K.K., Dinh, D.M., Groppi, V.E., Hajos, M. et al. (2005) Discovery and structure–activity relationship of quinuclidine benzamides as agonists of a7 nicotinic acetylcholine receptors. *Journal of Medicinal Chemistry*, 48, 905–908.

Böhme, T.M., Keim, C., Dannhardt, G., Mutschler, E. and Lambrecht, G. (2001) Design and pharmacology of quinuclidine derivatives as M2-selective muscarinic receptor ligands. *Bioorganic & Medicinal Chemistry Letters*, 11, 1241–1243.

Böhme, T.M., Keim, C., Kreutzmann, K., Linder, M., Dingermann, T., Dannhardt, G. et al. (2003) Structure–activity relationships of dimethindene derivatives as new M2-selective muscarinic receptor antagonists. *Journal of Medicinal Chemistry*, 46, 856–867.

Bohme, G.A., Letchworth, S.R., Piot-Grosjean, O., Gatto, G.J., Obinu, M.C., Caldwell, W.S. et al. (2004) In vitro and in vivo characterization of TC-1827, a novel brain a4b2 nicotinic receptor agonist with pro-cognitive activity. *Drug Development Research*, 62, 26–40.

Bolognesi, M.L., Andrisano, V., Bartolini, M., Minarini, A., Rosini, M., Tumiatti, V. and Melchiorre, C. (2001) Hexahydrochromeno[4,3-b]pyrrole derivatives as acetylcholinesterase inhibitors. *Journal of Medicinal Chemistry*, 44, 105.

Bon, S., Ayon, A., Leroy, J. and Massoulié, J. (2003) Trimerization domain of the collagen tail of acetylcholinesterase. *Neurochemical Research*, 28, 523–535.

Bourne, Y., Taylor, P., Radic, Z. and Marchot, P. (2003) Structural insights into ligand interactions at the acetylcholinesterase peripheral anionic site. *EMBO Journal*, 22, 1–12.

Briggs, C.A., Anderson, D.J., Brioni, J.D., Buccafusco, J.J., Buckley, M.J., Campbell, J.E. et al. (1997) Functional characterization of the novel neuronal nicotinic acetylcholine receptor ligand GTS-21 in vitro and in vivo. *Pharmacology, Biochemistry and Behavior*, 57, 231–241.

Broad, L.M., Zwart, R., Pearson, K.H., Lee, M., Wallace, L., McPhie, G.I. et al. (2006) Identification and pharmacological profile of a new class of selective nicotinic acetylcholine receptor potentiators. *Journal of Pharmacology and Experimental Therapeutics*, 318, 1108–1117.

Brodbeck, U., Schweikert, K., Gentinetta, R. and Rottenberg, M. (1979) Fluorinated aldehydes and ketones acting as quasi-substrate inhibitors of acetylcholinesterase. *Biochimica et Biophysica Acta*, 567, 357–369.

Brühlmann, C., Ooms, F., Carrupt, P.A., Testa, B., Catto, M., Leonetti, F. et al. (2001) Coumarins derivatives as dual inhibitors of acetylcholinesterase and monoamine oxidase. *Journal of Medicinal Chemistry*, 44, 3195.

Bullock, R., Touchon, J., Bergman, H., Gambina, G., He, Y., Rapatz, G. et al. (2005) Rivastigmine and donepezil treatment in moderate to moderately-severe Alzheimer's disease over a 2-year period. *Current Medical Research and Opinion*, 21, 1317.

Bunnelle, W.H., Dart, M.J. and Schrimpf, M.R. (2006) Design of ligands for the nicotinic acetylcholine receptors: the quest for selectivity. *Frontiers in Medicinal Chemistry*, 3, 195–247.

Burns, A., Friedhoff, L.T., Gauthier, S., Hecker, J., Moller, H.J., Petit, H. et al. (1999) The effects of donepezil in Alzheimer's disease—results from a multinational trial. *Dementia and Geriatric Cognitive Disorders*, 10, 237–244.

Camps, P. and Munoz-Torrero, D. (2002) Cholinergic drugs in pharmacotherapy of Alzheimer's disease. *Mini-Reviews in Medicinal Chemistry*, 2, 11–25.

Camps, P., Gomez, E, Munoz-Torrero, D, Badia, A, Vivas, N.M., Barril, X. et al. (2001) Synthesis, in vitro pharmacology, and molecular modeling of *syn*-huprines as acetylcholinesterase inhibitors. *Journal of Medicinal Chemistry*, 44, 4733.

Carbonell, T., Masip, I., Sanchez-Baeza, F., Delgado, M., Araya, E., Llorens, O. et al. (2002) Identification of selective and potent inhibitors of acetylcholinesterase from combinatorial libraries of 2,5-piperazinediones. *Molecular Diversity*, 5, 131.

Cardoso, C.L., Castro-Gamboa, I., Silva, D.H.S., Furlan, M., de Epifanio, R., Pinto, A.C. et al. (2004) Indole glucoalkaloids from *Chimarrhis turbinata* and their evaluation as antioxidant agents and acetylcholinesterase inhibitors. *Journal of Natural Products*, 67, 1882–1885.

Carlier, P.R., Chow, E.S.H., Han, Y., Liu, J., El Yazal, J. and Pang, Y.P. (1999) Heterodimeric tacrine-based acetylcholinesterase inhibitors: investigating ligand–peripheral site interactions. *Journal of Medicinal Chemistry*, 42, 4225.

Cassano, T., Carratu, M.R., Coluccia, A., Di Giovanni, V., Steardo, L., Cuomo, V. and Trabace, L. (2002) Preclinical progress with CHF2819, a novel orally active acetylcholinesterase inhibitor. *Drug Development Research*, 56, 354.

Cerasoli, D.M., Griffiths, E.M., Doctor, B.P., Saxena, A., Fedorko, J.M., Greig, N.H. et al. (2005) In vitro and in vivo characterization of recombinant human butyrylcholinesterase (Protexia) as a potential nerve agent bioscavenger. *Chemico-Biological Interactions*, 157–158, 363–365.

Changeux, J.P. (1966) Responses of acetylcholinesterase from *Torpedo marmorata* to salts and curarizing drugs. *Molecular Pharmacology*, 2, 369–392.

Chen, J.W., Luo, Y.L., Hwang, M.J., Peng, F.C. and Ling, K.H. (1999) Territrem B, a tremorgenic mycotoxin that inhibits acetylcholinesterase with a non-covalent yet irreversible binding mechanism. *Journal of Biological Chemistry*, 274, 34919.

Choudhary, M.I., Shahnaz, S., Parveen, S., Khalid, A., Ayatollahi, S.A.M., Atta-ur-Rahman and Parvez, M. (2003) New triterpenoid alkaloid cholinesterase inhibitors from *Buxus hyrcana*. *Journal of Natural Products*, 66, 739–742.

Choudhary, M.I., Devkota, K.P., Nawaz, S.A., Shaheen, F. and Atta-ur-Rahman (2004) Cholinesterase-inhibiting new steroidal alkaloids from *Sarcococca hookeriana* of Nepalese origin. *Helvetica Chimica Acta*, 87, 1099.

Chung, Y.K., Heo, H.J., Kim, E.K., Kim, H.K., Huh, T.L., Lim, Y. et al. (2001) Inhibitory effect of ursolic acid purified from *Origanum majorana* L. on the acetylcholinesterase. *Molecules and Cells*, 11, 137.

Clader, J.W., Billard, W., Binch, H., Chen, L.Y., Crosby, G., Duffy, R.A. et al. (2004) Muscarinic M2 antagonists: anthranilamide derivatives with exceptional selectivity and in vivo activity. *Bioorganic & Medicinal Chemistry*, 12, 319–326.

Cole, K.P. and Hsung, R.P. (2002) The first enantioselective total synthesis of (−)-arisugacin A. *Tetrahedron Letters*, 43, 8791.

Cole, K.P., Hsung, R.P. and Yang, X.F. (2002) The total synthesis of (±)-arisugacin A., *Tetrahedron Letters*, 43, 3341.

Colletier, J.P., Fournier, D., Greenblatt, H.M., Stojan, J. Sussman, J.L., Zaccai, J. et al. (2006) Structural insights into substrate traffic and inhibition in acetylcholinesterase. *EMBO Journal*, 25, 2746–2756.

Costagli, C. and Galli, A. (1998) Inhibition of cholinesterase-associated aryl acylamidase activity by anticholinesterase agents: focus on drugs potentially effective in Alzheimer's disease. *Biochemical Pharmacology*, 55, 1733.

Cummings, J.L., Vinters, H.V., Cole, G.M. and Khachaturian, Z.S. (1998) Alzheimer's disease: etiologies, pathophysiology, cognitive reserve, and treatment opportunities. *Neurology*, 51, S2–S17, discussion S65–S67.

Damodaran, T.V. and Abou-Donia, M.B. (2000) Alterations in levels of mRNAs coding for glial fibrillary acidic protein (GFAP) and vimentin genes in the central nervous system of hens treated with diisopropyl phosphorofluoridate (DFP). *Neurochemical Research*, 25, 809–816.

Dantoine, T., Auriacombe, S., Sarazin, M., Becker, H., Pere, J.J. and Bourdeix, I. (2006) Rivastigmine monotherapy and combination therapy with memantine in patients with moderately severe Alzheimer's disease who failed to benefit from previous cholinesterase inhibitor treatment. *International Journal of Clinical Practice*, 60, 110.

Davies, D.R. and Green, A.L. (1956) The kinetics of reactivation by oximes of cholinesterase inhibited by organophosphorus compounds. *Biochemical Journal*, 63, 520–535.

De Fiebre, C.M., Meyer, E.M., Henry, J.C., Muraskin, S.I., Kem, W.R. and Papke, R.L. (1995) Characterization of a series of anabaseine-derived compounds reveals that the 3-(4)-dimethylaminocinnamylidine derivative is a selective agonist at neuronal nicotinic a7/125I-a-bungarotoxin receptor subtypes. *Molecular Pharmacology*, 47, 164–171.

de Jong, R.H. (2003) Nerve gas terrorism: a grim challenge to anesthesiologists. *Anesthesia & Analgesia*, 96, 819–825.

de los Rios, C., Marco, J.L., Carreiras, M.D.C., Chinchon, P.M., Garcia, A.G. and Villarroya, M. (2002) Novel tacrine derivatives that block neuronal calcium channels. *Bioorganic & Medicinal Chemistry*, 10, 2077–2088.

Decker, M.W., Brioni, J.D., Sullivan, J.P., Buckley, M.J., Radek, R.J., Raskiewicz, J.L. et al. (1994) (*S*)-3-methyl-5-(1-methyl-2-pyrrolidinyl)isoxazole (ABT 418): a novel cholinergic ligand with cognition-enhancing and anxiolytic activities: II. In vivo characterization. *Journal of Pharmacology and Experimental Therapeutics*, 270, 319.

Desai, A.K. and Grossberg, G.T. (2005) Rivastigmine for Alzheimer's disease. *Expert Review of Neurotherapeutics*, 5, 563.

Di Angelantonio, S., Bernardi, G. and Mercuri, N.B. (2004) Donepezil modulates nicotinic receptors of substantia nigra dopaminergic neurones. *British Journal of Pharmacology*, 141, 644.

Dong, Y., Zheng, Z., Zhang, Q., Zhang, H., Lu, X., Shu, W. et al. (2002) N98-1021A, a selective acetylcholinesterase inhibitors derived from microorganisms. *Zhongguo Kangshengsu Zazhi*, 27, 260.

Doods, H.N. (1995) Lipophilic muscarinic M2 antagonists as potential drugs for cognitive disorders. *Drugs Future*, 20, 157.

Doody, R.S., Geldmacher, D.S., Gordon, B., Perdomo, C.A. and Pratt, R.D. (2001) Open-label, multicenter, phase 3 extension study of the safety and efficacy of donepezil in patients with Alzheimer disease. *Archives of Neurology*, 58, 427–433.

Dorronsoro, I., Alonso, D., Castro, A., Del Monte, M., Garcia-Palomero, E. and Martinez, A. (2005) Synthesis and biological evaluation of tacrine–thiadiazolidinone hybrids as dual acetylcholinesterase inhibitors. *Archiv der Pharmazie*, 338, 18–23.

Dudai, Y., Herzberg, M. and Silman, I. (1973) Molecular structures of acetylcholinesterase from electric organ tissue of the electric eel. *Proceedings of the National Academy of Sciences of the United States of America*, 70, 2473–2476.

Dvir, H., Harel, M., Bon, S., Liu, W.Q., Vidal, M., Garbay, C. et al. (2004), The synaptic acetylcholinesterase tetramer assembles around a polyproline II helix. *EMBO Journal*, 23, 4394–4405.

Earl, R.A., Zaczek, R., Teleha, C.A., Fisher, B.N., Maciag, C.M., Marynowski, M.E. et al. (1998) 2-Fluoro-4-pyridinylmethyl analogs of linopirdine as orally active acetylcholine release-enhancing agents with good efficacy and duration of action. *Journal of Medicinal Chemistry*, 41, 4615–4622.

Edwards, K.R., Hershey, L., Wray, L., Bednarczyk, E.M., Lichter, D., Farlow, M. and Johnson, S. (2004) Efficacy and safety of galantamine in patients with dementia with Lewy bodies: a 12-week interim analysis. *Dementia and Geriatric Cognitive Disorders*, 17, 40.

Ekstrom, F., Akfur, C., Tunemalm, A.K. and Lundberg, S. (2006) Structural changes of phenylalanine 338 and histidine 447 revealed by the crystal structures of tabun-inhibited murine acetylcholinesterase. *Biochemistry*, 45, 74–81.

Ellman, G.L., Courtney, K.D., Andres, V. and Featherstone, R.M. (1961) A new and rapid colorimetric determination of acetylcholinesterase activity. *Biochemical Pharmacology*, 7, 88–95.

Emmerling, M.R., Gregor, V.E., Schwarz, R.D., Scholten, J.D., Callahan, M.J., Lee, C. et al. (1995) CI-1002, a novel anticholinesterase and muscarinic antagonist. *Advances in Behavioral Biology*, 44, 483.

Engelhard, N., Prchal, K. and Nenner, M., (1967) Acetylcholinesterase. *Angewandte Chemie* (International Edition in English), 6, 615–626.

Erkinjuntti, T., Roman, G. and Gauthier, S. (2004) Treatment of vascular dementia—evidence from clinical trials with cholinesterase inhibitors. *Neurological Research*, 26, 603.

Ernst, G., Phillips, E. and Schmiesing, R.J. (2005) A preparation of diazabicyclononane derivatives, useful as nicotinic acetylcholine agonists. PCT Int. Appl., WO 05/030777 A1.

Farlow, M.R. (2003) Clinical pharmacokinetics of galantamine. *Clinical Pharmacokinetics*, 42, 1383.

Feldman, H., Gauthier, S., Hecker, J., Vellas, B., Subbiah, P. and Whalen, E. (2001) 24-week, randomized, double-blind study of donepezil in moderate to severe Alzheimer's disease. *Neurology*, 57, 613.

Gao, L.J., Waelbroeck, M., Hofman, S., Van Haver, D., Milanesio, M., Viterbo, D. and De Clercq, P.J. (2002) Synthesis and affinity studies of himbacine derived muscarinic receptor antagonists. *Bioorganic & Medicinal Chemistry Letters*, 12, 1909–1912.

Geula, C. (1998) Abnormalities of neural circuitry in Alzheimer's disease: hippocampus and cortical cholinergic innervation. *Neurology*, 51, S18–S29.

Ghelardini, C., Galeotti, N., Romanelli, M.N., Gualtieri, F., Bartolini, A. (2000) Pharmacological characterization of the novel ACh releaser a-tropanyl 2-(4-bromophenyl)propionate (PG-9). *CNS Drug Reviews*, 6, 63–78.

Gilson, M.K., Straatsma, T.P., McCammon, J.A., Ripoll, D.R., Faerman, C.H., Axelsen, P.H. et al. (1994) Open "back door" in a molecular dynamics simulation of acetylcholinesterase. *Science*, 263, 1276–1278.

Goličnik, M. and Stojan, J. (2002) Multi-step analysis as a tool for kinetic parameter estimation and mechanism discrimination in the reaction between tight-binding fasciculin 2 and electric eel acetylcholinesterase. *Biochimica et Biophysica Acta*, 1597, 164–172.

Goličnik, M., Fournier, D. and Stojan, J. (2002) Acceleration of *Drosophila melanogaster* acetylcholinesterase methanesulfonylation: peripheral ligand D-tubocurarine enhances the affinity for small methanesulfonylfluoride. *Chemico-Biological Interactions*, 139, 145–157.

Gracon, S.I., Knapp, M.J., Berghoff, W.G., Pierce, M., Dejong, R., Lobbestael, S.J. et al. (1998) Safety of tacrine: clinical trials, treatment IND, and postmarketing experience. *Alzheimer Disease and Associated Disorders*, 12, 93–101.

Greenberg, S.M., Tennis, M.K., Brown, L.B., Gomez-Isla, T., Hayden, D.L., Schoenfeld, D.A. et al. (2000) Donepezil therapy in clinical practice: a randomised crossover study. *Archives of Neurology*, 57, 94.

Greenlee, W., Clader, J., Asberom, T., McCombie, S., Ford, J., Guzik, H. et al. (2001) Muscarinic agonists and antagonists in the treatment of Alzheimer's disease. *Farmaco*, 56, 247–250.

Guandalini, L., Dei, S., Gualtieri, F., Romanelli, M.N., Scapecchi, S., Teodori, E. and Varani, K. (2002) Synthesis of hexahydro-2-pyrindine (= hexahydrocyclopenta[c]pyridine) derivatives as conformationally restricted analogs of the nicotinic ligands arecolone and isoarecolone. *Helvetica Chimica Acta*, 85, 96–107.

Harel, M., Sussman, J.L., Krejci, E., Bon, S., Chanal, P., Massoulie, J. et al. (1992) Conversion of acetylcholinesterase to butyrylcholinesterase: modeling and mutagenesis. *Proceedings of the National Academy of Sciences of the United States of America*, 89, 10827–10831.

Harel, M., Schalk, I., Ehret-Sabatier, L., Bouet, F., Goeldner, M., Hirth, C. et al. (1993) Quaternary ligand binding to aromatic residues in the active-site gorge of acetylcholinesterase. *Proceedings of the National Academy of Sciences of the United States of America*, 90, 9031–9035.

Harel, M., Kryger, G., Rosenberry, T.L., Mallender, W.D., Lewis, T., Fletcher, R.J. et al. (2000) Three-dimensional structures of *Drosophila melanogaster* acetylcholinesterase and of its complexes with two potent inhibitors. *Protein Science*, 9, 1063–1072.

Harel, M., Hyatt, J.L., Brumshtein, B., Morton, C.L., Wadkins, R.M., Silman, I. et al. (2005) The 3D structure of the anticancer prodrug CPT-11 with *Torpedo californica* acetylcholinesterase rationalizes its inhibitory action on AChE and its hydrolysis by butyrylcholinesterase and carboxylesterase. *Chemico-Biological Interactions*, 157–158, 153–157.

Heilbronn, E. (1963) In vitro reactivation and 'aging' of tabun-inhibited blood cholinesterases. Studies with *N*-methyl pyridinium-2-aloxime methane sulphonate and *N,N'*-trimethylenebis (pyridinium-4-aloxime) dibromide. *Biochemical Pharmacology*, 12, 25–36.

Heo, H.J., Hong, S.C., Cho, H.Y., Hong, B., Kim, H.K., Kim, E.K. and Shin, D.H. (2002) Inhibitory effect of zeatin, isolated from *Fatoua villosa*, on acetylcholinesterase activity from PC12 cells. *Molecules and Cells*, 13, 113.

Herrmann, N., Rabheru, K., Wang, J. and Binder, C. (2005) Galantamine treatment of problematic behavior in Alzheimer disease: post-hoc analysis of pooled data from three large trials. *The American Journal of Geriatric Psychiatry* (Official Journal of the American Association for Geriatric Psychiatry), 13, 527.

Hing, J.P., Piotrovsky, V., Kimko, H., Brashear, H.R. and Zhao, Q. (2005) Pharmacokinetic simulation for switching from galantamine immediate-release to extended-release formulation. *Current Medical Research and Opinion*, 21, 483.

Hirasawa, Y., Morita, H., Shiro, M. and Kobayashi, J. (2003) Sieboldine A, a novel tetracyclic alkaloid from *Lycopodium sieboldii*, inhibiting acetylcholinesterase. *Organic Letters*, 5, 3991–3993.

Hirasawa, Y., Kobayashi, J. and Morita, H. (2006) Lycoperine A, a novel C27N3-type pentacyclic alkaloid from *Lycopodium hamiltonii*, inhibiting acetylcholinesterase. *Organic Letters*, 8, 123–126.

Hobbiger, F. (1955) Effect of nicotinhydroxamic acid methiodide on human plasma cholinesterase inhibited by organophosphate containing a dialkylphosphate group. *British Journal of Pharmacology*, 10, 356–362.

Hoerr, R. and Noeldner, M. (2002) Ensaculin (KA-672 × HCl): a multitransmitter approach to dementia treatment. *CNS Drug Reviews*, 8, 143–158.

Holmes, C., Wilkinson, D., Dean, C., Vethanayagam, S., Olivieri, S., Langley, A. et al. (2004) The efficacy of donepezil in the treatment of neuropsychiatric symptoms in Alzheimer disease. *Neurology*, 63, 214–219.

Homma, A., Takeda, M., Imai, Y., Udaka, F., Hasegawa, K., Kameyama, M. et al. (2000) Clinical efficacy and safety of donepezil on cognitive and global function in patients with Alzheimer's disease: a 24-week, multicenter, double-blind, placebo-controlled study in Japan. *Dementia and Geriatric Cognitive Disorders*, 11, 299.

Hostettmann, K., Borloz, A., Urbain, A. and Marston, A. (2006) Natural product inhibitors of acetylcholinesterase. *Current Organic Chemistry*, 10, 825–847.

Houghton, P.J., Agbedahunsi, J.M. and Adegbulugbe, A. (2004) Choline esterase inhibitory properties of alkaloids from two Nigerian *Crinum* species. *Phytochemistry*, 65, 2893.

Hurst, R.S., Hajós, M., Raggenbass, M., Wall, T.W., Higdon, N.R., Lawson, J.A. et al. (2005) A novel positive allosteric modulator of the a7 neuronal nicotinic acetylcholine receptor: in vitro and in vivo characterization. *Journal of Neuroscience*, 25, 4396–4405.

Iimura, Y. and Kosasa, T. (2001) Acetylcholinesterase inhibitors containing 1-benzylpyridinium salts. PCT Int. Appl., WO 01/78728.

Imbibo, B.P. (2001) Pharmacodynamic-tolerability relationships of cholinesterase inhibitors for Alzheimer's disease. *CNS Drugs*, 15, 375.

Inestrosa, N.C., Roberts, W.L., Marshall, T.L. and Rosenberry, T.L. (1987) Acetylcholinesterase from bovine caudate nucleus is attached to membranes by a novel subunit distinct from those of acetylcholinesterases in other tissues. *Journal of Biological Chemistry*, 262, 4441–4444.

Inestrosa, N.C., Alvarez, A., Perez, C.A., Moreno, R.D., Vicente, M., Linker, C. et al. (1996) Acetylcholinesterase accelerates assembly of amyloid-b-peptides into Alzheimer's fibrils: possible role of the peripheral site of the enzyme. *Neuron*, 16, 881.

Ingkaninan, K., Changwijit, K. and Suwanborirux, K. (2006) Vobasinyl-iboga bisindole alkaloids, potent acetylcholinesterase inhibitors from *Tabernaemontana divaricata* root. *The Journal of Pharmacy and Pharmacology*, 58, 847–852.

Isomae, K., Morimoto, S., Hasegawa, H., Morita, K. and Kamei, J. (2003) Effects of T-82, a novel acetylcholinesterase inhibitor, on impaired learning and memory in passive avoidance task in rats. *European Journal of Pharmacology*, 465, 97–103.

Jain, K.K. (2004) Modulators of nicotinic acetylcholine receptors as analgesics. *Current Opinion in Investigational Drugs*, 5, 76–81.

Kaduszkiewicz, H., Zimmermann, T., Beck-Bornholdt, H.P. and van den Bussche, H. (2005) Cholinesterase inhibitors for patients with Alzheimer's disease: systematic review of randomised clinical trials. *British Medical Journal*, 331, 321.

Kalauni, S.K., Choudhary, M.I., Khalid, A., Manandhar, M.D., Shaheen, F., Atta-ur-Rahman and Gewali, M.B. (2002) New cholinesterase inhibiting steroidal alkaloids from the leaves of *Sarcococca corsiaceae* of Nepalese origin. *Chemical & Pharmaceutical Bulletin*, 50, 1423–1426.

Kang, S.Y., Lee, K.Y., Sung, S.H., Park, M.J. and Kim, Y.C. (2001) Coumarins isolated from *Angelica gigas* inhibit acetylcholinesterase: structure–activity relationships. *Journal of Natural Products*, 64, 683.

Kang, S.Y., Lee, K.Y., Park, M.J., Kim, Y.C., Markelonis, G.J., Oh, T.H. and Kim, Y.C. (2003) Decursin from *Angelica gigas* mitigates amnesia induced by scopolamine in mice. *Neurobiology of Learning and Memory*, 79, 11–18.

Kem, W.R. (2002) Anabaseine as a molecular model for design of a7 nicotinic receptor agonist drugs. *Perspectives in Molecular Toxinology*, 297, 297.

Kihara, T., Shimohama, S., Sawada, H., Kimura, J., Kume, T., Kochiyama, H. et al. (1997) Nicotinic receptor stimulation protects neurons against b-amyloid toxicity. *Annals of Neurology*, 42, 159–163.

Kim, W.G., Song, N.K. and Yoo, I.D. (2001) Quinolactacins A1 and A2, new acetylcholinesterase inhibitors from *Penicillium citrinum*. *Journal of Antibiotics*, 54, 831–835.

Kim, D.K., Lim, J.P., Yang, J.H., Eom, D.O., Eun, J.S. and Leem, K.H. (2002) Acetylcholinesterase inhibitors from the roots of *Angelica dahurica*. *Archives of Pharmacal Research*, 25, 856–859.

Kim, D.K., Lee, K.T., Baek, N.I., Kim, S.H., Park, H.W., Lim, J.P. et al. (2004) Acetylcholinesterase inhibitors from the aerial parts of *Corydalis speciosa*. *Archives of Pharmacal Research*, 27, 1127–1131.

Kitz, R.J., Ginsburg, S. and Wilson, I.B. (1967) The reaction of acetylcholinesterase with O-dimethylcarbamyl esters of quaternary quinolinium compounds. *Biochemical Pharmacology*, 16, 2201–2209.

Kozikowski, A.P., Fauq, A.H., Miller, J.H. and McKinney, M. (1992) Alzheimer's therapy: an approach to novel muscarinic ligands based upon the naturally occurring alkaloid himbacine. *Bioorganic & Medicinal Chemistry Letters*, 2, 797–802.

Kozlowski, J.A., Zhou, G., Tagat, J.R., Lin, S.I., McCombie, S.W., Ruperto, V.B. et al. (2002) Substituted 2-(R)-Methyl piperazines as muscarinic M2 selective ligands. *Bioorganic & Medicinal Chemistry Letters*, 12, 791–794.

Krupka, R.M. (1966a) Chemical structure and function of the active center of acetylcholinesterase. *Biochemistry*, 5, 1988–1997.

Krupka, R.M. (1966b) Hydrolysis of neutral substrates by acetylcholinesterase. *Biochemistry*, 5, 1983–1988.

Krupp, L.B., Christodoulou, C., Melville, P., Scherl, W.F., MacAllister, W.S. and Elkins, L.E. (2004) Donepezil improved memory in multiple sclerosis in a randomized clinical trial. *Neurology*, 63, 1579–1585.

Kuca, K., Bartosova, L., Kassa, J., Cabal, J., Bajgar, J., Kunesova, G. et al. (2005) Comparison of the potency of newly developed and currently available oximes to reactivate nerve agent-inhibited acetylcholinesterase in vitro and in vivo. *Chemico-Biological Interactions*, 157–158, 367–368.

Lachowicz, J.E., Duffy, R.A., Ruperto, V., Kozlowski, J., Zhou, G., Clader, J. et al. (2001) Facilitation of acetylcholine release and improvement in cognition by a selective M2 muscarinic antagonist, SCH 72788. *Life Sciences*, 68, 2585–2592.

Lanctôt, K.L., Herrmann, N., Yau, K.K., Khan, L.R., Liu, B.A., LouLou, M.M., and Einarson, T.R. (2003) Efficacy and safety of cholinesterase inhibitors in Alzheimer's disease: a meta-analysis. *Canadian Medical Association Journal*, 169, 557–564.

Lane, R.M., Kivipelto, M. and Greig, N.H. (2004) Acetylcholinesterase and its inhibition in Alzheimer disease. *Clinical Neuropharmacology*, 27, 141.

Levin, E.D., McClernon, F.J. and Rezvani, A.H. (2006) Nicotinic effects on cognitive function: behavioral characterization, pharmacological specification, and anatomic localization. *Psychopharmacology*, 184, 523–539.

Lin, N.H., Gunn, D.E., Ryther, K.B., Garvey, D.S., Donnelly-Roberts, D.L., Decker, M.W. et al. (1997) Structure–activity studies on 2-methyl-3-(2(*S*)-pyrrolidinylmethoxy)pyridine (ABT-089): an orally bioavailable 3-pyridyl ether nicotinic acetylcholine receptor ligand with cognition-enhancing properties. *Journal of Medicinal Chemistry*, 40, 385–390.

Lin, M.S., Hsin, L.W., Tsai, H.B., Tsai, M.C. and Cheng, C.Y. (2002) Synthesis of 1,4-dihydro-2-phenyl-4,4-bis(4-pyridinylmethyl)-2H-isoquinolin-3-one and related compounds as acetylcholine release enhancers. *Chinese Pharmaceutical Journal*, 54, 271–281.

Liu, Q.S., Kawai, H. and Berg, D.K. (2001) b-Amyloid peptide blocks the response of a7-containing nicotinic receptors on hippocampal neurons. *Proceedings of the National Academy of Sciences of the United States of America*, 98, 4734–4739.

Liu, H.C., Chi, C.W., Ko, S.Y., Wang, H.C., Hong, C.J., Lin, K.N. et al. (2005) Cholinesterase inhibitor affects the amyloid precursor protein isoforms in patients with Alzheimer's disease. *Dementia and Geriatric Cognitive Disorders*, 19, 345.

Lleo, A., Greenberg, S.M. and Growdon, J.H. (2006) Current pharmacotherapy for Alzheimer's disease. *Annual Review of Medicine*, 57, 513.

Loewenstein-Lichtenstein, Y., Schwarz, M., Glick, D., Norgaard-Pedersen, B., Zakut, H. and Soreq, H. (1995) Genetic predisposition to adverse consequences of anti-cholinesterases in 'atypical' BCHE carriers. *Nature Medicine*, 1, 1082–1085.

López, S., Bastida, J., Viladomat, F. and Codina, C. (2002) Acetylcholinesterase inhibitory activity of some Amaryllidaceae alkaloids and *Narcissus* extracts. *Life Sciences*, 71, 2521.

López-Pousa, S., Turon-Estrada, A., Garre-Olmo, J., Pericot-Nierga, I., Lozano-Gallego, M., Vilalta-Franch, M. et al. (2005) Differential efficacy of treatment with acetylcholinesterase inhibitors in patients with mild and moderate Alzheimer's disease over a 6-month period. *Dementia and Geriatric Cognitive Disorders*, 19, 189.

Luo, W., Yu, Q.S., Zhan, M., Parrish, D., Deschamps, J.R., Kulkarni, S.S. et al. (2005) Novel anticholinesterases based on the molecular skeletons of furobenzofuran and methanobenzodioxepine. *Journal of Medicinal Chemistry*, 48, 986–994.

Marchot, P., Bourne, Y., Prowse, C.N., Bougis, P.E. and Taylor, P. (1998) Inhibition of mouse acetylcholinesterase by fasciculin: crystal structure of the complex and mutagenesis of fasciculin. *Toxicon*, 36, 1613–1622.

Marco, J.L., de los Rios, C., Garcia, A.G., Villarroya, M., Carreiras, M.C., Martins, C. et al. (2004) Synthesis, biological evaluation and molecular modelling of diversely functionalized heterocyclic derivatives as inhibitors of acetylcholinesterase/butyrylcholinesterase and modulators of Ca^{2+} channels and nicotinic receptors. *Bioorganic & Medicinal Chemistry*, 12, 2199–2218.

Martinez, A., Fernandez, A.C., Santiago, C., Rodriguez-Franco, I., Baños, J.E. and Badia, A. (2000) *N*-benzylpiperidine derivatives of 1,2,4-thiadiazolidinone as new acetylcholinesterase inhibitors. *European Journal of Medicinal Chemistry*, 35, 913.

Mary, A., Rneko, D.Z., Guillou, C. and Thal, C. (1998) Potent acetylcholinesterase inhibitors: design, synthesis, and structure–activity relationships of bis-interacting ligands in the galanthamine series. *Bioorganic & Medicinal Chemistry*, 6, 1835.

Massoulié, J. (2002) The origin of the molecular diversity and functional anchoring of cholinesterases. *Neurosignals*, 11, 130–143.

Matsuyama, S. and Matsumoto, A. (2003) Epibatidine induces long-term potentiation (LTP) via activation of a7 and a4b2 nicotinic acetylcholine receptors (nAChRs) in vivo in the intact mouse dentate gyrus: both a7 and a4b2 nAChRs essential to nicotinic LTP. *Journal of Pharmacological Sciences*, 93, 180–187.

Matsuyama, S., Matsumoto, A., Enomoto, T. and Nishizaki, T. (2000) Activation of nicotinic acetylcholine receptors induces long-term potentiation in vivo in the intact mouse dentate gyrus. *European Journal of Neuroscience*, 12, 3741–3747.

Mazurov, A., Hauser, T. and Miller, C.H. (2006) Selective a7 nicotinic acetylcholine receptor ligands. *Current Medicinal Chemistry*, 13, 1567–1584.

McCombie, S.W., Lin, S.I., Tagat, J.R., Nazareno, D., Vice, S., Ford, J. et al. (2002) Synthesis and structure–activity relationships of M2-selective muscarinic receptor ligands in the 1-[4-(4-arylsulfonyl)-phenylmethyl]-4-(4-piperidinyl)-piperazine family. *Bioorganic & Medicinal Chemistry Letters*, 12, 795–798.

Metzger, H.P. and Wilson, I.B. (1967) The acceleration of the acetylcholinesterase catalyzed hydrolysis of acetyl fluoride. *Biochemical & Biophysical Research Communications*, 28, 263–269.

Meyer, M.D., Anderson, D.J., Campbell, J.E., Carroll, S., Marsh, K.C., Rodrigues, A.D. and Decker, M.W. (2000) Preclinical pharmacology of ABT-594: a nicotinic acetylcholine receptor agonist for the treatment of pain. *CNS Drug Reviews*, 6, 183–194.

Michel, H.O., Hackley, B.E. Jr., Berkowitz, L., List, G., Hackley, E.B., Gillilan, W. and Pankau, M. (1967) Ageing and dealkylation of Soman (pinacolylmethylphosphono fluoridate)-inactivated eel cholinesterase. *Archives of Biochemistry & Biophysics*, 121, 29–34.

Millard, C.B., Koellner, G., Ordentlich, A., Shafferman, A., Silman, I. and Sussman, J.L. (1999) Reaction products of acetylcholinesterase and VX reveal a mobile histidine in the catalytic triad. *Journal of American Chemical Society*, 121, 9883–9884.

Mohs, R.C., Doody, R.S., Morris, J.C., Ieni, J.R. and Rogers, S. (2001) A randomized, double-blind, placebo-controlled study of the efficacy and safety of donepezil in patients with Alzheimer's disease in the nursing home setting. *Journal of the American Geriatrics Society*, 49, 1590.

Moss, J.I. (2001) Many Gulf War illnesses may be autoimmune disorders caused by the chemical and biological stressors pyridostigmine bromide, and adrenaline. *Medical Hypotheses*, 56, 155–157.

Musilek, K., Kuca, K., Jun, D., Dohnal, V. and Dolezal, M. (2005) Synthesis of a novel series of bispyridinium compounds bearing a xylene linker and evaluation of their reactivation activity against chlorpyrifos-inhibited acetylcholinesterase. *Journal of Enzyme Inhibition and Medicinal Chemistry*, 20, 409–415.

Mustazza, C., Borioni, A., Del Giudice, M.R., Gatta, F., Ferretti, R., Meneguz, A. et al. (2002) Synthesis and cholinesterase activity of phenylcarbamates related to rivastigmine, a therapeutic agent for Alzheimer's disease. *European Journal of Medicinal Chemistry*, 37, 91.

Nicolet, Y., Lockridge, O., Masson, P., Fontecilla-Camps, J.C. and Nachon, F. (2003) Crystal structure of human butyrylcholinesterase and of its complexes with substrate and products. *Journal of Biological Chemistry*, 278, 41141–41147.

Nordberg, A. (2006) Mechanisms behind the neuroprotective actions of cholinesterase inhibitors in Alzheimer disease. *Alzheimer Disease and Associated Disorders*, 20, S12.

Ogura, H., Kosasa, T., Kuriya, Y. and Yamanishi, Y. (2000) Comparison of inhibitory activities of donepezil and other cholinesterase inhibitors on acetylcholinesterase and butyrylcholinesterase in vitro. *Methods and Findings in Experimental and Clinical Pharmacology*, 22, 609.

Omran, Z., Cailly, T., Lescot, E., Santos, J.S., Agondanou, J.H., Lisowski, V. et al. (2005) Synthesis and biological evaluation as AChE inhibitors of new indanones and thiaindanones related to donepezil. *European Journal of Medicinal Chemistry*, 40, 1222–1245.

Oosterbaan, R.A., Warringa, M.G.P.J., Jansz, H.S., Berends, F. and Cohen, J.A. (1958) The reaction of pseudocholinesterase with diisopropyl phosphorofluoridate (DFP). Abstracts of IV International Congress of Biochemistry, 4, 38, Vienna.

Orgogozo, J.M., Small, G.W., Hammond, G., Van Baelen, B. and Schwalen, S. (2004) Effects of galantamine in patients with mild Alzheimer's disease. *Current Medical Research and Opinion*, 20, 1815.

Orhan, I., Terzioglu, S. and Sener, B. (2003) A-onocerin: an acetylcholinesterase inhibitor from *Lycopodium clavatum*. *Planta Medica*, 69, 265–267.

Orozco, C., de los Rios, C., Arias, E., Leon, R., Garcia, A.G., Marco, J.L., Villarroya, M. and Lopez, M.G. (2004) ITH4012 (ethyl 5-amino-6,7,8,9-tetrahydro-2-methyl-4-phenylbenzol[1,8]naphthyridine-3-carboxylate), a novel acetylcholinesterase inhibitor with "calcium promotor" and neuroprotective properties. *Journal of Pharmacology and Experimental Therapeutics*, 310, 987–994.

Otoguro, K., Shiomi, K., Yamaguchi, Y., Arai, N., Sunazuka, T., Masuma, R. et al. (2000) Arisugacins C and D, novel acetylcholinesterase inhibitors and their related novel metabolites produced by *Penicillium* sp. FO-4259-11. *Journal of Antibiotics*, 53, 50.

Palani, A., Dugar, S., Clader, J.W., Greenlee, W.J., Ruperto, V., Duffy, R.A. and Lachowicz, J.E. (2004) Isopropyl amide derivatives of potent and selective muscarinic M2 receptor antagonists. *Bioorganic & Medicinal Chemistry Letters*, 14, 1791–1794.

Park, C.H., Choi, S.H., Koo, J.W., Seo, J.H., Kim, H.S., Jeong, S.J. and Suh, Y.H. (2002) Novel cognitive improving and neuroprotective activities of *Polygala tenuifolia* Willdenow extract, BT-11. *Journal of Neuroscience Research*, 70, 484.

Pavlič, M.R. (1973) On the nature of the acceleration of the methanesulfonylation of acetylcholinesterase by tetraethylammonium. *Biochimica et Biophysica Acta*, 327, 393–397.

Plosker, G.L. and Keating, G.M. (2004) Management of mild to moderate Alzheimer disease: defining the role of rivastigmine. *Disease Management & Health Outcomes*, 12, 55.

Pola, R., Flex, A., Ciaburri, M., Rovella, E., Valiani, A., Reali, G. et al. (2005) Responsiveness to cholinesterase inhibitors in Alzheimer's disease: a possible role for the 192 Q/R polymorphism of the *PON-1* gene. *Neuroscience Letters*, 382, 338.

Polinsky, R.J. (1998) Clinical pharmacology of rivastigmine: a new-generation acetylcholinesterase inhibitor for the treatment of Alzheimer's disease. *Clinical therapeutics*, 20, 634.

Potkin, S.G., Alva, G., Gunay, I., Koumaras, B., Chen, M. and Mirski, D. (2006) A pilot study evaluating the efficacy and safety of rivastigmine in patients with mixed dementia. *Drugs & Aging*, 23, 241.

Quirion, R., Wilson, A., Rowe, W.B., Aubert, I., Richard, J., Doods, H. et al. (1995) Facilitation of acetylcholine release and cognitive performance by an M(2)-muscarinic receptor antagonist in aged memory-impaired. *Journal of Neuroscience*, 15, 1455–1462.

Racchi, M., Porrello, E., Lanni, C., Lenzken, S.C., Mazzucchelli, M. and Govoni, S. (2006) Role of acetylcholinesterase inhibitors in pharmacological regulation of amyloid precursor protein processing. *Aging: Clinical and Experimental Research*, 18, 149.

Rampa, A., Piazzi, L., Belluti, F., Gobbi, S., Bisi, A., Bartolini, M. et al. (2001) Acetylcholinesterase inhibitors: SAR and kinetic studies on w-[N-methyl-N-(3-alkylcarbamoyloxyphenyl) methyl]aminoalkoxyaryl derivatives. *Journal of Medicinal Chemistry*, 44, 3810–3820.

Rangappa, K.S. and Basappa (2005) New cholinesterase inhibitors: synthesis and structure–activity relationship studies of 1,2-benzisoxazole series and novel imidazolyl-d2-isoxazolines. *Journal of Physical Organic Chemistry*, 18, 773–778.

Rao, K.C.S., Divakar, S., Karanth, N.G. and Sattur, A.P. (2001) 14-(2′,3′,5′-trihydroxyphenyl) tetradecan-2-ol, a novel acetylcholinesterase inhibitor from *Chrysosporium* sp. *Journal of Antibiotics*, 54, 848.

Raskind, M.A., Peskind, E.R., Wessel, T., Yuan, W., Allen, F.H. Jr., Aronson, S.M. et al. (2000) Galantamine in AD: a 6-month randomized, placebo-controlled trial with a 6-month extension. *Neurology*, 54, 2261.

Raskind, M.A, Peskind, E.R., Truyen, L., Kershaw, P. and Damaraju, C.V. (2004) The cognitive benefits of galantamine are sustained for at least 36 months: a long-term extension trial. *Archives of Neurology*, 61, 252.

Recanatini, M., Cavalli, A., Belluti, F., Piazzi, L., Rampa, A., Bisi, A. et al. (2000) SAR of 9-amino-1,2,3,4-tetrahydroacridine-based acetylcholinesterase inhibitors: synthesis, enzyme inhibitory activity, QSAR, and structure-based CoMFA of tacrine analogues. *Journal of Medicinal Chemistry*, 43, 2007–2018.

Ren, Y., Houghton, P.J., Hider, R.C. and Howes, M.J.R. (2004) Novel diterpenoid acetylcholinesterase inhibitors from *Salvia miltiorhiza*. *Planta Medica*, 70, 201–204.

Ripoll, D.R., Faerman, C.H., Axelsen, P.H., Silman, I. and Sussman, J.L. (1993) An electrostatic mechanism for substrate guidance down the aromatic gorge of acetylcholinesterase. *Proceedings of the National Academy of Sciences of the United States of America*, 90, 5128–5132.

Rockwood, K., Mintzer, J., Truyen, L., Wessel, T. and Wilkinson, D. (2001) Effects of a flexible galantamine dose in Alzheimer's disease: a randomized, controlled trial. *Journal of Neurology, Neurosurgery, and Psychiatry*, 71, 589.

Rodriguez-Franco, M.I., Fernandez-Bachiller, M.I., Perez, C., Hernandez-Ledesma, B. and Bartolome, B. (2006) Novel tacrine–melatonin hybrids as dual-acting drugs for Alzheimer disease, with improved acetylcholinesterase inhibitory and antioxidant properties. *Journal of Medicinal Chemistry*, 49, 459–462.

Rogers, S.L. and Friedhoff, L.T. (1996) The efficacy and safety of donepezil in patients with Alzheimer's disease: results of a US multicentre, randomized, double-blind, placebo-controlled trial. *Dementia*, 7, 293.

Rogers, S.L., Doody, R.S., Mohs, R.C., Friedhoff, L.T. and the Donepezil Study Group. (1998a) Donepezil improves cognition and global function in Alzheimer disease: a 15-week, double-blind, placebo-controlled study. *Archive of Internal Medicine*, 158, 1021–1031.

Rogers, S.L., Farlow, M.R., Doody, R.S., Mohs, R., Friedhoff, L.T. and the Donepezil Study Group. (1998b) A 24-week, double-blind, placebo-controlled trial of donepezil in patients with Alzheimer's disease. *Neurology*, 50, 136.

Rogers, S.L., Doody, R.S., Pratt, R.D. and Ieni, J.R. (2000) Long-term efficacy and safety of donepezil in the treatment of Alzheimer's disease: final analysis of a US multicentre open-label study. *European Neuropsychopharmacology*, 10, 195.

Roman, S., Vivas, N.M., Badia, A. and Clos, M.V. (2002) Interaction of a new potent anticholinesterasic compound ($\pm$) huprine X with muscarinic receptors in rat brain. *Neuroscience Letters*, 325, 103–106.

Roman, G.C., Wilkinson, D.G., Doody, R.S., Black, S.E., Salloway, S.P. and Schindler, R.J. (2005) Donepezil in vascular dementia: combined analysis of two large-scale clinical trials. *Dementia and Geriatric Cognitive Disorders*, 20, 338.

Rosenberry, T.L. (1975) Acetylcholinesterase. *Advances in Enzymology & Related Areas of Molecular Biology*, 43, 103–218.

Rosenberry, T.L., Rabl, C.R. and Neumann, E. (1996) Binding of the neurotoxin fasciculin 2 to the acetylcholinesterase peripheral site drastically reduces the association and dissociation rate constants for N-methylacridinium binding to the active site. *Biochemistry*, 35, 685–690.

Rowe, W.B., O'Donnell, J.P., Pearson, D., Rose, G.M., Meaney, M.J., Quirion, R. (2003) Long-term effects of BIBN-99, a selective muscarinic M2 receptor antagonist, on improving spatial memory performance in aged cognitively impaired rats. *Behavioural Brain Research*, 145, 171–178.

Satoru, M., Mori, E., Iseki, E. and Kosaka, K. (2006) Efficacy and safety of donepezil in patients with dementia with Lewy bodies: preliminary findings from an open-label study. *Psychiatry and Clinical Neurosciences*, 60, 190.

Savini, L., Gaeta, A., Fattorusso, C., Catalanotti, B., Campiani, G., Chiasserini, L. et al. (2003) Specific targeting of acetylcholinesterase and butyrylcholinesterase recognition sites. Rational design of novel, selective, and highly potent cholinesterase inhibitors. *Journal of Medicinal Chemistry*, 46, 1–4.

Scarpini, E., Scheltens, P. and Feldman, H. (2003) Treatment of Alzheimer's disease: current status and new perspectives. *Lancet Neurology*, 2, 539.

Šentjurc, M., Štalc, A. and Županičič, A.O. (1976) On the location of active serines of membrane acetylcholinesterase studied by the ESR method. *Biochimica et Biophysica Acta*, 438, 131–137.

Shen, Q., Peng, Q., Shao, J., Liu, X., Huang, Z., Pu, X. et al. (2005) Synthesis and biological evaluation of functionalized coumarins as acetylcholinesterase inhibitors. *European Journal of Medicinal Chemistry*, 40, 1307–1315.

Silman, I. and Futerman, A.H. (1987) Modes of attachment of acetylcholinesterase to the surface membrane. *European Journal of Biochemistry*, 170, 11–22.

Small, G., Erkinjuntti, T., Kurz, A. and Lilienfeld, S. (2003) Galantamine in the treatment of cognitive decline in patients with vascular dementia or Alzheimer's disease with cerebrovascular disease. *CNS Drugs*, 17, 905.

Sramek, J.J., Hourani, J., Jhee, S.S. and Cutler, N.R. (1999) NXX-066 in patients with Alzheimer's disease: a bridging study. *Life Sciences*, 64, 1215.

Stojan, J. (2005) Rational polynomial equation helps to select among homeomorphic kinetic models for cholinesterase reaction mechanism. *Chemico-Biological Interactions*, 157–158, 173–179.

Stojan, J. and Pavlič, M.R. (1991) On the inhibition of cholinesterase by D-tubocurarine. *Biochimica et Biophysica Acta*, 1079, 96–102.

Stojan, J., Brochier, L., Alies, C., Colletier, J.P. and Fournier, D. (2004) Inhibition of *Drosophila melanogaster* acetylcholinesterase by high concentrations of substrate. *European Journal of Biochemistry*, 271, 1364–1371.

Sugimoto, H., Yamanishi, Y., Limura, Y. and Kawakami, Y. (2000) Donepezil hydrochloride (E2020) and other acetylcholinesterase inhibitors. *Current Medicinal Chemistry*, 7, 303.

Sunazuka, T., Handa, M., Nagai, K., Shirahata, T., Harigaya, Y., Otoguro, K. et al. (2002) The first total synthesis of (±)-arisugacin A, a potent, orally bioavailable inhibitor of acetylcholinesterase. *Organic Letters*, 4, 367.

Sung, S.H., Kang, S.Y., Lee, K.Y., Park, M.J., Kim, J.H., Park, J.H. et al. (2002) (+)-α-Viniferin, a stilbene trimer from *Caragana chamlague*, inhibits acetylcholinesterase. *Biological & Pharmaceutical Bulletin*, 25, 125–127.

Sussman, J.L., Harel, M., Frolow, F., Oefner, C., Goldman, A., Toker, L. et al. (1991) Atomic structure of acetylcholinesterase from *Torpedo californica*: a prototypic acetylcholine-binding protein. *Science*, 253, 872–879.

Szegletes, T., Mallender, W.D. and Rosenberry, T.L. (1998) Nonequilibrium analysis alters the mechanistic interpretation of inhibition of acetylcholinesterase by peripheral site ligands. *Biochemistry*, 37, 4206–4216.

Takadoi, M., Katoh, T., Ishiwata, A. and, Terashima, S. (2002) Synthetic studies of himbacine, a potent antagonist of the muscarinic M2 subtype receptor 1. Stereoselective total synthesis and antagonistic activity of enantiomeric pairs of himbacine and (2'S,6'R)-diepihimbacine, 4-epihimbacine, and novel himbacine congeners. *Tetrahedron*, 58(50), 9903–9923.

Takeda, A., Loveman,. E., Clegg, A., Kirby, J., Picot, J., Payne, E. and Green, C. (2006) A systematic review of the clinical effectiveness of donepezil, rivastigmine and galantamine on cognition, quality of life and adverse events in Alzheimer's disease. *International Journal of Geriatric Psychiatry*, 21, 17.

Takeuchi, Y., Shibata, T., Suzuki, E., Iimura, Y., Kosasa, T., Yamanishi, Y. and Sugimoto, H. (2002) Preparation of 1-benzyl-4-[(5,6-dimethoxy-2-fluoro-1-indanon)-2-yl]methylpiperidine as acetylcholinesterase inhibitor. PCT Int. Appl., WO 02/20482.

Tang, X.C. and Han, Y.F. (1999) Pharmacological profile of huperzine A, a novel acetylcholinesterase inhibitor from Chinese herb. *CNS Drug Reviews*, 5, 281.

Tariot, P.N., Solomon, P.R., Morris, J.C., Kershaw, P., Lilienfeld, S. and Ding, C. (2000) A 5-month, randomized, placebo-controlled trial of galantamine in AD. *Neurology*, 54, 2269.

Tariot, P.N., Cummings, J.L., Katz, I.R., Mintzer, J., Perdomo, C.A., Schwam, E.M. and Whalen, E. (2001) A randomized, double-blind, placebo-controlled study of the efficacy and safety of donepezil in patients with Alzheimer's disease in the nursing home setting. *Journal of the American Geriatrics Society*, 49, 1590–1599.

Thatte, U. (2005) Phenserine (Axonyx). *Current Opinion in Investigational Drugs*, 6, 729–739.

Trabace, L., Cassano, T., Loverre, A., Steardo, L. and Cuomo, V. (2002) CHF2819: pharmacological profile of a novel acetylcholinesterase inhibitor. *CNS Drug Reviews*, 8, 53.

Urbain, A., Marston, A., Queiroz, E.F., Ndjoko, K. and Hostettmann, K. (2004) Xanthones from *Gentiana campestris* as new acetylcholinesterase inhibitors. *Planta Medica*, 70, 1011–1014.

Urbain, A., Marston, A. and Hostettmann, K. (2005) Coumarins from *Peucedanum ostruthium* as inhibitors of acetylcholinesterase. *Pharmaceutical Biology*, 43, 647–650.

Verhoeff, N.P.L.G. (2005) Acetylcholinergic neurotransmission and the b-amyloid cascade: implications for Alzheimer's disease. *Expert Review of Neurotherapeutics*, 5, 277.

Vernier, J.M., Holsenback, H., Cosford, N.D.P., Whitten, J.P., Menzaghi, F., Reid, R. et al. (1998) Conformationally restricted analogs of nicotine and anabasine. *Bioorganic & Medicinal Chemistry Letters*, 8, 2173–2178.

Vernier, J.M., El-Abdellaoui, H., Holsenback, H., Cosford, N.D.P., Bleicher, L., Barker, G. et al. (1999) 4-[[2-(1-Methyl-2-pyrrolidinyl)ethyl]thio]phenol hydrochloride (SIB-1553A): a novel cognitive enhancer with selectivity for neuronal nicotinic acetylcholine receptors. *Journal of Medicinal Chemistry*, 42, 1684–1686.

Viegas, C.J., Bolzani, V.S., Barreiro, E.J. and Fraga, C.A.M. (2005a) New anti-Alzheimer drugs from biodiversity: the role of the natural acetylcholinesterase inhibitors. *Mini-Reviews in Medicinal Chemistry*, 5, 915–926.

Viegas, C.J., Bolzani, V.S., Pimentel, L.S.B., Castro, N.G., Cabral, R.F., Costa, R.S. et al. (2005b) New selective acetylcholinesterase inhibitors designed from natural piperidine alkaloids. *Bioorganic & Medicinal Chemistry*, 13, 4184–4190.

Wagner-Jauregg, T. and Hackley, B.E. Jr. (1953) Model reactions of phosphorus-containing enzyme inactivators. III. Interaction of imidazole, pyridine and some of their derivates with dialkyl halogeno-phosphates. *Journal of American Chemical Society*, 75, 2125–2130.

Wang, H., Carlier, P.R., Ho, W.L., Wu, D.C., Lee, N.T.K., Li, C.P.L. et al. (1999) Effects of bis(7)-tacrine, a novel anti-Alzheimer's agent, on rat brain AChE. *NeuroReport*, 10, 789.

Wang, Y., Chackalamannil, S., Hu, Z., McKittrick, B.A., Greenlee, W., Ruperto, V. et al. (2002) Sulfide analogues as potent and selective M2 muscarinic receptor antagonists. *Bioorganic & Medicinal Chemistry Letters*, 12, 1087–1091.

Wang, Y., Sherwood, J.L. and Lodge, D. (2006) The a4b2 nicotinic acetylcholine receptor agonist TC-2559 impairs long-term potentiation in the dentate gyrus in vivo. *Neuroscience Letters*, 406, 183–188.

Weinstock, M., Poltyrev, T., Bejar, C. and Youdim, M.B.H. (2002) Effect of TV3326, a novel monoamine-oxidase cholinesterase inhibitor, in rat models of anxiety and depression. *Psychopharmacology*, 160, 318.

Wilcock, G.K., Lilienfeld, S., Gaens, E. et al. (2000) Efficacy and safety of galantamine in patients with mild to moderate Alzheimer's disease: multicentre randomised controlled trial. *British Medical Journal*, 321, 1.

Wilkinson, D. and Murray, J. (2001) Galantamine: a randomized, double-blind, dose comparison in patients with Alzheimer's disease. *International Journal of Geriatric Psychiatry*, 16, 852.

Wilson, I.B. (1951) Acetylcholinesterase. XI. Reversibility of tetraethyl pyrophosphate inhibition. *Journal of Biological Chemistry*, 190, 111–117.

Wilson, I.B. and Ginsburg, S. (1955) A powerful reactivator of alkylphosphate-inhibited acetylcholinesterase. *Biochimica et Biophysica Acta*, 18, 168–170.

Winblad, B., Engedal, K., Soininen, H., Verhey, F., Waldemar, G., Wimo, A. et al. (2001) A 1-year, randomized, placebo-controlled study of donepezil in patients with mild to moderate AD. *Neurology*, 57, 489.

Winblad, B., Kilander, L., Eriksson, S., Minthon, L., Batsman, S., Wetterholm, A.L., Jansson-Blixt, C. and Haglund, A. (2006) Donepezil in patients with severe Alzheimer's disease: double-blind, parallel-group, placebo-controlled study. *Lancet*, 367, 1057.

Windisch, M., Hutter-Paier, B., Jerkovic, L., Imbimbo, B. and Villetti, G. (2003) The protective effect of ganstigmine against amyloid beta25–35 neurotoxicity on chicken cortical neurons is independent from the cholinesterase inhibition. *Neuroscience Letters*, 341, 181–184.

Wong, D.M., Greenblat, H.M., Dvir, H., Carlier, P.R., Han, Y.F., Pang, Y.P. et al. (2003) Acetylcholinesterase complexed with bivalent ligands related to huperzine A: experimental evidence for species-dependent protein-ligand complementarity. *Journal of the American Chemical Society*, 125, 363.

Woodruff-Pak, D.S., Lander, C. and Geerts, H. (2002) Nicotinic cholinergic modulation: galantamine as a prototype. *CNS Drug Reviews*, 8, 405–426.

Wu, L.J., Hsiao, G., Yen, M.H. and Hu, M.K. (2002) Homodimeric tacrine congeners as acetylcholinesterase inhibitors. *Journal of Medicinal Chemistry*, 45, 2277.

Xie, W., Stribley, J.A., Chatonnet, A., Wilder, P.J., Rizzino, A., McComb, R.D. et al. (2000) Postnatal developmental delay and supersensitivity to organophosphate in gene-targeted mice lacking acetylcholinesterase. *Journal of Pharmacology & Experimental Therapeutics*, 293, 896–902.

Youdim, M.B.H., Weinstock, M. (2002) Molecular basis of neuroprotective activities of rasagiline and the anti-Alzheimer drug TV3326. *Cellular and Molecular Neurobiology*, 21, 555.

Yu, Q.S., Zhu, X., Holloway, H.W., Whittaker, N.F., Brossi, A. and Greig, N.H. (2002) Anticholinesterase activity of compounds related to geneserine tautomers. N-oxides and 1,2-oxazines. *Journal of Medicinal Chemistry*, 45, 3684.

Zaheer-ul-haq, Wellenzohn, B., Liedl, K.R. and Rode, B.M. (2003a) Molecular docking studies of natural cholinesterase-inhibiting steroidal alkaloids from *Sarcococca saligna*. *Journal of Medicinal Chemistry*, 46, 5087–5090.

Zaheer-ul-Haq, Wellenzohn, B., Tonmunphean, S., Khalid, A., Choudhary, M.I. and Rode, B.M. (2003b) 3D-QSAR studies on natural acetylcholinesterase inhibitors of *Sarcococca saligna* by comparative molecular field analysis (CoMFA). *Bioorganic & Medicinal Chemistry Letters*, 13(24), 4375–4380.

4 Reduction in Plaque Formation

Christian Czech, Helmut Jacobsen,
Celine Adessi, Bruno P. Imbimbo, Francesca
Speroni, Dario Cattaneo, Harry LeVine III,
and Corinne E. Augelli-Szafran

CONTENTS

4.1 SECRETASE AND APP PROCESSING

Christian Czech, Helmut Jacobsen, and Celine Adessi

4.1.1 AMYLOID CASCADE IS CAUSALLY INVOLVED IN ALZHEIMER'S DISEASE PATHOLOGY

Alzheimer's Disease (AD) is a progressive and neurodegenerative disorder of the brain characterized by loss of neurons and synapses, particularly in regions related to memory and cognition. The main neuropathological features of AD are accumulation of abnormal intracellular and extracellular proteinaceous deposits in the brain of the patients (Alois Alzheimer in 1907)—within neuronal cells, as accumulation of abnormally phosphorylated tau protein into paired helical filaments, known as neurofibrillary tangles (NFTs) (Terry, 1994) and outside cellular structures as proteinaceous aggregates in form of amyloid plaques and as amyloid in the wall of cerebral blood vessels (Masters et al., 1985; Selkoe, 1991). The sequence of pathological events leading to the observed deposits and finally to neurodegeneration is

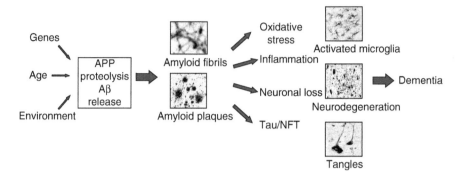

FIGURE 4.1 **(See color insert following page 296.)** The amyloid cascade hypothesis of Alzheimer's disease. Genes, environment, and age represent the main risk factors in the development of AD. The accumulation in the brain of Aβ peptide results in the deposition of plaques. The presence of amyloid plaques in the brain triggers a cascade of toxic events leading to dementia.

believed to start with the abnormal processing of the amyloid precursors protein (APP), leading to the generation of Aβ (Kang et al., 1987). The hypothesis states that the Aβ protein triggers through formation of toxic aggregates the subsequent formation of NFTs, loss of synapses, and finally loss of neuronal function. This sequence of events is now described as the amyloid cascade (Figure 4.1).

Central to the amyloid cascade is the proteolytic cleavage of APP by three proteolytic activities, which either destroy the Aβ region or lead by subsequent cleavage by two other proteases to the release of Aβ (Figure 4.2). These proteolytic activities consist of single proteins or protein complexes and have been named with the Greek letters α, β, and γ, whereby α cleaves APP within the Aβ region resulting in nonamyloidogenic fragments, β cleaves at the amino-terminal region of Aβ leading to the production of a intermediate carboxy-terminal fragment,

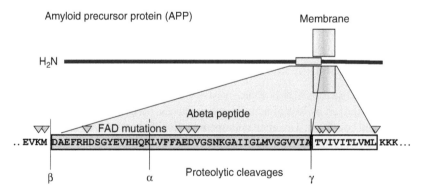

FIGURE 4.2 APP processing by secretases. Schematic representation of the amyloid precursor protein (APP) and the amino acid sequence of the Aβ peptide. Proteolytic cleavage sites by α-, β-, and γ-secretase are indicated. FAD, Familial Alzheimer's disease.

and γ-secretase finally releases Aβ (Figure 4.2). In the following sections, we discuss the biology of these proteolytic activities and their role in APP processing in more detail.

4.1.2 α-Secretase Cleavage

In the nonamyloidogenic pathway, APP is cleaved within the Aβ region at position 16, destroying Aβ and releasing α-secreted APP (sAPP-α). The remaining C-terminal stub of 83 amino acid length is subsequently cleaved by γ-secretase releasing the P3 fragment. These fragments do not contribute to the pathology of AD, and in case of sAPP-α the extracellular cleavage product is rather believed to have a neuroprotective activity (Turner et al., 2003). In the cell, the larger part of APP is sorted into the nonamyloidogenic pathway and cleaved by α-secretase. The proteolytic activity has been identified as belonging to the group of membrane-bound proteases of the ADAM family (a disintegrin and metalloprotease family. ADAM is a multifunctional gene family involved in a variety of biological processes (for review see Primakoff and Myles [2000]). At least three members of this family have been shown to cleave APP at the α-secretase site: ADAM-9, ADAM-10, and ADAM-17 (Asai et al., 2003). These proteases are activated by different pathways and contribute in different cell- and tissue-specific combinations to the cleavage of APP. Overexpression of ADAM-10 in APP-transgenic mice shifts APP processing into the nonamyloidogenic pathway and is able to reduce the formation of Aβ and to reduce its accumulation in amyloid plaques in the brain of these mice (Postina et al., 2004). These results suggest that activating the α-secretase pathway shifts APP into the nonamyloidogenic pathway and in parallel reduces Aβ release. This would make this mechanism an attractive target for pharmaceutical intervention to reduce the secretion of Aβ and increase the neurotropic sAPP-α generation. However, there are data showing that activation of the α-secretase does not always occur at the expense of β-secretase cleavage and does therefore not necessary lead to reduction in Aβ levels.

The G protein-coupled receptor PAC1 strongly activates APP α-secretase cleavage upon stimulation with its peptide ligands PACAP-27 and PACAP-38 without affecting β-secretase cleavage (Kojro et al., 2006). Similar results have been demonstrated upon activation of the 5HT-4 receptor. Activation of this receptor with its natural ligand serotonin or an artificial agonist leads to an increase of sAPP-α secretion in CHO cells transfected with APP695 without affecting Aβ production (Robert et al., 2001). Furthermore, activation of the muscarinic acetylcholine receptors M1 and M3 also increases secretion of sAPP-α (Nitsch et al., 1992), but in this case there is evidence that activation of M1 receptors does not only modify α-secretase cleavage but also has an effect on Aβ release. Transgenic mice treated with the M1 agonist AF267B show reduced Aβ deposits in the cortex and hippocampus (Caccamo et al., 2006).

The cellular signaling pathways downstream of the G protein-coupled receptor activation are only partially understood. The activation of intracellular second messenger systems is involved in regulation and targeting of APP to the secretory pathway. There is evidence that the intracellular cAMP is involved in this process. Treatment of cells with Forskolin, a direct activator of adenylate cyclase, leads to downstream activation of Protein Kinase A (PKA), followed by a large increase in α-secretase cleavage and release of sAPP-α in PC12 cells (Xu et al., 1996). In a second alternative pathway independent of PKA, the increase of cAMP triggers the cAMP guanine exchange factor Epac, which subsequently activates the small GTPase of the Rho family Rac via activation of Rap1 (Maillet et al., 2003). Another PKA-independent pathway was described for PAC1 receptor, and here the induction of α-secretase cleavage is independent of PKA but seems to be mediated by an increase in intracellular calcium and subsequent stimulation of PKC and activation of ERK1/2 MAP kinase pathway (Kojro et al., 2006). In any case, activation of these pathway changes sorting and trafficking only of APP destinated to α-secretase cleavage without affecting significantly the β-secretase pathway. However, even a small decrease of Aβ release over a longer period of time might significantly slow down the development of AD pathology.

4.1.3   β-Secretase Cleavage

APP is cleaved at the β site, leading to the release of a 99 amino acid C-terminal fragment by an enzyme termed BACE for β-amyloid cleavage enzyme. This protein, identified by an expression-cloning approach, was found to belong to the group of aspartic proteases (Vassar et al., 1999). BACE is a type-1 transmembrane protein composed of 501 amino acids with a prodomain that is cleaved by a furin-like protease upon maturation in the secretory pathway (Bennett et al., 2000). BACE is targeted to the plasma membrane, where it has been shown to cluster within lipid rafts; this is also where processing of APP occurs (Ehehalt et al., 2003). Furthermore, targeting BACE to lipid rafts by the addition of glycosylphosphatidylinositol (GPI) anchor, which replaces the transmembrane and C-terminal domains substantially upregulates the secretion of both sAPP-β and Aβ (Cordy et al., 2003). However, the plasma membrane is most likely not the only place where BACE cleaves APP. Since the optimum pH for BACE activity is around 4, APP cleavage will also occur after internalization of both proteins in the endosomic compartment (Gruninger-Leitch et al., 2002).

The gene for BACE1 is located on chromosome 11 (Saunders et al., 1999). There is a close homolog to the enzyme termed BACE2, which maps to chromosome 21 and together the two proteins form a new family of transmembrane aspartic proteases (Vassar, 2001). The amyloidogenic processing of APP is almost exclusively dependent on BACE1, as shown by knock-out mice in which Aβ generation is completely abolished (Cai et al., 2001; Luo et al., 2001). In cell culture experiments it has been shown that BACE2 cleaves within the Aβ region, closer to the α-secretase cleavage site (Farzan et al., 2000). However, it is not clear if this cleavage is relevant in vivo. BACE1 cleaves the APP substrate with the Swedish mutation much more efficiently than the wild-type APP sequence (Gruninger-Leitch et al., 2002) explaining the increased Aβ release in transfected cells and patients carrying the mutations. In addition to APP, the P-selectin glycoporotein ligand-1, the Sialyl-transferase, and lipoprotein receptor-related protein (LRP) have been identified as substrates for BACE1 (Kitazume et al., 2001; Lichtenthaler et al., 2003; von Arnim et al., 2005). BACE1 mRNA can be detected mainly in the brain and the pancreas and at lower levels in many other tissues (Vassar et al., 1999), but it seems that the amount of mRNA does not correlate with protein levels since it has been shown that expression of BACE is regulated on the level of mRNA translation. The $5'$ end of the BACE1 mRNA is suppressing effective translation of the protein (Lammich et al., 2004). There are reports showing that BACE protein but not mRNA levels are increased in sporadic AD patient compared with controls (Fukumoto et al., 2002; Holsinger et al., 2002; Yang et al., 2003), suggesting that regulation of BACE1 translation could be a contributing factor to the development of AD.

4.1.4 γ-SECRETASE CLEAVAGE

The γ-secretase catalyzes the final step in the release of the Aβ peptides from the APP. Its direct substrates are the APP C-terminal fragments that remain as membrane-bound stubs; after the prior removal of the extracellular domain by β-secretase, full-length APP is not cleaved by γ-secretase. The γ-secretase has several features that set it apart from most known proteases: (1) it is a complex of at least four proteins, (2) the complex is firmly integrated into cell membranes through a multitude of transmembrane domains, and (3) it cleaves its substrates within their transmembrane domain (Steiner and Haass, 2000; Wolfe, 2003; Haass, 2004).

The γ-secretase is formed by presenilin, nicastrin, anterior pharynx defective (aph-1), and presenilin enhancer (pen-2), which combine in a 1:1:1:1 ratio and form the basic unit of the complex (Edbauer et al., 2003; Kimberly et al., 2003; Fraering et al., 2004b). Dimers and higher orders of this basic unit have been described, but it is currently unclear what form represents the true physiological complex (Schroeter et al., 2003; Evin et al., 2005). Presenilin comes in two homologs, PS1 and PS2, encoded by genes on chromosomes 1 and 14. They were identified in 1995 in genetic studies on pedigrees with early onset familial AD (EOAD), and mutations in PS1 are the predominant reason for EOAD and more than 100 different mutations have been described (Rogaev et al., 1995; Sherrington et al., 1995). Presenilins are polytopic integral membrane proteins; there is substantial evidence that they form the catalytic subunit of γ-secretase. Nicastrin is a type-1 membrane protein that may have a role in substrate recognition by the complex (Shah et al., 2005). The precise roles of the two remaining partners, aph1 and pen-2, are less clear; however, assembly and maintenance of the active complex requires all four proteins. Aph1 is also represented by two homologs and splice variants, and thus a variety of γ-secretase complexes with different subunit composition exist (Shirotani et al., 2004; Serneels et al., 2005). If or how this structural diversity is linked to a corresponding functional diversity is currently incompletely understood.

All known presenilins have conserved Asp residues in transmembrane domains TMD6 and TMD7; the TMD7 Asp is embedded in a highly conserved GXGD motive. Mutation of any of the two Asp residues abolishes the γ-secretase activity (Wolfe et al., 1999). The same GXGD motif occurs also in the structurally related signal peptide peptidase (SPP) and in the distantly related prokaryotic type-4 prepilin peptidase family of proteases (Weihofen et al., 2000; Haass and Steiner, 2002; Martoglio and Golde, 2003). This observation and the inhibition by compounds known as inhibitors of typical Aspartyl proteases led to the suggestion that presenilins represent the prototype of a novel class of atypical Asp proteases (Wolfe and

Kopan, 2004). The mechanistic details of their peptidase activity, which occurs in the hydrophobic environment of the membrane, remain to be elucidated.

In contrast to β-secretase, the γ-secretase complex is expressed in all major tissues of the organism including the CNS. Genetic experiments and pharmacological intervention in different model organisms have clearly shown that it serves essential functions during embryonal development and also during adult life. This is likely due to its role in Notch receptor signaling (which will be discussed later). Within the cell, the primary assembly of the complex occurs in a highly ordered fashion in the ER compartment where it passes through several transient subcomplexes (LaVoie et al., 2003; Fraering et al., 2004a; Capell et al., 2005). The mature complex is transported through the Golgi and trans-Golgi network to the outer cell membrane but it is also localized in the endosomal compartments.

It was recognized early on that the Aβ peptides occur in various species in different lengths, Aβ 40 is the dominant peptide but Aβ 42 is the more pathogenic species. Peptides shorter then Aβ 40 are also found, for example, of 38 and 37 amino acid length. This led to the idea that the γ-cleavage is the result of different proteases working independently on the same substrate. However, later genetic experiments, that is, ablation of the presenilins or other components of the complex, and pharmacological interventions with specific inhibitors demonstrated that the γ-secretase complex produces all these Aβ peptides with their staggered C termini. These γ-cleavages occur more or less in the middle of the transmembrane domain of the APP substrate. However, to add another level of complexity, γ-secretase also cleaves approximately 10 residues C terminal to the γ-cleavage site close to the TMD's border to the cytoplasma. This latter cleavage site is called epsilon cleavage, and it occurs with high precision at the same residues (Weidemann et al., 2002). It releases the cytoplasmic domain of APP (AICD) which is short-lived, however, some of the AICD complexes with other proteins and translocates to the nucleus where it may serve some transcription-regulatory function (Cao and Sudhof, 2001). The role of γ-secretase in such signaling pathways is much clearer for the Notch pathways and will be discussed later.

It is not yet clear in what order γ and epsilon cleavage occur; so far, no intermediates like an Aβ peptide of 49 amino acid length or an AICD with an N terminus extending to the γ-site have been detected. It is assumed that the cleavages occur more or less in parallel and that the peptides corresponding to the short stretch between γ and epsilon cleavage are highly unstable and thus have so far escaped detection. This highly complicated cleavage pattern by γ-secretase is not restricted to APP but has also been observed for other substrates. The related SPP also cleave their substrates in their TMD, whereas presenilin or γ-secretase cleaves type-1 transmembrane proteins; the SPPs cleave type-2 substrates that fit with their inverted membrane topology compared with Presenilin (Weihofen et al., 2000).

It was mentioned earlier that mutated PS1 accounts for more than 50% of all cases of EOAD and that more than 100 different mutations have been described. All these mutations (plus the few PS2 mutants causing EOAD) have in common the ability to shift the ratio of Aβ 42 or Aβ 40 in favor of Aβ 42 (Citron et al., 1997). Although for wild-type PS the Aβ 42 constitutes approximately one-tenth of the newly formed Aβ peptides it may become 50% and more in some of the most aggressive forms of EOAD. This enhanced production of the more pathogenic Aβ 42 presumably causes the early onset of the disease in the most extreme cases as early as in the third or fourth decade of life.

A number of γ-secretase substrates have been described in addition to APP. They include the APP-like proteins, the Notch receptors, ErbB-4, some cadherins, p75 (NTR), and so on. Proteolytic shedding of the extracellular domain is always a prerequisite for the subsequent cleavage by γ-secretase. There are no clear sequence motives known that mark a type-1 membrane protein as substrate and it seems that many proteins can be cleaved when expressed to high levels in cell culture. However, it is more complicated to determine which of all these intracellular domains generated by γ-secretase cleavage serve a physiological function. The clearest evidence for such a function has been obtained for signaling by Notch receptors (Selkoe and Kopan, 2003). When these receptors bind one of their ligands their ectodomain is cleaved off by an ADAM-type protease. The truncated Notch proteins now become a substrate for γ-secretase, and its intracellular domain (NICD) is liberated. The NICD migrates to the nucleus where it activates the transcription factor RBP-kJ and thus regulates the expression of a set of genes, which are involved in cell lineage determination and cell differentiation. This pathway has been named signaling by regulated intramembranous proteolysis (RIP), a concept which in an analogous way was also found in the regulation of sterol metabolism (Brown et al., 2000). In the absence of γ-secretase, Notch signaling is blocked. The Notch receptors serve vital functions during embryonal development but also in maintenance of several tissues of the adult organism, which undergo continuous proliferation and differentiation like hematopoietic system and epithelia. The resulting consequences for the development of γ-secretase inhibitors as therapeutics for AD will be discussed later.

4.1.5 THERAPEUTIC APPROACHES FOR ALZHEIMER'S DISEASE BY INHIBITION OF SECRETASE

Due to its essential, nonredundant role in the production of the amyloidogenic Aβ peptides and the widespread acceptance of the amyloid cascade hypothesis of AD, γ-secretase became an interesting drug target for therapeutic intervention in AD (Beher and Graham, 2005; Churcher and Beher, 2005). Accordingly, many compounds that inhibit the γ-secretase activity have been described in the scientific literature and in patents, and many of the major pharmaceutical companies are active in this field. In the absence of any structural information (an X-ray structure of the γ-secretase complex is certainly a very long way off) and with only crude cell-free preparations of the complex available, the development of such inhibitors relies largely on cellular assays that measure Aβ production or other metabolites of APP (or Notch) proteolysis. Nevertheless, highly potent and selective inhibitors have been obtained which show nanomolar activity in in vitro assays but are also active in animal models like APP-transgenic mice or rat and guinea pig. At this point in time one compound has reportedly entered the clinic. Here we will focus mainly on compounds that have shown in vivo activity (Figure 4.3).

Early examples of specific and potent in vitro inhibitors are transition-state isosteres like the Merck compound L-685458 and the difluoroketone-type inhibitor MW167 described by M. Wolfe (Wolfe et al., 1998; Shearman et al., 2000).

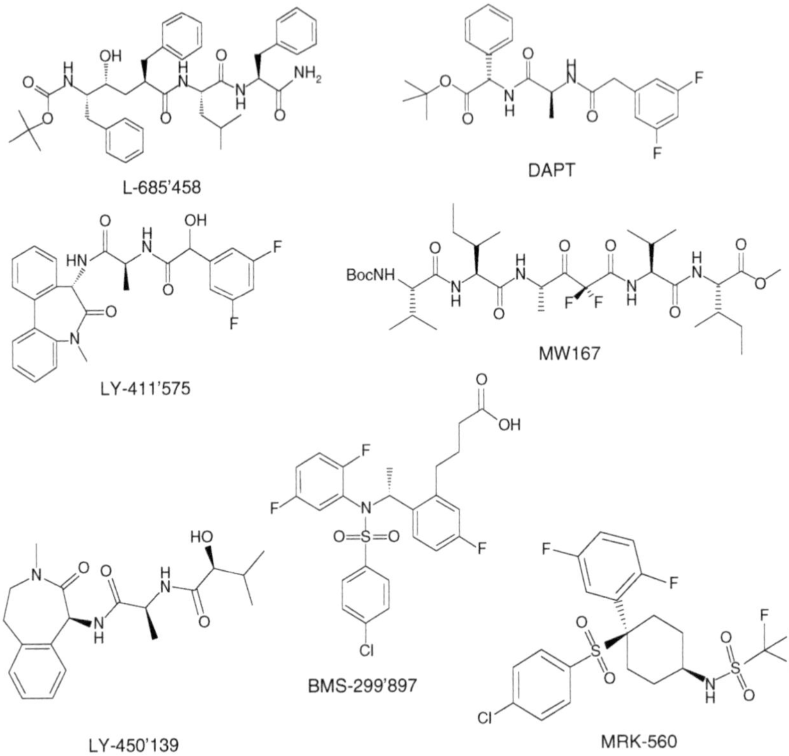

FIGURE 4.3 γ-Secretase inhibitors that have shown in vivo activity.

Derivatives of these inhibitors were used as affinity tags to capture the γ-secretase from cell lysates and contributed thus to the identification of presenilins as catalytic subunits of the γ-secretase complex. Since they inhibited the production of all Aβ peptides they also strongly supported the concept that the γ-secretase activity is due to one specific protease. However, these compounds lacked in vivo activity because of their limited pharmacological profile.

Preclinical evidence that an inhibitor of γ-secretase reduces brain Aβ was first obtained with the peptide-derived compound DAPT of Eli Lilly (Dovey et al., 2001). Oral application of DAPT to young, that is, preamyloid, PDAPP mice caused a dose-dependent reduction in brain Aβ. The studies also demonstrated, for the first time, that soluble Aβ has a relatively short half-life of only a few hours, a finding that was repeatedly confirmed with other inhibitors and in other models.

The benzodiazepine-type inhibitor LY-411575 has become a kind of preclinical benchmark compound for in vivo anti-amyloid potency. This compound is active in cellular Aβ-lowering assays at below nanomolar concentration and inhibits brain Aβ production after oral application to transgenic mice in a dose-dependent manner, starting from doses as low as 1 mg/kg (Lanz et al., 2004). This compound was also active in the rat, that is, a nontransgenic animal, at comparable doses (Best et al., 2005). Furthermore, in chronic application over several months to PDAPP mice it reduced the build-up of brain Aβ amyloid. These studies suggested that extended treatment with γ-secretase inhibitors is feasible and leads to the intended therapeutic effect. In vivo activity in an acute treatment model with a compound from a different structural class, biarylsulfonamides BMS-299897, was recently reported by a group from BMS using βAPP-YAC transgenic mice and guinea pigs (Anderson et al., 2005; Barten et al., 2005). Interestingly, the authors claimed that this compound was a more potent inhibitor of APP cleavage than of Notch, a very attractive and preferred profile for a γ-secretase inhibitor. A group from MSD reported on a cyclohexyl sulfone-type inhibitor, MRK-560, which showed strong Aβ-lowering activity in a rat model in acute treatment paradigms (Best et al., 2006).

Treatment with γ-secretase inhibitor reduces Aβ level in brain, CSF, and plasma, and the extent of reduction and the temporal order may differ between the various animal models and the compounds used. Changes of Aβ in plasma and CSF may be of relevance for the clinical development of secretase inhibitors since these compartments are accessible and can thus be used to follow the in vivo efficacy of the compound on trial. This was recently used by an Eli Lilly team to measure the efficacy of their clinical candidate LY-450139 during a phase 1 clinical trial (Siemers et al., 2005). A dose-dependent reduction in plasma Aβ was measured, which was followed by an unexpected, transient increase above the baseline level. However, despite Aβ lowering in plasma they failed to demonstrate significant changes in CSF Aβ level, and the reasons for this lack of efficacy are unclear.

Although preclinical studies with LY-411575 in mice and rats proved the feasibility of reducing brain Aβ through inhibition of γ-secretase they also showed its main liabilities, that is, mechanism-based side effects through inhibition of Notch signaling. These effects were largely predictable from genetic studies in the mouse, which had targeted the Notch system. The most dominant effects in the hematopoietic system are inhibition of thymic T-cell differentiation and depletion of marginal zone B cells from the spleen. Furthermore, γ-secretase inhibitor treatment affected the gut epithelia; it increased the amount of secretory goblet cells at the expense of enterocytes and enteroendocrine cells (Searfoss et al., 2003; Milano et al., 2004; Wong et al., 2004). These effects appeared after a few days of treatment and in the case of LY-411575 at doses similar to those that were needed to reduce brain Aβ. Additional, less obvious effects cannot be excluded because Notch receptors and ligands are expressed in a number of tissues, including the CNS. The Notch issue has not been resolved yet and is a clear obstacle for the clinical development of γ-secretase inhibitors. As mentioned earlier, there are indications that other structures like the sulfonamide BMS-299897 may be less problematic in this regard due to an apparent selectivity for APP over Notch processing, but a systematic and rigorous study to prove this point is still lacking.

There may be other options as inhibitors to use γ-secretase as a drug target for AD. A few years ago it was shown that some of the well-known NSAIDs affect the cleavage specificity of γ-secretase in a subtle way; they reduce the amount of Aβ 42 but increase the amount of the shorter Aβ 38 species. Aβ 40 production is not affected. Since Aβ 42 is considered to be the main driver of amyloid pathology, whereas Aβ 38 has no or very low amyloidogenic properties the net outcome of such a shift may be an inhibition of amyloid build-up. This activity of NSAIDs is unrelated to inhibition of their primary targets COX 1 and COX 2 since the effect can also be shown in cells lacking any COX expression and requires drug concentrations considerably higher than those required for COX inhibition, usually in the 10–100 μM range in cellular assays (see Section 4.3). Since the effect could also be demonstrated in cell-free Aβ-producing systems, and some NSAIDs can displace a full γ-secretase inhibitor in cell-free binding assay, it is assumed that this modulatory activity is a direct effect on the γ-secretase complex. Only a subgroup of the NSAIDs in clinical use have this activity, whereas other COX inhibitors like aspirin and naproxen lack this effect. Epidemiological evidence points to a protective effect of some NSAIDs, and those who take such drugs for extended periods for various indications have a significantly reduced risk to develop AD and there are even suggestions that those NSAIDs that have the Aβ 42-lowering activity are the most protective. However, considering the low potency of these compounds in the Aβ 42-lowering assays plus their low capacity for brain penetration it seems unlikely that in NSAIDs users such an effect in the brain can be achieved. The most interesting property of these so-called γ-secretase modulators is their lack of activity on Notch processing, which is not surprising since they do not affect the epsilon cleavage but only induce subtle changes at the γ-cleavage site. This makes them very interesting candidates for drug development without the liability of inhibiting crucial cell populations in the hematopoietic system or changing gut epithelia differentiation. Torrey Pines Pharmaceuticals reported such γ-secretase modulators with submicromolar activity. Myriad Pharmaceuticals announced a phase 3 study with R-flurbiprofen which, unlike its S-form, lacks COX inhibitory activity and thus has none of the dose-limiting gastrointestinal side effects. The future will tell which of the different approaches to target γ-secretase will lead to a clinically successful drug for treatment of AD.

4.2 MECHANISM OF PLAQUE FORMATION AND Aβ AGGREGATION INHIBITORS

Christian Czech, Helmut Jacobsen, and Celine Adessi

4.2.1 CHARACTERIZATION OF AMYLOID PLAQUES IN ALZHEIMER'S DISEASE BRAIN

Postmortem microscopic analysis of brain from AD-diagnosed patients reveals distinctive and widespread amyloid plaques. These types of deposits were first described in aged brain by neuropathologists already in the late nineteenth century, but in 1907 Alois Alzheimer was the first person to correlate such neuropathology features with the cognitive impairments, depression, and hallucinations observed in an early-onset Alzheimer's case. Amyloid plaques can be divided into two categories according to their structural appearance. Senile plaques comprise a dense core that can be seen both in electron and light microscopy (Puchtler et al., 1961). Senile plaques specifically bind Thioflavine-S and Congo-red stains, whereas Congo-red produces the typical green birefringence under polarized light, which correlates with the presence of β-sheet structure. Diffuse plaques are composed of homogenous deposit of fibrillar structures and can only be detected by immunohistochemistry using anti-Aβ peptide antibodies but not by Congo-red or Thioflavine-S. In AD brain amyloid plaques are extracellular deposits observed mainly in cortical regions such as neocortex, temporal cortex, enthorhinal cortex (CA1, CA4), and subiculum. The severity of the disease is correlated only weakly to the number of amyloid plaques but to the extent plaques have spread from the entorhinal cortex to the hippocampus and the neocortical regions. Anmyloid deposition seems to correlate with early stages of the disease and, in particular, their appearance in the hippocampus may coincide with onset of clinical symptoms (Tiraboschi et al., 2004). It was also proposed that senile plaque pathology begins first with diffuse plaques in the neocortex and extends hierarchically into further brain regions (for review see Thal et al. [2006b]).

The ultrastucture of amyloid plaques isolated from AD brain was studied early on by electron microscopy (EM) showing amyloid deposits as rigid, nonbranching fibrils with diameters of around 7–10 nm. Each fibril (diameter around 75Å–80Å) is likely composed of five or six smaller filaments, called protofilaments (diameter 25Å–30Å), arranged in parallel (Cohen and Calkins, 1959). The amyloid fibrils show an x-ray diffraction pattern known as cross-β (Kirschner et al., 1986). This pattern indicates that the strongest repeating feature of the fibrils is a set of β-sheets that are parallel to the fibril axis, with their strands perpendicular to the axis.

Many studies have been aimed at the isolation, identification, and characterization of plaque components. The main difficulty has been the insoluble nature of the cerebral AD amyloid plaques, both under physiological conditions or in the presence of detergents, and their resistance to proteolysis (Glenner and Wong, 1984). Later it was shown that only chaotropic agents and harsh acidic treatments led to partial or complete solubilization, allowing quantitative biochemical analysis (Permanne et al., 1995). The main constituent found is the amyloid β peptide which present in two major forms, Aβ 1–40 and Aβ 1–42. Several N- and C-terminal truncated isoforms have also been isolated from senile plaques such as Aβ peptides 3–40/42, 11–40/42, 17–40/42 and Aβ3(pE)-x-bearing pyroglutamate (pE) at residue position 3 (Tekirian et al., 1999). The abundance of these truncated isoforms in senile plaques suggests a role in plaque formation. Other molecules accumulate in senile plaques and have been proposed to participate in Aβ aggregation and plaque formation, for example, Apolipoprotein E and J, serum amyloid P (SAP), α2-macroglobulin receptor or low-density LRP, collagenous Alzheimer amyloid component or collagen XXV, interleukin-1α, interleukin-6, components of the complement system, heparan sulfate proteoglycans (HSPG), and enzymes such as acetylcholinesterase (AChE) and α-1 antichymotrypsin (ACT). In vitro studies have shown that some of these components, such as ApoE, bind to Aβ (Strittmatter et al., 1993; Soto et al., 1996a) and accelerate fibril formation (Wisniewski et al., 1994; Soto et al., 1995, 1997), but the role and mechanism by which these components contribute and affect amyloid plaque formation remain unknown. A more recent work combining laser capture microdissection of Thioflavine-S-positive plaque from postmortem AD brain tissues with liquid chromatography coupled with tandem mass spectrometry revealed a total of 488 proteins in amyloid plaques (Liao et al., 2004). Such novel technologies would be certainly useful to identify the proteins involved in the early stage of aggregate and to follow plaque evolution in AD transgenic mice.

4.2.2 In vitro Amyloid Fibrils Formation and Molecular Properties

Synthetic amyloid β peptide is used extensively to prepare and characterize fibrils in vitro and to study the amyloid formation process. Synthesis of Aβ peptide 1–40 and 1–42 amino acids is a challenge for peptide chemists (Sheppard, 2003), because of the high degree of hydrophobicity in the central part of the Aβ peptide chain (14 hydrophobic and bulky amino acids) and its tendency to aggregate rapidly even in weakly acidic or neutral media. Improvements in solid-phase peptide synthesis and purification techniques have led to improvements in yield and quality of synthetic Aβ. More recently, pure recombinant Aβ from *Escherichia coli* has been prepared on a milligram scale.

Amyloid fibrils obtained from the assembly of synthetic amyloid β peptide present strong structural similarity to the fibrils from senile plaques from AD brain. Thus, examinination of assembled full-length synthetic Aβ (1–40 and 1–42) by two-dimensional spectroscopy analysis, such as Fourier transform infrared spectroscopy (FTIR) and circular dichroism (CD), confirmed that Aβ fibrils adopt a β-sheet structure (Serpell, 2000). Negative stain transmission electron microscopy confirmed also that, like in amyloid deposits in AD brain, synthetic fibril average 70Å diameter and that the fibrils are composed of several protofilaments (Malinchik et al., 1998) wound around one another (Serpell et al., 2000; Stromer and Serpell, 2005). Hydrogen-bonding constraints from quenched hydrogen- or deuterium-exchange NMR analysis showed that at least two molecules of Aβ (1–42) are required to achieve the repeating structure of a protofilament (Luhrs et al., 2005). Liquid-state NMR, FTIR, and CD showed also that Aβ fibrils may have a turn at position 26–29 (for review see Irie et al. [2005]). However, high-resolution structural analysis of the Aβ fibrils is not possible since x-ray crystallography and liquid-state NMR cannot be applied to noncrystalline and insoluble Aβ fibrils.

The environment plays an important role in the process of amyloid peptide fibril formation. Peptide concentration, pH, ionic strength, metal ions content, temperature, type of Aβ species, and certainly quality of synthetic amyloid β peptide influence the assembly and final morphology of fibrils (Barrow et al., 1992; Exley et al., 1993; Wood et al., 1996; Stine et al., 2003). Thus, acidic pH increases dramatically the kinetics of fibril formation, resulting in shorter fibril fragments (Barrow et al., 1992). In the process of solubilization, the pH may pass through the isolelectric point of Aβ peptide, around 5.5, where precipitation and aggregation are maximal. In order to prepare in vitro Aβ aggregates of consistent quality different protocols have been developed. These comprise a first step in which amyloid β peptide is pretreated in an organic solvent with strong hydrogen bond donor properties such as dimethyl sulfoxide (DMSO), trifluoroethanol (TFE), hexafluoroethanol (HFIP), acidic solutions (trifluoroacetic acid, HCl), or basic solutions (NaOH, NH₄OH). This step is usually followed by filtration, size exclusion (SE) chromatography, and centrifugation techniques (Walsh et al., 1997; Dahlgren et al., 2002; Fezoui and Teplow, 2002). One of the advantages of presolubilizing Aβ peptide in fluoro-alcohol solvent is to eliminate any preformed peptide assemblies and that one obtains quasi stable conformers (mainly monomers) exhibiting predominately α-helical conformations, which do not form fibrils. As fluoro-alcohol solvent is highly volatile, a gentle evaporation procedure can be used to transfer the presolubilized Aβ peptide to the buffer or culture medium of choice (for review see Klein [2002]).

Protocols to produce Aβ aggregates of consistent quality and spectroscopic techniques have been combined to elucidate fundamental features of the early assembly of Aβ peptides. It was shown that fibril formation is a complex multistep process, which requires participation of several different types of metastable structures including oligomeric forms and protofibrils. Low- and high-molecular weight (LMW, HMW) soluble Aβ aggregates have been prepared, isolated, purified, and further characterized for their structural and biological properties. The distribution of size and structure of oligomers in solution, using synthetic Aβ peptide, is highly dependent on the method of preparation. Thus, SDS-stable LMW Aβ oligomers, from dimers to hexamers (LMW oligomers), are obtained by using pretreatment in fluoro-ethanol solvent, SE, or filtration (Walsh et al., 1997; Bitan et al., 2001; Dahlgren et al., 2002; Bernstein et al., 2005). By extending incubation time and adapting the centrifugation step, LMW oligomers can be grown further to produce HMW oligomers (3-mers to 24-mers), also referred to as amyloid-derived diffusible ligands (ADDLs) (Garzon-Rodriguez et al., 1997; Lambert et al., 1998). Interestingly, Aβ 1–42 has the tendency to form higher-MW oligomers than other isoforms, followed by a rapid transformation into protofibrils and mature fibrils, suggesting that Aβ 1–42 assemblies might predominate at the very early stages of AD (Bitan et al., 2003a, 2003b). In silico studies confirm that Aβ 1–42 monomer is more prone to form larger oligomers, with a significant amount of trimers and pentamers at the initial step, whereas Aβ 1–40 forms significantly more dimers (Urbanc et al., 2004). By atomic force microscopy (AFM) and transmission electron microscopy (TEM) analysis Aβ oligomers appear to be spheres of approximately 3–20 nm in diameter (Lashuel et al., 2002), described also as amylospheroid (ASPD) of 10–15 nm diameter for Aβ 1–40 and >10 nm for Aβ 1–42 (Hoshi et al., 2003). Nevertheless, the heterogeneity of in vitro oligomer preparations makes it difficult to reach a consensus about the true oligomeric state of Aβ and the structure of the individual aggregates.

4.2.3 CYTOTOXICITY PROPERTIES OF Aβ AGGREGATES

Another important question, which remains to be clarified, is the role of Aβ aggregates (oligomers and fibrils) in the pathology of AD and the mechanism by which these aggregates induce cellular dysfunction and injury, ultimately leading to the synaptic loss and neuronal death observed in AD brain. Synthetic Aβ fibrils have been demonstrated to be cytotoxic to different cell lines, primary neurons (Loo et al., 1993; Pike et al., 1993; Lorenzo and Yankner, 1994; Grace et al., 2002) and primary astrocytes in culture (Brera et al., 2000). Furthermore, Aβ peptides have been shown to impair synaptic plasticity in the form of long-term potentiation (LTP) in hippocampal slices and in vivo (Chen et al., 2000; Freir et al., 2003) and induce neuronal dystrophy (Grace et al., 2002). Differences in Aβ fibril morphology resulted in variable toxicity on primary neuronal cultures, suggesting that Aβ aggregates are not equivalent in mediating cytotoxicity (Petkova et al., 2005). In fact, several studies showed that not only fibrillar form but also soluble aggregation intermediates namely oligomers, produced from synthetic Aβ peptide, caused toxicity on cells and rat brain slice cultures (Lambert et al., 1998; Walsh et al., 1999; Dahlgren et al., 2002; Wang et al., 2002). Several potential neurotoxicity pathways were explored. Thus, Aβ-activated kinases such as c-Jun N-terminal kinase (JNK) (Shoji et al., 2000), p38 MAP kinase (MAPK) (McDonald et al., 1998), and cyclin-dependent kinase 5 (cdk5) (Alvarez et al., 1999) are known to be activated in AD brain. The high toxicity of Aβ amylospheroid on primary neuronal cultures was attributed to their ability to activate tau protein kinase I/glycogen synthase kinase-3β, suggesting a role in neurodegeneration (Demuro et al., 2005). LMW synthetic Aβ oligomers spontaneously enter the membrane of intact erythrocytes, producing permeable channels in the membrane, a key element in toxicity (Singer and Dewji, 2006). On the other hand, HMW oligomers have been reported to increase lipid layer conductance in the absence of channel or pore formation or ion selectivity (Kayed et al., 2004), resulting in increase of membrane permeability and depletion of intracellular Ca^{2+} in human neuroblastoma cells (Demuro et al., 2005) and leading to cell dysfunction and death. Using a specific anti-HMW Aβ oligomeric antibody, oligomers extracted from AD brain or made in vitro localized exclusively to synaptic terminals of cultured rat hippocampal neurons (Lacor et al., 2004). Specific immunodetection of oligomers showed that they tend to be found in the axon and axonal terminals in AD brain compared with nondemented brain, suggesting a role in synaptic dysfunction (Kokubo et al., 2005). It has therefore been proposed that one early toxic feature of Aβ is its ability to inhibit LTP and it is particularly interesting that specific inhibitors of JNK, p38 MAPK, cdk5 prevent naturally secreted Aβ oligomer-mediated inhibition of LTP in hippocampal slices (Wang et al., 2004). Selective blockage of metabotropic receptor subtype 5 (mGluR5) was also found to protect against Aβ-mediated toxicity in cortical cultures (Bruno et al., 2000) and to prevent Aβ-mediated inhibition of LTP in rat hippocampal slices (Wang et al., 2004). Aβ was also reported to promote endocytosis of the ionotropic glutamate NMDA receptor and to reduce its receptor levels at the cell surface (Snyder et al., 2005). However, despite these many evidences, the relevant cellular pathway that mediates synaptic Aβ toxicity in the AD brain is still unclear (for review see Selkoe [2002] and Tanzi [2005]).

Other studies have explored Aβ-mediated toxicity using Aβ oligomers isolated from AD brain (Roher et al., 1996) or naturally produced by cells. Culture medium of CHO cells overexpressing mutated APP751 (7PA2), containing significant amount of soluble LMW Aβ oligomers (Aβ diffusible ligands) but not fibrils, was able to impair LTP in rats and in hippocampal slices (Walsh et al., 2002b; Wang et al., 2004) and impaired memory functions after i.c.v administration in rats (Cleary et al., 2005). Experiments involving selective depletion of the 7PA2 culture medium of Aβ monomers by SE chromatography or protease degradation (using the insulin-degrading enzyme) add evidence that the Aβ-diffusible oligomers are indeed the cause of neuronal impairment, and the trimeric form may be particularly active in disrupting cognitive function. The work carried out more recently by Sylvain Lesné and Karen Ashe further investigated the time-dependent appearance of cerebral Aβ oligomers in Tg2576 mice (human APPSWE) and the correlation with cognitive decline (Lesne et al., 2006). A 56 kDa oligomers species (dodecamer), designated as Aβ star, appeared in the brain of the tg2576 at 6 months of age correlating with memory deficit in this model. Aβ star, purified by immunoprecipitation (IP) and SE and injected i.c.v. in healthy rats, induced impairment of long-term memory. Interestingly appearance of Aβ star preceded plaque formation in Tg2576 by 4 months. All these findings taken together suggest that oligomeric forms of Aβ are the toxic species that mediate neurotoxicity in the brain, contribute to synaptic dysfunction, neuronal loss, and plaque formation, and their appearance would correlate with early pathological events in AD. These findings would also explain the relative weak correlation between the severity of dementia and the density of amyloid plaques observed in AD brain (Terry et al., 1981; Braak and Braak, 1991), suggesting that amyloid plaques are relatively inert in AD pathology.

4.2.4 Aβ AGGREGATION INHIBITORS

Better understanding of the Aβ aggregation process in vitro and the identification and characterization of intermediate aggregates raised the interest in finding anti-aggregation molecules to inhibit the Aβ assembly into neurotoxic aggregate—not only to prevent Aβ fibrils or dissolve preformed fibrils but also to prevent formation and promote dissolution of oligomeric aggregates. The therapeutic benefit of an anti-Aβ aggregation strategy would therefore be to reduce the overall cerebral level of Aβ aggregates by promoting their dissolution into monomeric Aβ form, allowing more rapid degradation by proteases and clearance from the brain to the periphery. Molecules able to prevent Aβ-peptide aggregation and with the additional ability to dissolve preformed amyloid aggregates were intensively investigated. Two main strategies were followed: (1) rational design of short peptide sequences that inhibit amyloid aggregation and (2) random screening of chemical libraries for small-molecule inhibitors.

The self-recognition motif and the secondary structure of Aβ seem to determine several important properties of Aβ aggregation. Alanine-scanning studies revealed that the ability of Aβ to successively adopt a β-sheet conformation and to form fibrils is dependent on a specific hydrophobic region within the N-terminal domain of Aβ (amino acids 16–20: KLVFF) (Hilbich et al., 1992; Tjernberg et al., 1996). Therefore, peptide amyloid aggregation inhibitors bearing a KLVFF sequence were designed, and these showed activity in inhibiting amyloid formation and in preventing the formation of neurotoxic species of Aβ (Ghanta et al., 1996; Tjernberg et al., 1996). PPI-368 and PPI-1019 (a d-peptide analog) from Praecis Pharmaceuticals are examples of Aβ-derived peptides designed to prevent the polymerization process and which showed the ability to prevent Aβ nucleation and slowed fibril extension (Findeis, 2002; Wolfe, 2002). PPI-1019 was proposed as a candidate for clinical studies (patent WO 0052048). Another approach was to design peptides, called β-sheet breaker peptides, with high homology to the Aβ sequence responsible for the misfolding, but with a very low propensity to adopt a β-sheet conformation. Proline residues were added into the sequence to disrupt β-sheet formation, since incorporation of this amino acid within a β-pleated structure is highly unfavorable (Chou and Fasman, 1978; Kim and Berg, 1993; Wood et al., 1995). These inhibitors were able to inhibit the Aβ conformational change, dissolved preformed fibrils, and prevented neuronal death induced by Aβ in cell culture experiments (Soto et al., 1996b, 1998). A 5 residue peptide (iAβ5, Seq: LPFFD) was able to prevent deposition of amyloid aggregates to induce dissolution of preformed plaques in a rat brain model of amyloidosis and to reverse the associated cerebral histological changes, including neuronal shrinkage and microglial activation (Sigurdsson et al., 2000; Soto et al., 1998). End-protected iAβ5 (iAβ5p, Seq: Ac-LPFFD-NH2) caused a significant reduction of cerebral amyloid load, an increase in neuronal survival and a decrease in brain inflammation in a mouse model of AD (APP, V717I/PS1, A246E) (Permanne et al., 2002). iAβ5p was able to prevent Aβ amyloid-induced spatial memory impairment with partial reduction of amyloid deposits, reduction of neuronal loss, and astrocytic response (Chacon et al., 2004).

The peptides described earlier were designed to inhibit Aβ aggregation specifically, based on the self-recognition motif of Aβ. Interestingly anti-aggregation efficacy of a more general peptide motif has also been proven. In this case, a polylysine peptide was shown to potently dissolve preformed Aβ fibrils. This peptide has also been proposed as a universal dissolver of all types of oligomeric β-sheet peptides and which may globally retard protein aggregation (Nguyen et al., 2002).

The drug potential of peptide molecules is limited by their rapid degradation in vivo, their immunogenicity, and their poor bioavailability (Adessi and Soto, 2002). Therefore, development of nonpeptidic amyloid aggregation inhibitors is preferable. Several molecules have been shown to be effective inhibitors of Aβ aggregation, for example, Congo-red, small sulfonated anions, benzofuran-based compounds, rifampicin, melatonin, nicotine, oestrogen, nitrophenol, tetracycline, anthracyclinne 4'-iodo-4'deoxydoxorubicin, and curcumin. Certain inhibitors of synthetic Aβ fibrillogenesis were also able to inhibit the early oligomerization stages. Thus curcumin prevented formation of both oligomers and fibrils, and reduced amyloid level and plaque burden in transgenic mice (Yang et al., 2005). Pyridazine and related compounds inhibited early oligomerization of naturally secreted Aβ by 7PA2 cells and rescued LTP inhibition in rat hippocampal slices (Walsh et al., 2005). Unfortunately, the use of some of these molecules as therapeutic agents is limited by several factors; first their unknown mechanism of action which makes it difficult to improve potency, second by their weak selectivity for binding to Aβ aggregates rather than monomeric Aβ, and finally their toxicity.

The most clinically advanced anti-aggregation molecule is NC-531 (Alzhemed) from Neurochem. NC-531 is a sulfated glycosaminoglycan-mimetic able to inhibit Aβ fibrils formation in vivo. This compound is derived from a backbone-modified subunit of polyvinylsulfonate. It decreased amyloid burden in TgCRND8 transgenic mice after 8 weeks of treatment by 36% and 70% at 30 and 1000 mg/kg, respectively. Four phase I clinical trials involving a total of 117 healthy volunteers with three single dose and one multiple dose were completed and results showed good safety and tolerability with no serious adverse effects. In October 2002, a phase II trial started and reported results demonstrated again no safety concerns and presence of the compound in CSF (Tremblay et al., 2003). In November 2005, the compound entered a phase III trial with 930 mild to moderate AD patients receiving Alzhemed over a period of 18 months.

The prevention or dissolution of amyloid plaques has the advantage of directly targeting the removal of putative toxic elements from the brain. However, the prevention or dissolution of plaques may result in a significant increase of total cerebral soluble Aβ, which may trigger new formation or stabilization and stabilization of oligomers, which could be potentially neurotoxic. It is expected that the brain clearance machinery will overcome this increase of soluble Aβ by efficient enzymatic degradation and efflux transport.

Therefore, the ultimate validation of an anti-aggregation therapeutic approach will depend on well-designed clinical trials, which demonstrate that lowering of the amyloid burden in the brain is associated with a beneficial clinical outcome.

4.3 NONSTEROIDAL ANTI-INFLAMMATORY DRUGS

Bruno P. Imbimbo and Francesca Speroni

Epidemiological studies have shown that nonsteroidal anti-inflammatory drugs (NSAIDs) may protect from development of AD. Modulation of cyclooxygenase (COX), the classical biological target of NSAIDs, or of alternative targets like nuclear factor-κB and peroxisome proliferator-activated receptor-γ, could explain the NSAID effect on AD progression. However, recent studies indicate that some NSAIDs may have direct β-amyloid 1–42 (Aβ42)-lowering properties. This chapter reviews the pharmacology of NSAIDs and discusses the mechanisms underlying the protective effects of NSAIDs in AD and the potential therapeutic use of new NSAID derivatives in the treatment of AD.

4.3.1 INTRODUCTION

NSAIDs are small organic molecules with anti-inflammatory, analgesic, and antipyretic properties. Epidemiological studies have shown that prolonged use of NSAIDs may protect from development of AD. The classical biological target of NSAIDs is COX, the enzyme that converts arachidonic acid to prostanoids. Other important biological targets of NSAIDs are nuclear factor-κB and peroxisome proliferator-activated receptor-γ. Modulation of these pathways, all of which have been implicated in AD pathogenesis, could explain the NSAID effect on AD progression. However, recent studies indicate that the subset of NSAIDs, such as ibuprofen, indomethacin, and flurbiprofen, may have direct β-amyloid 1–42 (Aβ42)-lowering properties in cell cultures as well as in transgenic models of AD-like amyloidosis. A renewed interest in the old and the discovery of new pharmacological properties of these drugs are providing vital insight for future clinical trials in AD. This chapter reviews the pharmacology of NSAIDs and discusses the mechanisms underlying the protective effects of NSAIDs in AD and the potential therapeutic use of new NSAID derivatives in the treatment of AD.

4.3.2 STRUCTURE AND FUNCTIONS OF CYCLOOXYGENASE

COX, or prostaglandin H_2 synthase, is a membrane-bound glycoprotein, which is primarily located in the lumen of the nuclear envelope and endoplasmic reticulum.

COX consists of a homodimer with an associated heme group. COX has two enzymatic activities. The first is a bisoxygenase activity (COX) which catalyzes the oxidative cyclization of the central 5 carbons within the 20 carbon polyunsaturated fatty acid arachidonic acid, resulting in the hydroperoxy endoperoxide prostaglandin called PGG_2. The second is a peroxidase activity, which subsequently reduces PGG_2 to the hydroxy endoperoxide prostaglandin, called PGH_2 (Hamberg and Samuelsson, 1967a, 1967b; Kujubu et al., 1991; Xie et al., 1991). A range of enzymatic and nonenzymatic mechanisms transform PGH_2 into the primary prostanoids, for instance the prostaglandins PGE_2, $PGF_{2\alpha}$, PGD_2, PGI_2, and thromboxane A_2 (Figure 4.4).

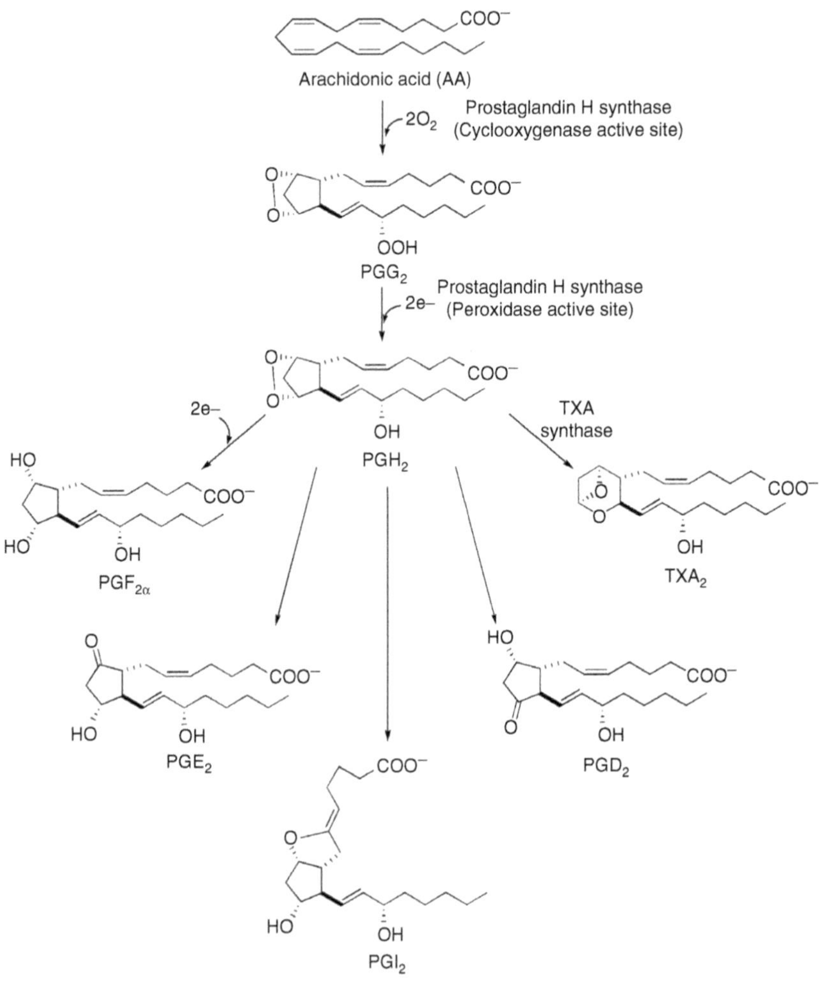

FIGURE 4.4 Scheme summarizing the formation of prostaglandins from arachidonic acid. Prostaglandin H synthase converts arachidonic acid to PGH_2, which acts as a substrate for different enzymes in the production of the principal prostaglandins and tromboxanes.

There are two main isoforms of COX, COX-1 and COX-2, which are coded by distinct genes located on human chromosomes 9 and 1, respectively (Kujubu et al., 1991; Xie et al., 1991). COX-1 is constitutively expressed and is mainly responsible for the production of prostanoids, which are involved in the maintenance of physiological functions such as gastric protection and renal function. The second isoform, COX-2, is regulated by growth factors, tumor promoters, cytokines, glucocorticoids, and bacterial endotoxins (Herschman, 1994, 1996). COX-2 is implicated in inflammatory responses and pathological changes in numerous diseases, including AD.

Recently, a number of studies have suggested the existence of another COX isoform named "COX-3," with a tissue-specific profile of expression and significantly present in the brain (Chandrasekharan et al., 2002; Dinchuk et al., 2003; Davies et al., 2004). The new enzyme could be a COX-1 splice variant (Chandrasekharan et al., 2002) selectively inhibited by paracetamol and phenacetin, and could represent a primary central mechanism by which these drugs decrease pain and possibly fever. Other studies suggest that this new isoform could be a COX-2 splice variant (Simmons et al., 1999). "COX-3" could be encoded by another gene even though there is currently no concrete scientific evidence for an actual third independent COX gene (Davies et al., 2004).

4.3.2.1 COX-1 and COX-2 Distribution and Physiological Role

COX-1 is constitutively expressed in platelets, endothelial cells, stomach, kidney, smooth muscle, and in other tissues. It is considered a "housekeeping" enzyme as the production of prostaglandins preserves the integrity of the stomach lining, maintains the normal renal function in a compromised kidney, and regulates gestation and parturition. In addition, COX-1, through the production of thromboxane A_2, causes the aggregation of platelets to prevent pathological bleeding.

COX-2 is constitutively expressed in brain and kidney (Yamagata et al., 1993; Komers and Epstein, 2002; Beher et al., 2004). The enzyme expression may be induced in synovial cells, fibroblasts, macrophages, and monocytes in response to growth factors, tumor promoters, and cytokines. COX-2 is thought to be involved in the production of prostaglandins involved in cell proliferation and differentiation. COX-2 expression may be transiently induced by specific inflammatory stimuli and is responsible for the synthesis of large quantities of prostanoids seen to occur during acute inflammation (Hla et al., 1993; Herschman, 1994).

Although both enzymes are membrane-bound proteins, COX-1 activity resides mainly in the endoplasmic reticulum, although it is present also in the nuclear envelope. COX-2 activity is located both in the endoplasmic reticulum and in the nuclear envelope where it appears to be more concentrated than COX-1 (Morita et al., 1995). This compartmentalization suggests that COX isoenzymes may represent two temporally and spatially separated prostanoid biosynthetic systems with COX-2 producing prostanoids for differential or replicative events and perhaps interacting with nuclear receptors (Smith, 1986; Smith and Dewitt, 1996; Lim et al., 1999). Although the gross kinetic properties (e.g., K_m, V_{max}) of the two enzymes are nearly identical, COX-1, but not COX-2, exhibits negative allosterism at low arachidonate concentrations. This difference may allow COX-2 to compete more effectively for newly released arachidonate when the isozymes are expressed in the same cell (Swinney et al., 1997; Chen et al., 1999).

4.3.2.2 Structure of COX-1 and COX-2

The primary structures of COX-1 and COX-2 have been determined for several species (Smith and Dewitt, 1996). Both enzymes contain signal peptides of varying lengths. Mature, processed COX-1 contains 576 amino acids, whereas the mature form of COX-2 contains 587 amino acids. Both isoforms are membrane-integrated proteins, and the major sequence differences occur in the membrane-binding domains. COX-2 has the same primary structure of COX-1 with 18 additional amino acids at the C terminus. The function of this insert is not established but may mark COX-2 for rapid proteolysis or provide a signal for subcellular trafficking (Xie et al., 1991; Hla et al., 1993; Smith et al., 2000).

The three-dimensional structure of both isoforms is virtually the same. Glycosylated COX-1 and COX-2 exist as homodimers with an apparent molecular mass of about 70 kDa per monomer. The three-dimensional structure of ovine COX-1 has been first determined at 3.5Å resolution by X-ray crystallography (Picot et al., 1994). Each monomer is composed of three distinct folding units: a short amino-terminal domain, a membrane-binding domain (MBD), and a C-terminal catalytic domain.

The N-terminal domain is formed by 50 amino acids held together by three disulfide bonds and is located at the dimer interface. The domain is covalently linked to the main body of the enzyme by another disulfide bridge and its conformation is similar to that of the epidermal growth factor (Picot et al., 1994). The MBD consists of a right-handed spiral of four α-helices along one side of the monomer. These helices are highly amphipathic and create a motif for insertion of the enzyme into the membrane bilayer. The helices are situated with their hydrophobic surfaces facing outward, away from the body of the protein, whereas the corresponding surfaces on the adjacent monomer face the same direction. This structure creates a large hydrophobic patch on the exterior of the protein, and the patch anchors the dimer in the membrane. Hence, COX-1 is a monotopic membrane protein (Picot and Garavito, 1994; Picot et al., 1994). The insertion into the membrane bilayer is important since arachidonic acid is hydrophobic and, once released from membrane phospholipids, can associate with the bilayer. As the domain helices form the entrance to the COX channel, their insertion into the membrane could allow a fatty-acid molecule to gain direct access to the hydrophobic channel from the interior of the bilayer (Picot et al., 1994; Kurumbail et al., 1996; Luong et al., 1996; Kiefer et al., 2000). The C-terminal catalytic domain is a globular α-helical structure containing both the COX and the peroxidase-active sites. It comprises two distinct lobes of polypeptide chain, a smaller and a larger one, organized in 17 helical segments (Picot et al., 1994).

4.3.2.3 Active Sites of COX-1 and COX-2

The final product of the arachidonic cascade, PGH_2, is formed by two sequential reactions in spatially distinct, but mechanistically coupled, active sites (Hamberg and Samuelsson, 1967b; Kulmacz et al., 1994). The peroxidase-active site lies on the exterior of the protein, on the side of the molecule opposite to the membrane-binding domain. The site is found at the interface between the large and the small lobes of the catalytic domain, in a shallow cleft containing the high-spin Fe(III)-protoporphyrin IX prosthetic group. His388, the proximal iron ligand that lies on Helix H8, coordinates the heme iron. On the distal side of the heme, Gln203 and the distal histidine His207 lie against the heme face but they do not coordinate the metal. Residues lining these two amino acids form a small shield that encloses the peroxidase-active site. Nevertheless, the shallow cleft exposes a large part of the heme to the solvent, possibly to accommodate a large hydroperoxide substrate as PGG_2. The peroxidase-active site is highly homologous to mammalian peroxidases such as myeloperoxidase (Picot et al., 1994). There are two significant differences between the peroxidase activities of COX-1 and COX-2. First, COX-1 catalyzes a two electron reduction of hydroperoxidase substrates almost exclusively, whereas COX-2 catalyzes 60% two electron and 40% one electron reductions (Landino et al., 1997). Second, the rate of formation of products is much faster with COX-2 (Lu et al., 1999).

The COX-active site consists of a long narrow channel extending from the outer surface of the membrane-binding motif through helices A, B, and C to the center of the monomer (Picot et al., 1994; Malkowski et al., 2000; Smith et al., 2000; Kurumbail et al., 2001). The long hydrophobic channel by which arachidonic acid and O_2 can gain access presents a broad entrance near the membrane domain and then a constriction composed of three residues, Arg120, Tyr355, and Glu524, which separates the active site located above. Hence, this site may be divided into the main substrate-binding channel, which is largely hydrophobic, and a smaller amphipathic side pocket (Kurumbail et al., 2001). Tyr385 is found at the top of the channel at the end of helix H8 and sits near the heme plan with its $C\gamma$ within 10Å of the heme iron. Ser530, which is known to be acetylated by aspirin, lies just below Tyr385, within a 6Å distance (Garavito, 1996), at a point where its acetylation could easily block the access to the upper part of the channel. The COX-active site is highly conserved between the two isoforms. The active site of COX-2 is approximately 20% larger and more accommodating than that of COX-1 (Figure 4.5). The difference in active site size and shape is due to three amino acids namely, Ile523 versus Val523 in the first shell, and Ile434 versus Val434 and His513 versus Arg513 in the surrounding second shell for COX-2 (Kurumbail et al., 1996, 2001; Smith et al., 2000). The substitution in position 523 with a valine residue may account for a minor steric hindrance (Gierse et al., 1996). This is likely to be the reason for the increased promiscuity of COX-2 for fatty acid substrates with different chain lengths (Laneuville et al., 1995) and the different selectivity profile of some classes of inhibitors. This size difference has been exploited in developing COX-2-specific NSAIDs.

4.3.2.4   COX–Substrates Complexes

COX-1 and COX-2 have similar kinetic parameters for arachidonate (Meade et al., 1993; Barnett et al., 1994; Laneuville et al., 1994). Both enzymes form tyrosyl radical (Hsi et al., 1994) and are inhibited by NSAIDs (Meade et al., 1993; Laneuville et al., 1994; Smith et al., 2000). There are differences between the two isoenzymes in selectivity toward fatty acid substrates (Laneuville et al., 1995). Although COX-1 can metabolize efficiently only arachidonic acid, COX-2 seems to be a more

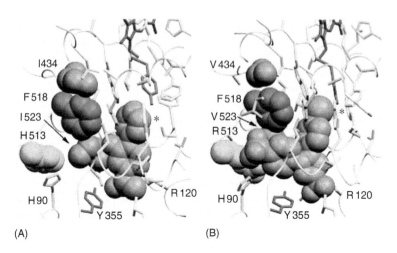

FIGURE 4.5 NSAID binding to the cyclooxygenase (COX)-active sites of prostaglandin synthase (PGHS). (A) Flurbiprofen bound in the COX-active site channel in ovine PGHS-1. Residues Ile434, His513, Phe518, and Ile523 are shown as space filling; Arg120 extends behind flurbiprofen to its carboxylate. (B) The COX-2 inhibitor SC-588 bound in the COX active site channel of mouse COX-2. Residues Val434, Arg513, Phe518, and Val523 are shown as space filling; the phenylsulfonamide group of SC-588 extends into the side pocket made accessible by Val523 (barely visible behind the drug) and interacts with Arg513. Access to the side pocket is made easier by the I434V change in PGHS-2, which allows Phe518 to move out of the way when drugs bind in this pocket. (From Smith, W.L., DeWitt, D.L., and Garavito, R.M., *Annu. Rev. Biochem.*, 69, 145, 2000. With permission.)

promiscuous enzyme as it can metabolize larger, neutral derivatives of arachidonic acid. Arachidonic acid can bind to COX-1-active site in four slightly different conformers, which are catalytically competent, whereas binding to COX-2 is achieved in three different conformers. Each different conformation leads to a different product (Xiao et al., 1997; Smith et al., 2000). Aspirin acetylation of COX-1 completely abolishes its COX activity, whereas acetylation of COX-2 shifts the product profile from primarily PGH_2 to exclusively 15-(R)-hydroxyeicosatetraenoic acid (HETE), a mono-oxygenated metabolite of arachidonic acid (Laneuville et al., 1995; Kurumbail et al., 2001).

The productive conformation of fatty acid substrates bound to prostaglandin synthase has been determined by the X-ray crystal structure of Co^{3+}-ovine COX-1 complexed with arachidonic acid (Malkowski et al., 2000). Arachidonic acid is bound in COX-active site and adopts an extended L-shaped conformation with two kinks in the center. Carbons of the arachidonic acid C-1 through C-3 bind in the channel with the carboxylate positioned to interact with the guanidinium group of Arg120 and the phenolic oxygen of Tyr355. C-7 through C-14 forms an S shape that weaves the substrate around the side chain of Ser530. C-13 is close to the phenolic oxygen of Tyr 385 and is oriented properly for abstraction of the pro*S* hydrogen. The S shape also positions C-11 above a small pocket into which O_2 could presumably migrate from the lipid bilayer. Thus, C-11 would be accessible to O_2 from the side opposite to hydrogen abstraction, a known aspect of the reaction (Thuresson et al., 2000). The ω-end of the substrate (from C-14 through C-20) binds in a *cul-de-sac* along helices 6 and 17 between Ser530 and Gly533. There are 19 residues lining the COX channel that are predicted to make a total of 50 contacts with the substrate (Malkowski et al., 2000). Two of these are hydrophilic and 48 involve hydrophobic interactions. The only hydrophilic interaction is that between the carboxylate moiety and Arg120 and Tyr355. The polar interaction between Arg120 and the carboxylate is a common feature in binding of fatty acid substrates and some classes of NSAIDs that possess this acid group. This ionic interaction plays a critical role in substrate binding to COX-1, whereas it is not critical for binding to COX-2. Moreover, Tyr355 lies across from Arg120 at the base of the COX channel such that its hydroxyl group makes a hydrogen bond to the carboxylate of arachidonic acid. Mutational analysis (Thuresson et al., 2001) allowed for the identification of five functional categories of residues lining COX-1-active site: (a) residues directly involved in hydrogen abstraction from C-14 of arachidonate (Tyr385); (b) residues essential for positioning C-13 of the arachidonate for hydrogen abstraction (Gly533 and Tyr348); (c) residues critical for high-affinity arachidonate binding (Arg120); (d) residues critical for positioning arachidonate in a conformation that yields the production of PGG_2 (Val349, Trp387, and Leu534); and (e) other active site residues, which individually make low but measurable contributions to catalytic efficiency.

The productive conformation of the fatty acid substrate observed in the crystal structures positions the 13-pro(*S*) hydrogen near the phenolic oxygen of Tyr385. It is, therefore, ideally located for the abstraction of the hydrogen by the radical derivative of tyrosine, which is the first step in arachidonic acid oxygenation. The side chain of Tyr348, which forms a hydrogen bond with Tyr385, makes numerous contacts to C-12 or C-14 of arachidonic acid. The tyrosyl radical (Figure 4.6) is produced by intramolecular oxidation by a ferryl-oxo derivative of the heme cofactor. The free radical thus generated on the C-13 of arachidonic acid then migrates to C-11 forming a pentadienyl radical. Molecular oxygen attacks the radical from the small pocket formed by Val349, Ala527, Ser530, and Leu531. The C-11 radical then attacks the C-9 to produce the endoperoxide bridge C-9 or C-11 leaving a radical centered on C-8: the cyclization can occur only after the reconfiguration of the fatty acid as its extended conformation could not permit the closure of the ring. These conformational transitions could also position C-15 optimally for the addition of the second oxygen molecule and for hydrogen donation by Tyr385 to the 15-peroxyl radical. This latter event completes the COX reaction and returns the radical to the catalytic tyrosine for the next turnover (Malkowski et al., 2000).

Nevertheless, this is not the only binding conformation for arachidonic acid in COX-active sites. The nonproductive conformation of arachidonic acid has been identified through the crystal structure of His207Ala mutant COX-2 cocrystallized with arachidonic acid (Kiefer et al., 2000). This enzyme is deficient in peroxidase activity. In this unexpected orientation of the fatty acids of COX-2, the side chain of Arg120 forms van der Waals interactions with the ω-half of the fatty acid, rather than the ion pair with the carboxylate that is seen in COX-1-active site. On the other hand, the carboxylate moiety is strongly hydrogen bound to the side chains of Tyr385 and

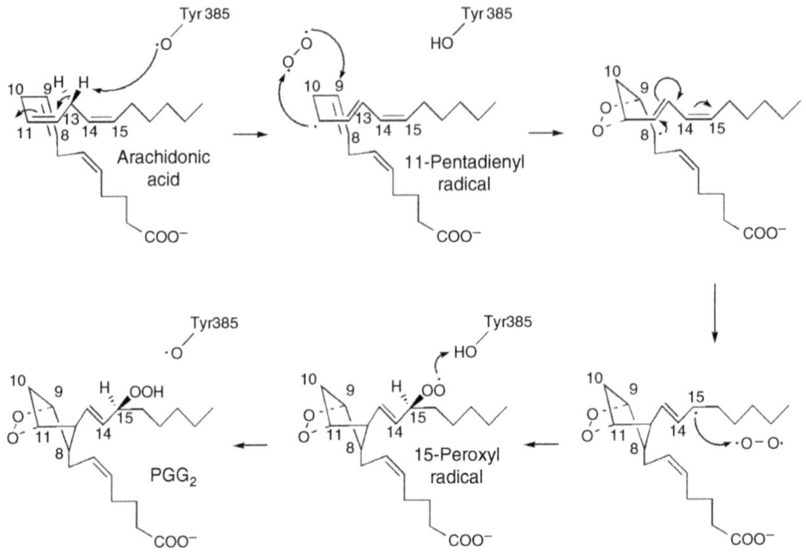

FIGURE 4.6 Reaction mechanisms underlying the transformation of arachidonic acid into PGG$_2$ at the COX-active site of prostaglandin synthase (see Section 4.2.3).

Ser 530. This conformation is not viable for COX-2 catalysis and might be an inhibitory substrate-binding mode even if so far there is no experimental evidence. This conformation underlines the importance of Tyr385 and Ser530 in chelating polar and negatively charged groups not only in arachidonic acid and aspirin; these groups could also be important for inhibition of COX-2 by clinically used NSAIDs such as diclofenac, piroxicam, and nimesulide (Rowlinson et al., 2003).

4.3.3 COX-1 AND COX-2 INHIBITORS

There are two classes of NSAIDs: classical NSAIDs and COX-2-selective inhibitors (Figure 4.7). Classical NSAIDs can be divided into (a) salicylate derivatives and diflunisal, (b) arylpropionic acids like the profens (**1** and **2**), naproxen (**3**) and zomepirac, (c) enol carboxamides, like piroxicam and meloxicam, (d) anthranilic acid, like the fenamates, and (e) indomethacin (**4**) and its prodrugs like sulindac (**5**). The relatively selective COX-2 inhibitors meloxicam, nimesulide, and etodolac were not designed specifically as COX-2 inhibitors, but were identified from pharmacological tests as potent anti-inflammatory drugs with low ulcerogenic activity in the rat stomach (Vane and Botting, 1998). After the discovery of COX-2, these three drugs were each found to preferentially inhibit COX-2, rather than COX-1. It was suggested that drugs with the highest potency for COX-2 and a better COX-2/COX-1 activity ratio would also have potent anti-inflammatory activity, with fewer side effects. The discovery of COX-2 has led to the development of highly selective COX-2 inhibitors (Figure 4.7), for instance celecoxib (**7**) and rofecoxib (**8**). The range of activities of NSAIDs against COX-1, when compared with COX-2, very well

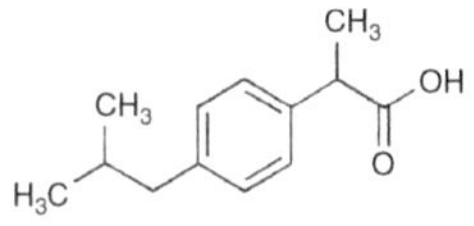

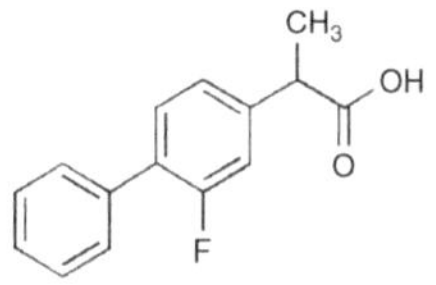

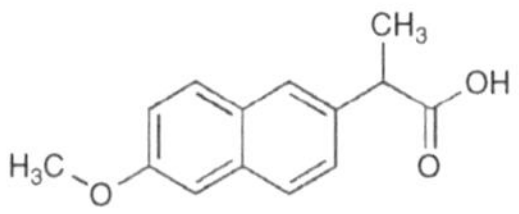

FIGURE 4.7 Most studied nonselective (ibuprofen (**1**), flurbiprofen (**2**), naproxen (**3**), indomethacin (**4**), sulindac sulfide (**5**), the sulindac active metabolite, and diclofenac (**6**)) and COX-2-selective (celecoxib (**7**), rofecoxib (**8**)) NSAIDs in AD.

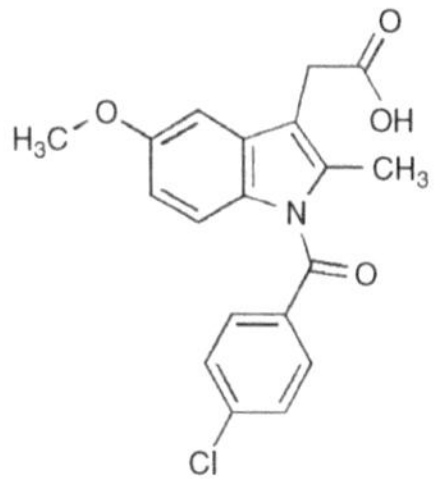

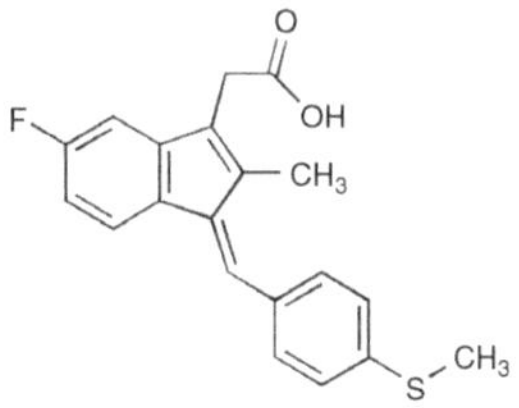

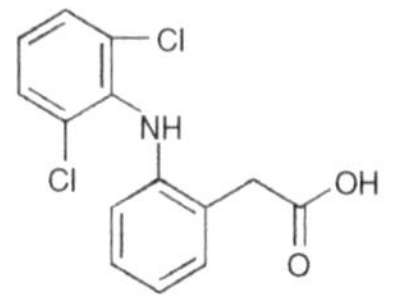

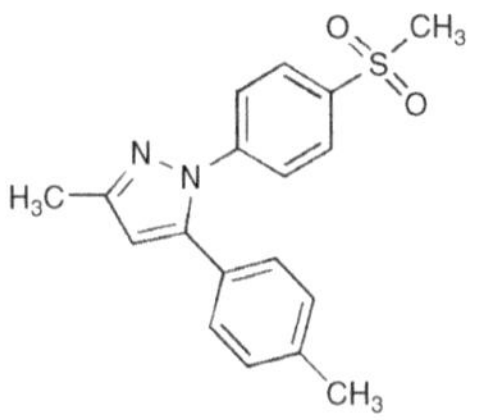

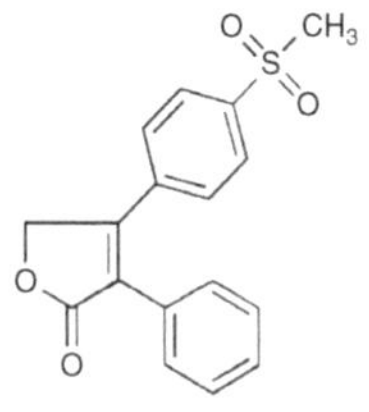

explain variations in the side effects of NSAIDs at their anti-inflammatory doses. The strongest inhibitors of COX-1 are the NSAIDs that cause the most damage to the stomach and kidney.

4.3.3.1 Type of Inhibition

NSAID interactions with COX-active sites of COX-1 and COX-2 have been studied extensively and a number of crystal structures of NSAID and COX complexes are available (Picot et al., 1994; Kurumbail et al., 1996; Loll et al., 1996; Luong et al., 1996; Selinsky et al., 2001). Although NSAIDs are chemically heterogeneous, they share the same capacity of competing with arachidonic acid for binding to the COX-active site (Smith et al., 2000). Their binding at the top of the COX channel in close proximity to Tyr385, a key residue for catalysis, blocks the access of the substrate and therefore inhibits the formation of prostaglandin H_2.

NSAIDs exhibit different types of kinetic inhibition (DeWitt, 1999; Marnett et al., 1999): (a) rapid, reversible binding followed by covalent modification (acetylation) of Ser530 (e.g., aspirin); (b) rapid, reversible binding (e.g., ibuprofen (**1**)); or (c) rapid, lower-affinity reversible binding followed by time-dependent, higher-affinity, slowly reversible binding (e.g., flurbiprofen (**2**)). Time-independent competitive inhibitors form reversible enzyme–inhibitor complexes. Time-dependent inhibitors initially form a reversible enzyme–inhibitor complex that is gradually converted to an inactive form in which the inhibitor is bound more tightly to the enzyme. The formation of this second complex may reflect a subtle protein conformational change (Copeland et al., 1994; Bhattacharyya et al., 1996; DeWitt, 1999; Marnett and Kalgutkar, 1999; Marnett et al., 1999). The structural basis for time-dependent inhibition of COX is not well understood and seems to vary among inhibitor classes and even between compounds belonging to the same class.

The prototype of NSAIDs is aspirin, the acetylated derivative of salicylic acid, which was introduced in 1897 (Dreser, 1899). Aspirin acetylates a serine residue in the arachidonic binding channel (Ser530 in COX-1 and Ser 516 in COX-2) and is not selective towards one of the two isoforms (Loll et al., 1995). When COX-1 acetylation creates a steric blockade that prevents the binding of substrate to the COX-active site, the aspirin-acetylated COX-2 retains COX activity, although the reaction produces a novel product, 15-(*R*)-hydroxyeicosatetraenoic acid (Mancini et al., 1994; Luong et al., 1996). The differential ability of acetylated COX-2 to use the substrate may be due to the additional volume available at the COX-active site. The potential use of a selective COX-2 acetylating agent led in 1998 to the design and synthesis of a class of irreversible COX-2 inhibitors useful for the treatment of inflammatory and proliferative disorders (Kalgutkar et al., 1998; Marnett and Kalgutkar, 1999).

4.3.3.2   Structure Requirements for COX-2 Inhibition Selectivity

The first identified COX-2-selective inhibitors were DuP697 (Gans et al, 1990) and NS-398 (Futaki et al., 1993). The structure of DuP697 led to the synthesis of the diarylheterocyclic family of selective inhibitors, which include celecoxib (**7**), rofecoxib (**8**), valdecoxib, and etoricoxib. Structure–activity studies suggested that a *cis*-stilbene moiety containing 4-methylsulfonyl or sulfonamide substituent in one of the phenyl rings, and the sulfur oxidation state for the methyl sulfone moiety are required for COX-2 specificity (Talley et al., 1999). Other classes of COX-2-selective inhibitors include acidic sulfonamides, indomethacin analogs, zomepirac analogs, and the diclofenac derivative lumiracoxib (Marnett and Kalgutkar, 1998; FitzGerald, 2003). Classical carboxylic acid containing NSAIDs ion pair with the guanidinium group of Arg120, a residue positioned at the mouth of the COX channel with Tyr355 and Glu524. Tyr355 sterically hinders the mouth of the channel, which accounts for the preferential inhibition exhibited by S-stereoisomers of 2-arylpropionc acids, like the profens (Bhattacharyya et al., 1996). On the other hand, the interaction with Arg120 seems to be unnecessary for binding to COX-2-active site as demonstrated by the crystal structure of the complex of diclofenac (**6**) with murine recombinant COX-2 (Rowlinson et al., 2003). COX-2 selectivity can be partly explained by the differences in the COX-active sites of COX-1 and COX-2. Luong and coworkers demonstrated in 1996 that NSAID-binding pocket in COX-2 is about 20% larger than that of COX-1. The larger binding site of COX-2 justifies the broad dimension of new selective COX-2 inhibitors. Substitution of Ile523 in COX-1 and Val523 in COX-2 results in a minor steric hindrance and produces a small pocket adjacent to the active-site channel, which increases the volume available for inhibitor binding. The substitution of two amino acids in the second shell lining the active-site channel of COX-2 (Ile434Val and His513Arg) further increases the volume available for inhibitor binding and may contribute to the local mobility of side chains within the pocket (Kurumbail et al., 1996; DeWitt, 1999). The substitution also alters the chemical environment in the side pocket (Wong et al., 1997). The side pocket could be the site for the specific interaction of the 4-methylsulfonylphenyl and 4-methylsulfamoylphenyl moieties that characterize diarylheterocyclic-selective COX-2 inhibitors. The interaction between 4-methylsulfonylphenyl and 4-methylsulfamoylphenyl substituents and Arg513 appears to be required for the time-dependent inhibition of COX-2 by these inhibitors (Kurumbail et al., 1996). Arg513 may also promote binding and time-dependent inhibition by the zomepirac analogs creating a hydrogen bond network with Glu524 and Tyr355 that seems to allow these inhibitors to bind more tightly to the enzyme (Luong et al., 1996). Site-directed mutagenesis studies elicited the important role of Tyr385 and Ser530. The two residues are involved in the chelation of electron-rich centers that are critical for acetylation of COX-2 by aspirin and for time-dependent inhibition by diclofenac (**6**) (Rowlinson et al., 2003). Nimesulide, an arylsulfonamide derivative that is modestly COX-2 selective, interacts with these two residues in generating time-dependent inhibition. Modeling of this compound into COX-2-active site showed that the sulfone moiety of this compound binds in the side pocket like sulfur-containing substituents in diarylheterocyclic inhibitors (Fabiola et al., 1998; Garcia-Nieto et al., 1999).

All COX-2-selective inhibitors cause time-dependent inhibition of the COX-2 isoenzyme but are time-independent competitive inhibitors of COX-1 (DeWitt, 1999; Smith et al., 2000). The new COX-2 inhibitors exhibit selectivity because they inhibit this isozyme by a time-dependent mechanism. The pharmacological consequence is that when the blood concentration of these drugs is below that required for half-maximal inhibition (IC_{50}) of COX-1, the activity of COX-1 will be minimally affected by the inhibitor, whereas COX-2 becomes effectively inactivated (Marnett and Kalgutkar, 1999; Smith et al., 2000).

4.3.3.3   Clinical Use of COX Inhibitors and Current Safety Issues

NSAIDs have been used to treat arthritis since 1899, when the analgesic and anti-inflammatory effects of aspirin were first recognized. Despite their potential gastrointestinal and renal toxicities, they are among the most widely used therapeutic classes of compounds primarily because they are generally effective for the relief of pain and inflammation. Aspirin is also used as a cardiovascular agent because of its protective effects against myocardial infarction and stroke (Dalen, 2006). COX-2 inhibitors and some other NSAIDs have received attention because of their protective effects against colon cancer (Clevers, 2006). Indeed, in late 1999, celecoxib (**7**) was approved by the FDA for the prevention of colon cancer in patients with familial adenomatous polyposis, which is a hereditary precancerous disease due to a loss of the adenomatous polyposis coli tumor suppressor gene.

Since the discovery of the COX-2 isoenzyme, COX-2-selective inhibitors have been developed with the idea that this isoform is inducible at the site of inflammation, whereas COX-1 is expressed constitutively in several tissues including gastric epithelium. This new class of NSAIDs was thought to be safer for ulcerations of the gastrointestinal mucosa observed with nonselective inhibitors. Nevertheless, in September 2004, Merck & Co. announced the voluntary withdrawal of rofecoxib (**8**) worldwide because of an increased risk of cardiovascular events. This decision and the unexpected findings of a colon cancer study, which has shown that celecoxib (**7**) might also increase the chance of heart attack and stroke in some patients, raised serious concerns about the safety of selective COX-2 inhibitors that are actively marketed today, and the ones currently under development (Finckh and Aronson, 2005). The mechanism with which COX-2 inhibitors may increase the risk of cardiovascular events is not clear but their depressive effects on prostaglandin I_2 formation has been advocated as the cause of blood pressure elevation, acceleration of atherogenesis, and exaggerated thrombotic response to the rupture of an atherosclerotic plaque (Fitzgerald, 2004). These findings represented a major drawback for the clinical use of COX-2 inhibitors in general (Couzin, 2004), and in December 2004 the Food and Drug Administration announced the suspension of a large, randomized, controlled, prevention trial of celecoxib in AD (Alzheimer's Disease Anti-inflammatory Prevention Trial or ADAPT). ADAPT was actually suspended prematurely because of an apparent increase in cardiovascular and cerebrovascular events in the naproxen arm compared with placebo. This event has revived the debate on the cardiovascular safety of NSAIDs, but this time, with a special emphasis on the impact of traditional NSAIDs on the incidence of cardiovascular events. The published data are quite discordant and one cannot conclude that there is clear evidence to support a cardiovascular hazard from the administration of traditional NSAIDs like ibuprofen (**1**) or naproxen (**3**) (Maillard and Burnier, 2006).

4.3.4   NSAIDs AND ALZHEIMER's DISEASE

4.3.4.1   Protective Effects of NSAIDs Used in Alzheimer's Disease

Epidemiological studies indicate that anti-inflammatory drugs, especially NSAIDs, decrease the risk of developing AD. Their beneficial effects may be due to interference of the chronic inflammatory reaction in AD (Hoozemans and O'Banion, 2005). The protective effects of NSAIDs appear to be linked to the duration of treatment. A population-based cohort study of 6989 subjects found a relative AD risk of 0.95 in subjects with short-term use of NSAIDs, 0.83 in those with intermediate-term use, and 0.20 in those with long-term use (in't Veld et al., 2001). Pooled data from 9 studies involving 14,654 subjects confirmed that the protective effects depend on the duration of NSAID use with relative risks of 0.95 among short-term (<1 month), 0.83 among intermediate-term (<24 months), and 0.27 among long term (>24 months) users (Etminan et al., 2003). The type of

NSAIDs used also appears to affect the magnitude of the protective effect. In a recent cohort study of 1301 dementia-free subjects at baseline and followed up for 6 years, no subjects who used nonaspirin NSAIDs for around 3 years developed AD 3 years later (Cornelius et al., 2004). More specifically, the reanalysis of the Rotterdam study indicates that risk decrease is restricted to NSAIDs that lower Aβ42 (ibuprofen (**1**), flurbiprofen (**2**), indomethacin (**4**) and sulindac (**5**), and diclofenac (**6**)) (Breteler et al., 2002). In a cross-sectional retrospective study involving 2708 community-dwelling elderly patients, a significantly decreased risk of cognitive impairment was found for patients using nonaspirin NSAIDs (Landi et al., 2003). The lowest relative risk for a single nonaspirin NSAID drug was observed for diclofenac (0.20). However, a further pooled analysis (Zandi et al., 2004b) did not find significant differences in the relative risk of AD among users of NSAIDs with or without Aβ42-lowering properties, thus arguing that the potential benefits of NSAIDs in AD still may rely on the anti-inflammatory properties of these drugs. In a cross-sectional study involving 526 nondemented elderly subjects, users of NSAIDs showed a nonsignificant trend to reduced Aβ42 plasma levels (Blasko et al., 2005).

4.3.4.2   Brain Inflammation in Alzheimer's Disease

Several studies indicate that Aβ deposition activates a series of inflammatory reactions that may mediate neuronal death. Indeed, in the AD brain there are elevated levels of a diverse range of proinflammatory molecules. These inflammatory molecules are produced principally by activated microglia and astrocytes, which are found to be clustered within and adjacent to the senile plaque (Blasko et al., 2004). The inflammatory products released by Ab-activated microglia include cytokines such as interleukin-1 (IL-1), interleukin-6 (IL-6), TNF-α, and transforming growth factor β (TGF-β) (Streit, 2004). Although their expression is induced by the presence of Aβ, these cytokines are also able to promote the accumulation of Aβ in a vicious circle that fuels the progression of the disease (Cacquevel et al., 2004). Activated microglia secretes also chemokines, a diverse group of small proteins that control the recruitment of cytotoxic and helper T lymphocyte to the sites of inflammation. These chemokines include interleukin-8 (IL-8), interferon-γ-inducible protein (IP-10), macrophage inflammatory protein-1α (MIP-1α), macrophage inflammatory protein-1β (MIP-1β), and monocyte chemoattractant protein-1 (MCP-1) (Xia and Hyman, 1999). Additionally, studies in glial cultures have shown that Aβ activated nuclear factor-kappaB (NF-κB), a transcription factor for several inflammatory mediators including IL-1β and IL-6 (Casal et al., 2004). In AD, the number of reactive astrocytes is increased and the expression of phospholipase A2 in these cells is upregulated, leading to increased arachidonic acid or prostaglandin inflammatory pathway activity. It has been shown that Aβ deposition is accompanied by activation of astrocyte to secrete chemokines, in particular MCP-1 and RANTES (regulated on activation normal T cell expressed and secreted), which serve as potent macrophage chemoattractants (Johnstone et al., 1999). Finally, there is also evidence that Aβ can bind and activate the classical complement cytolytic pathway (Rogers et al., 1992).

4.3.4.3 Clinical Trials of NSAIDs in Alzheimer's Disease

The hypothesis that suppression of these mechanisms will reduce the rate of disease progression and the evidence from epidemiological studies suggesting the protective effects of long-term use of NSAIDs against the onset of AD formed the rationale for a series of trials of different anti-inflammatory drugs in AD (Aisen and Davis, 1994). Specifically, NSAIDs are thought to inhibit microglial COX-1 activity, whereas selective COX-2 inhibitors are believed to act on COX-2 present in neurons. Indeed, both neuronal COX-2 and microglial COX-1 are overexpressed at a relatively early stage of AD (Hoozemans and O'Banion, 2005). The initial pilot studies with NSAIDs produced encouraging results. A 6 month, double-blind, placebo-controlled study in 44 AD patients showed a significant ($p = 0.003$) slower cognitive decline in patients receiving indomethacin (**4**) (100–150 mg/day) compared with those on placebo (Rogers et al., 1993). Unfortunately, 42% of patients abandoned the study because of adverse events, most of them being in the indomethacin-treated group (10 out of 16). Thus, the positive results of this study are difficult to interpret. In another 6 month, double-blind, placebo-controlled study in 41 AD patients, arithmetic trends were observed in patients receiving diclofenac (**6**) (50 mg/day) and misoprostol as gastroprotective agent (Scharf et al., 1999). Again, the withdrawal rate was high in the active treatment group (12 of 24 patients), indicating that AD patients poorly tolerate standard prescription doses of NSAIDs. Later, longer and larger studies with other anti-inflammatory agents produced completely negative results. A 1 year, double-blind, placebo-controlled trial of low doses of prednisone (10 mg daily) in 138 AD patients did not show any significant benefit of the glucocorticoid treatment on the rate of cognitive decline (Aisen et al., 2000). Indeed, patients treated with prednisone displayed a greater behavioral decline compared with those treated with placebo ($p = 0.003$). Another long-term, double-blind, placebo-controlled study evaluated the ability of hydroxychloroquine, a potent anti-inflammatory drug widely used in the treatment of rheumatoid arthritis and able to cross the blood–brain barrier, to delay progression of AD (Van Gool et al., 2001). The study involved 168 patients and was completed by 92% of participants. Unfortunately, at the end of the 18 month treatment period there were no significant differences in any of the efficacy outcome measures (activities of daily living, cognitive function, and behavioral abnormalities). A further large study was carried out with dapsone, an old antileprosy agent with strong anti-inflammatory properties and for which a preventive role in AD was advocated (McGeer et al., 1992). Dapsone (100 mg/day) and placebo were administered orally, once daily for 52 weeks in 201 patients with mild-to-moderate AD. At the end of treatment there were no significant differences between dapsone and placebo on either cognitive or other measures of efficacy.

In the brain, COX-2, the inducible isoform of COX is selectively expressed in neurons of the cerebral cortex, hippocampus, and amygdala (McGeer, 2000). COX-2 is upregulated in the AD brain and its expression in hippocampus increases as the disease progresses (Ho et al., 2001). Transgenic mice overexpressing COX-2 show memory dysfunction, neuronal apoptosis, and astrocytic activation in an age-dependent manner (Andreasson et al., 2001). These studies suggested that COX-2 may contribute to the neurodegeneration occurring in AD brains and that inhibition of COX-2 might be a useful therapeutic target. COX-2 inhibitors would appear to be preferred agents over classic NSAIDs, given their better tolerability at full anti-inflammatory doses. Unfortunately, a 1 year, double-blind placebo-controlled study with the COX-2 inhibitor celecoxib (**7**) failed to demonstrate efficacy in slowing cognitive decline of AD patients (Sainati et al., 2000). Another 1 year, double-blind, placebo-controlled study in 351 AD patients comparing the ability of rofecoxib (**8**) (25 mg/day), another COX-2 inhibitor, and naproxen (**3**) (220 mg/day) to slow cognitive deterioration produced negative results (Aisen et al., 2003) with cognitive decline being actually faster in rofecoxib-treated patients compared with placebo ($p = 0.044$). A further even larger study of rofecoxib (**8**) (25 mg/day) was recently completed in 692 patients and did not show any significant difference between rofecoxib (**8**) and placebo after 1 year of treatment (Reines et al., 2004).

These multiple failures have questioned the role of inflammation in AD (Launer, 2003). Even the use of prednisone, a glucocorticoid that suppresses the acute phase response and complement activation, resulted in negative results. It has been proposed that these negative effects may be duly ascribed to the dose of anti-inflammatory agents, which is insufficient to suppress brain inflammatory activity. Safety issues do not allow testing the hypothesis that higher doses of anti-inflammatory agents may be effective in AD patients. On the other hand, the initial positive results obtained with indomethacin (**4**) and partially with diclofenac (**6**) suggest that different NSAIDs may have different efficacy profiles in AD patients, depending on their specific chemical structure and ability to interfere with APP metabolism. It has been also argued that anti-inflammatory treatment does not prevent further deterioration after a diagnosis of AD has been established. However, the results of a recent 4 year study in 1457 patients with mild cognitive impairment showed that the annual conversion rate to AD was significantly higher in patients treated with rofecoxib (**8**) (25 mg/day) than in those treated with placebo (6.4% versus 4.5%, $p = 0.011$) (Thal et al., 2005). Finally, a large prevention study of naproxen (**3**) and celecoxib (**7**) in AD (ADAPT) was interrupted in December 2004, because of the wave of cerebrovascular safety concerns raised on long-term use of COX-2 inhibitors. Started in 2001, ADAPT was designed to investigate the preventive effect of naproxen (**3**) and celecoxib (**7**) in a 7 year follow-up period in 2625 cognitively normal individuals at risk for AD (age >70 years and first-degree relative with dementia). Subjects were randomized to receive celecoxib (200 mg bid) or naproxen (220 mg bid) or placebo. The major end point was conversion to dementia (Martin et al., 2002).

4.3.5 Effects of NSAIDs on β-Amyloid Secretion

4.3.5.1 In vitro Effects of NSAIDs on Aβ

The first report linking the protective role of NSAIDs against AD with their ability of lowering Aβ secretion was published in late 2001 (Weggen et al., 2001). In this seminal work, selected NSAIDs (ibuprofen (**1**), indomethacin (**4**), and sulindac sulfide (**5**)) were shown to preferentially inhibit, in a concentration-dependent manner, the secretion of the Aβ42 peptide in a variety of cultured cells. Marked Aβ42 inhibitions (>80%) were obtained with concentrations of 80–100 μM for

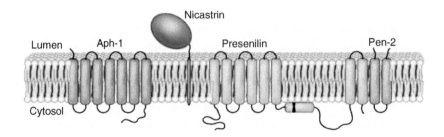

FIGURE 4.8 Schematic representation of the four components of the γ-secretase complex. (From Gandy, S., *J. Clin. Investig.*, 115, 1121, 2005. With permission.) The γ-secretase complex consists in a catalytic component, presenilin, and three cofactors, nicastrin, anterior pharynx defective 1 (Aph-1), and presenilin enhancer 2 (Pen-2). Presenilin is biologically activated by endoproteolysis that generates two heterodimers that each presents the key aspartate residue.

sulindac sulfide (**5**), 75–100 μM for indomethacin (**4**), and 500 μM for ibuprofen (**1**). The inhibitory effect was not seen with naproxen (**3**), a finding that fits well with the negative trial in AD patients (Aisen et al., 2003). The Aβ42-inhibiting activity of sulindac sulfide (**5**) appeared to be independent from anti-COX activity since it was still observed in fibroblasts deficient in COX-1 and COX-2. A parallel increase in another species of Aβ, Aβ38, was observed, thus suggesting that NSAIDs allosterically modulate the activity of γ-secretase, the enzymatic complex responsible for the final cleavage of APP with the generation of Aβ (Figure 4.8).

Other groups confirmed the in vitro Aβ42-lowering properties of selected NSAIDs (Morihara et al., 2002; Eriksen et al., 2003b; Takahashi et al., 2003; Gasparini et al., 2004b). Interestingly, *R*-enantiomers of both ibuprofen and flurbiprofen, known to have poor anti-COX activity, were also found to reduce Aβ42 production by human cells, thus confirming the independence of this effect from anti-COX activity (Morihara et al., 2002). Sulindac sulfide (**5**) was shown to inhibit Aβ42 production in a cell-free assay that employed membrane fractions of HeLa cells and the recombinant C-100 residual of APP as a substrate (Takahashi et al., 2003). In another large study, 20 commonly used NSAIDs were tested at 100 μM on Aβ42 or Aβ40 secretion in a human neuroglioma cell line (Eriksen et al., 2003b). Of the compounds tested, meclofenamic acid, (*R*)-flurbiprofen ((**9**) see Figure 4.9), and *S*-flurbiprofen lowered Aβ42 levels to the greatest extent. Interestingly, it was found that dapsone and naproxen were unable to decrease either Aβ42 or Aβ40 species, in agreement with the negative results obtained in AD patients (Aisen et al., 2003). Conversely, indomethacin (**4**) and diclofenac (**6**) significantly inhibited Aβ42 secretion and this again fits well with the initial encouraging results obtained in AD patients with these two NSAIDs (Rogers et al., 1993; Scharf et al., 1999). Flurbiprofen (**2**) and its enantiomers were shown to selectively lower Aβ42 levels also in a cell-free γ-secretase assay (Eriksen et al., 2003b).

A recent study tested both nonselective NSAIDs as well as COX-1 (sc-560) and COX-2 (celecoxib)-selective inhibitors in mouse neuroblastoma cells overexpressing

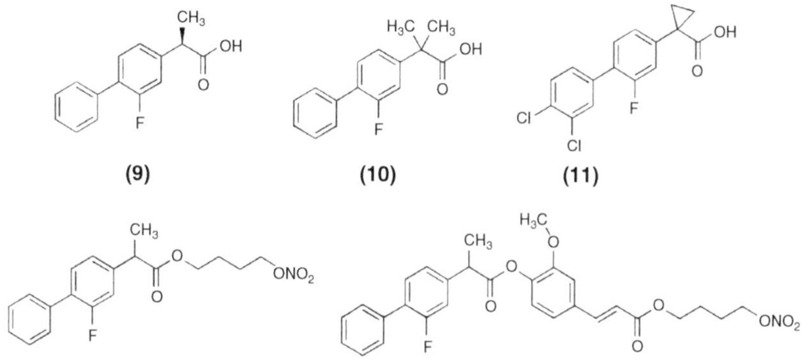

FIGURE 4.9 New NSAID derivatives under assessment as potential treatments of AD. (*R*)-flurbiprofen (**9**), methylflurbiprofen (**10**), CHF5074 (**11**), HCT 1026 (**12**), and NCX-2216 (**13**).

the Swedish mutant form of human APP (Gasparini et al., 2004b). This study confirmed the Aβ42-lowering properties of flurbiprofen (**2**) and sulindac sulfide (**5**) but surprisingly found that celecoxib (**7**) significantly increases Aβ42 secretion. The Aβ42 stimulating activity of celecoxib (**7**) agrees well with the pejorative effects compared with placebo found with COX-2 inhibitors in both AD patients (Sainati et al., 2000; Aisen et al., 2003; Reines et al., 2004) and in patients with mild cognitive impairment (Thal et al., 2005).

It is important to note that, although these observations in vitro were confirmed in vivo in both biochemical and behavioral studies (see Section 4.3.5.2), one should be very careful in explaining clinical results based on animal data. The efficacy of a drug in AD depends on a number of factors linked to both the complexity of the clinical condition (severity of the disease, age of onset, apolipoprotein E genotype, etc.) and the pharmacological properties of the compound (activity on the biological target, bioavailability, brain penetration, etc.).

4.3.5.2 In vivo Effects of NSAIDs on Aβ

A number of in vivo studies have evaluated the effects of selected NSAIDs on Aβ levels in brain of transgenic mice models of AD (Table 4.1).

The in vivo effects on cerebral Aβ were evaluated using transgenic mice models of AD that accumulate Aβ in the brain. Both animals overexpressing the Swedish mutant (K670N/M671L) form of human APP (Tg2567) and animals bearing a double mutation of human APP (K670N/M671L) and presenilin-1 (PS1) (M146L or A246E) were used. Studies can be divided according to the duration of NSAIDs administration in long-term (3–8 months) and short-term (3–7 days) experiments.

TABLE 4.1

Positive in vivo Studies on the Effects of Selected NSAIDs on Cerebral Aβ in Transgenic Animal Mouse Models of AD

Drug	Dose (mg/kg/day)	Administration Route	Treatment Duration	Transgenic Mouse	Age[a] (Months)	Aβ Reduction (%)	Reference
Ibuprofen	56	Diet	6 months	Tg2576	10	53	Lim et al. (2001)
Ibuprofen	46	Diet	3 months	Tg2576	14	33	Lim et al. (2001)
Ibuprofen	63	Diet	4 months	Tg2576	11	62[b]	Yan et al. (2003)
Ibuprofen	56	Diet	3 months	APP + PS1[c]	3	20[d]	Dedeoglu et al. (2003)
NCX-2216	63	Diet	5 months	APP + PS1[e]	7	42[f]	Jantzen et al. (2003)
HCT 1026	30	Diet	6 months	App + PS1[g]	8	45	van Groen and Kadish (2005)
Indomethacin	2.24	Drinking water	8 months	Tg2576	10	20[h]	Quinn et al. (2003)
Indomethacin	5	Drinking water	7 months	Tg2576	8	68[b]	Sung et al. (2004)
Ibuprofen	50	Oral gavage	3 days	Tg2576	3	39[b]	Weggen et al. (2001)
Ibuprofen	62.5	Diet	7 days	APP[i]	10	25[b]	Heneka et al. (2005)
Flurbiprofen	50	Oral gavage	3 days	Tg2576	3	70[b]	Eriksen et al. (2003a)

[a] Age at the beginning of treatment.
[b] Aβ42.
[c] APP (K670M/N671L + V717I) + PS1 (M146L).
[d] Aβ42/Aβ40 ratio.
[e] APP (K670N/M671L) + PS1 (M146L).
[f] In cerebral cortex.
[g] APP (K670N/M671L) + PS1 (A246E).
[h] In hippocampus.
[i] APP (V717I).

In the long-term studies, drugs were administered in diet or drinking water. Ibuprofen (**1**) is the best-documented compound with three independent groups showing quite marked decreasing effects (20%–62%) on brain Aβ42 or Aβ42/Aβ40 ratio (Lim et al., 2000, 2001; Dedeoglu et al., 2003; Yan et al., 2003). Two NO-donating derivatives of flurbiprofen (HCT1026 (**12**) and NCX-2216 (**13**) see Figure 4.9) were also studied and produced similar results (Jantzen et al., 2002; van Groen and Kadish, 2005). Consistently with in vitro data, long-term treatment with celecoxib (**7**) did not significantly affect brain Aβ levels (Jantzen et al., 2002; Dedeoglu et al., 2003).

An interesting study evaluated the effects of indomethacin (**4**) on both cerebral Aβ and prostaglandin concentrations (Quinn et al., 2003). Very low doses of indomethacin (**4**) (2.24 mg/kg/day) were administered in diet for 2 or 8 months. Although 90% reduction of brain prostaglandin concentrations were observed, significant reduction (20%) of Aβ was achieved only in hippocampus of animals treated for 8 months. The discordant effects on Aβ and prostaglandin suppression in this study is compatible with in vitro studies, indicating that Aβ42 production is suppressed by indomethacin (**4**) in a COX-independent manner (Weggen et al., 2001).

Short-term studies were mainly carried out by a research group from Mayo Clinic (Jacksonville, Florida), which described selective reductions of brain Aβ42 after only 3 days of treatment with NSAIDs administered by gavage four times a days (50 mg/kg/day) to young (3 month) transgenic animals (Weggen et al., 2001; Eriksen et al., 2003b). They evaluated the effects of 13 NSAIDs and the enantiomers of flurbiprofen on brain Aβ levels in APP-transgenic mice. Eight NSAIDs (ibuprofen (**1**), flurbiprofen (**2**), indomethacin (**4**), sulindac sulfide (**5**), diclofenac (**6**), diflunisal, fenoprofen, and meclofenamic acid) significantly lowered brain Aβ42 levels, meclofenamic acid and flurbiprofen (**2**) being the most active with 80% and 70% reductions, respectively. Other compounds, including naproxen (**3**), aspirin, nabumetone, and ketoprofen, did not lower brain Aβ42 levels in vivo. In general, there was a good correlation between the ability of an NSAID to lower Aβ42 levels in cell culture and in vivo activity. Aβ42 reductions were seen also in plasma with ibuprofen (**1**) and flurbiprofen (**2**) (30%–50%) but the correlation between plasma and brain Aβ42 inhibitions was poor. In vivo, brain Aβ42 lowering in mice occurred at drug plasma levels achievable in humans with standard doses. For ibuprofen (**1**), a significant correlation between brain Aβ42 lowering and brain drug levels was shown. However, these results were not fully confirmed by other groups. Lanz and colleagues (2005) found that after 3 days of treatment in APP-transgenic mice (Tg2576), flurbiprofen (**2**) (25 mg/kg/day) produced a 50%–60% reduction of both Aβ40 and Aβ42 in plasma and only a 30% reduction in Aβ40 in the cortex. Ibuprofen (**1**) and sulindac sulfide (**5**) at doses up to 50 mg/kg/day did not reduce significantly either brain or plasma Aβ40 or Aβ42. Similarly, Stock and colleagues (2006) have recently shown that a 3 day treatment of Tg2576 transgenic mice with ibuprofen (**1**) and a geminal dimethyl derivative of flurbiprofen ((**10**) see Figure 4.9) (84 mg/kg/day) produced a nonsignificant decrease in brain Aβ42 levels of 20% and 18%, respectively. Similarly, Peretto and colleagues (2005) were not able to show a significant inhibitory effect of a series of flurbiprofen analogs on brain Aβ42 levels after a 7 day treatment in transgenic mice. On the other hand, another study employing a 7 day treatment with ibuprofen (375 ppm in the diet corresponding to approximately 62.5 mg/kg/day) has shown a significant 27% decrease of Aβ42-positive amyloid deposits in the hippocampus of APPV717I transgenic mice (Heneka et al., 2005). The effects in cerebral cortex were lower (12%) and did not reach statistical significance. Similarly, the lowering effects of ibuprofen (**1**) on soluble brain Aβ42 (-22%) and Aβ40 (-30%) did not reach statistical significance (Heneka et al., 2005). Differences in extraction procedures and antibodies used for Aβ assay may explain these different results. One of the perplexing aspects of the Mayo Clinic data is the discrepancy between potency in vitro using cell cultures ($IC_{50} = 100$–300 μM) and potency in APP-transgenic mice with drug effective concentration in brain in the 1–3 μM (Eriksen et al., 2003b) or 20–40 μM (Stock et al., 2006) range, depending on the study. Several factors could account for this apparent discrepancy. First, it could be that NSAIDs accumulate within the neuronal membrane where γ-secretase is localized. Alternatively, it is possible that NSAIDs do not act centrally but instead reduce peripheral Aβ42 levels, which results in enhanced efflux of Aβ42 from the brain. This type of peripheral sink mechanism has been postulated to account for the Aβ-lowering effect of anti-Aβ immunotherapy (DeMattos et al., 2002) and statins (Sparks et al., 2002). On the other hand, it is possible that the Aβ-lowering properties of some NSAIDs are indirect due to their anti-inflammatory activity. Indeed, studies in transgenic mice have shown reduced microglial activation after both short-term and long-term treatment with ibuprofen (**1**) (Lim et al., 2000; Yan et al., 2003; Heneka et al., 2005) and after long-term administration of NCX-2216 (**13**) (Jantzen et al., 2002). A decrease in interleukin-1β brain levels after 3 month (-33%) (Lim et al., 2001) and 6 month (-65%) (Lim et al., 2000) ibuprofen (**1**) treatment (375 ppm in the diet) has been described in Tg2576 transgenic mice. Chronic ibuprofen (**1**) has been shown also to decrease the mRNA expression by 28% in transgenic mice cortex of α1-antichymotrypsin, an acute phase protein induced by inflammation and with proamyloidogenic properties (Morihara et al., 2005). Thus, there are evidences supporting the hypothesis that ibuprofen-dependent amyloid reduction is mediated by inhibition of interleukin-1β and its downstream target α1-antichymotrypsin.

The behavioral effects of some of the NSAIDs that lower Aβ42 in vivo (ibuprofen (**1**), sulindac (**5**), (*R*)-flurbiprofen (**9**), HCT 1026 (**12**), and mefenamic acid) were evaluated in different animal models of cognitive impairment (Table 4.2). Transgenic mice overexpressing mutated APP (Lim et al., 2001; Eriksen et al., 2003a), or rats with cognitive deficits induced by cerebral injection of lipopolysaccharide (Hauss-Wegrzyniak et al., 1999) or aggregated Aβ (Richardson et al., 2002; Joo et al., 2006) or rats with age-related deficits (Mesches et al., 2004) were used.

Chronic administration of these NSAIDs improved behavioral performance evaluated with different experimental paradigms (Morris water maze, radial arm water maze, two-lever operant chambers, open field) (Table 4.2).

4.3.5.3 Mechanism of Action of NSAIDs on Aβ

The initial study by Weggen and colleagues (2001) showed that sulindac sulfide (**5**) inhibited Aβ42 in fibroblasts deficient in both COX-1 and COX-2. Another group

TABLE 4.2
Studies on the Positive Behavioral Effects of NSAIDs in Different Animal Models of Cognitive Impairment

Drug	Dose (mg/kg/day)	Administration Route	Treatment Duration	Animal Species	Behavioral Paradigm	Reference
HCT 1026	15	s.c.	2 weeks	LPS-injected F-344 rats	Water maze	Hauss-Wegrzyniak et al. (1999)
Ibuprofen	46	p.o.	3 months	Tg2576 mice	Open field	Lim et al. (2001)
Ibuprofen	80	p.o.	3 months	Aβ-injected Wistar rats	Two-lever test	Richardson et al. (2002)
(R)-flurbiprofen	50	p.o.	3 months	Tg2576 mice	Water maze	Eriksen et al. (2003a)
Sulindac	61	p.o.	2 months	Aged Fischer rats	Radial maze	Mesches et al. (2004)
Mefenamic acid	5	i.p.	3 weeks	Aβ-injected Wistar rats	Water maze	Joo et al. (2006)

showed that R-enantiomers of flurbiprofen and ibuprofen, known to have poor anti-COX activity, reduced Aβ42 production in human embryonic kidney (HEK93) cells stably transfected with human mutated APP (Morihara et al., 2002). These studies clearly indicate that the reduction in Aβ42 by selected NSAIDs is not mediated by inhibition of COX activity.

The potential mediating role of other previously identified cellular targets of NSAIDs was also investigated. Using both pharmacological and genetic means it was demonstrated that alterations in activity of peroxisome proliferator-activated receptor-γ (PPAR-γ) do not affect Aβ42 production from APP-expressing cells or inhibit Aβ42-lowering properties of sulindac (Sagi et al., 2003). Nevertheless, other studies have shown that proinflammatory cytokines stimulate Aβ secretion by activating β-secretase and that ibuprofen (1) and indomethacin (4) may inhibit this effect through their PPAR-γ antagonism (Sastre et al., 2003). Similarly, it has been shown that genetic manipulation of nuclear factor κB (NK-κB) in embryonic fibroblasts do not alter the Aβ42-lowering properties of sulindac (Sagi et al., 2003). Another group confirmed that the Aβ42-lowering effect of (R)-flurbiprofen (9) and (R)-ibuprofen appear to be independent of NF-κB (Morihara et al., 2002). Nevertheless, a recent study indicated that reduction of Aβ pathology in transgenic mice induced by chronic administration of indomethacin (4) was coincidental with a reduction of NF-κB activity (Sung et al., 2004).

Finally, a potential mechanism for the Aβ42-lowering activity of NSAIDs through the inhibition of Rho, a small G protein, has been also proposed (Zhou et al., 2003). Indeed, the administration of Y-27632, a selective inhibitor of Rho-associated kinase (ROCK), preferentially lowers brain Aβ42 levels in a transgenic mouse model of AD. However, another group (Leuchtenberger et al., 2006) showed that both Y-27632 and HA1077, another selective ROCK inhibitor, do not select-ively reduce Aβ42 compared with total Aβ. In addition, ROCK inhibitors did not increase Aβ38 secretion in cell-based assays or reduce Aβ production in γ-secretase in vitro assays. Targeting ROCK by expression of dominant-negative or constitu-tively active ROCK mutants failed to modulate Aβ secretion. Taken together, these results seem to exclude a mechanistic involvement of ROCK in the Aβ42-lowering activity of NSAIDs.

An important clue concerning the real mechanism by which NSAIDs lower Aβ42 secretion comes from studies demonstrating that these compounds are active in cell-free assays using γ-secretase enzyme preparations. Indomethacin (4) and sulindac sulfide (5) and were shown to selectively decrease Aβ42 production in a cell-free assay of γ-secretase activity using membrane preparations of APP-transfected CHO cells (Weggen et al., 2003b). Another group, using membrane fractions of HeLa cells and the C-terminal fragment of APP (C100) as a substrate, showed that sulindac sulfide (5) noncompetitively inhibited γ-secretase activity for Aβ42 generation (Takahashi et al., 2003). Competitive binding studies indicated that NSAIDs compete for a site different from that of prototypical γ-secretase inhibitors (Beher et al., 2004). Interestingly, it was shown that ester and amide modifications of the Aβ42-lowering agents meclofenamic acid and indomethacin (4) converted these compounds into Aβ42-raising agents (Kukar et al., 2005). Using a fluorescence resonance energy transfer method, it has been shown that Aβ42-lowering NSAIDs

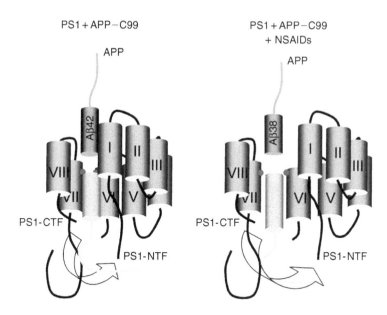

FIGURE 4.10 Model showing how certain NSAIDs may allosterically modulate the cleavage of APP-C99 by presenilin 1. (From Lleo, A., Berezovska, O., Herl, L., Raju, S., Deng, A., Bacskai, B.J., et al., *Nat. Med.*, 10, 1065, 2004. With permission.)

specifically affect the proximity between APP and presenilin 1 both in vitro and in vivo (Lleo et al., 2004) (Figure 4.10). This effect is opposite to that observed in familial AD-linked missense mutations located near the N and C termini, in the mid-region of presenilin 1, and the exon 9 deletion mutation changes the spatial relationship between presenilin 1 N and C termini in a similar way, increasing proximity of the two epitopes (Berezovska et al., 2005). Immunoprecipitation with mass spectrometry has demonstrated that the decrease in Aβ42 secretion induced by sulindac sulfide (**5**) in cultured cells is accompanied by an increase in the Aβ38 isoform (Weggen et al., 2001). Other groups extended this observation to other selected NSAIDs (Ellerbrock et al., 2003). All these studies support a model whereby Aβ42-lowering NSAIDs allosterically modulate presenilin 1 conformation. A model of the molecular mechanism with which NSAIDs allosterically inhibit γ-secretase has been recently proposed by Beher and Graham (2005). According to this model, the APP transmembrane domain assumes α-helical conformation so that the γ40-cleavage site between Val40 and Ile41 results in front of both the γ42-(Ala42-Thr43) and γ38-(Gly38-Val39) cleavage sites (Figure 4.11). A translation of γ-secretase along the same side of the substrate helix would lead to a decrease in Aβ42 production and a concomitant increase in Aβ38 production as observed on incubation of sulindac with purified enzyme (Fraering et al., 2004c). Translation of γ-secretase in the opposite sense may explain the increase in Aβ42 and the decrease Aβ38 observed with some ester amide derivatives of NSAIDs (Kukar et al., 2005).

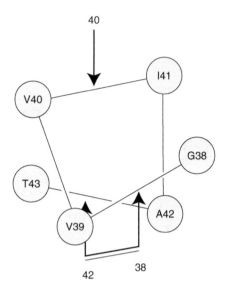

FIGURE 4.11 Model of the molecular mechanism with which certain NSAIDs may allosterically modulate γ-secretase. (From Beher, D. and Graham, S.L., *Expert Opin. Investig. Drugs*, 14, 1385, 2005. With permission.) Amino acids of APP are displayed in the single letter code and numbered according to their position in the Aβ peptide sequence.

4.3.5.4 Effects of NSAIDs on Other γ-Secretase Substrates

Several other substrates of γ-secretase have been identified in addition to APP. Notch-1, ErbB-4 (erythroblastic leukemia viral oncogene homolog 4), low-density LRP, E-cadherin, and CD44 have been demonstrated to be cleaved within their transmembrane domain by the γ-secretase complex. Human APP is cleaved by γ-secretase at different sites. Cleavage at Val40 and Ala42 (γ40 and γ42 sites) generate Aβ40 and Aβ42 peptides, respectively. Cleavage at Leu49 (ε site) generates the release of the 50 amino acid APP intracellular domain (AICD) into the cytoplasm with an important role in the regulation of gene transcription. Notch-1 is cleaved by γ-secretase at Ala1731 (S4 site) and Gly1743 (S3 sites) (Okochi et al., 2002). Cleavage at the S3 site generates the release of the NICD that translocates into the nucleus and regulates transcription of genes involved in cell development and in differentiation of adult self-renewing cells. A further cleavage occurs just outside the transmembrane domain and generates small Notch-1 fragments of 21 (Nβ21) and 25 (Nβ25) amino acids (Okochi et al., 2006). The exact γ-secretase cleavage site within the transmembrane domain of the ErbB-4, a growth factor receptor with tyrosine kinase activity, is currently unknown but the release of the ErbB-4 intracellular domain regulates cell proliferation and differentiation. Altered processing of these substrates by γ-secretase inhibitors could lead to toxic biological effects.

It has been shown that prototypical γ-secretase inhibitors do not discriminate between APP and Notch-1 cleavage sites (Lewis et al., 2003). The inhibitory effects of γ-secretase inhibitors on Notch activation in embryonic and fetal development may not be of concern for the treatment of AD elderly patients. However, it is known that Notch signaling plays an important role in the ongoing differentiation processes of the immune system (Maillard et al., 2003) and of the gastrointestinal tract (Stanger et al., 2005). Indeed, in vitro studies have showed that application of a peptidomimetic γ-secretase inhibitor to fetal thymus organ cultures results in inhibition of T-cell development in a manner consistent with loss or reduction of Notch-1 function (Hadland et al., 2001). An in vivo study in TgCRND8 APP-transgenic mice showed that LY-411,575, a nonpeptidic γ-secretase inhibitor, produces marked effects on lymphocyte development and on the intestine tissue morphology (Wong et al., 2004). By inhibiting Notch processing, LY-411,575 has been also shown to block peripheral inflammatory response associated to T-helper type-1 activation (Minter et al., 2005).

A 5 day treatment of rats with γ-secretase inhibitors with dibenzazepine and benzodiazepine structure causes dose-dependent increases in the size and number of mucosecreting goblet cells (Milano et al., 2004). These effects appear to be mediated by the lack of activation of the Notch-induced gene hairy/enhancer of split-1 (Hes-1), with the consequent increase of the serine protease adipsin (Searfoss et al., 2003).

NSAIDs seem to be more selective than prototypical γ-secretase inhibitors on substrates other than APP. In cell-based assays, sulindac sulfide (**5**) has been shown not altering either ε and S3-cleavage sites (Weggen et al., 2001). Other in vitro studies demonstrated that ibuprofen (**1**) and sulindac sulfide (**5**) preserve release of APP, Notch-1, and ErbB-4 intracellular domains (Weggen et al., 2003a). Although these observations were confirmed by other groups, at high concentrations of NSAIDs an inhibition of the S3-cleavage site of Notch-1 is observed, reminiscent of classical γ-secretase inhibition (Wrigley et al., 2003). Interestingly, indomethacin (**4**) and sulindac (**5**) preferentially decrease in vitro the formation of Nβ25, the Notch peptide that is predominantly formed in cell cultures expressing AD mutant form of presenilin (Okochi et al., 2006), a behavior reminiscent of the effects observed on Aβ42. Thus, it appears that the separation of the inhibitory effects on the various cleavage sites is concentration dependent leaving a "window of modulation." In vivo studies are needed to confirm the lack of biological effects of NSAIDs on Notch-1 processing.

4.3.5.5 New NSAID Derivatives as Aβ42-Lowering Agents

A number of NSAID derivatives are currently being developed for the treatment of AD (Figure 4.9). Unlike old NSAIDs, these new compounds do not retain either anti-COX activity or gastrointestinal toxicity potential in order to permit their chronic use at full doses in AD patients.

The most advanced compound is MPC-7869, the (*R*) enantiomer of flurbiprofen (**9**), presently in Phase 3 clinical development. This compound has poor or no anti-COX activity since this activity resides primarily in the (S)-isomer of NSAIDs. Different studies have extensively documented the Aβ42-lowering properties of (*R*)-flurbiprofen (**9**) in different cultured cell systems (Morihara et al., 2002; Eriksen et al., 2003b) as well as in cell-free assays (Eriksen et al., 2003b). Although a study on the effects of a 3 day treatment with (*R*)-flurbiprofen (**9**) on brain Aβ42 levels in

APP-transgenic mouse was published (Eriksen et al., 2003b), the effects of prolonged chronic treatment with (R)-flurbiprofen are not yet described. This is an
important piece of information since the Aβ42-lowering potency of (R)-flurbiprofen
(9) is quite low (IC$_{50}$ ≅ 300 μM). Nevertheless, a study on the behavioral effects of a
3 month treatment with (R)-flurbiprofen (9) in APP-transgenic mouse is also available but not fully published (Eriksen et al., 2003a). A 1 year, placebo-controlled
study in 207 AD patients employing two dose regimens of (R)-flurbiprofen (9) (400
and 800 mg twice a day) indicated that the drug is well tolerated and suggested that
the high dose regimen may significantly slow the cognitive and functional decay of
mild affected patients (Wilcock et al., 2005). However, in vivo racemization might
pose an issue on chronic administration of (R)-flurbiprofen (9) (Menzel-Soglowek
et al., 1992).

A conceptually similar strategy aimed to abolish the anti-COX activity from
NSAIDs was the synthesis of geminal dimethyl derivatives that inherently lack the
asymmetry in the α position of the carboxylic acid (Stock et al., 2006). Indeed, the
geminal dimethyl analog of flurbiprofen (10) showed reduced anti-COX activity
while maintaining Aβ42 inhibition (Stock et al., 2006). Unfortunately, a 3 day
treatment with this compound at the dose of 84 mg/kg/day produced a marginal
decrease (−18%) of cerebral Aβ42 levels in Tg2576 mice. In addition, the administration of 100 mg/kg/day for 5 days produced significant gastrointestinal damage
probably due to the high plasma levels reached with this dose regimen (430 μM) that
resulted in a significant residual anti-COX activity (Stock et al., 2006). Other NSAID
derivatives, in which the asymmetry α to the carboxylic moiety of flurbiprofen was
removed by substitution with a cycloalkyl group, were recently described by Peretto
et al. (2005). The appropriate substitution pattern at the α position of flurbiprofen
allowed for the complete removal of anti-COX activity, whereas modifications at the
terminal phenyl ring resulted in increased inhibitory potency on Aβ42 secretion. In
rats, some of the compounds appeared to be well absorbed after oral administration
and to penetrate into the central nervous system. Studies in a transgenic mice model
of AD showed that selected compounds (CHF5074 (11), Figure 4.9) significantly
decreased plasma Aβ42 concentrations (Imbimbo et al., 2006) and do not inhibit
NICD-mediated gene expression in cell culture (Moretto et al., 2006).

Another strategy to reduce gastrointestinal toxicity of NSAIDs was the preparation of nitric oxide-donating compounds. HCT 1026 (12) is the 4-(nitroxy)butyl
ester of flurbiprofen. The Phase 1 studies on this compound have been completed
(Scatena, 2004). The rationale underlying HCT 1026 (12) is based on the hypothesis
that nitric oxide release will decrease gastrointestinal toxicity (Perini et al., 2004).
Although the in vitro effects of HCT 1026 (12) on Aβ42 secretion are not described,
an in vivo study has shown that long-term treatment with HCT 1026 (12) significantly reduces Aβ deposition in the brain of mice bearing the mutated genes of APP
and PS1 (van Groen and Kadish, 2005). HCT 1026 (12) was administered as part of
the diet (66 and 200 ppm) for a period of 6 months, starting at 8 months of
age, before plaque formation. Treatment with the higher dose of HCT 1026 (12)
(30 mg/kg/day) produced a significant decrease in Aβ42 deposition in the dorsal
hippocampus. An endoscopic study in healthy volunteers has shown that HCT 1026
(12) has good gastric tolerability with 60%–80% dose dependent reduction of gastric

or duodenal ulcers in comparison with flurbiprofen (**2**) (Fiorucci et al., 2003). A multiple dose pharmacokinetic study in 24 healthy subjects showed that pharmacologically effective drug levels inhibiting prostaglandins were found in the cerebrospinal fluid after repeated oral administration of HCT 1026 (**12**) (75 and 150 mg bid) for 7 days (Tocchetti et al., 2003).

NCX-2216 (**13**), another NO-donating derivative of flurbiprofen, differs from HCT 1026 (**12**) in the presence of an antioxidant moiety, the ferulic acid (Gasparini et al., 2004a, 2005). NCX-2216 (**13**) administered in the diet (375 ppm corresponding to about 63 mg/kg/day) to double transgenic mice (APP + PS1) for 5 months determined a 42% reduction in Aβ load in the cortex (Jantzen et al., 2002). This reduction was associated with a robust microglial activation. In rats, single oral doses of 22 mg/kg of NCX-2216 (**13**) did not cause detectable gastric injury, whereas they significantly inhibited brain prostaglandin E_2 synthesis for up to 48 h (Wallace et al., 2004). In primary microglial cultures, NCX-2216 (**13**) was recently found to activate PPAR-γ (Bernardo et al., 2006) an effect claimed to explain the activity of the compound in the transgenic model of AD.

4.3.6 EFFECTS OF NSAIDS ON NUCLEAR FACTOR-κB

NF-κB is a transcription factor that plays crucial roles in inflammation, immunity, cell proliferation, and apoptosis (Delhalle et al., 2004). NF-κB resides in the cytoplasm as a heterodimer formed by two proteins, p50 and p65, physically linked to an inhibitory molecule, IκBα. Phosphorylation of IκBα by a kinase, IKKβ, causes the dissociation of NF-κB from IκBα and subsequent entrance into the nucleus, where it induces the expression of several immune and inflammatory genes (Viatour et al., 2005). NF-κB is widely expressed in the CNS and is present in both neurons and glial cells, where it plays a physiological role in learning and memory processes mediated by Ca^{2+} activation (Meffert and Baltimore, 2005). NF-κB expression is increased in AD brains compared with controls (Yoshiyama et al., 2001), mainly in association to Aβ diffuse plaques (Ferrer et al., 1998). In vitro studies have shown that Aβ formation activates NF-κB (Samuelsson et al., 2005) and that Aβ toxicity is mediated by activation of the NF-κB pathway (Jang and Surh, 2005). Increased activation of NF-κB is also observed in brains of mice overexpressing mutated human APP (Tg2576 animals) compared with wild-type littermates and this increase is associated with the deposition of Aβ (Sung et al., 2004).

Several NSAIDs can inhibit in vitro the NF-κB activation (Tegeder et al., 2001). Sulindac (**5**) inhibits NF-κB activation by decreasing IKKβ activity (Yamamoto et al., 1999). A prevention of NF-κB binding to DNA by NSAIDs has been also advocated to explain their inhibitory effect on NF-κB activation (Straus et al., 2000). Tg2576 transgenic animals treated with indomethacin (**4**) showed a reduced activation of NF-κB and decreased reactive astrocytosis in the same brain regions where the anti-amyloidogenic effect was observed (Sung et al., 2004). These observations support the hypothesis that the beneficial effect of NSAIDs in AD could be mediated by the suppression of the activation of NF-κB (Townsend and Pratico, 2005).

4.3.7 EFFECTS OF NSAIDs ON PEROXISOME PROLIFERATOR-ACTIVATED RECEPTOR-γ

Another mechanism that has been implicated in the anti-inflammatory action of NSAIDs is activation of the PPAR-γ, a member of the nuclear hormone receptor super family. PPAR-γ has classically been characterized for its implications in adipocyte differentiation and lipid and glucose metabolism, based on activation by endogenously secreted prostaglandins and fatty acids (Kota et al., 2005). Recently, PPAR-γ has been implicated in the pathophysiology of inflammatory and immune responses through its inhibitory action of the expression of inflammatory cytokines by monocytes (Jiang et al., 1998) and macrophages (Ricote et al., 1998). In the CNS, PPAR-γ activation regulates the inflammatory response of microglia (Kielian and Drew, 2003). In cytosolic homogenates from temporal cortex of AD patients, PPAR-γ levels appeared to be increased compared with age-matched controls (Kitamura et al., 1999). Pharmacological activation of PPAR-γ by synthetic agonists of the thiazolidinedione family (troglitazone, rosiglitazone, etc.) protects rat hippocampal neurons against Aβ-induced neurodegeneration (Inestrosa et al., 2005) and Aβ-mediated impairment of long-term potentiation (Costello et al., 2005), a possible mechanism being the stimulation of Aβ clearance (Camacho et al., 2004).

In vitro studies have shown that some NSAIDs, including ibuprofen (1) and indomethacin (4), bind and activate PPAR-γ at concentrations around 10–100 μM (Jaradat et al., 2001; Kojo et al., 2003) and, as a consequence, reduce cytokine production from monocytes (Jiang et al., 1998). HCT 1026 (12) (Bernardo et al., 2005), a nitroso-donating derivative of flurbiprofen, activated PPAR-γ in rat microglial cells at lower concentrations (1 μM). Another nitroso-derivative of flurbiprofen, NCX-2216, at 1 μM induced a long-lasting activation of PPAR-γ in rat primary microglial cultures (Bernardo et al., 2006). This effect was accompanied by an inhibition of LPS-induced release of TNF-α and IL-1β.

Both ibuprofen (1) and indomethacin (4) inhibited proinflammatory response and prevented neuronal death induced by Aβ-stimulated monocytes (Combs et al., 2000). In neuroblastoma cells, ibuprofen (1) and indomethacin (4) reversed the increase in Aβ production induced by proinflammatory cytokines (Sastre et al., 2003). A short-term (7 days) oral treatment of transgenic mice overexpressing human mutated APP (APPV717I) with ibuprofen (62.5 mg/kg/day) resulted in a reduction in the number of activated microglia and Ab42-positive amyloid deposits in the hippocampus (Heneka et al., 2005). Thus, it is possible that some of the beneficial effects from NSAID usage seen in AD could be ascribed to their action on PPAR-g and subsequent reduction of brain inflammation.

4.3.8 CONCLUSIONS

Over the last years considerable efforts have been undertaken to answer the crucial question of whether brain inflammation is a cause or a consequence of neuronal death in AD. Despite epidemiological evidences and an apparently robust biochemical rationale, all large, long-term, prospective, randomized, placebo-controlled studies aimed at reducing inflammation in the brain of AD patients have produced negative results. The failures include both traditional unselective NSAIDs and selective COX-2 inhibitors. These negative results apparently cannot be ascribed to the low doses used since negative results were obtained with potent anti-inflammatory drug like prednisone. It has also been argued that anti-inflammatory therapy cannot protect patients when dementia is fully established. However, a recent large trial in patients with MCI indicated that rofecoxib (**8**) could even accelerate the conversion to AD. Unfortunately, the large primary prevention trial with naproxen (**3**) and celecoxib (**7**) (ADAPT study) was interrupted because of emerging cerebrovascular safety concerns. The suspension of the ADAPT study represents a major setback in the efforts to test the inflammatory hypothesis of AD. However, it is likely that it would not have been very informative because it employed NSAIDs that do not lower Aβ42 production.

The discovery that some NSAIDs selectively decrease Aβ42 by allosterically inhibiting the γ-secretase complex sheds light on the apparent discrepancy between epidemiological studies and negative trials. The inhibition of Aβ42 production is independent of COX activity and depends on the chemical structure of the NSAIDs, some compounds being active (ibuprofen (**1**), flurbiprofen (**2**), indomethacin (**4**), sulindac sulfide (**5**) and diclofenac (**6**)) and others not (naproxen (**3**), aspirin, meloxicam, celecoxib (**7**)). Thus, it is not surprising that the results of the AD trial with naproxen (**3**), dapsone, prednisone, celecoxib (**7**), and rofecoxib (**8**) were negative, since none of these compounds lower Aβ42 production. Interestingly, the encouraging results observed in small AD trials with indomethacin (**4**) and diclofenac (**6**) are in agreement with their inhibitory activity on Aβ42 production. Even more interestingly, the pejorative results obtained with rofecoxib (**8**) in both AD and MCI agrees well with its ability to stimulate Aβ42 production. Thus, although previous negative trials suggest that the anti-inflammatory therapy in AD is over, clinical trials with specific NSAIDs active on β-amyloid production are still strongly warranted. Interestingly, NSAIDs target γ-secretase complex at a site different from that affected by prototypical inhibitors and responsible for the cleavage of other important substrates involved in cell development and hematopoiesis. This represents an opportunity for the development of safe Aβ42-lowering agents compared with classical γ-secretase inhibitors. However, significant gastrointestinal and renal toxicity associated with long-term COX-1 inhibition limit the clinical utility of current NSAIDS as Aβ42-lowering agents. Because the Aβ42-lowering effects appear independent from anti-COX activity, compounds with optimized Aβ42 reduction and little to no inhibition of COX activity have been identified. Such agents would represent a new generation of "anti-amyloid" drugs that selectively target production of the highly amyloidogenic Aβ42 species without inhibiting either COX activity or the vital physiological functions of γ-secretase. NSAIDs could provide a means for a pharmacological reversal of AD because most of the familial AD mutations actually elevate the production of Aβ42. Nevertheless, doubts still persist on the low potency of the present available Aβ42-lowering agents and their low brain penetration. Conversely, the ongoing trials of (R)-flurbiprofen (**9**) in mild AD will tell us if the Aβ42-lowering NSAIDs represent a new therapeutic class.

4.4   3-HYDROXY-3-METHYLGLUTARYL COENZYME A REDUCTASE INHIBITORS

Dario Cattaneo

3-Hydroxy-3-methylglutaryl coenzyme A reductase inhibitors, also referred as statins, are the most used cholesterol-lowering agents worldwide. Observations from epidemiological studies have suggested that these agents may be beneficial also for the treatment of AD, a chronic neurodegenerative disorder characterized by cognitive decline, neuropsychiatric symptoms, and diffuse structural abnormalities in the brain. Indeed, some but not all studies have shown that statin therapy—either by reducing cholesterol or by cholesterol-independent pleiotropic effects—was associated with a significant reduction in the risk of AD.

This chapter will provide the reader with perspective on the potential mechanisms by which cholesterol might affect the pathophysiology of AD and describing preliminary experiences with the use of HMG-CoA reductase inhibitors for the prevention or treatment of this disease.

4.4.1   INTRODUCTION

AD is a chronic neurodegenerative disorder that is manifested by cognitive decline, neuropsychiatric symptoms, and diffuse structural abnormalities in the brain. AD is characterized pathologically by deposition of extracellular β-amyloid and accumulation of NFTs. Therefore, strategies to delay the onset or biologic progression of AD largely have targeted the plaques formed by the deposition of β-amyloid. As a novel finding, increasing evidence indicates that important link exists between cholesterol and AD; notably, hypercholesterolemia occurs more often in AD patients than expected by chance. This evidence has provided the rationale for testing the potential beneficial effects of statins—also referred as 3-hydroxy-3-methylglutaryl coenzyme A (HMG-CoA) reductase inhibitors and actually considered the first-line treatment options for hypercholesterolemia—in reducing progression or preventing AD.

This chapter will summarize the actual knowledge on HMG-CoA reductase structure, physiologic function and its role in regulating cholesterol synthesis, to provide the reader with perspective on the potential mechanisms by which cholesterol might affect the pathophysiology of AD and describing preliminary experiences with the use of HMG-CoA reductase inhibitors for the prevention or treatment of this disease.

4.4.2   ENZYME STRUCTURE

The enzyme HMG-CoA reductase catalyzes the four-electron reduction of HMG-CoA to coenzyme A (CoA) and mevalonate (Figure 4.12), the precursor of cholesterol and isoprenoids (Istvan and Deisenhofer, 2000).

HMG-CoA reductase is a membrane-bound enzyme of the endoplasmic reticulum and is one of the most highly regulated enzymes in nature (Goldstein and Brown, 1990). Human HMG-CoA reductase consists of a single polypeptide chain of 888 amino acids with eight transmembrane domains that anchor the protein to the

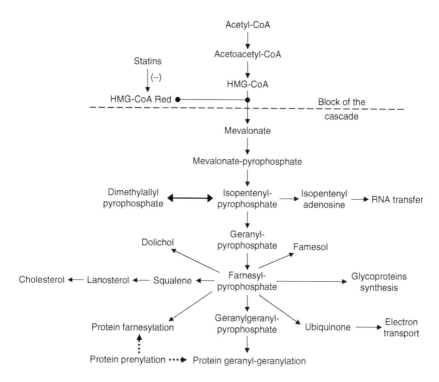

FIGURE 4.12 The mevalonate pathway. Statins, by inhibiting the conversion of 3-hydroxy-3-methyl-glutaryl coenzyme A to mevalonic acid, block not only the synthesis of cholesterol but also the production of several mediators, and the activation of prenylated proteins.

endoplasmic reticulum, and a cytosolic domain that contains the catalytic active site (Istvan et al., 2000). The amino-terminal 339 residues are membrane bound and reside in the endoplasmic reticulum membrane, whereas the catalytic activity of the protein resides in its cytoplasmic, soluble C-terminal portion (residues 460–888). A linker region (residues 340–459) connects the two portions of the protein. A sequence comparison of the human HMG-CoA reductase sequence with those from other organisms reveals two distinct classes: eukaryotic (class I) and prokaryotic (class II) HMG-CoA reductases. The amino acid sequences of catalytic portions are well conserved within each class. At variance, there are important differences, with sequence identities of less than 20%, between class I and class II (Goldstein and Brown, 1990).

The structure of the catalytic portion of human HMG-CoA reductase has recently been described (Istvan et al., 2000) and it has been shown that this enzyme forms tetramers, where the individual monomers wind around each other in an intricate fashion. The monomers are arranged in two dimers (called 1 and 2), each of which has two active sites. Residues from both monomers (called α and β) form the active sites. The HMG-CoA reductase monomer reveals a unique structure comprising three domains: an N-terminal helical domain (*N*-domain), a large domain (*L*-domain)—whose architecture resembles a prism with a 27 residue α-helix forming the central structural element—and a small domain (S-domain) that is inserted into the L-domain. The HMG-CoA binding site is located in the L-domain, whereas NADPH (the reduced form of nicotinamide adenine dinucleotide phosphate, where NADP$^+$ is the oxidized form) binds predominantly to the S-domain. S- and L-domains are connected by a loop that is stabilized by interactions with residues from neighboring monomer. This loop (residues 682–694) is essential in the formation of the HMG binding site.

At variance with data on the catalytic domain, available information on the membrane HMG-CoA reductase domain are scanty. It is known that the membrane domains of mammalian HMG-CoA reductase contain a 167 residue segment, termed the sterol sensing domain, which shares approximately 25% sequence identity with membrane regions of other proteins that are influenced by cholesterol (Brown and Goldstein, 1999). This domain is responsible for the enhanced degradation of HMG-CoA reductase in response to increased concentrations of oxysterols (Gil et al., 1985). Membrane sterol concentrations are not the only means to regulate the rate of HMG-CoA reductase protein degradation. Experiments in which monomeric soluble proteins were fused to the HMG-CoA reductase membrane domains illustrated that the protein is degraded faster when it is smaller than tetrameric (Lawrence et al., 1995). Therefore, it has been proposed that the soluble domains may initiate the tetramerization of the membrane domains, and that dissociation of the membrane domains increases the accessibility of HMG-CoA reductase for the protease cleavage, resulting in the inactivation of the enzyme.

4.4.3 PHYSIOLOGICAL FUNCTION

HMG-CoA reductase is generally referred as the enzyme catalyzing the rate-limiting step—the conversion from HMG-CoA to mevalonate—in the synthesis of cholesterol, a critical membrane lipid, precursor of bile acids and steroid hormones, as well as component of hedgehog protein, a signaling molecule involved in embryogenesis (Ness et al., 2000). Nevertheless, the catalysis of mevalonate formation influences also the production of several downstream intermediate products with important metabolic function (Figure 4.12). For instance, isoprenoids including farnesyl phosphates are important precursor for molecules involved in cell signaling, proliferation, and inflammatory response (Arnaud and Mach, 2005). Another molecule, ubiquinone (referred to as coenzyme Q10), an electron carrier essential for mitochondrial respiratory function, is synthesized from the intermediary metabolite of the cholesterol biosynthesis pathway (Nawarskas, 2005).

Although HMG-CoA reductase is found in virtually all tissues, the liver expresses one of the highest levels of this enzyme. HMG-CoA reductase expression or activity is mainly regulated, via a feedback mechanism, by the liver in response to dietary intake of cholesterol and to maintain the concentrations of mevalonate-derived products (Ness and Chambers, 2000). It has been suggested that HMG-CoA reductase may be regulated at multiple levels, including transcription, translation, protein stability or degradation, and phosphorylation or dephosphorylation (Goldstein and Brown, 1990; Ness and Chambers, 2000). The detailed mechanism of transcriptional regulation of HMG-CoA reductase has been studied extensively in recent years (Brown and

Goldstein, 1997), and the central role of sterol response element binding proteins (SREBP) has emerged from these studies. Overexpression of the active form of SREBP markedly increases hepatic mRNA levels of HMG-CoA reductase and several other enzymes involved in cholesterol and fatty acid biosynthesis (Shimano et al., 1996). In addition, it has been also shown that the feedback regulation of hepatic HMG-CoA reductase involves considerable posttranscriptional regulation, such as posttranscriptional regulatory sites in translation (Ness and Chambers, 2000). Transcription and translation of HMG-CoA reductase gene are low when products of the mevalonate pathway are abundant. The level of HMG-CoA reductase protein could also be determined in part by changes in its rate of degradation. This enzyme has a fairly short half-life, about 2.5 h (Keller et al., 1996). Therefore, modulation of this half-life mediated by exogenous cholesterol intake, endogenous sterol production, or pharmacological treatment markedly and rapidly changes the level of reductase protein (Ness and Chambers, 2000). As mentioned earlier, the catalytic activity efficiency of HMG-CoA reductase might be regulated also by phosphorylation or dephosphorylation. Indeed, it has been shown that the enzyme could be phosphorylated and inactivated by an AMP-activated protein-kinase, with serine 871 being identified as the phosphorylation site (Omkumar et al., 1994). All these regulatory mechanisms are effective in the short term. Long term control of HMG-CoA reductase activity is exerted primarily through control over the synthesis and degradation of the enzyme. When levels of cholesterol are high, the degree of expression of the HMG-CoA gene is reduced. Conversely, low levels of cholesterol activate expression of the gene. Of course, the full regulation of this enzyme cannot be considered alone but in the context of the mevalonate pathway.

4.4.4 ROLE OF THE MEVALONATE PATHWAY IN THE SYNTHESIS OF CHOLESTEROL

Cholesterol, either introduced with the diet or synthesized, is an extremely important biological molecule that has roles in membrane structure as well as being a precursor for the synthesis of the steroid hormones and bile acids. The synthesis and utilization of cholesterol must be tightly regulated in order to prevent overaccumulation and abnormal deposition within the body.

Less than half of the cholesterol in the body derives from the de novo biosynthesis, mainly in the liver and intestine. Cholesterol synthesis occurs in the cytoplasm and microsomes from the two-carbon acetate group of acetyl-CoA, and is mediated by sequential reactions, which are usually considered part of the "mevalonate pathway" (Bloch, 1965). Nevertheless, it must be remembered that mevalonate is also the precursor of isoprenoids, which are involved in the biosynthesis of many classes of compounds in addition to cholesterol, such as ubiquinone, dolichol, and prenylated proteins (Arnaud and Mach, 2005).

Cholesterol synthesis begins with the transport of acetyl-CoA units from mitochondrion to the cytosol. These units are converted by a series of reactions that begins with the formation of HMG-CoA. Two moles of acetyl-CoA are condensed in a reversal thiolase reaction, forming acetoacetyl-CoA that was converted to HMG-CoA in presence of a third mole of acetyl-CoA and the enzyme HMG-CoA synthase. HMG-CoA is converted to mevalonate by HMG-CoA reductase (see earlier). Mevalonate is then activated by three successive phosphorylations, yielding 5-pyrophosphate-mevalonate. In addition to activating mevalonate, the phosphorylations maintain its solubility, since otherwise it is insoluble in water. After phosphorylation, an ATP-dependent decarboxylation yields isopentenyl pyrophosphate (IPP), an activated isoprenoid molecule. IPP is in equilibrium with its isomer, dimethyl-allyl-pyrophosphate (DMPP). One molecule of IPP condenses with one molecule of DMPP to generate geranyl-pyrophosphate (GPP), GPP further condensates with another IPP molecule to generate farnesyl-pyrophosphate (FPP). The NADPH-requiring enzyme, squalene synthase catalyzes a 2 step reaction involving a "head-to-head" condensation of two molecules of FPP to form presqualene pyrophosphate, subsequently reduced to squalene. Squalene epoxidase catalyzes the epoxidation of squalene, which is then converted to lanosterol by oxido-squalene cyclase. There are finally, a series of multistep reactions, the order of which may vary, leading to the conversion of lanosterol to cholesterol. This conversion involves the oxidative removal of three methyl groups, reduction of double bonds, and migration of the lanosterol double bond to a new position in cholesterol (Bloch, 1965).

Normally healthy adults synthesize cholesterol at a rate of approximately 1 g/day and consume approximately 0.3 g/day. A relatively constant level of cholesterol in the body (150–200 mg/dL) is maintained primarily by controlling the level of de novo synthesis. The level of cholesterol synthesis is regulated in part by the dietary intake. The cellular supply of cholesterol is maintained at a steady level by three distinct mechanisms: (1) regulation of HMG-CoA reductase activity; (2) regulation of excess intracellular free cholesterol through the activity of acyl-CoA: cholesterol acyl transferase (ACAT); (3) regulation of plasma cholesterol levels via LDL receptor-mediated uptake and HDL-mediated reverse transport (Durrington, 2003). These physiological mechanisms guarantee the homeostasis of cholesterol.

After its synthesis, cholesterol is transported in the plasma predominantly as cholesterol esters associated with lipoproteins. Dietary cholesterol is transported from the small intestine to the liver within chylomicrons. Cholesterol is transported in the serum within low density lipoproteins (LDLs). The liver synthesizes very low density lipoproteins (VLDLs) and these are converted to LDLs through the action of endothelial cell-associated lipoprotein lipase. Cholesterol found in plasma membranes can be extracted by high density lipoproteins (HDLs) and esterified by the HDL-associated enzyme lecithin-cholesterol acyl transferase (LCAT). The cholesterol acquired from the peripheral tissues by HDLs can then be transferred to VLDLs and LDLs via the action of cholesteryl ester transfer protein (CETP) which is associated with HDLs. Reverse cholesterol transport allows peripheral cholesterol to be returned to the liver in LDLs. Ultimately it is excreted in the bile as free cholesterol or as bile salts following conversion to bile acids in the liver (Durrington, 2003).

Beside liver and intestine, cholesterol is synthesized also within the central nervous system (CNS). During development neurons synthesize most of the cholesterol needed for their growth and synaptogenesis, whereas mature neurons reduce their endogenous neuronal cholesterol synthesis and become dependent on cholesterol synthesized and secreted by glial cells, primarily astrocytes (Shobab et al., 2005). The transport of cholesterol from peripheral circulation into the brain is normally prevented by an intact blood–brain barrier (BBB), which separates the peripheral and central regulation of cholesterol. As a working hypothesis it has been proposed that excessive synthesis of cholesterol in the brain, or a vascular injury-mediated impairment in the BBB might lead to an accumulation of cholesterol in the CNS, ultimately leading to a wide range of neurological diseases, such as AD (Wolozin, 2004; Shobab et al., 2005).

4.4.5 INVOLVEMENT OF CHOLESTEROL IN ALZHEIMER'S DISEASE

AD is a chronic neurodegenerative disorder that is manifested by cognitive decline, neuropsychiatric symptoms, and diffuse structural abnormalities in the brain which account for more than 60% of all dementia cases in the world. AD is a highly specific system disease that is characterized primarily by the deposition of β-amyloid in the form of neuritic plaques and by accumulation of intracellular NFTs, which are insoluble deposition resulting from altered metabolism of the cytoskeletal tau protein. β-amyloid and tangles are cytotoxic and may lead to cell death. Other important pathologic changes associated with AD include region-specific neuronal death and deficits in neurotransmitters, such as acetylcholine, serotonin, noradrenalin, and glutamate (Carr et al., 1997).

The development of AD may be associated with various abnormalities, with the formation of plaques playing a pivotal role in the induction and progression of the disease. Plaques are observed in all patients with AD and are composed of altered metabolites of the APP, especially the 40 and 42 amino acid β-amyloid. These aggregates are formed by residue peptides released following proteolytic processing of the APP. The amyloidogenic pathway requires that APP be sequentially cleaved by β- and γ-secretases, two membrane associated proteins. β-Secretase cleaves APP close to the membrane to produce β-APPs (secreted), and a 12 kDa C100 transmembrane stub, subsequently cleaved by γ-secretase to produce the Aβ peptide. α-secretase cleaves APP within the Aβ sequence thus preventing its formation, producing the N-terminal α-PPS domain and the membrane-localized C-terminal stub, C83. Under normal conditions APP is cleaved by α-secretase just above the surface of the neuronal membrane, releasing soluble APP (Figure 4.13). In patients with AD, the APP molecule is cleaved further distal from the membrane surface by β-secretase and within the neuronal membrane by γ-secretase, releasing β-amyloid that can become deposited in the center of the amyloid plaque (Casserly and Topol, 2004). However, the mechanisms responsible for the shift from physiologic to pathologic activity of secretases have yet to be fully established.

A number of CVD risk factors have been shown to be associated with AD and dementia (Casserly and Topol, 2004). Furthermore, cardiovascular disease (CVD) and AD occur together more often than would be expected by chance, and neuritic plaques have been observed in patients with myocardial infarction without diagnosed dementia. These observations have led to the hypothesis that CVD and AD may share common or complementary antecedents. Hypercholesterolemia and inflammation have emerged as the dominant mechanism implicated in the development of CVD, and have important interactions (Casserly and Topol, 2004). Superimposed on these basic elements are several factors that seem to affect the disease process largely through effects on cholesterol homeostasis or the inflammatory response to injury.

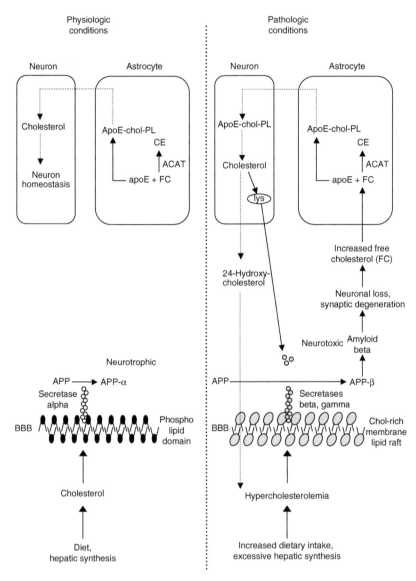

FIGURE 4.13 Transport of cholesterol in the central nervous system. High-cholesterol environment (either by increased dietary intake or excessive hepatic synthesis) favors the activity of β- and γ-secretases (right side: pathologic conditions), which increase the production of amyloid β from APP which, in turn, induces neuronal loss and synaptic degeneration. This pathologic condition leads to increased levels of free cholesterol, which is internalized in the astrocytes. After the internalization, cholesterol can either be stored as cholesterol esters (CE) via the action of acyl-CoA cholesterol acyl transferase (ACAT), or delivered to the neurons in a packaged form with phospholipids (PL) and apoliprotein E, or packaged with lysosomes (lys). Excess of cholesterol can be hydroxylated in the neurons and subsequently eliminated. At variance, in physiologic conditions (left side), the blood–brain barrier (BBB) favors the activity of α-secretase which, in turn, induces the formation of the neutrophic APP-α.

Inflammation has been implicated in AD pathogenesis for over a decade, and increasing evidence suggests that abnormalities in cholesterol homeostasis could have an important role. Similarly, many of the contributory factors in atherogenesis have emerged as potential contributors in AD.

Prompted largely by results of epidemiological studies, the concept of altered cholesterol homeostasis as an important factor in the pathogenesis of AD has emerged. Subsequently, in vitro studies (Casserly and Topol, 2004) have shown that increased or decreased cholesterol levels promote or inhibit the formation of β-amyloid peptide, respectively, demonstrating that cholesterol influences the activity of the enzymes involved in the metabolism of APP and in the production of β-amyloid. Indeed, animal studies have confirmed in-vitro observations, showing that high-cholesterol intake accelerates deposition of β-amyloid in the brain, whereas cholesterol-lowering strategies lower it (Wolozin, 2004; Shobab et al., 2005). The way by which cholesterol affects β-amyloid production and metabolism is not fully understood. Changes in membrane properties induced by cholesterol have been suggested to influence the activity of membrane-bound proteins, such a secretases. The high cholesterol content in the membrane regions where these enzymes are located facilitates the clustering of the β- and γ-secretases with their substrates into an optimum configuration, thereby promoting the undesirable pathogenetic cleavage of APP (Shobab et al., 2005). Although these studies suggest that cellular cholesterol modulates APP processing, other reports indicate that cholesterol esters, rather then free cholesterol, affect the secretase activities such that low cholesterol ester levels decrease β-amyloid formation (Cole and Vassar, 2006).

Increased exogenous cholesterol intake is only one of the several factors involved in promoting AD. As a complementary hypothesis, the homeostatic regulation of cholesterol metabolism may be altered in AD. As a matter of fact, several polymorphisms in genes involved in cholesterol transport and catabolism have been identified as risk factors for AD, the most important being those in the apolipoprotein E (apoE) gene (Bales et al., 2002; Poirier, 2005; Cole and Vassar, 2006). This apolipoprotein, in addition to playing crucial roles in the periphery, where it modulates the production and clearance of cholesterol-rich lipoproteins, can exert some important action also in the CNS.

The brain, together with the liver, is a major site for apoE expression in humans. This apolipoprotein is secreted by glial cells and acts as a ligand for members of the LDL receptors family in the brain. Of note, abnormal neurons containing NFTs in brains of individuals with AD were shown to contain apoE (Poirier, 2005). Polymorphisms within the human apoE gene account for the major apoE isoforms, designated apoE2, apoE3, and apoE4 arising from respective alleles epsilon 2, 3, and 4. In the nervous system, the importance of the polymorphic nature of apoE has recently been revealed, with particular regards to AD (Poirier, 2005). The alleles for apoE represent important genetic risk factors for the most common late-onset forms of AD, with the epsilon-4 and epsilon-2 alleles increasing and decreasing the risk for developing AD, respectively (Bales et al., 2002). Studies utilizing mouse models that mimic the neuropathology of AD have demonstrated an apoE isoform-dependent effect on β-amyloid peptide deposition, fibrillation, and neuritic plaque formation. Taken together, these data support an important role for apoE in the pathogenesis of AD, most likely by altering β-amyloid peptide clearance and metabolism.

Further elucidation of the exact cellular and molecular events mediating apoE isoform-dependent amyloid deposition and increased formation of Aβ by cholesterol could lead to novel therapeutic strategies for preventing or treating AD.

4.4.6 DISCOVERY OF STATINS

Cholesterol biosynthesis is a complex process involving more than 30 enzymes (Shobab et al., 2005). This pathway (Figure 4.12) was a natural target in the search for drugs to reduce plasma cholesterol concentrations, in the hope that these treatments would reduce the risk of cardiovascular disease. In contrast to other intermediates of the mevalonate pathway, HMG-CoA is water soluble and there are alternative metabolic pathways for its breakdown when HMG-CoA reductase is inhibited, so that there is no build-up of potentially toxic precursor. HMG-CoA reductase was, therefore, an attractive target.

The discovery of statins dates to the early 1970s when, in their search for antimicrobial agents, Endo and Kuroda identified a product from the mould *Penicillium citrinium* termed ML-236B, also referred as mevastatin or compactin (Endo et al., 1976). Additional investigations revealed that this compound, despite no useful antimicrobial activity, potently inhibited the rate-limiting enzyme in cholesterol biosynthesis, HMG-CoA reductase. Crystal-structure studies demonstrated that this inhibitory effect occurs by virtue of the ability of statins to occupy a portion of the HMG-CoA reductase binding site, thereby blocking access of this substrate to the enzyme's active site (Istvan, 2001). The affinity of statins for the active site of HMG-CoA reductase is several orders of magnitude higher than HMG-CoA, which leads to effective displacement of the natural substrate and inhibition of endogenous cholesterol synthesis. In addition, the inhibition of endogenous cholesterol synthesis leads to induction of LDL cholesterol receptor on hepatocytes, resulting in enhanced cholesterol clearance from the circulation. This dual mode of action accounts in large part for the potent cholesterol-lowering ability of statins.

Preliminary experimental studies showed that compactin significantly lowered plasma cholesterol levels in the rabbit, monkey, and dog, but not in the rat, due to massive induction of HMG-CoA reductase in rat liver by inhibitors of this enzyme (Tobert, 2003). Subsequently compactin was shown to be highly effective in reducing concentration of total and LDL cholesterol in the plasma of patients with heterozygous familial hypercholesterolemia. A few years later, in 1978, a new, potent inhibitor of HMG-CoA reductase was found in the fermentation broth of *Aspergillus terreus*. It was named lovastatin or mevinolin (Cattaneo et al., 2004). Similarly, two other statins were synthesized in 1982 and 1984, named pravastatin and simvastatin, respectively. All the above-mentioned statins share an HMG-CoA-like moiety, which may be present in an inactive lactone form. In vivo, these prodrugs are enzymatically hydrolyzed to their active hydroxyacid forms.

In the last few years, a number of strategies—both natural and synthetic—have been used to develop more potent statins that possess either a higher affinity for, or more prolonged binding to, HMG-CoA reductase. At present, nearly 20 molecules have been synthesized and tested as potent hypolipidemic agents (Cattaneo et al.,

2004). Some of these have been discontinued for safety reasons or for lack of ability to predict efficacy during registrative trials. However, 6–7 molecules (lovastatin, pravastatin, simvastatin, fluvastatin, atorvastatin, rosuvastatin, and, in the near future, pitavastatin) are now available in the clinical market (Figure 4.14), and constitute one of the most commonly prescribed class of medications worldwide.

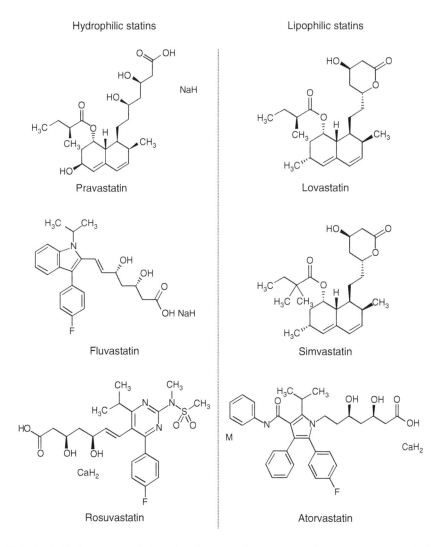

FIGURE 4.14 Protein prenylation. Prenylated proteins are a superfamily of GTPases actively involved in several physiological and pathological mechanisms. These proteins are post-translationally modified by the covalent attachment of mevalonate-derived isoprenoids groups (farnesyl- or geranylgeranly-pyrophosphate). Prenylation is a prerequisite for membrane association and protein activation. Statins, by blocking the production of mevalonate, could be considered as indirect inhibitors of protein prenylation.

4.4.7 CLINICAL USE OF STATINS

Observations from landmark cohort studies highlighted the fact that excess cholesterol is a risk factor for CVD. As such, it was logical to test whether reduction in cholesterol levels by statins could ameliorate the disease development and progression. Before the introduction of statins into clinical practice, the lipid-lowering armamentarium was limited to dietary changes, bile-acid sequestering resins and fibrates, all treatments with limited efficacy or tolerability.

The advent of statins has revolutionized the treatment of different forms of dyslipidemia. Today, these drugs are recommended by the National Educational Panel III (National Educational Cholesterol Panel, 2002) as the first-line agents for virtually all patients eligible for lipid modification. As a class, statins have substantially changed the approach to primary and secondary prevention of cardiovascular events. Indeed, large clinical trials such as 4S, WOSCOP, CARE, LIPID, AFCAPS, ASCOT-LLA, and HPS, involving over 100,000 patients with or without CVD, provided unequivocal evidence that statins reduced cardiovascular mortality and produce a greater absolute benefit in patients with higher baseline risk (Maron et al., 2000; Heart Protection Study, 2002). Recently, a multicenter clinical trial with over 20,000 high-risk patients enrolled (including 6,000 diabetics), has documented that allocation of 40 mg simvastatin daily reduced the rate of myocardial infarction, stroke, and revascularization by about 35%, irrespective of their initial blood cholesterol level (Heart Protection Study, 2002). The reduction in plasma cholesterol is crucial in reducing these events and has led to practice guidelines that emphasize the importance of achieving target goals for LDL levels. However, several of the above-mentioned studies also provided unanticipated results. For example, results of the CARE, LIPIS, and HPS trials suggested that the higher than expected degree of cardiovascular benefit did not correlate fully with the magnitude of cholesterol lowering, suggesting that the beneficial effects of statins did not relate exclusively to their ability to reduce cholesterol biosynthesis, by inhibiting the conversion from HMG-CoA to mevalonate.

The concept of statin benefit beyond lipid lowering is also supported by other observations. Mevalonic acid is the precursor not only of cholesterol, but also of many nonsteroidal isoprenoids compounds (Cattaneo and Remuzzi, 2005). Therefore, the inhibition of HMG-CoA reductase mediated by statins could potentially result in pleiotropic effects. Several proteins posttranslationally modified by the covalent attachment of mevalonate-derived isoprenoids have been identified. These proteins must be prenylated as a prerequisite for membrane association, which is required for their function (Figure 4.15). Members of this family are involved in a number of cell processes including cell signaling, cell differentiation and proliferation, cytoskeleton dynamics, production of inflammatory mediators, endothelial function, and many other biochemical pathways (Cattaneo and Remuzzi, 2005). As a matter of fact, statins can be considered not only hypocholesterolemic agents but also as indirect inhibitors of the prenylation pathways, providing the rationale, for their potential use in a wide range of disease conditions. Therefore, in the last few years, interest has been raised on novel applications and new therapeutic patents for the use of statins, alone or in combinations with other drugs, in several pathological conditions not necessarily associated with lipid abnormalities (Cattaneo et al., 2004).

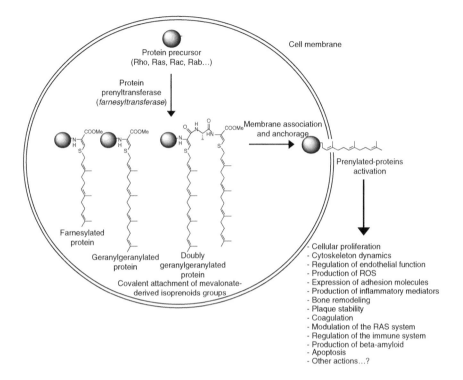

Protein precursor
(Rho, Ras, Rac, Rab…)

Cell membrane

Protein
prenyltransferase
(*farnesyltransferase*)

COOMe   COOMe   COOMe Membrane association
and anchorage

Prenylated-proteins
activation

Farnesylated
protein

Geranylgeranylated   Doubly
protein   geranylgeranylated
protein
Covalent attachment of mevalonate-
derived isoprenoids groups

- Cellular proliferation
- Cytoskeleton dynamics
- Regulation of endothelial function
- Production of ROS
- Expression of adhesion molecules
- Production of inflammatory mediators
- Bone remodeling
- Plaque stability
- Coagulation
- Modulation of the RAS system
- Regulation of the immune system
- Production of beta-amyloid
- Apoptosis
- Other actions…?

FIGURE 4.15 Structural formulas of statins available on the global market. Statins can be fairly divided in hydrophilic and lipophilic molecules. The physicochemical properties can significantly affect drug distribution within the central nervous system (see text).

A large proportion of patients have both hypercholesterolemia and hypertension, two well-known, possibly related, risk factors associated with increased CVD. Recently, preliminary studies have documented an effect of statins on blood pressure (van Dokkum et al., 2002). Blood pressure reduction has been reported in hypertensive patients given statins, independent of whether or not they were already on antihypertensive therapy with angiotensin-converting enzyme (ACE) inhibitors and calcium channel blockers. The hypotensive effect seems largely independent of the cholesterol-lowering properties of statins, and has been related to their impact on endothelial function or on the renin–angiotensin system (RAS) (Cattaneo and Remuzzi, 2005). Nevertheless, clinical evidence of the blood pressure-lowering effect of statins remains ill defined. Of note, in registration trials no beneficial effects of statins on blood pressure have been reported.

Recent in vitro experiments and in vivo animal studies have documented that inhibitors of HMG-CoA reductase increase the gene expression of bone morphogenetic protein-2 (BMP-2), an autocrine-paracrine factor for osteoblast differentiation (Mundy et al., 1999). Since BMP-2 has a strong bone-forming activity and accounts for the majority of the osteoinductive potential of bone extracts, statins could provide an important treatment for osteoporosis through this pathway.

Retrospective analyses from large clinical trials have found a positive relationship between statin use, bone mineral density and subsequent fractures, but other reports found no benefit. Some of these studies, however, are retrospective, the compliance of patients taking statins is unknown, and the dose of statin used varies considerably. Therefore, a definitive conclusion on the potential effect of statins on bone formation should be investigated in prospective clinical trials.

There is substantial evidence that statins may modulate the immune response by affecting recruitment, differentiation, proliferation, and secretory activity of a number of immune cells, including monocytes or macrophages and T cells (Palinski and Tsimikas, 2002). Statins inhibit major histocompatibility complex (MHC) class II expression on human macrophages, endothelial cells, and smooth muscle cells stimulated by interferon γ, downregulating T-cell proliferation and differentiation. In addition, statins have been shown to inhibit the expression of the β 2 integrin, leukocyte function antigen-1 (LFA-1), on the leukocyte cell surface. LFA-1 binds to its counter-receptor, intercellular adhesion molecule 1 on T cells and acts as costimulatory signal for T-cell activation. These findings suggest the possibility that statins may be beneficial for chronic inflammatory processes, autoimmune diseases as well as for organ transplantation.

Several studies have suggested a role for statins in cancer treatment based on preclinical evidence of their anti-proliferative, proapoptotic, anti-invasive properties (Chan et al., 2003). Recently, a case control study in more than 3,000 patients with incident cancer, matched with 20,000 controls, has shown that statins lowered the risk of cancer by 20% during a 4 year follow-up (Graaf et al., 2004). A group of investigators at the University of Michigan (Poynter et al., 2005) have reported that HMG CoA reductase inhibitors given to 267 of 3342 Israel patients resulted in 51% reduction in the risk of colorectal cancer, as compared with those who did not receive statins. These results, although preliminary and not conclusive, suggest that these drugs deserve further investigation in chemoprevention and therapeutic clinical trials.

Recent epidemiological reports suggest that statins could be of value for treatment of neuroinflammatory and neurodegenerative disorders (Wolozin, 2004). The beneficial effects may be mediated by reduction in cholesterol biosynthesis in the brain, lowering levels of apolipoprotein E-containing lipoproteins, or through pleiotropic activities, including reduction in β-amyloid or tumor necrosis factor α (TNFα) production, or modulation of the immune response.

4.4.8 Safety Profile of Statins

Given the growing widespread use of these agents, both as effective hypolipodemic agents as well as for their cholesterol-independent pleiotropic actions, it is extremely important to fully characterize their safety profile. Statins have been both safe and well tolerated in millions of patients over nearly 20 years of clinical use. The most important reported adverse effects are liver and muscle toxicity (Cattaneo et al., 2004). Myopathy associated with statin therapy—defined as muscle pain or weakness associated with creatine kinase levels 10 times higher than the normal upper limit—has been reported to occur in less than 1 over 1000 patients. Actually, one of the main concerns related to statin therapy is that myopathy may eventually progress to fatal or nonfatal rhabdomyolisis. In 2001, cerivastatin was withdrawn from the global market because of a high incidence of fatal rhabdomyolisis, raising doubts regarding the safety of the entire class of drugs. It should be pointed out; however, that a recent review of the FDA's entire spontaneously reported database has shown that the rate of fatal rhabdomyolysis associated with cerivastatin therapy is 16- to 80-fold higher than the rate for any other statins (Staffa et al., 2002). According to press reports, the use of cerivastatin was responsible for 31 fatalities in the United States and a further 21 deaths worldwide, particularly at the highest recommended dose (0.8 mg/day) or when given in combination with gemfibrozil, one of the most common lipid-lowering fibrates. Of note, this drug–drug interaction was implicated in 40% of the fatal events. Among the deaths associated with cerivastatin in the absence of gemfibrozil therapy, the majority (70%) occurred after use of the 0.8 mg daily dose, suggesting a strong relationship with drug dosing. As an additional explanation, cerivastatin shows the highest urinary or fecal excretion rate (40%); therefore it may accumulate in patients with mild to moderate renal insufficiency with significant increased toxicity as compared with the other statins, which have lower urinary excretion.

Hence, it can be reasonably speculated that the incidence of severe myopathy after statin therapy is dose-related, with a trend to increase when the drugs are used in combination with agents that share common metabolic pathways, such as fibrates, niacin, amiodarone, or immunosuppressive agents. As drug–drug interactions can increase plasma statin levels and drug-related adverse effects, avoiding the concomitant use of drugs with potential to inhibit cytochrome-dependent metabolism (i.e., amiodarone) or the elimination phase of statins may decrease the risk of statin-associated myopathy. Alternatively, if drug therapy with a potent cytochrome (CYP) inhibitor is inevitable, choosing a statin without relevant CYP metabolism should be considered.

In the last few years, during routine clinical monitoring in Phase III studies that compared different statins, an enhanced protein excretion rate, based on dipstick testing, was found in some subjects, most frequently with rosuvastatin (Wolfe, 2004). These episodes have been observed only at high rosuvastatin doses (80 mg). At lower doses (10–40 mg), the incidence of proteinuria was comparable with that found with comparator statins (Cattaneo et al., 2004). Although proteinuria was stated to be transient and not associated with worsening of renal function, adequate studies to get more insights to better define the renal effects and safety profile of this novel statin are required.

4.4.9 STATINS FOR THE TREATMENT OF AD: EARLY EXPERIENCE

Statins, in addition to reducing the risk of death or cardiovascular events in patients with or without CVD and being the therapy of choice for the treatment of hypercholesterolemia, might also have a protective effect against the risk of developing AD in humans. In this section, I will summarize actual clinical data supporting this concept.

In the past 5 years different epidemiological studies found an association between statin therapy and a reduced incidence of AD. In 2000, after a cross-sectional analysis of over 55,000 patients, Wolozin et al. (2000) showed, for the first time, a reduction in the occurrence of AD by approximately 67% in patients receiving statins. Similarly, data from the Canadian Study of Health and Aging revealed a 74% reduced risk of AD in statin users and in users of any other lipid-lowering agents compared with those not using lipid-lowering medications (Rockwood et al., 2002). More recently, an observational study in 342 AD patients, 129 of which were given lipid-lowering agents (mainly statins and fibrates), has confirmed that pharmacological correction of dyslipidemia may slow cognitive decline in AD and have a neuroprotective effect (Masse et al., 2005). All together, these analyses suggested that correction of dyslipidemia could reduce the incidence of AD. It was, however, unclear from these observations whether the beneficial effects were simply related to a reduction in the circulating lipid levels, and therefore shared by all lipid-lowering agents or, at variance, the neuroprotective properties were peculiar of statins.

Now, there is some evidence that statins in particular, rather than low cholesterol levels or lipid-lowering agents in general, are associated with the reduction in risk of AD. Indeed, a nested case-control study demonstrated a risk reduction of nearly 70% for dementia in statins users, independent of the presence or absence of untreated hyperlipidemia, or exposure to nonstatin lipid-lowering agents (Jick et al., 2000). Available data, however, did not distinguish between AD and other forms of dementia. In agreement with previous findings, an observational study in postmeno-pausal women with coronary artery diseases (CAD), documented that statins users scored higher on MMSE (Mini-Mental State Examination score) than nonusers of statins, independent of lipid levels (Yaffe et al., 2002). A trend for reduced risk of cognitive impairment in statin users was also observed. These results have led researchers to propose potential novel mechanisms for the possible benefits of statins in reducing the incidence of AD beside the hypocholesterolemic actions. From this point of view, it has been proposed that neuroprotective effects of statins may include upregulation of endothelial nitric oxide synthase, decreased platelet adhe-sion, suppression of inflammation, reduction of apoE levels, and, most importantly, inhibition of the formation of β amyloid (DeKosky, 2005).

Some studies have recently attempted to evaluate the effects of statins on β amyloid peptides in the human population. In one study, 10 mg/day of pravastatin given for at least 3 months in 46 hyperlipidemic patients without AD did not modify the levels of β-peptides (DeKosky, 2005). Similarly, a 12 month treatment with simvastatin did not change the β-peptide pattern in 19 patients characterized by a C-terminally truncated quintet of β-peptides (Hoglund et al., 2005). At variance, Friedhoff et al. (2001) showed that doses of 40 and 60 mg/day of controlled release lovastatin produced a significant dose-dependent decrease in serum β-amyloid peptides. These results suggest that, eventually, higher than conventional doses of statins should be used to obtain significant reduction in β-amyloid peptides.

Despite encouraging results from preliminary observations, other data, derived from observational or randomized, placebo-controlled prospective studies provided controversial conclusions, challenging the value of statins for the treatment of AD. Simons et al. (2002) have performed a randomized double-blind placebo-controlled trial of simvastatin 80 mg/day for 26 weeks in 44 patients with AD. Levels of sterols in the cerebro-spinal fluid (CSF) and serum levels of LDL were significantly decreased in patients on statin compared with placebo. However, these results did not parallel with improvement in the neurocognitive scores. On the same line, two community-based prospective cohort studies aimed at evaluating the effect of statins on dementia and AD failed to document an association between statin use and onset of neurological disease (Li et al., 2004; Zandi et al., 2005). Similarly, retrospective analysis of two large cardiological trials (Heart Protection Study, 2002; Packard et al., 2005) failed to observe a reduction in the incidence of new AD cases in patients given statins compared with those on placebo. It must be pointed out, however, that these studies, though, have been criticized because the primary measures focused on CVD, and the cognitive component was added on posthoc.

As a useful summary of the actual knowledge in this field, a systematic literature review including published human studies, which examined the cognitive effects of statins, has shown that although statins intuitively have appeal for the prevention or treatment of dementia, any final statement about their efficacy should await more definite evidence (Xiong et al., 2005). It can be, therefore, concluded that, at this time, there is insufficient evidence for proposing the use of statins for cognitive improvements in patients with AD, and prospective trials are urgently needed. Of note, preliminary results of a pilot, randomized trial with atorvastatin (80 mg/day) versus placebo which is involving nearly 100 patients with mild-to-moderate AD has shown that statin treatment reduced cholesterol levels and produced a positive signal on each of the neurological clinical outcome measures (Sparks et al., 2005). However, this trial is still ongoing and definite results are, therefore, pending.

4.4.10 PENETRATION OF STATINS IN THE CNS

Despite the common mechanism of action, statins differ in their physicochemical and pharmacokinetic properties. The characterization of their pharmacokinetics is important to understand not only their pharmacological action, but also the adverse effects associated with their long-term treatment. Comparative kinetic data among the clinically available statins have been provided by several authors (Garcia et al., 2003; Schachter, 2005) and are, therefore, beyond the scope of this chapter. We will focus only on difference in the physicochemical properties (Figure 4.14), which are relevant for the penetration of statins into the CNS.

With the exception of lovastatin and simvastatin—which are administered as lactone prodrugs and must be hydrolyzed in vivo to the corresponding β-hydroxy acid to achieve pharmacological activity—all statins are administered as the active β-hydroxy acid form, a fact that can influence their penetration in the CNS. Indeed, it was shown early on that statins in lactone form were transported via simple diffusion, whereas those having an acid form were transported across the BBB via a carrier-mediated transport mechanism for monocarboxylic acids, which is easily saturable (Tsuji et al., 1993). As a consequence the lipophilic statins lovastatin and simvastatin in their lactone form, but not in their acid form, crossed the BBB, whereas the more hydrophilic HMG-CoA reductase inhibitors pravastatin and fluvastatin did not reach the CNS in measurable concentrations (Christians et al., 1998). Subsequently, some differences in the ability to cross the BBB have also been documented for the novel statins. Rosuvastatin, which is hydrophilic, and unexpectedly atorvastatin, which is lipophilic, did not cross the BBB (Caballero and Nahata, 2004). As an additional confounding factor, it has been recently shown that a 24 h wash-out period was associated with low levels of statins in the cerebral cortex of rats chronically treated with these drugs (Johnson-Anuna et al., 2005), suggesting that statins were not accumulating in the brain but were being removed via active systems. Indeed, it has been recently proposed that distribution of statins into the CNS is dependent not only on lipophilicity but also on affinity for P-glycoproteins and multiple transporters, which constitute important efflux mechanisms for lipophilic drugs across the BBB (Kikuchi et al., 2004).

As additional confounding factors, the distribution of statins in the brain depends not only on their properties, but can also be affected by patient age and background. The BBB is a diffusion barrier, which impedes influx of most compounds from blood to brain. Dysfunction of the BBB complicates a number of neurologic diseases, allowing influx of substances potentially toxic for the brain. BBB impairment can be easily diagnosed using the CSF to serum albumin ratio (Miida et al., 1999). This ratio increases slightly after birth but falls to adult values by age 6 months and gradually increases again after age 40, being more than double in the older individual than that in children. Moreover, BBB permeability is likely to increase in AD as a result of hypoxia-ischemia and inflammation (Miida et al., 2005). In mice chronically treated with statins, both hydrophilic and hydrophobic molecules were detected in the brain, where cholesterol levels in the cerebral cortex were reduced and gene expression patterns were significantly altered (Miida et al., 2005). These results suggest that statins pass through the BBB to a greater extent in older AD patients than in young, nondemented individuals, and exert cholesterol-lowering and cholesterol-independent activity.

4.4.11 CONCLUSIONS

Advances in our understanding of how cholesterol and related products modulate APP processing provides potential insights into the mechanism by which abnormalities in brain cholesterol metabolism could affect the pathophysiology of AD. From this viewpoint, statins, by reducing cholesterol levels, regulating APP processing, microglial activation and possibly inhibiting atherosclerosis in the cerebral arteries, might beneficially affect AD. Up to now there is insufficient evidence to show that these drugs reduce the progression of the disease. Although some epidemiological studies have questioned the preventive efficacy of statins in patients with AD, recent evidence has given us some hope for the clinical use of statins in these patients, calling also for prospective, randomized trials.

Two major statin trials are currently under way to assess the effects of statins delaying the progression of AD in patients with serum cholesterol levels that do not require therapeutic intervention. The cholesterol-lowering agent to slow progression (CLASP) of AD study is a double-blind trial that randomized approximately 400 patients with AD to either simvastatin 20 mg/day or placebo for 6 weeks. Patients then received either simvastatin 40 mg/day or placebo for the remainder of the 18 month study. The Lipitor's effect in Alzheimer's dementia (LEADe) study is a double-blind trial that randomized approximately 600 patients with AD to either atorvastatin 80 mg/day or placebo for a period of 72 weeks. If these trials demon-strate that statin therapy confers symptomatic benefit in patients with AD, they will open up the possibility of AD prevention trials with statins.

4.5 Aβ POLYMERIZATION REDUCTION

Harry LeVine III and Corinne E. Augelli-Szafran

Multimers of the AD β-peptide have been postulated as an etiologic agent in disease progression. The process of assembly of the peptide into toxic species presents multiple avenues for attack as well as challenges for discovering and developing therapeutic agents that have the required specificity, efficacy, and safety. This chapter describes the current understanding of the pathways of Aβ assembly in vitro, identifies the species involved, and discusses the relevance of these species to their biological counterparts. The drug discovery strategies being used to success-fully identify inhibitors of Aβ polymerization are described, emphasizing the differ-ences between Aβ processes and the traditional enzyme and receptor drug targets. Strategies for clinical trials of polymerization inhibitors in AD are evolving while new technologies are being developed to determine the dynamics of different Aβ pools in the brain. The evaluation of the relationship of these polymerization inhibitors to clinical outcomes should assist in furthering successful drug discovery for this unmet medical need.

4.5.1 BACKGROUND AND RATIONALE

4.5.1.1 Protein Aggregation in Disease

Protein misfolding is increasingly being recognized as a likely driving force in disease pathology. The term originally described genetic disorders involving mis-folding of mutant unstable proteins in the endoplasmic reticulum such as in cystic fibrosis. Misfolding of otherwise normal proteins apparently triggered by aging or cellular stress is now being considered as the etiologic agent for multiple common sporadic forms of diseases. The most prevalent chronic neurodegenerative diseases, AD and Parkinson's disease (PD), are characterized by extensive deposition of specific proteins in the parenchyma of the brain. AD histopathology includes intra-cellular tau in NFTs and extracellular deposits of Aβ amyloid peptides in the parenchyma and cerebral vasculature.

Another intracellular deposit frequently found in AD subjects is Lewy bodies comprising the synaptic protein α-synuclein (Harrower et al., 2005). Lewy bodies can also occur in the absence of Aβ, but present a distinct clinical picture. PD is defined pathologically by α-synuclein inclusions (Eriksen et al., 2005; McNaught and Olanow, 2006). There is a controversy over whether the histologically visible deposits of protein in these diseases or the recently described soluble oligomeric forms of the same proteins are the proximal etiologic agent. This issue impacts the targets chosen for inhibition because, at least for the Aβ peptide, the pathways to formation of soluble oligomers and insoluble fibrils may be different. Other proteins may also differ in the relative contribution of soluble and insoluble material to disease.

Cellular toxicity studies suggest that both soluble and insoluble Aβ cause biological effects and hence, both forms of Aβ may be legitimate targets. Aβ oligomers in slices and in animal behavioral models seem to potently disturb fundamental synaptic mechanisms of memory formation such as long-term potentiation that could plausibly account for early clinical features of AD. Later stages of the disease could be exacerbated by amyloid fibril deposition, which engenders glial and immune responses as well as cellular stress. These processes further disrupt protective cellular mechanisms of dealing with protein misfolding, leading to neuronal dysfunction and death. Physical disruption of tissue organization by the masses of fibrillar and nonfibrillar Aβ deposits also disrupts neuronal connectivity in a manner analogous to the classical amyloidoses (Pepys, 2006).

4.5.1.2 Distinction of Intrinsically Disordered Proteins and Misfolding Proteins

The central dogma in protein chemistry is that the thermodynamics of primary amino acid sequences govern the folding of polypeptide chains. Since it was first articulated by Nobel laureate Christian Anfinsen in 1969, this principle has guided the prediction of the formation of protein structure. Nucleation of structure formation by local hydrophobic collapse has been central to the folding of both native and nonnative conformations, of which the latter often forms unstructured aggregates. Proteins destabilized by mutation or proteolysis can cause disease by a variety of mechanisms that involve loss of function and gain of (toxic) function. The misfolding process in these proteins follows a pathway from a destabilized native state. Under some conditions, the nonnative structures assemble into the regular and stable cross-β-sheet configuration of amyloid fibrils that are pathognomonic for the 25 WHO-recognized amyloid diseases (Sunde et al., 1997; Makin and Serpell, 2005). Under strongly destabilizing conditions, virtually all polypeptide chains have the capacity to form amyloid fibrils (Chiti et al., 1999; Calamai et al., 2005).

A number of diseased-related misfolding proteins have been recognized as being natively unstructured or unfolded (Gunasekaran et al., 2004; Fink, 2005). They are a subset of a growing number of proteins, whose tertiary structures appear to be disordered when assessed by circular dichroism, but do not normally misfold or form aggregates. Natively unstructured proteins exist as a mixture of disordered or rapidly interconverting loose protein conformations. Protein structure, like art, is in the eye of the beholder. Techniques such as nuclear magnetic resonance and fluorescence that are more sensitive than circular dichroism to population mixtures or to local transient structure formation in flexible chains can describe ensembles of conformations that interconvert on various timescales (Bernado et al., 2005).

There is purpose and function in the biological selection of these polar disordered sequences which may represent up to 30% of eukaryotic sequences (Fink, 2005). Upon encountering the right binding partner, these native disordered sequences fold into stable structures, usually devoting a significantly larger contact area to the interaction than two interacting proteins with stable folds. These disordered regions comprise an unusually high proportion of prolines and polar and charged amino acids, with relatively few hydrophobic amino acids dispersed throughout the sequence. The structures avoid nucleation via hydrophobic collapse by essentially diluting out the hydrophobic residues with polar residues that prefer to remain hydrated. They evade the cellular chaperone and proteasome defenses in eukaryotic cells that recognize regular unfolded proteins by the same mechanism. Long stretches of unfolded sequences in proteins are rarely (8%–12%) found in prokaryotic proteins because the major pathways of degradation do not recognize the same features as eukaryotes and the unfolded sequences are proteolyzed.

The formation of stable folded structures from exposing extended lengths of unfolded polypeptide in the absence of a binding partner does occur. This process is used by yeast in the formation of prions that encode a cytoplasmically heritable trait within an alternate protein conformation. For the yeast protein sup35, a translation terminator protein, it is a way to generate phenotypic diversity without genetic commitment. It has been suggested that long-term nongenetic changes in learning and memory in mammals promulgate through a stable prion-like conformational state in a neuronal member of the CPEB (cytoplasmic polyadenylation binding protein) family (Si et al., 2003).

Pathologic events can also occur by interactions between natively unfolded sequences. For reasons that are not completely clear, but may be related to particular vulnerabilities, these events seem to concentrate in the sporadically occurring forms of a number of chronic neurodegenerative diseases. These sequences include proteins with significant natively unfolded structures such as α-synuclein and tau. This group also includes the Alzheimer's β-peptide, which has both an unstructured N terminus and part of a hydrophobic transmembrane domain that is critical to its assembly and biological properties. Familial diseases precipitated by mutations in these proteins result in their accumulation in deposits accompanied by early onset neurodegenerative disease. Interestingly, the sporadic forms of PD and AD that involve the normal sequence proteins are more prevalent (~90% of cases), but are late-onset. This suggests that the process of aging, or sufficient time in nondividing cell types, lead to aberrant folding followed by stable amyloid structures. We will now focus on the progress and opportunities for therapeutic intervention in the assembly of Aβ into pathologically relevant forms.

4.5.2 Pathways of Aβ Misfolding

The elucidation of metabolic pathways and the enzymes that catalyzed the reactions have provided rationale, tools, and targets for pharmaceutical development. A similar detailed understanding of protein misfolding is essential for rational approaches that are likely to succeed in preventing, stopping, and curing protein misfolding diseases. The Aβ peptide that forms the conspicuous deposits that define AD attracted early attention because of the prevalence of AD and the relatively small size of the peptide (40–42 amino acids) that is amenable to solid phase chemical synthesis. Many of the principles discovered from working with this system, including organizational structure of the multimeric systems and biological toxicity, are applicable to the study of other misfolding protein systems.

4.5.2.1   Nomenclature—Amyloid-Speak

The vast majority of the Aβ peptide in the AD brain is (1–40) and (1–42). Modified (oxidation, glycation, isomerization, etc.) and truncated (x-40, 42; 1–38 etc.; pyroglu3–40, 42, etc.) Aβ species will not be discussed here because not enough information is available to draw conclusions. Under some circumstances, these modified species could be important.

Mutant forms of (1–40) and (1–42) Aβ, which are responsible for several types of familial amyloid angiopathy (E22Q, E22K, D23N) and an early-onset AD with atypical plaque morphology (E22G) have been instructive in elucidating the sequence dependence of Aβ aggregation. These substitutions have drastic effects on assembly and aggregation of the peptide. The affinity of the vasculotrophic peptides for the cerebral vasculature (Herzig et al., 2006) is unexplained. However, these rare forms present limited therapeutic targets.

Aβ assembles into a multitude of structures that have acquired names that are often used interchangeably. This has produced a great deal of confusion and rendered much work in the field uninterpretable. Table 4.3 illustrates a systematic utilitarian definition of the multimeric forms of Aβ. The generic term Aβ is used throughout this discussion when the short (40 residue) or long (42 residue) composition is not known. Useful criteria for defining a protein species include morphology, sedimentation characteristics (size and shape), biochemical or biophysical properties, reactivity with dyes commonly used to detect amyloid (Congo Red, Thioflavin T/S), and recently reactivity with oligomer-specific (Lambert et al., 2001; Kayed et al., 2003) or fibril-specific (O'Nuallain and Wetzel, 2002; Miller et al., 2003) antibodies. All oligomers of amyloidogenic proteins possess similar structural and immunologic characteristics despite differences in their primary amino acid sequence. The fibrils of all amyloidogenic proteins also share common characteristics that are distinct from those of the oligomers. For fibrils, this was a little less peculiar because all amyloid fibrils share similar X-ray fiber diffraction patterns and they are highly similar in appearance by electron microscopy and atomic force microscopy (Makin and Serpell, 2005).

These definitions and the mechanism by which the key intermediates are generated are vital information when considering a therapeutic target. The similarities among the assembled forms of misfolding proteins suggest that a common strategy could evolve if conformation-dependent rather than sequence-specific interventions are targeted. A potential downside is that a pan-therapeutic intervention against pathological assemblies may uncover important positive biological effects of the structures, or that similar protein folds are critical for the activity of some other proteins of which we are currently unaware. Given the generic toxicity of oligomers and fibrils, regardless of their peptide composition, the benefits of inhibiting assembly should outweigh possible side effects. Effects of compounds acting by a variety of mechanisms should help sort this out.

4.5.2.2   Pools of Aβ Peptide in the Brain

There is considerable genetic, biochemical, cellular biological, and animal model data supporting a pivotal role for the Aβ peptide in the etiology of AD, but the

TABLE 4.3
Properties of Aβ Assemblies

	Oligomer	Protofibrils	Fibrils	Amorphous Aggregates	Monomer
Sedimentability	>100 kg	15–100 kg	<15 kg	<15 kg	>100 kg
Proteolytic sensitivity	R	R	R	S	S
Structure	β-Sheet	β-Sheet	Parallel and antiparallel β-sheet	Mixture	Unstructured
Morphology	Globular 3–5 nm	Curvilinear, beaded	Twisted strands 8–10 nm × 1000	Amorphous varable	1–2 nm
Conformation-specific antibody	Linear epitope and nonsequence specific	Oligomer-like linear epitope	Linear epitope and nonsequence specific	Linear epitope (like monomer)	Linear epitope
Biological effects	Toxic	Toxic	Toxic	Nontoxic	Nontoxic
Thioflavine T fluorescence	–; +/–	+/–	+	–	–
Congo Red birefringence	–	–	+	–	–

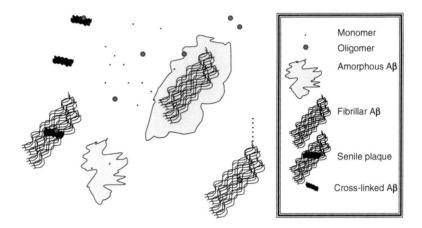

FIGURE 4.16 Forms of Aβ peptide in AD Brain.

connection is still not clear from the classical histological analyses. Recently, instead of considering Aβ as a monolithic agent, more detailed analysis of different forms of Aβ, such as temporal occurrence, localization, and biological effects are improving correlations with neuronal dysfunction or death and the clinical observations.

Figure 4.16 depicts pools of Aβ present in the AD brain. The Aβ peptide assembles into a variety of structures depending on conditions in the local environment. This polymorphism is controlled by physical chemical parameters such as pH, ionic strength, temperature, mechanical agitation, and interactions with membrane surfaces and particular proteins. Aβ pathology in AD has traditionally been defined by histologic methods that detect accumulations of insoluble structures or structures stabilized by chemical fixation. The in-life situation, however, is more complex than what is indicated by the histology. In vivo, the structures are subject to enzymatic processing (proteolytic, tranglutaminase), cellular remodeling (microglia, astrocytes), chemical modification (glycation, metabolite conjugation) (Reddy et al., 2002; Bieschke et al., 2005), and chemical isomerization of aspartyl residues with extended lifetimes (decades) (Roher et al., 1993; Shimizu et al., 2005). Radiochemical dating of plaques following isotopes incorporated during the years of nuclear testing in six cases of autopsy-diagnosed AD suggested that in some individuals, plaques predate development of clinical symptoms by up to 9 years (Lovell et al., 2002).

There are multiple species representing a gradient of increasing insolubility in the brain. The relevance of the individual species for disease progression remains to be established. Current dogma states that soluble species manifest potent cytotoxicity and that insoluble forms may be less toxic, but there is no consensus on how to define or quantify these species. A crude system of relative solubility based on extractability in solvents of increasing denaturing power PBS - > SDS - > formic acid or guanidine is used to report Aβ content of samples. However, the relationship of these fractions to histological structures or biological relevance has not been established.

Studies in βAPP-transgenic mice indicate that the soluble Aβ pools are in a complex dynamic equilibrium with different insoluble species (Bacskai et al., 2003; McLellan et al., 2003) and clearance or degradation processes (LeVine, 2004a; Tanzi et al., 2004; Bateman et al., 2006). Understanding the interrelationship between Aβ assemblies and monomeric material is important for defining pathology and the disease process that affects neuronal function and survival. The interrelationship among Aβ pools should be taken into account in the design of imaging studies and clinical trials that are used to evaluate the efficacy of new therapeutic regimens. This can have significant practical consequences in the assessment of the effectiveness of a test intervention (Craft et al., 2002; Shoghi-Jadid et al., 2006).

4.5.2.3 Forms of Aβ

A caveat to the following section is that it refers to the assembly of purified synthetic Aβ peptides in vitro under conditions controlled by the investigator. This is the only practical approach for early stage interventions in Aβ aggregation. Little is known about the details of the assembly of Aβ in vivo. The behavior of the peptide in cells, in the transgenic mouse models, and especially in the human brain, aged control and AD, is not well characterized. In vivo, Aβ is present as mixtures of the short and long peptides, in ratios ranging from 10:1 (1–40) or (1–42) to 1:1, depending on the system. Some of the details of this are being worked out (Lesne et al., 2006), but this work is in the early stages.

In vitro, soluble Aβ exists as monomeric peptide (~4 kDa) that fleetingly self-associates to multimers. The stability and structure of these multimers critically depend on whether the peptide is the short (1–40) or long (1–42) form (Bitan et al., 2003c). The biology of the short and long forms of Aβ may significantly differ from one another in ways that have only recently been appreciated. A major difference occurs at the level of the formation of small multimers (<10-mers) which do not appear in isolation in vitro but can be trapped by cross-linking (Vollers et al., 2005). The importance of the Aβ(1–42) peptide in AD has been supported by the fact that all known familial Alzheimer's disease (FAD) mutations increase Aβ(1–42) levels except for the Arctic mutation E22G. This difference may be explained by the observation that the mutant is hyperaggregation-prone and is deposited (Paivio et al., 2004).

Stable oligomers: Larger (~5 nm) spherical multimers (i.e., 100–150 kDa determined by gel permeation chromatography) of synthetic Aβ(1–42) are stable enough to be isolated and are relatively monodisperse (Stine et al., 2003; Kayed et al., 2004; LeVine, 2004b; Watson et al., 2005). Uncrosslinked SDS-stable trimer or tetramer and 50–60 kDa species are observed by SDS-PAGE, but rarely by size exclusion chromatography under native conditions. The relationship of these trimer or tetramers to the native species is undefined, but they may be kinetically stable species (Manning and Colon, 2004) that contain β-sheet and may represent building blocks of the larger forms. Biologically-derived soluble Aβ oligomers are similar in size to the synthetic species under nondenaturing conditions and decompose similarly in the presence of SDS (Walsh and Selkoe, 2004; Lesne et al., 2006). SDS-stable trimers are found intracellularly, whereas both trimers and tetramers occur extracellularly in the brains of young Tg 2576 human sequence βAPP-transgenic mice. As the animals age, nonamer and dodecamer species appear around the time behavioral deficits occur (Lesne et al., 2006; Oddo et al., 2006). The isolated dodecamer species, named Aβ56* from its migration on SDS-PAGE, reproduces the behavioral deficits when infused into the brains of naïve rats (Lesne et al., 2006). This suggests that at least this form of the peptide is capable of inducing physiological effects similar to those that occur in the mice. These effects may be analogous to deficits seen early in the progression of AD. Native soluble oligomers in AD brain are a heterodisperse population of which the most abundant species are 100–250 kDa by nondenaturing size exclusion chromatography. Some of this population may be Aβ associated with other protein complexes.

These oligomeric species are defined as soluble after centrifugation at 100,000g. The physical line dividing soluble species from insoluble species is not sharp. A certain fraction of this material is bound to extracellular matrix (likely in an exchangeable form) or to neuronal dendritic arbors (Lacor et al., 2004). Diffusion of a soluble species drastically increases its range of effectiveness and magnifies the biological consequences. Even a small fraction of toxic soluble oligomers can have biological effects that outweigh their proportion of the total material.

Protofibrils and Fibrils: Increases in polymerization number beyond stable oligomers (~150,000 Da) create sedimentable structures of decreased solubility. Small numbers of larger 100 kG-sedimentable species (>1.5 MDa) with beads-on-a-string morphology called protofibrils exist transiently, and then disappear. These species are succeeded by larger 15 kG-sedimentable twisted stranded fibrils of characteristic regular morphology that are microns in length and give characteristic fiber X-ray scattering patterns that reflect prominent cross-β-sheet secondary structure (Sunde et al., 1997; Makin and Serpell, 2005). Amorphous Aβ aggregates form under some conditions, but do not immediately mature into the quasicrystalline fibril morphology. These aggregates contain a mixture of secondary structure elements. Amorphous and fibrillar Aβ peptide also coexist, but demonstrate different stabilities to solubilization conditions and to proteolysis. The extended time (decades) over which Aβ fibrils can be present in AD (Lovell et al., 2002) leads to isomerization of amino acids in the peptide as well as chemical modifications, such as oxidation, nitration, transglutaminase and dityrosine crosslinking, and glycation. These modifications create a core of material that cannot be solubilized even under highly denaturing conditions. They are not seen in transgenic mouse brain Aβ, presumably because deposits are rarely more than 1 or 2 years of age (Kalback et al., 2002). Table 4.3 summarizes the distinguishing properties of Aβ forms that have been defined.

4.5.2.4 Aβ Folding Pathways—Where to Begin?

Most current molecular models of assembly consider a linear sequential pathway (Figure 4.17) progressing to the fibril, which is the thermodynamically lowest energy species. It is not entirely clear that a sequential model is appropriate for all amyloidogenic proteins, or even for the same protein under different conditions in the

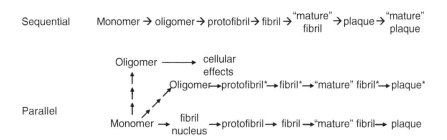

FIGURE 4.17 Aβ assembly pathways.

cellular milieu. The question of sequential versus parallel or branched pathways has practical consequences for therapeutic intervention. In a sequential scheme (Figure 4.17, *top*), blocking soluble oligomer formation should also block fibril formation. Attacking species early in the pathway would be most effective, especially since soluble oligomers are highly toxic. In parallel pathways (Figure 4.17, *bottom*), when oligomers and fibrils follow different routes via different intermediates, blocking oligomer formation may have little effect on fibrils. There is evidence for this dichotomy from atomic force microscopy (Mastrangelo et al., 2006) and scanning transmission electron microscopy (Losic et al., 2006). Furthermore, oligomers prepared from synthetic Aβ(1–42) by different procedures can have distinct biological activities (Deshpande et al., 2006). Depending on environmental conditions, oligomers may not progress to form protofibrils (Barghorn et al., 2005), or if they do form protofibrils and fibrils, they may be structural variants (Petkova et al., 2005) of fibrils produced by a different pathway. Dissimilar structural forms of fibrils may have distinct biological properties.

Although the spotlight has turned to soluble oligomeric Aβ species, fibrillar Aβ deposits do participate in pathology. Direct fibrillar cellular toxicity and physical properties that disrupt tissue structure, such as the weakening of blood vessel walls in cerebrovascular amyloid angiopathy are proposed pathologic mechanisms (Weller et al., 2000; Christie et al., 2001; Miao et al., 2005). The mechanism of fibril extension from preexisting fibrils may differ for different amyloid fibrils. Aβ fibrils and yeast prions extend by monomer addition rather than by addition of soluble oligomers (Tseng et al., 1999; Cannon et al., 2004; Collins et al., 2004). Other proteins may incorporate oligomers.

4.5.2.5   Targeting Individual Reactions in Aβ Assembly

In addition to the complexity of Aβ peptide assembly, the development of optimal screening assays is a challenge as well. Initially, screening assays were designed without the knowledge that multiple reactions were occurring during the Aβ assembly or that soluble oligomeric forms of the peptide were toxic. These issues increased the difficulty in the development of inhibitors of Aβ assembly. It is now recognized from antibody conformational specificity studies (Kayed et al., 2003) that assembly intermediates have distinct structures and hence, may interact differently with

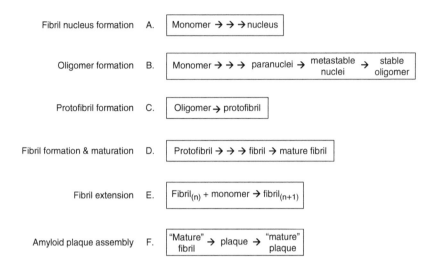

FIGURE 4.18 Aβ assembly reactions.

molecules designed to block further assembly by attaching to interfaces. If the assays are not designed to be rate limiting with respect to a particular target species, it is difficult to extract an informative structure–activity relationship (SAR). Some inhibitors may inhibit more than one reaction, and other inhibitors may inhibit one reaction, but accelerate another. This will be discussed further in Section 4.5.3.

Aβ assembly can be divided into subreactions (Figure 4.18): (1) fibril nucleus formation, (2) oligomer formation, (3) protofibril formation, (4) fibril formation and maturation, (5) fibril extension, and (6) plaque assembly. Plaque assembly cannot be reproduced in vitro because it is inherently a biological process. It is illustrated only for completeness. There may be manipulations of transgenic models that will give this information (Bacskai et al., 2001). Although it is difficult to formulate an assay that exclusively isolates one of these reaction steps, conditions can generally be arranged to favor a particular reaction to clarify the SAR. Some examples are available from the literature (Esler et al., 1999; LeVine, 1999; Stine et al., 2003).

4.5.2.6 Differences between Synthetic Aβ, Transgenic Mouse Aβ, and Aβ in AD

Biological Aβ displays some different properties than synthetic fibrils such as inducing amyloid deposition in hβAPP-overexpressing mice (Walker et al., 2006). Binding stoichiometries of Thioflavin-related dyes also indicate that the synthetic and biological forms are related but different. They also suggest that even the plaque material deposited in transgenic mouse brain differs in a significant way from the human brain. ^{11}C-6-OH-BTA-1 (PIB), an imaging agent for Aβ plaques based on the histologic amyloid dye Thioflavin T (ThT) binds to brain Aβ fibrils with high affinity and ~1:1 stoichiometry with respect to Aβ monomer in AD brain homogenates. It is being used clinically to quantify Aβ deposits in AD brain by PET imaging (Klunk et al., 2004; Price et al., 2005). Quantitative estimates of the high affinity binding stoichiometry of 6-OH-BTA-1 with respect to Aβ monomer is 1:370 (synthetic Aβ (1–40) or (1–42) fibrils), 1:100 (transgenic mouse brain), and 1:2 for affected areas of AD brain, but does not bind to the unaffected cerebellum of AD brain (Klunk et al., 2003, 2005). Aβ fibrils are clearly polymorphic in ThT binding. This difference could also translate into different biological effects. Oligomeric Aβ also exhibits polymorphic activities (Deshpande et al., 2006).

The importance of these observations for Aβ polymerization reduction is that polymerization inhibition strategies developed in vitro will require validation in animal models before clinical testing in humans. Proper endpoints for testing will be required, preferably ones that can be assessed rapidly and easily. For example, an oligomer formation inhibitor may not reverse amyloid load, but this type of inhibitor would be expected to alleviate behavioral manifestations caused by oligomers. Fortunately, in vitro fibrils and oligomers seem to share many of the qualitative properties of the biologically derived material even though they are not as quantitatively potent in producing biological effects. At a minimum, rank-order conclusions based on in vitro testing should be relevant.

From the brief analysis described so far, it appears that monomer reduction should be the approach that would address both the sequential and parallel pathway models. Targeting the APP-processing proteases that release Aβ monomer, immunological or scavenging approaches that deplete circulating and interstitial fluid Aβ (Jensen et al., 2005; Ma et al., 2006), or increasing degradation by Aβ metabolizing proteases (Saido and Iwata, 2006) are beyond the scope of this chapter. Even if monomer reduction is achieved without significant side effects, approaches that target assembly and disassembly of Aβ oligomers and fibrils are still useful for controlling the Aβ that escapes the extracellular antibody or secretase inhibition. In patients, there will be a significant amount of deposited exchangeable Aβ that will begin to clear and must be prevented from forming toxic species. Extending the scenario to other misfolding proteins in neurodegenerative diseases, such as α-synuclein, tau, and Huntingtin bearing an expanded polyglutamine segment, which are intracellular and not produced as a metabolic side reaction (e.g., no secretases), will also assure that controlling misfolded protein assembly will remain an opportunity for the development of therapeutics.

4.5.3 Challenges to Controlling Protein (Mis)Folding

Targeting the amyloid cascade at one particular step of this pathogenesis is a key focus for many researchers in the field of AD. Over the past several years, immunization clinical trials indicate that it may be beneficial to target early events in the Aβ cascade in order to produce a clinically observable effect in Alzheimer patients. However, due to the complexity of this "beast" of a protein, this surely is not an easy task.

There are many complications and factors to be considered when undertaking the challenge of discovering a drug to alter the progression of AD. Attacking misfolding proteins also poses a special set of challenges that go beyond those for the usual receptor and enzyme targets of the pharmaceutical industry. A useful summary of the drug discovery process can be found in Rang (2006).

4.5.3.1   What Is the Target?

A prominent approach since the identification of the processing β- and γ-secretases is targeting Aβ assembly at an early stage to prevent the production of the Aβ protein or formation of Aβ oligomers. This would accrue multiple downstream advantages such as decreases in toxicity and plaque formation and perhaps less tau neuropathology. Earlier discovery efforts that targeted prevention of fibril formation in vitro resulted in a decrease in the area occupied by amyloid plaques in transgenic mouse models of brain Aβ pathology (Augelli-Szafran et al., 2004). Because no animal other than *Homo sapiens* suffers large-scale neuronal death and degeneration as a result of Aβ accumulation, it is uncertain whether preventing plaque deposition will translate into a therapeutic benefit. The neuronal toxicity and electrophysiological effects of recently recognized early pathway soluble oligomeric Aβ may be a more proximal contributor to neurodegeneration. Many factors are part of the decision when choosing a drug target; the current state of knowledge, assay availability and reliability, quantitative interpretation of data, and potential chemical lead structures available to the effort. After all of these factors are taken into consideration, the target in the cascade is usually obvious.

4.5.3.2   Designing Informative Assays

Accurate, reliable, and quantitative assay readouts are essential to assist in the design of drug-like molecules and ultimately identify an effective drug. Consistent data significantly assist the strategy of optimization of chemical entities in a discovery program. In particular, from the medicinal chemistry aspect, test data must be quantitative, whether it is a repeatable IC_{50} or K_i value, a statistically significant in vivo result, or a definitive pharmacokinetic value. Otherwise, there is the risk of missing a minor structural change in a key chemical lead that could have had an impact on the design of future chemical targets. Another potential issue with assay readouts is false positive inhibition. It is critical to determine at an early stage of a discovery approach any potential interference between a compound and the assay components or conditions. False positives and false negatives can derail the search for a SAR. Alternative procedures to verify the effects of a compound are required.

Like any good screening program, testing for aggregation inhibitors is organized around a hierarchy of assays of increasing selectivity for a particular activity culminating in an animal model of CNS Aβ deposition. However, since the point of inhibition in the amyloid cascade that will translate to in vivo efficacy in animals as well as in humans is unknown, multiple primary assays are pursued in parallel in the screening paradigm. Testing compounds in several assays with different biological profiles identifies compounds active in one assay and not other assays. Different chemical classes of compounds are often effective in different assays and will likely display dissimilar profiles in vivo. Choices are then made to refine the testing strategy to be more efficient after identifying new chemical leads series that would have been overlooked. After compiling a substantial collection of in vitro data on representative compounds from various series, a correlation can be made between one or two selected in vitro assays and the in vivo outcome. This strategy provides further chemical series with different biological action as backups to the primary target or to launch a second generation of discovery. For the primary series, the focus narrows to a few key measures for decision-making.

4.5.3.3   Starting Points—Finding Inhibitors

Many factors should be considered to identify a chemical lead in a discovery program. Since high-resolution molecular structures of the target assemblies are not available for most misfolding proteins, accessing several sources of potential chemical leads at the early stages of the discovery process is the more conservative approach. This includes mining known inhibitors from the literature and collecting structural analogs. High throughput screening of large, structurally diverse chemical libraries can also pave the way for finding inhibitors. Executing these processes simultaneously will expedite the identification of chemical leads. Included in this process should be a definitive decision tree to evaluate the testing results to help narrow the field of leads to focus on a few key chemical series. The criteria should include the modification of drug-like properties, increasing efficacy and affinity, and the ease of synthesis and cost effectiveness for eventual production. If the biological assays are not as quantitative as desired, testing inhibitors across several assays that cause inhibition at the target and analyzing the data for rank-order trends among compounds as indicated in the previous section is a way to eliminate unpromising chemical leads and to prioritize the biological activity assays.

Another parameter that should be included in the decision-making process is the pharmacokinetic (PK) properties of representative compounds from a chemical series. Pharmacokinetic analysis has acquired increased importance in the early part of the drug discovery process. A practical series of compounds must be amenable to fine tuning of physical and chemical properties that have come to be accepted as predictive of absorption, disposition, metabolism, and elimination (ADME). This will increase the probability that the compounds will get to where they are required at high enough concentrations and stay there long enough to have the intended effect in a human subject. These properties are projected based on the outcome of a series of standard in vitro tests and animal pharmacokinetic (PK) studies. For compounds to be active against a neurodegenerative disease, they must also penetrate the blood–brain barrier. Building in these properties early in the discovery process reduces the attrition of drug candidates in the clinic and can save considerable time and money since clinical trials are a very expensive component of drug development. Hence, it is important to be selective in choosing the compounds to synthesize to make good progress in identifying a drug-like chemical lead or chemical series.

4.5.3.4   Small Molecules versus Peptides

Many targets for drug discovery involve some active site of a protein of which a peptide is a lead inhibitor. Peptide inhibitors are sometimes a favorite type of lead molecule of drug discovery endeavors due to potency. However, peptide inhibitors are usually high in molecular weight (>500 Da) and difficult to formulate into an acceptable dosage form, given the physical properties of peptides such as solubility and stability. Peptides are also expensive to produce on a commercial scale and usually have suboptimal PK properties such as rapid clearance and metabolism, poor bioavailability, and weak brain penetration. In addition, peptide-like analogs (peptoids) derived from the chiral peptide structure may involve a difficult chiral synthesis that requires intense efforts to fine-tune the inhibitor into an entity that is drug-like while retaining its inhibitory characteristics.

On the other hand, a small molecule inhibitor has advantages in the chemistry, formulation, overall cost, and in general, is much easier to study. A small molecule inhibitor is lower in molecular weight and a much more extensive series of structurally diverse analogs can be synthesized if an optimal synthetic route has been designed. The ability to evaluate several small molecule inhibitors greatly increases the probability of identifying drug-like candidates. Early in the search for Aβ aggregation inhibitors, there was concern that small molecules would not be able to interact with a large enough portion of the protein or peptide to exercise potent inhibition. This turned out to be unfounded as small critical sites on intermediate assemblies that bound a variety of small molecule ligands were critical for further growth into fibrils. These included peptides (Ghanta et al., 1996; Findeis, 2002; Soto and Estrada, 2005; Kokkoni et al., 2006), rifampicin derivatives, bi- and tricyclic pyridines, anthracyclines, acridinones, and naphthylazo compounds (Emmerling et al., 1999).

4.5.3.5 Interpreting an SAR

The SAR among different compounds designed for the inhibition of a specific protein is critical in driving the strategy toward the discovery of a therapeutic intervention. The biological response to the structural modifications of inhibitors of the protein assembly that increase or decrease affinity and efficacy are analyzed to evaluate the changes that are tolerated. A conceptual image can then be built, often called a pharmacophore model, which represents the interaction points between the small molecule and the target. Depending on what functional group is added or replaced on the small molecule inhibitor, the response of the biological system that correlates with the changes incorporated on the inhibitor directs the design of the next round of new inhibitors. The chemical and structural properties of the functional groups describe the type of inhibitors that may be preferred by the target protein. For example, a polar group on the inhibitor may increase potency and hence, could indicate that there is some type of interaction with a polar or H-bonding residue in the active site. If a bulky, lipophilic group is added, and potency is increased, this may indicate that there is a hydrophobic pocket in the active site that favors some form of steric interaction that translates into improved inhibition. The reverse would be true if a decrease in potency is observed with a modification. These are just a few simple examples of a minor change to a compound that translates into an effect on the observed potency.

It is important to note that modifications are not only necessary to improve potency of inhibitors, but they could also affect the pharmacokinetics and pharmacodynamics of the inhibitors. Unfortunately, there are many examples that a positive effect on one property of an inhibitor negatively affects other properties. Polar groups may be added to the molecule to increase solubility that is a key physical parameter to monitor when the formulation of the drug is being investigated. Too much of an increase in polarity may prevent required brain penetration or too much lipophilicity may cause the compound to accumulate in fat. Other changes may convert the molecule into a substrate for a metabolizing enzyme in the liver causing it to either be inactivated or be eliminated by the kidney. The properties of the molecule have to be fine-tuned to yield a complete successful druggable package. Persistence and strategy are key to achieving the right balance of all properties of the inhibitor. As one can imagine, this is a challenging, but very rewarding experience.

4.5.3.6   Current Approaches to Drug Design

For many drug targets, but not aggregation inhibitors, there is precedence for drug efficacy already on the market. Identifying new drugs that are better tolerated, have fewer side effects and perhaps are more effective is viewed as being a lower risk approach for a company than if the target does not have a known drug on the market that shows effectiveness in humans. There are currently no drugs in the market that prevent the onset of AD or halt its progression. This endeavor is high risk, but the potential benefit to patients and their families is very high reward. The complexity of this disease, and the lack of a demonstrated correlation between the human disease and an in vivo animal model requires the establishment of confidence that inhibition of Aβ amyloid aggregation as a therapeutic approach will be disease modifying. Endpoints of the in vitro and in vivo studies are particularly important for a nonprecedented (no currently marketed drugs) drug target such as amyloid for the treatment of AD. However, uncertainty over the translation of imperfect animal models to the human situation forces a combination of in vitro and in vivo data that are necessary for this targeted therapeutic intervention. The tendency, in this case, is to prefer the compounds that have the best pharmaceutical properties while still showing reasonable efficacy in the model.

4.5.3.7   Defining Efficacy and Specificity

Due to the incompleteness of current AD animal models, it is difficult to set rigid expectations from these models as decision-making tools before progressing compounds into clinical trials. Therefore, careful analysis of both in vitro and in vivo data plays a key role in realizing efficacy. Complete specificity for Aβ probably would not be necessary since other amyloids are associated with disease (except perhaps see Si et al. [2003] and Fowler et al. [2005]). In fact, a number of common features among amyloids may allow use of compounds that have been developed for AD to be used to also combat other amyloidoses. Of course, achieving specificity for amyloid proteins while not affecting other systems is also key to developing a viable therapeutic. This is a difficult issue to address a priori for amyloids. Nonspecific inhibitors are normally eliminated early on in the drug discovery process for well-characterized systems, that is, well before a drug candidate is being considered for clinical trials. This forces reliance on the detection of side effects in relevant safety animal models. Based on experience and collected data, a judgment is then made as to whether the undesired effects are significant enough to stop the compound from progressing into further development.

4.5.4 CURRENT APPROACHES

Aβ aggregation has been targeted at multiple levels. Fibril formation was the initial target, but the recognition of the toxic soluble oligomers has displaced much of that interest. The structural requirements for compounds inhibiting this early step of Aβ assembly appear to be significantly different. Outside of a smattering of publications, no drug discovery programs have published their work. Techniques for measuring inhibition of oligomer formation are still being developed, and a consensus has not been reached as to where interrupting the process is essential. A certain amount of target structure with at least quasistability is required to generate selective binding sites for compounds that will interrupt assembly.

4.5.4.1 Monomers

Removal of monomer is a logical starting point since the production of soluble toxic species and plaque deposition is directly related to the Aβ monomer concentration. The low fraction of persistent structure of the monomeric peptide (Lazo et al., 2005) made it an unlikely target for small molecule therapeutics. Other tactics such as passive and active immunization have proven effective in decreasing Aβ load in transgenic animals and in human clinical trials. However, unexpected side effects that are attributed to the immunization method have slowed development of this approach as noted elsewhere in this article. β- and γ-secretase inhibitors have attracted much interest because they are protease targets (Zimmermann et al., 2005), but side effects from inhibiting the activity of these multisubstrate enzymes could result in toxicological issues. Small molecule strategies that increase the level or activity of Aβ-degrading enzymes and destroy monomer are also being considered. Chronic glucocorticoid administration has been shown in macques to increase $Aβ(x–42)$, in line with increases seen during aging in corticosteroids (Kulstad et al., 2005). Angiotensin-converting-enzyme inhibitors also increase Aβ accumulation in transfected cultured cells (Hemming and Selkoe, 2005).

4.5.4.2 Fibril Formation

In other amyloid diseases the formation of insoluble deposits is central to the pathophysiology. Hence, blocking deposition of Aβ was a logical target. Aβ fibrils also possess a cellular toxicity that is enhanced during fibril formation and decreased after fibril formation is complete. This increased potency was later determined to be due to highly toxic intermediates that were generated when the fibrils were forming.

A wide variety of chemical structures prevent or delay fibril formation in vitro (Herbst and Wanker, 2006). A common theme among the compounds is the presence of flat aromatic ring structures such as polyphenols (Porat et al., 2006). Based on a core Aβ amyloidogenic sequence KVFFA, designed compounds such as "breaker peptides" (Estrada and Soto, 2006), N-methylated analogs (Kokkoni et al., 2006), and other modified peptides have been effective with Aβ on a stoichiometric basis. There are few reports of testing aggregation inhibitors using transgenic mouse models that overexpress human βAPP (except see Augelli-Szafran et al. [2004]). Since these mice fail to exhibit the large-scale neuronal death seen in AD, this crucial aspect of the efficacy of test compounds could not be evaluated in the plaque deposition model. Although little analysis was done on the mode of inhibition, substoichiometric inhibitors appeared to be acting at the level of a limiting nucleating species to prevent fibril formation. However, once frank fibrils are formed, these compounds are relatively ineffective against fibril growth that occurs at the end of the fibrils by monomer addition. This implies that the fibril surface, upon which growth takes place, or a solution conformation of the monomer, is different from early stage nucleation. A series of glycosaminoglycan mimics for amyloid diseases developed by Neurochem were designed to prevent fibril initiation. These compounds interfere with Aβ interactions and extracellular matrix components (Kisilevsky et al., 1995; Yang et al., 2001). They also interfere with monomer interactions. Phase III clinical trials for Alzhemed (3-amino-1-propanesulfonic acid) in AD were launched in September 2005.

Fibril dissolution with small molecules has the potential to destabilize or remove preexisting plaques. However, this approach has proven to be a difficult task. A few compounds have been reported to have this activity, but only with freshly formed fibrils. Fibril structure matures and stabilization against perturbation occurs even with synthetic peptides. This tactic is less likely to be effective in humans where the plaques can exist for decades and are chemically crosslinked, isomerized, and modified over time.

4.5.4.3 Oligomer and Protofibril Formation

One of the potential dangers of inhibiting fibril formation is the stabilization of soluble oligomeric species that account for the high potency of toxicity observed during fibril formation. Soluble oligomers and protofibrils are now recognized as the most cytotoxic species. These species are capable of recapitulating electrophysiologic effects on LTP (long-term potentiation) and behavioral perturbations seen in βAPP-overexpressing transgenic mice. It is hypothesized that these effects are mechanistically related to memory defects seen in early AD. The mechanism by which oligomers act is still under investigation, but explanations derived from oligomers prepared from synthetic Aβ(1–42) peptides range from nonselective permeability changes in membrane structure to receptor-dependent signaling cascades, depending on the method of preparation (Deshpande et al., 2006). The effects of biologically derived oligomers, especially those from AD brain, need to be evaluated in a comparable fashion before conclusions can be drawn.

Screening programs are now being run to identify small molecules that prevent oligomer formation. Although oligomer inhibition as a target is in its early stages, it is obvious that compounds blocking fibril formation are weak or totally ineffective in blocking oligomer formation. This complements the evidence from oligomer and fibril-specific antibodies that oligomer structures differ from fibril structures. Protofibrils have received little attention as a potential therapeutic intevention, partly because preventing protofibril assembly may cause the stabilization of highly toxic soluble oligomers. Calmidazolium has been shown to stabilize protofibrils and to slow the conversion of protofibrils into fibrils (Williams et al., 2005).

4.5.5 FUTURE Aβ APPROACHES

A number of epidemiological studies of AD suggested that the use of certain anti-inflammatory agents (NSAIDs), HMG CoA reductase inhibitors, and other drugs taken for relatively brief periods of time markedly reduced the incidence of AD (Zandi and Breitner, 2001; Zandi et al., 2004a,b; Rea et al., 2005; Zandi et al., 2005; Khachaturian et al., 2006). Surprisingly though, subsequent direct placebo-controlled clinical trials of those agents in subjects with AD unequivocally failed to provide benefit. Reanalysis of the epidemiological and clinical data indicated that the treatment effects were in prevention of whatever processes led to degeneration rather than in treatment of the degeneration once the disease was diagnosed. This is the equivalent of closing the barn door after the horse has left. Removing the initial cause of a problem is rarely beneficial once disease pathology is triggered if a different mechanism is involved in the disease. For some diseases, the distinction is more obvious; for AD it is not. This is a sobering thought and a reminder to carefully define disease stages and to consider the stage of the disease being targeted, particularly when clinical trials are involved. Additional approaches for therapeutic intervention designed to interrupt the cascade of Aβ-linked processes leading to disease are being considered.

4.5.5.1 Clearance

The accumulation of misfolded Aβ species appears related to decreased clearance of the multimeric species rather than an increase in the production of the monomer peptide (Zlokovic et al., 2000; Cirrito et al., 2003, 2005; Guenette, 2003). Physiological changes, attributable to normal processes of aging, account for a proportion of this decreasing capacity. These are exacerbated by processes set in motion by the accrual of chemically modified proteins and chronic upregulation of cellular and humoral elements of the immune response. β-amyloid peptide levels in the CSF begin to increase in early middle age, around age 50 (Fukuyama et al., 2000), which is a turning point for a number of aging changes in humans. Interestingly, similar aging changes occur in mice at an analogous age of 9–12 months (Kawarabayashi et al., 2001). The age dependence of Aβ accumulation in βAPP-transgenic mice, despite nearly constant expression levels, suggests that this is a physiological phenomenon that may in part account for the late-onset of sporadic AD. Familial forms of AD may exacerbate this process by increasing the ratio of Aβ42/Aβ40 produced from βAPP or by mutations in the Aβ peptide sequence that increase its propensity to aggregate or interact with vascular smooth muscle cells. The early onset seen in the human cases is paralleled by rapid development of deposits in transgenic animals carrying both βAPP and PS-1 mutations.

Most of Aβ is produced in the brain, mainly because the initial amyloidogenic proteolytic cleavage of βAPP is performed by the β-secretase BACE1, which is mainly found in brain. Clearance of Aβ from the brain interstitial tissue fluid through the blood–brain barrier occurs through periarterial pathways (Roher et al., 2003; Schley et al., 2005) and the CSF (Zlokovic et al., 2005) through the choroid plexus and the arachnoid granulations. Although the mechanisms involved are not entirely clear, studies that utilize peripheral antibody sinks for Aβ peptide clearly show in βAPP-transgenic mouse models that the equilibrium can be perturbed to move Aβ out of the brain (Hartman et al., 2005; Matsuoka et al., 2005; Morgan, 2005). Aβ deposits in the brains of these mice are in equilibrium with the soluble material, however there are distinct pools with different characteristics (Cirrito et al., 2003). Exchangeable deposits are cleared over time (Bacskai et al., 2001).

There is some question as to whether the same clearance processes can operate in humans. Unlike the transgenic mouse, deposited Aβ peptides in humans contain significant proportions of cross-linked material as well as chemically modified peptides, for example, oxidized or glycated (Lowenson et al., 1999; Kalback et al., 2002). Whether or not the extent of the modifications is due to the age of the human deposits (20–40 years) versus 1–2 years in the transgenic mice, the question remains as to whether the same clearance mechanisms are present in humans and whether that equilibrium is sufficient to significantly decrease the amyloid load. In addition, there is a question about the disposition of soluble or adsorbed nonfibrillar oligomeric Aβ that is increasingly being implicated in the biological effects of Aβ. Will greater amounts of toxic soluble oligomers be created by shifting the equilibrium if fibrils deplete soluble oligomers in vivo? In vitro studies with synthetic peptides suggest that fibrils grow primarily from monomer, not oligomer addition (Mastrangelo et al., 2006). The species produced by disassembly of fibrils requires more study. In particular, biological Aβ displays some different properties than synthetic fibrils, inducing amyloid deposition in hAPP-overexpressing mice (Kane et al., 2000; Walker and LeVine, 2000; Walker et al., 2002a, 2002b; Herzig et al., 2006) and binding stoichiometries of Thioflavin-related dyes (Klunk et al., 2005; LeVine, 2005; Ye et al., 2005).

Although there was no measurement of oligomers, cognitive improvements in patients in the AN-1792 Aβ immunization therapy trial might be due to decreases in soluble oligomeric Aβ. Synaptic physiology connected with memory processes has been shown to be perturbed by soluble Aβ oligomers and is restored by Aβ antibodies in cell and organotypic culture (Lambert et al., 2001; Rowan et al., 2005; Ma et al., 2006) and in transgenic mice (Frenkel et al., 2000; Lambert et al., 2001; Klyubin et al., 2005; Ma et al., 2006). ICV administration of soluble Aβ56* isolated from Tg2576 mice to naive rats induced behavioral abnormalities. These changes, believed to be due to the blocking of LTP by this oligomeric Aβ species, disappeared after a number of days. This observation suggests that at least the effects of the toxic species could be cleared or neutralized (Lesne et al., 2006). This is an important observation, since both synthetic and biologically produced Aβ oligomers are resistant to proteolysis and thus might be difficult to clear by the usual mechanisms. Newly synthesized biologic monomeric Aβ is turned over in the mouse brain within 2 h (Cirrito et al., 2003), and within 3 h in the guinea pig brain (Grimwood et al., 2005). In humans, approximately equal rates of synthesis and rate of turnover of ~8% of Aβ per hour is observed in the CSF (Bateman et al., 2006). This rapid turnover indicates that Aβ levels are dynamic and that decreased clearance is probably a major contributor to Aβ buildup in brain with aging.

4.5.5.2 Boosting Cellular Defenses Against Protein Misfolding

One of the processes that appear to be diminished in aging and particularly in AD is cellular protein quality control (de Vrij, 2004). The crowded intracellular milieu of the cell is poised for the aggregation of partially folded or misfolded proteins with exposed hydrophobic segments. Mammalian cell options for dealing with protein aggregates are much more restricted than for organisms such as bacteria, yeast, or plants. These organisms have to recover from sudden wide changes of temperature, osmotic, ionic, and other conditions in their environment. They can landfill aggregates into aggresome structures (Garcia-Mata et al., 2002), inclusion bodies, or use a process, such as autophagy (Massey et al., 2004; Rideout et al., 2004; Nixon et al., 2005) to create a lysosome to degrade them. The common cellular mechanism for protein turnover in multicellular organisms called ubiquitination, targets aggregates to the proteasome. However, these aggregates are poor substrates for the unfolding mechanism that feeds polypeptides into the proteasome's hydrolytic core. In addition, many misfolded forms of proteins, such as Aβ, tau, α-synuclein, expanded polyglutamine repeat proteins, and other aggregated proteins found in association with chronic neurodegenerative diseases form structured soluble oligomers and aggregates that are highly protease resistant which also favor their accumulation. Fibrillar and soluble oligomeric forms of Aβ and α-synuclein have been shown to interfere with proteasome function (Lindersson et al., 2004; Oh et al., 2005; Almeida et al., 2006; Lee et al., 2006).

The multiple molecular chaperone systems in mammalian cells provide surveillance for misfolded or unfolded proteins. They can sequester and refold these substrates and aid in processing them for ubiquitination and degradation. However, they are unable to deal with stable aggregated proteins. Overexpression of chaperone proteins, such as hsp70, in cellular models of misfolded protein accumulation is able to partially cope with small aggregates, but crosses of transgenic hsp70 overexpressors with transgenic overexpressors of expanded polyglutamine repeats, α-synuclein, Aβ, or tau (Chai et al., 1999; Hashimoto et al., 2003; Giffard et al., 2004; Klucken et al., 2004; Magrane et al., 2004) are much less effective. Accumulation of misfolded proteins elicits the unfolded protein response that induces increased chaperone production and several other adaptations to stress. However, these adaptations require short-term compromises for cellular function that cannot be sustained.

Nevertheless, there may be other ways suggested in this chapter of assisting endogenous chaperone and proteolytic processing systems to do their job. It is worth considering pharmacologically modifying soluble oligomer and aggregate structures to prevent their inhibition of disposal mechanisms and to make them susceptible to cellular removal systems.

4.5.6 ANIMAL MODELS

In order to be able to assess the efficacy of therapeutic measures for AD, models most closely related to the human disease would be the best choice. This is a difficult proposition since the true mechanism of AD has not been revealed. The most logical approach of overexpressing Aβ from neuronal-specific promotors in brain recapitulated the histopathological amyloid plaques, but not NFTs. It also did not promote neuronal cell death to the extent seen in AD. Multiple transgenes, including pathological tau, accounted for overlapping pathologies. However, the goal of pathology-dependent neuronal cell death has been elusive. Reviews comparing models of AD pathology are available (Davis and Laroche, 2003; Higgins and Jacobsen, 2003; Brandt et al., 2005; LeVine and Walker, 2006). Other species with human-sequence Aβ and reasonably long lifespans compared with humans (dogs, nonhuman primates) develop vascular and parenchymal Aβ fibrillar pathology and some apparent cognitive decline. However, no massive neuronal cell death develops as it occurs in AD. Thus, we are reduced to employing animal models that model potential disease mechanisms, but not AD itself. This dichotomy is not an uncommon scenario for animal models of disease (Morgan and Keller, 2002). The problems come when the limitations of the models are not taken into account in interpreting results. There is always the temptation to believe that the model is the same as the disease.

4.5.6.1 Utility of Current Models to Assess Modulation of Aβ Polymerization

Current transgenic mouse models of Aβ overexpression allow the assessment of whether a stratagem alters the deposition of Aβ peptide in the parenchyma and vasculature of the brain (reviewed in LeVine and Walker [2006] and Sankaranarayanan [2006]). Histological staining of plaques followed by quantification of the area of a brain region occupied by plaques using the unbiased stereology technique (Glaser and Glaser, 2000) is generally preferred, although some argue that the relevant measure is total brain volume occupied by plaques (Fiala and Harris, 2001). Several models produce behaviors in learning paradigms that are suspected to be due to synaptic effects of soluble forms of Aβ. This is supported by the observation that infusion of oligomeric Aβ into mice (Craft et al., 2004) or rats (Lesne et al., 2006) induces lasting learning behavioral changes. SDS-stable oligomers were decreased in a triple transgenic (APP/PS1/tau) mouse model by injection of an oligomer-specific antibody (Oddo et al., 2006). An antibody to Aβ(1–15) that interacted with multiple Aβ species also reduced oligomers, particularly dodecamers, in Tg2576 mice (Ma et al., 2006).

Soluble assembled forms of Aβ are only now being addressed in animal models of hAPP overexpression. Soluble oligomers have generally been detected in some cases as SDS-resistant multimers by Western blot (Walsh et al., 2002a,b) and more recently by oligomer-specific immunoreactivity (Chang et al., 2003; Oddo et al., 2006).

4.5.7 CLINICAL TRIALS

4.5.7.1 Current Trial Designs and Proof of Concept

The Gold Standard for clinical efficacy against AD is prevention of cognitive decline over a prolonged period. This was established both because the initial therapies being tested were symptomatic and because cognition is the clinical manifestation of AD. To date no surrogate markers equivalent to cholesterol levels for anti-atherosclerosis therapeutics have been established for AD. Although much evidence implicates Aβ in the etiology of AD, behavior remains the readout for trials. In a clinical trial there are two issues to be decided. First, does the treatment affect its intended target in the expected manner in humans, and second, does that effect on the target have the desired impact on the clinical readout for the disease.

For AD the field is still at the stage of proving that altering the target will impact the progression of the disease measured by clinical outcome. Because quantifying cognitive change requires months of slow, expensive longitudinal observation, markers reflecting the proposed mechanism of action of the therapeutic agent being tested are being developed. For therapies that target $A\beta$ levels or plaque deposition $A\beta$ measurements are being refined.

4.5.7.2   Aβ Levels

Therapeutics that target $A\beta$ production or clearance such as β- or γ-secretase inhibition or modulation, and immunization approaches exert their primary effect on $A\beta$ peptide levels in the brain, CSF, and plasma, which can be measured by sensitive and 42- or 40-$A\beta$ specific immunoassays of various types. Interpreting the results requires an appreciation of the equilibrium between insoluble forms of the peptide and the soluble material that is being measured. Quantification of soluble oligomeric forms of $A\beta$ with single-site immunoassays (El-Agnaf et al., 2003; LeVine, 2004b) or oligomer-specific antibody immunoassays (Georganopoulou et al., 2005; Haes et al., 2005; LeVine, 2006b) requires high sensitivity because the concentrations are low compared with the total $A\beta$ peptide content. There are also technical concerns with fidelity of the measures (Stenh et al., 2005). Nevertheless, in principle, measurements can be made and significant effects of therapies detected.

4.5.7.3   Amyloid Imaging

Amyloid-specific binding positron-emission (PET) radioligands that have the appropriate uptake, disposition, selectivity, and clearance characteristics have also been developed. These radioligands can provide in vivo images of the regional accumulation of amyloid deposits in the human brain (Klunk et al., 2004; Price et al., 2005; Archer et al., 2006). In combination with the standard diagnostic techniques of MRI and f-MRI for anatomical and blood flow, and with glucose metabolism by PET the relationship among these measures of pathology, structural change, and physiology is being established (Buckner et al., 2005). Longitudinal measurements of individuals can establish changes in these modalities. Unfortunately, the soluble species are probably too dilute to distinguish from background, even if specific small molecule reagents were developed.

4.5.8   FUTURE CHALLENGES

At the current time, the best hope for treatment of AD is in prevention. The next best option is to identify cognitive disturbances and intervene before they progress to dementia. Studies such as the Nun Study (Patzwald and Wildt, 2004) indicate that a significant proportion of the very early stages of cognitive decline do spontaneously reverse, but that this is an unstable equilibrium. This state, termed mild cognitive impairment (MCI) converts at a higher rate into AD than age-matched cognitive normals that never reached MCI. Reanalysis of retrospective studies that originally showed protection by therapeutic regimens turned out to exert their protective effect only in the predementia period, as was borne out by a lack of therapeutic effect in trials with patients diagnosed with early AD (Carlson et al., 2002; Zandi et al., 2002; Szekely et al., 2004; Zandi et al., 2004a, 2005; Khachaturian et al., 2006).

Thus, it appears that the maximum therapeutic benefit will be achieved by treating early in the disease process, before significant neuronal death or irreversible dysfunction occurs. Recreating missing neuronal pathways and establishing appropriate connections is a daunting prospect. Predictive biomarkers are required, and preferably ones that do not rely on cognitive function since enough damage may have already occurred to obviate therapeutic intervention. This has been an active area of endeavor, but has not yet yielded any single validated winners, although combinations of biomarkers show better predictive potential (Thal et al., 2006b).

In lieu of predictive biomarkers, a better definition of disease stages and tests that will provide reliable assessments, especially outside the tertiary care centers with their specialists, is needed (Thal et al., 2006a). Determining that someone has AD is relatively straightforward, but that is of little use to the patient, since only palliative treatments of limited efficacy are available at that stage of the disease. The key to both early diagnosis and possibly early effective treatment may lie in defining and monitoring the clinically relevant pathological species of Aβ. An encouraging start to this is exemplified by the suggestive work of Lesne et al. (2006) which isolates and demonstrates the biological activity of a particular aggregated but soluble form, Aβ56*. The extent to which this applies to AD remains to be demonstrated. However, this study broke a conceptual barrier, which could lead to future diagnostic and therapeutic progress.

REFERENCES

Adessi, C. and Soto, C. (2002) Converting a peptide into a drug: strategies to improve stability and bioavailability. *Current Medicinal Chemistry*, 9, 963–978.

Aisen, P.S. and Davis, K.L. (1994) Inflammatory mechanisms in Alzheimer's disease: implications for therapy. *American Journal of Psychiatry*, 151, 1105–1113.

Aisen, P.S., Davis, K.L., Berg, J.D., Schafer, K., Campbell, K., Thomas, R.G., et al. (2000) A randomized controlled trial of prednisone in Alzheimer's disease. *Alzheimer's Disease Cooperative Study Neurology*, 54, 588–593.

Aisen, P.S., Schafer, K.A., Grundman, M., Pfeiffer, E., Sano, M., Davis, K.L., et al. (2003) Effects of rofecoxib or naproxen vs placebo on Alzheimer disease progression: a randomized controlled trial. *JAMA*, 289, 2819–2826.

Almeida, C.G., Takahashi, R.H., and Gouras, G.K. (2006) Beta-amyloid accumulation impairs multivesicular body sorting by inhibiting the ubiquitin-proteasome system. *Journal of Neuroscience*, 26, 4277–4288.

Alvarez, A., Toro, R., Caceres, A., and Maccioni, R.B. (1999). Inhibition of tau phosphorylating protein kinase cdk5 prevents beta-amyloid-induced neuronal death. *FEBS Letters*, 459, 421–426.

Anderson, J.J., Holtz, G., Baskin, P.P., Turner, M., Rowe, B., Wang, B., Kounnas, M.Z., Lamb, B.T., Barten, D., Felsenstein, K., et al. (2005). Reductions in beta-amyloid concentrations in vivo by the g-secretase inhibitors BMS-289948 and BMS-299897. *Biochemistry Pharmacology*, 69, 689–698.

Andreasson, K.I., Savonenko, A., Vidensky, S., Goellner, J.J., Zhang, Y., Shaffer, A., et al. (2001) Age-dependent cognitive deficits and neuronal apoptosis in cyclooxygenase-2 transgenic mice. *Journal of Neuroscience*, 21, 8198–8209.

Archer, H.A., Edison, P., Brooks, D.J., Barnes, J., Frost, C., Yeatman, T., et al. (2006) Amyloid load and cerebral atrophy in Alzheimer's disease: an (11)C-PIB positron emission tomography study. *Annals of Neurology*, 60, 145–147.

Arnaud, C. and Mach, F. (2005) Pleiotropic effects of statins in atherosclerosis: role on endothelial function, inflammation and immunomodulation. *Archives des maladies du coeur et des vaisseaux*, 98, 661–666.

Asai, M., Hattori, C., Szabo, B., Sasagawa, N., Maruyama, K., Tanuma, S., and Ishiura, S. (2003). Putative function of ADAM9, ADAM10, and ADAM17 as APP alpha-secretase. *Biochemical and Biophysical Research Communications*, 301, 231–235.

Augelli-Szafran, C.E., Bian, F., Callahan, M.J., Feng, R., Iwai, A., Lai, J., et al. (2004) PD 0118057 and PD 0202091: Small Molecule Inhibitors of Beta Amyloid Peptide Aggregation. 228th ACS National Meeting, August 22–26 (2004) Philadelphia, Pennsylvania, MEDI-321.

Bacskai, B.J., Kajdasz, S.T., Christie, R.H., Carter, C., Games, D., Seubert, P., et al. (2001) Imaging of amyloid-beta deposits in brains of living mice permits direct observation of clearance of plaques with immunotherapy. *Nature Medicine*, 7, 369–372.

Bacskai, B.J., Hickey, G.A., Skoch, J., Kajdasz, S.T., Wang, Y., Huang, G.F., et al. (2003) Four-dimensional multiphoton imaging of brain entry, amyloid binding, and clearance of an amyloid-{beta} ligand in transgenic mice. *Proceedings of the National Academy of Sciences of the United States of America*, 100, 12462–12467.

Bales, K.R., Dodart, J.C., DeMattos, R.B., Holtzman, D.M., and Paul, S.M. (2002) Apolipoprotein E, amyloid, and Alzheimer disease. *Molecular Interventions*, 2, 363–375.

Barghorn, S., Nimmrich, V., Striebinger, A., Krantz, C., Keller, P., Janson, B., et al. (2005) Globular amyloid beta-peptide oligomers—a homogenous and stable neuropathological protein in Alzheimer's disease. *Journal of Neurochemistry*, 95, 834–847.

Barnett, J., Chow, J., Ives, D., Chiou, M., Mackenzie, R., Osen, E., et al. (1994) Purification, characterization and selective inhibition of human prostaglandin G/H synthase 1 and 2 expressed in the baculovirus system. *Biochimica et Biophysica Acta*, 1209, 130–139.

Barrow, C.J., Yasuda, A., Kenny, P.T., and Zagorski, M.G. (1992) Solution conformations and aggregational properties of synthetic amyloid beta-peptides of Alzheimer's disease. Analysis of circular dichroism spectra. *Journal of Molecular Biology*, 225, 1075–1093.

Barten, D.M., Guss, V.L., Corsa, J.A., Loo, A., Hansel, S.B., Zheng, M., Munoz, B., Srinivasan, K., Wang, B., Robertson, B.J., et al. (2005). Dynamics of {beta}-amyloid reductions in brain, cerebrospinal fluid, and plasma of {beta}-amyloid precursor protein transgenic mice treated with a {g-}-secretase inhibitor. *Journal of Pharmacology and Experimental Therapeutics*, 312, 635–643.

Bateman, R.J., Munsell, L.Y., Morris, J.C., Swarm, R., Yarasheski, K.E., and Holtzman, D.M. (2006) Human amyloid-beta synthesis and clearance rates as measured in cerebrospinal fluid in vivo. *Nature Medicine*, 12, 856–861.

Beher, D., and Graham, S.L. (2005) Protease inhibitors as potential disease-modifying therapeutics for Alzheimer's disease. *Expert Opinion on Investigational Drugs*, 14, 1385–1409.

Beher, D., Clarke, E.E., Wrigley, J.D., Martin, A.C., Nadin, A., Churcher, I., et al. (2004) Selected non-steroidal anti-inflammatory drugs and their derivatives target g-secretase at a novel site. Evidence for an allosteric mechanism. *Journal of Biological Chemistry*, 279, 43419–43426.

Bennett, B.D., Denis, P., Haniu, M., Teplow, D.B., Kahn, S., Louis, J.C., Citron, M., and Vassar, R. (2000) A furin-like convertase mediates propeptide cleavage of BACE, the Alzheimer's beta-secretase. *Journal of Biological Chemistry*, 275, 37712–37717.

Berezovska, O., Lleo, A., Herl, L.D., Frosch, M.P., Stern, E.A., Bacskai, B.J., et al. (2005) Familial Alzheimer's disease presenilin 1 mutations cause alterations in the conformation of presenilin and interactions with amyloid precursor protein. *Journal of Neuroscience*, 25, 3009–3017.

Bernado, P., Blanchard, L., Timmins, P., Marion, D., Ruigrok, R.W., and Blackledge, M. (2005) A structural model for unfolded proteins from residual dipolar couplings and small-angle x-ray scattering. *Proceedings of the National Academy of Sciences of the United States of America*, 102, 17002–17007.

Bernardo, A., Ajmone-Cat, M.A., Gasparini, L., Ongini, E., and Minghetti, L. (2005) Nuclear receptor peroxisome proliferator-activated receptor-g is activated in rat microglial cells by the anti-inflammatory drug HCT1026, a derivative of flurbiprofen. *Journal of Neurochemistry*, 92, 895–903.

Bernardo, A., Gasparini, L., Ongini, E., and Minghetti, L. (2006) Dynamic regulation of microglial functions by the non-steroidal anti-inflammatory drug NCX 2216: implications for chronic treatments of neurodegenerative diseases. *Neurobiology of Disease*, 22, 25–32.

Bernstein, S.L., Wyttenbach, T., Baumketner, A., Shea, J.E., Bitan, G., Teplow, D.B., and Bowers, M.T. (2005) Amyloid beta-protein: monomer structure and early aggregation states of Abeta42 and its Pro19 alloform. *Journal of the American Chemical Society*, 127, 2075–2084.

Best, J.D., Jay, M.T., Otu, F., Ma, J., Nadin, A., Ellis, S., Lewis, H.D., Pattison, C., Reilly, M., Harrison, T., et al. (2005) Quantitative measurement of changes in amyloid-beta(40) in the rat brain and cerebrospinal fluid following treatment with the g-secretase inhibitor LY-411575 [N2-[(2S)-2-(3,5-difluorophenyl)-2-hydroxyethanoyl]-N1-[(7S)-5-methyl-6-ox o-6,7-dihydro-5H-dibenzo[b,d]azepin-7-yl]-L-alaninamide]. *Journal of Pharmacology and Experimental Therapeutics*, 313, 902–908.

Best, J.D., Jay, M.T., Otu, F., Churcher, I., Reilly, M., Morentin-Gutierrez, P., Pattison, C., Harrison, T., Shearman, M.S., and Atack, J.R. (2006). In vivo characterisation of A{beta}(40) changes in brain and CSF using the novel {g-}-secretase inhibitor MRK-560 (*N*-[*cis*-4-[(4-chlorophenyl)sulfonyl]-4-(2,5-difluorophenyl)cyclohexyl]-1,1,1- trifluoromethanesulfonamide) in the rat. *Journal of Pharmacology and Experimental Therapeutics*, 317, 786–790.

Bhattacharyya, D.K., Lecomte, M., Rieke, C.J., Garavito, M., and Smith, W.L. (1996) Involvement of arginine 120, glutamate 524, and tyrosine 355 in the binding of arachidonate and 2-phenylpropionic acid inhibitors to the cyclooxygenase active site of ovine prostaglandin endoperoxide H synthase-1. *Journal of Biological Chemistry*, 271, 2179–2184.

Bieschke, J., Zhang, Q., Powers, E.T., Lerner, R.A., and Kelly, J.W. (2005) Oxidative metabolites accelerate Alzheimer's amyloidogenesis by a two-step mechanism, eliminating the requirement for nucleation. *Biochemistry*, 44, 4977–4983.

Bitan, G., Lomakin, A., and Teplow, D.B. (2001) Amyloid beta-protein oligomerization: prenucleation interactions revealed by photo-induced cross-linking of unmodified proteins. *Journal of Biological Chemistry*, 276, 35176–35184.

Bitan, G., Kirkitadze, M.D., Lomakin, A., Vollers, S.S., Benedek, G.B., and Teplow, D.B. (2003a) Amyloid beta-protein (Abeta) assembly: Abeta 40 and Abeta 42 oligomerize through distinct pathways. *Proceedings of the National Academy of Sciences of the United States of America*, 100, 330–335.

Bitan, G., Vollers, S.S., and Teplow, D.B. (2003b) Elucidation of primary structure elements controlling early amyloid beta-protein oligomerization. *Journal of Biological Chemistry*, 278, 34882–34889.

Bitan, G., Kirkitadze, M.D., Lomakin, A., Vollers, S.S., Benedek, G.B., and Teplow, D.B. (2003c) Amyloid beta-protein (Abeta) assembly: Abeta 40 and Abeta 42 oligomerize through distinct pathways. *Proceedings of the National Academy of Sciences of the United States of America*, 100, 330–335.

Blasko, I., Stampfer-Kountchev, M., Robatscher, P., Veerhuis, R., Eikelenboom, P., and Grubeck-Loebenstein, B. (2004) How chronic inflammation can affect the brain and support the development of Alzheimer's disease in old age: the role of microglia and astrocytes. *Aging Cell*, 3, 169–176.

Blasko, I., Kemmler, G., Krampla, W., Jungwirth, S., Wichart, I., Jellinger, K., et al. (2005) Plasma amyloid b protein 42 in non-demented persons aged 75 years: effects of concomitant medication and medial temporal lobe atrophy. *Neurobiology of Aging*, 26, 1135–1143.

Bloch, K. (1965) The biological synthesis of cholesterol. *Science*, 150, 19–28.

Braak, H. and Braak, E. (1991). Neuropathological staging of Alzheimer-related changes. *Acta Neuropatholgica* (Berl.), 82, 239–259.

Brandt, R., Hundelt, M., and Shahani, N. (2005) Tau alteration and neuronal degeneration in tauopathies: mechanisms and models. *Biochimica et Biophysica Acta*, 1739, 331–354.

Brera, B., Serrano, A., and de Cellabos, L. (2000) Beta-amyloid peptides are cytotoxic to astrocytes in culture: a role for oxidative stress. *Neurobiology of Disease*, 7, 395–405.

Breteler, M.M., in t' Veld, B.A., Breteler, M.M., and Stricker, B.H. (2002) Ab42 peptide lowering NSAIDs and Alzheimer's disease, *Eighth International Congress on Alzheimer's Disease and Related Disorders*, Stockholm, Sweden.

Brown, M.S. and Goldstein, J.L. (1997) The SREBP pathway: regulation of cholesterol metabolism by proteolysis of a membrane-bound transcription factor. *Cell*, 89, 331–340.

Brown, M.S. and Goldstein, J.L. (1999) A proteolytic pathway that controls the cholesterol content of membranes, cells, and blood. *Proceedings of the National Academy of Sciences of the United States of America*, 96, 11041–11048.

Brown, M.S., Ye, J., Rawson, R.B., and Goldstein, J.L. (2000) Regulated intramembrane proteolysis: a control mechanism conserved from bacteria to humans. *Cell*, 100, 391–398.

Bruno, V., Ksiazek, I., Battaglia, G., Lukic, S., Leonhardt, T., Sauer, D., Gasparini, F., Kuhn, R., Nicoletti, F., and Flor, P.J. (2000) Selective blockade of metabotropic glutamate receptor subtype 5 is neuroprotective. *Neuropharmacology*, 39, 2223–2230.

Buckner, R.L., Snyder, A.Z., Shannon, B.J., LaRossa, G., Sachs, R., Fotenos, A.F. et al. (2005) Molecular, structural, and functional characterization of Alzheimer's disease: evidence for a relationship between default activity, amyloid, and memory. *Journal of Neuroscience*, 25, 7709–7717.

Caballero, J. and Nahata, M. (2004) Do statins slow down Alzheimer's disease? A review. *Journal of Clinical Pharmacy and Therapeutics*, 29, 209–213.

Caccamo, A., Oddo, S., Billings, L.M., Green, K.N., Martinez-Coria, H., Fisher, A., and Laferla, F.M. (2006) M1 receptors play a central role in modulating AD-like pathology in transgenic mice. *Neuron*, 49, 671–682.

Cacquevel, M., Lebeurrier, N., Cheenne, S., and Vivien, D. (2004) Cytokines in neuro-inflammation and Alzheimer's disease. *Current Drug Targets*, 5, 529–534.

Cai, H., Wang, Y., McCarthy, D., Wen, H., Borchelt, D.R., Price, D.L., and Wong, P.C. (2001) BACE1 is the major beta-secretase for generation of Abeta peptides by neurons. *Nature Neuroscience*, 4, 233–234.

Calamai, M., Chiti, F., and Dobson, C.M. (2005) Amyloid fibril formation can proceed from different conformations of a partially unfolded protein. *Biophysics Journal*, 89, 4201–4210.

Camacho, I.E., Serneels, L., Spittaels, K., Merchiers, P., Dominguez, D., and De Strooper, B. (2004) Peroxisome-proliferator-activated receptor g induces a clearance mechanism for the amyloid-b peptide. *Journal of Neuroscience*, 24, 10908–10917.

Cannon, M.J., Williams, A.D., Wetzel, R., and Myszka, D.G. (2004) Kinetic analysis of beta-amyloid fibril elongation. *Analytical Biochemistry*, 328, 67–75.

Cao, X. and Sudhof, T.C. (2001) A transcriptionally [correction of transcriptively] active complex of APP with Fe65 and histone acetyltransferase Tip60. *Science*, 293, 115–120.

Capell, A., Beher, D., Prokop, S., Steiner, H., Kaether, C., Shearman, M.S., and Haass, C. (2005) G-secretase complex assembly within the early secretory pathway. *Journal of Biological Chemistry*, 280, 6471–6478.

Carlson, M.C., Tschanz, J.T., Norton, M.C., Welsh-Bohmer, K., Martin, B.K., and Breitner, J.C. (2002) H2 histamine receptor blockade in the treatment of Alzheimer disease: a randomized, double-blind, placebo-controlled trial of nizatidine. *Alzheimer Disease and Associated Disorders*, 16, 24–30.

Carr, D.B., Goate, A., Phil, D., and Morris, J.C. (1997) Current concepts in the pathogenesis of Alzheimer's disease. *American Journal of Medicine*, 103, 3S–10S.

Casal, C., Serratosa, J., and Tusell, J.M. (2004) Effects of b-AP peptides on activation of the transcription factor NF-kappaB and in cell proliferation in glial cell cultures. *Neuroscience Research*, 48, 315–323.

Casserly, I. and Topol, E. (2004) Convergence of atherosclerosis and Alzheimer's disease: inflammation, cholesterol, and misfolded proteins. *Lancet*, 363, 1139–1146.

Cattaneo, D. and Remuzzi, G. (2005) Lipid oxidative stress and the anti-inflammatory properties of statins and ACE inhibitors. *Journal of Renal Nutrition*, 15, 71–76.

Cattaneo, D., Baldelli, S., Merlini, S., Zenoni, S., Perico, N., and Remuzzi, G. (2004) Therapeutic use of HMG-CoA reductase inhibitors: current practice and future perspectives. *Expert Opinion on Therapeutic Patents*, 14, 1553–1566.

Chacon, M.A., Barria, M.I., Soto, C., and Inestrosa, N.C. (2004) Beta-sheet breaker peptide prevents Abeta-induced spatial memory impairments with partial reduction of amyloid deposits. *Molecular Psychiatry*, 9, 953–961.

Chai, Y., Koppenhafer, S.L., Bonini, N.M., and Paulson, H.L. (1999) Analysis of the role of heat shock protein (Hsp) molecular chaperones in polyglutamine disease. *Journal of Neuroscience*, 19, 10338–10347.

Chan, K.K., Oza, A.M., and Siu, L.L. (2003) The statins as anticancer agents. *Clinical Cancer Research*, 9, 10–19.

Chandrasekharan, N.V., Dai, H., Roos, K.L., Evanson, N.K., Tomsik, J., Elton, T.S., et al. (2002) COX-3, a cyclooxygenase-1 variant inhibited by acetaminophen and other analgesic/antipyretic drugs: cloning, structure, and expression. *Proceedings of the National Academy of Sciences of the United States of America*, 99, 13926–13931.

Chang, L., Bakhos, L., Wang, Z., Venton, D.L., and Klein, W.L. (2003) Femtomole immunodetection of synthetic and endogenous amyloid-beta oligomers and its application to Alzheimer's disease drug candidate screening. *Journal of Molecular Neuroscience*, 20, 305–313.

Chen, W., Pawelek, T.R., and Kulmacz, R.J. (1999) Hydroperoxide dependence and cooperative cyclooxygenase kinetics in prostaglandin H synthase-1 and -2. *Journal of Biological Chemistry*, 274, 20301–20306.

Chen, Q.S., Kagan, B.L., Hirakura, Y., and Xie, C.W. (2000) Impairment of hippocampal long-term potentiation by Alzheimer amyloid beta-peptides. *Journal of Neuroscience Research*, 60, 65–72.

Chiti, F., Webster, P., Taddei, N., Clark, A., Stefani, M., Ramponi, G., et al. (1999) Designing conditions for in vitro formation of amyloid protofilaments and fibrils [see comments]. *Proceedings of the National Academy of Sciences of the United States of America*, 96, 3590–3594.

Chou, P.Y. and Fasman, G.D. (1978) Empirical predictions of protein conformation. *Annual Reviews in Biochemistry*, 47, 251–276.

Christians, U., Jacobsen, W., and Floren, L.C. (1998) Metabolism and drug interactions of 3-hydroxy-3-methylglutaryl coenzyme A reductase inhibitors in transplant patients: are the statins mechanistically similar? *Pharmacology & Therapeutics*, 80, 1–34.

Christie, R., Yamada, M., Moskowitz, M., and Hyman, B. (2001) Structural and functional disruption of vascular smooth muscle cells in a transgenic mouse model of amyloid angiopathy. *American Journal of Pathology*, 158, 1065–1071.

Churcher, I. and Beher, D. (2005) G-secretase as a therapeutic target for the treatment of Alzheimer's disease. *Current Pharmacology Design*, 11, 3363–3382.

Cirrito, J.R., May, P.C., O'Dell, M.A., Taylor, J.W., Parsadanian, M., Cramer, J.W., et al. (2003) In vivo assessment of brain interstitial fluid with microdialysis reveals plaque-associated changes in amyloid-beta metabolism and half-life. *Journal of Neuroscience*, 23, 8844–8853.

Cirrito, J.R., Deane, R., Fagan, A.M., Spinner, M.L., Parsadanian, M., Finn, M.B. et al. (2005) P-glycoprotein deficiency at the blood–brain barrier increases amyloid-beta deposition in an Alzheimer disease mouse model. *Journal of Clinical Investigation*, 115, 3285–3290.

Citron, M., Westaway, D., Xia, W., Carlson, G., Diehl, T., Levesque, G., Johnson-Wood, K., Lee, M., Seubert, P., Davis, A., et al. (1997) Mutant presenilins of Alzheimer's disease increase production of 42-residue amyloid beta-protein in both transfected cells and transgenic mice. *Nature Medicine*, 3, 67–72.

Cleary, J.P., Walsh, D.M., Hofmeister, J.J., Shankar, G.M., Kuskowski, M.A., Selkoe, D.J., and Ashe, K.H. (2005) Natural oligomers of the amyloid-beta protein specifically disrupt cognitive function. *Nature Neuroscience*, 8, 79–84.

Clevers, H. (2006) Colon cancer—understanding how NSAIDs work. *New England Journal of Medicine*, 354, 761–763.

Cohen, A.S. and Calkins, E. (1959) Electron microscopic observations on a fibrous component in amyloid of diverse origins. *Nature*, 183, 1202–1203.

Cole, S.L., and Vassar, R. (2006) Isoprenoids and Alzheimer's disease: a complex relationship. *Neurobiology of Disease*, 22, 209–222.

Collins, S.R., Douglass, A., Vale, R.D., and Weissman, J.S. (2004) Mechanism of prion propagation: amyloid growth occurs by monomer addition. PLoS Biol, 2, E321.

Combs, C.K., Johnson, D.E., Karlo, J.C., Cannady, S.B., and Landreth, G.E. (2000) Inflammatory mechanisms in Alzheimer's disease: inhibition of b-amyloid-stimulated proinflammatory responses and neurotoxicity by PPARg agonists. *Journal of Neuroscience*, 20, 558–567.

Copeland, R.A., Williams, J.M., Giannaras, J., Nurnberg, S., Covington, M., Pinto, D., et al. (1994) Mechanism of selective inhibition of the inducible isoform of prostaglandin G/H synthase. *Proceedings of the National Academy of Sciences of the United States of America*, 91, 11202–11206.

Cordy, J.M., Hussain, I., Dingwall, C., Hooper, N.M., and Turner, A.J. (2003) Exclusively targeting beta-secretase to lipid rafts by GPI-anchor addition up-regulates beta-site processing of the amyloid precursor protein. *Proceedings of the National Academy of Sciences of the United States of America*, 100, 11735–11740.

Cornelius, C., Fastbom, J., Winblad, B., and Viitanen, M. (2004) Aspirin, NSAIDs, risk of dementia, and influence of the apolipoprotein E epsilon 4 allele in an elderly population. *Neuroepidemiology*, 23, 135–143.

Costello, D.A., O'Leary, D.M., and Herron, C.E. (2005) Agonists of peroxisome proliferator-activated receptor-g attenuate the Ab-mediated impairment of LTP in the hippocampus in vitro. *Neuropharmacology*, 49, 359–366.

Couzin, J. (2004) Clinical trials. Nail-biting time for trials of COX-2 drugs. *Science*, 306, 1673–1675.

Craft, D.L., Wein, L.M., and Selkoe, D.J. (2002) A mathematical model of the impact of novel treatments on the Abeta burden in the Alzheimer's brain, CSF and plasma. *Bulletin of Mathematical Biology*, 64, 1011–1031.

Craft, J.M., Van Eldik, L.J., Zasadzki, M., Hu, W., and Watterson, D.M. (2004) Aminopyridazines attenuate hippocampus-dependent behavioral deficits induced by human beta-amyloid in a Murine model of neuroinflammation. *Journal of Molecular Neuroscience*, 24, 115–122.

Dahlgren, K.N., Manelli, A.M., Stine, W.B., Jr., Baker, L.K., Krafft, G.A., and LaDu, M.J. (2002). Oligomeric and fibrillar species of amyloid-beta peptides differentially affect neuronal viability. *Journal of Biological Chemistry*, 277, 32046–32053.

Dalen, J.E. (2006) Aspirin to prevent heart attack and stroke: what's the right dose? *American Journal of Medicine*, 119, 198–202.

Davies, N.M., Good, R.L., Roupe, K.A., and Yanez, J.A. (2004) Cyclooxygenase-3: axiom, dogma, anomaly, enigma or splice error?—Not as easy as 1, 2, 3. *Journal of Pharmacy and Pharmaceutical Sciences*, 7, 217–226.

Davis, S. and Laroche, S. (2003) What can rodent models tell us about cognitive decline in Alzheimer's disease? *Molecular Neurobiology*, 27, 249–276.

Dedeoglu, A., Choi, J., Cormier, K.S.A.M., Ferrante, R.J., Jenkins, B.G., et al. (2003) Ibuprofen reduces Ab1–42/Ab1–40 ratio in Alzheimer mice cortex at ages where no metabolic changes are noted by magnetic resonance spectroscopy, *33rd Annual Meeting of Society of Neuroscience*, New Orleans, LA, Abstract 295.2.

DeKosky, S.T. (2005) Statin therapy in the treatment of Alzheimer disease: what is the rationale? *American Journal of Medicine*, 118 (Suppl 12A), 48–53.

Delhalle, S., Blasius, R., Dicato, M., and Diederich, M. (2004) A beginner's guide to NF-kappaB signaling pathways. *Annals of the New York Academy of Sciences*, 1030, 1–13.

DeMattos, R.B., Bales, K.R., Cummins, D.J., Paul, S.M., and Holtzman, D.M. (2002) Brain to plasma amyloid-b efflux: a measure of brain amyloid burden in a mouse model of Alzheimer's disease. *Science*, 295, 2264–2267.

Demuro, A., Mina, E., Kayed, R., Milton, S.C., Parker, I., and Glabe, C.G. (2005) Calcium dysregulation and membrane disruption as a ubiquitous neurotoxic mechanism of soluble amyloid oligomers. *Journal of Biological Chemistry*, 280, 17294–17300.

Deshpande, A., Mina, E., Glabe, C., and Busciglio, J. (2006) Different conformations of amyloid beta induce neurotoxicity by distinct mechanisms in human cortical neurons. *Journal of Neuroscience*, 26, 6011–6018.

DeWitt, D.L. (1999) Cox-2-selective inhibitors: the new super aspirins. *Molecular Pharmacology*, 55, 625–631.

Dinchuk, J.E., Liu, R.Q., and Trzaskos, J.M. (2003) COX-3: in the wrong frame in mind. *Immunology Letters*, 86, 121.

Dovey, H.F., John, V., Anderson, J.P., Chen, L.Z., de Saint Andrieu, P., Fang, L.Y., Freedman, S.B., Folmer, B., Goldbach, E., Holsztynska, E.J., et al. (2001) Functional g-secretase inhibitors reduce beta-amyloid peptide levels in brain. *Journal of Neurochemistry*, 76, 173–181.

Dreser, H. (1899) Pharmakologisches über aspirin (acetylsalicylsäure). Pflugers Archiv *European Journal of Physiology*, 76, 306–318.

Durrington, P. (2003) Dyslipidaemia. *Lancet*, 362, 717–731.

Edbauer, D., Winkler, E., Regula, J.T., Pesold, B., Steiner, H., and Haass, C. (2003) Reconstitution of g-secretase activity. *Nature Cell Biology*, 5, 486–488.

Ehehalt, R., Keller, P., Haass, C., Thiele, C., and Simons, K. (2003) Amyloidogenic processing of the Alzheimer beta-amyloid precursor protein depends on lipid rafts. *Journal of Cell Biology*, 160, 113–123.

El-Agnaf, O.M., Walsh, D.M., and Allsop, D. (2003) Soluble oligomers for the diagnosis of neurodegenerative diseases. *Lancet Neurology*, 2, 461–462.

Ellerbrock, B.R., Fleck, T.J., Landrum, C.S., Pauley, A.M., Carter, D.B., Peach, M.L., et al. (2003) NSAID modulators of the g-secretase show similar potencies at inhibition of Ab1–42 and induction of Ab1–38 as measured by a novel neoepitope-specific ELISA, *33rd Annual Meeting of Society of Neuroscience*, New Orleans, LA, Abstract 667.10.

Emmerling, M., Spiegel, K., Hall, E.D., LeVine, H. III, Walker, L., Schwarz, R.D., et al. (1999) Emerging strategies for the treatment of Alzheimer's disease at the millennium. *Emerging Drugs, The Prospect for Improved Medicines*, 4, 35–86.

Endo, A., Kuroda, M., and Tsujita, Y. (1976) ML-236A, ML-236B, and ML-236C, new inhibitors of cholesterogenesis produced by *Penicillium citrinium*. *Journal of Antibiotics*, 29, 1346–14348.

Eriksen, J.L., Nicolle, M.M., Prescott, S., Ozols, V.V., Monnier, T.E., Beard, J.L., et al. (2003a) Chronic treatment of transgenic APP mice with R-flurbiprofen, *33rd Annual Meeting of Society of Neuroscience*, New Orleans. LA, Program No. 295.22.

Eriksen, J.L., Sagi, S.A., Smith, T.E., Weggen, S., Das, P., McLendon, D.C., et al. (2003b) NSAIDs and enantiomers of flurbiprofen target g-secretase and lower Ab 42 in vivo. *Journal of Clinical Investigation*, 112, 440–449.

Eriksen, J.L., Wszolek, Z., and Petrucelli, L. (2005) Molecular pathogenesis of Parkinson disease. *Archives of Neurology*, 62, 353–357.

Esler, W.P., Stimson, E.R., Mantyh, P.W., and Maggio, J.E. (1999) Deposition of soluble amyloid-beta onto amyloid templates: with application for the identification of amyloid fibril extension inhibitors. *Methods in Enzymology*, 309, 350–374.

Estrada, L.D. and Soto, C. (2006) Inhibition of protein misfolding and aggregation by small rationally-designed peptides. *Current Pharmaceutical Design*, 12, 2557–2567.

Etminan, M., Gill, S., and Samii, A. (2003) Effect of non-steroidal anti-inflammatory drugs on risk of Alzheimer's disease: systematic review and meta-analysis of observational studies. *British Medical Journal*, 327, 128.

Evin, G., Canterford, L.D., Hoke, D.E., Sharples, R.A., Culvenor, J.G., and Masters, C.L. (2005). Transition-state analogue g-secretase inhibitors stabilize a 900 kDa presenilin/nicastrin complex. *Biochemistry*, 44, 4332–4341.

Exley, C., Price, N.C., Kelly, S.M., and Birchall, J.D. (1993) An interaction of beta-amyloid with aluminium in vitro. *FEBS Letters*, 324, 293–295.

Fabiola, G.F., Pattabhi, V., and Nagarajan, K. (1998) Structural basis for selective inhibition of COX-2 by nimesulide. *Bioorganic & Medicinal Chemistry*, 6, 2337–2344.

Farzan, M., Schnitzler, C.E., Vasilieva, N., Leung, D., and Choe, H. (2000) BACE2, a beta-secretase homolog, cleaves at the beta site and within the amyloid-beta region of the amyloid-beta precursor protein. *Proceedings of the National Academy of Sciences of the United States of America*, 97, 9712–9717.

Ferrer, I., Marti, E., Lopez, E., and Tortosa, A. (1998) NF-kB immunoreactivity is observed in association with b A4 diffuse plaques in patients with Alzheimer's disease. *Neuropathology and Applied Neurobiology*, 24, 271–277.

Fezoui, Y. and Teplow, D.B. (2002) Kinetic studies of amyloid beta-protein fibril assembly. Differential effects of alpha-helix stabilization. *Journal of Biological Chemistry*, 277, 36948–36954.

Fiala, J.C. and Harris, K.M. (2001) Extending unbiased stereology of brain ultrastructure to three-dimensional volumes. *Journal of the American Medical Informatics Association*, 8, 1–16.

Finckh, A. and Aronson, M.D. (2005) Cardiovascular risks of cyclooxygenase-2 inhibitors: where we stand now. *Annals of Internal Medicine*, 142, 212–214.

Findeis, M.A. (2002) Peptide inhibitors of beta amyloid aggregation. *Current Topics in Medicinal Chemistry*, 2, 417–423.

Fink, A.L. (2005) Natively unfolded proteins. *Current Opinion on Structural Biology*, 15, 35–41.

Fiorucci, S., Santucci, L., Sardina, M., Santus, G., Fransioni, A., Del Soldato, P., et al. (2003) Effect of HCT 1026, a nitric oxide (NO) releasing derivative of flurbiprofen on gastrointestinal mucosa: a double blind placebo-controlled endoscopic study, *The Digestive Disease Week 2003 Conference*, Orlando, FL.

FitzGerald, G.A. (2003) COX-2 and beyond: approaches to prostaglandin inhibition in human disease. Nature Reviews. *Drug Discovery*, 2, 879–890.

Fitzgerald, G.A. (2004) Coxibs and cardiovascular disease. *New England Journal of Medicine*, 351, 1709–1711.

Fowler, D.M., Koulov, A.V., Alory-Jost, C., Marks, M.S., Balch, W.E., and Kelly, J.W. (2005) Functional amyloid formation within Mammalian tissue. *PLoS Biol*, 4, e6.

Fraering, P.C., LaVoie, M.J., Ye, W., Ostaszewski, B.L., Kimberly, W.T., Selkoe, D.J., and Wolfe, M.S. (2004a) Detergent-dependent dissociation of active g-secretase reveals an interaction between Pen-2 and PS1-NTF and offers a model for subunit organization within the complex. *Biochemistry*, 43, 323–333.

Fraering, P.C., Ye, W., Strub, J.M., Dolios, G., LaVoie, M.J., Ostaszewski, B.L., van Dorsselaer, A., Wang, R., Selkoe, D.J., and Wolfe, M.S. (2004b) Purification and characterization of the human g-secretase complex. *Biochemistry*, 43, 9774–9789.

Fraering, P.C., Ye, W., Strub, J.M., Dolios, G., LaVoie, M.J., Ostaszewski, B.L., et al. (2004c) Purification and characterization of the human g-secretase complex *Biochemistry*, 43, 9774–9789.

Freir, D.B., Costello, D.A., and Herron, C.E. (2003) A beta 25–35-induced depression of long-term potentiation in area CA1 in vivo and in vitro is attenuated by verapamil. *Journal of Neurophysiology*, 89, 3061–3069.

Frenkel, D., Solomon, B., and Benhar, I. (2000) Modulation of Alzheimer's beta-amyloid neurotoxicity by site-directed single-chain antibody. *Journal of Neuroimmunology*, 106, 23–31.

Friedhoff, L.T., Cullen, E.I., Geoghagen, N.S., and Buxbaum, J.D. (2001) Treatment with controlled-release lovastatin decreases serum concentrations of human beta-amyloid (A beta) peptide. *International Journal of Neuropsychopharmacology*, 4, 127–130.

Fukumoto, H., Cheung, B.S., Hyman, B.T., and Irizarry, M.C. (2002) Beta-secretase protein and activity are increased in the neocortex in Alzheimer disease. *Archive in Neurology*, 59, 1381–1389.

Fukuyama, R., Mizuno, T., Mori, S., Nakajima, K., Fushiki, S., and Yanagisawa, K. (2000) Age-dependent change in the levels of Abeta40 and Abeta42 in cerebrospinal fluid from control subjects, and a decrease in the ratio of Abeta42 to Abeta40 level in cerebrospinal fluid from Alzheimer's disease patients. *European Neurology*, 43, 155–160.

Futaki, N., Yoshikawa, K., Hamasaka, Y., Arai, I., Higuchi, S., Iizuka, H., et al. (1993) NS-398, a novel non-steroidal anti-inflammatory drug with potent analgesic and antipyretic effects, which causes minimal stomach lesions. *General Pharmacology*, 24, 105–110.

Gandy, S. (2005) The role of cerebral amyloid b accumulation in common forms of Alzheimer disease. *Journal of Clinical Investigation*, 115, 1121–1129.

Gans, K.R., Galbraith, W., Roman, R.J., Haber, S.B., Kerr, J.S., Schmidt, W.K., et al. (1990) Anti-inflammatory and safety profile of DuP 697, a novel orally effective prostaglandin synthesis inhibitor. *Journal of Pharmacology and Experimental Therapeutics*, 254, 180–187.

Garavito, R.M. (1996) The cyclooxygenase-2 structure: new drugs for an old target? *Nature Structural Biology*, 3, 897–901.

Garcia-Mata, R., Gao, Y.S., and Sztul, E. (2002) Hassles with taking out the garbage: aggravating aggresomes. *Traffic*, 3, 388–396.

Garcia, M.J., Reinoso, R.F., Sanchez Navarro, A., and Prous, J.R. (2003) Clinical pharmacokinetics of statins. *Methods and Findings in Experimental and Clinical Pharmacology*, 25, 457–481.

Garcia-Nieto, R., Perez, C., Checa, A., and Gago, F. (1999) Molecular model of the interaction between nimesulide and human cyclooxygenase-2. *Rheumatology* (Oxford), 38 (Suppl 1), 14–18.

Garzon-Rodriguez, W., Sepulveda-Becerra, M., Milton, S., and Glabe, C.G. (1997) Soluble amyloid Abeta-(1–40) exists as a stable dimer at low concentrations. *Journal of Biological Chemistry*, 272, 21037–21044.

Gasparini, L., Ongini, E., and Wenk, G. (2004a) Non-steroidal anti-inflammatory drugs (NSAIDs) in Alzheimer's disease: old and new mechanisms of action. *Journal of Neurochemistry*, 91, 521–536.

Gasparini, L., Rusconi, L., Xu, H., del Soldato, P., and Ongini, E. (2004b) Modulation of b-amyloid metabolism by non-steroidal anti-inflammatory drugs in neuronal cell cultures. *Journal of Neurochemistry*, 88, 337–348.

Gasparini, L., Ongini, E., Wilcock, D., and Morgan, D. (2005) Activity of flurbiprofen and chemically related anti-inflammatory drugs in models of Alzheimer's disease. Brain Research. *Brain Research Reviews*, 48, 400–408.

Georganopoulou, D.G., Chang, L., Nam, J.M., Thaxton, C.S., Mufson, E.J., Klein, W.L. et al. (2005) From the cover: nanoparticle-based detection in cerebral spinal fluid of a soluble pathogenic biomarker for Alzheimer's disease. *Proceedings of the National Academy of Sciences of the United States of America*, 102, 2273–2276.

Ghanta, J., Shen, C.L., Kiessling, L.L., and Murphy, R.M. (1996) A strategy for designing inhibitors of beta-amyloid toxicity. *Journal of Biological Chemistry*, 271, 29525–29528.

Gierse, J.K., McDonald, J.J., Hauser, S.D., Rangwala, S.H., Koboldt, C.M., and Seibert, K. (1996) A single amino acid difference between cyclooxygenase-1 (COX-1) and -2 (COX-2) reverses the selectivity of COX-2 specific inhibitors. *Journal of Biological Chemistry*, 271, 15810–15814.

Giffard, R.G., Xu, L., Zhao, H., Carrico, W., Ouyang, Y., Qiao, Y. et al. (2004) Chaperones, protein aggregation, and brain protection from hypoxic/ischemic injury. *Journal of Experimental Biology*, 207, 3213–3220.

Gil, G., Faust, J.R., Chin, D.J., Goldstein, J.L., and Brown, M.S. (1985) Membrane-bound domain of HMG CoA reductase is required for sterol-enhanced degradation of the enzyme. *Cell*, 41, 249–258.

Glaser, J.R. and Glaser, E.M. (2000) Stereology, morphometry, and mapping: the whole is greater than the sum of its parts. *Journal of Chemical Neuroanatomy*, 20, 115–126.

Glenner, G.G. and Wong, C.W. (1984) Alzheimer's disease: initial report of the purification and characterization of a novel cerebrovascular amyloid protein. *Biochemical and Biophysical Research Communications*, 120, 885–890.

Goldstein, J.L. and Brown, M.S. (1990) Regulation of the mevalonate pathway. *Nature*, 343, 425–430.

Graaf, M.R., Beiderbeck, A.B., Egberts, A.C., Richel, D.J., and Guchelaar, H.J. (2004) The risk of cancer in users of statins. *Journal of Clinical Oncology*, 22, 2388–2394.

Grace, E.A., Rabiner, C.A., and Busciglio, J. (2002) Characterization of neuronal dystrophy induced by fibrillar amyloid beta: implications for Alzheimer's disease. *Neuroscience*, 114, 265–273.

Grimwood, S., Hogg, J., Jay, M.T., Lad, A.M., Lee, V., Murray, F. et al. (2005) Determination of guinea-pig cortical gamma-secretase activity ex vivo following the systemic administration of a gamma-secretase inhibitor. *Neuropharmacology*, 48, 1002–1011.

Gruninger-Leitch, F., Schlatter, D., Kung, E., Nelbock, P., and Dobeli, H. (2002) Substrate and inhibitor profile of BACE (beta-secretase) and comparison with other mammalian aspartic proteases. *Journal of Biological Chemistry*, 277, 4687–4693.

Guenette, S.Y. (2003) Mechanisms of Abeta clearance and catabolism. *Neuromolecular Medicine*, 4, 147–160.

Gunasekaran, K., Tsai, C.J., and Nussinov, R. (2004) Analysis of ordered and disordered protein complexes reveals structural features discriminating between stable and unstable monomers. *Journal of Molecular Biology*, 341, 1327–1341.

Haass, C. (2004) Take five-BACE and the g-secretase quartet conduct Alzheimer's amyloid beta-peptide generation. *EMBO Journal*, 23, 483–488.

Haass, C. and Steiner, H. (2002) Alzheimer disease g-secretase: a complex story of GxGD-type presenilin proteases. *Trends in Cell Biology*, 12, 556–562.

Hadland, B.K., Manley, N.R., Su, D., Longmore, G.D., Moore, C.L., Wolfe, M.S., et al. (2001) g-Secretase inhibitors repress thymocyte development. *Proceedings of the National Academy of Sciences of the United States of America*, 98, 7487–7491.

Haes, A.J., Chang, L., Klein, W.L., and Van Duyne, R.P. (2005) Detection of a biomarker for Alzheimer's disease from synthetic and clinical samples using a nanoscale optical biosensor. *Journal of the American Chemical Society*, 127, 2264–2271.

Hamberg, M. and Samuelsson, B. (1967a) On the mechanism of the biosynthesis of prostaglandins E-1 and F-1-alpha. *Journal of Biological Chemistry*, 242, 5336–5343.

Hamberg, M. and Samuelsson, B. (1967b) Oxygenation of unsaturated fatty acids by the vesicular gland of sheep. *Journal of Biological Chemistry*, 242, 5344–5354.

Harrower, T.P., Michell, A.W., and Barker, R.A. (2005) Lewy bodies in Parkinson's disease: protectors or perpetrators? *Experimental Neurology*, 195, 1–6.

Hartman, R.E., Izumi, Y., Bales, K.R., Paul, S.M., Wozniak, D.F., and Holtzman, D.M. (2005) Treatment with an amyloid-beta antibody ameliorates plaque load, learning deficits, and hippocampal long-term potentiation in a mouse model of Alzheimer's disease. *Journal of Neuroscience*, 25, 6213–6220.

Hashimoto, M., Rockenstein, E., and Masliah, E. (2003) Transgenic models of alpha-synuclein pathology: past, present, and future. *Annals of the New York Academy of Science*, 991, 171–188.

Hauss-Wegrzyniak, B., Vraniak, P., and Wenk, G.L. (1999) The effects of a novel NSAID on chronic neuroinflammation are age dependent. *Neurobiology of Aging*, 20, 305–313.

Heart Protection Study Collaborative Group (2002) MRC/BHF heart protection study of cholesterol lowering with simvastatin in 2,0536 high-risk individuals: a randomised placebo-controlled trial. *Lancet*, 360, 7–22.

Hemming, M.L. and Selkoe, D.J. (2005) Amyloid beta-protein is degraded by cellular angiotensin-converting enzyme (ACE) and elevated by an ACE inhibitor. *Journal of Biological Chemistry*, 280, 37644–37650.

Heneka, M.T., Sastre, M., Dumitrescu-Ozimek, L., Hanke, A., Dewachter, I., Kuiperi, C., et al. (2005) Acute treatment with the PPARg agonist pioglitazone and ibuprofen reduces glial inflammation and Ab1–42 levels in APPV717I transgenic mice. *Brain*, 128, 1442–1453.

Herbst, M. and Wanker, E.E. (2006) Therapeutic approaches to polyglutamine diseases: combating protein misfolding and aggregation. *Current Pharmaceutical Design*, 12, 2543–2555.

Herschman, H.R. (1994) Regulation of prostaglandin synthase-1 and prostaglandin synthase-2. *Cancer Metastasis Reviews*, 13, 241–256.

Herschman, H.R. (1996) Prostaglandin synthase 2. *Biochimica et Biophysica Acta*, 1299, 125–140.

Herzig, M.C., Van Nostrand, W.E., and Jucker, M. (2006) Mechanism of cerebral beta-amyloid angiopathy: murine and cellular models. *Brain Pathology*, 16, 40–54.

Higgins, G.A. and Jacobsen, H. (2003) Transgenic mouse models of Alzheimer's disease: phenotype and application. *Behavioural Pharmacology*, 14, 419–438.

Hilbich, C., Kisters-Woike, B., Reed, J., Masters, C.L., and Beyreuther, K. (1992) Substitutions of hydrophobic amino acids reduce the amyloidogenicity of Alzheimer's disease beta A4 peptides. *Journal of Molecular Biology*, 228, 460–473.

Hla, T., Ristimaki, A., Appleby, S., and Barriocanal, J.G. (1993) Cyclooxygenase gene expression in inflammation and angiogenesis. *Annals of the New York Academy of Sciences*, 696, 197–204.

Ho, L., Purohit, D., Haroutunian, V., Luterman, J.D., Willis, F., Naslund, J., et al. (2001) Neuronal cyclooxygenase 2 expression in the hippocampal formation as a function of the clinical progression of Alzheimer disease. *Archives of Neurology*, 58, 487–492.

Hoglund, K., Syversen, S., Lewczuk, P., Wallin, A., Wiltfang, J., and Blennow, K. (2005) Statin treatment and a disease-specific pattern of beta-amyloid peptides in Alzheimer's disease. *Experimental Brain Research*, 164, 205–214.

Holsinger, R.M., McLean, C.A., Beyreuther, K., Masters, C.L., and Evin, G. (2002) Increased expression of the amyloid precursor beta-secretase in Alzheimer's disease. *Annals in Neurology*, 51, 783–786.

Hoozemans, J.J. and O'Banion, M.K. (2005) The role of COX-1 and COX-2 in Alzheimer's disease pathology and the therapeutic potentials of non-steroidal anti-inflammatory drugs. *Current Drug Targets. CNS and Neurological Disorders*, 4, 307–315.

Hoshi, M., Sato, M., Matsumoto, S., Noguchi, A., Yasutake, K., Yoshida, N., and Sato, K. (2003) Spherical aggregates of beta-amyloid (amylospheroid) show high neurotoxicity and activate tau protein kinase I/glycogen synthase kinase-3beta. *Proceedings of the National Academy of Sciences of the United States of America*, 100, 6370–6375.

Hsi, L.C., Hoganson, C.W., Babcock, G.T., and Smith, W.L. (1994) Characterization of a tyrosyl radical in prostaglandin endoperoxide synthase-2. *Biochemical and Biophysical Research Communications*, 202, 1592–1598.

Imbimbo, B.P., Del Giudice, E., Cenacchi, V., Villetti, G., Facchinetti, F., and Leon, A. (2006) Pharmacokinetics and pharmacodynamics of CHF5022 and CHF5074, two new b-amyloid 1–42 lowering agents, after multiple doses in Tg2576 transgenic mice, *10th International Conference on Alzheimer Disease and Related Disorders*, Madrid, Spain.

in't Veld, B.A., Ruitenberg, A., Hofman, A., Launer, L.J., van Duijn, C.M., Stijnen, T., et al. (2001) Nonsteroidal anti-inflammatory drugs and the risk of Alzheimer's disease. *New England Journal of Medicine*, 345, 1515–1521.

Inestrosa, N.C., Godoy, J.A., Quintanilla, R.A., Koenig, C.S., and Bronfman, M. (2005) Peroxisome proliferator-activated receptor g is expressed in hippocampal neurons and its activation prevents b-amyloid neurodegeneration: role of Wnt signaling. *Experimental Cell Research*, 304, 91–104.

Irie, K., Murakami, K., Masuda, Y., Morimoto, A., Ohigashi, H., Ohashi, R., Takegoshi, K., Nagao, M., Shimizu, T., and Shirasawa, T. (2005) Structure of beta-amyloid fibrils and its relevance to their neurotoxicity: implications for the pathogenesis of Alzheimer's disease. *Journal of Bioscience and Bioengineering*, 99, 437–447.

Istvan, E.S. and Deisenhofer, J. (2000) The structure of the catalytic portion of human HMG-CoA reductase. *Biochimica & Biophysica Acta*, 1529, 9–18.

Istvan, E.S. and Deisenhofer, J. (2001) Structural mechanism for statin inhibition of HMG-CoA reductase. *Science*, 292, 1160–1164.

Istvan, E.S., Palnitkar, M., Buchanan, S.K., and Deisenhofer, J. (2000) Crystal structure of the catalytic portion of human HMG-CoA reductase: insights into regulation of activity and catalysis. *EMBO Journal*, 19, 819–830.

Jang, J.H., and Surh, Y.J. (2005) b-amyloid-induced apoptosis is associated with cyclooxygenase-2 up-regulation via the mitogen-activated protein kinase-NF-kappaB signaling pathway. *Free Radical Biology & Medicine*, 38, 1604–1613.

Jantzen, P.T., Connor, K.E., DiCarlo, G., Wenk, G.L., Wallace, J.L., Rojiani, A.M., et al. (2002) Microglial activation and b-amyloid deposit reduction caused by a nitric oxide-releasing nonsteroidal anti-inflammatory drug in amyloid precursor protein plus presenilin-1 transgenic mice. *Journal of Neuroscience*, 22, 2246–2254.

Jaradat, M.S., Wongsud, B., Phornchirasilp, S., Rangwala, S.M., Shams, G., Sutton, M., et al. (2001) Activation of peroxisome proliferator-activated receptor isoforms and inhibition of prostaglandin H(2) synthases by ibuprofen, naproxen, and indomethacin. *Biochemical Pharmacology*, 62, 1587–1595.

Jensen, M.T., Mottin, M.D., Cracchiolo, J.R., Leighty, R.E., and Arendash, G.W. (2005) Lifelong immunization with human beta-amyloid (1–42) protects Alzheimer's transgenic mice against cognitive impairment throughout aging. *Neuroscience*, 130, 667–684.

Jiang, C., Ting, A.T., and Seed, B. (1998) PPAR-g agonists inhibit production of monocyte inflammatory cytokines. *Nature*, 391, 82–86.

Jick, H., Zornberg, G.L., Jick, S.S., Seshadri, S., and Drachman, D.A. (2000) Statins and the risk of dementia. *Lancet*, 356, 1627–1631.

Johnson-Anuna, L.N., Eckert, G.P., Keller, J.H., Igbavboa, U., Franke, C., Fechner, T., et al. (2005) Chronic administration of statins alters multiple gene expression patterns in mouse cerebral cortex. *Journal of Pharmacology and Experimental Therapeutics*, 312, 786–793.

Johnstone, M., Gearing, A.J., and Miller, K.M. (1999) A central role for astrocytes in the inflammatory response to b-amyloid; chemokines, cytokines and reactive oxygen species are produced. *Journal of Neuroimmunology*, 93, 182–193.

Joo, Y., Kim, H.S., Woo, R.S., Park, C.H., Shin, K.Y., Lee, J.P., et al. (2006) Mefenamic acid shows neuroprotective effects and improves cognitive impairment in in vitro and in vivo Alzheimer's disease models. *Molecular Pharmacology*, 69, 76–84.

Kalback, W., Watson, M.D., Kokjohn, T.A., Kuo, Y.-M., Weiss, N., Luehrs, D.C. et al. (2002) APP transgenic mice Tg2576 accumulate Abeta peptides that are distinct from the chemically modified and insoluble peptides deposited in Alzheimer's disease senile plaques. *Biochemistry*, 41, 922–928.

Kalgutkar, A.S., Crews, B.C., Rowlinson, S.W., Garner, C., Seibert, K., and Marnett, L.J. (1998) Aspirin-like molecules that covalently inactivate cyclooxygenase-2. *Science*, 280, 1268–1270.

Kane, M.D., Lipinski, W.J., Callahan, M.J., Bian, F., Durham, R.A., Schwarz, R.D. et al. (2000) Evidence for seeding of beta-amyloid by intracerebral infusion of Alzheimer brain extracts in beta-amyloid precursor protein-transgenic mice. *Journal of Neuroscience*, 20, 3606–3611.

Kang, J., Lemaire, H.G., Unterbeck, A., Salbaum, J.M., Masters, C.L., Grzeschik, K.H., Multhaup, G., Beyreuther, K., and Muller, H.B. (1987) The precursor of Alzheimer's disease amyloid A4 protein resembles a cell-surface receptor. *Nature*, 325, 733–736.

Kawarabayashi, T., Younkin, L.H., Saido, T.C., Shoji, M., Ashe, K.H., and Younkin, S.G. (2001) Age-dependent changes in brain, CSF, and plasma amyloid (beta) protein in the Tg2576 transgenic mouse model of Alzheimer's disease. *Journal of Neuroscience*, 21, 372–381.

Kayed, R., Head, E., Thompson, J.L., McIntire, T.M., Milton, S.C., Cotman, C.W. et al. (2003) Common structure of soluble amyloid oligomers implies common mechanism of pathogenesis. *Science*, 300, 486–489.

Kayed, R., Sokolov, Y., Edmonds, B., MacIntire, T.M., Milton, S.C., Hall, J.E. et al. (2004) Permeabilization of lipid bilayers is a common conformation-dependent activity of soluble amyloid oligomers in protein mis-folding diseases. *Journal of Biological Chemistry*, 279, 46363–46366.

Keller, R.K., Zhao, Z., Chambers, C., and Ness, G.C. (1996) Farnesol is not the nonsterol regulator mediating degradation of HMG-CoA reductase in rat liver. *Archives of Biochemistry & Biophysics*, 328, 324–330.

Khachaturian, A.S., Zandi, P.P., Lyketsos, C.G., Hayden, K.M., Skoog, I., Norton, M.C. et al. (2006) Antihypertensive medication use and incident Alzheimer disease: the Cache County Study. *Archives of Neurology*, 63, 686–692.

Kiefer, J.R., Pawlitz, J.L., Moreland, K.T., Stegeman, R.A., Hood, W.F., Gierse, J.K., et al. (2000) Structural insights into the stereochemistry of the cyclooxygenase reaction. *Nature*, 405, 97–101.

Kielian, T. and Drew, P.D. (2003) Effects of peroxisome proliferator-activated receptor-g agonists on central nervous system inflammation. *Journal of Neuroscience Research*, 71, 315–325.

Kikuchi, R., Kusuhara, H., Abe, T., Endou, H., and Sugiyama, Y. (2004) Involvement of multiple transporters in the efflux of 3-hydroxy-3-methylglutaryl-CoA reductase inhibitors across the blood-brain barrier. *Journal of Pharmacology and Experimental Therapeutics*, 311, 1147–1153.

Kim, C.A. and Berg, J.M. (1993) Thermodynamic beta-sheet propensities measured using a zinc-finger host peptide. *Nature*, 362, 267–270.

Kimberly, W.T., LaVoie, M.J., Ostaszewski, B.L., Ye, W., Wolfe, M.S., and Selkoe, D.J. (2003) g-secretase is a membrane protein complex comprised of presenilin, nicastrin, Aph-1, and Pen-2. *Proceedings of the National Academy of Sciences of the United States of America*, 100, 6382–6387.

Kirschner, D.A., Abraham, C., and Selkoe, D.J. (1986) X-ray diffraction from intraneuronal paired helical filaments and extraneuronal amyloid fibers in Alzheimer disease indicates cross-beta conformation. *Proceedings of the National Academy of Sciences of the United States of America*, 83, 503–507.

Kisilevsky, R., Lemieux, L.J., Fraser, P.E., Kong, X., Hultin, P.G., and Szarek, W.A. (1995) Arresting amyloidosis in vivo using small-molecule anionic sulphonates or sulphates: implications for Alzheimer's disease [see comments]. *Nature Medicine*, 1, 143–148.

Kitamura, Y., Shimohama, S., Koike, H., Kakimura, J., Matsuoka, Y., Nomura, Y., et al. (1999) Increased expression of cyclooxygenases and peroxisome proliferator-activated receptor-g in Alzheimer's disease brains. *Biochemical and Biophysical Research Communications*, 254, 582–586.

Kitazume, S., Tachida, Y., Oka, R., Shirotani, K., Saido, T.C., and Hashimoto, Y. (2001) Alzheimer's beta-secretase, beta-site amyloid precursor protein-cleaving enzyme, is responsible for cleavage secretion of a Golgi-resident sialyltransferase. *Proceedings of the National Academy of Sciences of the United States of America*, 98, 13554–13559.

Klein, W.L. (2002) Amyloid beta toxicity in Alzheimer's Disease: globular oligomers (ADDLs) a new vaccine and drug targets. *Neurochemistry International*, 41, 345–352.

Klucken, J., Shin, Y., Masliah, E., Hyman, B.T., and McLean, P.J. (2004) Hsp70 reduces alpha-synuclein aggregation and toxicity. *Journal of Biological Chemistry*, 279, 25497–25502.

Klunk, W.E., Wang, Y., Huang, G.F., Debnath, M.L., Holt, D.P., Shao, L. et al. (2003) The binding of 2-(4¢-methylaminophenyl)benzothiazole to postmortem brain homogenates is dominated by the amyloid component. *Journal of Neuroscience*, 23, 2086–2092.

Klunk, W.E., Engler, H., Nordberg, A., Wang, Y., Blomqvist, G., Holt, D.P. et al. (2004) Imaging brain amyloid in Alzheimer's disease with Pittsburgh compound-B. *Annals of Neurology*, 55, 306–319.

Klunk, W.E., Lopresti, B.J., Ikonomovic, M.D., Lefterov, I.M., Koldamova, R.P., Abrahamson, E.E. et al. (2005) Binding of the positron emission tomography tracer Pittsburgh compound-B reflects the amount of amyloid-beta in Alzheimer's disease brain but not in transgenic mouse brain. *Journal of Neuroscience*, 25, 10598–10606.

Klyubin, I., Walsh, D.M., Lemere, C.A., Cullen, W.K., Shankar, G.M., Betts, V. et al. (2005) Amyloid beta protein immunotherapy neutralizes Abeta oligomers that disrupt synaptic plasticity in vivo. *Nature Medicine*, 11, 556–561.

Kojo, H., Fukagawa, M., Tajima, K., Suzuki, A., Fujimura, T., Aramori, I., et al. (2003) Evaluation of human peroxisome proliferator-activated receptor (PPAR) subtype selectivity of a variety of anti-inflammatory drugs based on a novel assay for PPAR d(b). *Journal of Pharmacological Sciences*, 93, 347–355.

Kojro, E., Postina, R., Buro, C., Meiringer, C., Gehrig-Burger, K., and Fahrenholz, F. (2006) The neuropeptide PACAP promotes the alpha-secretase pathway for processing the Alzheimer amyloid precursor protein. *FASEB Journal*, 20, 512–514.

Kokkoni, N., Stott, K., Amijee, H., Mason, J.M., and Doig, A.J. (2006) N-methylated peptide inhibitors of beta-amyloid aggregation and toxicity. Optimization of the inhibitor structure. *Biochemistry*, 45, 9906–9918.

Kokubo, H., Kayed, R., Glabe, C.G., and Yamaguchi, H. (2005) Soluble Abeta oligomers ultrastructurally localize to cell processes and might be related to synaptic dysfunction in Alzheimer's disease brain. *Brain Research*, 1031, 222–228.

Komers, R. and Epstein, M. (2002) Cyclooxygenase-2 expression and function in renal pathophysiology. *Journal of Hypertension* (Supplement: official journal of the International Society of Hypertension), 20, S11–S15.

Kota, B.P., Huang, T.H., and Roufogalis, B.D. (2005) An overview on biological mechanisms of PPARs. *Pharmacological Research*, 51, 85–94.

Kujubu, D.A., Fletcher, B.S., Varnum, B.C., Lim, R.W., and Herschman, H.R. (1991) TIS10, a phorbol ester tumor promoter-inducible mRNA from Swiss 3T3 cells, encodes a novel prostaglandin synthase/cyclooxygenase homologue. *Journal of Biological Chemistry*, 266, 12866–12872.

Kukar, T., Murphy, M.P., Eriksen, J.L., Sagi, S.A., Weggen, S., Smith, T.E., et al. (2005) Diverse compounds mimic Alzheimer disease-causing mutations by augmenting Ab42 production. *Nature Medicine*, 11, 545–550.

Kulmacz, R.J., Pendleton, R.B., and Lands, W.E. (1994) Interaction between peroxidase and cyclooxygenase activities in prostaglandin-endoperoxide synthase. Interpretation of reaction kinetics. *Journal of Biological Chemistry*, 269, 5527–5536.

Kulstad, J.J., McMillan, P.J., Leverenz, J.B., Cook, D.G., Green, P.S., Peskind, E.R., et al. (2005) Effects of chronic glucocorticoid administration on insulin-degrading enzyme and amyloid-beta peptide in the aged macaque. *Journal of Neuropathology and Experimental Neurology*, 64, 139–146.

Kurumbail, R.G., Stevens, A.M., Gierse, J.K., McDonald, J.J., Stegeman, R.A., Pak, J.Y., et al. (1996) Structural basis for selective inhibition of cyclooxygenase-2 by anti-inflammatory agents. *Nature*, 384, 644–648.

Kurumbail, R.G., Kiefer, J.R., and Marnett, L.J. (2001) Cyclooxygenase enzymes: catalysis and inhibition. *Current Opinion in Structural Biology*, 11, 752–760.

Lacor, P.N., Buniel, M.C., Chang, L., Fernandez, S.J., Gong, Y., Viola, K.L., et al. (2004) Synaptic targeting by Alzheimer's-related amyloid beta oligomers. *Journal of Neuroscience*, 24, 10191–10200.

Lambert, M.P., Barlow, A.K., Chromy, B.A., Edwards, C., Freed, R., Liosatos, M., Morgan, T.E., Rozovsky, I., Trommer, B., Viola, K.L., et al. (1998) Diffusible, nonfibrillar ligands derived from Abeta1–42 are potent central nervous system neurotoxins. *Proceedings of the National Academy of Sciences of the United States of America*, 95, 6448–6453.

Lambert, M.P., Viola, K.L., Chromy, B.A., Chang, L., Morgan, T.E., Yu, J., et al. (2001) Vaccination with soluble Abeta oligomers generates toxicity-neutralizing antibodies. *Journal of Neurochemistry*, 79, 595–605.

Lammich, S., Schobel, S., Zimmer, A.K., Lichtenthaler, S.F., and Haass, C. (2004) Expression of the Alzheimer protease BACE1 is suppressed via its 5¢-untranslated region. *EMBO Reports*, 5, 620–625.

Landi, F., Cesari, M., Onder, G., Russo, A., Torre, S., and Bernabei, R. (2003) Non-steroidal anti-inflammatory drug (NSAID) use and Alzheimer disease in community-dwelling elderly patients. *American Journal of Geriatric Psychiatry*, 11, 179–185.

Landino, L.M., Crews, B.C., Gierse, J.K., Hauser, S.D., and Marnett, L.J. (1997) Mutational analysis of the role of the distal histidine and glutamine residues of prostaglandin-endoperoxide synthase-2 in peroxidase catalysis, hydroperoxide reduction, and cyclooxygenase activation. *Journal of Biological Chemistry*, 272, 21565–21574.

Laneuville, O., Breuer, D.K., Dewitt, D.L., Hla, T., Funk, C.D., and Smith, W.L. (1994) Differential inhibition of human prostaglandin endoperoxide H synthases-1 and -2 by nonsteroidal anti-inflammatory drugs. *Journal of Pharmacology and Experimental Therapeutics*, 271, 927–934.

Laneuville, O., Breuer, D.K., Xu, N., Huang, Z.H., Gage, D.A., Watson, J.T., et al. (1995) Fatty acid substrate specificities of human prostaglandin-endoperoxide H synthase-1 and -2. Formation of 12-hydroxy-(9Z, 13E/Z, 15Z)-octadecatrienoic acids from alpha-linolenic acid. *Journal of Biological Chemistry*, 270, 19330–19336.

Lanz, T.A., Hosley, J.D., Adams, W.J., and Merchant, K.M. (2004) Studies of Abeta pharmacodynamics in the brain, cerebrospinal fluid, and plasma in young (plaque-free) Tg2576 mice using the g-secretase inhibitor N2-[(2S)-2-(3,5-difluorophenyl)-2-hydroxyethanoyl]-N1-[(7S)-5-methyl-6-oxo-6,7-dihydro-5H-dibenzo[b,d]azepin-7-yl]-L-alaninamide (LY-411575). *Journal of Pharmacology and Experimental Therapeutics*, 309, 49–55.

Lanz, T.A., Fici, G.J., and Merchant, K.M. (2005) Lack of specific amyloid-b1–42 suppression by nonsteroidal anti-inflammatory drugs in young, plaque-free Tg2576 mice and in guinea pig neuronal cultures. *Journal of Pharmacology and Experimental Therapeutics*, 312, 399–406.

Lashuel, H.A., Hartley, D., Petre, B.M., Walz, T., and Lansbury, P.T., Jr. (2002) Neurodegenerative disease: amyloid pores from pathogenic mutations. *Nature*, 418, 291.

Launer, L.J. (2003) Nonsteroidal anti-inflammatory drugs and Alzheimer disease: what's next? *Journal of American Medical Association*, 289, 2865–2867.

LaVoie, M.J., Fraering, P.C., Ostaszewski, B.L., Ye, W., Kimberly, W.T., Wolfe, M.S., and Selkoe, D.J. (2003) Assembly of the g-secretase complex involves early formation of an intermediate subcomplex of Aph-1 and nicastrin. *Journal of Biological Chemistry*, 278, 37213–37222.

Lawrence, C.M., Rodwell, V.W., and Stauffacher, C.V. (1995) Crystal structure of *Pseudomonas mevalonii* HMG-CoA reductase at 3.0 angstrom resolution. *Science*, 268, 1758–1762.

Lazo, N.D., Grant, M.A., Condron, M.C., Rigby, A.C., and Teplow, D.B. (2005) On the nucleation of amyloid {beta}-protein monomer folding. *Protein Science*, 14, 1581–1596.

Lee, E.K., Park, Y.W., Shin, D.Y., Mook-Jung, I., and Yoo, Y.J. (2006) Cytosolic amyloid-beta peptide 42 escaping from degradation induces cell death. *Biochemical and Biophysical Research Communications*, 344, 471–477.

Lesne, S., Koh, M.T., Kotilinek, L., Kayed, R., Glabe, C.G., Yang, A., Gallagher, M., and Ashe, K.H. (2006) A specific amyloid-beta protein assembly in the brain impairs memory. *Nature*, 440, 352–357.

Leuchtenberger, S., Kummer, M.P., Kukar, T., Czirr, E., Teusch, N., Sagi, S.A., et al. (2006) Inhibitors of Rho-kinase modulate amyloid-b (Ab) secretion but lack selectivity for Ab42. *Journal of Neurochemistry*, 96, 355–365.

LeVine, H., III (1999) Screening for pharmacologic inhibitors of amyloid fibril formation. *Methods in Enzymology*, 309, 467–476.

LeVine, H., III (2004a) The amyloid hypothesis and the clearance and degradation of Alzheimer's beta-peptide. *Journal of Alzheimer's Disease*, 6, 303–314.

LeVine, H., III (2004b) Alzheimer's beta-peptide oligomer formation at physiologic concentrations. *Analytical Biochemistry*, 335, 81–90.

LeVine, H., III (2005) Multiple ligand binding sites on A beta(1–40) fibrils. *Amyloid*, 12, 5–14.

LeVine, H., III, and Walker, L.C. (2006) Chapter 11. Models of Alzheimer's Disease, in *Handbook of Models for Human Aging*, (Conn, M.P., 121–134, Academic Press/Elsevier, New York, NY).

Lewis, H.D., Perez Revuelta, B.I., Nadin, A., Neduvelil, J.G., Harrison, T., Pollack, S.J., et al. (2003) Catalytic site-directed g-secretase complex inhibitors do not discriminate pharmacologically between Notch S3 and b-APP cleavages. *Biochemistry*, 42, 7580–7586.

Li, G., Higdon, R., Kukull, W.A., Peskind, E., Van Valen Moore, K., et al. (2004) Statin therapy and risk of dementia in the elderly: a community-based prospective cohort study. *Neurology*, 63, 1624–1628.

Liao, L., Cheng, D., Wang, J., Duong, D.M., Losik, T.G., Gearing, M., Rees, H.D., Lah, J.J., Levey, A.I., and Peng, J. (2004) Proteomic characterization of postmortem amyloid plaques isolated by laser capture microdissection. *Journal of Biological Chemistry*, 279, 37061–37068.

Lichtenthaler, S.F., Dominguez, D.I., Westmeyer, G.G., Reiss, K., Haass, C., Saftig, P., De Strooper, B., and Seed, B. (2003) The cell adhesion protein P-selectin glycoprotein ligand-1 is a substrate for the aspartyl protease BACE1. *Journal of Biological Chemistry*, 278, 48713–48719.

Lim, H., Gupta, R.A., Ma, W.G., Paria, B.C., Moller, D.E., Morrow, J.D., et al. (1999) Cyclooxygenase-2-derived prostacyclin mediates embryo implantation in the mouse via PPARdelta. *Genes & Development*, 13, 1561–1574.

Lim, G.P., Yang, F., Chu, T., Chen, P., Beech, W., Teter, B., et al. (2000) Ibuprofen suppresses plaque pathology and inflammation in a mouse model for Alzheimer's disease. *Journal of Neuroscience*, 20, 5709–5714.

Lim, G.P., Yang, F., Chu, T., Gahtan, E., Ubeda, O., Beech, W., et al. (2001) Ibuprofen effects on Alzheimer pathology and open field activity in APPsw transgenic mice. *Neurobiology of Aging*, 22, 983–991.

Lindersson, E., Beedholm, R., Hojrup, P., Moos, T., Gai, W., Hendil, K.B., et al. (2004) Proteasomal inhibition by alpha-synuclein filaments and oligomers. *Journal of Biological Chemistry*, 279, 12924–12934.

Lleo, A., Berezovska, O., Herl, L., Raju, S., Deng, A., Bacskai, B.J., et al. (2004) Nonsteroidal anti-inflammatory drugs lower Ab42 and change presenilin 1 conformation. *Nature Medicine*, 10, 1065–1066.

Loll, P.J., Picot, D., and Garavito, R.M. (1995) The structural basis of aspirin activity inferred from the crystal structure of inactivated prostaglandin H2 synthase. *Nature Structural Biology*, 2, 637–643.

Loll, P.J., Picot, D., Ekabo, O., and Garavito, R.M. (1996) Synthesis and use of iodinated nonsteroidal anti-inflammatory drug analogs as crystallographic probes of the prostaglandin H2 synthase cyclooxygenase active site. *Biochemistry*, 35, 7330–7340.

Loo, D.T., Copani, A., Pike, C.J., Whittemore, E.R., Walencewicz, A.J., and Cotman, C.W. (1993) Apoptosis is induced by beta-amyloid in cultured central nervous system neurons. *Proceedings of the National Academy of Sciences of the United States of America*, 90, 7951–7955.

Lorenzo, A., and Yankner, B.A. (1994) Beta-amyloid neurotoxicity requires fibril formation and is inhibited by Congo red. *Proceedings of the National Academy of Sciences of the United States of America*, 91, 12243–12247.

Losic, D., Martin, L.L., Mechler, A., Aguilar, M.I., and Small, D.H. (2006) High resolution scanning tunnelling microscopy of the beta-amyloid protein (Abeta1–40) of Alzheimer's disease suggests a novel mechanism of oligomer assembly. *Journal of Structural Biology*, 155, 104–110.

Lovell, M.A., Robertson, J.D., Buchholz, B.A., Xie, C., and Markesbery, W.R. (2002) Use of bomb pulse carbon-14 to age senile plaques and neurofibrillary tangles in Alzheimer's disease. *Neurobiology of Aging*, 23, 179–186.

Lowenson, J.D., Clarke, S., and Roher, A.E. (1999) Chemical modifications of deposited amyloid-beta peptides. *Methods in Enzymology*, 309, 89–105.

Lu, G., Tsai, A.L., Van Wart, H.E., and Kulmacz, R.J. (1999) Comparison of the peroxidase reaction kinetics of prostaglandin H synthase-1 and -2. *Journal of Biological Chemistry*, 274, 16162–16167.

Luhrs, T., Ritter, C., Adrian, M., Riek-Loher, D., Bohrmann, B., Dobeli, H., Schubert, D., and Riek, R. (2005) 3D structure of Alzheimer's amyloid-beta(1–42) fibrils. *Proceedings of the National Academy of Sciences of the United States of America*, 102, 17342–17347.

Luo, Y., Bolon, B., Kahn, S., Bennett, B.D., Babu-Khan, S., Denis, P., Fan, W., Kha, H., Zhang, J., Gong, Y., et al. (2001) Mice deficient in BACE1, the Alzheimer's beta-secretase, have normal phenotype and abolished beta-amyloid generation. *Nature Neuroscience*, 4, 231–232.

Luong, C., Miller, A., Barnett, J., Chow, J., Ramesha, C., and Browner, M.F. (1996) Flexibility of the NSAID binding site in the structure of human cyclooxygenase-2. *Nature Structural Biology*, 3, 927–933.

Ma, Q.L., Lim, G.P., Harris-White, M.E., Yang, F., Ambegaokar, S.S., Ubeda, O.J. et al. (2006) Antibodies against beta-amyloid reduce Abeta oligomers, glycogen synthase kinase-3beta activation and tau phosphorylation in vivo and in vitro. *Journal of Neuroscience Research*, 83, 374–384.

Magrane, J., Smith, R.C., Walsh, K., and Querfurth, H.W. (2004) Heat shock protein 70 participates in the neuroprotective response to intracellularly expressed beta-amyloid in neurons. *Journal of Neuroscience*, 24, 1700–1706.

Maillard, I., Adler, S.H., and Pear, W.S. (2003) Notch and the immune system. *Immunity*, 19, 781–791.

Maillard, M., and Burnier, M. (2006) Comparative cardiovascular safety of traditional nonsteroidal anti-inflammatory drugs. *Expert Opinion on Drug Safety*, 5, 83–94.

Maillet, M., Robert, S.J., Cacquevel, M., Gastineau, M., Vivien, D., Bertoglio, J., Zugaza, J.L., Fischmeister, R., and Lezoualc'h, F. (2003) Crosstalk between Rap1 and Rac regulates secretion of sAPPalpha. *Nature Cell Biology*, 5, 633–639.

Makin, O.S., and Serpell, L.C. (2005) Structures for amyloid fibrils. *FEBS Journal*, 272, 5950–5961.

Malinchik, S.B., Inouye, H., Szumowski, K.E., and Kirschner, D.A. (1998) Structural analysis of Alzheimer's beta(1–40) amyloid: protofilament assembly of tubular fibrils. *Biophysical Journal*, 74, 537–545.

Malkowski, M.G., Ginell, S.L., Smith, W.L., and Garavito, R.M. (2000) The productive conformation of arachidonic acid bound to prostaglandin synthase. *Science*, 289, 1933–1937.

Mancini, J.A., O'Neill, G.P., Bayly, C., and Vickers, P.J. (1994) Mutation of serine-516 in human prostaglandin G/H synthase-2 to methionine or aspirin acetylation of this residue stimulates 15-R-HETE synthesis. *FEBS Letters*, 342, 33–37.

Manning, M. and Colon, W. (2004) Structural basis of protein kinetic stability: resistance to sodium dodecyl sulfate suggests a central role for rigidity and a bias toward beta-sheet structure. *Biochemistry*, 43, 11248–11254.

Marnett, L.J. and Kalgutkar, A.S. (1998) Design of selective inhibitors of cyclooxygenase-2 as nonulcerogenic anti-inflammatory agents. *Current Opinion in Chemical Biology*, 2, 482–490.

Marnett, L.J. and Kalgutkar, A.S. (1999) Cyclooxygenase 2 inhibitors: discovery, selectivity and the future. *Trends in Pharmacological Sciences*, 20, 465–469.

Marnett, L.J., Rowlinson, S.W., Goodwin, D.C., Kalgutkar, A.S., and Lanzo, C.A. (1999) Arachidonic acid oxygenation by COX-1 and COX-2. Mechanisms of catalysis and inhibition. *The Journal of Biological Chemistry*, 274, 22903–22906.

Maron, D.J., Fazio, S., and Linton, M.F. (2000) Current perspectives on statins. *Circulation*, 101, 207–213.

Martin, B.K., Meinert, C.L., and Breitner, J.C. (2002) Double placebo design in a prevention trial for Alzheimer's disease. *Controlled Clinical Trials*, 23, 93–99.

Martoglio, B., and Golde, T.E. (2003) Intramembrane-cleaving aspartic proteases and disease: presenilins, signal peptide peptidase and their homologs. *Human Molecular Genetics*, 12 Spec No. 2, R201–R206.

Masse, I., Bordet, R., Deplanque, D., Al Khedr, A., Richard, F., Libersa, C., et al. (2005) Lipid lowering agents are associated with a slower cognitive decline in Alzheimer's disease. *Journal of Neurology Neurosurgery and Psychiatry*, 76, 1624–1629.

Massey, A., Kiffin, R., and Cuervo, A.M. (2004) Pathophysiology of chaperone-mediated autophagy. *International Journal of Biochemistry and Cell Biology*, 36, 2420–2434.

Masters, C., Simms, G., Weinmann, N.A., Multhaup, G., McDonald, B.L., and Beyreuther, K. (1985) Amyloid plaque core protein in Alzheimer's disease and Down syndrome. *Proceedings of the National Academy of Sciences of the United States of America*, 82, 4245–4249.

Mastrangelo, I.A., Ahmed, M., Sato, T., Liu, W., Wang, C., Hough, P., et al. (2006) High-resolution atomic force microscopy of soluble Abeta42 oligomers. *Journal of Molecular Biology*, 358, 106–119.

Matsuoka, Y., Shao, L., Debnath, M., Lafrancois, J., Becker, A., Gray, A. et al. (2005) An Abeta sequestration approach using non-antibody Abeta binding agents. *Current Alzheimer Research*, 2, 265–268.

McDonald, D.R., Bamberger, M.E., Combs, C.K., and Landreth, G.E. (1998) Beta-amyloid fibrils activate parallel mitogen-activated protein kinase pathways in microglia and THP1 monocytes. *Journal of Neuroscience*, 18, 4451–4460.

McGeer, P.L. (2000) Cyclo-oxygenase-2 inhibitors: rationale and therapeutic potential for Alzheimer's disease. *Drugs & Aging*, 17, 1–11.

McGeer, P.L., Harada, N., Kimura, H., McGeer, E.G., and Schulzer, M. (1992) Prevalence of dementia amongst elderly Japanese with leprosy—Apparent effect of chronic drug-therapy. *Dementia*, 3, 146–149.

McLellan, M.E., Kajdasz, S.T., Hyman, B.T., and Bacskai, B.J. (2003) In vivo imaging of reactive oxygen species specifically associated with thioflavine s-positive amyloid plaques by multiphoton microscopy. *Journal of Neuroscience*, 23, 2212–2217.

McNaught, K.S. and Olanow, C.W. (2006) Protein aggregation in the pathogenesis of familial and sporadic Parkinson's disease. *Neurobiology of Aging*, 27, 530–545.

Meade, E.A., Smith, W.L., and DeWitt, D.L. (1993) Differential inhibition of prostaglandin endoperoxide synthase (cyclooxygenase) isozymes by aspirin and other non-steroidal anti-inflammatory drugs. *Journal of Biological Chemistry*, 268, 6610–6614.

Meffert, M.K. and Baltimore, D. (2005) Physiological functions for brain NF-kappaB. *Trends in Neurosciences*, 28, 37–43.

Menzel-Soglowek, S., Geisslinger, G., Beck, W.S., and Brune, K. (1992) Variability of inversion of (R)-flurbiprofen in different species. *Journal of Pharmaceutical Sciences*, 81, 888–891.

Mesches, M.H., Gemma, C., Veng, L.M., Allgeier, C., Young, D.A., Browning, M.D., et al. (2004) Sulindac improves memory and increases NMDA receptor subunits in aged Fischer 344 rats. *Neurobiology of Aging*, 25, 315–324.

Miao, J., Xu, F., Davis, J., Otte-Holler, I., Verbeek, M.M., and Van Nostrand, W.E. (2005) Cerebral microvascular amyloid {beta} protein deposition induces vascular degeneration and neuroinflammation in transgenic mice expressing human vasculotropic mutant amyloid {beta} precursor protein. *American Journal of Pathology*, 167, 505–515.

Miida, T., Yamazaki, F., Sakurai, M., Wada, R., Yamadera, T., Asami, K., et al. (1999) The apolipoprotein E content of HDL in cerebrospinal fluid is higher in children than in adults. *Clinical Chemistry*, 45, 1294–1296.

Miida, T., Takahashi, A., Tanabe, N., and Ikeuchi, T. (2005) Can statin therapy really reduce the risk of Alzheimer's disease and slow its progression? *Current Opinion Lipidology*, 16, 619–623.

Milano, J., McKay, J., Dagenais, C., Foster-Brown, L., Pognan, F., Gadient, R., Jacobs, R.T., Zacco, A., Greenberg, B., and Ciaccio, P.J. (2004) Modulation of notch processing by g-secretase inhibitors causes intestinal goblet cell metaplasia and induction of genes known to specify gut secretory lineage differentiation. *Toxicological Sciences*, 82, 341–358.

Miller, D.L., Currie, J.R., Mehta, P.D., Potempska, A., Hwang, Y.W., and Wegiel, J. (2003) Humoral immune response to fibrillar beta-amyloid peptide. *Biochemistry*, 42, 11682–11692.

Minter, L.M., Turley, D.M., Das, P., Shin, H.M., Joshi, I., Lawlor, R.G., et al. (2005) Inhibitors of g-secretase block in vivo and in vitro T helper type 1 polarization by preventing Notch upregulation of Tbx21. *Nature Immunology*, 6, 680–688.

Moretto, N., Grassi, F., Imbimbo, B.P., Villetti, G., Riccardi, B., Puccini, P., et al. (2006) Substrate selectivity of three novel g-secretase modulators analyzed by target gene expression profiling, *10th International Conference on Alzheimer Disease and Related Disorders*, Madrid, Spain.

Morgan, D. (2005) Mechanisms of Abeta plaque clearance following passive Abeta immunization. *Neurodegenerative Disease*, 2, 261–266.

Morgan, D. and Keller, R.K. (2002) What evidence would prove the amyloid hypothesis? Towards rational drug treatments for Alzheimer's disease. *Journal of Alzheimer's Disease*, 4, 257–260.

Morihara, T., Chu, T., Ubeda, O., Beech, W., and Cole, G.M. (2002) Selective inhibition of Ab42 production by NSAID R-enantiomers. *Journal of Neurochemistry*, 83, 1009–1012.

Morihara, T., Teter, B., Yang, F., Lim, G.P., Boudinot, S., Boudinot, F.D., et al. (2005) Ibuprofen suppresses interleukin-1b induction of pro-amyloidogenic a1-antichymotrypsin to ameliorate b-amyloid (Ab) pathology in Alzheimer's models. *Neuropsychopharmacology*, 30, 1111–1120.

Morita, I., Schindler, M., Regier, M.K., Otto, J.C., Hori, T., DeWitt, D.L., et al. (1995) Different intracellular locations for prostaglandin endoperoxide H synthase-1 and -2. *Journal of Biological Chemistry*, 270, 10902–10908.

Mundy, G., Garrett, R., Harris, S., Chan, J., Chen, D., Rossini, G., et al. (1999) Stimulation of bone formation in vitro and in rodents by statins. *Science*, 286, 1946–1949.

National Cholesterol Education Program (NCEP) (2002) Expert panel on detection, evaluation, and treatment of high blood cholesterol in adults (Adult Treatment Panel III). *Circulation*, 106, 3143–3151.

Nawarskas, J.J. (2002) HMG-CoA reductase inhibitors and coenzyme Q10. *Cardiology in Review*, 13, 76–79.

Ness, G.C. and Chambers, C.M. (2000) Feedback and hormonal regulation of hepatic 3-hydroxy-3-methylglutaryl coenzyme A reductase: the concept of cholesterol buffering capacity. *Proceedings of the Society for Experimental Biology and Medicine*, 224, 8–19.

Nguyen, K.V., Gendrault, J.L., and Wolff, C.M. (2002) Poly-l-lysine dissolves fibrillar aggregation of the Alzheimer beta-amyloid peptide in vitro. *Biochemical and Biophysical Research Communications*, 291, 764–768.

Nitsch, R.M., Slack, B.E., Wurtman, R.J., and Growdon, J.H. (1992) Release of Alzheimer amyloid precursor derivatives stimulated by activation of muscarinic acetylcholine receptors. *Science*, 258, 304–307.

Nixon, R.A., Wegiel, J., Kumar, A., Yu, W.H., Peterhoff, C., Cataldo, A. et al. (2005) Extensive involvement of autophagy in Alzheimer disease: an immuno-electron microscopy study. *Journal of Neuropathology and Experimental Neurology*, 64, 113–122.

O'Nuallain, B. and Wetzel, R. (2002) Conformational Abs recognizing a generic amyloid fibril epitope, *Proceedings of the National Academy of Sciences of the United States of America*, 99, 1485–1490.

Oddo, S., Caccamo, A., Tran, L., Lambert, M.P., Glabe, C.G., Klein, W.L., et al. (2006) Temporal profile of Abeta oligomerization in an in vivo model of Alzheimer's disease: a link between Abeta and tau pathology. *Journal of Biological Chemistry*, 281, 1599–1604.

Oh, S., Hong, H.S., Hwang, E., Sim, H.J., Lee, W., Shin, S.J., et al. (2005) Amyloid peptide attenuates the proteasome activity in neuronal cells. *Mechanisms of Ageing and Development*, 126, 1292–1299.

Okochi, M., Steiner, H., Fukumori, A., Tanii, H., Tomita, T., Tanaka, T., et al. (2002) Presenilins mediate a dual intramembranous g-secretase cleavage of Notch-1. *EMBO Journal*, 21, 5408–5416.

Okochi, M., Fukumori, A., Jiang, J., Itoh, N., Kimura, R., Steiner, H., et al. (2006) Secretion of the Notch-1 Ab-like peptide during Notch signaling. *Journal of Biological Chemistry*, 281, 7890–7898.

Omkumar, R.V. and Rodwell, V.W. (1994) Phosphorylation of Ser871 impairs the function of His865 of Syrian hamster 3-hydroxy-3-methylglutaryl-CoA reductase. *Journal of Biological Chemistry*, 269, 16862–16866.

Packard, C.J., Ford, I., Robertson, M., Shepherd, J., Blauw, G.J., Murphy, M.B., et al. (2005) Plasma lipoproteins and apolipoproteins as predictors of cardiovascular risk and treatment benefit in the prospective study of Pravastatin in the elderly at risk (PROSPER). *Circulation*, 112, 3058–3065.

Paivio, A., Jarvet, J., Graslund, A., Lannfelt, L., and Westlind-Danielsson, A. (2004) Unique physicochemical profile of beta-amyloid peptide variant Abeta1–40E22G protofibrils: conceivable neuropathogen in Arctic mutant carriers. *Journal of Molecular Biology*, 339, 145–159.

Palinski, W. and Tsimikas, S. (2002) Immunomodulatory effects of statins: mechanisms and potential impact on arteriosclerosis. *Journal of the American Society of Nephrology*, 13, 1673–1681.

Patzwald, G.A. and Wildt, S.C. (2004) The use of convent archival records in medical research: the School Sisters of Notre Dame archives and the nun study. *American Archives*, 67, 86–106.

Pepys, M.B. (2006) Amyloidosis. *Annual Review of Medicine*, 57, 223–241.

Peretto, I., Radaelli, S., Parini, C., Zandi, M., Raveglia, L.F., Dondio, G., et al. (2005) Synthesis and biological activity of flurbiprofen analogues as selective inhibitors of b-amyloid1–42 secretion. *Journal of Medicinal Chemistry*, 48, 5705–5720.

Perini, R., Fiorucci, S., and Wallace, J.L. (2004) Mechanisms of nonsteroidal anti-inflammatory drug-induced gastrointestinal injury and repair: a window of opportunity for cyclooxygenase-inhibiting nitric oxide donors. *Canadian Journal of Gastroenterology*, 18, 229–236.

Permanne, B., Buee, L., David, J.P., Fallet-Bianco, C., Di Menza, C., and Delacourte, A. (1995) Quantitation of Alzheimer's amyloid peptide and identification of related amyloid proteins by dot-blot immunoassay. *Brain Research*, 685, 154–162.

Permanne, B., Adessi, C., Saborio, G.P., Fraga, S., Frossard, M.J., Van Dorpe, J., Dewachter, I., Banks, W.A., Van Leuven, F., and Soto, C. (2002) Reduction of amyloid load and cerebral damage in a transgenic mouse model of Alzheimer's disease by treatment with a beta-sheet breaker peptide. *FASEB Journal*, 16, 860–862.

Petkova, A.T., Leapman, R.D., Guo, Z., Yau, W.M., Mattson, M.P., and Tycko, R. (2005) Self-propagating, molecular-level polymorphism in Alzheimer's beta-amyloid fibrils. *Science*, 307, 262–265.

Picot, D. and Garavito, R.M. (1994) Prostaglandin H synthase: implications for membrane structure. *FEBS Letters*, 346, 21–25.

Picot, D., Loll, P.J., and Garavito, R.M. (1994) The X-ray crystal structure of the membrane protein prostaglandin H2 synthase-1. *Nature*, 367, 243–249.

Pike, C.J., Burdick, D., Walencewicz, A.J., Glabe, C.G., and Cotman, C.W. (1993) Neurodegeneration induced by beta-amyloid peptides in vitro: the role of peptide assembly state. *Journal of Neuroscience*, 13, 1676–1687.

Poirier, J. (2005) Apolipoprotein E, cholesterol transport and synthesis in sporadic Alzheimer's disease. *Neurobiology of Aging*, 226, 355–361.

Porat, Y., Abramowitz, A., and Gazit, E. (2006) Inhibition of amyloid fibril formation by polyphenols: structural similarity and aromatic interactions as a common inhibition mechanism. *Chemistry of Biology and Drug Design*, 67, 27–37.

Postina, R., Schroeder, A., Dewachter, I., Bohl, J., Schmitt, U., Kojro, E., Prinzen, C., Endres, K., Hiemke, C., Blessing, M., et al. (2004) A disintegrin-metalloproteinase prevents amyloid plaque formation and hippocampal defects in an Alzheimer disease mouse model. *Journal of Clinical Investigation*, 113, 1456–1464.

Poynter, J.N., Gruber, S.B., Higgins, P.D., Almog, R., Bonner, J.D., Rennert, H.S., et al. (2005) Statins and the risk of colorectal cancer. *New England Journal of Medicine*, 352, 2184–2192.

Price, J.C., Klunk, W.E., Lopresti, B.J., Lu, X., Hoge, J.A., Ziolko, S.K., et al. (2005) Kinetic modeling of amyloid binding in humans using PET imaging and Pittsburgh Compound-B. *Journal of Cerebral Blood Flow and Metabolism*, 25, 1528–1547.

Primakoff, P. and Myles, D.G. (2000) The ADAM gene family: surface proteins with adhesion and protease activity. *Trends in Genetics*, 16, 83–87.

Puchtler, H., Chandler, A.B., and Sweat, F. (1961) Demonstration of fibrin in tissue sections by the Rosindole method. *Journal of Histochemistry and Cytochemistry*, 9, 340–341.

Quinn, J., Montine, T., Morrow, J., Woodward, W.R., Kulhanek, D., and Eckenstein, F. (2003) Inflammation and cerebral amyloidosis are disconnected in an animal model of Alzheimer's disease. *Journal of Neuroimmunology*, 137, 32–41.

Rang, H. (2006) *Drug Discovery and Development—Technology in Transition*, Churchill Livingstone, London, England, 346.

Rea, T.D., Breitner, J.C., Psaty, B.M., Fitzpatrick, A.L., Lopez, O.L., Newman, A.B., et al. (2005) Statin use and the risk of incident dementia: the cardiovascular health study. *Archives of Neurology*, 62, 1047–1051.

Reddy, V.P., Obrenovich, M.E., Atwood, C.S., Perry, G., and Smith, M.A. (2002) Involvement of Maillard reactions in Alzheimer disease. *Neurotoxicity Research*, 4, 191–209.

Reines, S.A., Block, G.A., Morris, J.C., Liu, G., Nessly, M.L., Lines, C.R., et al. (2004) Rofecoxib: no effect on Alzheimer's disease in a 1-year, randomized, blinded, controlled study. *Neurology*, 62, 66–71.

Richardson, R.L., Kim, E.M., Shephard, R.A., Gardiner, T., Cleary, J., and O'Hare, E. (2002) Behavioural and histopathological analyses of ibuprofen treatment on the effect of aggregated Ab1–42 injections in the rat. *Brain Research*, 954, 1–10.

Ricote, M., Li, A.C., Willson, T.M., Kelly, C.J., and Glass, C.K. (1998) The peroxisome proliferator-activated receptor-g is a negative regulator of macrophage activation. *Nature*, 391, 79–82.

Rideout, H.J., Lang-Rollin, I., and Stefanis, L. (2004) Involvement of macroautophagy in the dissolution of neuronal inclusions. *International Journal of Biochemistry and Cell Biology*, 36, 2551–2562.

Robert, S.J., Zugaza, J.L., Fischmeister, R., Gardier, A.M., and Lezoualc'h, F. (2001) The human serotonin 5-HT4 receptor regulates secretion of non-amyloidogenic precursor protein. *Journal of Biological Chemistry*, 276, 44881–44888.

Rockwood, K., Kirkland, S., Hogan, D.B., MacKnight, C., Merry, H., Verreault, R., et al. (2002) Use of lipid-lowering agents, indication bias, and the risk of dementia in community-dwelling elderly people. *Archives of Neurology*, 59, 223–227.

Rogaev, E.I., Sherrington, R., Rogaeva, E.A., Levesque, G., Ikeda, M., Liang, Y., Chi, H., Lin, C., Holman, K., Tsuda, T., et al. (1995) Familial Alzheimer's disease in kindreds with missense mutations in a gene on chromosome 1 related to the Alzheimer's disease type 3 gene. *Nature*, 376, 775–778.

Rogers, J., Cooper, N.R., Webster, S., Schultz, J., McGeer, P.L., Styren, S.D., et al. (1992) Complement activation by b-amyloid in Alzheimer disease. *Proceedings of the National Academy of Sciences of the United States of America*, 89, 10016–10020.

Rogers, J., Kirby, L.C., Hempelman, S.R., Berry, D.L., McGeer, P.L., Kaszniak, A.W., et al. (1993) Clinical trial of indomethacin in Alzheimer's disease. *Neurology*, 43, 1609–1611.

Roher, A.E., Lowenson, J.D., Clarke, S., Wolkow, C., Wang, R., Cotter, R.J., et al. (1993) Structural alterations in the peptide backbone of beta-amyloid core protein may account for its deposition and stability in Alzheimer's disease. *Journal of Biological Chemistry*, 268, 3072–3083.

Roher, A.E., Chaney, M.O., Kuo, Y.-M., Webster, S.D., Stine, W.B., Haverkamp, L.J., Woods, A.S., Cotter, R.J., Tuohy, J.M., Krafft, G.A., et al. (1996) Morphology and toxicity of Abeta-(1–42) dimer derived from neuritic and vascular amyloid deposits of Alzheimer's disease. *Journal of Biological Chemistry*, 271, 20631–20635.

Roher, A.E., Kuo, Y.M., Esh, C., Knebel, C., Weiss, N., Kalback, W. et al. (2003) Cortical and leptomeningeal cerebrovascular amyloid and white matter pathology in Alzheimer's disease. *Molecular Medicine*, 9, 112–122.

Rowan, M.J., Klyubin, I., Wang, Q., and Anwyl, R. (2005) Synaptic plasticity disruption by amyloid beta protein: modulation by potential Alzheimer's disease modifying therapies. *Biochemical Society Transactions*, 33, 563–567.

Rowlinson, S.W., Kiefer, J.R., Prusakiewicz, J.J., Pawlitz, J.L., Kozak, K.R., Kalgutkar, A.S., et al. (2003) A novel mechanism of cyclooxygenase-2 inhibition involving interactions with Ser-530 and Tyr-385. *Journal of Biological Chemistry*, 278, 45763–45769.

Sagi, S.A., Weggen, S., Eriksen, J., Golde, T.E., and Koo, E.H. (2003) The non-cyclo-oxygenase targets of non-steroidal anti-inflammatory drugs, lipoxygenases, peroxisome proliferator-activated receptor, inhibitor of kappa B kinase, and NF kappa B, do not reduce amyloid b 42 production. *Journal of Biological Chemistry*, 278, 31825–31830.

Saido, T.C. and Iwata, N. (2006) Metabolism of amyloid beta peptide and pathogenesis of Alzheimer's disease towards presymptomatic diagnosis, prevention and therapy. *Neuroscience Research*, 54, 235–253.

Sainati, S.M., Ingram, D., Talwalker, S., and Geis, G. (2000) Results of a double-blind, randomized, placebo-controlled study of celecoxib in the treatment of progression of Alzheimer's disease, *Sixth International Stockholm-Springfield Symposium of Advances in Alzheimer's Therapy*, Stockholm, Sweden, Abstract p. 180.

Samuelsson, M., Fisher, L., and Iverfeldt, K. (2005) b-Amyloid and interleukin-1b induce persistent NF-kB activation in rat primary glial cells. *International Journal of Molecular Medicine*, 16, 449–453.

Sankaranarayanan, S. (2006) Genetically modified mice models for Alzheimer's disease. *Current Topics in Medicinal Chemistry*, 6, 609–627.

Sastre, M., Dewachter, I., Landreth, G.E., Willson, T.M., Klockgether, T., van Leuven, F., et al. (2003) Nonsteroidal anti-inflammatory drugs and peroxisome proliferator-activated receptor-g agonists modulate immunostimulated processing of amyloid precursor protein through regulation of b-secretase. *Journal of Neuroscience*, 23, 9796–9804.

Saunders, A., Kim, T.-W., and Tanyi, R. (1999) BACE maps to chromosome 11 and a BACE homolog, BACE2 reside in the obligate Downs syndrome region of chromosome 21. *Science*, 286, 1255a.

Scatena, R. (2004) Nitroflurbiprofen (NicOx). *Current Opinion in Investigational Drugs*, 5, 551–556.

Schachter, M. (2005) Chemical, pharmacokinetic and pharmacodynamic properties of statins: an update. *Fundamental & Clinical Pharmacology*, 19, 117–125.

Scharf, S., Mander, A., Ugoni, A., Vajda, F., and Christophidis, N. (1999) A double-blind, placebo-controlled trial of diclofenac/misoprostol in Alzheimer's disease. *Neurology*, 53, 197–201.

Schley, D., Carare-Nnadi, R., Please, C.P., Perry, V.H., and Weller, R.O. (2006) Mechanisms to explain the reverse perivascular transport of solutes out of the brain. *Journal of Theoretical Biology*, 8, 151–162.

Schroeter, E.H., Ilagan, M.X., Brunkan, A.L., Hecimovic, S., Li, Y.M., Xu, M., Lewis, H.D., Saxena, M.T., De Strooper, B., Coonrod, A., et al. (2003) A presenilin dimer at the core of the g-secretase enzyme: insights from parallel analysis of Notch 1 and APP prote-olysis. *Proceedings of the National Academy of Sciences of the United States of America*, 100, 13075–13080.

Searfoss, G.H., Jordan, W.H., Calligaro, D.O., Galbreath, E.J., Schirtzinger, L.M., Berridge, B.R., Gao, H., Higgins, M.A., May, P.C., and Ryan, T.P. (2003) Adipsin, a biomarker of gastrointestinal toxicity mediated by a functional g-secretase inhibitor. *Journal of Biological Chemistry*, 278, 46107–46116.

Selinsky, B.S., Gupta, K., Sharkey, C.T., and Loll, P.J. (2001) Structural analysis of NSAID binding by prostaglandin H2 synthase: time-dependent and time-independent inhibitors elicit identical enzyme conformations. *Biochemistry*, 40, 5172–5180.

Selkoe, D.J. (1991) The molecular pathology of Alzheimer's disease. *Neuron*, 6, 487–498.

Selkoe, D.J. (2002) Alzheimer's disease is a synaptic failure. *Science*, 298, 789–791.

Selkoe, D., and Kopan, R. (2003) Notch and presenilin: regulated intramembrane proteolysis links development and degeneration. *Annual Reviews in Neuroscience*, 26, 565–597.

Serneels, L., Dejaegere, T., Craessaerts, K., Horre, K., Jorissen, E., Tousseyn, T., Hebert, S., Coolen, M., Martens, G., Zwijsen, A., et al. (2005) Differential contribution of the three Aph1 genes to g-secretase activity in vivo. *Proceedings of the National Academy of Sciences of the United States of America*, 102, 1719–1724.

Serpell, L.C. (2000) Alzheimer's amyloid fibrils: structure and assembly. *Biochimica et Biophysica Acta*, 1502, 16–30.

Serpell, L.C., Sunde, M., Benson, M.D., Tennent, G.A., Pepys, M.B., and Fraser, P.E. (2000) The protofilament substructure of amyloid fibrils. *Journal of Molecular Biology*, 300, 1033–1039.

Shah, S., Lee, S.F., Tabuchi, K., Hao, Y.H., Yu, C., LaPlant, Q., Ball, H., Dann, C.E., III, Sudhof, T., and Yu, G. (2005) Nicastrin functions as a g-secretase-substrate receptor. *Cell*, 122, 435–447.

Shearman, M.S., Beher, D., Clarke, E.E., Lewis, H.D., Harrison, T., Hunt, P., Nadin, A., Smith, A.L., Stevenson, G., and Castro, J.L. (2000) L-685,458, an aspartyl protease transition state mimic, is a potent inhibitor of amyloid beta-protein precursor g-secretase activity. *Biochemistry*, 39, 8698–8704.

Sheppard, R. (2003) The fluorenylmethoxycarbonyl group in solid phase synthesis. *Journal of Peptide Sciences*, 9, 545–552.

Sherrington, R., Rogaev, E.I., Liang, Y., Rogaeva, E.A., Levesque, G., Ikeda, M., Chi, H., Lin, C., Li, G., Holman, K., et al. (1995) Cloning of a gene bearing missense mutations in early-onset familial Alzheimer's disease. *Nature*, 375, 754–760.

Shimano, H., Horton, J.D., Hammer, R.E., Shimomura, I., Brown, M.S., and Goldstein, J.L. (1996) Overproduction of cholesterol and fatty acids causes massive liver enlargement in transgenic mice expressing truncated SREBP-1a. *Journal of Clinical Investigations*, 98, 1575–1584.

Shimizu, T., Matsuoka, Y., and Shirasawa, T. (2005) Biological significance of isoaspartate and its repair system. *Biological and Pharmaceutical Bulletin*, 28, 1590–1596.

Shirotani, K., Edbauer, D., Prokop, S., Haass, C., and Steiner, H. (2004) Identification of distinct g-secretase complexes with different APH-1 variants. *Journal of Biological Chemistry*, 279, 41340–41345.

Shobab, L.A., Hsiung, G.Y., and Feldman, H.H. (2005) Cholesterol in Alzheimer's disease. *Lancet Neurology*, 4, 841–852.

Shoghi-Jadid, K., Barrio, J.R., Kepe, V., and Huang, S.C. (2006) Exploring a mathematical model for the kinetics of beta-amyloid molecular imaging probes through a critical analysis of plaque pathology. *Molecular Imaging and Biology*, 238, 962–974.

Shoji, M., Iwakami, N., Takeuchi, S., Waragai, M., Suzuki, M., Kanazawa, I., Lippa, C.F., Ono, S., and Okazawa, H. (2000) JNK activation is associated with intracellular beta-amyloid accumulation. *Brain Research Molecular Brain Research*, 85, 221–233.

Si, K., Lindquist, S., and Kandel, E.R. (2003) A neuronal isoform of the aplysia CPEB has prion-like properties. *Cell*, 115, 879–891.

Siemers, E., Skinner, M., Dean, R.A., Gonzales, C., Satterwhite, J., Farlow, M., Ness, D., and May, P.C. (2005) Safety, tolerability, and changes in amyloid beta concentrations after administration of a g-secretase inhibitor in volunteers. *Clinical Neuropharmacology*, 28, 126–132.

Sigurdsson, E.M., Permanne, B., Soto, C., Wisniewski, T., and Frangione, B. (2000) In vivo reversal of amyloid-beta lesions in rat brain. *Journal of Neuropathology and Experimental Neurology*, 59, 11–17.

Simmons, D.L., Botting, R.M., Robertson, P.M., Madsen, M.L., and Vane, J.R. (1999) Induction of an acetaminophen-sensitive cyclooxygenase with reduced sensitivity to nonsteroid antiinflammatory drugs. *Proceedings of the National Academy of Sciences of the United States of America*, 96, 3275–3280.

Simons, M., Schwarzler, F., Lutjohann, D., von Bergmann, K., Beyreuther, K., Dichgans, J., et al. (2002) Treatment with simvastatin in normocholesterolemic patients with Alzheimer's disease: a 26-week randomized, placebo-controlled, double-blind trial. *Annals of Neurology*, 52, 346–350.

Singer, S.J. and Dewji, N.N. (2006) Evidence that Perutz's double-beta-stranded subunit structure for beta-amyloids also applies to their channel-forming structures in membranes. *Proceedings of the National Academy of Sciences of the United States of America*, 103, 1546–1550.

Smith, W.L. (1986) Prostaglandin biosynthesis and its compartmentation in vascular smooth muscle and endothelial cells. *Annual Review of Physiology*, 48, 251–262.

Smith, W.L. and Dewitt, D.L. (1996) Prostaglandin endoperoxide H synthases-1 and -2. *Advances in Immunology*, 62, 167–215.

Smith, W.L., DeWitt, D.L., and Garavito, R.M. (2000) Cyclooxygenases: structural, cellular, and molecular biology. *Annual Review of Biochemistry*, 69, 145–182.

Snyder, E.M., Nong, Y., Almeida, C.G., Paul, S., Moran, T., Choi, E.Y., Nairn, A.C., Salter, M.W., Lombroso, P.J., Gouras, G.K., and Greengard, P. (2005) Regulation of NMDA receptor trafficking by amyloid-beta. *Nature Neuroscience*, 8, 1051–1058.

Soto, C. and Estrada, L. (2005) Amyloid inhibitors and beta-sheet breakers. *Subcellular Biochemistry*, 38, 351–364.

Soto, C., Castano, E.M., Prelli, F., Kumar, R.A., and Baumann, M. (1995) Apolipoprotein E increases the fibrillogenic potential of synthetic peptides derived from Alzheimer's, gelsolin and AA amyloids. *FEBS Letters*, 371, 110–114.

Soto, C., Golabek, A., Wisniewski, T., and Castano, E.M. (1996a) Alzheimer's beta-amyloid peptide is conformationally modified by apolipoprotein E in vitro. *Neuroreport*, 7, 721–725.

Soto, C., Kindy, M.S., Baumann, M., and Frangione, B. (1996b) Inhibition of Alzheimer's amyloidosis by peptides that prevent beta-sheet conformation. *Biochemical and Biophysical Research Communications*, 226, 672–680.

Soto, C., Ghiso, J., and Frangione, B. (1997) Alzheimer's amyloid beta aggregation is modulated by the interaction of multiple factors. *Alzheimer's Research*, 3, 3, 215–222.

Soto, C., Sigurdsson, E.M., Morelli, L., Kumar, R.A., Castano, E.M., and Frangione, B. (1998) Beta-sheet breaker peptides inhibit fibrillogenesis in a rat brain model of amyloidosis: implications for Alzheimer's therapy. *Nature Medicine*, 4, 822–826.

Sparks, D.L., Connor, D.J., Browne, P.J., Lopez, J.E., and Sabbagh, M.N. (2002) HMG-CoA reductase inhibitors (statins) in the treatment of Alzheimer's disease and why it would be ill-advise to use one that crosses the blood–brain barrier. *Journal of Nutrition, Health & Aging*, 6, 324–331.

Sparks, D.L., Sabbagh, M.N., Connor, D.J., Lopez, J., Launer, L.J., Browne, P., et al. (2005) Atorvastatin for the treatment of mild to moderate Alzheimer disease: preliminary results. *Archives of Neurology*, 62, 753–757.

Staffa, J.A., Chang, J., and Green, L. (2002) Cerivastatin and reports of fatal rhabdomyolysis. *New England Journal of Medicine*, 346, 539–540.

Stanger, B.Z., Datar, R., Murtaugh, L.C., and Melton, D.A. (2005) Direct regulation of intestinal fate by Notch. *Proceedings of the National Academy of Sciences of the United States of America*, 102, 12443–12448.

Steiner, H. and Haass, C. (2000) Intramembrane proteolysis by presenilins. *Nature Review in Molecular Cell Biology*, 1, 217–224.

Stenh, C., Englund, H., Lord, A., Johansson, A.S., Almeida, C.G., Gellerfors, P., et al. (2005) Amyloid-beta oligomers are inefficiently measured by enzyme-linked immunosorbent assay. *Annals of Neurology*, 58, 147–150.

Stine, W.B., Jr., Dahlgren, K.N., Krafft, G.A., and LaDu, M.J. (2003) In vitro characterization of conditions for amyloid-beta peptide oligomerization and fibrillogenesis. *Journal of Biological Chemistry*, 278, 11612–11622.

Stock, N., Munoz, B., Wrigley, J.D., Shearman, M.S., Beher, D., Peachey, J., et al. (2006) The geminal dimethyl analogue of Flurbiprofen as a novel Ab42 inhibitor and potential Alzheimer's disease modifying agent. *Bioorganic & Medicinal Chemistry Letters*, 16, 2219–2223.

Straus, D.S., Pascual, G., Li, M., Welch, J.S., Ricote, M., Hsiang, C.H., et al. (2000) 15-deoxy-delta 12,14-prostaglandin J2 inhibits multiple steps in the NF-kappa B signaling pathway. *Proceedings of the National Academy of Sciences of the United States of America*, 97, 4844–4849.

Streit, W.J. (2004) Microglia and Alzheimer's disease pathogenesis. *Journal of Neuroscience Research*, 77, 1–8.

Strittmatter, W.J., Weisgraber, K.H., Huang, D.Y., Dong, L.M., Salvesen, G.S., Pericak-Vance, M., Schmechel, D., Saunders, A.M., Goldgaber, D., and Roses, A.D. (1993) Binding of human apolipoprotein E to synthetic amyloid beta peptide: isoform-specific effects and implications for late-onset Alzheimer disease. *Proceedings of the National Academy of Sciences of the United States of America*, 90, 8098–8102.

Stromer, T., and Serpell, L.C. (2005) Structure and morphology of the Alzheimer's amyloid fibril. *Microscopy Research and Technique*, 67, 210–217.

Sunde, M., Serpell, L.C., Bartlam, M., Fraser, P.E., Pepys, M.B., and Blake, C.C. (1997) Common core structure of amyloid fibrils by synchrotron X-ray diffraction. *Journal of Molecular Biology*, 273, 729–739.

Sung, S., Yang, H., Uryu, K., Lee, E.B., Zhao, L., Shineman, D., et al. (2004) Modulation of nuclear factor-kappa B activity by indomethacin influences Ab levels but not Ab precursor protein metabolism in a model of Alzheimer's disease. *American Journal of Pathology*, 165, 2197–2206.

Swinney, D.C., Mak, A.Y., Barnett, J., and Ramesha, C.S. (1997) Differential allosteric regulation of prostaglandin H synthase 1 and 2 by arachidonic acid. *Journal of Biological Chemistry*, 272, 12393–12398.

Szekely, C.A., Thorne, J.E., Zandi, P.P., Ek, M., Messias, E., Breitner, J.C., et al. (2004) Nonsteroidal anti-inflammatory drugs for the prevention of Alzheimer's disease: a systematic review. *Neuroepidemiology*, 23, 159–169.

Takahashi, Y., Hayashi, I., Tominari, Y., Rikimaru, K., Morohashi, Y., Kan, T., et al. (2003) Sulindac sulfide is a noncompetitive g-secretase inhibitor that preferentially reduces Ab42 generation. *Journal of Biological Chemistry*, 278, 18664–18670.

Talley, J.J., Bertenshaw, S.R., Brown, D.L., Carter, J.S., Graneto, M.J., Koboldt, C.M., et al. (1999) 4,5-Diaryloxazole inhibitors of cyclooxygenase-2 (COX-2). *Medicinal Research Reviews*, 19, 199–208.

Tanzi, R.E. (2005) The synaptic Abeta hypothesis of Alzheimer disease. *Nature Neuroscience*, 8, 977–979.

Tanzi, R.E., Moir, R.D., and Wagner, S.L. (2004) Clearance of Alzheimer's abeta Peptide; the many roads to perdition. *Neuron*, 43, 605–608.

Tegeder, I., Pfeilschifter, J., and Geisslinger, G. (2001) Cyclooxygenase-independent actions of cyclooxygenase inhibitors. *FASEB Journal*, 15, 2057–2072.

Tekirian, T.L., Yang, A.Y., Glabe, C., and Geddes, J.W. (1999) Toxicity of pyroglutaminated amyloid beta-peptides 3(pE)-40 and -42 is similar to that of A beta1-40 and -42. *Journal of Neurochemistry*, 73, 1584–1589.

Terry, R.D. (1994) Neuropathological changes in Alzheimer disease. *Progress in Brain Research*, 101, 383–390.

Terry, R.D., Peck, A., DeTeresa, R., Schechter, R., and Horoupian, D.S. (1981) Some morphometric aspects of the brain in senile dementia of the Alzheimer type. *Annals in Neurology*, 10, 184–192.

Thal, L.J., Ferris, S.H., Kirby, L., Block, G.A., Lines, C.R., Yuen, E., et al. (2005) A randomized, double-blind, study of rofecoxib in patients with mild cognitive impairment. *Neuropsychopharmacology*, 30, 1204–1215.

Thal, L.J., Kantarci, K., Reiman, E.M., Klunk, W.E., Weiner, M.W., Zetterberg, H., et al. (2006a) The role of biomarkers in clinical trials for Alzheimer disease. *Alzheimer Disease and Associated Disorders*, 20, 6–15.

Thal, D.R., Capetillo-Zarate, E., Del Tredici, K., and Braak, H. (2006b) The development of amyloid beta protein deposits in the aged brain. *Science Aging Knowledge Environment*, 2006, re1.

Thuresson, E.D., Lakkides, K.M., and Smith, W.L. (2000) Different catalytically competent arrangements of arachidonic acid within the cyclooxygenase active site of prostaglandin endoperoxide H synthase-1 lead to the formation of different oxygenated products. *Journal of Biological Chemistry*, 275, 8501–8507.

Thuresson, E.D., Lakkides, K.M., Rieke, C.J., Sun, Y., Wingerd, B.A., Micielli, R., et al. (2001) Prostaglandin endoperoxide H synthase-1: the functions of cyclooxygenase active site residues in the binding, positioning, and oxygenation of arachidonic acid. *Journal of Biological Chemistry*, 276, 10347–10357.

Tiraboschi, P., Hansen, L.A., Thal, L.J., and Corey-Bloom, J. (2004) The importance of neuritic plaques and tangles to the development and evolution of AD. *Neurology*, 62, 1984–1989.

Tjernberg, L.O., Naslund, J., Lindqvist, F., Johansson, J., Karlstrom, A.R., Thyberg, J., Terenius, L., and Nordstedt, C. (1996) Arrest of beta-amyloid fibril formation by a pentapeptide ligand. *Journal of Biological Chemistry*, 271, 8545–8548.

Tobert, J.A. (2003) Lovastatin and beyond: the history of the HMG-CoA reductase inhibitors. *Nature Reviews Drug Discovery*, 2, 517–526.

Tocchetti, P., Sardina, M., Santus, G., and Del Soldato, P. (2003) Cerebrospinal fluid concentration of flurbiprofen after multiple administrations of HCT 1026, a NO-donating derivative of flurbiprofen, in humans, *33rd Annual Meeting of Society of Neuroscience*, New Orleans, LA, Abstract 525.12.

Townsend, K.P. and Pratico, D. (2005) Novel therapeutic opportunities for Alzheimer's disease: focus on nonsteroidal anti-inflammatory drugs. *FASEB Journal*, 19, 1592–1601.

Tremblay, P., Yu, M., Aman, A., Paquette, J., Krzywkowski, P., Morissette, C., King, X., and Gervais, F. (2003) GAG mimetics as a therapeutic approach to AD and CAA: reduction of parenchymal and vascular amyloid deposition in APP transgenic mice Paper presented at 6th international conference AD/PD (Sevilla, Spain).

Tseng, B.P., Esler, W.P., Clish, C.B., Stimson, E.R., Ghilardi, J.R., Vinters, H.V., et al. (1999) Deposition of monomeric, not oligomeric, Abeta mediates growth of Alzheimer's disease amyloid plaques in human brain preparations. *Biochemistry*, 38, 10424–10431.

Tsuji, A., Saheki, A., Tamai, I., and Terasaki, T. (1993) Transport mechanism of 3-hydroxy-3-methylglutaryl coenzyme A reductase inhibitors at the blood-brain barrier. *Journal of Pharmacology and Experimental Therapeutics*, 267, 1085–1090.

Turner, P.R., O'Connor, K., Tate, W.P., and Abraham, W.C. (2003) Roles of amyloid precursor protein and its fragments in regulating neural activity, plasticity and memory. *Progress in Neurobiology*, 70, 1–32.

Urbanc, B., Cruz, L., Yun, S., Buldyrev, S.V., Bitan, G., Teplow, D.B., and Stanley, H.E. (2004) In silico study of amyloid beta-protein folding and oligomerization. *Proceedings of the National Academy of Sciences of the United States of America*, 101, 17345–17350.

van Dokkum, R.P., Henning, R.H., and de Zeeuw, D. (2002) Statins: cholesterol, blood pressure and beyond. *Journal of Hypertension*, 20, 2351–2353.

van Groen, T. and Kadish, I. (2005) Transgenic AD model mice, effects of potential anti-AD treatments on inflammation and pathology. Brain Research. *Brain Research Reviews*, 48, 370–378.

Van Gool, W.A., Weinstein, H.C., Scheltens, P. and Walstra, G.J. (2001) Effect of hydroxychloroquine on progression of dementia in early Alzheimer's disease: an 18-month randomised, double-blind, placebo-controlled study. *Lancet*, 358, 455–460.

Vane, J.R., and Botting, R.M. (1998) Anti-inflammatory drugs and their mechanism of action. *Inflammation Research*, 47 (Suppl 2), S78–S87.

Vassar, R. (2001) The beta-secretase, BACE: a prime drug target for Alzheimer's disease. *Journal of Molecular Neuroscience*, 17, 157–170.

Vassar, R., Bennett, B.D., Babu-Khan, S., Kahn, S., Mendiaz, E.A., Denis, P., Teplow, D.B., Ross, S., Amarante, P., Loeloff, R., et al. (1999) Beta-secretase cleavage of Alzheimer's amyloid precursor protein by the transmembrane aspartic protease BACE. *Science*, 286, 735–741.

Viatour, P., Merville, M.P., Bours, V., and Chariot, A. (2005) Phosphorylation of NF-kappaB and IkappaB proteins: implications in cancer and inflammation. *Trends in Biochemical Sciences*, 30, 43–52.

Vollers, S.S., Teplow, D.B., and Bitan, G. (2005) Determination of peptide oligomerization state using rapid photochemical crosslinking. *Methods in Molecular Biology*, 299, 11–18.

von Arnim, C.A., Kinoshita, A., Peltan, I.D., Tangredi, M.M., Herl, L., Lee, B.M., Spoelgen, R., Hshieh, T.T., Ranganathan, S., Battey, F.D., et al. (2005) The low density lipoprotein receptor-related protein (LRP) is a novel beta-secretase (BACE1) substrate. *Journal of Biological Chemistry*, 280, 17777–17785.

Walker, L.C. and LeVine, H. (2000) The cerebral proteopathies: neurodegenerative disorders of protein conformation and assembly. *Molecular Neurobiology*, 21, 83–95.

Walker, L.C., Bian, F., Callahan, M.J., Lipinski, W.J., Durham, R.A., and LeVine, H. (2002a) Modeling Alzheimer's disease and other proteopathies in vivo: is seeding the key? *Amino Acids*, 23, 87–93.

Walker, L.C., Callahan, M.J., Bian, F., Durham, R.A., Roher, A.E., and Lipinski, W.J. (2002b) Exogenous induction of cerebral beta-amyloidosis in betaAPP-transgenic mice. *Peptides*, 23, 1241–1247.

Walker, L.C., Levine, H., III, Mattson, M.P. and Jucker, M. (2006) Inducible proteopathies. *Trends in Neuroscience*, 29, 438–443.

Wallace, J.L., Muscara, M.N., de Nucci, G., Zamuner, S., Cirino, G., del Soldato, P., et al. (2004) Gastric tolerability and prolonged prostaglandin inhibition in the brain with a nitric oxide-releasing flurbiprofen derivative, NCX-2216 [3-[4-(2-fluoro-alpha-methyl-[1,1¢-biphenyl]-4-acetyloxy)-3-methoxyphenyl]-2-propenoic acid 4-nitrooxy butyl ester]. *Journal of Pharmacology and Experimental Therapeutics*, 309, 626–633.

Walsh, D.M. and Selkoe, D.J. (2004) Deciphering the molecular basis of memory failure in Alzheimer's disease. *Neuron*, 44, 181–193.

Walsh, D.M., Lomakin, A., Benedek, G.B., Condron, M.M., and Teplow, D.B. (1997) Amyloid beta-protein fibrillogenesis. Detection of a protofibrillar intermediate. *Journal of Biological Chemistry*, 272, 22364–22372.

Walsh, D.M., Hartley, D.M., Kusumoto, Y., Fezoui, Y., Condron, M.M., Lomakin, A., Benedek, G.B., Selkoe, D.J., and Teplow, D.B. (1999) Amyloid beta-protein fibrillogenesis. Structure and biological activity of protofibrillar intermediates. *Journal of Biological Chemistry*, 274, 25945–25952.

Walsh, D.M., Klyubin, I., Fadeeva, J.V., Rowan, M.J., and Selkoe, D.J. (2002a) Amyloid-beta oligomers: their production, toxicity and therapeutic inhibition. *Biochemistry Society Transactions*, 30, 552–557.

Walsh, D.M., Klyubin, I., Fadeeva, J.V., Cullen, W.K., Anwyl, R., Wolfe, M.S., Rowan, M.J., and Selkoe, D.J. (2002b) Naturally secreted oligomers of amyloid beta protein potently inhibit hippocampal long-term potentiation in vivo. *Nature*, 416, 535–539.

Walsh, D.M., Townsend, M., Podlisny, M.B., Shankar, G.M., Fadeeva, J.V., Agnaf, O.E., Hartley, D.M., and Selkoe, D.J. (2005) Certain inhibitors of synthetic amyloid beta-peptide (Abeta) fibrillogenesis block oligomerization of natural Abeta and thereby rescue long-term potentiation. *Journal of Neuroscience*, 25, 2455–2462.

Wang, H.W., Pasternak, J.F., Kuo, H., Ristic, H., Lambert, M.P., Chromy, B., Viola, K.L., Klein, W.L., Stine, W.B., Krafft, G.A., and Trommer, B.L. (2002) Soluble oligomers of beta amyloid (1–42) inhibit long-term potentiation but not long-term depression in rat dentate gyrus. *Brain Research*, 924, 133–140.

Wang, Q., Walsh, D.M., Rowan, M.J., Selkoe, D.J., and Anwyl, R. (2004) Block of long-term potentiation by naturally secreted and synthetic amyloid beta-peptide in hippocampal slices is mediated via activation of the kinases c-Jun *N*-terminal kinase, cyclin-dependent kinase 5, and p38 mitogen-activated protein kinase as well as metabotropic glutamate receptor type 5. *Journal of Neuroscience*, 24, 3370–3378.

Watson, D., Castano, E., Kokjohn, T.A., Kuo, Y.M., Lyubchenko, Y., Pinsky, D., et al. (2005) Physicochemical characteristics of soluble oligomeric Abeta and their pathologic role in Alzheimer's disease. *Neurology Research*, 27, 869–881.

Weggen, S., Eriksen, J.L., Das, P., Sagi, S.A., Wang, R., Pietrzik, C.U., et al. (2001) A subset of NSAIDs lower amyloidogenic Ab42 independently of cyclooxygenase activity. *Nature*, 414, 212–216.

Weggen, S., Eriksen, J.L., Sagi, S.A., Pietrzik, C.U., Golde, T.E., and Koo, E.H. (2003a) Ab42-lowering nonsteroidal anti-inflammatory drugs preserve intramembrane cleavage of the amyloid precursor protein (APP) and ErbB-4 receptor and signaling through the APP intracellular domain. *Journal of Biological Chemistry*, 278, 30748–30754.

Weggen, S., Eriksen, J.L., Sagi, S.A., Pietrzik, C.U., Ozols, V., Fauq, A., et al. (2003b) Evidence that nonsteroidal anti-inflammatory drugs decrease amyloid b 42 production by direct modulation of g-secretase activity. *Journal of Biological Chemistry*, 278, 31831–31837.

Weidemann, A., Eggert, S., Reinhard, F.B., Vogel, M., Paliga, K., Baier, G., Masters, C.L., Beyreuther, K., and Evin, G. (2002) A novel epsilon-cleavage within the transmembrane domain of the Alzheimer amyloid precursor protein demonstrates homology with Notch processing. *Biochemistry*, 41, 2825–2835.

Weihofen, A., Lemberg, M.K., Ploegh, H.L., Bogyo, M., and Martoglio, B. (2000) Release of signal peptide fragments into the cytosol requires cleavage in the transmembrane region by a protease activity that is specifically blocked by a novel cysteine protease inhibitor. *Journal of Biological Chemistry*, 275, 30951–30956.

Weller, R.O., Massey, A., Kuo, Y.M., and Roher, A.E. (2000) Cerebral amyloid angiopathy: accumulation of A beta in interstitial fluid drainage pathways in Alzheimer's disease. *Annals of the New York Academy of Science*, 903, 110–117.

Wilcock, G., Haworth, J., Hendrix, S., Zavitz, K., Binger, M.H., Roch, J.M., et al. (2005) A placebo-controlled, double-blind trial of the selective Aß42 lowering agent, MPC-7869 (R-flurbiprofen) in patients with mild to moderate Alzheimer's diseases, *Neuroscience 2005 Meeting*, Washington, DC.

Williams, A.D., Sega, M., Chen, M., Kheterpal, I., Geva, M., Berthelier, V., et al. (2005) Structural properties of Abeta protofibrils stabilized by a small molecule. *Proceedings of the National Academy of Sciences of the United States of America*, 102, 7115–7120.

Wisniewski, T., Castano, E.M., Golabek, A., Vogel, T., and Frangione, B. (1994) Acceleration of Alzheimer's fibril formation by apolipoprotein E in vitro. *American Journal of Pathology*, 145, 1030–1035.

Wolfe, M.S. (2002) Therapeutic strategies for Alzheimer's disease. *Nature Reviews in Drug Discovery*, 1, 859–866.

Wolfe, M.S. (2003) g-secretase-intramembrane protease with a complex. *Science Aging Knowledge Environment*, 2003, PE7.

Wolfe, M.S., and Kopan, R. (2004) Intramembrane proteolysis: theme and variations. *Science*, 305, 1119–1123.

Wolfe, M.S., Citron, M., Diehl, T.S., Xia, W., Donkor, I.O., and Selkoe, D.J. (1998) A substrate-based difluoro ketone selectively inhibits Alzheimer's g-secretase activity. *Journal of Medicinal Chemistry*, 41, 6–9.

Wolfe, M.S., Xia, W., Ostaszewski, B.L., Diehl, T.S., Kimberly, W.T., and Selkoe, D.J. (1999) Two transmembrane aspartates in presenilin-1 required for presenilin endoproteolysis and g-secretase activity. *Nature*, 398, 513–517.

Wolozin, B. (2004) Cholesterol, statins and dementia. *Current Opinion in Lipidology*, 15, 667–672.

Wolozin, B., Kellman, W., Ruosseau, P., Celesia, G.G., and Siegel, G. (2000) Decreased prevalence of Alzheimer disease associated with 3-hydroxy-3-methyglutaryl coenzyme A reductase inhibitors. *Archives of Neurology*, 57, 1439–1443.

Wong, E., Bayly, C., Waterman, H.L., Riendeau, D., and Mancini, J.A. (1997) Conversion of prostaglandin G/H synthase-1 into an enzyme sensitive to PGHS-2-selective inhibitors by a double His513 – > Arg and Ile523 – > Val mutation. *Journal of Biological Chemistry*, 272, 9280–9286.

Wong, G.T., Manfra, D., Poulet, F.M., Zhang, Q., Josien, H., Bara, T., Engstrom, L., Pinzon-Ortiz, M., Fine, J.S., Lee, H.J., et al. (2004) Chronic treatment with the g-secretase inhibitor LY-411,575 inhibits beta-amyloid peptide production and alters lymphopoiesis and intestinal cell differentiation. *Journal of Biological Chemistry*, 279, 12876–12882.

Wood, S.J., Wetzel, R., Martin, J.D., and Hurle, M.R. (1995) Prolines and amyloidogenicity in fragments of the Alzheimer's peptide beta/A4. *Biochemistry*, 34, 724–730.

Wood, S.J., Maleeff, B., Hart, T., and Wetzel, R. (1996) Physical, morphological and functional differences between ph 5.8 and 7.4 aggregates of the Alzheimer's amyloid peptide Abeta. *Journal of Molecular Biology*, 256, 870–877.

Wrigley, J.D., Martin, A.C., Clarke, E.E., Lewis, H.D., Pollack, S.J., Shearman, M.S., et al. (2003) Modulation versus inhibition - selected NSAIDs can act as secretase inhibitors in vitro, *33rd Annual Meeting of Society of Neuroscience*, New Orleans, LA, Abstract 295.2.

Wolfe, S.M. (2004) Dangers of rosuvastatin identified before and after FDA approval. *Lancet*, 363, 2189–2190.

Xia, M.Q. and Hyman, B.T. (1999) Chemokines/chemokine receptors in the central nervous system and Alzheimer's disease. *Journal of Neurovirology*, 5, 32–41.

Xiao, G., Tsai, A.L., Palmer, G., Boyar, W.C., Marshall, P.J., and Kulmacz, R.J. (1997) Analysis of hydroperoxide-induced tyrosyl radicals and lipoxygenase activity in aspirin-treated human prostaglandin H synthase-2. *Biochemistry*, 36, 1836–1845.

Xie, W.L., Chipman, J.G., Robertson, D.L., Erikson, R.L., and Simmons, D.L. (1991) Expression of a mitogen-responsive gene encoding prostaglandin synthase is regulated by mRNA splicing. *Proceedings of the National Academy of Sciences of the United States of America*, 88, 2692–2696.

Xiong, G.L., Benson, A., and Doraiswamy, P.M. (2005) Statins and cognition: what can we learn from existing randomized trials? *CNS Spectrums*, 10, 867–874.

Xu, H., Sweeney, D., Greengard, P., and Gandy, S. (1996) Metabolism of Alzheimer beta-amyloid precursor protein: regulation by protein kinase A in intact cells and in a cell-free system. *Proceedings of the National Academy of Sciences of the United States of America*, 93, 4081–4084.

Yaffe, K., Barrett-Connor, E., Lin, F., and Grady, D. (2002) Serum lipoprotein levels, statin use, and cognitive function in older women. *Archives of Neurology*, 59, 378–384.

Yamagata, K., Andreasson, K.I., Kaufmann, W.E., Barnes, C.A., and Worley, P.F. (1993) Expression of a mitogen-inducible cyclooxygenase in brain neurons: regulation by synaptic activity and glucocorticoids. *Neuron*, 11, 371–386.

Yamamoto, Y., Yin, M.J., Lin, K.M., and Gaynor, R.B. (1999) Sulindac inhibits activation of the NF-kappaB pathway. *Journal of Biological Chemistry*, 274, 27307–27314.

Yan, Q., Zhang, J., Liu, H., Babu-Khan, S., Vassar, R., Biere, A.L., et al. (2003) Anti-inflammatory drug therapy alters b-amyloid processing and deposition in an animal model of Alzheimer's disease. *Journal of Neuroscience*, 23, 7504–7509.

Yang, D.S., Serpell, L.C., Yip, C.M., McLaurin, J., Chrishti, M.A., Horne, P. et al. (2001) Assembly of Alzheimer's amyloid-beta fibrils and approaches for therapeutic intervention. *Amyloid*, 8 (Suppl 1), 10–19.

Yang, F., Lim, G.P., Begum, A.N., Ubeda, O.J., Simmons, M.R., Ambegaokar, S.S., Chen, P.P., Kayed, R., Glabe, C.G., Frautschy, S.A., and Cole, G.M. (2005) Curcumin inhibits formation of amyloid beta oligomers and fibrils, binds plaques, and reduces amyloid in vivo. *Journal of Biological Chemistry*, 280, 5892–5901.

Yang, L.B., Lindholm, K., Yan, R., Citron, M., Xia, W., Yang, X.L., Beach, T., Sue, L., Wong, P., Price, D., et al. (2003) Elevated beta-secretase expression and enzymatic activity detected in sporadic Alzheimer disease. *Nature Medicine*, 9, 3–4.

Ye, L., Morgenstern, J.L., Gee, A.D., Hong, G., Brown, J., and Lockhart, A. (2005) Delineation of PET imaging agent binding sites on beta-amyloid peptide fibrils. *Journal of Biological Chemistry*, 280, 23599–23604.

Yoshiyama, Y., Arai, K., and Hattori, T. (2001) Enhanced expression of I-kappaB with neurofibrillary pathology in Alzheimer's disease. *Neuroreport*, 12, 2641–2645.

Zandi, P.P., and Breitner, J.C. (2001) Do NSAIDs prevent Alzheimer's disease? And, if so, why? The epidemiological evidence. *Neurobiology of Aging*, 22, 811–817.

Zandi, P.P., Carlson, M.C., Plassman, B.L., Welsh-Bohmer, K.A., Mayer, L.S., Steffens, D.C. et al. (2002) Hormone replacement therapy and incidence of Alzheimer disease in older women: the Cache County Study. *Journal of the American Medical Association*, 288, 2123–2129.

Zandi, P.P., Anthony, J.C., Khachaturian, A.S., Stone, S.V., Gustafson, D., Tschanz, J.T. et al. (2004a) Reduced risk of Alzheimer disease in users of antioxidant vitamin supplements: the Cache County Study. *Archives of Neurology*, 61, 82–88.

Zandi, P.P., Szekely, C.A., Green, R.C., Breitner, J.C. and Welsh-Bohmer, K.A. (2004b) S1–03–03 Pooled analysis of the association between different NSAIDS and AD: preliminary findings. *Neurobiology of Aging*, 25, S5–S6.

Zandi, P.P., Sparks, D.L., Khachaturian, A.S., Tschanz, J., Norton, M., Steinberg, M. et al. (2005) Do statins reduce risk of incident dementia and Alzheimer disease? The Cache County Study. *Archives of General Psychiatry*, 62, 217–224.

Zhou, Y., Su, Y., Li, B., Liu, F., Ryder, J.W., Wu, X., et al. (2003) Nonsteroidal anti-inflammatory drugs can lower amyloidogenic Ab42 by inhibiting Rho. *Science*, 302, 1215–1217.

Zimmermann, M., Gardoni, F., and Di Luca, M. (2005) Molecular rationale for the pharmacological treatment of Alzheimer's disease. *Drugs & Aging*, 22 (Suppl 1), 27–37.

Zlokovic, B.V., Yamada, S., Holtzman, D., Ghiso, J., and Frangione, B. (2000) Clearance of amyloid beta-peptide from brain: transport or metabolism? *Nature Medicine*, 6, 718–719.

Zlokovic, B.V., Deane, R., Sallstrom, J., Chow, N., and Miano, J.M. (2005) Neurovascular pathways and Alzheimer amyloid beta-peptide. *Brain Pathology*, 15, 78–83.

5 Carbonic Anhydrase Activators as Potential Anti-Alzheimer's Disease Agents

Claudiu T. Supuran and Andrea Scozzafava

CONTENTS

Activators of the zinc enzymes carbonic anhydrases (CAs) accelerate the hydration of carbon dioxide (CO_2) to bicarbonate and a proton by favoring the rate-determining step in the catalytic cycle. CA activators (CAAs) bind within the enzyme active site cavity, at a site different to the inhibitor- or substrate-binding sites, facilitating the proton-transfer processes between the active site and the reaction medium. Adducts of isoforms CA I and II with various activators have been characterized by means of X-ray crystallography, allowing thus for a rationale in the design of such agents. Solution studies were performed on the activation of isoforms CA I, II, IV, VA, VII, and XIV with a variety of activators. Such studies led to the discovery of strong CAAs belonging to several chemical classes. Structure–activity correlations for such activators are discussed for the various isozymes for which the phenomenon has been studied. In the human brain, at least seven CA isoforms are present (CA I, II, IV, VII, VIII, XII, and XIV), most of which are activated by these types of modulators of activity. The physiologic relevance of brain CA activation and the possible application of activators in the management of Alzheimer's disease and other memory therapies are also reviewed, although research in this field is in its early phase.

5.1 INTRODUCTION

At least 16 different isoforms of the zinc enzyme carbonic anhydrase (CA, EC 4.2.1.1) have been characterized up to now in mammals, including *Homo sapiens* (Hewett-Emmett, 2000; Supuran et al., 2004). Most of them act as efficient catalysts for the reversible hydration of CO_2 to bicarbonate and a proton, under physiological conditions. Detailed studies based on kinetics, spectroscopy, X-ray crystallography, and site-specific mutagenesis have undoubtedly revealed that some of these isoforms (such as CA II, VII, and IX) are among the fastest enzymes known, possessing turnover numbers $k_{cat} \geq 10^6\ s^{-1}$ and second-order steady-state rate constants close to the diffusion-controlled processes (hCA II has $k_{cat}/K_M \approx 10^8\ M^{-1}s^{-1}$) (Stams and Christianson, 2000; Supuran et al., 2004).

Activators of CAs constituted a rather controversial issue for a long time (Supuran et al., 2004). Although CA activation was reported (Leiner, 1940) simultaneously with its inhibition (Mann and Keilin, 1940), the research development in these two related fields had completely different fates (Supuran and Puscas, 1994; Supuran and Scozzafava, 2000a). Although CA inhibitors (CAIs) have been constantly and extensively studied, as well as fruitfully exploited clinically for the treatment or prevention of a multitude of diseases (Supuran and Scozzafava, 2000b, 2001), CA activation studies have registered important developments only in the last decade (Supuran et al., 2004).

Starting with the early 1990s, this field was relaunched by several relevant discoveries, such as (1) the report of CA III anionic activators (Shelton and Chegwidden, 1988), shortly followed by the report of a large variety of many structural classes of organic compounds with good activating effects against isozymes I, II, and IV (Supuran and Puscas, 1994; Supuran and Scozzafava, 2000a; Supuran et al., 2003); (2) the elucidation of the activation mechanism at the molecular level by means of kinetic, spectroscopic, and X-ray crystallographic data (Briganti et al., 1997; Supuran and Scozzafava, 2000a; Supuran et al., 2004); (3) provision of compelling evidences that CA genetic deficiency, although much less encountered than the enzyme excessive activity, causes severe physiological/pathological disorders, which might be corrected/treated through an appropriate activation of the involved isozymes (Roth et al., 1992; De Coursey, 2000; Alper, 2002; Rousselle and Heymann, 2002; Sun and Alkon, 2002); (4) the report of Sun and Alkon (2001) and Sun et al. (2001) on the possible applications of CAAs as memory therapy agents, and probably also as anti-Alzheimer's disease agents, which provided new impetus to research in the field.

5.2 CA THREE-DIMENSIONAL STRUCTURE
AND ACTIVATION MECHANISM

CAs are globular metalloproteins containing one polypeptide chain and a catalytically critical zinc ion (Supuran et al., 2004) (Figure 5.1). The metal ion is coordinated by three conserved histidine residues (His94, His96, and His119) and a water

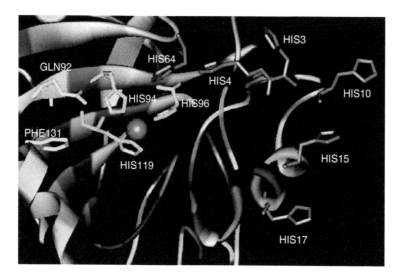

FIGURE 5.1 Carbonic anhydrase isoform II (hCA II) active site. Zinc ion (central sphere), its three histidine ligands—His94, His96, His119, and the histidine cluster—His64, His4, His3, His17, His15, His10 are evidenced. The figure was generated from the X-ray crystallographic coordinates reported by Briganti et al. (1997) (PDB entry 4TST). (Reproduced from Scozzafava, A., and Supuran, C.T., *Bioorg. Med. Chem. Lett.*, 12, 1177, 2002a. With permission.)

molecule, which by deprotonation at neutral pH leads to a potent nucleophile (the zinc-bound hydroxide ion), which then converts CO_2 to bicarbonate with very high efficiency (Supuran et al., 2004). Thus, the catalytic cycle implies two reactions (Equations 5.1 and 5.2) shown below:

$$EZn^{2+} - OH^- + CO_2 \overset{H_2O}{\Leftrightarrow} EZn^{2+} - HCO_3^- \Leftrightarrow EZn^{2+} - OH_2 + HCO_3^- \quad (5.1)$$

$$EZn^{2+} - OH_2 \Leftrightarrow EZn^{2+} - HO^- + H^+ \quad (5.2)$$

In Equation 5.1, the interconversion between CO_2 and bicarbonate is achieved by means of the nucleophilic attack of the zinc-bound hydroxide ion to CO_2 bound in a hydrophobic pocket in its neighborhood. Equation 5.2, the rate-determining step for the entire catalytic turnover, involves regeneration of the basic form of the enzyme, by deprotonation of the water molecule coordinated to Zn(II), a process generally assisted by a histidine residue situated in the middle of the active site cavity (usually His64), shown schematically in Figures 5.1 and 5.2 (Supuran et al., 2004).

Crystallographic data (Nair and Christianson, 1991) have shown that the imidazole side chain of His64 exhibits a pH-dependent conformational mobility, changing gradually its orientation relative to the metal site through a 64° ring flipping. Thus, at pH 8.5 conformer "in" predominates, the imidazole moiety being directed toward zinc ion and linked to it through hydrogen-bond bridges. After the proton catching the "out" conformation, in which the heterocyclic ring points toward the enzyme

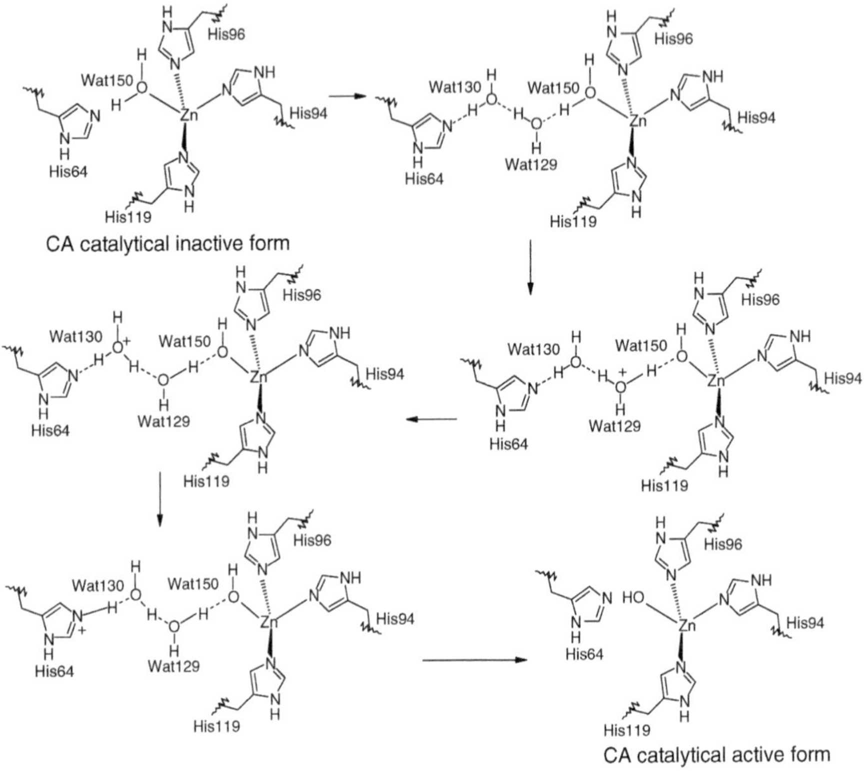

FIGURE 5.2 Schematic representation of proton transfer steps through the hydrogen-bond network Wat150—Wat129—Wat130—His64 in the hCA II active site (without activators).

surface and engages no hydrogen bond contacts, becomes increasingly predominant as the pH decreases, reaching the maximum conformational percentage at the value of 5.7. This high flexibility between the two-mentioned conformations is crucial for the catalytic proton shuttling (Christianson and Fierke, 1996).

The CA activation mechanism (Supuran, 1992; Briganti et al., 1997, 1998; Supuran et al., 2004) is closely connected with that of the catalytic cycle of the enzyme, in the absence of activators, mentioned earlier (Ilies et al., 2004). It has been shown that a CAA interferes directly in the proton transfer step of the catalytic cycle, facilitating the intramolecular proton transfer by means of transient enzyme–activator complexes (Equation 5.3). Because intramolecular reactions are faster than the intermolecular ones (Page and Williams, 1989), the result is a significantly enhanced catalytic rate.

$$EZn^{2+} - OH_2 + A \rightleftharpoons [EZn^{2+} - OH_2 - A \rightleftharpoons EZn^{2+} - OH^- - AH^+] \rightleftharpoons EZn^{2+} - OH^- + AH^+$$
$$\text{enzyme–activator complexes}$$

$$(5.3)$$

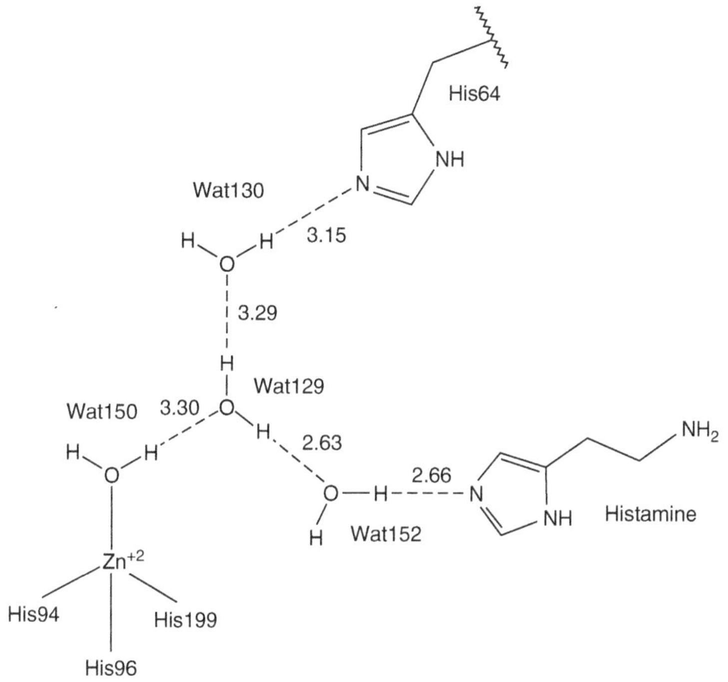

FIGURE 5.3 Binding of histamine to hCA II. Scheme of hydrogen bonds between the activator, His64, and the zinc-bound water molecule (Wat150) (distances are in Å). (Reproduced from Briganti, F., Mangani, S., Orioli, P., Scozzafava, A., Vernaglione, G., and Supuran, C.T., *Biochemistry*, 36, 10384, 1997. With permission.)

Several X-ray crystallographic structures of the most studied and ubiquitous iso-zyme, hCA II, demonstrated the formation of CA–activator complexes (Briganti et al., 1997; Temperini et al., 2005, 2006a, 2006b). The first one was obtained for the adduct of hCA II with histamine (Briganti et al., 1997). This structure has confirmed the previous hypothesis on the activation mechanism (Supuran, 1992), revealing that histamine binds in the hydrophilic region located at the entrance of the active site, establishing through the nitrogen atoms of the imidazole ring several hydrogen bonds with water molecules and polar amino acids residues, whereas the aliphatic amino group of the activator fluctuates free into the solvent (Figure 5.3). The supplementary hydrogen-bond pathways generated by the binding of the activator have two key consequences for the rate-determining step of catalysis: (1) stabilize the His64 "in" conformation—a steric requirement for the proton shuttling; (2) offer adjacent routes for the proton transport from the zinc-bound water molecule to the external medium. Furthermore, histamine shows few contacts with the enzyme, limited to the imidazole ring, which interacts with three amino acids residues Asn62, Asn67, and Gln92. This interaction is favorable for the CA–histamine complex dissociation in the last step of the activation mechanism, in the same way in which the conformational flexibility of His64 confers to this residue the ability to easily oscillate between the deep and the opening of the active site cavity during the proton-transfer processes. It can be thus concluded that the activator actually acts as an efficient second proton shuttle, apart from the native one, His64.

Activation of six human isoforms, hCA I, II, IV, VA, VII, and XIV, with L- and D-histidine compounds closely related structurally to histamine was also investigated kinetically and by X-ray crystallography more recently (Temperini et al., 2005, 2006a). L-His was shown to be a potent activator of isozymes hCA I, VA, VII, and XIV, and a weaker activator of isozymes hCA II and IV. D-His showed good hCA I, VA, and VII activatory properties, being a moderate activator of hCA XIV and a weak activator of hCA II and IV. X-ray crystallographic structures of the hCA II–L-His/D-His adducts showed the activators to be anchored at the entrance of active site (in the same region where histamine also binds), participating in extended networks of hydrogen bonds with amino acid residues/water molecules present in the cavity, explaining their different potency and interaction patterns with various isozymes. Residues involved in L-His recognition were His64, Asn67, and Gln92, whereas three water molecules connected the activator to the zinc-bound hydroxide (Figure 5.4). Only the imidazole moiety of L-His interacted with these amino acids. For the D-His adduct, residues involved in recognition of the activator were Trp5, His64, and Pro201, whereas two water molecules connected the zinc-bound water to the activator. Only the COOH and NH_2 moieties of D-His participated in hydrogen bonds with these residues (Figure 5.4). This was the first study showing different binding modes of stereoisomeric activators within the hCA II active site (see Figure 5.5 where the two structures are superimposed), with consequences for the overall proton-transfer processes. This also points out that differences of activation efficiency between various isozymes with structurally related activators are exploitable for designing alternative proton-transfer pathways, useful both for a better understanding of the catalytic mechanism and for obtaining pharmacologically useful derivatives, for example, for the management of Alzheimer's disease (Temperini et al., 2006a).

In another recent study, the activation of six human brain carbonic anhydrases, hCA I, II, IV, VA, VII, and XIV, with L-/D-phenylalanine was investigated kinetically and by X-ray crystallography. L-Phe was a potent activator of isozymes I, II, and XIV (K_{AS} of 13–240 nM), a weaker activator of hCA VA and VII (K_{AS} of 9.8–10.9 μM), and quite an inefficient activator of hCA IV (K_A of 52 μM). D-Phe showed good hCA II activatory properties (K_A of 35 nM), being a moderate hCA VA, VII, and XIV (K_{AS} of 4.6–9.7 μM) and a weak hCA I and IV activator (K_{AS} of 63–86 μM). X-ray crystallography of the hCA II–L-Phe/D-Phe adducts showed the activators to be anchored at the entrance of the active site, participating in numerous hydrogen bonds and hydrophobic interactions with amino acid residues His64, Thr200, Trp5, and Pro201 (Figure 5.6A and B). This study also showed the different binding modes of stereoisomeric activators within the hCA II active site, similarly to the previous one (Temperini et al., 2006a, 2006b). CAAs may thus lead to the design of pharmacologically useful derivatives for the enhancement of synaptic efficacy, which may represent a conceptually new approach for the treatment of Alzheimer's disease, aging, and other conditions in which spatial learning and memory therapy must be enhanced (see later in the text). As the blood and brain concentrations of L-Phe

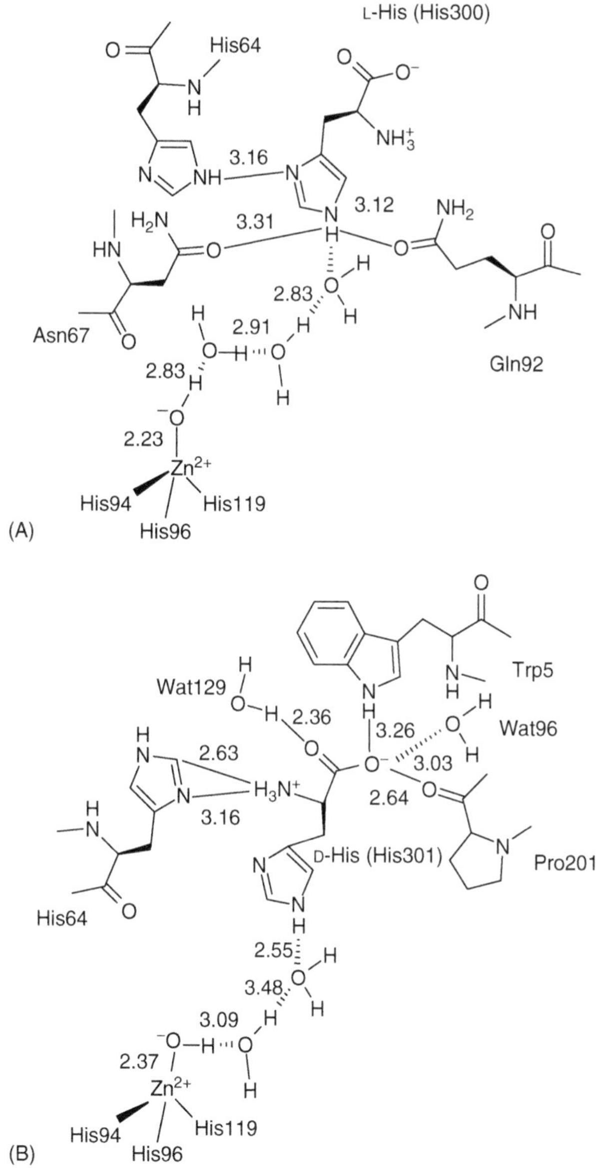

FIGURE 5.4 Schematic representation for the binding of L-His (numbered as His300) (A) and D-His (numbered as His301) (B) within the hCA II active site. The Zn(II) ligands and hydrogen bonds connecting the Zn(II) ion and the activator molecules through a network of several water molecules are shown, as well as the hydrogen bonds (*dotted lines*) between the activator molecules and amino acid residues involved in their binding (figures represent distances in Å).

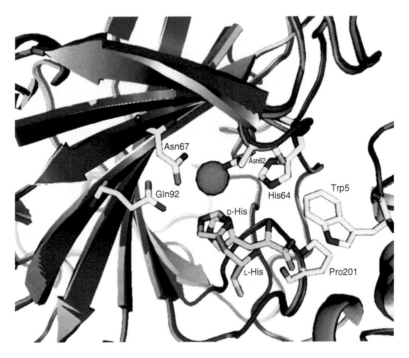

FIGURE 5.5 Superimposition of the two hCA II adducts with ʟ-His and ᴅ-His, with the zinc ion (central sphere) and amino acid residues present in the activator-binding site evidenced. (From Temperini, C., Scozzafava, A., Vullo, D., and Supuran, C.T., *Chemistry*, 12, 7057, 2006a. With permission.)

are quite variable (30–73 μM), activity of some brain CAs may strongly be influenced by the level of activator(s) present in such tissues.

A ternary adduct of hCA II with the activator ʟ-Phe and the inhibitor azide was also reported earlier (Briganti et al., 1998). Analyzing the refined atomic model of this adduct, it appeared obvious that azide replaces the hydroxide anion from the Zn(II) native enzyme coordination sphere, binding directly to the metal site (with the maintenance of the overall tetrahedral geometry), and extending into the hydrophobic half of the active site cavity. Unlike the inhibitor molecule and similar to histamine, phenylalanine experienced no contacts with the zinc ion, being anchored through hydrogen-bond bridges in the proximity of the active site entrance, with the phenyl ring oriented toward this entrance (Figure 5.6C). It should be also noted that a water molecule (Wat73) linked simultaneously the inhibitor and activator molecules by means of two strong hydrogen bonds involving the zinc-bound nitrogen atom and the amino group, respectively (Figure 5.6C). The very different binding of Phe in the normal adduct with hCA II and the ternary adduct reported above must be observed (Briganti et al., 1998; Temperini et al., 2006b).

It is worth stressing that, although belonging to diverse classes of compounds, all hCA II activators discussed up to now bind in a similar location of the enzyme active site, at its entrance. It has been assumed that this particular location of the activators

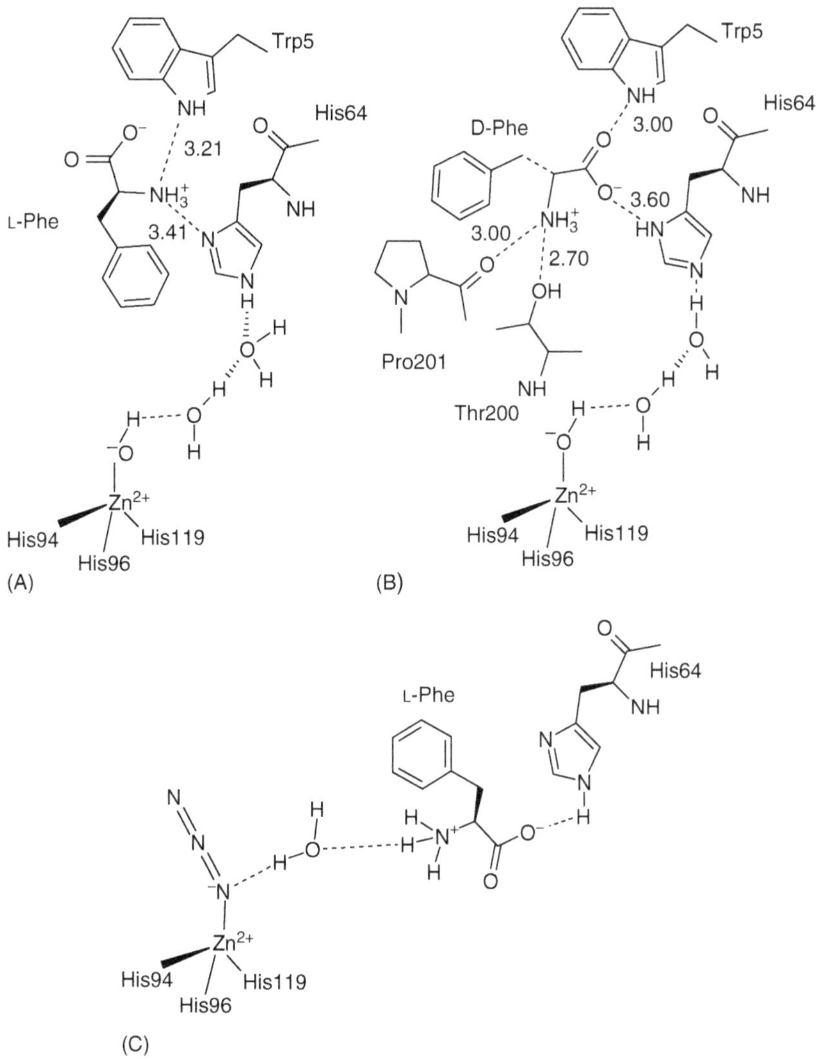

FIGURE 5.6 Schematic representation of the active site in the hCA II–L-Phe adduct (A), hCA II–D-Phe adduct (B), and the hCA II–azide-L-Phe ternary complex (C) (figures represent distances in Å).

is also a consequence of their weak Lewis base character, which makes them unable to replace the stronger HO^- zinc ligand (case in which they should act as CAIs). All the above findings led to the conclusion that in addition to the CO_2 substrate-binding site (the hydrophobic pocket consisting of the amino acid residues Val121, Val143, Leu198, Thr199, Val207, and Trp209) and the inhibitor-binding site (the zinc ion), CAs possess a third modulators-binding site—*the activator-binding site*, located at the entrance of the active site cavity.

Very recently, the first X-ray crystal structure of an hCA I activator adduct has been resolved, that of L-His (Temperini et al., 2006c). On the one hand, the binding affinity of L-His for hCA I is high, with an affinity constant of 30 nM for this isozyme, on the other hand it is a weaker activator of hCA II, with an affinity constant of around 10 μM (Temperini et al., 2006c). Kinetic measurements lead to the observation that the activation is because of an enhancement of k_{cat}, which for hCA I in the absence of activators (for the physiological reaction, at 25°C and pH 7.5) is of 2.0×10^5 s^{-1}, whereas in the presence of 10 μM L-His, this parameter is $k_{cat} = 13.4 \times 10^5$ s^{-1}. In all these experiments, the Michaelis–Menten constant K_M was the same, in the absence or in the presence of activators, the measured values being: $K_M = 25$ mM (Temperini et al., 2006c). It was observed that 10 μM L-His (concentrations much higher than this one, in the range of 60–120 mM, are present in many tissues, including the brain) led to an enhancement of 670% of the hCA I catalytic activity. Such a phenomenon surely translates into important physiological consequences, probably correlated with pH homeostasis or ion transport processes in which hCA I is known to be involved (Temperini et al., 2006c).

The activator-binding site of hCA I is rather different from that of hCA II (Figures 5.7 and 5.8), where the same activator, L-His, was shown to bind at the entrance of the cavity, being anchored by several strong hydrogen bonds to His64, Asn67, and Gln92 (Figure 5.7A). In the case of the hCA I–L-His adduct, it may be observed that the activator was bound much deeper within the enzyme active site, bridging by means of two hydrogen bonds (of 3.15 and 2.85 Å, respectively) the zinc-bound hydroxide ion and the Nε imidazole atom of His200 (an amino acid residue characteristic for this isozyme, as in hCA II there is a Thr in position 200). Furthermore, the orientation of the activator molecule is quite different in the two adducts, as it is the carboxylate moiety of L-His participating in the main interactions with the hCA I active site (in fact the COO$^-$ points toward the zinc ion, Figure 5.7B), whereas the imidazole moiety of L-His points toward the Zn(II) ion in the hCA II adduct, and also participates in a network of four hydrogen bonds with three amino acid residues and a water molecule bound within the cavity (Figure 5.7A). In addition, L-His is in van der Waals contact with Thr199, His64, His67, and His200 in the hCA I adduct (Figure 5.7B). The zinc-bound hydroxide is directly hydrogen bonded to the activator molecule in the hCA I adduct, but is bridged by three different water molecules in the hCA II adduct (Figure 5.7A and B). These important differences in binding of L-His to hCA I and II are also clearly observed in Figure 5.8, where the two adducts were superimposed. It may be observed that the globular shape and active site architecture of the two CA isozymes is quite similar, but the activator binds at the entrance of the active site cavity for hCA II, and quite deep within this site for hCA I. This may also explain the rather different affinity of this activator for the two isozymes, which is in the nanomolar range for hCA I and in the micromolar range for hCA II.

5.3 CAAs DRUG DESIGN

Searching for structural particularities which may determine CA activation, many authors (Puscas et al., 1990; Supuran et al., 1991; Supuran, 1992) investigated CA

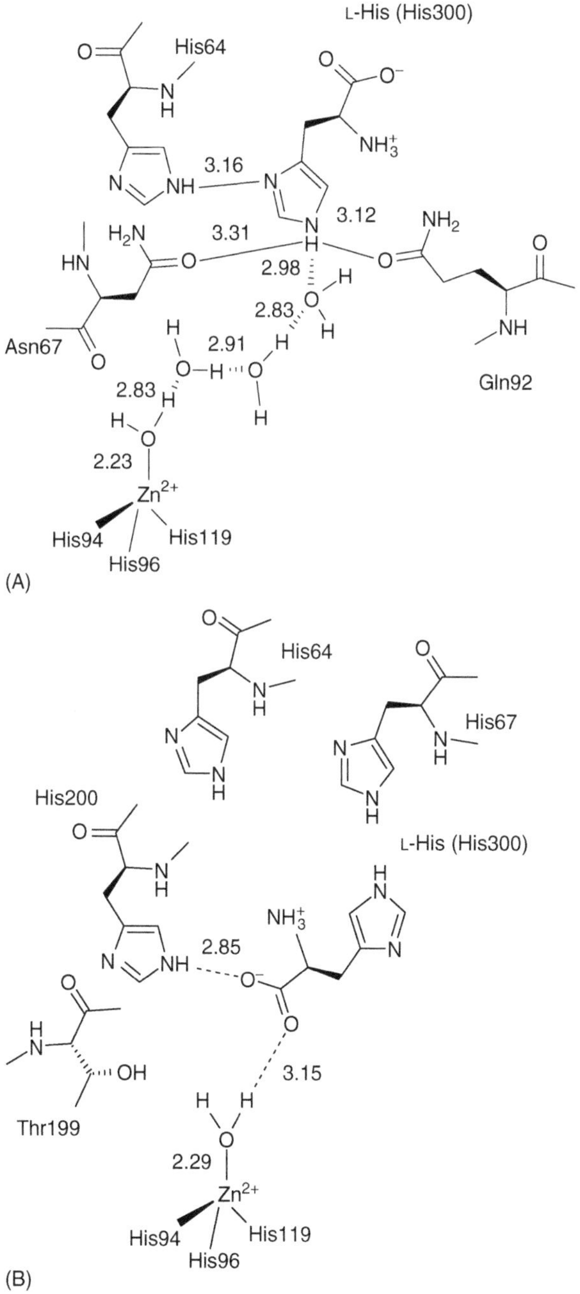

FIGURE 5.7 Schematic representation for the binding of L-His (numbered as His300) to the hCA II (A) and hCA I (B) active sites. The Zn(II) ligands and hydrogen bonds connecting the Zn (II) ion and the activator molecules through a network of water molecules are shown, as well as the hydrogen bonds (*dotted lines*) between the activator molecules and amino acid residues involved in their binding (figures represent distances in Å). (From Temperini, C., Scozzafava, A., and Supuran, C.T., *Bioorg. Med. Chem. Lett.*, 16, 5152, 2006c. With permission.)

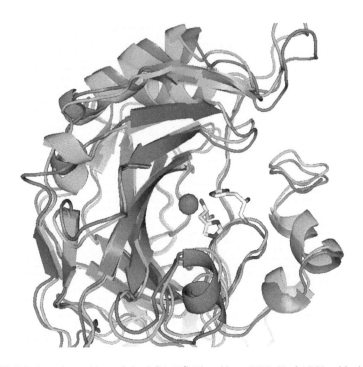

FIGURE 5.8 Superimposition of the hCA II/l-His adduct (PDB ID 2ABE) with the hCA I/l-His adduct (PDB ID 2FW4), clearly showing the deep binding of the activator to hCA I and the external binding to hCA II. (From Temperini, C., Scozzafava, A., and Supuran, C.T., *Bioorg. Med. Chem. Lett.*, 16, 5152, 2006c. With permission.)

activity in the presence of various natural or synthetic compounds with biological relevance, such as biogenic amines, amino acids and their derivatives, oligopeptides, and several pharmacological agents (vide infra). It was shown (Supuran, 1992) that all these compounds fit well into a general structure of type (**5.1**), incorporating as main structural elements a protonable moiety (a primary or secondary amino group) attached to a bulky aromatic/heterocyclic ring through an aliphatic carbon chain linker. Activation data for some of these compounds are shown in Table 5.1.

$$
\begin{array}{c}
\text{R2} \\
\text{Ar}\diagdown\diagup\diagdown\diagup\text{N}\diagdown\text{R1} \\
\overset{|}{\text{R3}}\qquad\overset{|}{\text{H}}
\end{array}
$$

5.1

The data of Table 5.1 afforded the first structure–activity relationships (SAR) for CAAs. Thus, consistent with the activation mechanism previously presented, substitution of the parent structure (**5.1**) with polar moieties able to engage in hydrogen bonds (such as hydroxyl and carboxyl groups) increases the activator potency. The

TABLE 5.1
hCA I and II Activation by Some Biologically Active Compounds, for the CO$_2$ Hydration at 10^{-5} M of Activator

Compound[a]	Ar	R^1	R^2	R^3	% CA Activity[b]	
					hCA I	hCA II
Phenethylamine	Ph	H	H	H	114	110
Dopamine	3,4-Di(OH)Ph	H	H	H	138	141
Noradrenaline	3,4-Di(OH)Ph	H	H	OH	140	143
Adrenaline	3,4-Di(OH)Ph	Me	H	OH	145	153
Isoprotenerole	3,4-Di(OH)Ph	i-Pr	H	OH	143	146
Histamine	4-Imidazolyl	H	H	H	180	173
2-Pyridyl-ethylamine	2-Pyridyl	H	H	H	134	120
Serotonin	5-OH-indol-3-yl	H	H	H	128	115
Phenylalanine	Ph	H	COOH	H	170	196
4-Hydroxyphenylalanine	4-OH-Ph	H	COOH	H	174	202
3,4-Dihydroxyphenylalanine	3,4-Di(OH)Ph	H	COOH	H	164	142
4-Fluorophenylalanine	4-F-Ph	H	COOH	H	169	175
3-Amino-4-hydroxyphenylalanine	3-NH$_2$-4-OH-Ph	H	COOH	H	171	177
4-Aminophenylalanine	4-NH$_2$-Ph	H	COOH	H	152	163
Histidine	4-Imidazolyl	H	COOH	H	153	149
Tryptophan	3-Indolyl	H	COOH	H	129	124

[a] Amino acids were L-enantiomers.
[b] CA activity without activator added is taken as 100%.

Source: Adapted from Supuran, C.T. and Puscas, I., *Carbonic Anhydrase and Modulation of Physiologic and Pathologic Processes in the Organism,* Helicon, Timisoara, 1994.

major part of the arylalkylamines and all the amino acids from Table 5.1 were also included in a quantitative structure–activity relationships (QSAR) study (Clare and Supuran, 1994). This is the only QSAR investigation on hCA II activators available up to now. By means of regression analysis and partial least squares method, good equations of quantitative correlation were derived, which translated into mathematical terms the interdependence between biological activity and two main physicochemical parameters—the molecular charge distribution and the molecular volume. The activating properties were expressed in the form of logA, where A is the hCA II activation percentage. The electronic density and the size of the activator molecule were expressed in the form of a large number of different descriptors calculated by complete neglect of differential overlap (CNDO) approximation. Such optimized equations have shown that CAAs efficiency is determined both by steric and electronic factors. Thus, considering a molecular model projection on a three-dimensional axis, it clearly appeared that decreasing the two smaller orthogonal dimensions of the CAA molecule yielded an intensification of its biological activity. This fact has presented for the first time the idea that CAAs bind to the enzyme in a site with limited size, and this has been subsequently confirmed through x-ray crystallographic studies (Briganti et al., 1997). On the other hand, the higher the charge of the most electronegative atom in the molecule, the more active that compound as a CAA. This second theoretical result explained experimental data showing that molecules bearing highly charged oxygen or nitrogen atoms (such as amino acids and amino-azole derivatives) were more powerful CAAs than analogous compounds in which the amino group is attached to a molecular skeleton containing only carbon atoms (Supuran et al., 1991, 1993a). This first QSAR of CAAs afforded two main conclusions: (1) the activating properties are conditioned by a relatively compact structure of the molecule, strictly depending on the activator-binding site dimensions; (2) the activating properties are significantly increased by the presence in the molecule of some strongly polarized oxygen/nitrogen-containing moieties. These conclusions, together with the chemical structure of compounds from Table 5.1, constituted the starting point for different directions in the design and synthesis of CAAs.

5.3.1   CA ACTIVATION WITH AMINES

Three premises have been considered in the development of this type of CAA: (1) aliphatic amines with the general formula (**5.2**) were shown to have no activatory effect against CA II (Puscas et al., 1990); (2) 2-amino-5-(2-aminoethyl)-1,3,4-thiadiazole (**5.3**), reported as a histamine agonist and proved to possess gastric acid-stimulating influences, has been recognized as a powerful CAA (Supuran and Puscas, 1994); (3) the pyridinium ring presents several features (i.e., aromatic character, lasting positive charge, strong chemical stability—especially when it is alkyl/aryl substituted), which recommend it as a candidate moiety for obtaining biologically active compounds with pharmacological applications. Efficient positively charged CAAs of type (**5.4**) were thus designed and synthesized (Supuran et al., 1993b, 1996a), by condensing different 2,4,6-trisubstituted pyrylium salts with the corresponding amine derivatives in the classical Baeyer–Piccard reaction conditions. Such pyridinium compounds generally showed excellent CA-activating properties (Supuran and Puscas, 1994).

(R = alkyl, OH, SH, NH$_2$, etc.)

   The following SAR conclusions were derived: (1) The polymethylene bridge between the amino group and the heterocyclic moiety may incorporate either two or three carbon atoms; a greater number leads to a decrease in the activating properties. (2) The derivatives bearing methyl or t-butyl groups in the 2- and 6-positions of the pyridinium cycle are much more potent CAAs than the 2,6-diphenyl substituted analogs, probably because of steric impairment of activator access into the enzymatic site. On the other hand, 2,6-dialkyl-4-phenyl derivatives were highly active, supporting the previous QSAR finding according to which molecule elongation along the axis passing through the enzyme Zn(II) ion, the CAA molecule itself, and the active site exit is the most beneficial substitution pattern for the activating properties. (3) The complete derivatization of the ω-amino alkyl group eliminates the activatory effect. Furthermore, these studies (Supuran et al., 1993b, 1996a) provided the first suggestion for obtaining selective CAAs. Because of their cationic nature, pyridinium derivatives are membrane-impermeant, therefore, theoretically they should discriminate between the membrane-bound CA IV and the very similar cytosolic CA II (Ilies et al., 2002).

5.3.2 CA ACTIVATION WITH AZOLES

Imidazole is widely used as buffer, whereas the imidazole moiety is present in both histamine (the prototypical CAA) and histidine (an amino acid of crucial importance for the CA catalytic cycle and also a CAA). Consequently, the research into the field started with investigations of CA activation with azoles. A major difference between isozymes I and II was thus unveiled: imidazole is a unique competitive inhibitor with CO_2 as substrate for hCA I (Khalifah, 1971), whereas it behaves as a very efficient activator for isozyme II (Parkes and Coleman, 1989; Supuran, 1992). Subsequent studies (Supuran et al., 1993a, 1996b) afforded CA II activation data with a large series of mono-, bis- and tris-azole derivatives **5.5–5.7** as well as with pyridinium azoles (Ilies et al., 1997, 2002; Supuran et al., 1998; Scozzafava et al., 2000a).

$n = 1, 2$; R, R', R" = H, Me, Et

5.3.3 CA ACTIVATION WITH HISTAMINE DERIVATIVES

Histamine belongs to the amino-azole class of CAAs. Therefore, one may consider the CAAs derived from this lead molecule as an independent and particular category of such compounds. The rational for designing novel activators of this type was based on structural aspects of the hCA II–histamine interaction, revealed by x-ray crystallography (discussed above): (1) the unique histidine cluster of the hCA II active site surrounds its opening and consists of six residues, some of them possessing flexible conformations, which could easily participate in hydrogen bonds with activators/inhibitors bound within the active site (Briganti et al., 1997, 1998); (2) the histamine molecule binds at the hCA II active site entrance, being anchored to amino acid side chains and to water molecules through hydrogen bonds involving only the nitrogen atoms of the imidazole moiety, whereas its aliphatic amino group extends into the solvent, making no contact with the enzyme (Briganti et al., 1997). Thus, it appeared of interest to derivatize the amino aliphatic end of the histamine molecule with highly polarized moieties. Such groups might interfere in a favorable way with polar amino acid residues at the rim of the active site. Decreasing the whole energy of binding of the activators, these additional contacts would lead to more stable enzyme–activator adducts, hence to more efficient activation processes. This approach, based on the improvement in the energy of binding of certain compounds to the enzyme through chemically modified units of a given parent structure, was also reported both by Whitesides' group (Gao et al., 1995) and Supuran's group (Scozzafava et al., 1999) for the design of tight-binding sulfonamide CAIs.

R = alkyl, aryl, substituted aryl; X = O, S

To obtain a much better affinity for the protein as compared with histamine, the following ideas have been considered in the choice of the polar derivatizing groups: (1) as it has been proved by earlier QSAR calculations (Clare and Supuran, 1994), oxygen and nitrogen atoms stimulate considerably the activating properties; (2) a peptide type side chain, fulfilling the previous condition, would additionally confer to the compound a specific embedded structure through which it could interact better with amino acid residues from the active site (Supuran and Scozzafava, 1999); (3) according to the activation mechanism previously described in this chapter, a free amino or imidazole moiety acting as a second proton-transferring group (besides the imidazole ring from the lead histamine) might influence positively the CAAs efficacy (Supuran and Scozzafava, 1999, 2000c). Some of the best activators obtained considering all these facts possessed the structures **5.8–5.11**, and were much more effective CAAs as compared with the lead compound histamine (Briganti et al., 1999; Supuran and Scozzafava, 1999, 2000c; Scozzafava and Supuran, 2000a, 2000b, 2002b).

5.3.4 CA ACTIVATION WITH AMINO ACID DERIVATIVES

Supuran et al. (1991) reported the first systematic study of CA activation with amino acids and related compounds. Various such compounds, including natural/synthetic amino acids, their esters, N-alkyl or N-acyl derivatives, as well as some pyridinium derivatives were assayed against isozymes I and II, for the reversible CO_2 hydration reaction.

The following SAR emerged: (1) The most powerful CAAs in these series were proline, homoproline, histidine, and many of the aromatic amino acids, especially those possessing the general formula (**5.1**), such as phenylalanine. X-ray crystallographic data (Briganti et al., 1997b, 1998) clearly demonstrated that the aryl/hetaryl moieties of these compounds increases the stability of the enzyme–activator complexes. (2) The amino or carboxyl groups derivatization strongly diminished the activating profile, mainly because of the decrease in charge on the most electronegative atom of the molecule (through the induced electronic effects), as it was proved later by QSAR calculations (Clare and Supuran, 1994). (3) Similar to azole derivatives (Supuran et al., 1996b) and pyridinium amine derivatives of type (**5.4**) (Supuran and Balaban, 1994), a strong correlation between the pK_a value and the activator strength was also found for the investigated amino acids (Supuran and Balaban, 1994).

It should be stressed that no isozyme specificity was detected for this CAAs class, some of the investigated amino acids possessing an increased affinity for CA I, whereas others for CA II. In summary, the above-mentioned studies (Supuran et al., 1991; Supuran and Balaban, 1994) delineated two major factors that modulate the activating behavior of the amino acid derivatives: (1) the presence of cyclic structural units in the molecule is favorable for stabilizing enzyme–activator adducts; (2) the pK_a value of the proton shuttle moiety strongly influences the activator's efficiency.

The interesting results of CA activation with amino acids (Supuran et al., 1991), as well as the report of pentagastrin (Puscas et al., 1996) and hemoglobin (Parkes and Coleman, 1989) as CAAs, motivated further search for similar properties in polypeptide derivatives, such as angiotensins (Puscas et al., 1994). The activatory efficacy of these compounds against bovine CA II (bCA II) for the CO_2 hydration varied in wide limits. CA activation with tri- and tetrapeptides incorporating alanine and aspartic acid residues has recently been investigated in detail (Scozzafava and Supuran, 2002a). These novel activators exhibited a significantly increased affinity for hCA I, hCA II, and bCA IV, respectively, as compared with the parent amino acids. SAR conclusions of this study were (Scozzafava and Supuran, 2002a) (1) the first CAAs for which hCA II possesses the highest sensitivity among the three investigated isozymes have been reported. The activation pattern has been hCA II > hCA I > bCA IV, whereas the major part of the activators tested up to now (including the classical histamine) have shown a general range of isozyme affinity in the order: hCA I > bCA IV > hCA II (Ilies et al., 1997; Briganti et al., 1999; Supuran and Scozzafava, 1999, 2000c; Scozzafava and Supuran, 2000a, 2000b, 2002b; Scozzafava et al., 2000b; Ilies et al., 2002). Thus, the positively charged imidazolium moieties present in the hCA II histidine cluster presumably engage appropriate contacts with the carboxylate groups of the above-mentioned oligopeptides, this favored interaction not being available for isozymes I and IV, which do not have such a cluster. (2) Aspartyl-alanyl-aspartyl-aspartic acid (DADD), the best CAA in this new series, possesses the sequence recently evidenced (Vince and Reithmeier, 2000) to be essential for the binding of hCA II to the C-terminal region of the bicarbonate/chloride anion exchanger AE1. Consequently, this study

(Scozzafava and Supuran, 2002a) has advanced the idea that in the metabolon formed by these two proteins (Reithmeier, 2001) the AE1 acts actually as a natural activator of isozyme II. In other words, it seems that AE1 facilitates the formation/transport of bicarbonate in erythrocytes through an efficient promotion of the hCA II catalysis for CO_2 hydration.

5.4   CAAs ENHANCE SYNAPTIC EFFICACY, MEMORY, AND SPATIAL LEARNING

Many naturally occurring compounds found in the human body activate CAs at physiological concentrations (Supuran et al., 1991; Supuran and Balaban, 1994; Scozzafava and Supuran, 2002b). Such examples are histamine, serotonin, dopamine, adrenaline, noradrenaline, and other catecholamines, which are important autacoids, as well as some amino acids and oligopeptides, some of which are hormones (Supuran and Puscas, 1994), whereas others (such as β-alanyl-histidine) have been proved to possess different physiological roles (Quinn et al., 1992; Hipkiss and Chana, 1998). Furthermore, their mechanism of action at the molecular level has been clearly explained (Briganti et al., 1997, 1998; Supuran and Scozzafava, 2000a). It has been also shown very recently that the AE1 Cl^-/HCO_3^- exchanger is a natural hCA II activator (Scozzafava and Supuran, 2002a). These results support the assumption (Supuran and Puscas, 1994) that CAAs possess important physiological functions as metabolic modulators and intra- or intercellular signal-transducing systems, but obviously more detailed studies are needed for clarifying in detail the in vivo role of such compounds.

The recent proposal (Sun et al., 2001; Sun and Alkon, 2002) of CAAs use as therapeutic targets in cognitive disorders represents a new direction in this field. The following experimental conclusions strongly support this suggestion: (1) A significant decline of several CA isozymes present in the brain was found to constantly accompany the memory and learning alterations occurring in Alzheimer's disease or during aging (Meyer-Ruge et al., 1984). (2) The synaptic switch of the γ-aminobutyric acid (GABA)ergic responses from inhibitory to excitatory (which amplify input signals) depends on the increased HCO_3^- conductance through the $GABA_A$ receptor–channel complex, hence it directly depends on the activity of the CA present within the intracellular compartments of the hippocampus pyramidal cells (Sun et al., 2001). (3) CAAs, such as phenylalanine (Clare and Supuran, 1994) or imidazole (Parkes and Coleman, 1989), administered to experimental rats directly into the brain, increase HCO_3^- concentrations in memory-related neural structures and thereby stimulate the different groups of GABA-releasing interneurons to change their function from filters to amplifiers, with a clear improvement of learning and memory abilities (Sun and Alkon, 2001, 2002). Furthermore, in the absence of the CAAs, the same stimuli are insufficient to trigger the previously mentioned synaptic switch (Sun and Alkon, 2001), whereas CA inhibition suppressed it (Sun et al., 2001).

Such investigations (Sun and Alkon, 2001, 2002; Sun et al., 2001) have shown that CA and its activators act as an effective gate-control, which modulates the signal transfer through neural networks. CAAs may have pharmacological applications in the enhancement of synaptic efficacy, which in turn may constitute an excellent new approach for the treatment of Alzheimer's disease and other conditions in need of achieving spatial learning and memory therapy (Sun and Alkon, 2001; Scozzafava and Supuran, 2002b). An even more selective approach has been further suggested, that is, phenylalanine could be used in the majority of individuals who do not have genetic lack of phenylalanine hydroxylase, whereas the development of more potent nonphenylalanine activators (such as derivatives of histamine or imidazole) could be applied to those with hydroxylase dysfunction (Sun and Alkon, 2001). It must be stressed that at least seven different CA isoforms (CA I, II, IV, VII, VIII, XII, and XIV) are present normally in the human brain (Halmi et al., 2006), although the exact role of most of them is not clearly understood at this point. However, as we showed recently, most of them are activatable by many CAAs reviewed here (Temperini et al., 2006a, 2006b, 2006c). Thus, the hypothesis that modulating the activity of such CAs may lead to an enhancement of synaptic efficacy and a possible new approach for the treatment of Alzheimer's disease should be investigated in greater detail.

Last but not least, it should be mentioned that CAAs derived from β-alanyl-histidine as lead molecule have been proposed very recently as precursors for another particular class of putative pharmacological agents (Scozzafava and Supuran, 2002b). The clinical value of such agents would be related to the physiological roles of carnosine (i.e., antioxidant, free radical, and aldehyde scavenger, or heavy metal ion complexing agent), to its enhanced metabolic survival in vivo, as well as to the fact that its protective functions have already put this natural dipeptide into the medical practice as a potential antiaging agent (Quinn et al., 1992; Hipkiss and Chana, 1998).

5.5 CONCLUSIONS

At present, the CA activation phenomenon seems to have been clarified in great detail. Once the CAAs existence was undoubtedly proved, hundreds of such new compounds were synthesized and fully characterized, their biological activity being evaluated against different CA isozymes and investigated for potential pharmacological applications. These compounds belong to well-defined structural classes, that is, anions and amines, azoles, amino-azoles, or amino acid derivatives. The largest majority of the reported potent activators consist of histamine or histidine derivatives, most of them possessing a peptidomimetic backbone. This focused approach has been motivated by: (1) the experimental evidences for an efficient CA activation with biogenic amines and natural amino acids; and (2) the detailed x-ray analysis of the hCA II–histamine adduct, which has provided useful insights for the design of novel CAAs.

Although belonging to diverse categories of organic derivatives, all the investigated CAAs share a common mechanism of action, well understood now at the molecular level by means of kinetic, spectroscopic, and crystallographic studies.

Important SAR in all CAAs series (and also QSAR for a limited number of compounds) were established and thereby some fundamental conclusions regarding an improved general strategy for CAAs design might be drawn. Thus, an efficient CAA molecule should incorporate two main structural elements: (1) at least one moiety able to shuttle protons between the enzyme active site and the external medium (generally, an amino group or a NH imidazole ring), preferably with a pK_a value in the range 6–8; and (2) additional polar moieties that ensure a tight binding to the enzyme, of arylsulfonyl, carboxamido, ureido/thioureido, or substituted pyridinium type. The substitution pattern of such a parent scaffold is an efficient tool for fine-tuning both the energy of binding of the activator molecule to the protein and the pK_a of the proton-transferring group, hence for modulating the intensity of the activating properties. A certain type of in vitro/ex vivo isozyme specificity was also identified for some of the various investigated CAAs.

CAAs are considered to be essential for the investigation of these widely spread enzymes, which are deeply involved in major physiological processes in living organisms. Thus, CAAs are critical for: (1) clarifying some aspects of the catalytic mechanism; (2) explaining the great differences in activity between the many CA isozymes; (3) making it easier to understand the still unknown physiological function of certain isoforms. Different natural CAAs present in the human body possess significant physiological roles. In addition to all these, CAAs may also possess useful medical applications in the development of drugs/diagnostic agents for the management of both CA deficiency syndrome, for which no clinical treatment is available at present, and neurological disorders related to learning and memory impairments, such as Alzheimer's disease or aging. This field comprising potential pharmacological uses of CAAs is largely unexplored for the moment.

ACKNOWLEDGMENTS

Research from the author's laboratory was financed in part by a grant of the sixth Framework Programme of the European Union (EUROXY project) and in part by an Italian FIRB project (MIUR/FIRB RBNE03PX83_001).

REFERENCES

Alper, S.L. (2002) Genetic diseases of acid–base transporters. *Annual Review of Physiology*, **64**, 899–923.

Briganti, F., Mangani, S., Orioli, P., Scozzafava, A., Vernaglione, G. and Supuran, C.T. (1997) Carbonic anhydrase activators: X-ray crystallographic and spectroscopic investigations for the interaction of isozymes I and II with histamine. *Biochemistry*, **36**, 10384–10392.

Briganti, F., Iaconi, V., Mangani, S., Orioli, P., Scozzafava, A., Vernaglione, G. and Supuran, C.T. (1998) A ternary complex of carbonic anhydrase: X-ray crystallographic structure of the adduct of human carbonic anhydrase II with the activator phenylalanine and the inhibitor azide. *Inorganica Chimica Acta*, **275–276**, 295–300.

Briganti, F., Scozzafava, A. and Supuran, C.T. (1999) Novel carbonic anhydrase isozymes I, II and IV activators incorporating sulfonyl-histamino moieties. *Bioorganic and Medicinal Chemistry Letters*, **9**, 2043–2048.

Christianson, D.W. and Fierke, C.A. (1996) Carbonic anhydrase: evolution of the zinc binding site by nature and by design. *Accounts of Chemical Research*, **29**, 331–339.

Clare, B.W. and Supuran, C.T. (1994) Carbonic anhydrase activators. Part 3. Structure–activity correlations for a series of isozyme II activators. *Journal of Pharmaceutical Sciences*, **83**, 768–773.

De Coursey, T.E. (2000) Hypothesis: do voltage-gated H^+ channels in alveolar epithelial cells contribute to CO_2 elimination by the lung? *American Journal of Physiology. Cell Physiology*, **278**, C1–C10.

Gao, J.M., Qiao, S. and Whitesides, G.M. (1995) Increasing binding constants of ligands to carbonic anhydrase by using "greasy tails." *Journal of Medicinal Chemistry*, **38**, 2292–2301.

Halmi, P., Parkkila, S. and Honkaniemi, J. (2006) Expression of carbonic anhydrases II, IV, VII, VIII and XII in rat brain after kainic acid induced status epilepticus. *Neurochemistry International*, **48**, 24–30.

Hewett-Emmett, D. (2000) Evolution and distribution of the carbonic anhydrase gene families. In *The Carbonic Anhydrases—New Horizons,* Chegwidden, W.R., Edwards, Y. and Carter, N. (Eds.), Birkhäuser Verlag, Basel, pp. 29–76.

Hipkiss, A.R. and Chana, H. (1998) Carnosine protects against methyl-glyoxal-mediated modifications. *Biochemical and Biophysical Research Communications*, **248**, 28–32.

Ilies, M.A., Banciu, M.D., Ilies, M., Chiraleu, F., Briganti, F., Scozzafava, A. and Supuran, C.T. (1997) Carbonic anhydrase activators. Part 17. Synthesis and activation study of a series of 1-(1,2,4-triazole-(1H)-3-yl)-2,4,6-trisubstituted-pyridinium salts against isozymes I, II and IV. *European Journal of Medicinal Chemistry*, **32**, 911–918.

Ilies, M., Banciu, M.D., Ilies, M.A., Scozzafava, A., Caproiu, M.T. and Supuran, C.T. (2002) Carbonic anhydrase activators: design of high affinity isozymes I, II, and IV activators, incorporating tri-/tetrasubstituted-pyridinium-azole moieties. *Journal of Medicinal Chemistry*, **45**, 504–510.

Ilies, M., Scozzafava, A. and Supuran, C.T. (2004) Carbonic anhydrase activators. In *Carbonic Anhydrase—Its Inhibitors and Activators*, Supuran, C.T., Scozzafava, A. and Conway, J. (Eds.) CRC Press, Boca Raton 2004, pp. 317–352.

Khalifah, R.G. (1971) The carbonic dioxide hydration activity of carbonic anhydrase. *Journal of Biological Chemistry*, **246**, 2561–2573.

Leiner, M. (1940) Das Ferment Kohlensäureanhydrase in Tierkörper. *Naturwiss*, **28**, 316–317.

Mann, T. and Keilin, D. (1940) Sulphanilamide as a specific inhibitor of carbonic anhydrase. *Nature*, **146**, 164–165.

Meyer-Ruge, W., Iwangoff, P. and Reichlmeier, K. (1984) Neurochemical enzyme changes in Alzheimer's disease and Pick's disease. *Archives of Gerontology and Geriatrics*, **3**, 161–165.

Nair, S.K. and Christianson, D.W. (1991) Unexpected pH-dependent conformation of His-64, the proton shuttle of carbonic anhydrase II. *Journal of American Chemical Society*, **113**, 9455–9458.

Page, M.I. and Williams, A. (1989) *Enzyme Mechanism,* Royal Society of Chemistry, London, pp. 1–13.

Parkes, J.L. and Coleman, P. (1989) Enhancement of carbonic anhydrase activity by erythrocyte membrane. *Archives of Biochemistry and Biophysics*, **275**, 459–468.

Puscas, I., Supuran, C.T. and Manole, G. (1990) Carbonic anhydrase inhibitors. Part 3. Relations between chemical structure of inhibitors and activators. *Revue Roumaine de Chimie*, **35**, 683–689.

Puscas, I., Coltau, M., Puscas, I.C. and Supuran, C.T. (1994) Carbonic anhydrase activators. Part 9. Kinetic parameters for isozyme II activation by histamine prove a non-competitive type of interaction. *Revue Roumaine de Chimie*, **39**, 114–119.

Puscas, I., Coltau, M., Supuran, C.T., Moisi, F., Lazoc, L. and Farcau, D. (1996) Carbonic anhydrase activators. Part 10. Kinetic studies of isozyme II activation with pentagastrin. *Revue Roumaine de Chimie*, **41**, 159–163.

Quinn, P.J., Boldyrev, A.A. and Formazuyk, V.E. (1992) Carnosine: its properties, functions and potential therapeutic applications. *Molecular Aspects of Medicine*, **13**, 379–444.

Reithmeier, R.A. (2001) A membrane metabolon linking carbonic anhydrase with chloride/bicarbonate anion exchangers. *Blood Cells, Molecules and Diseases*, **27**, 85–89.

Roth, D.E., Venta, P.J., Tashian, R.E. and Sly, W.S. (1992) Molecular basis of human carbonic anhydrase II deficiency. *Proceedings of the National Academy of Sciences of the United States of America*, **89**, 1804–1808.

Rousselle, A.-V. and Heymann, D. (2002) Osteoclastic acidification pathways during bone resorption. *Bone*, **30**, 533–540.

Scozzafava, A. and Supuran, C.T. (2000a) Carbonic anhydrase activators. Part 21. Novel activators of isozymes I, II and IV incorporating carboxamido and ureido histamine moieties. *European Journal of Medicinal Chemistry*, **35**, 31–39.

Scozzafava, A. and Supuran, C.T. (2000b) Carbonic anhydrase activators. Part 24. High affinity isozymes I, II and IV activators, derivatives of 4-(4-chlorophenylsulfonylureido-amino acyl)ethyl-1*H*-imidazole. *European Journal of Pharmaceutical Sciences*, **10**, 29–41.

Scozzafava, A. and Supuran, C.T. (2002a) Carbonic anhydrase activators: human isozyme II is strongly activated by oligopeptides incorporating the carboxyterminal sequence of the bicarbonate anion exchanger AE1. *Bioorganic Medicinal Chemistry Letters*, **12**, 1177–1180.

Scozzafava, A. and Supuran, C.T. (2002b) Carbonic anhydrase activators: high affinity isozymes I, II, and IV activators, incorporating a β-alanyl-histidine scaffold. *Journal of Medicinal Chemistry*, **45**, 284–291

Scozzafava, A., Menabuoni, L., Mincione, F., Briganti, F., Mincione, G. and Supuran, C.T. (1999) Carbonic anhydrase inhibitors. Synthesis of water-soluble, topically effective intraocular pressure lowering aromatic/heterocyclic sulfonamides containing cationic or anionic moieties. Is the tail more important than the ring? *Journal of Medicinal Chemistry*, **42**, 2641–2650.

Scozzafava, A., Briganti, F., Ilies, M.A. and Supuran, C.T. (2000a) Carbonic anhydrase inhibitors. Synthesis of membrane-impermeant low molecular weight sulfonamides possessing in vivo selectivity for the membrane-bound versus the cytosolic isozymes. *Journal of Medicinal Chemistry*, **43**, 292–300.

Scozzafava, A., Iorga, B. and Supuran, C.T. (2000b) Carbonic anhydrase activators: synthesis of high affinity isozymes I, II and IV activators, derivatives of 4-(4-tosylureido-amino acyl)ethyl-1*H*-imidazole (histamine derivatives). *Journal of Enzyme Inhibition*, **15**, 139–161.

Shelton, J.B. and Chegwidden, W.R. (1988) Activation of carbonic anhydrase III by phosphate. *Biochemical Society Transactions*, **16**, 853–854.

Stams, T. and Christianson, D.W. (2000) X-ray crystallographic studies of mammalian carbonic anhydrase isozymes. In *The Carbonic Anhydrases—New Horizons,* Chegwidden, W.R., Edwards, Y. and Carter, N. (Eds.), Birkhäuser Verlag, Basel, pp. 159–174, and references cited therein.

Sun, M.-K. and Alkon, D.L. (2001) Pharmacological enhancement of synaptic efficacy, spatial learning and memory through carbonic anhydrase activation in rats. *Journal of Pharmacology and Experimental Therapeutics*, **297**, 961–967.

Sun, M.-K. and Alkon, D.L. (2002) Carbonic anhydrase gating of attention: memory therapy and enhancement. *Trends in Pharmacological Sciences*, **23**, 83–89.

Sun, M.K., Dahl, D. and Alkon, D.L. (2001) Heterosynaptic transformation of GABAergic gating in the hippocampus and effects of carbonic anhydrase inhibition. *Journal of Pharmacology and Experimental Therapeutics*, **296**, 811–817.

Supuran, C.T. (1992) Carbonic anhydrase activators. Part 4. A general mechanism of action for activators of isozymes I, II and III. *Revue Roumaine de Chimie*, **37**, 411–421.

Supuran, C.T. and Balaban, A.T. (1994) Carbonic anhydrase activators. Part 8. pK$_a$–activation relationship in a series of amino acid derivatives activators of isozyme II. *Revue Roumaine de Chimie*, **39**, 107–113.

Supuran, C.T. and Puscas, I. (1994) Carbonic anhydrase activators. In *Carbonic Anhydrase and Modulation of Physiologic and Pathologic Processes in the Organism*, Puscas, I. (Ed.), Helicon, Timisoara, pp. 113–145.

Supuran, C.T. and Scozzafava, A. (1999) Carbonic anhydrase activators. Part 23. Aminoacyl/dipeptidyl histamine derivatives bind with high affinity to isozymes I, II and IV and act as efficient activators. *Bioorganic Medicinal Chemistry*, **7**, 2397–2406.

Supuran, C.T. and Scozzafava, A. (2000a) Activation of carbonic anhydrase isozymes. In *The Carbonic Anhydrases—New Horizons*, Chegwidden, W.R., Edwards, Y. and Carter, N. (Eds.), Birkhäuser Verlag, Basel, pp. 197–219.

Supuran, C.T. and Scozzafava, A. (2000b) Carbonic anhydrase inhibitors and their therapeutic potential. *Expert Opinion on Therapeutic Patents*, **10**, 575–600.

Supuran, C.T. and Scozzafava, A. (2000c) Carbonic anhydrase activators. Synthesis oh high affinity isozymes I, II and IV activators, derivatives of 4-(arylsulfonylureido-amino acyl)ethyl-1*H*-imidazole. *Journal of Enzyme Inhibition*, **15**, 471–486.

Supuran, C.T. and Scozzafava, A. (2001) Carbonic anhydrase inhibitors. *Current Medicinal Chemistry—Immunologic, Endocrine and Metabolic Agents*, **1**, 61–97.

Supuran, C.T., Dinculescu, A., Manole, G., Savan, F., Puscas, I. and Balaban, A.T. (1991) Carbonic anhydrase activators. Part 2. Amino acids and some of their derivatives may act as potent enzyme activators. *Revue Roumaine de Chimie*, **36**, 937–946.

Supuran, C.T., Balaban, A.T., Cabildo, P., Claramunt, R.M., Lavandera, J.L. and Elguero, J. (1993a) Carbonic anhydrase activators. Part 7. Isozyme II activation with bis-azolylmethanes,-ethanes and related azoles. *Biological and Pharmaceutical Bulletin*, **16**, 1236–1239.

Supuran, C.T., Dinculescu, A. and Balaban, A.T. (1993b) Carbonic anhydrase activators. Part 5. CA II activation by 2,4,6-trisubstituted pyridinium cations with 1-(ω-aminoalkyl)side chains. *Revue Roumaine de Chimie*, **38**, 343–349.

Supuran, C.T., Barboiu, M., Luca, C., Pop, E., Brewster, M.E. and Dinculescu, A. (1996a) Carbonic anhydrase activators. Part 14. Synthesis of mono- and bis-pyridinium salt derivatives of 2-amino-5-(2-aminoethyl)- and 2-amino-5-(3-aminopropyl)-1,3,4-thiadiazole, and their interaction with isozyme II. *European Journal of Medicinal Chemistry*, **31**, 597–606.

Supuran, C.T., Claramunt, R.M., Lavandera, J.L. and Elguero, J. (1996b) Carbonic anhydrase activators. Part 15. A kinetic study of the interaction of bovine isozyme II with pyrazoles, bis- and tris-azolylmethanes. *Biological and Pharmaceutical Bulletin*, **19**, 1417–1422.

Supuran, C.T., Scozzafava, A., Ilies, M.A., Iorga, B., Cristea, T., Briganti, F., Chiraleu, F. and Banciu, M.D. (1998) Carbonic anhydrase inhibitors. Part 53. Synthesis of substituted-pyridinium derivatives of aromatic sulfonamides: The first non polymeric membrane-impermeable inhibitors with selectivity for isozyme IV. *European Journal of Medicinal Chemistry*, **33**, 577–594.

Supuran, C.T., Scozzafava, A. and Casini, A. (2003) Carbonic anhydrase inhibitors. *Medicinal Research Reviews*, **23**, 146–189.

Supuran, C.T., Scozzafava, A. and Conway, J. (Eds.). (2004) *Carbonic Anhydrase—Its Inhibitors and Activators*, CRC Press, Boca Raton, pp. 1–368.

Temperini, C., Scozzafava, A., Puccetti, L. and Supuran, C.T. (2005) Carbonic anhydrase activators: X-ray crystal structure of the adduct of human isozyme II with L-histidine as a platform for the design of stronger activators. *Bioorganic and Medicinal Chemistry Letters*, **15**, 5136–5141.

Temperini, C., Scozzafava, A., Vullo, D. and Supuran, C.T. (2006a) Carbonic anhydrase activators. Activation of isozymes I, II, IV, VA, VII, and XIV with L- and D-histidine and crystallographic analysis of their adducts with isoform II: engineering proton-transfer processes within the active site of an enzyme. *Chemistry*, **12**, 7057–7066.

Temperini, C., Scozzafava, A., Vullo, D. and Supuran, C.T. (2006b) Carbonic nhydrase activators. Activation of isoforms I, II, IV, VA, VII, and XIV with L- and D-phenylalanine and crystallographic analysis of their adducts with isozyme II: stereospecific recognition within the active site of an enzyme and its consequences for the drug design. *Journal of Medicinal Chemistry*, **49**, 3019–3027.

Temperini, C., Scozzafava, A. and Supuran, C.T. (2006c) Carbonic anhydrase activators. The first X-ray crystallographic study of an adduct of isoform I. *Bioorganic and Medicinal Chemistry Letters*, **16**, 5152–5156.

Vince, J.W. and Reithmeier, R.A. (2000) Identification of the carbonic anhydrase II binding site in the Cl^-/HCO_3^- anion exchanger AE1. *Biochemistry*, **39**, 5527–5533.

6 Detection and Reduction of Neurofibrillary Lesions

Jeff Kuret

CONTENTS

6.1 INTRODUCTION

Alzheimer's disease (AD) is a dementing illness of the elderly defined by the hierarchical emergence of intra- and extracellular brain lesions in distinct cortical and subcortical regions (Chapter 1). It is definitively diagnosed and staged on the basis of postmortem neuropathology. Neurofibrillary lesions, which consist of neurofibrillary tangles (NFTs) occupying neuronal cell bodies, neuropil threads within neuronal processes, and dystrophic neurites associated with neuritic plaques (Buee et al., 2000), have special utility in this regard. Each manifestation of neurofibrillary pathology is composed of tau, a microtubule-associated protein that normally functions to promote cytoskeletal integrity. Tau that accumulates in neurofibrillary lesions differs from normal microtubule-associated protein in its states of aggregation and posttranslational modification. Aggregation is accompanied by conformational changes in tau protomers that can be sensed by selective monoclonal antibodies and by small molecules such as thioflavin dyes. Thus, tau-bearing lesions present novel binding sites that could be exploited to develop ligands capable of selectively detecting their presence. Such agents could have practical usage as contrast agents capable of staging disease in living cases.

In addition to its utility as a surrogate marker for neurodegeneration in AD, abnormalities in tau biology also may contribute directly or indirectly to neuronal death. If so, then pharmacological modulation of normal tau function (e.g., microtubule binding) or misfunction (e.g., inappropriate posttranslational modification and aggregation) could have therapeutic utility. For these reasons, small-molecule ligands that cross the blood–brain barrier and modulate tau biology are being investigated in two contexts: as a means of leveraging an established marker of neurodegeneration for premortem diagnosis of AD and other tauopathic neurodegenerative diseases, and as a potential avenue to therapeutics that attack a root cause of neurodegeneration.

6.1.1 Basic Concepts

6.1.1.1 Normal Tau Structure and Function

Six tau isoforms varying in length from 352 to 441 amino acid residues are found in the human central nervous system (CNS) as a result of alternative splicing of a single gene on chromosome 17 (Figure 6.1). The N-terminal third of each isoform is hydrophilic and contains a high proportion of amino acids whose side chains are negatively charged at physiological pH. This region also contains two inserts of 29 residues each encoded by alternatively spliced exons 2 and 3 of the tau gene that serve an unknown function. The middle third of the tau molecule is enriched in proline residues and carries a positive charge at neutral pH. This is followed by a segment containing 31 or 32 residue imperfect repeat sequences that mediate the association of tau with tubulin. This microtubule-binding repeat region contains the final alternatively spliced sequence (31 residues), encoded by exon 10 of the tau gene, so that three tau isoforms contain four repeats whereas the remainder contain only three. Although the function of the additional repeat sequence is not fully known, at a minimum it confers increased binding affinity for tubulin in vitro. Sequences homologous to the microtubule repeats are found in other microtubule-associated proteins, including MAP2 and MAP4, suggesting that these proteins compose a distinct family of microtubule-associated proteins with a common mechanism of binding (Dehmelt and Halpain, 2005). Tau binds tubulin, promotes microtubule assembly, copurifies with tubulin during cycles of assembly–disassembly, and colocalizes extensively (but not exclusively) with tubulin in vivo.

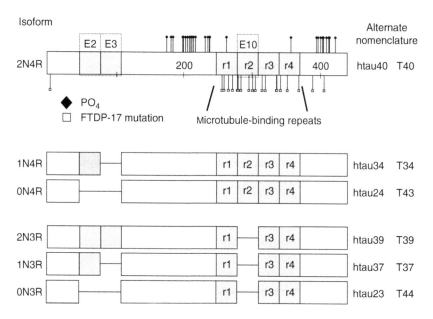

FIGURE 6.1 Schematic representation of tau primary structure. Six tau isoforms exist in the human central nervous system owing to alternative splicing of exons 2, 3, and 10 in transcripts derived from a single tau-encoding gene. Isoforms are named according to their complement of N-terminal segments and C-terminal microtubule-binding repeats, but alternative nomenclature is common. (From Goedert, M., Spillantini, M.G., Jakes, R., Rutherford, D. and Crowther, R.A., *Neuron*, 3, 519, 1989; Giasson, B.I., Forman, M.S., Higuchi, M., Golbe, L.I., Graves, C.L., Kotzbauer, P.T. et al., *Science*, 300, 636, 2003.) For example, isoform 2N4R contains both N-terminal segments (each 29 residues in length) and all four microtubule binding repeats (labeled r1–r4). The microtubule binding domain mediates association with tubulin and also forms the core of filamentous aggregates in disease. AD-derived tau proteins contain covalently bound phosphate distributed across at least 27 sites (marked by *diamonds*) clustered primarily on each side of tau-microtubule binding region. (From Morishima-Kawashima, M., Hasegawa, M., Takio, K., Suzuki, M., Yoshida, H., Watanabe, A. et al., *Neurobiol. Aging*, 16, 365, 1995; Hanger, D.P., Betts, J.C., Loviny, T.L., Blackstock, W.P. and Anderton, B.H., *J. Neurochem.*, 71, 2465, 1998; Derkinderen, P., Scales, T.M., Hanger, D.P., Leung, K.Y., Byers, H.L., Ward, M.A. et al., *J. Neurosci.*, 25, 6584, 2005.) Tau phosphorylation modulates its affinity for tubulin and may promote tau aggregation and other toxic reactions. In contrast, exonic mutations that cause FTDP-17 cluster primarily in the microtubule-binding region (marked by *squares*). FTDP-17 mutations can modulate affinity for tubulin, promote spontaneous tau aggregation, and modify susceptibility to phosphorylation. As a result, they are frequently employed to generate biological models of neurofibrillary lesion formation.

Alone in solution, tau proteins behave hydrodynamically as natively unfolded monomers, consistent with complete absence of tertiary structure (Schweers et al., 1994). This state is compatible with microtubule binding, which proceeds with submicromolar affinity (Makrides et al., 2004). Because tubulin dimer concentrations are ~20 μM and brain tau levels are low micromolar, most tau is thought to bind microtubules under normal conditions. However, tau–tubulin binding affinity is modulated by posttranslational modifications, including phosphorylation, and so the percentage of tau that is microtubule associated cannot be predicted solely on the basis of bulk concentrations.

In summary, normal tau binds microtubules, with the portion that is unbound presenting a disorganized structure that is unsuited for high-affinity binding of small molecule ligands.

6.1.1.2 Tau Structure in Disease

In AD brain, tau dissociates from microtubules and aggregates to form filamentous inclusions. Although some lesions develop predominantly straight morphology (Perry et al., 1991), late-stage disease is dominated by paired helical filaments (PHF) displaying 80 nm periodicity. Filamentous morphology reflects the cross-β-sheet structure (Berriman et al., 2003) adopted by a ~93 residue segment located in the microtubule-binding repeat regions of all six naturally occurring CNS tau isoforms. This conformation, which is characterized by parallel, in-register β-sheets that run approximately perpendicular to the long axes of the filaments (Margittai and Langen, 2004), makes up the protease-resistant core of each filament (Figure 6.2). Other proteins that adopt cross-β-sheet conformation in complex with glycosaminoglycans in the extracellular space are termed amyloid (Westermark et al., 2005). Thus, despite differences in molecular composition, significant conformational similarity exists between tau aggregates formed within degenerating neurons and the protein aggregates found in other compartments of AD tissue (such as β-amyloid plaques in the extracellular space) and also in other diseases. Although the specific repeats involved in aggregate formation vary among tau isoforms (depending on whether they contain three or four repeats), self-association minimally involves residues spanning three repeats as demonstrated by proteolysis experiments (Novak et al., 1993).

Sequences outside the core region also may adopt limited ordered structure. For example, tau-specific monoclonal antibodies such as Tau2, Alz50, and MC1 bind to tau filaments with 1–2 orders of magnitude higher affinity than to random coil tau monomer (Carmel et al., 1996). Because the epitope for each of these antibodies at least partially resides in the N-terminal third of tau, these data suggest that the aggregation of the microtubule repeat region may contact or influence the conformation of sequences well beyond the ~93 residue filament core. Nonetheless, sequences outside the core are not required for filament stability and are gradually lost through proteolysis as disease progresses.

Tau lesion formation is accompanied by four- to eightfold increases in tissue tau levels, which eventually rise to ~120 pmol/mg protein in late stage AD frontal cortex (Khatoon et al., 1992). This greatly exceeds the concentration of all plaque-associated β-amyloid peptides (3–4 pmol/mg protein in midfrontal, parietal, or temporal cortex in late stage AD; Wang et al., 1999; Li et al., 2004b). The causes of this increase are not fully understood, but both increases in tau translation rate owing to activation of signal transduction pathways and decreases in tau degradation rate owing to sequestration of tau from the neuron's protein turnover machinery may be involved. The latter seems to make a major contribution. In fact, late-stage filaments become difficult to solubilize in strong denaturants including ionic detergents (Iqbal et al.,

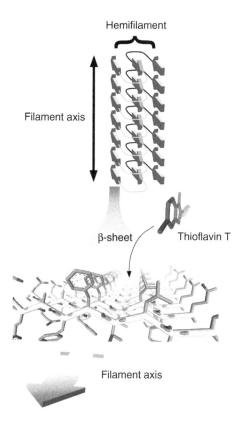

FIGURE 6.2 Hypothetical model of filamentous tau substructure. *Top*: Three β-sheets are drawn to symbolize the three microtubule binding repeats that compose the core of a PHF hemifilament, although their relative orientation is speculative. Each β-sheet consists of parallel, in-register β-stands aligned perpendicular to the long axis of the filament. *Bottom*: Atomic model of a representative segment of solvent-exposed β-sheet resident in β-amyloid filaments (From Petkova, A.T., Yau, W.M. and Tycko, R., *Biochemistry*, 45, 498, 2006), with the α-carbon trace overlayed and the parallel β-sheet hydrogen-bonding pattern depicted with dashed lines. Solvent exposed β-sheets contain channels between amino acid sidechains that parallel the long axis of the filament and bind aromatic heterocycles such as thioflavin dyes with high affinity. Dye-binding blocks free rotation of the bond bridging the benzothiazole and benzaminic rings, leading to altered fluorescence spectra and a slightly twisted dye conformation. (From Dzwolak, W. and Pecul, M., *FEBS Lett.*, 579, 6601, 2005.) These interactions are presumably common to proteins in cross-β-sheet conformation, including tau, although the relative affinity of small molecule binding depends on the amino acid side chains that form the binding channels.

1984), suggesting that stability and resistance to degradation stems from covalent cross-links among tau protomers. Some neurofibrillary lesions survive cell death. The resultant "ghost tangles" retain the ability to bind small molecule fluorophores such as thioflavin S (Defossez and Delacourte, 1987), Congo red (Schwab et al., 1995), and Thiazin red (Uchihara et al., 2001a) indicating that cross-β-sheet conformation is stably maintained in these lesions.

Together these data suggest that the organized folding of tau associated with aggregation creates novel ligand-binding sites in AD lesions not present in normal tissue, that these attain relatively high concentrations in late-stage disease, and that they survive destruction of the cell. The observed stability and cross-linking of filaments also suggests the unlikelihood of reducing neurofibrillary lesions in the late stages of disease.

6.1.1.3 Fibrillization Pathways

Although the structure of tau in normal and diseased states is relatively well characterized, the pathway through which the transition occurs is not. It appears that filament formation follows a primary nucleation pathway (i.e., the phase transition proceeds from nonfibrillar tau without the participation of preexisting filaments because the latter do not form as part of normal cell biology). Primary nucleation may be homogeneous, where tau molecules spontaneously self-associate to nucleate and extend filaments without the contribution of foreign particles. For example, β-amyloid peptide can spontaneously aggregate by homogeneous nucleation both in vitro and in biological models. However, results with tau protein in vitro, in cell culture, and in transgenic animals suggest that this pathway is kinetically disfavored in the case of tau, even under conditions where tau concentration exceeds the concentration of endogenous tubulin (Chirita and Kuret, 2004; Terwel et al., 2005). As a result, it has been difficult to model neurofibrillary lesion formation in the absence of exogenous inducers or aggregation-promoting tau mutations. Both appear to stabilize assembly competent conformations that, once attained, result in spontaneous tau aggregation. Assembly competence is associated with adoption of limited β-sheet structure, and this conformational change may represent a crucial step in disease pathogenesis. The pathophysiological inducer of conformational change in disease is unknown, but at least a portion of tau molecules begin the transition from natively unfolded to cross-β-sheet conformation in association with intracellular membranes (Galvan et al., 2001).

Together these observations suggest that heterogeneous nucleation and growth of tau filaments may be a dominant pathway in vivo, where foreign bodies supply a surface which selectively binds tau and facilitates the adoption of extended β-sheet structure at levels of supersaturation well below those needed to support homogeneous nucleation (Kuret et al., 2005). The process of heterogeneous nucleation, combined with the presence of partially folded, assembly competent intermediates in the reaction pathway, implies that tau aggregation pathways may present multiple small-molecule binding sites for modulation of the aggregation reaction. Because spontaneous aggregation of tau is slow, most in vitro assays of tau aggregation, including screens for interacting ligands, rely on exogenous inducers of aggregation (heparin, fatty acids, and others) to create high-affinity binding sites from natively unfolded recombinant tau.

6.1.1.4 Posttranslational Modifications

Tau is the target of multiple posttranslational modifications beyond the proteolysis and cross-linking reactions summarized above. Phosphorylation, which normally attains ~2–3 mol/mol stoichiometry, is an established modulator of tau–tubulin binding affinity through which tau effects on microtubule stability and dynamics are controlled. In AD brain regions undergoing neurofibrillary lesion formation, tau phosphorylation rises to 7–8 mol/mol, representing a three- to fourfold increase relative to normal tau (Ksiezak-Reding et al., 1992; Kopke et al., 1993). The phosphate is distributed into at least 27 sites, most of which are clustered into two regions flanking the microtubule-binding domain (Figure 6.1). Because few of these sites are unique to filamentous tau, hyperphosphorylation may reflect increased occupancy of the phosphorylation sites found in normal tau (Matsuo et al., 1994; Seubert et al., 1995). Some phosphorylation sites fill in hierarchical fashion as neurofibrillary lesions mature, suggesting a sequential correlation between site occupancy and neuritic lesion development (Kimura et al., 1996; Augustinack et al., 2002). Very high stoichiometry phosphorylation (>10 mol/mol) promotes tau aggregation in vitro (Alonso et al., 2001). The mechanism may involve stabilization of filaments through decreased protomer dissociation rates (Necula and Kuret, 2005). Nonetheless, it is possible that some phosphorylation silently accompanies rather than directly drives filament formation in vivo. For example, another posttranslational modification, detyrosination, accumulates in stably polymerized microtubules, but is not the cause of microtubule stability (Khawaja et al., 1988).

At a minimum, therefore, hyperphosphorylation contributes to AD pathogenesis by neutralizing the microtubule-binding activity of tau, yielding a cytosolic population available for aggregation (Biernat et al., 1993; Bramblett et al., 1993; Patrick et al., 1999; Li et al., 2004a). It may also increase the rate and extent of aggregation, potentially by slowing the rate of protomer dissociation and regulating tau degradation as summarized below.

6.1.1.5 Tau Lesions as Markers for AD

Despite the high prevalence of AD in the elderly population, it is not synonymous with dementia. For example, ~20% of American dementia cases result from vascular cognitive impairment arising from blood vessel occlusion (Roman et al., 2004). In countries such as Japan, these cases comprise up to 50% of dementia cases and exceed AD in prevalence. Another portion of cases correlate with Lewy body disease. Lewy bodies are intracellular aggregates of α-synuclein protein in cross-β-sheet conformation (Cummings, 2004). Many of these cases also have neurofibrillary and amyloid plaque lesions. A third group of dementia cases, termed frontotemporal dementias, present with focal atrophy and NFTs but not amyloid plaques (Munoz et al., 2003). The differential premortem diagnosis of dementia subtype is important because each may have different underlying causes and may respond differently to therapy. With the many new potential treatments for halting or slowing AD under development comes the need for identifying the subpopulations of dementia cases capable of benefiting from them.

Selective detection of neurofibrillary lesions may address this need because they develop according to a common spatiotemporal pattern reflecting the sequence, type, and severity of cognitive decline and neuronal loss (Ghoshal et al., 2002; Royall et al., 2002). Their appearance has been used to stage AD (Braak and Braak, 1991) and distinguish it from other forms of dementia in postmortem specimens (Feany and Dickson, 1996). In addition, certain brain areas (such as transentorhinal region of the temporal lobe cortex, locus coeruleus in brainstem, and magnocellular nuclei of basal forebrain) develop neurofibrillary lesions well before neocortical deposition of β-amyloid plaques or signs of dementia can be detected (Figure 6.3). Thus, the spatial distribution of neurofibrillary lesions may reveal very early stage disease.

Because tau protein is the principal component of neurofibrillary lesions, the first-generation diagnostic strategy focused on its detection in cerebral spinal fluid (CSF) using sensitive immunochemical technique. Although CSF reflects the CNS complement of most small-molecule substances, tau presents special challenges for detection owing to its aggregation and sequestration in AD brain. As a result, increased levels of tau that appear in CSF depend on proteolytic release of soluble fragments from insoluble tau aggregates. Thus, only certain regions of the tau molecule appear in CSF. Assaying for posttranslationally modified tau has increased

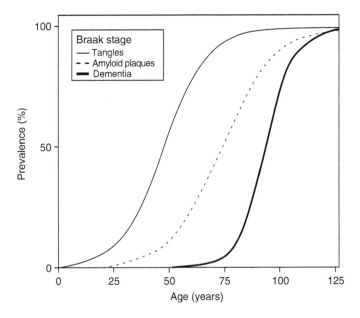

FIGURE 6.3 Estimated prevalence of cortical amyloid and cortical neurofibrillary pathologies relative to dementia during AD progression made on the basis of Braak staging. (From Braak, H. and Braak, E., *Acta Neuropathol.*, 82, 239, 1991; Duyckaerts, C. and Hauw, J.J., *Neurobiol. Aging*, 18, 362, 1997.) Six Braak stages mark the progression of neurofibrillary lesions from transentorhinal cortex and subcortical nuclei in early-stage disease (Braak stages 1–2) through destruction of most neocortical association areas (Braak stages 5–6). A second Braak staging scheme describes the progression of neocortical amyloid lesions from low density (Braak stage A) through high density (Braak stage C) distributions. Neurofibrillary lesion formation (Braak stages 1–6; *solid line*) precede deposition of cortical β-amyloid plaques (Braak stages A–C; *dashed line*) and the onset of dementia (*heavy solid line*). Most of the early prevalence results from neurofibrillary lesion formation in the transentorhinal cortex, basal forebrain, and brainstem (Braak stages 1–2).

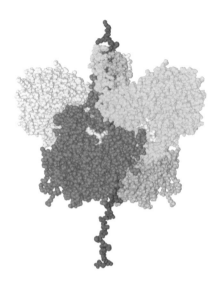

COLOR FIGURE 3.6 (a) The AChE hetero-pentamer (red: part of the proline-rich anchoring protein; white: rim of the active-site gorge).

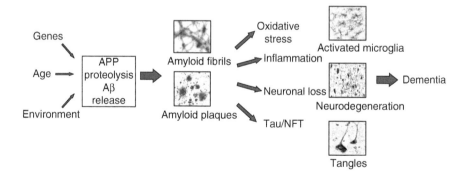

COLOR FIGURE 4.1 The amyloid cascade hypothesis of Alzheimer's disease. Genes, environment, and age represent the main risk factors in the development of AD. The accumulation in the brain of Aβ peptide results in the deposition of plaques. The presence of amyloid plaques in the brain triggers a cascade of toxic events leading to dementia.

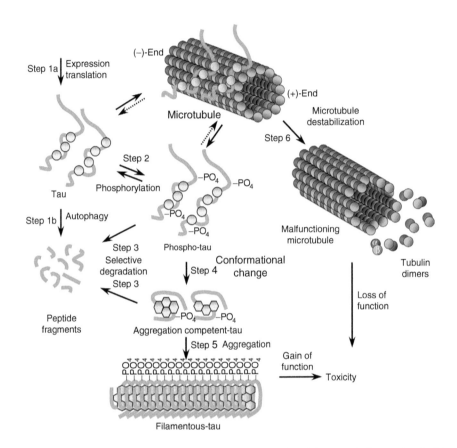

COLOR FIGURE 6.4 Targets for pharmacological control of toxicity potentially associated with tau misfunction in AD. Both three and four-repeat tau isoforms are depicted. Toxicity arising from gain of function effects may be controlled at Step 1, modulation of tau levels (by selective depression of tau expression or translation or by increasing the rate of bulk tau degradation); Step 2, reversal of tau hyperphosphorylation through modulation of phosphotransferase activities; Step 3, selective destruction of aggregation-competent or hyperphosphorylated species; and Steps 4 and 5, direct interference with conformational changes and aggregation. Toxicity associated with loss of microtubule function may be controlled through microtubule stabilization, Step 6.

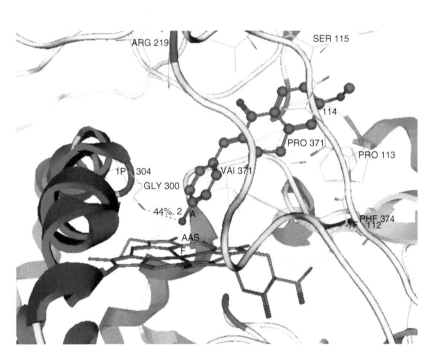

COLOR FIGURE 8.8 Interaction of tetralone inhibitor with the CYP26A1 active site; Hydrogen bonding (red dashed lines), transition metal interaction (purple dashed line), labeled amino acid residues are involved in hydrophobic interactions.

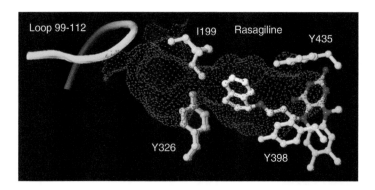

COLOR FIGURE 10.10 Structure of the active site of human MAO B in complex with rasagiline (pdb code: 1S2Q). The surface of the cavity is shown as a purple dot mesh, containing rasagiline (white) covalently linked to the isoalloxazine ring system of FAD (green). Key amino acid residues are shown in cyan, including the "gating switch" loop 99–112 (in tube representation). The ligand-binding cavity (right) is separated from the smaller entrance cavity (left) by a ring of amino acid side chains, including Ile199 and Tyr326. The conformation of these side chains may vary in different crystal structures depending on the bound inhibitor. This is exemplified with the side chain conformations taken from the crystallographic complex of MAO B with the small inhibitor isatin, (pdb code: 1OJA), as superimposed on the current structure. (From Binda, C., Li, M., Hubalek, F., Restelli, N., Edmondson, D.E., and Mattevi, A., *Proc. Natl. Acad. Sci. U.S.A.*, 100, 9750, 2003. With permission.)

detection sensitivity and specificity. Although this approach captures the increase in tau levels associated with AD, it loses all spatial resolution and therefore does not fully take advantage of the power of neuritic lesions to stage AD (de Leon et al., 2004; Hampel et al., 2004). Moreover, because tau aggregation appears in fronto-temporal and Lewy body dementias as well, the differential diagnosis of these conditions can be problematic.

Second-generation diagnostic tests seek to overcome these limitations by using whole brain imaging. Toward this end, small-molecule ligands (see Section 6.2) radiolabeled with short-lived isotopes for positron emission tomography (PET; ^{18}F, ^{11}C) and single photon emission computed tomography (SPECT; ^{123}I) that selectively bind proteins in cross-β-sheet conformation are under development (Mathis et al., 2004). The principal advantage of this approach is that it has the potential to capture the spatial distribution of AD lesions in situ (at low resolution) while being insensitive to proteolytic degradation occurring outside the core region of tau. Nonetheless, the approach has two limitations in the context of AD. First, ligands identified to date react with cross-β-sheet structure common to both tau and β-amyloid bearing lesions, suggesting that the staging information inherent in neuritic lesions will be confounded by cross-reactivity with amyloid plaques. Densities of β-amyloid plaques, especially diffuse plaques (i.e., those that lack neuritic involvement), correlate poorly with AD progression (Braak and Braak, 1991; Arriagada et al., 1992). Thus, while a promising approach for following amyloid burden, it is suboptimal as an AD diagnostic. Second, because α-synuclein also adopts extended β-sheet conformation in Lewy bodies (Serpell et al., 2000), which also appear in AD, these agents may detect lesions unrelated to amyloid plaques and tau-bearing tangles. In contrast, selective binding agents have the potential to distinguish the composition of protein aggregates char-acteristic of AD, frontotemporal, and Lewy body dementias. Ligand-binding sites that accompany tau filament formation are summarized in Section 6.2.

6.1.1.6 Tau as Toxic Mediator

Because only palliative treatments are available for AD, new approaches to directly suppress neurodegeneration are being sought. Tau protein is one such target. In addition to marking degenerating neurons, tau may directly contribute to neuro-degeneration as suggested by the synaptic deficits that accompany neurofibrillary lesion formation (Callahan and Coleman, 1995) and by the toxicity associated with the misexpression of tau in animal models (Ishihara et al., 1999; Spittaels et al., 1999; Hall and Yao, 2000; Hall et al., 2001). Above all, the involvement of various tau missense mutations in certain genetic forms of frontotemporal dementia (fronto-temporal dementia with Parkinsonism linked to chromosome 17; FTDP-17) demon-strates that misfunction of tau protein is sufficient to cause neurodegeneration and dementia (Ingram and Spillantini, 2002).

Two general hypotheses have been put forward to rationalize a direct relation-ship between tau misfunction and disease. The first argues that abnormally phos-phorylated or aggregated tau is directly toxic to cells that harbor them. For example, tau aggregates are capable of inhibiting the ubiquitin–proteasome system (Keck et al., 2003), which plays important roles in cell homeostasis. Loss of proteasome activity can lead to cellular stress (Bence et al., 2001; Bennett et al., 2005). In addition, aggregates composed of hyperphosphorylated tau could also sequester binding proteins such as PIN1, a phosphoprotein-specific prolyl isomerase important for cell viability (Lim and Lu, 2005). High cytoplasmic levels of phospho-tau may also be directly toxic (Kosik and Shimura, 2005), potentially by binding and activating tyrosine phosphorylation-mediated signal transduction pathways (Lee, 2005). In each of these examples, tau phosphorylation or aggregation leads to a "gain of function" that does not predominate in normal tissue. Potential targets for antagonizing gain-of-function toxicity are summarized in Figure 6.4, and include modulation of tau levels by selective depression of tau expression or by increasing

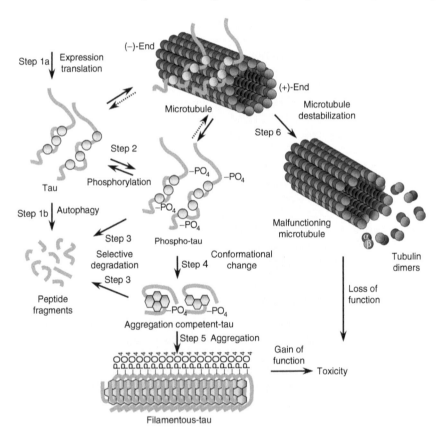

FIGURE 6.4 (See color insert following page 296.) Targets for pharmacological control of toxicity potentially associated with tau misfunction in AD. Both three and four-repeat tau isoforms are depicted. Toxicity arising from gain of function effects may be controlled at Step 1, modulation of tau levels (by selective depression of tau expression or translation or by increasing the rate of bulk tau degradation); Step 2, reversal of tau hyperphosphorylation through modulation of phosphotransferase activities; Step 3, selective destruction of aggregation-competent or hyperphosphorylated species; and Steps 4 and 5, direct interference with conformational changes and aggregation. Toxicity associated with loss of microtubule function may be controlled through microtubule stabilization, Step 6.

bulk tau degradation rates (Step 1), reversal of tau hyperphosphorylation through modulation of phosphotransferase activities (Step 2), selective destruction of aggregation competent or hyperphosphorylated species (Step 3), and direct interference with conformational changes and aggregation (Step 4 and Step 5).

A second hypothesis argues that sequestration of tau from microtubules destabilizes the cytoskeleton leading to a loss-of-function toxicity (Alonso et al., 1996). For example, loss of microtubule-binding activity may make microtubules more susceptible to microtubule-severing enzymes, leading to destabilization of the microtubule array (Qiang et al., 2006). In fact, microtubule density and total length decreases in AD brain in a process not necessarily correlating with or caused by tau aggregation (Cash et al., 2003). Although loss of tau is well tolerated over both long (Harada et al., 1994; Kimura et al., 1996) and short (Tint et al., 1998) time periods in model systems, simultaneous loss of both tau and other microtubule-associated proteins can lead to severe phenotypes and neurodegeneration (DiTella et al., 1996). For example, imbalances that lead to tau hyperphosphorylation may also affect the phosphorylation state and activity of other microtubule-associated proteins irrespective of effects on tau aggregation. Alternatively, it has been proposed that phospho-tau may directly bind and sequester other normal microtubule-associated proteins leading to loss of function phenotype (Alonso et al., 1997). Potential targets for antagonizing loss of tau function include modulation of tau phosphorylation as described above and also direct stabilization of microtubules (Figure 6.4, Step 6). Progress in developing ligands capable of interrogating fundamental aspects of tau biology are summarized in Section 6.3.

6.2 TAU FILAMENT BINDING AGENTS

Small molecules that bind cross-β-sheet conformation with submicromolar affinity have been characterized in most detail in the context of β-amyloid filaments. These filaments are composed of two protofilaments, each of which contributes two parallel, in-register β-sheets to the mature filament (Petkova et al., 2006). The parallel orientation of β-strands creates narrow channels between amino acid sidechains that extend along the length of the filament (Krebs et al., 2005). Ligands capable of binding the channels are long, thin, and mostly planar aromatic heterocycles and include thioflavin dyes such as thioflavins S and T (**6.1–6.2**). Both of these are benzothiazole derivatives that exhibit altered fluorescence upon binding amyloid fibrils.

(6.1) Thioflavin S (component of mixture)   **(6.2)** Thioflavin T

Bound dyes adopt the orientation of the channels so that their long axes parallel the filament long axis (Figure 6.2). It has been estimated that filaments must contain at least four β-strands (i.e., be ≥ 1.6 nm in length) to completely envelop a molecule having the length of thioflavin T (Krebs et al., 2005). It is possible, however, that multiple channels exist on each protofibril so that the stoichiometry of ligand to β-amyloid protomer at maximum binding (B_{max}) exceeds the predicted 1:4 molar ratio, or that additional or overlapping sites appear with lower frequency. In fact, benzothiazole derivatives can differentially bind at least three independent and partially overlapping sites on synthetic β-amyloid filaments, with affinities inversely dependent on binding stoichiometry (Lockhart et al., 2005). The lowest affinity sites ($K_D \sim 1$ µM) are present at 1:4 stoichiometry, with the highest affinity sites ($K_D \sim$ low nM) present at only ~1:300. Balance between binding affinity and stoichiometry is of critical importance in the development of radioligands for whole brain imaging applications (Ichise et al., 2001). As an additional complication, cross-β-sheet conformation can be polymorphic, with different molecular packings potentially leading to distinct ligand-binding sites (Petkova et al., 2005). For example, polymorphism may underlie the failure of certain benzothiazole derivatives to detect the β-amyloid plaques of certain transgenic mouse models while avidly binding β-amyloid filaments derived from authentic AD brain (Klunk et al., 2005). Overall, experience with β-amyloid aggregates indicates that successful development of imaging agents will depend on balancing ligand affinity with binding stoichiometry while retaining avidity for filament morphologies appearing early in the course of disease.

Unlocking the potential of neurofibrillary lesions for diagnosis and staging of AD will depend on applying these lessons to tau protein (as well as on gaining access to the intracellular compartment and maintaining blood–brain barrier permeability). First, ligands must bind neurofibrillary lesions at higher stoichiometry and affinity than amyloid plaques. Although the molar concentration of fibrillar tau protein exceeds that of β-amyloid by an order of magnitude (Section 6.1.1.2), neither the accessibility of the tau nor the binding stoichiometry of candidate ligands is established. Under conditions where B_{max} for each lesion is similar, mere selectivity of binding rather than absolute specificity should be adequate for this purpose. Second, ligands must bind tau filament morphologies that appear early in disease. In some cases, for example, straight filament morphology precedes paired helical morphology (Perry et al., 1991). The organization of tau molecules in the two morphologies appears to be similar but not identical (Crowther, 1991). Moreover, the morphology of tau filaments in other tauopathies, such as corticobasal degeneration, differs from the paired-helical and straight morphologies found in AD (Ksiezak-Reding and Wall, 2005). Therefore, it may be challenging to identify a single agent that reacts equally well with all tau aggregates that appear in tauopathic neurodegenerative diseases.

(6.3) IMSB

(6.4) FDDNP

Adoption of cross-β-sheet conformation by the ~93 residue core region of tau creates binding sites similar to those residing on filaments composed of β-amyloid and other proteins. Not surprisingly, agents developed to image β-amyloid, such as the bis-styryl benzene derivative IMSB (**6.3**), the substituted naphthalene FDDNP (**6.4**), and the benzothiazole aniline Pittsburgh compound B (**6.5**; PIB), also bind to NFTs in authentic AD tissue (Agdeppa et al., 2003, Klunk et al., 2003, Kung et al., 2003). Nonetheless, the amino acid side chains that form ligand-binding channels differ between tau and β-amyloid, and it is reasonable to expect at least some pharmacological selectivity among lesions. For example, some benzothiazole derivatives have up to two orders of magnitude selectivity for sites on filaments composed of insulin relative to those composed of β-amyloid (Caprathe et al., 1999). The benzothiazole scaffold may also offer a route to selective tau-binding agents, although to date only the opposite has been demonstrated. For example, halogenation of benzothiazoles at position 6 (e.g., TZDM; **6.6**) facilitates binding to high-density sites on β-amyloid plaques (Lockhart et al., 2005) but not to sites on NFTs (Kung et al., 2003).

(6.5) PIB R_1 = OH; R_2 = H
(6.6) TZDM R_1 = I; R_2 = CH_3

In fact, only one selective tau-binding agent, termed Thiazin red, is in regular use (Mena et al., 1995). Thiazin red is a 1-naphthol-2-phenylazo derivative that exists as a mixture of azo and hydrazone tautomers (**6.7**). The latter may be most important for filament binding owing to intramolecular hydrogen bonding which further stabilizes planarity (Antonov et al., 1998). The reagent has poor reactivity with tau inclusions in corticobasal degeneration, consistent with the altered morphology of these fila-ments having differing ligand-binding sites (Uchihara et al., 2001b). In contrast to binding selectivity seen in human tissue, Thiazin red has been reported to bind β-amyloid plaques and cerebral amyloid angiopathy in transgenic mouse models (Bacskai et al., 2003). As noted above, however, these aggregates appear to differ from authentic AD-derived β-amyloid at least with respect to benzothiazole affinity.

Information on the affinity and stoichiometry of Thiazin red binding to tau filaments is still lacking.

(6.7) Thiazine red R

Compounds with the selectivity of Thiazin red and with suitable pharmacokinetics may offer a route to neurofibrillary lesion-selective contrast agents. Other compounds that are planar or can adopt planar structure with conjugated π-systems and limited electrostatic interactions may emerge from high-throughput screens. Because many of these compounds can organize cross-β-sheet structure in natively unfolded tau protein, candidates should be examined for the ability to nucleate tau filament formation so that neurofibrillary lesion formation is not exacerbated (Chirita et al., 2005).

6.3   MODULATION OF BULK TAU LEVELS

Because the rate and extent of filament formation directly depends on tau concentration, lowering the rate of tau expression or translation (Figure 6.4, Step 1a) or increasing its rate of degradation (Figure 6.4, Step 1b) could potentially decrease neurofibrillary lesion formation. Although total tau mRNA levels do not change substantially in AD (Chambers et al., 1999), bulk tau levels still depend on rates of tau mRNA translation and protein turnover.

6.3.1   mTOR INHIBITORS

In fact, a protein kinase involved in the regulation of protein translation and turnover in response to nutrient levels, mTOR (mammalian target of rapamycin), is upregulated in AD (Li et al., 2005). The mTOR-dependent protein translation pathway, which increases translation of mRNAs containing oligopyrimidine tracts at their 5' ends (including tau mRNA), acts through the phosphorylation of multiple substrates including p70 S6 kinase (Figure 6.5). Activation of the pathway also inhibits autophagy-mediated protein degradation (Raught et al., 2001). Therefore, upregulation of

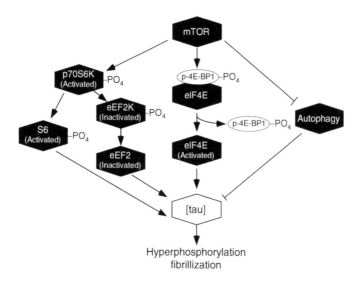

FIGURE 6.5 Hypothetical role for mTOR protein kinase in balancing tau protein synthesis and degradation. mTOR, which is activated in AD tissue, could potentially promote neurofibrillary lesion formation by increasing bulk tau levels. First, mTOR increases translation of tau and other mRNAs containing 5′ terminal oligopyrimidine tracts through its phosphorylation and activation of the p70S6 kinase and ribosomal protein S6. Activation of the p70S6 kinase may also increase elongation rates by depressing the phosphorylation of eEF2 (eukaryotic translation elongation factor 2). Second, Tor can activate eIF4E (eukaryotic translation initiation factor 4E) by phosphorylating and dissociating its inhibitor, 4E-BP1. Finally, mTOR activity depresses autophagy, a major pathway for tau turnover. Restoration of normal mTOR activity in neurons may reverse these posttranscriptional effects on tau levels.

the mTOR system may contribute to the elevated levels of bulk tau seen in AD through increased rates of translation and decreased rates of degradation. Conversely, inhibition of the pathway could potentially decrease neurofibrillary lesion formation by reducing tau levels. For example, in mouse and Drosophila model systems, inhibition of mTOR decreases formation of lesions from multiple aggregation-prone proteins (Ravikumar et al., 2004).

Inhibitors of the mTOR system are under investigation in many biological contexts and include rapamycin (sirolimus), a naturally occurring lipophilic macro-lide isolated from *Streptomyces hygroscopicus* (a soil bacterium). Rapamycin binds members of the immunophilin protein family, forming a complex that then inhibits mTOR-mediated signalling. Derivitization of rapamycin at position 42 yields compounds with superior water solubility such as temsirolimus (CCI-779; a dihydroxy-methyl propionic acid ester derivative), everolimus (a hydroxyethyl derivative), and AP23573 (a phospho-derivative), which also are under investigation (**6.8–6.10**). These compounds appear to cross the blood–brain barrier and should be useful for assessing the feasibility of minimizing neurofibrillary lesion formation through modulation of bulk tau protein levels.

(6.8) Rapamycin R = H
(6.9) CCL-779 R = 3-hydroxy-2-(hydroxymethyl)-2-methylpropanoate
(6.10) AP23573 R = PO$_3$

6.4 INHIBITORS OF TAU HYPERPHOSPHORYLATION

Tau hyperphosphorylation (Figure 6.4, Step 2) results from an imbalance among cellular protein kinase and phosphoprotein phosphatase activities. Hundreds of protein kinases catalyze the Mg–ATP dependent incorporation of phosphate into proteins, and many of these recognize tau as protein cosubstrate in vitro. Characterization of the pathological phosphorylation sites on PHF–tau has not conclusively identified responsible protein kinases because multiple enzymes fulfill the deduced substrate selectivity requirements. For example, some tau phosphorylation sites correspond to the –Ser/Thr–Pro– motif recognized by mitogen-activated protein kinases (MAPKs), cyclin-dependent kinases (CDKs), and glycogen synthase kinase-3 (GSK3). Still other protein kinases, such as those that recognize basic amino acid residues N-terminal to the phosphorylatable hydroxyamino acid as part of their substrate selectivity signature (e.g., the cAMP-dependent protein kinase), phosphorylate pathological tau sites in vitro. In fact, multiple protein kinases may contribute to tau hyperphosphorylation in neuritic lesions. Major targets are summarized below.

6.4.1 PROTEIN KINASE TARGETS

6.4.1.1 GSK3

Human GSK3 consists of homologous ~50 kDa α- and β-isoforms encoded by genes on chromosomes 3 and 19. GSK3 phosphorylates tau in vitro and also in biological models on sites occupied in AD, mediates the tau phosphorylation promoting activity of β-amyloid peptide applied to cultured neurons, and localizes weakly with neuritic lesions in authentic AD brain (Bhat et al., 2004). GSK3 also modulates processing of the amyloid precursor protein to β-amyloid peptide by an unknown mechanism (Phiel et al., 2003). Thus, GSK3 may be involved in the formation of the two preeminent lesions of AD.

GSK3 participates in signal transduction networks including the insulin and Wnt pathways, and so it may be possible to modulate its activity indirectly through ligands acting on cell-surface receptors or other pathway components. GSK3 may also be inhibited directly by compounds that interfere with its catalytic activity. The first widely used GSK3 inhibitor, lithium, inhibits GSK3 directly by acting as a competitive inhibitor of Mg^{2+} cosubstrate (IC_{50} ~ low millimolar) and indirectly by interfering with a regulatory phosphorylation on GSK3 (Chalecka-Franaszek and Chuang, 1999; Ryves et al., 2002). Direct inhibitory activity is selective (but not specific) for GSK3 within the protein kinase family (Davies et al., 2000). Specificity is further limited by its interaction with other enzymes such as inositol monophosphatase (Phiel and Klein, 2001). Lithium treatment decreases tau hyperphosphorylation and aggregation in a mouse model over-expressing the P301L FTDP-17 mutation in a four-repeat tau background (Noble et al., 2005), consistent with GSK3 mediating tau phosphorylation in this model.

(6.11) AR-A04418

R_1 = Bulky aromatic

(6.12) 2,4-Disubstituted thiadiazolidinones

Synthetic direct GSK3 inhibitors with greater potency have been prepared (Meijer et al., 2004). Many of these act as competitive inhibitors of ATP cosubstrate and have limited selectivity for GSK3. Thiazole derivatives, such as AR-A04418 (**6.11**), inhibit GSK3 selectively relative to CDKs and with $K_i = 38$ nm (Bhat et al., 2003). AR-A04418 is able to recapitulate many of the effects of lithium treatment in the transgenic mouse model of tauopathy described above, including a decrease in the amount of insoluble tau accumulating in brain (Noble et al., 2005). Compounds that inhibit GSK3 noncompetitively with respect to MgATP and cross the blood–brain barrier, including 2,4-disubstituted thiadiazolidinones (**6.12**), are now entering clinical trials (Martinez et al., 2005).

Despite these advances, no direct inhibitor capable of selectively antagonizing GSK3α or GSK3β has been identified. Selective inhibitors may be desirable since individual GSK3 isoforms appear to mediate key AD-related processes such as the processing of amyloid precursor protein (Phiel et al., 2003).

6.4.1.2 MAPK

MAPKs are monomeric protein kinases that function as part of phosphorylation cascades to transduce diverse cellular signals and stresses (Johnson and Lapadat, 2002).

Three well-characterized MAPK subfamilies include the extracellular signal-regulated kinases (ERKs), the c-jun NH_2-terminal kinases (JNKs), and the p38 MAPKs. Each MAPK family member is regulated by phosphorylation catalyzed by an upstream "kinase kinase." Direct and selective inhibitors of the JNK (SP600125) and p38 MAPK (SB203580) subfamilies have been developed, whereas indirect inhibition of ERK-mediated pathways has been achieved at the level of its upstream kinase (PD098059) (**6.13–6.15**).

(6.13) SP600125

(6.14) SB203580

(6.15) PD098059

Application of these inhibitors to tau biology has revealed that SP600125 antagonizes the phosphorylation of tau catalyzed in response to mitosis (Tatebayashi et al., 2006), whereas PD098059 inhibits tau phosphorylation and neurite degeneration catalyzed in response to β-amyloid treatment of culture models (Rapoport and Ferreira, 2000). Efforts to confirm these findings with direct ERK inhibitors have been hampered by the lack of selective inhibitors. Nonselective inhibitors include SRN-003-556 (**6.16**), an orally bioavailable CNS-penetrating derivative of K252a (**6.17**). K252a is a naturally occurring indolo[2,3α]carbazole derivative related to the staurosporin family of microbial alkaloid toxins (**6.18**) (Omura et al., 1995). SRN-003-556 inhibits ERK2 with submicromolar affinity but like most staurosporin-related compounds, inhibits a broad range of protein kinases. Nonetheless, SRN-003-556 inhibits tau hyperphosphorylation and slows the onset of behavioral deficits observed in transgenic mice overexpressing the P301L tauopathy mutation in a four-repeat tau background (Le Corre et al., 2006). Together these data implicate MAPKs as mediating tau hyperphosphorylation in response to stresses related to mitotis, expression of tauopathy mutations, and exposure to exogenous β-amyloid.

(6.16) SRN-003-556   **(6.17)** K252a   **(6.18)** Staurosporin

6.4.1.3 CDK5

Although a member of the cyclin-dependent kinase family, CDK5 neither binds cyclin nor regulates cell-cycle progression. Rather, CDK5 associates with activator proteins (p35 and p39) and is enriched in postmitotic neurons where it has been implicated in the regulation of the cytoskeleton. The ability of CDK5 to directly phosphorylate tau has been demonstrated in vitro and in cell culture, and activation of CDK5 in transgenic mice leads to tau hyperphosphorylation and aggregation (Cruz et al., 2003; Noble et al., 2003).

Multiple inhibitors have been developed against CDKs, most of which act competitively with ATP cosubstrate and are nonspecific among CDK family members (Sridhar et al., 2006). Considering the large number of CDKs and their importance in controlling cell-cycle progression and transcription, the issue of selectivity is a major hurdle in assessing the potential of CDK5 antagonists. Nonetheless, CDKs have been extensively characterized with respect to three-dimensional structure, and features unique to the ATP binding site of CDK5 have been identified. These may lead to more selective inhibitors in future studies (Ahn et al., 2005).

6.4.1.4 CK1

Members of the CK1 (Casein kinase-1) branch of the protein kinase superfamily are candidate tau kinases because they phosphorylate tau in vitro and in cell culture, copurify with PHF out of AD brain, and display a striking colocalization with both neuritic and granulovacuolar degeneration lesions in intact AD brain sections (Ghoshal et al., 1999; Kannanayakal et al., 2006). Moreover, these colocalizations correlate strongly with cognitive decline in longitudinal studies of AD progression (Ghoshal et al., 2002). CK1 activity is encoded by at least six genes (Ckiα, γ1, γ2, γ3, δ, ε) with further diversity generated by alternative splicing (Knippschild et al., 2005). On the basis of sequence similarity, all isoforms are believed to share a phosphate-directed substrate selectivity, where priming phosphorylation of a protein substrate three-to-four residues N-terminal to the target hydroxyamino acid improves phosphorylation kinetics. CK1 isoforms are, therefore, well suited to catalyze the clustered phosphorylation pattern shown in Figure 6.1.

(6.19) IC261   **(6.20)** D4476

CK1 isoforms can be selectively inhibited with either IC261 (**6.19**), a 3-substituted indolin-2-one derivative, or by D4476 (**6.20**) (Mashhoon et al., 2000, Rena et al., 2004). IC261 inhibits CK1 by binding at the ATP substrate site and stabilizing a partially open conformation. Selectivity among protein kinases arises from the novel axis of rotation relating the N- and C-terminal subdomains of CK1. The three-dimensional structure of the IC261/CK1 complex suggests the possibility of creating more potency while retaining selectivity.

6.4.2 PHOSPHOPROTEIN PHOSPHATASE TARGETS

In contrast to the diversity of candidate protein kinases potentially involved in pathological tau phosphorylation, only one protein phosphatase, PP2A, is responsible for tau dephosphorylation (Goedert et al., 1995). PP2A functions as a heterotrimer of scaffolding (A), regulatory (B), and catalytic (C) subunits (Figure 6.6a). Formation of active trimer, which depends on reversible association with diverse B subunits, is modulated by methylation of the C subunit. Methylation catalyzed by PP2A methyltransferase (PPMT) and its cosubstrate S-adenosylmethionine stabilizes active trimer, whereas demethylation catalyzed by the specific PP2A methylesterase PME-1 (Lee et al., 1996; Ogris et al., 1999) leads to B-subunit dissociation and loss of protein phosphatase activity. Hypomethylation, therefore, is predicted to lead to reduced PP2A activity and tau hyperphosphorylation. In fact, PP2A activity is downregulated in AD, as are levels of methylation and PPMT expression (Gong et al., 1995; Sontag et al., 2004a, 2004b). In addition, elevated levels of homocysteine, an intermediate in S-adenosylmethionine

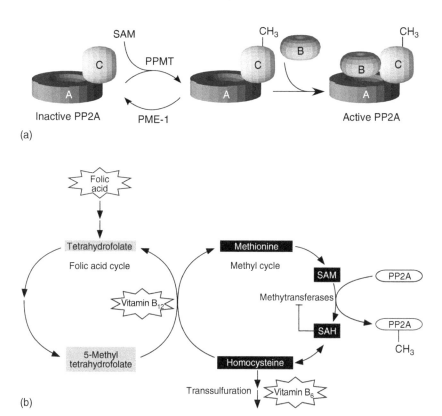

FIGURE 6.6 (a) Phosphoprotein phosphatase 2A structure. The dimer core of PP2A, composed of A (scaffold) and C (catalytic) subunits, reversibly associates with a family of regulatory B subunits to yield fully active holoenzyme. Association requires methylation of the C subunit catalyzed by the PP2A methyltransferase (PPMT) and its cosubstrate S-adenosylmethionine (SAM). Demethylation is catalyzed by a specific PP2A methylesterase (PME-1), a member of the α/β hydrolase family. Tau hyperphosphorylation in AD may reflect diminished phosphatase activity secondary to depressed PP2A methylation. (b) The methyl cycle and risk factors for AD. Methylation cosubstrate SAM is a product of both folic acid and methyl metabolic cycles and is itself an allosteric regulator of cycle enzymes. Both cycles, along with the biodegradative transsulfuration pathway, depend on the B vitamins folic acid, pyridoxal phosphate (vitamin B_6), and cobalamin (vitamin B_{12}) for normal function. Thus, B vitamin insufficiency can reduce cellular methylation by lowering SAM levels and by depressing breakdown of homocysteine, a metabolite associated with increased risk for developing AD. The methyl cycle may represent the nexus between homocysteine levels and tau hyperphosphorylation through its control of PP2A activity.

metabolism associated with inhibition of cellular methylation, is a risk factor for developing AD. Thus, tau hyperphosphorylation of tau may in part reflect depressed activity of PP2A secondary to deficiencies in cellular methylation (Vafai and Stock, 2002).

These observations suggest that tau hyperphosphorylation could be minimized through restoration of methylation pathways. For example, it may be possible to counter methylation imbalance engendered by downregulation of PPMT activity or elevated levels of homocysteine in AD by inhibiting PME-1. In addition, nutritional regimens that fully support the methyl cycle while minimizing homocysteine levels, such as one rich in folate and B vitamins, may serve a protective function by minimizing PP2A-mediated tau hyperphosphorylation (Figure 6.6b). These hypotheses are currently under investigation in model systems. Preliminary evidence demonstrates that diet modulates the onset of behavioral deficits observed in transgenic mice overexpressing the P301L tauopathy mutation in a four-repeat tau background (Michaelis, 2006).

6.5   TURNOVER OF MODIFIED TAU

6.5.1   Hsp90 Inhibitors

Tau degradation is mediated in part by the ubiquitin–proteasome system, which can selectively detect and degrade modified tau species (Figure 6.4, Step 3). For example, site-specific phosphorylation of tau renders it a substrate for CHIP (carboxyl terminus of the Hsp70-interacting protein), a ubiquitin ligase that functions as a cochaperone with Hsp70 (Petrucelli et al., 2004; Shimura et al., 2004). The balance between proper folding and degradation of proteins is controlled in part by the activity of chaperones such as Hsp70 and Hsp90 and their cochaperones (including CHIP). Because overexpression of Hsp70 shifts this balance toward degradation for phospho-tau, strategies for increasing Hsp70 levels are under investigation as a means for lowering cellular levels of abnormal tau species associated with AD. Induction of the Hsp70 chaperone system can be achieved by inhibition of Hsp90, which regulates the folding and conformation of a number of proteins including transcription factors involved in regulating Hsp70 expression. Hsp90 is organized into three domains (Ali et al., 2006): an N-terminal ATP-binding domain (N), a middle domain (M) that regulates the ATPase activity of the N domain, and a C-terminal dimerization domain (C). Ligands that bind the ATP-binding site located in the N domain and inhibit ATPase activity include geldanamycin, a naturally occurring product of *S. hygroscopicus* that is in clinical trials as a potential anticancer agent. Synthetic products with nearly the potency of geldanamycin analogs but with greater solubility and oral activity have been developed on the basis of the purine pharmacophore that interacts with the ATP-binding domain of Hsp90 (Zhang et al., 2006). Preliminary studies indicate that synthetic Hsp90 inhibitors can selectively clear both phosphorylated and conformationally altered species of tau in cellular models (Dickey et al., 2006).

6.6   TAU AGGREGATION INHIBITORS

The ability of fluorescent benzothiazole dyes such as thioflavin S to detect cross-β-sheet conformation provides a convenient primary screen for antagonists of tau fibrillization (Figure 6.4, Step 5). These assays typically employ anionic inducers to initiate tau aggregation and are conducted either under reducing (i.e., in the presence of sulfhydryl reducing agent) or nonreducing conditions. Because loss of the thioflavin S signal could result from nonspecific effects, including competitive displacement of the thioflavin S probe or interference with its fluorescence, secondary screens employing electron microscopy or other physical methods are employed to confirm aggregation antagonist activity (Pickhardt et al., 2005b).

(6.21) 1,3,8-trihydroxy-6-methylanthraquinone **(6.22)** 2,3,4-trihydroxybenzophenone

(6.23) Baicalein

The largest primary screen of this kind employed a library of 200,000 small molecules and was conducted using heparin as anionic fibrillization inducer under nonreducing conditions (Pickhardt et al., 2005a; Mandelkow et al., 2006). Over 400 inhibitor candidates belonging to different classes of heterocycle scaffold have been disclosed (Mandelkow et al., 2006), with anthraquinone (e.g., 1,3,8-trihydroxy-6-methylanthraquinone; **6.21**) and polyphenol derivatives (e.g., 2,3,4-trihydroxybenzophenone; **6.22**) being characterized in greatest detail. Many members of this class inhibit the aggregation of β-amyloid and α-synuclein as well as tau (Masuda et al., 2006), and act by forming covalent adducts with protein (Conway et al., 2001; Zhu et al., 2004). In each case, the active species appears to be a quinone, which then reacts with protein nucleophiles (e.g., the sidechains of Cys, Tyr, or Met residues) via the Michael addition (Figure 6.7).

Polyphenolic compounds appear as hits in these screens owing to their rapid oxidation to quinone under nonreducing conditions. For example 5,6,7-trihydroxy-flavone (baicalein; **6.23**) oxidizes at pH 7.5 with $t_{1/2} \sim 15$ h, so that incubations conducted over a period of days in the absence of reducing agent (i.e., typical tau aggregation conditions used in screens) generate substantial amounts of protein-reactive baicalein quinone (Zhu et al., 2004). Covalent modification of tau effectively lowers the concentration of assembly-competent species, resulting in inhibition of aggregation and promotion of disaggregation of mature filaments (Figure 6.8).

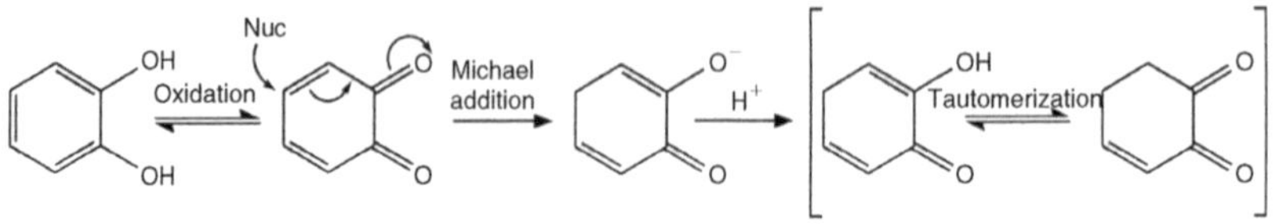

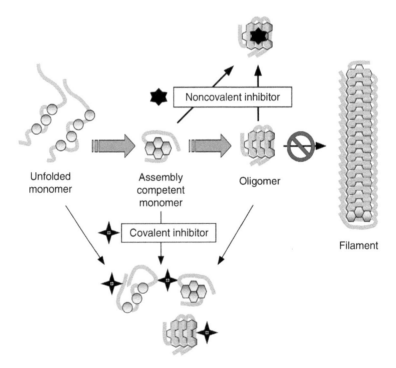

FIGURE 6.7 Proposed mechanism for polyphenol inhibitors of tau aggregation. A simple catechol is shown as an example. Under nonreducing conditions, polyphenols oxidize to form quinones, which can act as Michael acceptors for protein nucleophiles (Nuc) to form covalent adducts. (From Bolton, J.L., Trush, M.A., Penning, T.M., Dryhurst, G., and Monks, T.J., *Chem. Res. Toxicol.*, 13, 135, 2000.)

One advantage of this mechanism is that aggregation can be inhibited at substoichiometric concentrations of antagonist. For example, some polyphenolic compounds inhibit the aggregation of 10–23 μM full-length tau isoforms with low-micromolar IC_{50}s (Pickhardt et al., 2005a; Taniguchi et al., 2005). On the other hand, the reaction

FIGURE 6.8 Hypothetical mechanisms for inhibition of tau aggregation by small molecules. Covalent inhibitors, such as quinones, can react with and sequester multiple species in the reaction pathway including unfolded monomer. Noncovalent inhibitors, such as phenothiazines, may preferentially sequester protein oligomers and partially folded monomers owing to their affinity for β-sheet structure. Both mechanisms would be expected to raise the minimal concentration of protein required to support aggregation and also potentially slow rates of nucleation. They may also destabilize mature filaments by shifting the equilibrium toward monomers and small oligomers.

is sensitive to the presence of reducing agent, which can greatly diminish inhibitory potency against tau and other protein substrates (Conway et al., 2001; Pickhardt et al., 2005a). It is also associated with cellular damage through alkylation of macromolecules (Bolton et al., 2000). Although further studies will be required to determine whether this mechanism can be useful in vivo, preliminary evidence suggests that it is possible to decrease tau aggregation in tissue culture cell models (Pickhardt et al., 2005a).

A second antagonist class identified under nonreducing conditions consists of nitro-piperazine derivatives (Mandelkow et al., 2006) (**6.24**). Characterization of these compounds and their mechanism of action has not been elaborated. It is conceivable, however, that the strong electron-withdrawing nitro moiety combined with good leaving groups at positions R1 and R2 facilitates covalent modification of tau much like the quinone derivatives described above (Aptula et al., 2005).

A third structural class identified in small screens under nonreducing conditions includes planar aromatic heterocycles such as acridine and phenothiazine derivatives (Wischik et al., 1996; Taniguchi et al., 2005). Some members of this class inhibit the aggregation of β-amyloid and α-synuclein in addition to tau (Masuda et al., 2006), and also inhibit misfolding of prion protein to the disease-associated scrapie conformation (PrP^{sc}) (Korth et al., 2001). It has been proposed that these agents act by stabilizing and trapping soluble protein oligomers (Masuda et al., 2006). The mechanism resembles that of polyphenols and quinones except that it may involve noncovalent binding to β-sheet or other secondary structure rather than covalent modification of protein (Figure 6.8).

X and Y = N or C
R$_1$ and R$_2$ = Electron withdrawing groups
R$_3$ = Aromatic ring system

(6.24) Nitro-piperazine derivatives

(6.25) Cyanine N744

Tau aggregation reactions conducted under reducing conditions and anionic surfactant inducers differ qualitatively from those described earlier. Because they are more efficient, ligand screens are conducted at low micromolar tau protein concentrations over a period of hours instead of days. The resultant aggregates have predominantly hemifilament morphology, with half the mass per unit length of authentic PHF (King et al., 2001). Together, these differences dramatically change the behavior of ligands on aggregation. For example, ligands that require oxidation to quinone are inactive under these conditions, and some planar aromatic heterocycles that inhibit aggregation under nonreducing conditions, such as Congo red, become aggregation inducers (Chirita et al., 2005). In contrast, cyanine derivatives (**6.25**) inhibit tau fibrillization under reducing conditions with submicromolar $IC_{50}s$ and promote endwise disaggregation of mature filaments (Chirita et al., 2004; Necula et al., 2005). Cyanines act by raising the critical monomer concentration supporting aggregation. This behavior could arise from noncovalent sequestration of soluble oligomers or other species as proposed for acridine and phenothiazine derivatives. Why these planar scaffolds display antagonist activity is not clear but may be related to their propensity to aggregate. In fact, cyanines form noncovalent dimers in the presence of tau protein under aggregation conditions, suggesting a potential source of cooperative concentration effect relationships (Necula et al., 2005). If so, then macrocyclic cyanine and other scaffolds may be a route toward more potent aggregation inhibitors. For example, bis-acridine derivatives (i.e., two acridine heterocycles tethered by a linker) block PrP^{sc} misfolding with nearly an order of magnitude greater than potent acridine monomers such as quinocrine (May et al., 2003). This effect was attributed to multivalency of the inhibitory moiety in open conformation interacting with multiple PrP^{sc} protomers in complexes, but it is also possible that bis-acridines collapse in aqueous solution so that the acridine nuclei form covalent, face-to-face dimers (Figure 6.9). This reaction would also be consistent with the

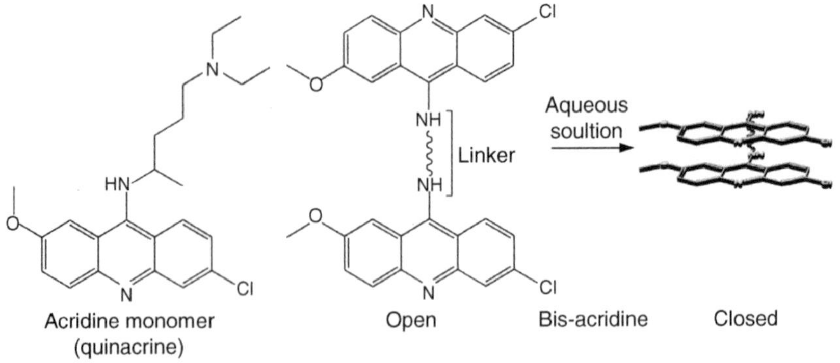

FIGURE 6.9 Structures of acridine derivatives. Bis-acridines inhibit PrP^{sc} misfolding with up to an order of magnitude greater potency than simple acridines such as quinacrine. Although the bis-acridine open conformation should predominate in organic solvents, the closed conformation may predominate in aqueous solution. Inhibitory activity may reside with dimeric species formed noncovalently from monomers or by the closed conformation of covalent dimer species.

observed dependence of potency on linker length (May et al., 2003). In the case of cyanine dyes, a structure–activity relationship will help identify inhibitory species. Bis-cyclo-cyanines (which collapse to mimic face-to-face dimer formation at all concentrations in aqueous solution) (Herz, 1974) may be useful in this effort.

6.7 MICROTUBULE STABILIZATION

Microtubules play crucial roles in adult neurons despite the absence of mitotic spindles found in dividing cells. For example, extended neuronal processes, termed axons and dendrites, contain microtubule arrays that function as tracks for motor-driven transport of cellular material between the cell body and distal segments. Drug-induced disassembly of these microtubules leads to rapid retraction of processes, nuclear collapse, and cell death. Because tau stabilizes total microtubule mass and modulates its dynamic behavior (i.e., the rates of association and dissociation of tubulin dimers at microtubule ends), loss of tau–tubulin binding affinity resulting from tau hyperphosphorylation could contribute to cytoskeletal collapse. On the other hand, levels of normal, microtubule-binding competent tau do not change markedly in authentic AD tissue (Khatoon et al., 1992; Khatoon et al., 1994), suggesting that abnormal tau represents an accumulation rather than complete conversion and depletion of normal species. To distinguish these possibilities and assess the potential microtubule stabilization as a therapeutic target in AD, agents that preserve or restore microtubule stability and dynamics are under investigation for their ability to overcome the effects of tau aggregation and hyperphosphorylation on the cytoskeleton.

Neuronal microtubules are polymers composed of 12–13 protofilaments aligned in parallel (Figure 6.4). Each protofilament consists of globular dimers of primarily α and β-tubulins polymerized in a head-to-tail arrangement. Therefore microtubules are polar structures, with the β-tubulin end ("+" end) being the principal locus of microtubule elongation. Microtubules contain at least three pharmacologically distinct ligand-binding sites that modulate microtubule stability and dynamics (Jordan and Wilson, 2004). The Vinca alkaloid site lies primarily on the β-subunit at an interface between two tubulin αβ heterodimers (Gigant et al., 2005). As a result, ligands such as vinblastine bind with high affinity to microtubules along their lengths and also their + ends but not with free tubulin dimers. Ligand binding inhibits microtubule dynamics at low concentration and induces microtubule dissociation at high concentration.

The colchicine-binding site also lies primarily on the β-subunit, but at the interface between the αβ subunits of a single tubulin dimer (Ravelli et al., 2004). As a result, colchicine binds to microtubules along their lengths but also binds free tubulin dimer. Binding inhibits polymerization and dynamics at low concentration and promotes depolymerization at high concentration.

Finally, the binding site for taxanes such as paclitaxel (**6.26**), a natural product originally isolated from the bark of yew trees, lies within the β-subunit (Löwe et al., 2001). Occupancy of this site along the length of the microtubule induces conformational changes that stabilize microtubules against depolymerization (Xiao et al., 2006).

(6.26) Paclitaxel; R = Acetate
(6.27) TX-67;   R = Butanedioate

Although all three classes of microtubule-binding agents modulate dynamics at low concentration, efforts to antagonize the effects of tau hyperphosphorylation have focused on the microtubule-stabilizing activity of taxanes. Preliminary data indicates that paclitaxel treatment protects cultured neurons against Aβ induced cell death (Michaelis et al., 1998). Because paclitaxel has poor blood–brain barrier permeability owing to its interaction with the P-glycoprotein efflux system, derivatives that gain access to the CNS have been prepared by modifications at position 10. One such blood–brain barrier permeable paclitaxel analog, TX-67 (**6.27**), has a similar effect on neuronal protection against Aβ-induced cell death as taxol (Li et al., 2003, Rice et al., 2005). Compounds such as TX-67 are currently under investigation in mouse models of tauopathy (Michaelis, 2006).

6.8   CONCLUSIONS

Although tau misfunction initiates neurodegeneration in familial tauopathies linked to mutations in the tau gene (FTDP-17), it does not play this role in AD. Rather, in sporadic AD, it follows upstream events that are poorly understood. Nonetheless, misfunction of genetically normal tau may make a significant contribution to neurodegeneration as disease progresses. Further development of compounds useful for interrogating tau biology will clarify the feasibility of antagonizing tau lesion formation and the potential benefits of doing so. In the end, only clinical studies can validate the targets identified on the basis of biological models. Regardless of its role as a potential mediator of toxicity, tau is a powerful marker of neurodegeneration that can be leveraged for improved premortem diagnosis and staging of AD and other tauopathic neurodegenerative diseases.

ACKNOWLEDGMENT

The author thanks Lauren Crissman for assistance with figure preparation and the National Institutes of Health and the Alzheimer's Association for financial support.

REFERENCES

Agdeppa, E.D., Kepe, V., Liu, J., Small, G.W., Huang, S.C., Petric, A. et al. (2003). 2-Dialkylamino-6-acylmalononitrile substituted naphthalenes (DDNP analogs): novel diagnostic and therapeutic tools in Alzheimer's disease. *Molecular Imaging and Biology*, **5**, 404–417.

Ahn, J.S., Radhakrishnan, M.L., Mapelli, M., Choi, S., Tidor, B., Cuny, G.D. et al. (2005). Defining Cdk5 ligand chemical space with small molecule inhibitors of tau phosphorylation. *Chemistry and Biology*, **12**, 811–823.

Ali, M.M., Roe, S.M., Vaughan, C.K., Meyer, P., Panaretou, B., Piper, P.W. et al. (2006). Crystal structure of an Hsp90-nucleotide-p23/Sba1 closed chaperone complex. *Nature*, **440**, 1013–1017.

Alonso, A.C., Grundke-Iqbal, I. and Iqbal, K. (1996). Alzheimer's disease hyperphosphorylated tau sequesters normal tau into tangles of filaments and disassembles microtubules. *Nature Medicine*, **2**, 783–787.

Alonso, A.D., Grundke-Iqbal, I., Barra, H.S. and Iqbal, K. (1997). Abnormal phosphorylation of tau and the mechanism of Alzheimer neurofibrillary degeneration: sequestration of microtubule-associated proteins 1 and 2 and the disassembly of microtubules by the abnormal tau. *Proceedings of the National Academy of Sciences of the United States of America*, **94**, 298–303.

Alonso, A., Zaidi, T., Novak, M., Grundke-Iqbal, I. and Iqbal, K. (2001). Hyperphosphorylation induces self-assembly of tau into tangles of paired helical filaments/straight filaments. *Proceedings of the National Academy of Sciences of the United States of America*, **98**, 6923–6928.

Antonov, L., Kawauchi, S., Satoh, M. and Komiyama, J. (1998). Theoretical investigations on the tautomerism of 1-phenylazo-4-naphthol and its isomers. *Dyes and Pigments*, **38**, 157–164.

Aptula, A.O., Roberts, D.W., Cronin, M.T. and Schultz, T.W. (2005). Chemistry-toxicity relationships for the effects of di- and trihydroxybenzenes to *Tetrahymena pyriformis*. *Chemical Research in Toxicology*, **18**, 844–854.

Arriagada, P.V., Growdon, J.H., Hedley-Whyte, E.T. and Hyman, B.T. (1992). Neurofibrillary tangles but not senile plaques parallel duration and severity of Alzheimer's disease. *Neurology*, **42**, 631–639.

Augustinack, J.C., Schneider, A., Mandelkow, E.M. and Hyman, B.T. (2002). Specific tau phosphorylation sites correlate with severity of neuronal cytopathology in Alzheimer's disease. *Acta Neuropathologica*, **103**, 26–35.

Bacskai, B.J., Hickey, G.A., Skoch, J., Kajdasz, S.T., Wang, Y., Huang, G.F. et al. (2003). Four-dimensional multiphoton imaging of brain entry, amyloid binding, and clearance of an amyloid-beta ligand in transgenic mice. *Proceedings of the National Academy of Sciences of the United States of America*, **100**, 12462–12467.

Bence, N.F., Sampat, R.M. and Kopito, R.R. (2001). Impairment of the ubiquitin-proteasome system by protein aggregation. *Science*, **292**, 1552–1555.

Bennett, D.A., Schneider, J.A., Bienias, J.L., Evans, D.A. and Wilson, R.S. (2005). Mild cognitive impairment is related to Alzheimer disease pathology and cerebral infarctions. *Neurology*, **64**, 834–841.

Berriman, J., Serpell, L.C., Oberg, K.A., Fink, A.L., Goedert, M. and Crowther, R.A. (2003). Tau filaments from human brain and from in vitro assembly of recombinant protein show cross-beta structure. *Proceedings of the National Academy of Sciences of the United States of America*, **100**, 9034–9038.

Bhat, R., Xue, Y., Berg, S., Hellberg, S., Ormo, M., Nilsson, Y. et al. (2003). Structural insights and biological effects of glycogen synthase kinase 3-specific inhibitor AR-A014418. *Journal of Biological Chemistry*, **278**, 45937–45945.

Bhat, R.V., Budd Haeberlein, S.L. and Avila, J. (2004). Glycogen synthase kinase 3: a drug target for CNS therapies. *Journal of Neurochemistry*, **89**, 1313–1317.

Biernat, J., Gustke, N., Drewes, G., Mandelkow, E.M. and Mandelkow, E. (1993). Phosphorylation of Ser262 strongly reduces binding of tau to microtubules: distinction between PHF-like immunoreactivity and microtubule binding. *Neuron*, **11**, 153–163.

Bolton, J.L., Trush, M.A., Penning, T.M., Dryhurst, G. and Monks, T.J. (2000). Role of quinones in toxicology. *Chemical Research in Toxicology*, **13**, 135–160.

Braak, H. and Braak, E. (1991). Neuropathological stageing of Alzheimer-related changes. *Acta Neuropathologica*, **82**, 239–259.

Bramblett, G.T., Goedert, M., Jakes, R., Merrick, S.E., Trojanowski, J.Q. and Lee, V.M. (1993). Abnormal tau phosphorylation at Ser396 in Alzheimer's disease recapitulates development and contributes to reduced microtubule binding. *Neuron*, **10**, 1089–1099.

Buee, L., Bussiere, T., Buee-Scherrer, V., Delacourte, A. and Hof, P.R. (2000). Tau protein isoforms, phosphorylation and role in neurodegenerative disorders. *Brain Research. Brain Research Reviews*, **33**, 95–130.

Callahan, L.M. and Coleman, P.D. (1995). Neurons bearing neurofibrillary tangles are responsible for selected synaptic deficits in Alzheimer's disease. *Neurobiology of Aging*, **16**, 311–314.

Caprathe, B.W., Gilmore, J.L., Hays, S.J., Jaen, J.C. and LeVine, H. (1999). Method of imaging amyloid deposits. *United States Patent*, 6,001,331.

Carmel, G., Mager, E.M., Binder, L.I. and Kuret, J. (1996). The structural basis of monoclonal antibody Alz50's selectivity for Alzheimer's disease pathology. *Journal of Biological Chemistry*, **271**, 32789–32795.

Cash, A.D., Aliev, G., Siedlak, S.L., Nunomura, A., Fujioka, H., Zhu, X. et al. (2003). Microtubule reduction in Alzheimer's disease and aging is independent of tau filament formation. *American Journal of Pathology*, **162**, 1623–1627.

Chalecka-Franaszek, E. and Chuang, D.M. (1999). Lithium activates the serine/threonine kinase Akt-1 and suppresses glutamate-induced inhibition of Akt-1 activity in neurons. *Proceedings of the National Academy of Sciences of the United States of America*, **96**, 8745–8750.

Chambers, C.B., Lee, J.M., Troncoso, J.C., Reich, S. and Muma, N.A. (1999). Overexpression of four-repeat tau mRNA isoforms in progressive supranuclear palsy but not in Alzheimer's disease. *Annals of Neurology*, **46**, 325–332.

Chirita, C.N. and Kuret, J. (2004). Evidence for an intermediate in tau filament formation. *Biochemistry*, **43**, 1704–1714.

Chirita, C.N., Necula, M. and Kuret, J. (2004). Ligand-dependent inhibition and reversal of tau filament formation. *Biochemistry*, **43**, 2879–2887.

Chirita, C.N., Congdon, E.E., Yin, H. and Kuret, J. (2005). Triggers of full-length tau aggregation: a role for partially folded intermediates. *Biochemistry*, **44**, 5862–5872.

Conway, K.A., Rochet, J.C., Bieganski, R.M. and Lansbury, P.T., Jr. (2001). Kinetic stabilization of the alpha-synuclein protofibril by a dopamine-alpha-synuclein adduct. *Science*, **294**, 1346–1349.

Crowther, R.A. (1991). Straight and paired helical filaments in Alzheimer disease have a common structural unit. *Proceedings of the National Academy of Sciences of the United States of America*, **88**, 2288–2292.

Cruz, J.C., Tseng, H.C., Goldman, J.A., Shih, H. and Tsai, L.H. (2003). Aberrant Cdk5 activation by p25 triggers pathological events leading to neurodegeneration and neurofibrillary tangles. *Neuron*, **40**, 471–483.

Cummings, J.L. (2004). Dementia with lewy bodies: molecular pathogenesis and implications for classification. *Journal of Geriatric Psychiatry and Neurology*, **17**, 112–119.

Davies, S.P., Reddy, H., Caivano, M. and Cohen, P. (2000). Specificity and mechanism of action of some commonly used protein kinase inhibitors. *Biochemical Journal*, **351**, 95–105.

de Leon, M.J., DeSanti, S., Zinkowski, R., Mehta, P.D., Pratico, D., Segal, S. et al. (2004). MRI and CSF studies in the early diagnosis of Alzheimer's disease. *Journal of Internal Medicine*, **256**, 205–223.

Defossez, A. and Delacourte, A. (1987). Transformation of degenerating neurofibrils into amyloid substance in Alzheimer's disease: histochemical and immunohistochemical studies. *Journal of the Neurological Sciences*, **81**, 1–10.

Dehmelt, L. and Halpain, S. (2005). The MAP2/Tau family of microtubule-associated proteins. *Genome Biology*, **6**, 204.

Derkinderen, P., Scales, T.M., Hanger, D.P., Leung, K.Y., Byers, H.L., Ward, M.A. et al. (2005). Tyrosine 394 is phosphorylated in Alzheimer's paired helical filament tau and in fetal tau with c-Abl as the candidate tyrosine kinase. *Journal of Neuroscience*, **25**, 6584–6593.

Dickey, C.A., Dunmore, J., Lu, B., Wang, J.W., Lee, W.C., Kamal, A. et al. (2006). HSP induction mediates selective clearance of tau phosphorylated at proline-directed Ser/Thr sites but not KXGS (MARK) sites. *FASEB Journal*, **20**, 753–755.

DiTella, M.C., Feiguin, F., Carri, N., Kosik, K.S. and Caceres, A. (1996). MAP-1B/TAU functional redundancy during laminin-enhanced axonal growth. *Journal of Cell Science*, **109**, 467–477.

Duyckaerts, C. and Hauw, J.J. (1997). Prevalence, incidence and duration of Braak's stages in the general population: can we know? *Neurobiology of Aging*, **18**, 362–369.

Dzwolak, W. and Pecul, M. (2005). Chiral bias of amyloid fibrils revealed by the twisted conformation of Thioflavin T: an induced circular dichroism/DFT study. *FEBS Letters*, **579**, 6601–6603.

Feany, M.B. and Dickson, D.W. (1996). Neurodegenerative disorders with extensive tau pathology: a comparative study and review. *Annals of Neurology*, **40**, 139–148.

Galvan, M., David, J.P., Delacourte, A., Luna, J. and Mena, R. (2001). Sequence of neurofibrillary changes in aging and Alzheimer's disease: a confocal study with phospho-tau antibody, AD2. *Journal of Alzheimer's Disease*, **3**, 417–425.

Ghoshal, N., Smiley, J.F., DeMaggio, A.J., Hoekstra, M.F., Cochran, E.J., Binder, L.I. et al. (1999). A new molecular link between the fibrillar and granulovacuolar lesions of Alzheimer's disease. *American Journal of Pathology*, **155**, 1163–1172.

Ghoshal, N., Garcia-Sierra, F., Wuu, J., Leurgans, S., Bennett, D.A., Berry, R.W. et al. (2002). Tau conformational changes correspond to impairments of episodic memory in mild cognitive impairment and Alzheimer's disease. *Experimental Neurology*, **177**, 475–493.

Giasson, B.I., Forman, M.S., Higuchi, M., Golbe, L.I., Graves, C.L., Kotzbauer, P.T. et al. (2003). Initiation and synergistic fibrillization of tau and α-synuclein. *Science*, **300**, 636–640.

Gigant, B., Wang, C., Ravelli, R.B., Roussi, F., Steinmetz, M.O., Curmi, P.A. et al. (2005). Structural basis for the regulation of tubulin by vinblastine. *Nature*, **435**, 519–522.

Goedert, M., Spillantini, M.G., Jakes, R., Rutherford, D. and Crowther, R.A. (1989). Multiple isoforms of human microtubule-associated protein tau: sequences and localization in neurofibrillary tangles of Alzheimer's disease. *Neuron*, **3**, 519–526.

Goedert, M., Jakes, R., Qi, Z., Wang, J.H. and Cohen, P. (1995). Protein phosphatase 2A is the major enzyme in brain that dephosphorylates tau protein phosphorylated by proline-directed protein kinases or cyclic AMP-dependent protein kinase. *Journal of Neurochemistry*, **65**, 2804–2807.

Gong, C.X., Shaikh, S., Wang, J.Z., Zaidi, T., Grundke-Iqbal, I. and Iqbal, K. (1995). Phosphatase activity toward abnormally phosphorylated tau: decrease in Alzheimer disease brain. *Journal of Neurochemistry*, **65**, 732–738.

Hall, G.F. and Yao, J. (2000). Neuronal morphology, axonal integrity, and axonal regeneration in situ are regulated by cytoskeletal phosphorylation in identified lamprey central neurons. *Microscopy Research and Technique*, **48**, 32–46.

Hall, G.F., Lee, V.M., Lee, G. and Yao, J. (2001). Staging of neurofibrillary degeneration caused by human tau overexpression in a unique cellular model of human tauopathy. *American Journal of Pathology*, **158**, 235–246.

Hampel, H., Mitchell, A., Blennow, K., Frank, R.A., Brettschneider, S., Weller, L. et al. (2004). Core biological marker candidates of Alzheimer's disease—perspectives for diagnosis, prediction of outcome and reflection of biological activity. *Journal of Neural Transmission*, **111**, 247–272.

Hanger, D.P., Betts, J.C., Loviny, T.L., Blackstock, W.P. and Anderton, B.H. (1998). New phosphorylation sites identified in hyperphosphorylated tau (paired helical filament-tau) from Alzheimer's disease brain using nanoelectrospray mass spectrometry. *Journal of Neurochemistry*, **71**, 2465–2476.

Harada, A., Oguchi, K., Okabe, S., Kuno, J., Terada, S., Ohshima, T. et al. (1994). Altered microtubule organization in small-calibre axons of mice lacking tau protein. *Nature*, **369**, 488–491.

Herz, A.H. (1974). Dye–dye interactions of cyanines in solution and at silver bromide surfaces. *Photographic Science and Engineering*, **18**, 323–335.

Ichise, M., Meyer, J.H. and Yonekura, Y. (2001). An introduction to PET and SPECT neuroreceptor quantification models. *The Journal of Nuclear Medicine*, **42**, 755–763.

Ingram, E.M. and Spillantini, M.G. (2002). Tau gene mutations: dissecting the pathogenesis of FTDP-17. *Trends in Molecular Medicine*, **8**, 555–562.

Iqbal, K., Zaidi, T., Thompson, C.H., Merz, P.A. and Wisniewski, H.M. (1984). Alzheimer paired helical filaments: bulk isolation, solubility, and protein composition. *Acta Neuropathologica*, **62**, 167–177.

Ishihara, T., Hong, M., Zhang, B., Nakagawa, Y., Lee, M.K., Trojanowski, J.Q. et al. (1999). Age-dependent emergence and progression of a tauopathy in transgenic mice over-expressing the shortest human tau isoform. *Neuron*, **24**, 751–762.

Johnson, G.L. and Lapadat, R. (2002). Mitogen-activated protein kinase pathways mediated by ERK, JNK, and p38 protein kinases. *Science*, **298**, 1911–1912.

Jordan, M.A. and Wilson, L. (2004). Microtubules as a target for anticancer drugs. *Nature Reviews. Cancer*, **4**, 253–265.

Kannanayakal, T.J., Tao, H., Vandre, D.D. and Kuret, J. (2006). Casein kinase-1 isoforms differentially associate with neurofibrillary and granulovacuolar degeneration lesions. *Acta Neuropathologica*, **111**, 413–421.

Keck, S., Nitsch, R., Grune, T. and Ullrich, O. (2003). Proteasome inhibition by paired helical filament-tau in brains of patients with Alzheimer's disease. *Journal of Neurochemistry*, **85**, 115–122.

Khatoon, S., Grundke-Iqbal, I. and Iqbal, K. (1992). Brain levels of microtubule-associated protein tau are elevated in Alzheimer's disease: a radioimmuno-slot-blot assay for nanograms of the protein. *Journal of Neurochemistry*, **59**, 750–753.

Khatoon, S., Grundke-Iqbal, I. and Iqbal, K. (1994). Levels of normal and abnormally phosphorylated tau in different cellular and regional compartments of Alzheimer disease and control brains. *FEBS Letters*, **351**, 80–84.

Khawaja, S., Gundersen, G.G. and Bulinski, J.C. (1988). Enhanced stability of microtubules enriched in detyrosinated tubulin is not a direct function of detyrosination level. *Journal of Cell Biology*, **106**, 141–149.

Kimura, T., Ono, T., Takamatsu, J., Yamamoto, H., Ikegami, K., Kondo, A. et al. (1996). Sequential changes of tau-site-specific phosphorylation during development of paired helical filaments. *Dementia*, **7**, 177–181.

King, M.E., Ghoshal, N., Wall, J.S., Binder, L.I. and Ksiezak-Reding, H. (2001). Structural analysis of Pick's disease-derived and in vitro-assembled tau filaments. *American Journal of Pathology*, **158**, 1481–1490.

Klunk, W.E., Wang, Y., Huang, G.F., Debnath, M.L., Holt, D.P., Shao, L. et al. (2003). The binding of 2-(4'-methylaminophenyl)benzothiazole to postmortem brain homogenates is dominated by the amyloid component. *Journal of Neuroscience*, **23**, 2086–2092.

Klunk, W.E., Lopresti, B.J., Ikonomovic, M.D., Lefterov, I.M., Koldamova, R.P., Abrahamson, E.E. et al. (2005). Binding of the positron emission tomography tracer Pittsburgh compound-B reflects the amount of amyloid-beta in Alzheimer's disease brain but not in transgenic mouse brain. *Journal of Neuroscience*, **25**, 10598–10606.

Knippschild, U., Gocht, A., Wolff, S., Huber, N., Lohler, J. and Stoter, M. (2005). The casein kinase 1 family: participation in multiple cellular processes in eukaryotes. *Cellular Signalling*, **17**, 675–689.

Kopke, E., Tung, Y.C., Shaikh, S., Alonso, A.C., Iqbal, K. and Grundke-Iqbal, I. (1993). Microtubule-associated protein tau. Abnormal phosphorylation of a non-paired helical filament pool in Alzheimer disease. *Journal of Biological Chemistry*, **268**, 24374–24384.

Korth, C., May, B.C., Cohen, F.E. and Prusiner, S.B. (2001). Acridine and phenothiazine derivatives as pharmacotherapeutics for prion disease. *Proceedings of the National Academy of Sciences of the United States of America*, **98**, 9836–9841.

Kosik, K.S. and Shimura, H. (2005). Phosphorylated tau and the neurodegenerative foldopathies. *Biochimica et Biophysica Acta*, **1739**, 298–310.

Krebs, M.R., Bromley, E.H. and Donald, A.M. (2005). The binding of thioflavin-T to amyloid fibrils: localisation and implications. *Journal of Structural Biology*, **149**, 30–37.

Ksiezak-Reding, H. and Wall, J.S. (2005). Characterization of paired helical filaments by scanning transmission electron microscopy. *Microscopy Research and Technique*, **67**, 126–140.

Ksiezak-Reding, H., Liu, W.K. and Yen, S.H. (1992). Phosphate analysis and dephosphorylation of modified tau associated with paired helical filaments. *Brain Research*, **597**, 209–219.

Kung, M.P., Skovronsky, D.M., Hou, C., Zhuang, Z.P., Gur, T.L., Zhang, B. et al. (2003). Detection of amyloid plaques by radioligands for Aβ40 and Aβ42: potential imaging agents in Alzheimer's patients. *Journal of Molecular Neuroscience*, **20**, 15–24.

Kuret, J., Congdon, E.E., Li, G., Yin, H., Yu, X. and Zhong, Q. (2005). Evaluating triggers and enhancers of tau fibrillization. *Microscopy Research and Technique*, **67**, 141–155.

Le Corre, S., Klafki, H.W., Plesnila, N., Hubinger, G., Obermeier, A., Sahagun, H. et al. (2006). An inhibitor of tau hyperphosphorylation prevents severe motor impairments in tau transgenic mice. *Proceedings of the National Academy of Sciences of the United States of America*, **103**, 9673–9678.

Lee, G. (2005). Tau and src family tyrosine kinases. *Biochimica et Biophysica Acta*, **1739**, 323–330.

Lee, J., Chen, Y., Tolstykh, T. and Stock, J. (1996). A specific protein carboxyl methylesterase that demethylates phosphoprotein phosphatase 2A in bovine brain. *Proceedings of the National Academy of Sciences of the United States of America*, **93**, 6043–6047.

Li, G., Faibushevich, A., Turunen, B.J., Yoon, S.O., Georg, G., Michaelis, M.L. et al. (2003). Stabilization of the cyclin-dependent kinase 5 activator, p35, by paclitaxel decreases beta-amyloid toxicity in cortical neurons. *Journal of Neurochemistry*, **84**, 347–362.

Li, G., Yin, H. and Kuret, J. (2004a) Casein kinase 1 delta phosphorylates Tau and disrupts its binding to microtubules. *Journal of Biological Chemistry*, **279**, 15938–15945.

Li, R., Lindholm, K., Yang, L.B., Yue, X., Citron, M., Yan, R. et al. (2004b) Amyloid beta peptide load is correlated with increased beta-secretase activity in sporadic Alzheimer's disease patients. *Proceedings of the National Academy of Sciences of the United States of America*, **101**, 3632–3637.

Li, X., Alafuzoff, I., Soininen, H., Winblad, B. and Pei, J.J. (2005). Levels of mTOR and its downstream targets 4E-BP1, eEF2, and eEF2 kinase in relationships with tau in Alzheimer's disease brain. *FEBS Journal*, **272**, 4211–4220.

Lim, J. and Lu, K.P. (2005). Pinning down phosphorylated tau and tauopathies. *Biochimica et Biophysica Acta*, **1739**, 311–322.

Lockhart, A., Ye, L., Judd, D.B., Merritt, A.T., Lowe, P.N., Morgenstern, J.L. et al. (2005). Evidence for the presence of three distinct binding sites for the thioflavin T class of Alzheimer's disease PET imaging agents on beta-amyloid peptide fibrils. *Journal of Biological Chemistry*, **280**, 7677–7684.

Löwe, J., Li, H., Downing, K.H. and Nogales, E. (2001). Refined structure of alpha beta-tubulin at 3.5 A resolution. *Journal of Molecular Biology*, **313**, 1045–1057.

Makrides, V., Massie, M.R., Feinstein, S.C. and Lew, J. (2004). Evidence for two distinct binding sites for tau on microtubules. *Proceedings of the National Academy of Sciences of the United States of America*, **101**, 6746–6751.

Mandelkow, E., Mandelkow, E.-M., Biernat, J., Bergen, M.V. and Pickhardt, M. (2006). Treating neurodegenerative conditions. *PCT International Application* WO 2006/007864 A1.

Margittai, M. and Langen, R. (2004). Template-assisted filament growth by parallel stacking of tau. *Proceedings of the National Academy of Sciences of the United States of America*, **101**, 10278–10283.

Martinez, A., Alonso, M., Castro, A., Dorronsoro, I., Gelpi, J.L., Luque, F.J. et al. (2005). SAR and 3D-QSAR studies on thiadiazolidinone derivatives: exploration of structural requirements for glycogen synthase kinase 3 inhibitors. *Journal of Medicinal Chemistry*, **48**, 7103–7112.

Mashhoon, N., DeMaggio, A.J., Tereshko, V., Bergmeier, S.C., Egli, M., Hoekstra, M.F. et al. (2000). Crystal structure of a conformation-selective casein kinase-1 inhibitor. *Journal of Biological Chemistry*, **275**, 20052–20060.

Masuda, M., Suzuki, N., Taniguchi, S., Oikawa, T., Nonaka, T., Iwatsubo, T. et al. (2006). Small molecule inhibitors of alpha-synuclein filament assembly. *Biochemistry*, **45**, 6085–6094.

Mathis, C.A., Wang, Y. and Klunk, W.E. (2004). Imaging beta-amyloid plaques and neurofibrillary tangles in the aging human brain. *Current Pharmaceutical Design*, **10**, 1469–1492.

Matsuo, E.S., Shin, R.W., Billingsley, M.L., Van deVoorde, A., O'Connor, M., Trojanowski, J.Q. et al. (1994). Biopsy-derived adult human brain tau is phosphorylated at many of the same sites as Alzheimer's disease paired helical filament tau. *Neuron*, **13**, 989–1002.

May, B.C., Fafarman, A.T., Hong, S.B., Rogers, M., Deady, L.W., Prusiner, S.B. et al. (2003). Potent inhibition of scrapie prion replication in cultured cells by bis-acridines. *Proceedings of the National Academy of Sciences of the United States of America*, **100**, 3416–3421.

Meijer, L., Flajolet, M. and Greengard, P. (2004). Pharmacological inhibitors of glycogen synthase kinase 3. *Trends in Pharmacological Sciences*, **25**, 471–480.

Mena, R., Edwards, P., Perez-Olvera, O. and Wischik, C.M. (1995). Monitoring pathological assembly of tau and beta-amyloid proteins in Alzheimer's disease. *Acta Neuropathologica*, **89**, 50–56.

Michaelis, M.L. (2006). Ongoing in vivo studies with cytoskeletal drugs in tau transgenic mice. *Current Alzheimer Research*, **3**, 215–219.

Michaelis, M.L., Ranciat, N., Chen, Y., Bechtel, M., Ragan, R., Hepperle, M. et al. (1998). Protection against beta-amyloid toxicity in primary neurons by paclitaxel (Taxol). *Journal of Neurochemistry*, **70**, 1623–1627.

Morishima-Kawashima, M., Hasegawa, M., Takio, K., Suzuki, M., Yoshida, H., Watanabe, A. et al. (1995). Hyperphosphorylation of tau in PHF. *Neurobiology of Aging*, **16**, 365–371.

Munoz, D.G., Dickson, D.W., Bergeron, C., Mackenzie, I.R., Delacourte, A. and Zhukareva, V. (2003). The neuropathology and biochemistry of frontotemporal dementia. *Annals of Neurology*, **54**, S24-S28.

Necula, M. and Kuret, J. (2005). Site-specific pseudophosphorylation modulates the rate of tau filament dissociation. *FEBS Letters*, **579**, 1453–1457.

Necula, M., Chirita, C.N. and Kuret, J. (2005). Cyanine dye n744 inhibits tau fibrillization by blocking filament extension: implications for the treatment of tauopathic neurodegenerative diseases. *Biochemistry*, **44**, 10227–10237.

Noble, W., Olm, V., Takata, K., Casey, E., Mary, O., Meyerson, J. et al. (2003). Cdk5 is a key factor in tau aggregation and tangle formation in vivo. *Neuron*, **38**, 555–565.

Noble, W., Planel, E., Zehr, C., Olm, V., Meyerson, J., Suleman, F. et al. (2005). Inhibition of glycogen synthase kinase-3 by lithium correlates with reduced tauopathy and degeneration in vivo. *Proceedings of the National Academy of Sciences of the United States of America*, **102**, 6990–6995.

Novak, M., Kabat, J. and Wischik, C.M. (1993). Molecular characterization of the minimal protease resistant tau unit of the Alzheimer's disease paired helical filament. *EMBO Journal*, **12**, 365–370.

Ogris, E., Du, X., Nelson, K.C., Mak, E.K., Yu, X.X., Lane, W.S. et al. (1999). A protein phosphatase methylesterase (PME-1) is one of several novel proteins stably associating with two inactive mutants of protein phosphatase 2A. *Journal of Biological Chemistry*, **274**, 14382–14391.

Omura, S., Sasaki, Y., Iwai, Y. and Takeshima, H. (1995). Staurosporine, a potentially important gift from a microorganism. *Journal of Antibiotics*, **48**, 535–548.

Patrick, G.N., Zukerberg, L., Nikolic, M., de la Monte, S., Dikkes, P. and Tsai, L.H. (1999). Conversion of p35 to p25 deregulates Cdk5 activity and promotes neurodegeneration. *Nature*, **402**, 615–622.

Perry, G., Kawai, M., Tabaton, M., Onorato, M., Mulvihill, P., Richey, P. et al. (1991). Neuropil threads of Alzheimer's disease show a marked alteration of the normal cytoskeleton. *Journal of Neuroscience*, **11**, 1748–1755.

Petkova, A.T., Leapman, R.D., Guo, Z., Yau, W.M., Mattson, M.P. and Tycko, R. (2005). Self-propagating, molecular-level polymorphism in Alzheimer's beta-amyloid fibrils. *Science*, **307**, 262–265.

Petkova, A.T., Yau, W.M. and Tycko, R. (2006). Experimental Constraints on Quaternary Structure in Alzheimer's β-Amyloid Fibrils. *Biochemistry*, **45**, 498–512.

Petrucelli, L., Dickson, D., Kehoe, K., Taylor, J., Snyder, H., Grover, A. et al. (2004). CHIP and Hsp70 regulate tau ubiquitination, degradation and aggregation. *Human Molecular Genetics*, **13**, 703–714.

Phiel, C.J. and Klein, P.S. (2001). Molecular targets of lithium action. *Annual Review of Pharmacology and Toxicology*, **41**, 789–813.

Phiel, C.J., Wilson, C.A., Lee, V.M. and Klein, P.S. (2003). GSK-3α regulates production of Alzheimer's disease amyloid-beta peptides. *Nature*, **423**, 435–439.

Pickhardt, M., Gazova, Z., von Bergen, M., Khlistunova, I., Wang, Y., Hascher, A. et al. (2005a) Anthraquinones inhibit tau aggregation and dissolve alzheimer paired helical filaments in vitro and in cells. *Journal of Biological Chemistry*, **280**, 3628–3635.

Pickhardt, M., von Bergen, M., Gazova, Z., Hascher, A., Biernat, J., Mandelkow, E.M. et al. (2005b) Screening for inhibitors of tau polymerization. *Current Alzheimer Research*, **2**, 219–226.

Qiang, L., Yu, W. Andreadis, A., Luo, M. and Baas, P.W. (2006). Tau protects microtubules in the axon from severing by katanin. *Journal of Neuroscience*, **26**, 3120–3129.

Rapoport, M. and Ferreira, A. (2000). PD98059 prevents neurite degeneration induced by fibrillar beta-amyloid in mature hippocampal neurons. *Journal of Neurochemistry*, **74**, 125–133.

Raught, B., Gingras, A.C. and Sonenberg, N. (2001). The target of rapamycin (TOR) proteins. *Proceedings of the National Academy of Sciences of the United States of America*, **98**, 7037–7044.

Ravelli, R.B., Gigant, B., Curmi, P.A., Jourdain, I., Lachkar, S., Sobel, A. et al. (2004). Insight into tubulin regulation from a complex with colchicine and a stathmin-like domain. *Nature*, **428**, 198–202.

Ravikumar, B., Vacher, C., Berger, Z., Davies, J.E., Luo, S., Oroz, L.G. et al. (2004). Inhibition of mTOR induces autophagy and reduces toxicity of polyglutamine expansions in fly and mouse models of Huntington disease. *Nature Genetics*, **36**, 585–595.

Rena, G., Bain, J., Elliott, M. and Cohen, P. (2004). D4476, a cell-permeant inhibitor of CK1, suppresses the site-specific phosphorylation and nuclear exclusion of FOXO1a. *EMBO Reports*, **5**, 60–65.

Rice, A., Liu, Y., Michaelis, M.L., Himes, R.H., Georg, G.I. and Audus, K.L. (2005). Chemical modification of paclitaxel (Taxol) reduces P-glycoprotein interactions and increases permeation across the blood-brain barrier in vitro and in situ. *Journal of Medicinal Chemistry*, **48**, 832–838.

Roman, G.C., Sachdev, P., Royall, D.R., Bullock, R.A., Orgogozo, J.M., Lopez-Pousa, S. et al. (2004). Vascular cognitive disorder: a new diagnostic category updating vascular cognitive impairment and vascular dementia. *Journal of the Neurological Sciences*, **226**, 81–87.

Royall, D.R., Palmer, R., Mulroy, A.R., Polk, M.J., Roman, G.C., David, J.P. et al. (2002). Pathological determinants of the transition to clinical dementia in Alzheimer's disease. *Experimental Aging Research*, **28**, 143–162.

Ryves, W.J., Dajani, R., Pearl, L. and Harwood, A.J. (2002). Glycogen synthase kinase-3 inhibition by lithium and beryllium suggests the presence of two magnesium binding sites. *Biochemical and Biophysical Research Communications*, **290**, 967–972.

Schwab, C., Steele, J.C., Akiyama, H., McGeer, E.G. and McGeer, P.L. (1995). Relationship of amyloid beta/A4 protein to the neurofibrillary tangles in Guamanian parkinsonism-dementia. *Acta Neuropathologica*, **90**, 287–298.

Schweers, O., Schonbrunn-Hanebeck, E., Marx, A. and Mandelkow, E. (1994). Structural studies of tau protein and Alzheimer paired helical filaments show no evidence for beta-structure. *Journal of Biological Chemistry*, **269**, 24290–24297.

Serpell, L.C., Berriman, J., Jakes, R., Goedert, M. and Crowther, R.A. (2000). Fiber diffraction of synthetic alpha-synuclein filaments shows amyloid-like cross-beta conformation. *Proceedings of the National Academy of Sciences of the United States of America*, **97**, 4897–4902.

Seubert, P., Mawal-Dewan, M., Barbour, R., Jakes, R., Goedert, M., Johnson, G.V. et al. (1995). Detection of phosphorylated Ser262 in fetal tau, adult tau, and paired helical filament tau. *Journal of Biological Chemistry*, **270**, 18917–18922.

Shimura, H., Schwartz, D., Gygi, S.P. and Kosik, K.S. (2004). CHIP-Hsc70 complex ubiquitinates phosphorylated tau and enhances cell survival. *Journal of Biological Chemistry*, **279**, 4869–4876.

Sontag, E., Hladik, C., Montgomery, L., Luangpirom, A., Mudrak, I., Ogris, E. et al. (2004a) Downregulation of protein phosphatase 2A carboxyl methylation and methyltransferase may contribute to Alzheimer disease pathogenesis. *Journal of Neuropathology and Experimental Neurology*, **63**, 1080–1091.

Sontag, E., Luangpirom, A., Hladik, C., Mudrak, I., Ogris, E., Speciale, S. et al. (2004b) Altered expression levels of the protein phosphatase 2A ABαC enzyme are associated with Alzheimer disease pathology. *Journal of Neuropathology and Experimental Neurology*, **63**, 287–301.

Spittaels, K., Van den Haute, C., Van Dorpe, J., Bruynseels, K., Vandezande, K., Laenen, I. et al. (1999). Prominent axonopathy in the brain and spinal cord of transgenic mice overexpressing four-repeat human tau protein. *American Journal of Pathology*, **155**, 2153–2165.

Sridhar, J., Akula, N. and Pattabiraman, N. (2006). Selectivity and potency of cyclin-dependent kinase inhibitors. *The AAPS Journal*, **8**, E204–E221.

Taniguchi, S., Suzuki, N., Masuda, M., Hisanaga, S., Iwatsubo, T., Goedert, M. et al. (2005). Inhibition of heparin-induced tau filament formation by phenothiazines, polyphenols, and porphyrins. *Journal of Biological Chemistry*, **280**, 7614–7623.

Tatebayashi, Y., Planel, E., Chui, D.H., Sato, S., Miyasaka, T., Sahara, N. et al. (2006). c-jun N-terminal kinase hyperphosphorylates R406W tau at the PHF-1 site during mitosis. *FASEB Journal*, **20**, 762–764.

Terwel, D., Lasrado, R., Snauwaert, J., Vandeweert, E., Van Haesendonck, C., Borghgraef, P. et al. (2005). Changed conformation of mutant Tau-P301L underlies the moribund tauopathy, absent in progressive, nonlethal axonopathy of Tau-4R/2N transgenic mice. *Journal of Biological Chemistry*, **280**, 3963–3973.

Tint, I., Slaughter, T., Fischer, I. and Black, M.M. (1998). Acute inactivation of tau has no effect on dynamics of microtubules in growing axons of cultured sympathetic neurons. *Journal of Neuroscience*, **18**, 8660–8673.

Uchihara, T., Nakamura, A., Yamazaki, M. and Mori, O. (2001a) Evolution from pretangle neurons to neurofibrillary tangles monitored by thiazin red combined with Gallyas method and double immunofluorescence. *Acta Neuropathologica*, **101**, 535–539.

Uchihara, T., Nakamura, A., Yamazaki, M., Mori, O., Ikeda, K. and Tsuchiya, K. (2001b) Different conformation of neuronal tau deposits distinguished by double immunofluorescence with AT8 and thiazin red combined with Gallyas method. *Acta Neuropathologica*, **102**, 462–466.

Vafai, S.B. and Stock, J.B. (2002). Protein phosphatase 2A methylation: a link between elevated plasma homocysteine and Alzheimer's Disease. *FEBS Letters*, **518**, 1–4.

Wang, J., Dickson, D.W., Trojanowski, J.Q. and Lee, V.M. (1999). The levels of soluble versus insoluble brain Abeta distinguish Alzheimer's disease from normal and pathologic aging. *Experimental Neurology*, **158**, 328–337.

Westermark, P., Benson, M.D., Buxbaum, J.N., Cohen, A.S., Frangione, B., Ikeda, S. et al. (2005). Amyloid: toward terminology clarification. Report from the Nomenclature Committee of the International Society of Amyloidosis. *Amyloid*, **12**, 1–4.

Wischik, C.M., Edwards, P.C., Lai, R.Y., Roth, M. and Harrington, C.R. (1996). Selective inhibition of Alzheimer disease-like tau aggregation by phenothiazines. *Proceedings of the National Academy of Sciences of the United States of America*, **93**, 11213–11218.

Xiao, H., Verdier-Pinard, P., Fernandez-Fuentes, N., Burd, B., Angeletti, R., Fiser, A. et al. (2006). Inaugural article: insights into the mechanism of microtubule stabilization by Taxol. *Proceedings of the National Academy of Sciences of the United States of America*, **103**, 10166–10173.

Zhang, L., Fan, J., Vu, K., Hong, K., Le Brazidec, J.Y., Shi, J. et al. (2006). 7′-substituted benzothiazolothio- and pyridinothiazolothio-purines as potent heat shock protein 90 inhibitors. *Journal of Medicinal Chemistry*, **49**, 5352–5362.

Zhu, M., Rajamani, S., Kaylor, J., Han, S., Zhou, F. and Fink, A.L. (2004). The flavonoid baicalein inhibits fibrillation of α-synuclein and disaggregates existing fibrils. *Journal of Biological Chemistry*, **279**, 26846–26857.

7 Protein Misfolding in Alzheimer's Disease: Pathogenic or Protective?

Rudy J. Castellani, Hyoung-Gon Lee, Akihiko Nunomura, Xiongwei Zhu, George Perry, and Mark A. Smith

CONTENTS

For nearly a century, the pathological hallmarks of Alzheimer's disease (AD), namely senile plaques and neurofibrillary tangles (NFT), have been suspected to play a major role in disease pathogenesis. This, not surprisingly, has led the field to focus on the biochemistry of amyloid-β deposition as senile plaques, or the phosphorylation and aggregation of tau as NFT. In this review we take a contrary view, where, rather than remaining initiators of disease pathogenesis, we suspect that the lesions function as a primary line of antioxidant defense. If amyloid-β and tau accumulations reflect a physiological reaction to chronic stress, the rationale of current therapeutic efforts targeted toward lesion removal would obviously be misguided. Expanding beyond AD, we suspect that this concept of protection is also true for misfolded proteins and pathological lesions in other neurodegenerative diseases.

7.1 INTRODUCTION

Amyloid-β and tau protein are among the best studied proteins in all of neurobiology and figure centrally into much of the research dedicated to AD. While not surprising since the pathological diagnosis of AD is dependent on the quantity of amyloid-β and tau deposits within cortical gray matter (Mirra et al., 1991; Hyman and

Trojanowski, 1997), we suggest that this strict linkage of diagnostic and mechanistic views is misleading, particularly in the case of neurodegenerative disease. Nevertheless, since amyloid-β and tau are crucial proteins that are exploited for diagnostic purposes, this focus has led many to conclude that these same proteins, corresponding to stereotypical AD-type lesions, are signatures of an aberration that speaks directly to disease etiology. Unfortunately, it now appears that such an inductive leap is not warranted on objective review of available data. Rather, we believe the mechanistic importance of senile plaques or amyloid-β and NFT or tau has far less to do with their consequences than with the factors that led to their formation. Indeed, it is important to recognize that, unlike the pathological diagnosis of most other processes, the pathological diagnosis of AD brains merely represents an association of presumed pathological findings together with a given clinical disease. Therefore, since AD-type pathological changes in cognitively intact elderly patients (Davis et al., 1999), and, conversely, AD-like dementia in the absence of AD-type pathology (Tiraboschi et al., 2004), the diagnosis of AD requires positive data for both clinical and pathological aspects. In this review, we present an alternative hypothesis for the role of amyloid-β and tau deposition in AD that may herald a paradigm shift in our view of not only AD but also many other neurodegenerative diseases characterized by lesions.

7.2   AMYLOID-β

The scientific literature has literally dogmatized the fundamental concept that amyloid-β causes disease (Hardy and Selkoe, 2002). Genetic data are often suggested as a priori evidence of this fact since, for example, amyloid-β protein precursor (AβPP) mutations lead to familial, early-onset AD (autosomal dominant), and since patients with Down's syndrome, who carry an extra copy of the AβPP gene, consistently develop AD changes with prolonged survival. Clinicopathological data may also be cited, as amyloid-β deposits are increased in the AD brain (Knowles et al., 1998). On the other hand, AD kindreds with AβPP mutations are exceedingly rare and it remains to be determined whether these kindreds are only tangentially representative of sporadic AD (Nunomura et al., 2000). Indeed, it is notable that oxidative stress precedes amyloid-β deposition by decades in Down's syndrome, sporadic AD, and familial AD (Odetti et al., 1998; Nunomura et al., 1999, 2000, 2001, 2004). Similarly, the genetic aberration in Down's syndrome clearly leads to a cascade of pathophysiology over and above amyloid-β deposits and sporadic AD (Nunomura et al., 2000). Moreover, the notion of amyloid-β deposits per se as primary neurotoxic lesions in AD may be called into question simply by the early appearance of oxidative stress sequelae relative to amyloid-β deposits in both sporadic (Nunomura et al., 2000, 2001; Smith et al., 2000) as well as genetic (Nunomura et al., 2004) cases of disease.

Of paramount importance, neurons respond to oxidative stress by increasing amyloid-β production (Yan et al., 1995) and such increases in amyloid-β are associated with a consequent reduction in oxidative stress (Nunomura et al., 2000, 2001). In fact, amyloid-β is a genuine antioxidant that can act as a potent superoxide dismutase (Cuajungco et al., 2000). By this logic, therefore, AD kindreds with AβPP mutations lose effective antioxidant capacity (due to mutation-driven protein dysfunction), while the extensive amyloid-β deposits themselves are signatures not of neurotoxicity per se but of oxidative imbalance and an oxidative stress response. This is consistent with the data that virtually everyone over the age of 40 contain detectable amyloid-β deposits (Nunomura et al., 2001), and manifestly more logical than the alternative view that everyone at midlife is on the verge of developing AD, a view also directly contradicted by the fact that a large percentage of cognitively intact, aged individuals contain amyloid-β loads equivalent to patients with AD (Davis et al., 1999).

Fibrillar or aggregated forms of amyloid-β, such as in senile plaque cores, in the obviously artificial cell culture environment are toxic to cultured neurons in vitro (Pike et al., 1991; Rottkamp et al., 2001). However, in vivo, the presence and density of amyloid-β correlate weakly with the onset and severity of AD (Arriagada et al., 1992; Giannakopoulos et al., 2003), while recent data suggest that the presence of the soluble form of amyloid-β in the brain may be a better predictor of the disease (McLean et al., 1999). Specifically, SDS-stable oligomers, and not monomers, of this form of amyloid-β seem to play an important role, as shown by the augmented presence of these oligomers during the expression of mutations in AβPP or pre-senilin (Xia et al., 1997), as well as by their capacity to inhibit neuronal plasticity parameters (LTP) in vivo when microinjected into the brains of rodents (Walsh et al., 2002). Whether oligomeric amyloid-β is fundamental to disease pathogenesis is controversial and is discussed later.

Conversely, amyloid-β is not always present in the brains of cognitively normal elderly people. Whether this indicates that some individuals have efficient endogenous antioxidant defense systems and thus age more effectively, or whether such individuals may have supplemented their diets with antioxidants throughout their lifespan, compensating for age-related declines in antioxidant defenses, remains to be elucidated (Joseph et al., 1998, 1999; Bickford et al., 2000). If amyloid-β deposits serve an antioxidant function, this process will be recruited during times when oxidative stress is high and the endogenous antioxidant-defenses are compromised. On the other hand, if this system is efficient and is supported by exogenous antioxidant supplementation, the antioxidant effects of amyloid-β may not be necessary.

Some studies have suggested that there may be little or no neuronal loss during normal aging despite, as pointed out earlier, the presence of an increasing number of amyloid-β plaques (Long et al., 1999). Interestingly, even the hyperphysiologic levels of amyloid-β in mouse models of AD (Hsiao et al., 1996) only lead to senile plaque formation in middle-aged mice and, like their human counterparts, these mice show evidence of oxidative stress that precedes the amyloid-β deposits (Pappolla et al., 1998; Smith et al., 1998; Pratico et al., 2001; Drake et al., 2003). Taken together, these findings indicate that amyloid-β is a consequence of the pathogenesis that serves an antioxidant function.

The idea that amyloid-β is protective should not necessarily be surprising. Neuronal degeneration is associated with a number of responses including the induction of heat shock proteins (Smith et al., 1994; Anthony et al., 2003) that, like amyloid-β, show a relationship with cognitive decline. Yet only amyloid-β is considered pathogenic since amyloid-β is neurotoxic in vitro. On the other hand, as alluded to earlier, neurotoxicity in cultured cells may be an artifact of in vitro conditions (Rottkamp et al., 2001), since neither isolated senile plaques nor immobilized amyloid-β elicits neurotoxicity in vivo or in vitro (Frautschy et al., 1992; Canning et al., 1993; DeWitt et al., 1998). Thus, the capacity of amyloid-β to induce oxidative stress remains controversial (Walter et al., 1997) but may be akin to the known pro-oxidant effect of all antioxidants that are dependent on environmental conditions.

7.3 THERAPEUTIC STRATEGIES: AMYLOID-β

As detailed earlier, in the field of AD, the predominant hypothesis is the Amyloid Cascade Hypothesis, the original version of which proposed insoluble fibrillar amyloid-β as central to disease pathogenesis (Hardy and Higgins, 1992). Supporting this hypothesis, amyloid-β fibrils were found to be toxic in vitro (Lorenzo and Yankner, 1994) and, since then, considerable efforts have, and continue to be, spent on developing therapeutic modalities that target amyloid-β fibrils. In this regard, transgenic mice, that overexpress mutant forms of the amyloid-β protein precursor (AβPP) and develop extensive fibrillar amyloid-β senile plaque deposits (Hsiao et al., 1996), have become invaluable despite the fact that such mice, by not provoking neurodegeneration or other features of AD, would seem to disprove the hypothesis. That withstanding such transgenic mice have become mainstay screening tools for potential therapeutics and, of the many agents found to treat these mice (i.e., reduce amyloid-β plaque deposits and rescue cognitive deficits), none has raised as much enthusiasm and expectations as the vaccination strategy (Schenk et al., 1999; Buttini et al., 2005).

Around the same time that the vaccination approach was taken into human clinical trials, the original amyloid hypothesis (Hardy and Higgins, 1992) underwent a slight modification where the emphasis switched to oligomeric, rather than fibrillar, forms of amyloid-β (Hardy and Selkoe, 2002). Today, oligomeric amyloid-β is viewed, almost universally in the field, as the most toxic and, therefore, more important species (Lesne et al., 2006). What did this mean for the clinical trial that targeted fibrillar amyloid? Before this answer became apparent, the trial was suspended due to encephalitis in a small percentage of individuals, which was subsequently blamed on the adjuvant (Birmingham and Frantz, 2002). However, recently an alternate explanation may be more relevant. Specifically, while initial studies in mice supported vaccination-mediated decreases in both fibrillar and oligomeric species (Schenk et al., 1999), in human patients who received the vaccination, oligomeric amyloid may in fact increase (Patton et al., 2006). These findings essentially remove any "doubt" (sic) that amyloid is indeed responsible for AD (Lee et al., 2006). Simply, if patients improve, it is because amyloid-β (fibrils) are removed. On the other hand, if patients deteriorate, it is because of increases in (oligomeric) amyloid-β. Success or failure will equally support the amyloid-β hypothesis.

7.4 TAU

Like amyloid-β, tau accumulation in the form of neurofibrillary pathology has been predominantly viewed as detrimental. Yet, a recent transgenic mouse study from Karen Ashe and colleagues (Santacruz et al., 2005) goes a long way toward bringing light to this specific issue. These studies quite convincingly demonstrate that NFT-like accumulations of phospho-tau are not associated with neurodegeneration, echoing a recent similar conclusion from the Peter Davies group (Andorfer et al., 2005). While these studies implicate NFTs as beneficial, since they are disconnected with cell death, better evidence for the exoneration of phospho-tau or NFTs in the disease process is provided by human studies [reviewed in Lee et al., (2005)]. For example, in patients with AD, phospho-tau and NFTs are unlikely to be factors in the neurodegenerative process since, while both the total number and length of microtubules are significantly and selectively reduced in pyramidal neurons, such decrements are unrelated to NFTs (Cash et al., 2003). This poses the question as to what causes the neurodegeneration. One suggestion is that NFTs sequester abnormal phospho-tau microaggregates, thereby rendering them harmless (Binder et al., 2005). Therefore, much like microaggregates of amyloid-β (ADDLs) (Klein, 2002), the tau story has shifted to microaggregates of phospho-tau, which we have termed tau-derived diffusible ligands (TADDLs or TAuDDLs). The evidence for such TADDLs, much like their counterpart ADDLs (Lee et al., 2004), is scant but at least it keeps the prime suspects front and center. However, as we recently reviewed (Lee et al., 2005), there is far more evidence that phospho-tau, as TADDLs or NFTs, is a protective compensatory response mounted by neurons in an effort to stave off another driving pathogenic force. Indeed, oxidative damage not only chronologically precedes phospho-tau (Nunomura et al., 2000; Liu et al., 2005), but, supporting a protective response, such oxidative stress is significantly reduced in AD in neurons containing phospho-tau (Nunomura et al., 2001).

7.5 PARALLELS TO OTHER NEURODEGENERATIVE DISORDERS

Aberrantly (sic) folded proteins are common to a great number of neurodegenerative diseases and are, for the most part, vilified. The focus of Parkinson disease, Pick disease, and amyotrophic lateral sclerosis, for example, has been on Lewy bodies, Pick bodies, and spheroids, their respective protein components. However, the concept that such intracellular inclusions are manifestations of cell survival may be a common feature of all neurodegenerative diseases. Such a notion, while heretical to most, recently found support in a Huntington's disease model (Arrasate et al., 2004). In this neuronal model, cell death was mutant-huntingtin-dose- and polyglutamine-dependent; however, huntingtin inclusion formation correlated with cell survival. Thus, in this model, as in AD, inclusion formation represents adaptation, or a productive, beneficial response to the otherwise neurodegenerative process. Taken together with our studies, this represents a fundamental and necessary change in which pathological manifestations of neurodegenerative disease are interpreted.

7.6 SUMMARY

For most investigators in the field of AD, pathology has become equated with pathogenesis. As such, the two leading theories concerning the disease revolve around the amyloid-β of senile plaques and the phospho-tau of NFTs. Contrasting this viewpoint, a growing contingency suggests that pathology is not central to pathogenesis; rather, pathology may even be the antipathogenesis and, rather than causing the disease, the pathology may be protecting from the disease (Rottkamp et al., 2002; Smith et al., 2002; Lee et al., 2005). Given this, the classical notion of neurodegenerative disease pathology as signifying disease per se should be reorganized into a modern framework that recognizes the difference between cause and effect (Lee et al., 2003). Only through such an effort will the greatest potential for continued diagnostic and therapeutic advances be realized.

ACKNOWLEDGMENT

Work in the authors' laboratories is supported by the National Institutes of Health, the Alzheimer's Association, and by Philip Morris USA, Inc., and Philip Morris International.

REFERENCES

Andorfer, C., Acker, C.M., Kress, Y., Hof, P.R., Duff, K., and Davies, P. (2005) Cell-cycle reentry and cell death in transgenic mice expressing nonmutant human tau isoforms. *Journal of Neuroscience*, 25, 5446.

Anthony, S.G., Schipper, H.M., Tavares, R., Hovanesian, V., Cortez, S.C., Stopa, E.G. et al. (2003) Stress protein expression in the Alzheimer-diseased choroid plexus. *Journal of Alzheimer's Disease*, 5, 171.

Arrasate, M., Mitra, S., Schweitzer, E.S., Segal, M.R., and Finkbeiner, S. (2004) Inclusion body formation reduces levels of mutant huntingtin and the risk of neuronal death. *Nature*, 431, 805.

Arriagada, P.V., Growdon, J.H., Hedley-Whyte, E.T., and Hyman, B.T. (1992) Neurofibrillary tangles but not senile plaques parallel duration and severity of Alzheimer's disease. *Neurology*, 42, 631.

Bickford, P.C., Gould, T., Briederick, L., Chadman, K., Pollock, A., Young, D. et al. (2000) Antioxidant-rich diets improve cerebellar physiology and motor learning in aged rats. *Brain Research*, 866, 211.

Binder, L.I., Guillozet-Bongaarts, A.L., Garcia-Sierra, F., and Berry, R.W. (2005) Tau, tangles, and Alzheimer's disease. *Biochimica et Biophysica Acta*, 1739, 216.

Birmingham, K. and Frantz, S. (2002) Set back to Alzheimer vaccine studies. *Nature Medicine*, 8, 199.

Buttini, M., Masliah, E., Barbour, R., Grajeda, H., Motter, R., Johnson-Wood, K. et al. (2005) Beta-amyloid immunotherapy prevents synaptic degeneration in a mouse model of Alzheimer's disease. *Journal of Neuroscience*, 25, 9096.

Canning, D.R., McKeon, R.J., DeWitt, D.A., Perry, G., Wujek, J.R., Frederickson, R.C. et al. (1993) Beta-Amyloid of Alzheimer's disease induces reactive gliosis that inhibits axonal outgrowth. *Experimental Neurology*, 124, 289.

Cash, A.D., Aliev, G., Siedlak, S.L., Nunomura, A., Fujioka, H., Zhu, X. et al. (2003) Microtubule reduction in Alzheimer's disease and aging is independent of tau filament formation. *American Journal of Pathology*, 162, 1623.

Cuajungco, M.P., Goldstein, L.E., Nunomura, A., Smith, M.A., Lim, J.T., Atwood, C.S. et al. (2000) Evidence that the beta-amyloid plaques of Alzheimer's disease represent the redox-silencing and entombment of Abeta by zinc. *Journal of Biological Chemistry*, 275, 19439.

Davis, D.G., Schmitt, F.A., Wekstein, D.R., and Markesbery, W.R. (1999) Alzheimer neuropathologic alterations in aged cognitively normal subjects. *Journal of Neuropathology and Experimental Neurology*, 58, 376.

DeWitt, D.A., Perry, G., Cohen, M., Doller, C., and Silver, J. (1998) Astrocytes regulate microglial phagocytosis of senile plaque cores of Alzheimer's disease. *Experimental Neurology*, 149, 329.

Drake, J., Link, C.D., and Butterfield, D.A. (2003) Oxidative stress precedes fibrillar deposition of Alzheimer's disease amyloid beta-peptide (1–42) in a transgenic *Caenorhabditis elegans* model. *Neurobiology of Aging*, 24, 415.

Frautschy, S.A., Cole, G.M., and Baird, A. (1992) Phagocytosis and deposition of vascular beta-amyloid in rat brains injected with Alzheimer beta-amyloid. *American Journal of Pathology*, 140, 1389.

Giannakopoulos, P., Herrmann, F.R., Bussiere, T., Bouras, C., Kovari, E., Perl, D.P. et al. (2003) Tangle and neuron numbers, but not amyloid load, predict cognitive status in Alzheimer's disease. *Neurology*, 60, 1495.

Hardy, J.A. and Higgins, G.A. (1992) Alzheimer's disease: the amyloid cascade hypothesis. *Science*, 256, 184.

Hardy, J. and Selkoe, D.J. (2002) The amyloid hypothesis of Alzheimer's disease: progress and problems on the road to therapeutics. *Science*, 297, 353.

Hsiao, K., Chapman, P., Nilsen, S., Eckman, C., Harigaya, Y., Younkin, S. et al. (1996) Correlative memory deficits, Abeta elevation, and amyloid plaques in transgenic mice. *Science*, 274, 99.

Hyman, B.T. and Trojanowski, J.Q. (1997) Consensus recommendations for the postmortem diagnosis of Alzheimer disease from the National Institute on Aging and the Reagan Institute Working Group on diagnostic criteria for the neuropathological assessment of Alzheimer disease. *Journal of Neuropathology and Experimental Neurology*, 56, 1095.

Joseph, J.A., Shukitt-Hale, B., Denisova, N.A., Prior, R.L., Cao, G., Martin, A. et al. (1998) Long-term dietary strawberry, spinach, or vitamin E supplementation retards the onset of age-related neuronal signal-transduction and cognitive behavioral deficits. *Journal of Neuroscience*, 18, 8047.

Joseph, J.A., Shukitt-Hale, B., Denisova, N.A., Bielinski, D., Martin, A., McEwen, J.J. et al. (1999) Reversals of age-related declines in neuronal signal transduction, cognitive, and motor behavioral deficits with blueberry, spinach, or strawberry dietary supplementation. *Journal of Neuroscience*, 19, 8114.

Klein, W.L. (2002) Abeta toxicity in Alzheimer's disease: globular oligomers (ADDLs) as new vaccine and drug targets. *Neurochemistry International*, 41, 345.

Knowles, R.B., Gomez-Isla, T., and Hyman, B.T. (1998) Abeta associated neuropil changes: correlation with neuronal loss and dementia. *Journal of Neuropathology and Experimental Neurology*, 57, 1122.

Lee, H.G., Petersen, R.B., Zhu, X., Honda, K., Aliev, G., Smith, M.A. et al. (2003) Will preventing protein aggregates live up to its promise as prophylaxis against neurodegenerative diseases? *Brain Pathology*, 13, 630.

Lee, H.G., Casadesus, G., Zhu, X., Joseph, J.A., Perry, G., and Smith, M.A. (2004) Perspectives on the amyloid-beta cascade hypothesis. *Journal of Alzheimer's Disease*, 6, 137.

Lee, H.G., Perry, G., Moreira, P.I., Garrett, M.R., Liu, Q., Zhu, X. et al. (2005) Tau phosphorylation in Alzheimer's disease: pathogen or protector? *Trends in Molecular Medicine*, 11, 164.

Lee, H.G., Zhu, X., Nunomura, A., Perry, G., and Smith, M.A. (2006) Amyloid-beta vaccination: testing the amyloid hypothesis: heads we win, tails you lose! *American Journal of Pathology*, 169, 738.

Lesne, S., Koh, M.T., Kotilinek, L., Kayed, R., Glabe, C.G., Yang, A. et al. (2006) A specific amyloid-beta protein assembly in the brain impairs memory. *Nature*, 440, 352.

Liu, Q., Smith, M.A., Avila, J., DeBernardis, J., Kansal, M., Takeda, A. et al. (2005) Alzheimer-specific epitopes of tau represent lipid peroxidation-induced conformations. *Free Radical Biology and Medicine*, 38, 746.

Long, J.M., Mouton, P.R., Jucker, M., and Ingram, D.K. (1999) What counts in brain aging? Design-based stereological analysis of cell number. *Journals of Gerontology. Series A, Biological Sciences and Medical Sciences*, 54, B407.

Lorenzo, A. and Yankner, B.A. (1994) Beta-amyloid neurotoxicity requires fibril formation and is inhibited by congo red. *Proceedings of the National Academy of Sciences of the United States of America*, 91, 12243.

McLean, C.A., Cherny, R.A., Fraser, F.W., Fuller, S.J., Smith, M.J., Beyreuther, K. et al. (1999) Soluble pool of Abeta amyloid as a determinant of severity of neurodegeneration in Alzheimer's disease. *Annals of Neurology*, 46, 860.

Mirra, S.S., Heyman, A., McKeel, D., Sumi, S.M., Crain, B.J., Brownlee, L.M. et al. (1991) The consortium to establish a registry for Alzheimer's disease (CERAD). Part II. Standardization of the neuropathologic assessment of Alzheimer's disease. *Neurology*, 41, 479.

Nunomura, A., Perry, G., Pappolla, M.A., Wade, R., Hirai, K., Chiba, S. et al. (1999) RNA oxidation is a prominent feature of vulnerable neurons in Alzheimer's disease. *Journal of Neuroscience*, 19, 1959.

Nunomura, A., Perry, G., Pappolla, M.A., Friedland, R.P., Hirai, K., Chiba, S. et al. (2000) Neuronal oxidative stress precedes amyloid-beta deposition in Down syndrome. *Journal of Neuropathology and Experimental Neurology*, 59, 1011.

Nunomura, A., Perry, G., Aliev, G., Hirai, K., Takeda, A., Balraj, E.K. et al. (2001) Oxidative damage is the earliest event in Alzheimer disease. *Journal of Neuropathology and Experimental Neurology*, 60, 759.

Nunomura, A., Chiba, S., Lippa, C.F., Cras, P., Kalaria, R.N., Takeda, A. et al. (2004) Neuronal RNA oxidation is a prominent feature of familial Alzheimer's disease. *Neurobiology of Disease*, 17, 108.

Odetti, P., Angelini, G., Dapino, D., Zaccheo, D., Garibaldi, S., Dagna-Bricarelli, F. et al. (1998) Early glycoxidation damage in brains from Down's syndrome. *Biochemical and Biophysical Research Communications*, 243, 849.

Pappolla, M.A., Chyan, Y.J., Omar, R.A., Hsiao, K., Perry, G., Smith, M.A. et al. (1998) Evidence of oxidative stress and in vivo neurotoxicity of beta-amyloid in a transgenic mouse model of Alzheimer's disease: a chronic oxidative paradigm for testing antioxidant therapies in vivo. *American Journal of Pathology*, 152, 871.

Patton, R.L., Kalback, W.M., Esh, C.L., Kokjohn, T.A., Van Vickle, G.D., Luehrs, D.C. et al. (2006) Abeta peptide remnants in AN-1792-immunized Alzheimer's disease patients: a biochemical analysis. *American Journal of Pathology*, 169, 1048–1063.

Pike, C.J., Walencewicz, A.J., Glabe, C.G., and Cotman, C.W. (1991) Aggregation-related toxicity of synthetic beta-amyloid protein in hippocampal cultures. *European Journal of Pharmacology*, 207, 367.

Pratico, D., Uryu, K., Leight, S., Trojanoswki, J.Q., and Lee, V.M. (2001) Increased lipid peroxidation precedes amyloid plaque formation in an animal model of Alzheimer amyloidosis. *Journal of Neuroscience*, 21, 4183.

Rottkamp, C.A., Raina, A.K., Zhu, X., Gaier, E., Bush, A.I., Atwood, C.S. et al. (2001) Redox-active iron mediates amyloid-beta toxicity. *Free Radical Biology and Medicine*, 30, 447.

Rottkamp, C.A., Atwood, C.S., Joseph, J.A., Nunomura, A., Perry, G., and Smith, M.A. (2002) The state versus amyloid-beta: the trial of the most wanted criminal in Alzheimer disease. *Peptides*, 23, 1333.

Santacruz, K., Lewis, J., Spires, T., Paulson, J., Kotilinek, L., Ingelsson, M. et al. (2005) Tau suppression in a neurodegenerative mouse model improves memory function. *Science*, 309, 476.

Schenk, D., Barbour, R., Dunn, W., Gordon, G., Grajeda, H., Guido, T. et al. (1999) Immunization with amyloid-beta attenuates Alzheimer-disease-like pathology in the PDAPP mouse. *Nature*, 400, 173.

Smith, M.A., Kutty, R.K., Richey, P.L., Yan, S.D., Stern, D., Chader, G.J. et al. (1994) Heme oxygenase-1 is associated with the neurofibrillary pathology of Alzheimer's disease. *American Journal of Pathology*, 145, 42.

Smith, M.A., Hirai, K., Hsiao, K., Pappolla, M.A., Harris, P.L., Siedlak, S.L. et al. (1998) Amyloid-beta deposition in Alzheimer transgenic mice is associated with oxidative stress. *Journal of Neurochemistry*, 70, 2212.

Smith, M.A., Joseph, J.A., and Perry, G. (2000) Arson. Tracking the culprit in Alzheimer's disease. *Annals of the New York Academy of Sciences*, 924, 35.

Smith, M.A., Casadesus, G., Joseph, J.A., and Perry, G. (2002) Amyloid-beta and tau serve antioxidant functions in the aging and Alzheimer brain. *Free Radical Biology and Medicine*, 33, 1194.

Tiraboschi, P., Sabbagh, M.N., Hansen, L.A., Salmon, D.P., Merdes, A., Gamst, A. et al. (2004) Alzheimer disease without neocortical neurofibrillary tangles: "a second look." *Neurology*, 62, 1141.

Walsh, D.M., Klyubin, I., Fadeeva, J.V., Cullen, W.K., Anwyl, R., Wolfe, M.S. et al. (2002) Naturally secreted oligomers of amyloid beta protein potently inhibit hippocampal long-term potentiation in vivo. *Nature*, 416, 535.

Walter, M.F., Mason, P.E., and Mason, R.P. (1997) Alzheimer's disease amyloid beta peptide 25–35 inhibits lipid peroxidation as a result of its membrane interactions. *Biochemical and Biophysical Research Communications*, 233, 760.

Xia, W., Zhang, J., Kholodenko, D., Citron, M., Podlisny, M.B., Teplow, D.B. et al. (1997) Enhanced production and oligomerization of the 42-residue amyloid beta-protein by Chinese hamster ovary cells stably expressing mutant presenilins. *Journal of Biological Chemistry*, 272, 7977.

Yan, S.D., Yan, S.F., Chen, X., Fu, J., Chen, M., Kuppusamy, P. et al. (1995) Non-enzymatically glycated tau in Alzheimer's disease induces neuronal oxidant stress resulting in cytokine gene expression and release of amyloid beta-peptide. *Nature Medicine*, 1, 693.

8 Enhancement of Brain Retinoic Acid Levels

Ann B. Goodman, Peter McCaffery, Joana A. Palha, Claire Simons, and Arthur B. Pardee

CONTENTS

8.1 INTRODUCTION

The purpose of this chapter is to review the developing body of evidence suggesting the involvement of vitamin A and derivatives (retinoids, Figure 8.1) in the causality of late onset Alzheimer's disease (LOAD). Further, we suggest treatment strategies, including the administration of retinol or retinyl esters, retinoic acid (RA) receptor agonists, and engineering the increase of RA in the brain through targeting of the RA-inactivating enzymes, the CYP26s.

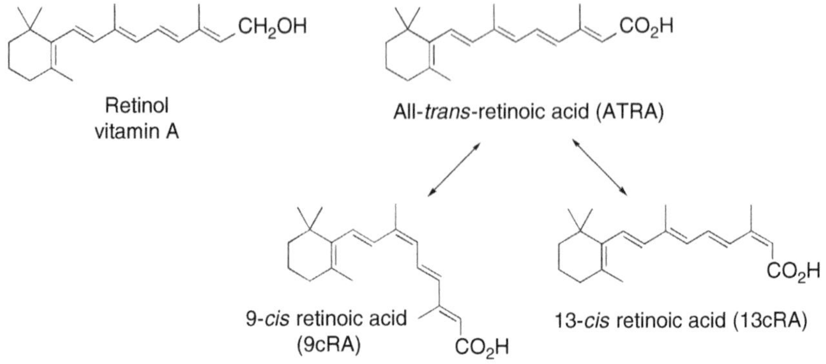

FIGURE 8.1 Retinoids—Vitamin A (retinol) and isomers of retinoic acid (RA), RA, 9cRA, and 13cRA.

8.1.1 ALZHEIMER'S DISEASE (AD) AND RETINOIDS

LOAD, the most common cause of dementia in later life, is a worldwide problem for those affected, for their families, and for society at large. LOAD is essentially a disease of neuronal insufficiency and loss. The loss of neurons creates defects in brain functions including deficiencies in memory, and the cell death is connected to the creation of extracellular amyloid protein plaques and intracellular tau protein tangles. (Amyloid plaques are discussed in Chapter 4 and tangles in Chapter 6.) Effective treatments for preventing the disease, slowing its progression, or alleviating its symptoms are sorely needed, but this is hampered by the lack of complete understanding of the underlying molecular mechanisms of the disease (Hardy and Selkoe, 2002).

LOAD involves both genetic risk factors (Cacabelos, 2005) and environmental influences (Hendrie et al., 2001). The hypothesis of this chapter is that LOAD is influenced by the availability in the brain of RA, the final product of the retinoid metabolic cascade that originates from ingested vitamin A (Figure 8.2). RA combines with its receptor proteins and activates transcription of genes that modulate amyloid plaque formation and brain cell death. RA carries out the main neurological functions of retinoids (Mey and McCaffery, 2004; Lane and Bailey, 2005), regulates cell proliferation and migration, and causes cell death at high concentrations (Wang et al., 2005a). RA also mediates apoptosis, involving release by phospholipases of arachidonic acid (Farooqui et al., 2004). RA promotes apoptosis of some hNT neurons (Zigova et al., 2001) and increases survival of cholinergic neurons but not GABAergic neurons (Wuarin and Sidell, 1991). Disruption of the retinoid-signaling pathway causes a deposition of amyloid beta (Aβ) in the adult rat brain (Corcoran et al., 2004). Transcription of amyloid precursor protein (APP) is subject to regulation by RA (Konig et al., 1990; Fukuchi et al., 1992; Hung et al., 1992; Lahiri and Nall, 1995; Yang et al., 1998). Upregulation of retinoid receptor expression alleviates performance deficits in aged mice, supporting the role of retinoids in the cognitive decline associated with aging (Etchamendy et al., 2001). A specific decline (20%–30%) in levels of mRNAs of the retinoid receptors RARβ and RXRγ in the

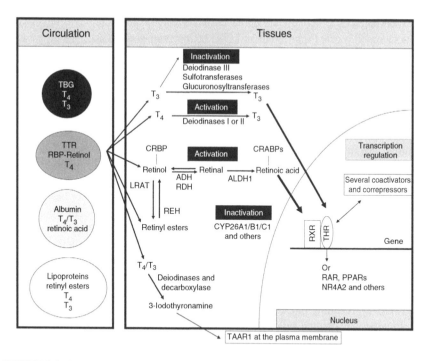

FIGURE 8.2 Overview of thyroid hormone and retinoid transport, metabolism, and mechanism of action. (From Palha, J.A. and Goodman, A.B., *Brain Res. Rev.*, 51, 61, 2006. With permission.)

hippocampus of aged mice was reported (Etchamendy et al., 2001). Described below is the way in which environmental, dietary, genetic, or metabolic induced changes in RA may influence LOAD.

8.1.2 EARLY AND LATE ONSET AD

Early onset AD (EOAD) primarily has a genetic cause. Mutations determine plaque formation and cell death. To date, three mutated genes, APP, presenilin-1 (PS-1), and presenilin-2 (PS-2), have been identified as having direct effects in EOAD (Cacabelos, 2005). Aβ accumulates extracellularly from cleavage of APP by secretases. Insulin degrading enzyme (IDE) catalyzes plaque degradation, as reviewed (Vekrellis et al., 2000; De Strooper and Wodgett, 2003). There is also intracell accumulation of phosphorylated tau protein. Although beta-site APP-cleaving enzyme (BACE) (Cai et al., 2001) is required in the molecular processing to produce amyloid plaques and neurofibrillary tangles, with rare exceptions, other genes have not yet been generally found to be directly mutated in classical EOAD or LOAD. However, the apolipoprotein E4 (APOE4) variant increases susceptibility to both early and late diseases (Mahley and Rall, 2000; Cedazo-Minguez and Cowburn, 2001; Poirier, 2005).

In contrast to EOAD, both genetic and environmental (dietary) factors are involved in LOAD. This difference is similar to heart disease in which early onset is due to mutations that increase cholesterol accumulation, and late onset heart attack is exacerbated by excess dietary cholesterol and may be prevented by drugs. Experiments support a role for retinoids in LOAD (Corcoran et al., 2004). We propose that both genetic and environmental causes can result in a reduction in RA signaling in the brain. This decline can be brought about by mutation of genes that are involved in retinoid uptake from the diet, retinoid transport to the brain, metabolism of retinol to RA, binding of RA to its receptors, as well as epigenetic changes that modify RA-activated gene transcriptions. Environmental factors that may reduce RA signaling include diet, drugs, and chemicals that influence the enzymes of RA production and removal. All of these factors are modified in the course of natural aging, leading to a reduction in retinoid levels with age (Connor and Sidell, 1997; Jimenez-Jimenez et al., 1999; Quadro et al., 1999; Bourdel-Marchasson et al., 2001; Mecocci et al., 2002). Although mutations of the genes identified in EOAD have not been found in LOAD, RA and the retinoid nuclear receptors directly or indirectly regulate the transcription of genes identified as causal in EOAD, including APP, PS-1, and PS-2 (reviewed in Goodman and Pardee [2003]). There we proposed that allelic variants of genes of the retinoid cascade will cause pathological modulation of the transcriptional expression of such EOAD genes and of other genes now thought to be involved in Alzheimer's pathology, including BACE, IDE, and microtubule-associated protein tau (MAPT), resulting in LOAD.

In what way could a deficiency in RA influence the brain in a way that promotes AD? Much of what is known about the influence of RA on the CNS comes, not from studies of the aging brain but from analysis of the embryonic CNS. Surprisingly though these investigations at the opposite pole of the life span may have provided some important clues as to its function in the aging adult brain.

8.2   RA, NEURONAL PLASTICITY, AND AD

8.2.1   FUNCTIONS FOR RA IN THE EMBRYONIC CNS

An abundance of studies have shown that retinoids control differentiation of the brain in the embryo. Insights into the function of RA in the adult brain, and possibly how RA may be involved in AD, have come from the study of its function in the developing brain. Since the 1930s it has been recognized that vitamin A is necessary for embryonic development. CNS abnormalities in vitamin A-deprived embryos have signified the reliance of the developing brain on this vitamin (Hale, 1933). Extensive studies over the decades since have shown that RA is the predominant metabolite of vitamin A necessary to control two crucial functions in the embryonic brain—morphogenesis and differentiation. Morphogenesis is the creation of shape and pattern in the embryo. RA can organize pattern in the embryo by the formation of gradients. The varying concentration of RA at different points within the gradient will induce different assemblages of gene expression. For instance, for the developing hindbrain, RA is synthesized by the synthetic enzyme RALDH2 in the mesenchyme around the spinal cord and posterior to the hindbrain. RA diffuses from this site into the hindbrain creating a gradient from posterior to anterior. This gradient is essential for setting up the correct segmentation of the hindbrain, which subdivides into compartments known as rhombomeres (Gavalas, 2002). RA helps guide specification of the posterior rhombomeres through induction of the *hox* group of homeobox containing transcription factors (Krumlauf et al., 1993) specifically *hoxb* genes (Gavalas and Krumlauf, 2000; Oosterveen et al., 2003). The transcription of Hox is regulated by retinoids and the *hoxb1* gene contains an RA response element (Huang et al., 2002). Hox clusters are colinear, suggesting a role for chromosome structure in gene regulation.

The RA catabolic enzymes of the CYP26 family limit the spread of RA in the anterior hindbrain (Maden, 1999) as well as control the timing of RA action (Sirbu et al., 2005). Alternatively to RA's action within a gradient, it can be limited to zones of synthesis, regions often separated by bands of catabolism. This is the case for the retina where two regions of synthesis, in the dorsal and ventral retina, each determined by a different synthetic enzyme, are separated by an equatorial stripe of CYP26. This divides the retina into sections of high, low, and moderate RA levels controlling expression of different genes depending on the concentration of RA (McCaffery et al., 1999). RA synthesis in the adult brain may create both patterns of RA distribution, generating gradients by diffusion of RA synthesized by the meninges (Wagner et al., 2002; Sakai et al., 2004) and choroid plexus (Yamamoto et al., 1998) and also creating localized regions of relatively high RA concentration such as in a subpopulation of periglomerular cells in the olfactory bulb (Wagner et al., 2002).

The second function for RA in the embryonic CNS is regulation of neuronal differentiation. Differentiation occurs through steps in which, at each new cell division, the daughter cell is progressively more committed toward a mature fate. Different growth factors act at different steps of this series. RA acts at the beginning of the sequence, initiating differentiation and inducing growth factor receptors such as those for neurotrophins (Kaplan et al., 1993; Xie et al., 1997) as well as TGFβ (Yoshizawa et al., 1998) and EGF (Joh et al., 1992; Sizemore et al., 1998). The cells are then responsive to such growth factors, which can direct later events of differentiation.

Studies with embryonic cell lines have indicated that RA induces the expression of several genes thought to be involved in Alzheimer pathology. Teratocarcinoma cell lines induced to differentiate into neurons by RA show induction of PS-1 or PS-2 (Ren et al., 1999; Culvenor et al., 2000). RA also promotes the expression of the TGF-β2 receptor in NT2 cells, allowing TGF-β2 to protect the differentiated cells from the 25–35 Aβ peptide fragment (Ren et al., 1997). RA induces TGF-β2 expression from a very wide range of murine embryonal carcinoma and embryonic stem cells (Mummery et al., 1990) whereas early embryonic deprivation of RA results in a reduction of TGF-β1 (Bavik et al., 1996). Further indicative of the interrelationship between the RA and TGF-β-signaling pathways during development is the finding that TGF-β regulates expression of cellular RA binding protein type I and II (CRABPI and II) as well as the RA receptor RARβ (Nugent et al., 1995).

As a result of increased responsiveness to growth factors, RA will promote later steps of neuronal differentiation and trigger neurite outgrowth in progenitors (Clagett-Dame et al., 2006). RA may also act more directly to promote neurite outgrowth via activation of a number of kinase pathways including Cdk5 (Lee and Kim, 2004) and phosphatidylinositol 3-kinase (Lopez-Carballo et al., 2002; Pan et al., 2005). As such, RA can act at several points of neurogenesis, promoting the initial steps of neuronal birth but also guiding later stages of differentiation. As described below, this points to a role for RA in the adult brain in regulating neuronal plasticity.

8.2.2 Regions of RA Signaling in the Adult Brain

Much less is known about the function of RA in the adult brain. The regulatory roles of particular genes are often revealed from the abnormalities or diseases that result from gene mutation in humans or animal models. Little genetic evidence though has pointed to a role for RA in any neural disorder. However, any such disorder is most likely hidden because of the absolute requirement of the embryo for RA; any large change in RA signaling will inevitably result in profound embryonic defects in neural development, masking any effects on the mature brain. In several examples where RA has been associated with an adult disorder, the effects of RA are likely to occur primarily during embryonic development. In the case of schizophrenia and autism, both disorders are believed to include an essential defect that occurs in the embryo (Akshoomoff et al., 2002; McCaffery and Deutsch, 2005; Rehn and Rees, 2005), whereas in the case of Parkinson's disease, where a link with RA has been suggested from the motor defects in RA receptor null mutants (Krezel et al., 1999), these defects may result from a decline in dopamine receptor expression in the embryonic striatum (Valdenaire et al., 1998; Wang and Liu, 2005).

It is clear that RA has a much less widespread role in its control of adult brain function as compared with its function in the embryo. This is revealed by the decline in RA signaling seen in transgenic RA reporter mice. These reporter mice indicate regions of RA signaling and either use a lacZ reporter linked to an RA response element (RARE/lacZ) (Rossant et al., 1991; Colbert et al., 1993) or make use of an RA receptor ligand binding domain linked to a Gal-4 DNA-binding domain acting on an appropriately controlled reporter gene (Mata de Urquiza et al., 1999). There is a gradual decline in the strength of RA signaling as indicated by the reduction in RA reporter response as development proceeds. The regions of RA signaling in the adult brain become progressively more restricted with two main regions of RA signaling being the olfactory bulb (Thompson et al., 2002) and the hippocampus (Misner et al., 2001; Sakai et al., 2004). This is despite the continued expression of RA receptors (RARs) in the adult (Krezel et al., 1999) suggesting that other factors result in this decline in RA signaling. The decline in RA is not due to limiting amounts of ligand for even if RA is injected into the reporter animals there is still no increase in RA reporter activity outside the regions in which signaling already occurs. This loss of signal may be due to a decrease in coactivators or increase in corepressors necessary to control RA receptor mediated transcription (Bastien and Rochette-Egly, 2004). In the absence of ligand, the RARs can become active repressors (Weston et al., 2003) and transcriptional repression may be an important function for the RARs in the adult brain.

The repressor activity of the RARs in the absence of RA means that the distribution of these receptors is not a straightforward indicator of where RA acts. Better indicators are the binding proteins that retain these lipids in the cytoplasm. CRABPI and II are the cytoplasmic binding proteins for RA (Figure 8.2). Although there is no absolute requirement for these proteins, because their mutation in mice results in no obvious phenotype (Lampron et al., 1995) they enter the nucleus (Gustafson et al., 1996; Gaub et al., 1998; Sessler and Noy, 2005) and, at least for CRABPII, associate with the RA receptor transcriptional complex (Delva et al., 1999). One function for these binding proteins may then be to transport RA into the nucleus and they presumably modulate the strength of the RA signal. CRABPI may also assist in delivery of RA to the catabolizing enzymes that degrade this lipid (Boylan and Gudas, 1992; Chen et al., 2003). This type of carrier function may also be true for the cellular retinol binding proteins I and II (CRBP I and II) that carry retinol to either the retinol dehydrogenases to promote RA synthesis or to the retinol esterases for retinol storage (Napoli, 1999). Again, although null mutation of CRBP does not result in gross changes in CNS development (Ghyselinck et al., 1999), these binding proteins exist to provide a fine level of control for RA signaling. Both CRABPs and CRBPs are strongly expressed in local regions of the adult brain and are particularly strongly expressed in the olfactory bulb, striatum, and hippocampus (Zetterstrom et al., 1994; Zetterstrom et al., 1999).

Further indicators of regions of RA function are the sites of RA synthesis that express the RA synthesizing enzyme retinaldehyde dehydrogenase (RALDH). In the adult brain three such enzymes have been identified—RALDH1, 2, and 3. Subsets of dopaminergic neurons of the ventral tegmental area express RALDH1, which is transported along axons to terminals in the striatum (McCaffery and Dräger, 1994), whereas RALDH3 is strongly expressed in the olfactory bulb (Wagner et al., 2002). The hippocampus, a third region of high RA signaling, receives RA by diffusion from the meninges and the choroid plexus, which strongly express RALDH2 and lie alongside the dentate gyrus (Sakai et al., 2004; McCaffery et al., 2006).

8.2.3 RA AND NEURONAL PLASTICITY IN THE BRAIN AND IN AD

Thus, a sophisticated system has been developed to meticulously control the regional intensity of RA signaling in the adult brain. Several studies have noted that these regions, including the hippocampus, olfactory bulb, and striatum, are areas high in "neural plasticity" (Zetterstrom et al., 1999; Thompson et al., 2002; Husson et al., 2004; Mey and McCaffery, 2004). Plasticity is defined as the brain's capacity for self-reorganization via its ability to modulate its neural connections. Plastic changes in the adult brain include processes similar to the events that fashion the embryonic brain such as modification of synapses, remodeling of the dendritic tree, and birth of new neurons (neurogenesis). The maintenance of RA in regions of high plasticity is not surprising because RA can recapitulate its role in the developing brain in these areas of neural flexibility. For instance, vitamin A is necessary for the changes in synaptic efficiency measured as long-term potentiation (LTP) and long-term depression (LTD) (Misner et al., 2001; Crandall et al., 2004) promoting both dendritic outgrowth (Clagett-Dame et al., 2006) and influencing adult neurogenesis (Crandall et al., 2004; Wang et al., 2005b). Reduction in RA signaling will impair neural plasticity, which in turn will result in higher susceptibility to any disorder that also acts to degrade this quality. AD has been suggested to be such a disorder and it has been noted that impairment of synaptic plasticity may be a key event that results in cognitive difficulties in AD (Flood and Coleman, 1990; DeWitt and Silver, 1996; Trinchese et al., 2004). Arendt (2003) correlated the degeneration in AD with those neurons that exhibit a high degree of plasticity and proposed that neurons with a high potential for plasticity might be predisposed to the development of neurofibrillary tangles. Mesulam (2000) proposed a "plasticity burden" on neurons in AD. Such a burden may derive from changes in the proteins that predispose to AD; including APP and PS both of which regulate plasticity in the adult brain (Turner et al., 2003; Wines-Samuelson and Shen, 2005). Interestingly, both are also inducible by RA; APP directly (Konig et al., 1990; Lahiri and Nall, 1995) and PS indirectly (Ren et al., 1999; Culvenor et al., 2000) perhaps as part of RA's control of plasticity. A reduction in RA would be predicted to be a risk factor for AD by adding a "plasticity burden" to neurons already sensitized due to damaging mutations in, for instance, APP or PS. One mechanism by which this senstitization could take place is through loss of RA driven growth factor receptor expression such as the reduction in the normally protective TGF-β2 receptor (Ren et al., 1997). Supporting the idea that RA decreases in AD are the multiple reports of lowered levels of antioxidants, including retinoids, in serum or plasma of LOAD patients compared with controls (Jimenez-Jimenez et al., 1999; Bourdel-Marchasson et al., 2001; Mecocci et al., 2002). These lowered levels appear to be specific to LOAD, because no such differences are found in a general population of aged humans with memory impairment (Perkins et al., 1999).

AD has also been reported to include an aberrant *increase* in neuritic outgrowth (Geddes et al., 1985; Cotman and Anderson, 1988; Ihara, 1988; Geddes and Cotman, 1991; Bowser et al., 1997; Mikkonen et al., 2001) with an increase in some proteins involved in plasticity, such as polysialylated neural cell adhesion molecule (Mikkonen et al., 1999). Recently, AD has been shown to lead to an increase in neurogenesis in the hippocampus (Jin et al., 2004). Such changes may represent an attempt by the brain to self-repair and may include a regional upregulation of RA. Connor and Sidell (1997) reported a local increase in RALDH in the hippocampus but not in the frontal cortex and proposed that the RALDH increase may be in response to an initial decline in RA levels. Although no reduction in vitamin A could be measured in the study of Connor and Sidell, we propose here that the key factor may be a decline in RA signaling, which may result from a mutation in RARs or binding proteins just as much as a decline in retinoid levels themselves. This attempt to restore RA levels may be part of the disease process itself and may help to explain why this disorder can show a complex picture of altered neural plasticity with some features of decreased and other features of increased plasticity.

In adults, dietary retinoid status has marked effects on neuronal functioning, memory, and neuronal plasticity (Chiang et al., 1998; Takahashi et al., 1999; Misner et al., 2001). During differentiation of embryonal carcinoma cells, two cycles after RA addition epigenetic expression of serotonin receptor 2C (Htr2c) is strongly increased, involving DNA duplication and a chromatin remodeling process (Bancescu et al., 2004). Lowered levels of retinoid in LOAD may thus alter the normal processes of neuronal repair and remodeling.

8.3 GENETICS OF LOAD

We have hypothesized that transcription of genes that are mutated in EOAD are modulated in LOAD by allelic variants of genes that alter the availability of retinoid in target tissues (Goodman and Pardee, 2003). This suggestion is supported by other work (Corcoran et al., 2004). These genes are involved in all aspects of RA signaling, including the transport of retinoids to the brain, metabolism of retinoids, whereas others are retinoid target genes. RARs plus their ligands regulate the direct or indirect expression of genes in a variety of cell types in the brain. Genes relevant to AD include MAPT (Fukuchi et al., 1992; Heicklen-Klein et al., 2000), APP (Konig et al., 1990; Fukuchi et al., 1992; Hung et al., 1992; Lahiri and Nall, 1995; Yang et al., 1998), PS2 (Culvenor et al., 2000), and BACE (Satoh and Kuroda, 2000) among others. The IDE gene contains an RARα response element in its promoter and transcription is regulated by RA (Melino et al., 1996).

APP contributes to Aβ synthesis, but is not itself neurotoxic and promotes neurite outgrowth (Konig et al., 1990; Yang et al., 1998). APP is also activated by TGF-β, a cytokine centrally involved in LOAD plaque formation (Burton et al., 2002), brain injury, and inflammatory responses (Grammas and Ovase, 2002). Terminally differentiated neurons expressing TGF-β2 receptors appear to be protected from Aβ toxicity by the administration of TGF-β2 (Ren et al., 1997), which reduces plaque burden in transgenic mice (Wyss-Coray et al., 2001). TGF-β2 pathways are upregulated by RA and diminished under conditions of retinoid deficiency (Freemantle et al., 2002). In retinoid-deficient rats, TGF-β2 expression declines but is restored in tissue-specific fashion by the application of RA (Glick et al., 1991). Thus, RA may increase production of APP in normal neurons in part via the TGF-β pathway involving SMAD4, which is stained strongly in AD brain (Burton et al., 2002).

RA has been shown to regulate the MAPT promoter (Heicklen-Klein et al., 2000). Inspection of the nucleic acid sequence of MAPT (Goedert et al., 1989) indicates that it contains an RAR response element TGAACxxTGAAC (Duester et al., 1991), beginning at 59,216. There are also four different TGACC motifs throughout this gene, which may confer retinoid response (Vasios et al., 1991; Hall et al., 1992). This suggests that the transcriptional activation of these elements may be modified by variants of one or another of the RARs or by the availability of RA.

8.3.1 Genomic Colocalizations Suggest a Role of Retinoids in LOAD

Of the several chromosomal loci identified by genome scans, chromosomes 10q23 and 12q13 are the most frequently associated with LOAD (Bertram and Tanzi, 2001; Myers and Goate, 2001; Sorbi et al., 2001). Genome screens at LOAD loci have not unequivocally identified any particular genes; however, remarkably, important gene (s) related to retinoids are found at each of these loci (Table 8.1). The functions of these genes are discussed below.

Significant genetic linkages to LOAD are demonstrated for markers close to four of the six RARs: RA receptor gamma (RARγ) at 12q13.13, retinoid X receptor β (RXRβ) at 6p21.3, RXRγ at 1q23, and RARα at 17q21. Three of the four retinol-binding proteins (RBPs) at 3q23 and 10q23 and the RA-degrading cytochrome P450 enzymes at 10q23 and 2p13 map to AD linkages.

Chromosome 12q13.13 presents strong evidence for LOAD linkage (Schellenberg et al., 2000; Tanzi and Bertram, 2001; D'Introno et al., 2005; Liang et al., 2006). Low-density lipoprotein receptor-related protein 1 (LRP1) is a proposed candidate gene (Zerbinatti and Bu, 2005), but its role has not been clearly established (Pritchard et al., 2005), suggesting that another gene in the vicinity, as yet unspecified and unknown, is causal (Scott et al., 1998; Lambert et al., 1999; D'Introno et al., 2005).

TABLE 8.1
Chromosomal Positions of Retinoid Cascade Genes and AD Linkages or Associations

Retinoid Locus		LOAD Locus				
Band Gene	kb	Band Gene or Marker	kb	cM	LOD Significance	Reference
12q13		12q13				
RARγ	53,518–53,528	D12S345–D12S78	32,352–104,143	55.25–111.87	LOD 6.374	Poduslo and Yin (2001)
RDH5	56,187–56,192	D12S96	53,147	67.16	$P = 0.001$	Rogaeva et al. (1998)
RODH	57,370–57,405	D12S390	52,901	67.17	LOD 2.3	Pericak-Vance et al. (1998)
		D12S1632	56,979	71.7	LOD 2.43 APOE4–	Scott et al. (2000)
		LRP1	57,747–57,764	—	O.R. 1.8; $P = 0.01$	Niederreither et al. (2002)
					No association	Boussaha et al. (2002)
10q23		10q23				
CYP26A1	94,067–94,071	IDE	93,434–93,555	—	$P = 0.04$	Bertram et al. (2000)
					Disequilibrium No linkage	Abraham et al. (2001)
CYP26C1	94,811–94,818					
RBP4	94,585–94,595	D10S583	93,590	115.27	Z max 2.8	Bertram et al. (2000)
CYP2C9	95,932–95,982				$P = 0.008$	Ait-Ghezala et al. (2002)
CYP2C8	96,030–96,062	D10S1239	102,430	121.81	LOD 2.62	Li et al. (2002)
2p12		2p13				
CYP26A2	72,567–72,586	D2S1356	43,542	64.29	MLS 3.52	Bacanu et al. (2002)
P450RAI-2						

Gene (band)	kb	Marker (band)	kb	cM	Statistics	Reference
17q12–q21 RARα	40,640–40,688	17q12–q21 D17S1787	41.835	70.08	MLOD 5.51	Rademakers et al. (2002)
1q21.3–q23 CRABP2	152,434–152,441	1q21.3–q23 D1S1595–D1S2844	151.436–160.583	148.85–160.43	No mutations or SNPs	Yu et al. (2000)
RXRγ	161,021–161,065			170	MLS 2.67	Kehoe et al. (1999)
		D1S518	182.978	188.02	P = 0.03	Zubenko et al. (1998)
6p21.3 RXRβ	33,158–33,165	6p21.3 TNFα	31,597–31,600	51.31	NPL 2.3	Collins et al. (2000)
					P = 0.03 APOE4+	McCusker et al. (2001)
		D6S1019	38.975	—	LOD 1.3	Pericak-Vance et al. (1998)
3q23 RBP2 RBP1	140,000–140,028 140,065–140,088	3q23 D3S3554–D3S1569	140.435–144.219	146.07–150.8	MLOD 4.17 plaque only	Poduslo et al. (1999)
9q34.3 RXRα	129,062–129,102	9q34.2	129,062–129,102		MLS 1.96	Pericak-Vance et al. (2000)

Source: From Goodman, A.B. and Pardee, A.B., *Proc. Natl. Acad. Sci. U.S.A.,* 100, 2901, 2003. With permission.

Chromosomal positions as band, kilobase (kb), centiMorgan (cM) for selected genes of the retinoid cascade and genes or markers linked to AD at those loci. Markers and genes were located using the Locus Link database (http://www.ncbi.nlm.nih.gov/LocuLink) posted as of January 22, 2003. Where available, the DeCode position was reported.

The RARγ at 12q13.13 position 51,890–51,912 kb lies within 1 Mb of LOAD-linked markers D12S368, D12S96, and D12S390 (Pericak-Vance et al., 1998; Rogaeva et al., 1998; Liang et al., 2006) and between two extensively researched candidate genes (D'Introno et al., 2005), specifically CP2 at position 49,774–49,852 kb and LRP1 at position 55,808–55,893 (Human Genome Browser Gateway http://genome.ucsc.edu/cgi-bin/hgGateway). Also in the 12q13 band are clustered five of the seven retinol dehydrogenases, which reversibly convert retinol to retinal.

Chromosome 10q23 harbors the marker D10S583 within the IDE gene (Bertram et al., 2000). This marker is significantly linked to LOAD, and IDE is a potential candidate as it degrades Aβ (Edbauer et al., 2002). Although extreme linkage disequilibrium is evident at the IDE locus, at least three groups have found no evidence for linkage of IDE itself to LOAD, and no polymorphisms in IDE have been significantly associated with LOAD (Abraham et al., 2001; Boussaha et al., 2002; Feuk et al., 2005; Nowotny et al., 2005), suggesting that transcriptional regulation rather than translation of the IDE protein may increase vulnerability to LOAD. CYP26A1 is 417 kb from IDE. CYP26A1 causes the hydroxylation and degradation of all-*trans* (AT)RA (Niederreither et al., 2002), and thus controls the availability of RA. A second gene, CYP26C1 is located just distal to CYP26A1. Also included in the linkage region is plasma RBP4, the primary retinol transporter (Goodman, 1984). RBP4 is increased in type 2 diabetes and obesity (Yang et al., 2005), two conditions associated with increased risk for LOAD.

Chromosome 2p13 has been linked to AD with psychosis (Bacanu et al., 2002). A third RA-inactivating enzyme, CYP26B1, is at chromosome 2p13. This CYP is most strongly expressed in the adult cerebellum and pons, and also elsewhere in the brain (Trofimova-Griffin and Juchau, 2002). Importantly, this recent report now establishes genetic links to both CYP26 RA-degrading enzyme chromosomal genetic loci. We suggest CYP26A1, CYP26B1, and CYP26C1 as candidates in LOAD.

Chromosome 17q21 is the locus of RARα immediately upstream of the anonymous marker D17S1787, which has been recently linked to MAPT-negative frontal lobe dementia in a single family with a multiple logarithm of odds score of 5.51. This score is among the highest obtained for any dementia linkage. LOAD could not be excluded in 4 of the 12 cases within this family. Extensive mutation analysis at 17q21 of MAPT and Saitohin, another AD candidate gene, excluded these two genes, leading the authors to suggest that an unknown gene in the region is responsible (Rademakers et al., 2002).

Chromosome 1q21–23 locus is linked to LOAD in two genome scanning studies (Zubenko et al., 1998; Kehoe et al., 1999). CRABPII and RXRγ are within the linked region. Both are highly expressed in brain (Chiang et al., 1998; Yamamoto et al., 1998). In a LOAD search, no mutations or polymorphisms were detected in an interval including CRABP2 (Yu et al., 2000), but RXRG lies just outside of the 14 centiMorgan region sequenced in this study.

Chromosome 6p21.3 is associated with LOAD in at least three studies (Pericak-Vance et al., 1998; Collins et al., 2000; McCusker et al., 2001). Within this band and close to the linked markers is RXRβ.

Chromosome 3q23 is strongly linked to LOAD in one study (Poduslo et al., 1999). Cellular RBP1 and RBP2 map to the region.

The above finding of colocalization of LOAD loci and retinoid-related genes suggests that retinoids have a role in the disease. How could mutations of these retinoid-related genes be involved in AD, as suggested by their colocalizations with LOAD loci? We propose that rare mutations of the retinoid receptors RARγ at 12q13, RXRγ at 1q23, RARA at 17q21, and RXRβ at 6p21.3 can coordinately dysregulate LOAD candidate genes, according to the availability of RA, because these receptors' genes have been repeatedly linked to AD loci (Table 8.1). For example, the RARγ/RXRγ heterodimer activated by RA upregulates the expression of the RA-degrading enzyme CYP26A1 (Loudig et al., 2000, 2005). RA thereby influences its own removal.

As Table 8.1 shows, there is a consistent relationship between areas in the genome repeatedly linked to LOAD and the loci of genes in the retinoid metabolic cascade, retinoid transporters, the RBPs, and the retinoid nuclear receptors. However, none of the loci connected to familial EOAD is near loci of genes of the retinoid cascade or to the retinoid nuclear receptors, with the exception of one or two rare mutations in single pedigrees. These data show that the colocalizations found in LOAD are not random. We propose these retinoid genes at LOAD-linked loci as specific candidates for LOAD.

The above genetic results support our hypothesis that hypofunctioning of retinoid is a key factor in development of Aβ toxicity. This is particularly relevant in light of the age-related decline in retinoid supply or retinoid signaling in both normal and LOAD samples (Jimenez-Jimenez et al., 1999; Bourdel-Marchasson et al., 2001; Mecocci et al., 2002). To determine whether genetic variations in the retinoid genes at loci linked to AD increase vulnerability, it will be necessary to sequence these specific candidates in coding (Jimenez-Jimenez et al., 1999; Werner and Deluca, 2002) and promoter regions. Even lacking the discovery of mutations in these genes, relative expression of the genes may be dysregulated in brain regions relevant to LOAD. For example, specific upregulation of RARα has been found in granule cells of the dentate gyrus in schizophrenia (Rioux and Arnold, 2005).

8.4 SUPPLY OF RA TO THE BRAIN

The retinoid metabolic cascade (Figure 8.2) determines the amount of RA in brain. The amounts of RA in brain are determined by three stepwise factors; (1) the supply of retinoids in the diet, (2) their transport to the brain, and (3) production and removal of the functional end product of the cascade, RA. Mutations that affect the retinoid cascade could alter the level of RA in the brain, and mutations in genes coding for the RARs and RXRs could modify transcriptional functioning of genes activated by the retinoid receptor complexes. In this section, a description of the components of the retinoid cascade is provided with special reference to those at AD loci.

8.4.1 Diet and RA

Vitamin A, which is essential for life, cannot be synthesized by humans, and therefore must be supplied in the diet, either as ingested beta-carotene from plant products, preformed retinyl esters and retinol from animal origin (liver and fish oil)

(Blomhoff et al., 1992), or as chemically synthesized retinol (Goodman et al., 1966; Goodman, 1984). Retinoid supplementation is prominent in the rescue of function of retinoid-deficient cells and tissues (Glick et al., 1991; Chiang et al., 1998; Takahashi et al., 1999; Misner et al., 2001; Etchamendy et al., 2001).

A diet high in fat correlated with increase of AD and increased dietary RA was associated with a reduced risk of LOAD (Smith et al., 1999). All-*trans*, 9-*cis*, and 13-*cis* RA protected primary cultures of hippocampal neurons from Aβ-induced apoptotic neurodegeneration (Sahin et al., 2005). The mechanism was suggested to involve prevention of oxidative stress through scavenging of free radicals by retinoids or blocking the cell cycle in G1 phase through activation of p21 which is an inhibitor of cyclin-dependent kinase (Webber et al., 2005), or by preventing mitotic catastrophe which is one of the earliest events in neural degeneration (Zhu et al., 2004).

A remarkable epidemiological finding that provides a clue for etiology and risk factors for AD is that the age-standardized rate of LOAD is more than double among African Americans in Indianapolis, USA than among a comparison sample of Africans in Ibadan, Nigeria (Ogunniyi et al., 2000; Hendrie et al., 2001). No explanation has been offered. The researchers reported that the diet of the Ibadans consists mainly of red palm oil and yams (Hendrie as quoted in *The New York Times*, February 14, 2001). This diet, high in provitamin A (Fawzi et al., 1997), should maximize the retinoid available for adequate storage in target tissues (Blomhoff, 1994) and transport to brain by APOE (Norum and Blomhoff, 1992) and other retinoid transporters, for example, apolipoprotein D (APOD) (Drayna et al., 1987) and transthyretin (TTR) (Herbert et al., 1986). Consistent with this hypothesis, APOE4 is not a risk factor for AD in Ibadan (Ogunniyi et al., 2000). The findings from this initial cross-national epidemiologic study are well worth replication with other samples.

8.4.2 PROTEINS INVOLVED IN RETINOID TRANSPORT

Retinoid transport from the intestine is necessary for its storage in target tissues including brain (Blomhoff, 1994), and this transport system (RBP, CRABP, APOE, etc.) appears to be modified in LOAD. Retinoids are carried through the body circulation by means of a complex cascade (Goodman, 1987). RBPs are the major carriers of retinol, and they titrate the availability of retinol throughout the body (Goodman et al., 1966; Werner and Deluca, 2002). Retinyl esters transport retinoid from intestine to stellate cells in liver for storage, and from there to target tissues by a highly regulated process dependent on the assembly and secretion of chylomicrons (Nayak et al., 2001). APOE is a transport protein for retinyl esters in chylomicrons (Blomhoff, 1994). This is one of the alternative and redundant pathways by which necessary retinoid is made available to various target tissues. Several properties place the retinyl ester–APOE complex in a pivotal position for impacting the transcription of retinoid-regulated target genes. The APOE2 allele clears postprandial chylomicron remnants containing retinyl esters more slowly than do APOE3 or APOE4 (Boerwinkle et al., 1994). APOE4 is strongly associated with increased risk for AD of both early and late onset in genetic and clinical studies (for reviews see Mahley and Rall, 2000; Poirier, 2005). APOE2 apparently is protective against LOAD (Chartier-Harlin et al., 1994; Corder et al., 1994) and has been shown to protect against memory impairment in rats (Hashimoto et al., 2002). In feedback fashion, the transcriptional expression of APOE in brain astrocytes is strongly upregulated by RA (Cedazo-Minguez and Cowburn, 2001).

Diffusion at high retinyl ester concentrations of retinol into cells in the nervous system is enabled by transporters other than APOE (Norum and Blomhoff, 1992), for example, the lipocalin APOD (Drayna et al., 1987). APOD expression is regulated by RARA (Lopez-Boado et al., 1996), and increased in stressed neurons of LOAD patients (Terrisse et al., 1998), as well as in normal aging, before the accumulation of neurofibrillary tangles (Belloir et al., 2001). We suggest that the increased expression of APOD may be the result of feedback mechanisms dependent on the reduced amounts of retinol in aging individuals (Quadro et al., 1999), particularly those at risk for LOAD. This lipoprotein transport system markedly affects APP processing and Aβ plaque formation (Poirier, 2005). We further suggest that APOE2, which gradually clears retinyl esters may be a preferred carrier of retinol to brain, with APOE4 rapidly clearing retinyl esters and preferentially transporting LDL cholesterol. In the absence of retinoid, the feedback regulation of APOD could be obstructed creating elevated levels of APOD, a preferred carrier of cholesterol.

8.4.3 TRANSTHYRETIN

The RBPs are able to carry retinol alone as well as complexed with TTR (Figure 8.2). TTR is thought to be the major carrier of retinol bound to RBP from liver stores through plasma (Goodman, 1987) and across the choroid plexus to target tissues in the brain (Soprano et al., 1985; Herbert et al., 1986). An increased interest in TTR in the brain has emerged from studies in AD. TTR has been described to bind the Alzheimer's Aβ peptide. By sequestration of Aβ TTR prevents Aβ deposition as amyloid, a landmark in AD pathology, both in vitro and in vivo. More precisely, while in certain in vitro conditions Aβ peptide solutions readily form amyloid, this process is inhibited in the presence of TTR (Schwarzman et al., 1994). Similarly, Aβ amyloid formation is abolished in the presence of cerebrospinal fluid (CSF), which was attributed to the ability of TTR and apolipoproteins E and J to bind Aβ (Golabek et al., 1995; Matsubara et al., 1996; Mazur-Kolecka et al., 1997). In vivo, while a transgenic *Caenorhabditis elegans* model for human mutated APP showed amyloid deposition, the coexpression of human TTR decreased the amyloid load (Link, 1995). Three human studies implicated the decreased TTR CSF levels found in elderly AD patients as contributing to AD pathology (Serot et al., 1997; Merched et al., 1998; Riisoen, 1998), suggesting decreased TTR expression or mutations in TTR that render it less able to bind the peptide. These observations suggest that TTR may be an important sequesterer for Aβ in the CSF, decreasing Aβ availability for amyloid plaque formation, thus modulating susceptibility to the disease. The implication of TTR in AD could then result from decreased TTR affinity to Aβ due to the presence of mutations or decreased TTR circulating levels. A small study failed in identifying mutations in the coding regions of the TTR gene associated with AD (Palha et al., 1996). This observation suggests, as referred to previously in the case of IDE, that regulation of transcription or translation rather than mutated TTR forms might be implicated in AD.

Recently, using microarrays and RT-PCR methods, increased TTR expression in several regions of the brain parenchyma has been implicated as neuroprotective in AD. Specifically, nutritional supplementation of omega-3 polyunsaturated fatty acids (PUFAs), a diet shown to decrease impairment of learning ability in an AD rat model, resulted in elevated TTR expression in the hippocampus (Puskas et al., 2003). A similar effect was observed with the use of *Ginko biloba*, a medicinal plant extract used to counteract age-related neurological disorders (Watanabe et al., 2001). In addition, the lack of full AD pathology in a transgenic mouse model overexpressing a mutated form of APP was related to augmented TTR expression in the hippocampus (Stein and Johnson, 2002). These studies suggested novel sites of TTR synthesis in the brain where, until now, TTR had solely been found expressed in the choroid plexus and the meninges (Dickson et al., 1986; Herbert et al., 1986). Using laser dissection microscopy, a detailed analysis of TTR expression of individual groups of cells in the hippocampus and cortex in wild type and in two validated human APP-expressing transgenic mouse models of AD, the APP-V717I (Moechars et al., 1999; Dewachter et al., 2000), and the Tg2576 (K670N, M671L) (Hsiao et al., 1996) excluded TTR expression from the brain parenchyma (Sousa et al., 2007). These studies refocus attention on the involvement of TTR in AD to the choroid plexus, meninges, and the CSF rather than to the brain parenchyma. In accordance, a recent study showed a protective effect of estrogen in AD as being mediated by increased TTR expression in the choroid plexus (Tang et al., 2004). Very striking is the observation that human APP transgenic mice display a remarkable decrease in amyloid load after treatment with insulin growth factor I (IGF-I) (Carro et al., 2002, 2006a). This effect was attributed to the ability of IGF-I to increase the levels of Aβ sequesterers in the CSF; specifically TTR, albumin, and apolipoprotein J (Carro et al., 2006b). Further dissection of this clearance pathway suggests megalin, a multiligand receptor present at the blood–brain barrier and in the choroid plexus (Zlokovic et al., 1996), as an export protein for Aβ from the brain into the blood (Carro et al., 2005). By binding to these receptors, increased levels of TTR might influence Aβ clearance out of the brain and, therefore, protect against neurodegeneration.

Although TTR might be implicated in AD as a carrier for Aβ, it is also possible that it influences the brain and behavior given its properties as a carrier for vitamin A and thyroid hormones, particularly across the choroid plexus–CSF barrier (Herbert et al., 1986). Although, thus far, studies in TTR-null mice have not shown impaired thyroid hormone or retinoid metabolisms, (Palha et al., 1994, 1997, 2000, 2002; Wei et al., 1995; Gottesman et al., 2001; Van Bennekum et al., 2001), TTR-null mice display decreased depression- and anxiety-like behaviors (Sousa et al., 2004). Interestingly, TTR levels were found decreased in the CSF of depressed patients after a long period of treatment (Sullivan et al., 1999). The effect of medication could, therefore, result in the decrease of TTR CSF levels observed in these patients. Recently, decreased expression and synthesis of TTR in the choroid plexus following chronic administration of lithium chloride, a mood-stabilizing agent has been reported (Pulford et al., 2002). However, the molecular mechanism underlying this effect or the precise role of TTR remains to be clarified. TTR-null dams have decreased levels of circulating total thyroxine (T_4) and vitamin A, even though these are sufficient to meet tissue demands in the adult mice. We cannot exclude, however, that hypothyroxinemia and hypovitaminosis A during development might interfere with the normal development of the CNS with impact on adult behavior, as referred to in the previous section. Further studies are, therefore, needed to understand how TTR, as a carrier for both thyroxine and retinol, or through a novel function to be identified, influences behavior.

In the context of this book, it should be mentioned that most of the studies on TTR relate to its own ability to form extracellular amyloid depositions mainly in the peripheral nervous system. A primary example is the autosomal dominant neurodegenerative disorder designated as familial amyloidotic polyneuropathy (FAP). FAP has an onset of clinical symptoms at around age 20–30 years, and is characterized by early temperature and pain sensation impairment in the feet and autonomic dysfunction, ultimately leading to death (Saraiva, 2001). A single nucleotide mutation (Sasaki et al., 1984), resulting in a methionine for valine substitution in position 30, was identified as an abnormal TTR form in the FAP amyloid deposits (Saraiva et al., 1984). To date, up to 70 mutations have been identified in TTR, most of which are associated with amyloidosis with diverse tissue preferences (Saraiva, 2001). Interestingly, despite the fact that TTR is the major protein synthesized by the choroid plexus, no amyloid load has ever been observed in any region of the brain parenchyma, including those cases with mutated TTR associated with FAP.

8.5 BIOCHEMISTRY AND MOLECULAR BIOLOGY

8.5.1 RA SYNTHESIS

Dietary intake of retinoids results in transport of retinol–retinyl ester complex to the liver for storage and mobilization. The retinyl ester hydroxylase, lipoprotein lipase (LPL) are expressed not only in the liver, but also at high levels in the hippocampus (Blaner et al., 1994), suggesting a role for release of retinol from retinyl ester in brain. Retinol is bound to the CRBPs and serum transport is accomplished by RBP4 bound to TTR. The resulting CRBP–retinol complex serves as a substrate for two different microsomal enzymes: lecithin:retinol acyl transferase (LRAT) and retinol dehydrogenase. LRAT catalyses the esterification of retinol to retinyl esters the storage form of retinol in many tissues (Blomhoff et al., 1992; Napoli, 1996) (Figure 8.3 (1)). RA can promote strong induction of LRAT and hence a reduction in the conversion of retinol to RA (Kurlandsky et al., 1996) (Figure 8.3 (2)). Retinol dehydrogenase catalyses the oxidation of retinol to retinaldehyde, the rate limiting step in the oxidative formation of RA (Blomhoff et al., 1992; Blaner and Olson, 1994; Napoli, 1996; Chen et al., 2000). The metabolism of retinaldehyde to RA is mediated mainly by dehydrogenase but it can also be metabolized by cytochrome P450 enzymes, namely the CYPs 1A1, 1A2, and 3A4 in humans (Zhang et al., 2000). CRABP facilitates RA uptake, and allows RA to bind to cytochrome P450 hydroxylase for inactivation (Fiorella and Napoli, 1991). Thus, the binding of RA to CRABP decreases the elimination half-life of RA (Figure 8.3 (3)). RA is highly efficiently synthesized in the adult hippocampus in rats (Werner and Deluca, 2002).

Three cytochrome P450 RA inducible enzymes, CYP26A1, CYP26B1, and CYP26C1 (designated CYP26) appear to be novel cytochrome P450 hydroxylase

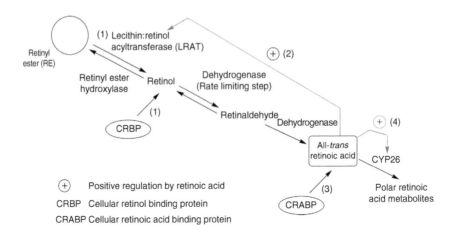

FIGURE 8.3 Regulation of retinoic acid homeostasis; LRAT and CYP26 play a crucial role in the downregulation of all-*trans* RA biosynthesis.

enzymes expressed in numerous tissues, and rapidly induced by RA (White et al., 1997; Taimi et al., 2004) (Figure 8.3 (4)).

The hydroxylation of ATRA by CYP26 produces polar metabolites, that is, hydroxyl or keto metabolites of ATRA, which are less bioactive than ATRA. ATRA is the preferred substrate for all three members of CYP26, except for CYP26C1 which can metabolize both ATRA and 9-*cis* RA (Taimi et al., 2004). In the brain, 4-hydroxylation of ATRA is catalyzed by enzymes CYP26A1 and CYP26B1 (White et al., 2000; Niederreither et al., 2002) and other circulating P450s (McSorley and Daly, 2000; Njar, 2002). CYP26A1 limits programmed cell death (Osanai and Petkovich, 2005). CYP26C1 has not been studied in adult brain, but in the normally developing quail brain CYP26C1 is highly expressed in the cranial mesenchyme and downregulated the presence of excess RA (Reijntjes et al., 2005). How CYP26C1 would respond to a deficit of RA is a question to be investigated in conditions in which lowered levels of retinoid are proposed as pathologic, for example, LOAD.

8.5.2 NUCLEAR RECEPTORS OF RA

The RARs and retinoid X receptors (RXRs) are each composed of three subtypes (α, β, and γ). The RAR family (RARα, RARβ, and RARγ) is activated by both ATRA and by 9-*cis* RA (Mangelsdorf et al., 1995), whereas, the RXR family (RXRα, RXRβ, and RXRγ) is activated by 9-*cis* RA (Heyman et al., 1992) (Figure 8.2) and other small molecules, including docosahexaenoic acid and arachidonic acid (Lengqvisst et al., 2004). The RXRs are involved in all retinoid-signaling cascades (Kliewer et al., 1992), and all are expressed in the adult brain (Krezel et al., 1999). RA controls transcriptions by binding to RAR–RXR dimers, which locally modifies chromatin structure. Synthesis and degradation of RAR and RXR by proteasomes control amount and timing of the retinoid responses. These receptors also modulate phosphorylations (Bastien and Rochette-Egly, 2004).

8.5.3 FUNCTIONS OF RETINOID-RELATED GENES

Unliganded receptors can act as repressors of gene expression by changing chromatin structure through activities of histone deacetylases and silencing mediators for retinoid and thyroid hormone receptors (Weston et al., 2003). The main function of RA is to modulate gene transcriptions by liganding nuclear receptors that are bound to their DNA response element motifs in promoters of target genes (Mangelsdorf and Evans, 1995). RA binds to dimeric RAR–RXR receptors and induces transcription of numerous genes (Bastien and Rochette-Egly, 2004). PUFAs such as docosahexanoic acid also bind to RAR–RXR (de Urquiza et al., 2000). Some of the many genes activated are involved in cognition and the sustainability of LTP and LTD necessary in learning, memory, brain formation, and neurotransmitter functioning (Chiang et al., 1998; Goodman, 1998; Krezel et al., 1999; Takahashi et al., 1999; Etchamendy et al., 2001; Misner et al., 2001; Thompson et al., 2002). Docosahexanoic acid in diet maintains and improves neural membrane function (Horrocks and Farooqui, 2004). APOE regulates the transport and synthesis in neuronal membranes of both docosahexanoic acid and retinoids (Igbavboa et al., 2002).

8.6 APPLICATIONS TO PREVENTION, THERAPY, AND DETECTION

Biochemical results support lower retinoid supply, synthesis, impaired transport, and hypofunction as contributing factors to LOAD. These findings suggest testable experiments to determine whether increasing the availability of retinoid in brain, possibly through pharmacologic targeting of the RARs and the cytochrome P450 RA-inactivating enzymes, can prevent loss of neurons or decrease amyloid plaque formation. Small molecule approaches have been suggested for promoting adult neural processes and synaptic plasticity in LOAD (Longo et al., 2006). Many authors (Etchamendy et al., 2003; Goodman and Pardee, 2003; Corcoran et al., 2004; Farooqui et al., 2004; Prinzen et al., 2005; Sahin et al., 2005; Jacobs et al., 2006) have suggested that the small molecule RA analogs should be considered. However, because too much RA decreases neurogenesis (Crandall et al., 2004), there may be problems of toxicity with RA analogs or deactivating enzymes. Alternatively, we have suggested supplementation through RA precursors, for example, retinol or retinyl esters carried in chylomicrons (Goodman and Pardee, 2003). An established model is prevention of heart disease by statins, which block cholesterol-synthesizing enzymes. A similar paradigm suggests prevention of LOAD by increasing RA supply through modulation of the RA synthesizing or degrading enzymes.

8.6.1 RAR/RXR AGONISTS

Thousands of retinoids have been synthesized and some of these might be effective against AD (Etchamendy et al., 2001; Brand et al., 2002; Prinzen et al., 2005). However, retinoid therapy is limited by toxicity, and only a few retinoids are licensed for clinical use in the treatment of acne, psoriasis, acute promyelocytic leukemia

TABLE 8.2

Structure and Therapeutic Application of Retinoids Used Clinically

Retinoid	Structure	Target	Application
Tretinoin (RA)		RAR agonist	APL, acne
Isotretinoin (13cRA)		RAR agonist	Acne
Alitretinoin (9cRA)		RAR and RXR agonist	Psoriasis, APL, KS
Acitretin Etretinate	Acitretin, R = H; Etretinate, R = Et	RAR agonist	Psoriasis, cancer
Adapalene		RARβ agonist	Acne, psoriasis
Tazarotene		RARα agonist	Psoriasis, acne, cancer

(continued)

TABLE 8.2 (Continued)
Structure and Therapeutic Application of Retinoids Used Clinically

Retinoid	Structure	Target	Application
Bexarotene		RXR agonist	Psoriasis, cancer, keratosis, eczema Schizophrenia

(APL), Kaposi's sarcoma (KS) lesions, and various cancers (Table 8.2). The most widely used and studied RA analog, Accutane (Isotretinoin, 13cRA), has been shown to increase depression in mice (O'Reilly et al., 2006) and possibly human psychiatric disorders (Mey and McCaffery, 2004), increase use of mental health services (Friedman et al., 2006), increase risk factors for atherosclerosis (De Marchi et al., 2006), cause altered brain function, and decreased brain metabolism (Bremner et al., 2005), and result in decrease in neurogenesis and learning ability in exposed mice (Crandall et al., 2004). An open label clinical trial of the RXR agonist Bexarotene reports significant improvement in symptoms among a small group of persons with chronic schizophrenia (Lerner et al., 2007).

RAR-selective ligands, retinoids have been developed to improve chemical stability and pharmacological properties. Replacement of the unstable polyene chain with aromatic rings (TTNPB) and heteroatoms (Tamibarotene, Am80) has resulted in potent and chemically stable RAR-selective agonists (Kagechika, 2002). Am80 rescued memory deficit in rats treated with scopolamine (Shudo et al., 2004). The availability of crystal structures for RARβ (Germain et al., 2004) and RARγ (Renaud et al., 1995) has revealed the ligand-binding domains allowing development of subtype-selective RAR agonists (Figure 8.4) (Kagechika and Shudo, 2005), for example, AGN-193836 has greater RARα subtype selectivity (Teng et al., 1996) whereas BMS-961 is an RARγ-selective retinoid (Klaholz et al., 1998).

An RAR agonist of interest (Figure 8.4) is fenretinide (4-HPR), an atypical retinoid with retinoid and retinoid nuclear receptor-independent activities, currently in Phase III clinical trials for application in breast, prostate, and CNS cancers. Fenretinide has been shown to ameliorate glucose intolerance in diabetes (Yang et al., 2005), both of which conditions are risk factors in LOAD. Fenretinide has been suggested as an adjuvant or preventive therapy for AD (Goodman, 2006). Other RAR agonists include the adamantyl derivative adapalene (Table 8.2), and the unusual carborane derivative BR403, which possesses RAR agonist activity comparable with RA (Figure 8.4).

RXRs are the silent partners of RARs. RXR agonists (rexinoids) alone cannot activate the RXR–RAR heterodimers. However, as a result of retinoid synergism RXR agonists can increase the potencies of RAR ligands (Forman et al., 1995).

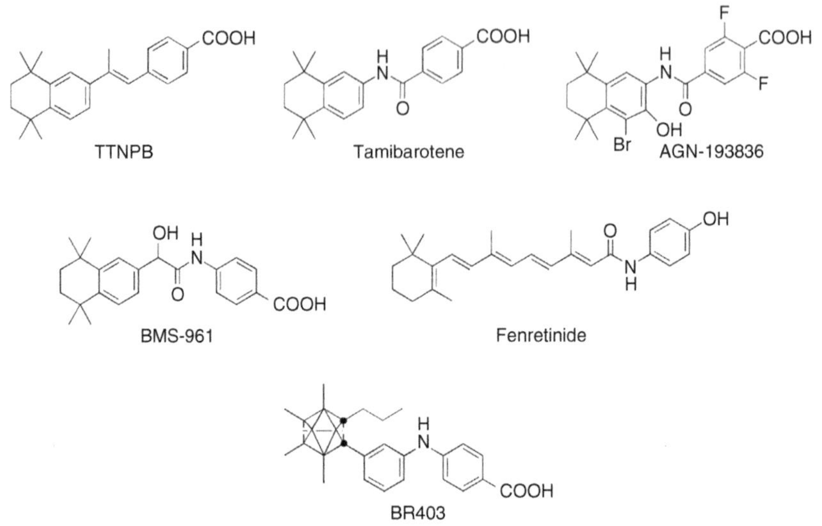

FIGURE 8.4 RAR agonists (retinoids).

RXRs are the heterodimeric partners of other nuclear receptors (vitamin D_3 receptor, thyroid hormone receptor, peroxisome proliferators-activated receptor, nuclear receptor 4A2, liver X receptors, and farnesoid X receptors). RXR agonists alone can activate these heterodimers and owing to the complex protein–protein interactions different RXR agonists can display varying biological activities. A recent report suggests development of RXR heterodimers to mediate APOE expression, secretion, and cholesterol homeostasis in LOAD (Liang et al., 2004).

RXR-selective agonists have been designed, but RXR subtype specificity has not been achieved (Kagechika and Shudo, 2005). The first RXR-selective agonist was SR11,237 (BMS-649), further development led to HX630 and PA024 (Umemiya et al., 1997) (Figure 8.5). Combination of PA024 with the RAR agonist tamibarotene (Figure 8.4) produced a gene expression profile similar to 9cRA with increased induction of cell apoptosis suggesting combinations of RAR–RXR agonists may be an option in retinoid therapy (Ishida et al., 2003). The agonist and antagonist activities of rexinoids change dependent on the heterodimeric partner, for example, LG100754 displayed RXR agonistic activity in RXR–RAR and PPAR–RXR heterodimers with subsequent antidiabetic activity associated with PPAR–RXR activation (Cesario et al., 2001).

The continued development of retinoids or rexinoids with improved selectivity and toxicity profiles and the options for targeting RXR heterodimers provides considerable scope for therapeutic intervention and application in AD.

8.6.2 CYP26 Inhibitors

Owing to the rapid metabolism of RA in cells, RA shows a decrease in plasma concentrations after repeated dosage (Muindi et al., 1994) therefore inhibition of the

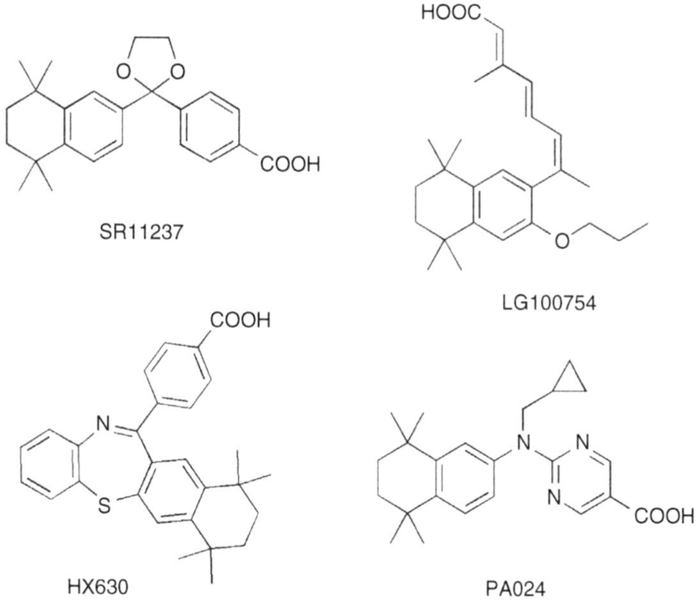

FIGURE 8.5 RXR agonists (rexinoids).

CYP26 RA-degrading enzymes at the 10q23 and 2p13 loci linked to AD is an approach. Many research groups are searching for inhibitors of these CYPs that mediate the metabolism of RA, that is, RA metabolism blocking agents (RAMBAs). The use of RAMBAs should delay in vivo RA metabolism thus resulting in increased endogenous levels (Njar, 2002).

Liarozole and ketoconazole (Figure 8.6) are capable of inhibiting the CYP-dependent metabolism of RA by hamster (Van Wauwe et al., 1990; Njar et al., 2000) and rat (Kirby et al., 2003) liver microsomes, respectively. Moreover, it has been shown that liarozole with RA potentiates the antiproliferative effect in in vitro tumor cell lines (Wouters et al., 1992; Djikman et al., 1994). In vivo studies using

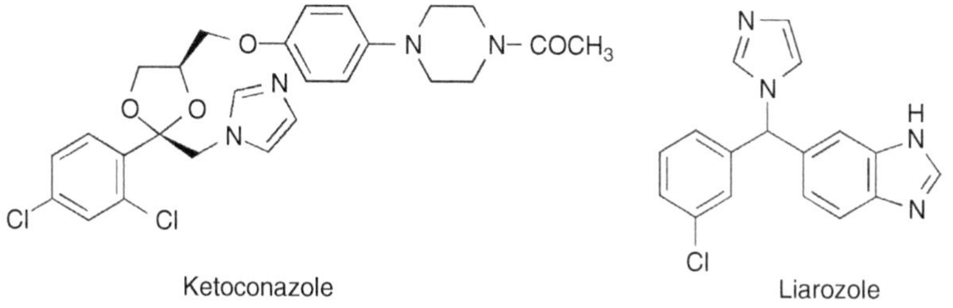

FIGURE 8.6 CYP26 inhibitors: ketoconazole and liarozole.

liarozole showed enhanced plasma levels of RA following oral dosing of liarozole in rat (Van Wauwe et al., 1992).

Liarozole and ketoconazole are not selective inhibitors and target other P450 enzymes; therefore, further development of CYP26 inhibitors has concentrated on the preparation of compounds with greater enzyme selectivity as well as increased potency. Potent inhibitors such as the 2-benzothiazolamine compounds, R115866 and R110610 (Figure 8.7) (Stoppie et al., 2000; Aelterman et al., 2001; Van Heusden et al., 2002) have been described. In an in vivo study, R115866 was able to enhance endogenous RA levels and mimic the effects of RA (Stoppie et al., 2000). Potent and selective CYP26 inhibitors, with a naphthyl-based backbone structure (Figure 8.7) have been described with $IC_{50} = 3.3$ nM for the most potent compound (OSI Pharma) (Mulvihill et al., 2005). Azolyl retinoid compounds showed interesting activities against RA metabolism enzyme(s) in hamster liver microsomes ($IC_{50} = 0.68-1.6$ µM) (Njar et al., 2000) and in breast and prostate cancer cell lines, which demonstrated enhanced antiproliferative action compared with RA and liarozole (Njar, 2002; Patel et al., 2004). The majority of CYP26 inhibitors contain an azolyl group, either an imidazole or triazole heterocycle, which coordinates with the Fe^{3+} of the porphyrin haem (Figure 8.7). Nonazoyl CYP26 inhibitors have been described such as the tetralone derivatives (Yee et al., 2005), designed using a computer-generated homology model of CYP26A1 (Gomaa et al., 2006). These tetralone derivatives had inhibitory activity comparable with liarozole and were shown to bind within the ligand-binding hydrophobic tunnel (Yee et al., 2005) (Figure 8.8).

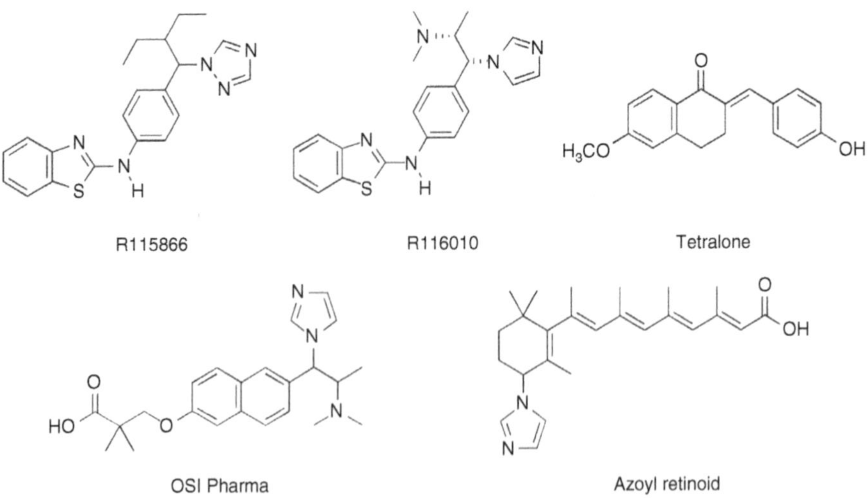

FIGURE 8.7 Retinoic acid metabolism blocking agents (RAMBAs).

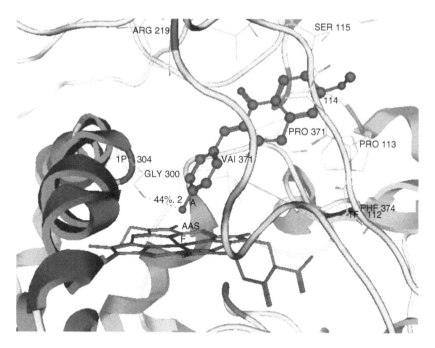

FIGURE 8.8 (See color insert following page 296.) Interaction of tetralone inhibitor with the CYP26A1 active site; Hydrogen bonding (red dashed lines), transition metal interaction (purple dashed line), labeled amino acid residues are involved in hydrophobic interactions.

8.7 CONCLUSIONS

A variety of potential therapies come to mind for future testing. Drugs could induce enzymes that increase RA synthesis, in particular, in the hippocampus (Mey and McCaffery, 2004) and other areas involved in AD pathology. In contrast, homocysteine, the level of which is increased in AD, has the opposite effect of inhibiting RA synthesis (Limpach et al., 2000; Seshadri et al., 2002). Specific CYP26 inhibitors, for example, liarazole (Van Wauwe et al., 1992), could be applied to increase RA; low concentrations may be necessary because xenobiotics frequently induce proteins involved in their detoxification (Schuetz et al., 1996). Alternatively, supplementation by means of the RA precursors retinol or retinyl esters has been suggested (Goodman and Pardee, 2003).

The increased specific ligation of RA to the RARγ–RARα heterodimer regulating IDE transcription (Wyss-Coray et al., 2001) might cause plaque degradation by increasing IDE (Bertram et al., 2000; Edbauer et al., 2002). Ligands that increase TTR should result in increased clearance of Aβ from brain (Carro et al., 2002). Alternately, suggestions involve targeting RXRγ. The RARγ–RXRγ heterodimer regulates the atypical expression of CYP26A1 (Loudig et al., 2000), which could be manipulated by receptor ligand agonists. The retinoid acid analog, fenretinide, could reduce glucose intolerance, thus mediating known risk factors (Goodman, 2006; Lerner et al., 2007).

Without further basic and pharmacological knowledge, the complexity of retinoid regulation in the CNS and the remarkably wide range of regulatory and signaling processes in which retinoids are involved could make successful implementation of these strategies difficult to accomplish safely. Although pharmacologic strategies to increase retinoid activation in brain may in this early stage seem far-fetched, recall the remarkable success against heart disease when dietary and drug measures were developed and applied to lower cholesterol after this molecule was identified as a major risk factor (Goodman, 1991).

REFERENCES

Abraham, R., Myers, A., Wavrant-DeVrieze, F., Hamshere, M.L., Thomas, H.V., Marshall, H. et al. (2001). Substantial linkage disequilibrium across the insulin-degrading enzyme locus but no association with late-onset Alzheimer's disease. *Human Genetics*, **109**, 646–652.

Aelterman, W., Lang, Y., Willemsens, B., Vervest, I., Leurs, S. and De Knaep, F. (2001). Conversion of the laboratory synthetic route of the *N*-aryl-2-benzothiazolamine R116010 to a manufacturing method. *Organic Process Research & Development*, **5**, 467–471.

Ait-Ghezala, G., Abdullah, L., Crescentini, R., Crawford, F., Town, T., Singh, S. et al. (2002). Confirmation of association between D10S583 and Alzheimer's disease in a case–control sample. *Neuroscience Letters*, **325**, 87–90.

Akshoomoff, N., Pierce, K. and Courchesne, E. (2002). The neurobiological basis of autism from a developmental perspective. *Developmental Psychopathology*, **14**, 613–634.

Arendt, T. (2003). Synaptic plasticity and cell cycle activation in neurons are alternative effector pathways: the 'Dr. Jekyll and Mr. Hyde concept' of Alzheimer's disease or the yin and yang of neuroplasticity. *Progress in Neurobiology*, **71**, 83–248.

Bacanu, S.A., Devlin, B., Chowdari, K.V., DeKosky, S.T., Nimgaonkar, V.L. and Sweet, R.A. (2002). Linkage analysis of Alzheimer disease with psychosis. *Neurology*, **59**, 118–120.

Bancescu, D.L., Glatt-Deeley, H. and Lalande, M. (2004). Epigenetic activation of the 5-hydroxytryptamine (serotonin) receptor 2C in embryonal carcinoma cells is DNA replication-dependent. *Experimental Cell Research*, **298**, 262–267.

Bastien, J. and Rochette-Egly, C. (2004). Nuclear retinoid receptors and the transcription of retinoid-target genes, *Gene*, **328**, 1–16.

Bavik, C., Ward, S.J. and Chambon, P. (1996). Developmental abnormalities in cultured mouse embryos deprived of retinoic by inhibition of yolk-sac retinol binding protein synthesis. *Proceedings of the National Academy of Sciences of the United States of America*, **93**, 3110–3114.

Belloir, B., Kovari, E., Surini-Demiri, M. and Savioz, A. (2001). Altered apolopoprotein D expression in the brain of patients with Alzheimer disease. *Journal of Neuroscience Research*, **64**, 61–69.

Bertram, L. and Tanzi, R.E. (2001). Dancing in the dark? The status of late-onset Alzheimer's disease genetics. *Journal of Molecular Neuroscience*, **17**, 127–136.

Bertram, L., Blacker, D., Mullin, K., Keeney, D., Jones, J., Basu, S. et al. (2000). Evidence for genetic linkage of Alzheimer's disease to chromosome 10q. *Science*, **290**, 2302–2303.

Blaner, W.S. and Olson, J.A. (1994). Retinol and retinoic acid metabolism, In *The Retinoids: Biology, Chemistry and Medicine* (Sporn, M.B., Roberts, A.B. and Goodman, D.S. eds, 2nd Ed.) pp. 229–255, Raven Press, Ltd., New York.

Blaner, W.S., Obunike, J.C., Kurlandsky, S.B., Al-Haideri, M., Piantedosi, R., Deckelbaum, R.J. et al. (1994). Lipoprotein-lipase hydrolysis of retinyl ester—possible implications for retinoid uptake by cells. *Journal of Biological Chemistry*, **269**, 16559–16565.

Blomhoff, R. (1994). Transport and metabolism of vitamin A. *Nutrition Reviews*, **52**, 13–23.

Blomhoff, R., Green, M.H. and Norum, K.R. (1992). Vitamin A: physiological and biochemical processing. *Annual Review of Nutrition*, **12**, 37–57.

Boerwinkle, E., Brown, S., Sharrett, A.R., Heiss, G. and Patsch, W. (1994). Apolipoprotein-E polymorphism influences postprandial retinyl palmitate but not triglyceride concentrations. *American Journal of Human Genetics*, **54**, 341–360.

Bourdel-Marchasson, I., Delmas-Beauvieux, M.C., Peuchant, E., Richard-Harston, S. Decamps, A., Reignier, B. et al. (2001). Antioxidant defences and oxidative stress markers in erythrocytes and plasma from normally nourished elderly Alzheimer patients. *Age and Ageing*, **30**, 235–241.

Boussaha, M., Hannequin, D., Verpillat, P., Brice, A., Frebourg, T. and Campion, D. (2002). Polymorphisms of insulin degrading enzyme gene are not associated with Alzheimer's disease. *Neuroscience Letters*, **329**, 121–123.

Bowser, R., Kordower, J.H. and Mufson, E.J. (1997). A confocal microscopic analysis of galaninergic hyperinnervation of cholinergic basal forebrain neurons in Alzheimer's disease. *Brain Pathology*, **7**, 723–730.

Boylan, J.F. and Gudas, L.J. (1992). The level of CRABP-I expression influences the amounts and types of all-trans-retinoic acid metabolites in F9 teratocarcinoma stem cells. *Journal of Biological Chemistry*, **267**, 21486–21491.

Brand, C., Segard, P., Plouvier, P., Formstecher, P., Danze, P.-M. and Lefebvre, P. (2002). Selective alteration of gene expression in response to natural and synthetic retinoids. *BMC Pharmacology*, **2**, 13–27.

Bremner, J.D., Fani, N., Ashraf, A., Votaw, J.R., Brummer, M.E., Cummins, T. et al. (2005). Functional brain imaging alterations in acne patients treated with isotretinoin. *American Journal of Psychiatry*, **162**, 983–991.

Burton, T., Liang, B., Dibrov, A. and Amara, F. (2002). Transcriptional activation and increase in expression of Alzheimer's beta-amyloid precursor protein gene is mediated by TGF-beta in normal human astrocytes. *Biochemical and Biophysical Research Communications*, **295**, 702–712.

Cacabelos, R. (2005). Pharmacogenomics and therapeutic prospects in Alzheimer's disease. *Expert Opinion in Pharmacotherapy*, **6**, 1967–1987.

Cai, H.B, Wang, Y.S., McCarthy, D., Wen, H.J., Borchelt, D.R., Price, D.L. et al. (2001). BACE1 is the major beta-secretase for generation of A beta peptides by neurons. *Nature Neuroscience*, **4**, 233–234.

Carro, E., Trejo, J.L., Gomez-Isla, T., LeRoith, D. and Torres-Aleman, I. (2002). Serum insulin-like growth factor I regulates brain amyloid-beta levels. *Nature Medicine*, **8**, 1390–1397.

Carro, E., Spuch, C., Trejo, J.L., Antequera, D. and Torres-Aleman, I. (2005). Choroid plexus megalin is involved in neuroprotection by serum insulin-like growth factor I. *Journal of Neuroscience*, **25**, 10884–10893.

Carro, E., Trejo, J.L., Gerber, A., Loetscher, H., Torrado, J., Metzger, F. et al. (2006a). Therapeutic actions of insulin-like growth factor I on APP/PS2 mice with severe brain amyloidosis. *Neurobiology of Aging*, **27**, 1250–1257.

Carro, E., Trejo, J.L., Spuch, C., Bohl, D., Heard, J.M. and Torres-Aleman, I. (2006b). Blockade of the insulin-like growth factor I receptor in the choroid plexus originates Alzheimer's-like neuropathology in rodents: new cues into the human disease? *Neurobiology of Aging*, **27**, 1618–1631.

Cedazo-Minguez, A. and Cowburn, R.F. (2001). Apolipoprotein E: a major piece in the Alzheimer's disease puzzle. *Journal of Cellular and Molecular Medicine*, **5**, 254–266.

Cesario, R.M., Klausing, K., Razzaghi, H., Crombie, D., Rungta, D., Heyman, R.A. et al. (2001). The rexinoid LG100754 is a novel RXR:PPAR gamma agonist and decreases glucose levels in vivo. *Molecular Endocrinology*, **15**, 1360–1369.

Chartier-Harlin, M.C., Parfitt, M., Legrain, S., Perez-Tur, J., Brousseau, T., Evans, A. et al. (1994). Apolipoprotein-E, epsilon-4 allele as a major risk factor for sporadic early and late-onset forms of Alzheimers-disease—analysis of the 19q13.2 chromosomal region. *Human Molecular Genetics*, **3**, 569–574.

Chen, H., Howald, W.N. and Juchau, M.R. (2000). Biosynthesis of all-*trans*-retinoic acid from all-*trans*-retinol: catalysis of all-*trans*-retinol oxidation by human P-450 cytochromes. *Drug Metabolism and Disposition*, **28**, 315–322.

Chen, A.C., Yu, K., Lane, M.A. and Gudas, L.J. (2003). Homozygous deletion of the CRABPI gene in AB1 embryonic stem cells results in increased CRABPII gene expression and decreased intracellular retinoic acid concentration. *Archives of Biochemistry and Biophysics*, **411**, 159–173.

Chiang, M.Y., Misner, D., Kempermann, G., Schikorski, T., Giguere, V., Sucov, H.M. et al. (1998). An essential role for retinoid receptors RARβeta and RXRgamma in long-term potentiation and depression. *Neuron*, **21**, 1353–1361.

Clagett-Dame, M., McNeill, E.M. and Muley, P.D. (2006). The role of all-trans retinoic acid in neurite outgrowth and axonal elongation, *Journal of Neurobiology*, **66**, 739–756.

Colbert, M.C., Linney, E. and LaMantia, A.S. (1993). Local sources of retinoic acid coincide with retinoid-mediated transgene activity during embryonic development. *Proceedings of the National Academy of Sciences of the United States of America*, **90**, 6572–6576.

Collins, J.S., Perry R.T., Watson. B. Jr., Harrell, L.E., Acton, R.T., Blacker, D. et al. (2000). Association of a haplotype for tumor necrosis factor in siblings with late-onset Alzheimer disease: the NIMH Alzheimer disease genetics initiative. *American Journal of Medical Genetics*, **96**, 823–830.

Connor, M.J. and Sidell, N. (1997). Retinoic acid synthesis in normal and Alzheimer diseased brain and human neural cells. *Molecular and Chemical Neuropathology*, **30**, 239–252.

Corcoran, J.P.T., So, P.L. and Maden, M. (2004). Disruption of the retinoid signalling pathway causes a deposition of amyloid beta in the adult rat brain. *European Journal of Neuroscience*, **20**, 896–902.

Corder, E.H., Saunders, A.M., Risch, N.J., Strittmatter, W.J., Schmechel, D.E., Gaskell, P.C. Jr. et al. (1994). Protective effect of apolipoprotein-E type-2 allele for late-onset Alzheimer-disease. *Nature Genetics*, **7**, 180–184.

Cotman, C.W. and Anderson, K.J. (1988). Synaptic plasticity and functional stabilization in the hippocampal formation: possible role in Alzheimer's disease. *Advances in Neurology*, **47**, 313–335.

Crandall, J., Sakai, Y., Zhang, J., Koul, O., Mineur, Y., Crusio, W.E. and McCaffery, P. (2004). 13-cis-retinoic acid suppresses hippocampal cell division and hippocampal-dependent learning in mice. *Proceedings of the National Academy of Sciences of the United States of America*, **101**, 5111–5116.

Culvenor, J.G., Evin, G., Cooney, M.A., Wardan, H., Sharples, R.A., Maher, F. et al. (2000). Presenilin 2 expression in neuronal cells: induction during differentiation of embryonic carcinoma cells. *Experimental Cell Research*, **255**, 192–206.

Delva, L., Bastie, J.N., Rochette-Egly, C., Kraiba, R., Balitrand, N., Despouy, G. et al. (1999). Physical and functional interactions between cellular retinoic acid binding protein II and the retinoic acid-dependent nuclear complex. *Molecular Cell Biology*, **19**, 7158–7167.

De Marchi, M.A., Maranhao, R.C., Brandizzi, L.I. and Souza, D.R. (2006). Effects of isotretinoin on the metabolism of triglyceride-rich lipoproteins and on the lipid profile in patients with acne. *Archives of Dermatological Research*, **297**, 403–408.

De Strooper, B. and Wodgett, J. (2003). Mental plaque removal. *Nature*, **423**, 392–393.

de Urquiza, A.M., Liu, S., Sjoberg, M., Zetterstrom, R.H., Griffiths, W., Sjovall, J. et al. (2000). Docosahexaenoic acid, a ligand for the retinoid X receptor in mouse brain. *Science*, **290**, 2140–2144.

Dewachter, I., Van Dorpe, J., Smeijers, L., Gilis, M., Kuiperi, C., Laenen, I. et al. (2000). Aging increased amyloid peptide and caused amyloid plaques in brain of old APP/V717I transgenic mice by a different mechanism than mutant presenilin1. *Journal of Neuroscience*, **20**, 6452–6458.

DeWitt, D.A. and Silver, J. (1996). Regenerative failure: a potential mechanism for neuritic dystrophy in Alzheimer's disease. *Experimental Neurology*, **142**, 103–110.

Dickson, P.W., Aldred, A.R., Marley, P.D., Bannister, D. and Schreiber, G. (1986). Rat choroid plexus specializes in the synthesis and the secretion of transthyretin (prealbumin). Regulation of transthyretin synthesis in choroid plexus is independent from that in liver. *Journal of Biological Chemistry*, **261**, 3475–3478.

D'Introno, A., Solfrizzi, V., Colacicco, A.M., Capurso, C., Amodio, M., Todarello, O. et al. (2005). Current knowledge of chromosome 12 susceptibility genes for late-onset Alzheimer's disease. *Neurobiology of Aging*, 2005 Oct 26 [Epub ahead of print].

Djikman, G.A., Van Moorselaar, R.J.A., Van Ginckel, R., Van Stratum, P., Wouters, L., Debruyne, F.M.J. et al. (1994). Antitumoral effects of liarozole in androgen-dependent and independent R3327-Dunning prostate adenocarcinomes. *Journal of Urology*, **151**, 217–222.

Drayna, D.T., McLean, J.W., Wion, K.L., Trent, J.M., Drabkin, H.A. and Lawn, R. (1987). Human apolipoprotein-D gene—gene sequence, chromosome localization, and homology to the alpha 2U-globulin superfamily. *DNA – A Journal of Molecular and Cellular Biology*, **6**, 199–204.

Duester, G., Shean, M.L., McBridge, M.S. and Stewart, M.J. (1991). Retinoic acid response element in the human alcohol-dehydrogenase gene ADH3—implications for regulation of retinoic acid synthesis. *Molecular Cell Biology*, **11**, 1638–1646.

Edbauer, D., Willem, M., Lammich, S., Steiner, H. and Haass, C. (2002). Insulin-degrading enzyme rapidly removes the beta-amyloid precursor protein intracellular domain (AICD). *Journal of Biological Chemistry*, **277**, 13389–13393.

Etchamendy, N., Enderlin, V., Marighetto, A., Vouimba, R.M., Pallet, V., Jaffard, R. et al. (2001). Alleviation of a selective age-related relational memory deficit in mice by pharmacologically induced normalization of brain retinoid signaling. *Journal of Neuroscience*, **21**, 6423–6429.

Etchamendy, N., Enderlin, V., Marighetto, A., Pallet, V., Higueret, P. and Jaffard, R. (2003). Vitamin A deficiency and relational memory deficit in adult mice: relationships with changes in brain retinoid signalling. *Behavioural Brain Research*, **145**, 37–49.

Farooqui, A.A., Antony, P., Ong, W.Y., Horrocks, L.A. and Freysz, L. (2004). Retinoic acid-mediated phospholipase A2 signaling in the nucleus. *Brain Research Reviews*, **45**, 179–195.

Fawzi, W.W., Herrera, M.G., Willett, W.C., Nestel, P., el Amin, A. and Mohamed, K.A. (1997). Dietary vitamin A intake in relation to child growth. *Epidemiology*, **4**, 402–407.

Feuk, L., McCarthy, S., Andersson, B., Prince, J.A. and Brookes, A.J. (2005). Mutation screening of a haplotype block around the insulin degrading enzyme gene and association with Alzheimer's disease. *American Journal of Medical Genetics - Part B Neuropsychiatric Genetics*, **136**, 69–71.

Fiorella, P.D. and Napoli, J.L. (1991). Expression of cellular retinoic acid binding protein (CRABP) in *Escherichia coli*. *Journal of Biological Chemistry*, **266**, 16572–16579.

Flood, D.G. and Coleman, P.D. (1990). Hippocampal plasticity in normal aging and decreased plasticity in Alzheimer's disease. *Progress in Brain Research*, **83**, 435–443.

Forman, M., Umesono, K., Chen, J. and Evans, R.M. (1995) Unique response pathways are established by allosteric interactions among nuclear hormone receptors. *Cell*, **81**, 541–550.

Freemantle, S.J., Kerley, J.S., Olsen, S.L., Gross, R.H. and Spinella, M.J. (2002). Developmentally-related candidate retinoic acid target genes regulated early during neuronal differentiation of human embryonal carcinoma. *Oncogene*, **21**, 2880–2889.

Friedman, T., Wohl, Y., Knobler, H.Y., Lubin, G., Brenner, S., Levi, Y., Barak, Y. (2006). Increased use of mental health services related to isotretinoin treatment: A 5-year analysis. *European Neuropsychopharmacology*, **16**, 413–416.

Fukuchi, K., Deeb, S.S., Kamino, K., Ogburn, C.E., Snow, A.D., Sekiguchi, R.T. et al. (1992). Increased expression of beta-amyloid protein-precursor and microtubule-associated protein tau during the differentiation of murin embryonal carcinoma-cells. *Journal of Neurochemistry*, **58**, 1863–1873.

Gaub, M.P., Lutz, Y., Ghyselinck, N.B., Scheuer, I., Pfister, V., Chambon, P. et al. (1998). Nuclear detection of cellular retinoic acid binding proteins I and II with new antibodies. *Journal of Histochemistry & Cytochemistry*, **46**, 1103–1111.

Gavalas, A. (2002). Arranging the hindbrain. *Trends in Neuroscience*, **25**, 61–64.

Gavalas, A. and Krumlauf, R. (2000). Retinoid signalling and hindbrain patterning. *Current Opinion in Genetics & Development*, **10**, 380–386.

Geddes, J.W. and Cotman, C.W. (1991). Plasticity in Alzheimer's disease: too much or not enough? *Neurobiology of Aging*, **12**, 330–333; discussion 352–355.

Geddes, J.W., Monaghan, D.T., Cotman, C.W., Lott, I.T., Kim, R.C. and Chui, H.C. (1985). Plasticity of hippocampal circuitry in Alzheimer's disease. *Science*, **230**, 1179–1181.

Germain, P., Kammerer, S., Peluso-Iltis, C., Tortolani, D., Zusi, F.C., Starrett, J. et al. (2004). Rational design of RAR-selective ligands revealed by RARβ crystal structure. *EMBO Reports*, **5**, 877–882.

Ghyselinck, N.B., Bavik, C., Sapin, V., Mark, M., Bonnier, D., Hindelang, C. et al. (1999). Cellular retinol-binding protein I is essential for vitamin A homeostasis. *EMBO Journal*, **18**, 4903–4914.

Glick, A.B., McCune, B.K., Abdulkarem, N., Flanders, K.C., Lumadue, J.A., Smith, J.M. et al. (1991). Complex regulation of TGF-beta expression by retinoic acid in the vitamin-A-deficient rat. *Development*, **111**, 1081–1086.

Goedert, M., Spillantni, M.G., Jakes, R., Rutherford, D. and Crowther, R.A. (1989). Multiple isoforms of human microtubule-associated protein-tau—sequences and localization in neurofibrillary tangles of Alzheimers-disease. *Neuron*, **3**, 519–526.

Golabek, A., Marques, M.A., Lalowski, M. and Wisniewski, T. (1995). Amyloid beta binding proteins in vitro and in normal human cerebrospinal fluid. *Neuroscience Letters*, **191**, 79–82.

Gomaa, M.S., Yee, S.W., Milbourne, C.E., Barbara, M.-C., Brancale, A. and Simons, C. (2006). Homology model of human retinoic acid metabolising enzyme cytochrome P450 26A1 (CYP26A1): active site architecture and ligand binding. *Journal of Enzyme Inhibition and Medicinal Chemistry*, **21**, 361–369.

Goodman, A.B. (1998). Three independent lines of evidence suggest retinoids as causal to schizophrenia. *Proceedings of the National Academy of Sciences of the United States of America*, **95**, 7240–7244.

Goodman, A.B. (2006). Retinoid receptors, transporters, and metabolizers as therapeutic targets in late onset Alzheimer disease. *Journal of Cellular Physiology*, **209**, 598–603.

Goodman, A.B. and Pardee, A.B. (2003). Evidence for defective retinoid transport and function in late onset Alzheimer's disease. *Proceedings of the National Academy of Sciences of the United States of America*, **100**, 2901–2905.

Goodman, D.S. (1984). Plasma retinol-binding protein, In *The Retinoids*. (Sporn, M.B., Roberts, A.B. and Goodman, D.S. eds, Vol. 2) pp. 41–88, Raven Press Ltd., New York.

Goodman, D.S. (1987). Retinoids and retinol-binding proteins, In *The Harvey Lectures 1987*, Series **81**, 111–132.

Goodman, D.S. (1991). The national cholesterol education program—guidelines, status, and issues. *American Journal of Medicine*, **90**, 32S–35S.

Goodman, D.S., Huang, H.S. and Shiratori, T. (1966). Mechanism of the biosynthesis of vitamin A from beta-carotene. *Journal of Biological Chemistry*, **241**, 1929–1932.

Gottesman, M.E., Quadro, L. and Blazer, W.S. (2001). Studies of vitamin A metabolism in mouse model systems. *Bioessays*, **23**, 409–419.

Grammas, P. and Ovase, R. (2002). Cerebrovascular transforming growth factor-beta contributes to inflammation in the Alzheimer's disease brain. *American Journal of Pathology*, **160**, 1583–1587.

Gustafson, A.L., Donovan, M., Annerwall, E., Dencker, L. and Eriksson, U. (1996). Nuclear import of cellular retinoic acid-binding protein type I in mouse embryonic cells. *Mechanisms of Development*, **58**, 27–38.

Hale, F. (1933). Pigs born without eyeballs. *Journal of Heredity*, **24**, 105–106.

Hall, R.K., Scott, D.K., Noisin, E.L., Lucas, P.C. and Granner, D.K. (1992). Activation of the phosphoenolpyruvate carboxykinase gene retinoic acid response element is dependent on a retinoic acid receptor coregulator complex. *Molecular Cell Biology*, **12**, 5527–5535.

Hardy, J. and Selkoe, D.J. (2002). Medicine—the amyloid hypothesis of Alzheimer's disease: progress and problems on the road to therapeutics. *Science*, **297**, 353–356.

Hashimoto, M., Hossain, S., Shimada, T., Sogioka, K., Yamasaka, H., Fujii, Y. et al. (2002). Docosahexaenoic acid provides protection from impairment of learning ability in Alzheimer's disease model rats. *Journal of Neurochemistry*, **81**, 1084–1091.

Heicklen-Klein, A., Aronov, S. and Ginzburg, I. (2000). Tau promoter activity in neuronally differentiated P19 cells. *Brain Research*, **874**, 1–9.

Hendrie, H.C., Ogunniyi, A., Hall, K.S., Baiyewu, O., Unverzagt, F.W., Gureje, O. et al. (2001). Incidence of dementia and Alzheimer disease in 2 communities—Yoruba residing in Ibadan, Nigeria, and African Americans residing in Indianapolis, Indiana. *Journal of the American Medical Association*, **285**, 739–747.

Herbert, J., Wilcox, J.N., Pham, K.T., Fremeau, R.T. Jr., Zeviani, M., Dwork, A. et al. (1986). Transthyretin: a choroid plexus-specific transport protein in human brain. *Neurology*, **36**, 900–911.

Heyman, R.A., Mangelsdorf, D.J., Dyck, J.A., Stein, R.B., Eichele, G., Evans, R.M. et al. (1992). 9-*Cis* retinoic acid is a high affinity ligand for the retinoid X receptor. *Cell*, **68**, 397–406.

Horrocks, L.A. and Farooqui, A.A. (2004). Docosahexaenoic acid in the diet: its importance in maintenance and restoration of neural membrane function. *Prostaglandins Leukotrienes and Essential Fatty Acids*, **70**, 361–372.

Hsiao, K., Chapman, P., Nilsen, S., Eckman, C., Harigaya, Y., Younkin, S. et al. (1996). Correlative memory deficits, Aβ elevation, and amyloid plaques in transgenic mice. *Science*, **274**, 99–102.

Huang, D., Chen, S.W. and Gudas, L.J. (2002). Analysis of two distinct retinoic acid response elements in the homeobox gene Hoxb1 in transgenic mice. *Developmental Dynamics*, **223**, 353–370.

Hung, A.Y., Koo, E.H., Haass, C. and Selkoe, D.J. (1992). Increased expression of beta-amyloid precursor protein during neuronal differentiation is not accompanied by secretory cleavage. *Proceedings of the National Academy of Sciences of the United States of America*, **89**, 9439–9443.

Husson, M., Enderlin, V., Alfos, S., Boucheron, C., Pallet, V. and Higueret, P. (2004). Expression of neurogranin and neuromodulin is affected in the striatum of vitamin A-deprived rats. *Molecular Brain Research*, **123**, 7–17.

Igbavboa, U., Hamilton, J., Kim, H.Y., Sun, G.Y. and Wood, W.G. (2002). A new role for apolipoprotein E: modulating transport of polyunsaturated phospholipid molecular species in synaptic plasma membranes. *Journal of Neurochemistry*, **80**, 255–261.

Ihara, Y. (1988). Massive somatodendritic sprouting of cortical neurons in Alzheimer's disease. *Brain Research*, **459**, 138–144.

Ishida, S., Shigemoto-Mogami, Y., Kagechika, H., Shudo, K., Ozawa, S., Sawada, J. et al. (2003). Clinically potential subclasses of retinoid synergists revealed by gene expression profiling. *Molecular Cancer Therapeutics*, **3**, 49–58.

Jacobs, S., Lie, D.C., Decicco, K.L., Shi, Y., Deluca, L.M., Gage, F.H. et al. (2006). Retinoic acid is required early during adult neurogenesis in the dentate gyrus. *Proceedings of the National Academy of Sciences of the United States of America*, **103**, 3902–3907.

Jimenez-Jimenez, F.J., Molina, J.A., de Bustos, F., Orti-Pareja, M., Benito-Leon, J., Tallon-Barranco, A. et al. (1999). Serum levels of beta-carotene, alpha-carotene and vitamin A in patients with Alzheimer's disease. *European Journal of Neurology*, **6**, 495–497.

Jin, K., Peel, A.L., Mao, X.O., Xie, L., Cottrell, B.A., Henshall, D.C. et al. (2004). Increased hippocampal neurogenesis in Alzheimer's disease. *Proceedings of the National Academy of Sciences of the United States of America*, **101**, 343–347.

Joh, T., Darland, T., Samuels, M., Wu, J.X. and Adamson, E.D. (1992). Regulation of epidermal growth factor receptor gene expression in murine embryonal carcinoma cells. *Cell Growth & Differentiation*, **3**, 315–325.

Kagechika, H. (2002). Novel synthetic retinoids and separation of the pleiotropic retinoidal activities. *Current Medicinal Chemistry*, **9**, 591–608.

Kagechika, H. and Shudo, K. (2005). Synthetic retinoids: recent developments concerning structure and clinical utility. *Journal of Medicinal Chemistry*, **48**, 1–9.

Kaplan, D.R., Matsumoto, K., Lucarelli, E. and Thiele, C.J. (1993). Induction of TrkB by retinoic acid mediates biologic responsiveness to BDNF and differentiation of human neuroblastoma cells. *Neuron*, **11**, 321–331.

Kehoe, P., Wavrant-De Vrieze, F., Crook, R., Wu, W.S., Holmans, P., Fenton, I. et al. (1999). A full genome scan for late onset Alzheimer's disease. *Human Molecular Genetics*, **8**, 237–245.

Kirby, A.J., Le Lain, R., Maharlouie, F., Mason, P., Nicholls, P.J., Smith, H.J. and Simons, C. (2003). Inhibition of retinoic acid metabolising enzymes by 2-(4-aminophenylmethyl)-6-hydroxy-3,4-dihydronaphthalene-1(2*H*)-one and related compounds. *Journal of Enzyme Inhibition and Medicinal Chemistry*, **18**, 27–33.

Klaholz, B.P., Renaud, J.-P., Mitschler, A., Zusi, C., Chambon, P., Gronemeyer, H. et al. (1998). Conformational adaptation of agonists to the human nuclear receptor RARγ. *Nature Structural Biology*, **5**, 199–202.

Kliewer, S.A., Umesono, K., Heyman, R.A., Mangelsdorf, D.J., Dyck, J.A. and Evans, R.M. (1992). Retinoid X-receptor coup-tf interactions modulate retinoic acid signaling. *Proceedings of the National Academy of Sciences of the United States of America*, **89**, 1448–1452.

Konig, G., Masters, C.L. and Beyreuther, K. (1990). Retinoic acid induced differentiated neuroblastoma cells show increased expression of the beta A4 amyloid gene of Alzheimer's disease and an altered splicing pattern. *FEBS Letters*, **269**, 305–310.

Krezel, W., Kastner, P. and Chambon, P. (1999). Differential expression of retinoid receptors in the adult mouse central nervous system. *Neuroscience*, **89**, 1291–1300.

Krumlauf, R., Marshall, H., Studer, M., Nonchev, S., Sham, M.H. and Lumsden, A. (1993). Hox homeobox genes and regionalisation of the nervous system. *Journal of Neurobiology*, **24**, 1328–1340.

Kurlandsky, S.B., Duell, E.A., Kang, S., Voorhees, J.J. and Fisher, G.J. (1996). Autoregulation of retinoic acid biosynthesis through regulation of retinol esterification in human keratinocytes. *Journal of Biological Chemistry*, **271**, 15346–15352.

Lahiri, D.K. and Nall, C. (1995). Promoter activity of the gene encoding the beta-amyloid precursor protein is up-regulated by growth factors, phorbol ester, retinoic acid and interleukin-1. *Molecular Brain Research*, **32**, 233–240.

Lambert, J.C., Chartier-Harlin, M.C., Cottel, D., Richard, F., Neuman, E., Guez, D. et al. (1999). Is the LDL receptor-related protein involved in Alzheimer's disease? *Neurogenetics*, **2**, 109–113.

Lampron, C., Rochette-Egly, C., Gorry, P., Dollé, P., Mark, M., Lufkin, T. et al. (1995). Mice deficient in cellular retinoic acid binding protein II (CRABPII) or in both CRABPI and CRABPII are essentially normal. *Development*, **121**, 539–548.

Lane, M.A. and Bailey, S.J. (2005). Role of retinoid signalling in the adult brain. *Progress in Neurobiology*, **75**, 275–293.

Lee, J.H. and Kim, K.T. (2004). Induction of cyclin-dependent kinase 5 and its activator p35 through the extracellular-signal-regulated kinase and protein kinase A pathways during retinoic-acid mediated neuronal differentiation in human neuroblastoma SK-N-BE(2)C cells. *Journal of Neurochemistry*, **91**, 634–647.

Lengqvist, J., Mata De Urquiza, A., Bergman, A.C., Willson, T.M., Sjovall, J., Perlmann, T. et al. (2004). Polyunsaturated fatty acids including docosahexaenoic and arachidonic acid bind to the retinoid X receptor alpha ligand-binding domain. *Molecular Cell Proteomics*, **7**, 692–703.

Lerner, V., Miodownik, C., Gibel, A., Kovalyonok, E., Shleifer, T., Goodman, A.B. and Ritsner, M.S. (2007). Bexarotene as add- on to antipsychotic treatment in schizophrenia patients: A pilot open-label trial. *Clinical Neuropharmacology*, in press.

Li, Y.J., Scott, W.K., Hedges, D.J., Zhang, F., Gaskell, P.C., Nance, M.A. et al. (2002). Age at onset in two common neurodegenerative diseases is genetically controlled. *American Journal of Human Genetics*, **70**, 985–993.

Liang, X., Schnetz-Boutaud, N., Kenealy, S.J., Jiang, L., Bartlett, J., Lynch, B. et al. (2006). Covariate analysis of late-onset Alzheimer disease refines the chromosome 12 locus. *Molecular Psychiatry*, **11**, 280–285.

Liang, Y., Lin, S., Beyer, T.P., Zhang, Y., Wu, X., Bales, K.R. et al. (2004). A liver X receptor and retinoid X receptor heterodimer mediates apolipoprotein E expression, secretion and cholesterol homeostasis in astrocytes. *Journal of Neurochemistry*, **88**, 623–634.

Limpach, A., Dalton, M., Miles, R. and Gadson, P. (2000). Homocysteine inhibits retinoic acid synthesis: a mechanism for homocysteine-induced congenital defects. *Experimental Cell Research*, **260**, 166–174.

Link, C.D. (1995). Expression of human beta-amyloid peptide in transgenic *Caenorhabditis elegans*. *Proceedings of the National Academy of Sciences of the United States of America*, **92**, 9368–9372.

Longo, F.M., Yang, T., Xie, Y. and Massa, S.M. (2006). Small molecule approaches for promoting neurogenesis. *Current Alzheimer Research*, **3**, 5–10.

Lopez-Boado, Y.S., Klaus, M., Dawson, M.I. and Lopez-Otin, C. (1996). Retinoic acid-induced expression of apolipoprotein D and concomitant growth arrest in human breast cancer cells are mediated through a retinoic acid receptor RAR alpha-dependent signaling pathway. *Journal of Biological Chemistry*, **271**, 32105–32111.

Lopez-Carballo, G., Moreno, L., Masia, S., Perez, P. and Barettino, D. (2002). Activation of the phosphatidylinositol 3-kinase/Akt signaling pathway by retinoic acid is required for neural differentiation of SH-SY5Y human neuroblastoma cells. *Journal of Biological Chemistry*, **277**, 25297–25304.

Loudig, O., Babichuk, C., White, J., Abu-Abed, S., Mueller, C. and Petkovich, M. (2000). Cytochrome P450RAI(CYP26) promoter: a distinct composite retinoic acid response element underlies the complex regulation of retinoic acid metabolism. *Molecular Endocrinology*, **14**, 1483–1497.

Loudig, O., Maclean, G.A., Dore, N.L., Luu, L. and Petkovich, M. (2005). Transcriptional co-operativity between distant retinoic acid response elements in regulation of Cyp26A1 inducibility. *Biochemistry Journal*, **392**, 241–248.

Maden, M. (1999). Heads or tails? Retinoic acid will decide. *Bioessays*, **21**, 809–812.

Mahley, R.W. and Rall, S.C. Jr. (2000). Apolipoprotein E: far more than a lipid transport protein. *Annual Review of Genomics and Human Genetics*, **1**, 507–537.

Mangelsdorf, D.J. and Evans, R.M. (1995). The RXR heterodimers and orphan receptors. *Cell*, **83**, 841–850.

Mangelsdorf, D.J., Thummel, C., Beato, M., Herrlich, P., Schutz, G., Umesono, K. et al. (1995). The nuclear receptor superfamily—the 2nd decade. *Cell*, **83**, 835–839.

Matsubara, E., Soto, C., Governale, S., Frangione, B. and Ghiso, J. (1996). Apolipoprotein J and Alzheimer's amyloid beta solubility. *Biochemical Journal*, **316**, 671–679.

Mata De Urquiza, A., Solomin, L. and Perlmann, T. (1999). Feedback-inducible nuclear-receptor-driven reporter gene expression in transgenic mice. *Proceedings of the National Academy of Sciences of the United States of America*, **96**, 13270–13275.

Mazur-Kolecka, B., Frackowiak, J., Carroll, R.T. and Wisniewski, H.M. (1997). Accumulation of Alzheimer amyloid-beta peptide in cultured myocytes is enhanced by serum and reduced by cerebrospinal fluid. *Journal of Neuropathology and Experimental Neurology*, **56**, 263–272.

McCaffery, P. and Deutsch, C.K. (2005). Macrocephaly and the control of brain growth in autistic disorders. *Progress in Neurobiology*, **77**, 38–56.

McCaffery, P. and Dräger, U.C. (1994). High levels of a retinoic-acid generating dehydrogenase in the meso-telencephalic dopamine system. *Proceedings of the National Academy of Sciences of the United States of America*, **91**, 7772–7776.

McCaffery, P., Zhang, J. and Crandall, J.E. (2006). Retinoic acid signaling and function in the adult hippocampus. *Journal of Neurobiology*, **66**, 780–791.

McCaffery, P., Wagner, E., O'Neil, J., Petkovich, M. and Dräger, U.C. (1999). Dorsal and ventral retinoic territories defined by retinoic acid synthesis, break-down and nuclear receptor expression. *Mechanisms of Development*, **85**, 203–214.

McCusker, S.M., Curran, M.D., Dynan, K.B., McCullagh, C.D., Urquhart, D.D., Middleton, D. et al. (2001). Association between polymorphism in regulatory region of gene encoding tumour necrosis factor alpha and risk of Alzheimer's disease and vascular dementia: a case–control study. *Lancet*, **357**, 436–439.

McSorley, L.C. and Daly, A.K. (2000). Identification of human cytochrome P450 isoforms that contribute to all-trans-retinoic acid 4-hydroxylation. *Biochemical Pharmacology*, **60**, 517–526

Mecocci, P., Polidori, M.C., Cherubini, A., Ingegni, T., Mattioli, P., Catani, M. et al. (2002). Lymphocyte oxidative DNA damage and plasma antioxidants in Alzheimer disease. *Archives of Neurology*, **59**, 794–798.

Melino, G., Draoui, M., Bernardini, S., Bellincampi, L., Reichert, U. and Cohen, P. (1996). Regulation by retinoic acid of insulin-degrading enzyme and of a related endoprotease in human neuroblastoma cell lines. *Cell Growth & Differentiation*, **7**, 787–796.

Merched, A., Serot, J.M., Visvikis, S., Aguillon, D., Faure, G. and Siest, G. (1998). Apolipoprotein E, transthyretin and actin in the CSF of Alzheimer's patients: relation with the senile plaques and cytoskeleton biochemistry. *FEBS Letters*, **425**, 225–228.

Mesulam, M.M. (2000). A plasticity-based theory of the pathogenesis of Alzheimer's disease. *Annals of the New York Academy of Science*, **924**, 42–52.

Mey, J. and McCaffery, P. (2004). Retinoic acid signaling in the nervous system of adult vertebrates. *Neuroscientist*, **10**, 409–421.

Mikkonen, M., Soininen, H., Tapiola, T., Alafuzoff, I. and Miettinen, R. (1999). Hippocampal plasticity in Alzheimer's disease: changes in highly polysialylated NCAM immunoreactivity in the hippocampal formation. *European Journal of Neuroscience*, **11**, 1754–1764.

Mikkonen, M., Soininen, H., Alafuzof, I. and Miettinen, R. (2001). Hippocampal plasticity in Alzheimer's disease. *Reviews in Neuroscience*, **12**, 311–325.

Misner, D.L., Jacobs, S., Shimizu, Y., de Urquiza, A.M., Solomin, L., Perlmann, T. et al. (2001). Vitamin A deprivation results in reversible loss of hippocampal long-term synaptic plasticity. *Proceedings of the National Academy of Sciences of the United States of America*, **98**, 11714–11719.

Moechars, D., Dewachter, I., Lorent, K., Reverse, D., Baekelandt, V., Naidu, A. et al. (1999). Early phenotypic changes in transgenic mice that overexpress different mutants of amyloid precursor protein in brain. *Journal of Biological Chemistry*, **274**, 6483–6492.

Muindi, J.R., Young, C.W. and Warrell, J.R.P. (1994). Clinical pharmacology of all-*trans* retinoic acid. *Leukemia*, **8**, 1807–1812.

Mulvihill, M.J., Kan, J.L., Beck, P., Bittner, M., Cesario, C., Cooke, A. et al. (2005). Potent and selective [2-imidazol-1-yl-2-(6-alkoxy-naphthalen-2-yl)-1-methyl-ethyl]-dimethylamines as retinoic acid metabolic blocking agents (RAMBAs). *Bioorganic & Medicinal Chemistry Letters*, **15**, 1669–1673.

Mummery, C.L., Slager, H., Kruijer, W., Feijen, A., Freund, E., Koornneef, I. et al. (1990). Expression of transforming growth factor beta 2 during the differentiation of murine embryonal carcinoma and embryonic stem cells. *Developmental Biology*, **137**, 161–170.

Myers, A.J. and Goate, A.M. (2001). The genetics of late-onset Alzheimer's disease. *Current Opinion in Neurology*, **14**, 433–440.

Napoli, J.L. (1996). Retinoic acid biosynthesis and metabolism. *The FASEB Journal*, **10**, 993–1001.

Napoli, J.L. (1999). Interactions of retinoid binding proteins and enzymes in retinoid metabolism. *Biochimica et Biophysica Acta*, **1440**, 139–162.

Nayak, N., Harrison, E.H. and Hussain, M.M. (2001). Retinyl ester secretion by intestinal cells: a specific and regulated process dependent on assembly and secretion of chylomicrons. *Journal of Lipid Research*, **42**, 272–280.

Niederreither, K., Abu-Abed, S., Schuhbaur, B, Petkovich, M., Chambon, P. and Dolle, P. (2002). Genetic evidence that oxidative derivatives of retinoic acid are not involved in retinoid signaling during mouse development. *Nature Genetics*, **3**, 84–88.

Njar, V.C.O. (2002). Cytochrome p450 retinoic acid 4-hydroxylase inhibitors: potential agents for cancer therapy. *Mini Reviews in Medicinal Chemistry*, **2**, 261–269.

Njar, V.C.O., Nnane, I.P. and Brodie, A.M.H. (2000). Potent inhibition of retinoic acid metabolism enzyme(s) by novel azolyl retinoids. *Bioorganic & Medicinal Chemistry Letters*, **10**, 1905–1908.

Nugent, P., Potchinsky, M., Lafferty, C. and Greene, R.M. (1995). TGF-beta modulates the expression of retinoic acid-induced RAR-beta in primary cultures of embryonic palate cells. *Experimental Cell Research*, **220**, 495–500.

Norum, K.R. and Blomhoff, R. (1992). McCollum award lecture, 1992—vitamin-A absorption, transport, cellular uptake, and storage. *American Journal of Clinical Nutrition*, **56**, 735–744.

Nowotny, P., Hinrichs, A.L., Smemo, S., Kauwe, J.S., Maxwell, T., Holmans, P. et al. (2005). Association studies between risk for late-onset Alzheimer's disease and variants in insulin degrading enzyme. *American Journal of Medical Genetics - Part B Neuropsychiatric Genetics*, **136**, 62–68.

Ogunniyi, A., Baiyewu, O., Gureje, O., Hall, K.S., Unverzagt, F., Siu, S.H. et al. (2000). Epidemiology of dementia in Nigeria: results from the Indianapolis–Ibadan study. *European Journal of Neurology*, **7**, 485–490.

O'Reilly, K.C., Shumake, J., Gonzalez-Lima, F., Lane, M.A. and Bailey, S.J. (2006). Chronic administration of 13-cis-retinoic acid increases depression-related behavior in mice. *Neuropsychopharmacology*, **31**, 1919–1927.

Oosterveen, T., Niederreither, K., Dolle, P., Chambon, P., Meijlink, F. and Deschamps, J. (2003). Retinoids regulate the anterior expression boundaries of 5′ Hoxb genes in posterior hindbrain. *EMBO Journal*, **22**, 262–269.

Osanai, M. and Petkovich, M. (2005). Expression of the retinoic acid-metabolizing enzyme CYP26A1 limits programmed cell death. *Molecular Pharmacology*, **67**, 1808–1817.

Palha, J.A. and Goodman, A.B. (2006). Thyroid hormones and retinoids: a possible link between genes and environment in schizophrenia. *Brain Research Reviews*, **51**, 61–71.

Palha, J.A., Episkopou, V., Maeda, S., Shimada, K., Gottesman, M.E. and Saraiva, M.J.M. (1994). Thyroid hormone metabolism in a transthyretin-null mouse strain. *Journal of Biological Chemistry*, **269**, 33135–33139.

Palha, J.A., Moreira, P., Wisniewski, T., Frangione, B. and Saraiva, M.J. (1996). Transthyretin gene in Alzheimer's disease patients. *Neuroscience Letters*, **204**, 212–214.

Palha, J.A., Hays, M.T., Morreale de Escobar, G., Episkopou, V., Gottesman, M. and Saraiva, M.J.M. (1997). Transthyretin is not essential for thyroxine to reach the brain and other tissues in a transthyretin null mouse strain. *American Journal of Physiology*, **272**, E485–E493.

Palha, J.A., Fernandes, R., Morreale de Escobar, G., Episkopou, V., Gottesman, M. and Saraiva, M.J.M. (2000). Transthyretin regulates thyroid hormone levels in the choroid plexus but not in the brain: study in a transthyretin-null mouse model. *Endocrinology*, **141**, 3267–3272.

Palha, J.A., Nissanov, J., Fernandes, R., Sousa, J.C., Bertrand, L., Dratman, M.B. et al. (2002). Thyroid hormone distribution in the mouse brain: the role of transthyretin. *Neuroscience*, **113**, 837–842.

Pan, J., Kao, Y.L., Joshi, S., Jeetendran, S., Dipette, D. and Singh, U.S. (2005). Activation of Rac1 by phosphatidylinositol 3-kinase in vivo: role in activation of mitogen-activated protein kinase (MAPK) pathways and retinoic acid-induced neuronal differentiation of SH-SY5Y cells. *Journal of Neurochemistry*, **93**, 571–583.

Patel, J.B., Huynh, C.K., Handratta, V.D., Gediya, L.K., Brodie, A.M.H., Goloubeva, O.G. et al. (2004). Novel retinoic acid metabolism blocking agents endowed with multiple biological activities are efficient growth inhibitors of human breast and prostate cancer cells in vitro and a human breast tumor xenograft in nude mice. *Journal of Medicinal Chemistry*, **47**, 6716–6729.

Pericak-Vance, M.A., Bass, M.L., Yamaoka, L.H., Gaskell, P.C., Scott, W.K., Terwedow, H.A. et al. (1998). Complete genomic screen in late-onset familial Alzheimer's disease. *Neurobiology of Aging*, **19** (Suppl. 1), S39–S42.

Pericak-Vance, M.A., Grubber, J., Bailey, L.R., Hedges, D., West, S., Santoro, L. et al. (2000). Identification of novel genes in late-onset Alzheimer's disease. *Experimental Gerontology*, **35**, 1343–1352.

Perkins, A.J., Hendrie, H.C., Callahan, C.M., Gao, S., Unverzagt, F.W., Xu, Y. et al. (1999). Association of antioxidants with memory in a multiethnic elderly sample. Using the Third National Health and Nutrition Examination Survey. *American Journal of Epidemiology*, **150**, 37–44.

Poduslo, S.E. and Yin, X. (2001). Chromosome 12 and late-onset Alzheimer's disease. *Neuroscience Letters*, **310**, 188–190.

Poduslo, S.E., Yin, X., Hargis, J., Brumback, R.A., Mastrianni, J.A. and Schwankhaus, J. (1999). A familial case of Alzheimer's disease without tau pathology may be linked with chromosome 3 markers. *Human Genetics*, **105**, 32–37.

Poirier, J. (2005). Apolipoprotein E, cholesterol transport and synthesis in sporadic Alzheimer's disease. *Neurobiology of Aging*, **26**, 355–361.

Prinzen, C., Muller, U., Endres, K., Fahrenholz, F. and Postina, R. (2005). Genomic structure and functional characterization of the human ADAM10 promoter. *FASEB Journal*, **19**, 1522–1524.

Pritchard, A., Harris, J., Pritchard, C.W., St Clair, D., Lemmon, H., Lambert, J.C. et al. (2005). Association study and meta-analysis of low-density lipoprotein receptor related protein in Alzheimer's disease. *Neuroscience Letters*, **382**, 221–226.

Pulford, D.J., Henry, B., Adams, F., Harries, D.N., Mallinson, D.J., Reid, I.C. et al. (2002). Chronic administration of lithium chloride down-regulates transthyretin mRNA expression in rat brain. In *Society for Neuroscience 32nd Annual Meeting. Abstract Viewer/Itinerary Planner Program No. 308.2*. Society for Neuroscience, Washington, DC. Online.

Puskas, L.G., Kitajka, K., Nyakas, C., Barcelo-Coblijn, G., and Farkas, T. (2003). Short-term administration of omega 3 fatty acids from fish oil results in increased transthyretin transcription in old rat hippocampus. *Proceedings of the National Academy of Sciences of the United States of America*, **100**, 1580–1585.

Quadro, L., Blaner, W.S., Salchow, D.J., Vogel, S., Piantedosi, R., Gouras, P. et al. (1999). Impaired retinal function and vitamin A availability in mice lacking retinol-binding protein. *EMBO Journal*, **18**, 4633–4644.

Rademakers, R., Cruts, M., Dermaut, B., Sleegers, K., Rosso, S.M., Van den Broeck, M. et al. (2002). Tau negative frontal lobe dementia at 17q21: significant finemapping of the candidate region to a 4.8 cM interval. *Molecular Psychiatry*, **7**, 1064–1074.

Ren, R.F., Hawver, D.B., Kim, R.S. and Flanders, K.C. (1997). Transforming growth factor-beta protects human hNT cells from degeneration induced by beta-amyloid peptide: involvement of the TGF-beta type II receptor. *Molecular Brain Research*, **48**, 315–322.

Ren, R.F., Lah, J.J., Diehlmann, A., Kim, E.S., Hawver, D.B., Levey, A.I. et al. (1999). Differential effects of transforming growth factor-beta(s) and glial cell line-derived neurotrophic factor on gene expression of presenilin-1 in human post-mitotic neurons and astrocytes. *Neuroscience*, **93**, 1041–1049.

Rehn, A.E. and Rees, S.M. (2005). Investigating the neurodevelopmental hypothesis of schizophrenia. *Clinical and Experimental Pharmacology and Physiology*, **32**, 687–696.

Reijntjes, S., Blentic, A., Gale, E. and Maden, M. (2005). The control of morphogen signalling: regulation of the synthesis and catabolism of retinoic acid in the developing embryo. *Developmental Biology*, **285**, 224–237.

Renaud, J.P., Rochel, N., Ruff, M., Vivat, V., Chambon, P., Gronemeyer, H. et al. (1995). Crystal structure of the RAR-γ ligand-binding domain bound to all-*trans* retinoic acid. *Nature*, **378**, 681–689.

Riisoen, H. (1988). Reduced prealbumin (transthyretin) in CSF of severely demented patients with Alzheimer's disease. *Acta Neurology Scandinavia*, **78**, 455–459.

Rioux, L. and Arnold, S.E. (2005). The expression of retinoic acid receptor alpha is increased in the granule cells of the dentate gyrus in schizophrenia. *Psychiatry Research*, **133**, 13–21.

Rogaeva, E., Premkumar, S., Song, Y., Sorbi, S., Brindle, N., Paterson, A. et al. (1998). Evidence for an Alzheimer disease susceptibility locus on chromosome 12 and for further locus heterogeneity. *Journal of the American Medical Association*, **280**, 614–618.

Rossant, J., Zirngibl, R., Cado, D., Shago, M. and Giguère, V. (1991). Expression of a retinoic acid response element-hsplacZ transgene defines specific domains of transcriptional activity during mouse embryogenesis. *Genes and Development*, **5**, 1333–1344.

Sahin, M., Karauzum, S.B., Perry, G., Smith, M.A. and Aliciguzel, Y. (2005). Retinoic acid isomers protect hippocampal neurons from amyloid-beta induced neurodegeneration. *Neurotoxicology Research*, **7**, 243–250.

Sakai, Y., Crandall, J.E., Brodsky, J. and McCaffery, P. (2004). 13-*cis* retinoic acid (accutane) suppresses hippocampal cell survival in mice. *Annals of the New York Academy of Science*, **1021**, 436–440.

Saraiva, M.J. (2001). Transthyretin amyloidosis: a tale of weak interactions. *FEBS Letters*, **498**, 201–203.

Saraiva, M.J., Birken, S., Costa, P.P. and Goodman, D.S. (1984). Amyloid fibril protein in familial amyloidotic polyneuropathy, Portuguese type. Definition of molecular abnormality in transthyretin (prealbumin). *Journal of Clinical Investigation*, **74**, 104–119.

Sasaki, H., Sakaki, Y., Matsuo, H., Goto, I., Kuroiwa, Y., Sahashi, I. et al. (1984). Diagnosis of familial amyloidotic polyneuropathy by recombinant DNA techniques. *Biochemical and Biophysical Research Communications*, **125**, 636–642.

Satoh, J. and Kuroda, Y. (2000). Amyloid precursor protein beta-secretase (BACE) mRNA expression in human neural cell lines following induction of neuronal differentiation and exposure to cytokines and growth factors. *Neuropathology*, **20**, 289–296.

Schellenberg, G.D., D'Souza, I. and Poorkaj, P. (2000). The genetics of Alzheimer's disease. *Current Psychiatry Reports*, **2**, 158–164.

Schuetz, E.G., Beck, W.T. and Schuetz, J.D. (1996). Modulators and substrates of P-glycoprotein and cytochrome P4503A coordinately up-regulate these proteins in human colon carcinoma cells. *Molecular Pharmacology*, **49**, 311–318.

Schwarzman, A.L., Gregori, L., Vitek, M.P., Lyubski, S., Strittmatter, W.J., Enghilde, J.J. et al. (1994). Transthyretin sequesters amyloid-beta protein and prevents amyloid formation. *Proceedings of the National Academy of Sciences of the United States of America*, **91**, 8368–8372.

Scott, W.K., Yamaoka, L.H., Bass, M.P., Gaskell, P.C., Conneally, P.M., Small, G.W. et al. (1998). No genetic association between the LRP receptor and sporadic or late-onset familial Alzheimer disease. *Neurogenetics*, **1**, 179–183.

Scott, W.K., Grubber, J.M., Conneally, P.M., Small, G.W., Hulette, C.M., Rosenberg, C.K. et al. (2000). Fine mapping of the chromosome 12 late-onset Alzheimer disease locus: potential genetic and phenotypic heterogeneity. *American Journal of Human Genetics*, **66**, 922–932.

Serot, J.M., Christmann, D., Dubost, T. and Couturier, M. (1997). Cerebrospinal fluid transthyretin: aging and late onset Alzheimer's disease. *Journal of Neurology Neurosurgery and Psychiatry*, **63**, 506–508.

Seshadri, S., Beiser, A., Selhub, J., Jacques, P.F., Rosenberg, I.H., D'Agostino, R.B. et al. (2002). Plasma homocysteine as a risk factor for dementia and Alzheimer's disease. *New England Journal of Medicine*, **346**, 476–483.

Sessler, R.J. and Noy, N. (2005). A ligand-activated nuclear localization signal in cellular retinoic acid binding protein-II. *Molecular Cell*, **18**, 343–353.

Shudo, K., Kagechika, H., Yamazaki, N., Igarashi, M. and Tateda, C. (2004). A synthetic retinoid Am80 (Tamibarotene) rescues the memory deficit caused by scopolamine in a passive avoidance paradigm. *Biological and Pharmaceutical Bulletin*, **27**, 1887–1889.

Sirbu, I.O., Gresh, L., Barra, J. and Duester, G. (2005). Shifting boundaries of retinoic acid activity control hindbrain segmental gene expression. *Development*, **132**, 2611–2622.

Sizemore, N., Choo, C.K., Eckert, R.L. and Rorke, E.A. (1998). Transcriptional regulation of the EGF receptor promoter by HPV16 and retinoic acid in human ectocervical epithelial cells. *Experimental Cell Research*, **244**, 349–356.

Smith, M.A., Petot, G.J. and Perry, G. (1999). Diet and oxidative stress: a novel synthesis of epidemiological data on Alzheimer's Disease. *Journal of Alzheimer's Disease*, **1**, 203–206.

Soprano, D.R., Herbert, J., Soprano, K.J., Schon, E.A. and Goodman, D.S. (1985). Demonstration of transthyretin messenger-RNA in the brain and other extrahepatic tissues in the rat. *Journal of Biological Chemistry*, **260**, 1793–1798.

Sorbi, S., Forleo, P., Tedde, A., Cellini, E., Ciantelli, M., Bagnoli, S. et al. (2001). Genetic risk factors in familial Alzheimer's disease. *Mechanisms of Aging and Development*, **122**, 1951–1960.

Sousa, J.C., Grandela, C., Fernandez-Ruiz, J., de Miguel, R., de Sousa, L., Magalhães, A.I. et al. (2004). Transthyretin is involved in depression-like behaviour and exploratory activity. *Journal of Neurochemistry*, **88**, 1052–1058.

Sousa, J.C., Cardoso, I., Marques, F., Saraiva, M.J. and Palha, J.A. (2007). Transthyretin and Alzheimer's disease: where in the brain? *Neurobiology of Aging*, 28, 713–718.

Stein, T.D. and Johnson, J.A. (2002). Lack of neurodegeneration in transgenic mice overexpressing mutant amyloid precursor protein is associated with increased levels of transthyretin and the activation of cell survival pathways. *Journal of Neuroscience*, **22**, 7380–7388.

Stoppie, P., Borgers, M., Borghgraef, P., Dillen, L., Goossens, J., Sanz, G. et al. (2000). R115866 inhibits all-*trans*-retinoic acid metabolism and exerts retinoidal effects in rodents. *Journal of Pharmacology and Experimental Therapeutics*, **293**, 304–312.

Sullivan, G.M., Hatterer, J.A., Herbert, J., Chen, X., Roose, S.P., Attia, E. et al. (1999). Low levels of transthyretin in the CSF of depressed patients. *American Journal of Psychiatry*, **156**, 710–715.

Taimi, M., Helvig, C., Wisniewski, J., Ramshaw, H., White, J., Amad, M. et al. (2004). A novel human cytochrome P450, CYP26C1, involved in metabolism of 9-*cis* and all-*trans* isomers of retinoic acid. *Journal of Biological Chemistry*, **279**, 77–85.

Takahashi, J., Palmer, T.D. and Gage, F.H. (1999). Retinoic acid and neurotrophins collaborate to regulate neurogenesis in adult-derived neural stem cell cultures. *Journal of Neurobiology*, **38**, 65–81.

Tang, Y.P., Haslam, S.Z., Conrad, S.E. and Sisk, C.L. (2004). Estrogen increases brain expression of the mRNA encoding transthyretin, an amyloid beta scavenger protein. *Journal of Alzheimers Disease*, **6**, 413–420.

Tanzi, R.E. and Bertram, L. (2001). New frontiers in Alzheimer's disease genetics. *Neuron*, **32**, 181–184.

Teng, M., Duong, T.T., Klein, E.S., Pino, M.E. and Chandraratna, R.A. (1996). Identification of a retinoic acid receptor r subtype specific agonist. *Journal of Medicinal Chemistry*, **39**, 3035–3038.

Terrisse, L., Poirier, J., Bertrand, P., Merched, A., Visvikis, S., Siest, G. et al. (1998). Increased levels of apolipoprotein D in cerebrospinal fluid and hippocampus of Alzheimer's patients. *Journal of Neurochemistry*, **71**, 1643–1650.

Thompson Haskell, G., Maynard, T.M., Shatzmiller, R.A. and Lamantia, A.S. (2002). Retinoic acid signaling at sites of plasticity in the mature central nervous system. *Journal of Comparative Neurology*, **452**, 228–241.

Trinchese, F., Liu, S., Battaglia, F., Walter, S., Mathews, P.M. and Arancio, O. (2004). Progressive age-related development of Alzheimer-like pathology in APP/PS1 mice. *Annals of Neurology*, **55**, 801–814.

Trofimova-Griffin, M.E. and Juchau, M.R. (2002). Developmental expression of cytochrome CYP26B1 (P450RAI-2) in human cephalic tissues. *Developmental Brain Research*, **136**, 175–178.

Turner, P.R., O'Connor, K., Tate, W.P. and Abraham, W.C. (2003). Roles of amyloid precursor protein and its fragments in regulating neural activity, plasticity and memory. *Progress in Neurobiology*, **70**, 1–32.

Umemiya, H., Fukasawa, H., Ebisawa, M., Eyrolles, L., Kawachi, E., Eisenmann, G. et al. (1997). Regulation of retinoidal actions by diazepinylbenzoic acids. Retinoid synergists which activate the RXR-RAR heterodimers. *Journal of Medicinal Chemistry*, **40**, 4222–4234.

Valdenaire, O., Maus-Moatti, M., Vincent, J.D., Mallet, J. and Vernier, P. (1998). Retinoic acid regulates the developmental expression of dopamine D2 receptor in rat striatal primary cultures. *Journal of Neurochemistry*, **71**, 929–936.

Van Bennekum, A.M., Wei, S., Gamble, M.V., Vogel, S., Piantedosi, R., Gottesman, M. et al. (2001). Biochemical basis for depressed serum retinol levels in transthyretin-deficient mice. *Journal of Biological Chemistry*, **276**, 1107–1113.

Van Heusden, J., Van Ginckel, R., Bruwiere, H., Moelans, P., Janssen, B., Floren, W. et al. (2002). Inhibition of all-*trans*-retinoic acid metabolism by R116010 induces antitumour activity. *British Journal of Cancer*, **86**, 605–611.

Van Wauwe, J., Coene, M.C., Goossens, J., Cools, W. and Monbaliu, J. (1990). Effects of cytochrome P450 inhibitors on the *in vivo* metabolism of all-*trans*-retinoic acid in rats. *Journal of Pharmacology and Experimental Therapeutics*, **252**, 365–369.

Van Wauwe, J., Van Nyen, G., Coene, M.C., Stoppie, P., Cools, W., Goossens, J. et al. (1992). Liarozole, an inhibitor of retinoic acid metabolism, exerts retinoid-mimetic effects *in vivo*. *Journal of Pharmacology and Experimental Therapeutics*, **261**, 773–779.

Vasios, G., Mader, S., Gold, J.D., Leid, M., Lutz, Y., Gaub, M.P. et al. (1991). The late retinoic acid induction of laminin-B1 gene-transcription involves RAR binding to the responsive element. *EMBO Journal*, **19**, 1149–1159.

Vekrellis, K., Ye, Z., Qiu, W.Q., Walsh, D., Hartley, D., Chesneau, V. et al. (2000). Neurons regulate extracellular levels of amyloid beta-protein via proteolysis by insulin-degrading enzyme. *Journal of Neuroscience*, **20**, 1657–1665.

Wagner, E., Luo, T. and Drager, U.C. (2002). Retinoic acid synthesis in the postnatal mouse brain marks distinct developmental stages and functional systems. *Cerebral Cortex*, **12**, 1244–1253.

Wang, H.F. and Liu, F.C. (2005). Regulation of multiple dopamine signal transduction molecules by retinoids in the developing striatum. *Neuroscience*. **134**, 97–105.

Wang, L., Mear, J.P., Kuan, C.Y. and Colbert, M.C. (2005a). Retinoic acid induces CDK inhibitors and growth arrest specific (Gas) genes in neural crest cells. *Development Growth and Differentiation*, **47**, 119–130.

Wang, T.W., Zhang, H. and Parent, J.M. (2005b). Retinoic acid regulates postnatal neurogenesis in the murine subventricular zone-olfactory bulb pathway. *Development*, **132**, 2721–2732.

Watanabe, C.M.H., Wolffram, S., Ader, P., Rimbach, G., Packer, L., Maguire, J.J. et al. (2001). The in vivo neuromodulatory effects of the herbal medicine ginkgo biloba. *Proceedings of the National Academy of Sciences of the United States of America*, **98**, 6577–6580.

Webber, K.M., Raina, A.K., Marlatt, M.W., Zhu, X., Prat, M.I., Morelli, L. et al. (2005). The cell cycle in Alzheimer disease: a unique target for neuropharmacology. *Mechansims of Ageing and Development*, **126**, 1019–1025.

Wei, S., Episkopou, V., Piantedosi, R., Maeda, S., Shimada, K., Gottesman, M.E. et al. (1995). Studies on the metabolism of retinol and retinol-binding protein in transthyretin-deficient mice produced by homologous recombination. *Journal of Biological Chemistry*, **270**, 866–870.

Werner, E.A. and Deluca, H.F. (2002). Retinoic acid is detected at relatively high levels in the CNS of adult rats. *American Journal of Physiology Endocrinology and Metabolism*, **282**, E672–E678.

Weston, A.D., Blumberg, B. and Underhill, T.M. (2003). Active repression by unliganded retinoid receptors in development: less is sometimes more. *Journal of Cell Biology*, **161**, 223–228.

White, J.A., Beckett-Jones, B., Guo, Y.D., Dilworth, J., Bonasoro, J., Jones, G. et al. (1997). cDNA cloning of human retinoic acid-metabolizing enzyme (hP450RAI) identifies a novel family of cytochromes P450 (CYP26). *Journal of Biological Chemistry*, **272**, 18538–18541.

White, J.A., Ramshaw, H., Taimi, M., Stangle, W., Zhang, A., Everingham, S. et al. (2000). Identification of the human cytochrome P450, P450RAI-2, which is predominantly expressed in the adult cerebellum and is responsible for all-trans-retinoic acid metabolism. *Proceedings of the National Academy of Sciences of the United States of America*, **97**, 6403–6408.

Wines-Samuelson, M. and Shen, J. (2005). Presenilins in the developing, adult, and aging cerebral cortex. *Neuroscientist*, **11**, 441–451.

Wouters, W., van Dun, J., Dillen, A., Coene, M.-C., Cools, W. and De Coster, R. (1992). Effects of liarozole, a new antitumoral compound, on retinoic acid-induced metabolism in MCF-7 human breast cancer cells. *Cancer Research*, **52**, 2841–2846.

Wuarin, L. and Sidell, N. (1991). Differential susceptibilities of spinal cord neurons to retinoic acid-induced survival and differentiation. *Developmental Biology*, **144**, 429–435.

Wyss-Coray, T., Lin, C., Yan, F., Yu, G.Q., Rohde, M., McConlogue, L. et al. (2001). TGF-beta 1 promotes microglial amyloid-beta clearance and reduces plaque burden in transgenic mice. *Nature Medicine*, **7**, 612–618.

Xie, P., Cheung, W.M., Ip, F.C., Ip, N.Y. and Leung, M.F. (1997). Induction of TrkA receptor by retinoic acid in leukaemia cell lines. *Neuroreport*, **8**, 1067–1070.

Yamamoto, M., Drager, U.C., Ong, D.E. and McCaffery, P. (1998). Retinoid-binding proteins in the cerebellum and choroid plexus and their relationship to regionalized retinoic acid synthesis and degradation. *European Journal of Biochemistry*, **257**, 344–350.

Yang, Q., Graham, T.E., Mody, N., Preitner, F., Peroni, O.D., Zabolotny, J.M. et al. (2005). Serum retinol binding protein 4 contributes to insulin resistance in obesity and type 2 diabetes. *Nature*, **436**, 356–362.

Yang, Y.X., Quitschke, W.W. and Brewer, G.J. (1998). Upregulation of amyloid precursor protein gene promoter in rat primary hippocampal neurons by phorbol ester, IL-1 and retinoic acid, but not by reactive oxygen species. *Molecular Brain Research*, **60**, 40–49.

Yee, S.W., Jarno, L., Gomaa, M.S., Elford, C., Ooi, L.-L., Coogan, M.P. et al. (2005). Novel tetralone-derived retinoic acid metabolism blocking agents: synthesis and *in vitro* evaluation with liver microsomal and MCF-7 CYP26A1 cell assays. *Journal of Medicinal Chemistry*, 2005, **48**, 7123–7131.

Yoshizawa, M., Miyazaki, H. and Kojima, S. (1998). Retinoids potentiate transforming growth factor-beta activity in bovine endothelial cells through up-regulating the expression of transforming growth factor-beta receptors. *Journal of Cell Physiology*, **176**, 565–573.

Yu, G., Nishimura, M., Arawaka, S., Levitan, D., Zhang, L.L., Tandon, A. et al. (2000). Nicastrin modulates presenilin-mediated notch/glp-1 signal transduction and beta APP processing. *Nature*, **407**, 48–54.

Zhang, Q.Y., Dunbar, D. and Kaminsky, L. (2000). Human cytochrome P-450 metabolism of retinals to retinoic acids. *Drug Metabolism and Disposition*, **28**, 292–297.

Zerbinatti, C.V. and Bu, G. (2005). LRP and Alzheimer's disease. *Reviews in Neuroscience*, **16**, 123–135.

Zetterstrom, R.H., Simon, A., Giacobini, M.M.J., Ericksson, U. and Olson, L. (1994). Localization of cellular retinoid-binding proteins suggests specific roles for retinoids in the adult central nervous system. *Neuroscience*, **62**, 899–918.

Zetterstrom, R.H., Lindqvist, E., Mata de Urquiza, A., Tomac, A., Eriksson, U., Perlmann, T. et al. (1999). Role of retinoids in the CNS: differential expression of retinoid binding proteins and receptors and evidence for presence of retinoic acid. *European Journal of Neuroscience*, **11**, 407–416.

Zhu, X., Webber, K.M., Casadesus, G., Raina, A.K., Lee, H.G., Marlatt, M. et al. (2004). Mitotic and gender parallels in Alzheimer disease: therapeutic opportunities. *Current Drug Targets*, **5**, 559–563.

Zigova, T., Willing, A.E., Saporta, S., Daadi, M.M., McGrogan, M.P., Randall, T.S. et al. (2001). Apoptosis in cultured hNT neurons. *Brain Research Development Brain Research*, **127**, 63–70.

Zlokovic, B.V., Martel, C.L., Matsubara, E., McComb, J.G., Zheng, G., McCluskey, R.T. et al. (1996). Glycoprotein 330/megalin: probable role in receptor-mediated transport of apolipoprotein J alone and in a complex with Alzheimer disease amyloid beta at the blood-brain and blood-cerebrospinal fluid barriers. *Proceedings of the National Academy of Sciences of the United States of America*, **93**, 4229–4234.

Zubenko, G.S., Hughes, H.B., Stiffler, J.S., Hurtt, M.R. and Kaplan, B.B. (1998). A genome survey for novel Alzheimer disease risk loci: results at 10-cM resolution. *Genomics*, **50**, 121–128.

Part II

Parkinson's Disease

9 Parkinson's Disease: What Is It? What Causes It? And How Can It Be Cured?

Tom Foltynie, Andrew W. Michell, and Roger A. Barker

CONTENTS

Parkinson's disease (PD) is a common neurodegenerative disorder of the brain and typically presents with a disorder of movement. The core pathological event under-lying the condition is the loss of the dopaminergic nigrostriatal pathway with the formation of α-synuclein positive Lewy bodies. As a result, drugs that target the degenerating dopaminergic network within the brain work well, at least in the early stages of the disease. Unfortunately with time, these therapies fail and produce their own unique side-effect profile and this, coupled to the more diffuse pathological and clinical findings in advancing disease, has led to a search for more effective therap-ies. In this review, we discuss the epidemiology of PD with an emphasis on risk factors and then discuss the clinical features and pathogenic pathways before con-sidering some of the newer more curative therapies.

9.1 INTRODUCTION

PD is a common neurodegenerative condition, which takes its name from James Parkinson (Parkinson, 1817) and is characterized by a triad of motor symptoms: bradykinesia, rigidity, resting tremor. The pathological hallmark of the condition is degeneration of the dopaminergic nigrostriatal tract with the formation of intracellular Lewy Bodies (Jellinger, 1987), although cell loss is more widespread than this and involves a host of sites and neurotransmitters. The cause of PD is not known, but as more has been learnt regarding the etiological risk factors for PD, it has become clear that this question is too simplistic and we should perhaps be attempting to identify a range of etiological risk factors—both genetic and environmental that together pre-dispose to the development of *various forms* of Parkinson-type diseases.

The complexity of the subject will become apparent but at the very outset of the discussion of Parkinson's epidemiology, it must first be made clear that there must be differentiation between primary forms of Parkinson's disease as opposed to second-ary forms of "parkinsonism" such as the well recognized vascular and drug induced forms of the disease. Secondary forms of parkinsonism, although important, will not be the subject of discussion in this chapter.

Within the concept of *primary* forms of Parkinson-type diseases, academics and clinicians consider a disease process that is *neurodegenerative* and has at its core a constellation of clinical features namely bradykinesia, that is, poverty or slowness of movements, muscle rigidity, and in many but not all cases, the presence of a tremor that is typically more obvious at rest and classically has a large amplitude and a 3–4 Hz frequency. Along with these typical motor features, there are frequently also cognitive, affective, and autonomic features all of which will be discussed in more detail in subsequent sections of this chapter. For ease of reference, we will refer to primary neurodegenerative forms of Parkinson-type diseases as PD and do not include other clinically and pathologically distinct conditions such as the Parkinson plus disorders—Progressive Supranuclear Palsy (PSP), Multiple Systems Atrophy (MSA), and Corticobasal degeneration (CBD)—although it is acknowledged that mistakes in clinical diagnoses occur and may only come to light in the few patients who have postmortem pathological examination of their brains.

It can thus be recognized that a major limitation in our search for a cause for Parkinson's disease is the lack of precision of the clinical definition. On one hand, it is tempting to disregard the clinical phenotype and redefine the disease based on a more precise definition such as pathological appearance. On the other hand, the major limitation of pathology is a practical one, in that performing a brainstem biopsy during life is impossible and implementation of pathological criteria can only occur postmortem.

Specific biomarkers for neurodegenerative diseases including PD would assist the field considerably and are being actively sought, partly driven by the observation that Lewy bodies (the pathological hallmark of PD) are predominantly composed of aggregated α-synuclein (Spillantini et al., 1997). Whether α-synuclein pathology may be detectable in tissues outside the CNS such as skin, muscle, GI tract, or platelets is also a focus of research (Ferrarese et al., 2001), and again we will discuss this later in this chapter.

9.2 EPIDEMIOLOGY OF PD

9.2.1 DESCRIPTIVE EPIDEMIOLOGY

The worldwide frequency of PD has been the subject of a multitude of studies. Comparisons are limited due to variable case definition, methods of case ascertainment, and also whether figures are presented as crude estimates, age standardized to a reference population, or as figures specific for varying age strata. Overall, in studies using screening methodologies, the *crude* prevalence of the disease ranges from 7–43 per 100,000 in Africa (Tekle-Haimanot et al., 1990; Attia et al., 1993) to 137–270 per 100,000 in Europe and the United States (Schoenberg et al., 1985; Acosta et al., 1989; Morgante et al., 1992; de Rijk et al., 1997; Milanov et al., 2000), with crude annual incidence figures ranging from 4.5 every 100,000 per year in Africa (records based only study) (Ashok et al., 1986) to 10–20 every 100,000 per year in Asia, Europe, and the United States (Morens et al., 1996; Chen et al., 2001; Van Den Eeden et al., 2003). However, comparing blacks and whites living under similar environmental conditions showed virtually the same prevalence ratios of PD (Schoenberg et al., 1985). Without doubt, it can be seen from epidemiological studies that the most important risk factor for PD is age (the incidence increasing with age), and there is a trend for a small excess in the numbers of PD cases among men than women, in both caucasian and noncaucasian populations. Although the explanation for this is not yet clear, gender differences in PD risk may be related to differential exposure to environmental risks, either in the workplace or in the home, genetic risk or protective factors on the X chromosome, or hormonal influences.

9.2.2 ANALYTIC EPIDEMIOLOGY

Previously, PD was considered primarily to be due to an environmental toxin; however, we now know of two genes, α-synuclein (Polymeropoulos et al., 1997) and LRRK2 (leucine rich repeat kinase 2) (Paisan-Ruiz et al., 2004; Zimprich et al., 2004), mutations of which cause autosomal dominant PD and three genes, parkin (Matsumine et al., 1997), PINK1 (Valente et al., 2004), and DJ-1 (Bonifati et al., 2003), that lead to autosomal recessive PD. Inevitably, a distinction arises between patients with obvious genetic forms of PD and the more common apparently sporadic forms of the disease. The previously identified genes do not account for the majority of cases of sporadic PD patients but have been hugely instructive on possible pathophysiological mechanisms that may underlie the sporadic forms of the disease. Before discussing these important genetic discoveries, the current knowledge regarding the role of environmental exposure and PD will be presented.

9.2.2.1 Environmental Risks for PD

A number of environmental factors have been implicated in the etiology of PD, through ecological, case–control and cohort studies. No single exposure has been consistently associated with PD although several candidate exposures warrant further study. If a single exposure is responsible for the majority of cases of sporadic PD, then it should be compatible with observed age, sex, and geographical patterns and fit in with observations of the temporal trends of the disease. Based on the clear difficulty in identifying such an agent, it is more likely that a single exposure is not the cause of all cases of sporadic PD (Tanner and Ben Shlomo, 1999).

The factors most associated with PD are:

Pesticides: A meta-analysis of studies published prior to 2000, was performed to evaluate rural living, well-water drinking, and pesticide exposure as PD risks (Priyadarshi et al., 2000). Significant heterogeneity was detected among the studies, and there was no attempt to evaluate different subtypes of pesticides or quantify exposure. Overall however, this meta-analysis suggested that the risk of PD is indeed increased following rural residence and exposure to pesticides but not from well-water drinking (Priyadarshi et al., 2000). Evidence that well-water drinking itself is a genuine risk factor for all forms of PD is therefore lacking. It remains possible however that regional contamination of well-water supplies occurs or has previously occurred, and a genuine association exists in specific areas.

Dietary constituents: Overall, dietary studies of PD risks have not been conclusive although it seems that elevated saturated fat intake may increase risk (Chen et al., 2003), and high caffeine intake (Ascherio et al., 2001) and foods rich in vitamin E (but not supplements) (Zhang et al., 2002) may be protective. Nonnutrient dietary components, where there is linkage of food items with a naturally occurring toxin, may be important and require further study. Furthermore, having said all this, it is also possible that development of subclinical PD may influence consuming habits and observed associations do not have a causal role.

Smoking: Apart from age, the most consistent epidemiological association is that cigarette smoking appears to afford protection from PD. This may be an important key to understanding the disease. In those studies that show statistically significant protection from smoking, the association appears to be strong (Odds Ratio (OR) ~0.5), consistent across multiple countries, and appears to have a dose–response relationship (Hernan et al., 2001; Paganini-Hill, 2001; Abbott et al., 2003). There are several possible confounders that may explain the observed relationship including the possibility that naturally occurring high levels of dopamine perhaps lead to both addiction and smoking behavior (Paulson, 1992), and it may be that these high levels confer protection from PD. Confounding by coffee or alcohol drinking (that might instead be the protective agents) is possible as these are both used more commonly by smokers. A third possible confounder is that certain personality types are more likely to develop PD, and are less likely to smoke. If the association between smoking and protection from PD is indeed a causal one, there are several potential mechanisms—inhibition of the activity of MAO-B (Baron, 1986), competitive binding at nicotinic receptors (Paganini-Hill, 2001), or by increasing levels of neurotrophic factors (Maggio et al., 1998).

The overall evidence suggests that environmental neurotoxins possibly including certain pesticides, combined with lack of protection from dietary antioxidants increase the risk of PD with advancing age. Putative mechanisms of cell toxicity have been elucidated through the identification of some of the genetic risk factors for rare kindreds with PD.

9.2.2.2   Genetic Epidemiology

There have been numerous kindreds identified among whom PD is inherited in a Mendelian pattern, although in the majority of cases it is truly sporadic. Nevertheless, it seems reasonable to conclude that familial clustering of the disease is genuine, and among the community-based studies results in a 1.3-fold to 3-fold elevated familial relative risk (Marder et al., 1996; Elbaz et al., 1999; Kuopio et al., 2001). The precise contribution of genetics to this observed familial clustering of PD can be elucidated by studying disease concordance rates in monozygotic (MZ) and dizygotic (DZ) twins. The early twin studies (Duvoisin et al., 1981; Ward et al., 1983; Marsden, 1987; Marttila et al., 1988), found low concordance rates (5%–8%) in both MZ and DZ twins with little evidence for an excess concordance in MZ twins.

However, clinical concordance may actually only become apparent after an interval of up to 26 years (Johnson et al., 1990), and so more recent studies (Tanner et al., 1999) have found similar concordance rates in MZ twins (16%) and DZ twins (11%) when twins with any age at disease onset were included, but a concordance rate for the disease of 100% in MZ twins and 16% in DZ twins if age at onset was below 50 years (relative risk (RR) of 6.0). This strongly supported a primarily inherited cause of *early onset* PD. Although this study included examination of all cases and cotwins by a qualified neurologist, the presence of subclinical disease among the "unaffected" twins could not be excluded. The development of ^{18}F-dopa PET scans as a research tool, has enabled imaging of the surviving nigrostriatal dopaminergic neurons, and thereby enabled the identification of preclinical cases of PD (see Section 9.6.4.) Using decline in ^{18}F-dopa uptake over 4 years as a marker for preclinical disease, concordance was found in 75% of MZ twins compared with 22% of DZ twins (Burn et al., 1992; Piccini et al., 1999b). This was not confined to young onset pairs of twins or cases with other affected family members and establishes that genetic component is extremely important even in late-onset PD patients although does not diminish the possibility of important concomitant environmental factors, either interacting with genetic risks or acting independently. On the basis of epidemiological and twin studies all suggesting a genetic contribution to PD susceptibility, attempts to map the position of responsible genes have been made.

α-Synuclein: The first PD gene locus was discovered within a large Italian family (Contursi kindred) who exhibited a dominant inheritance pattern with linkage identified to Chromosome 4q (Polymeropoulos et al., 1996). The gene at this site codes for α-synuclein, which is a major component of Lewy bodies (Spillantini et al., 1997). Subsequent explorations have revealed single base pair changes within the gene (missense mutations) which result in amino acid substitutions in the α-synuclein protein (A53T) (Polymeropoulos et al., 1997) and (A30P) (Kruger et al., 1998) in several unrelated kindreds. In addition there appears to be a gene dosage effect in that families with either a duplication or a triplication of α-synuclein also develop early onset PD and inherit this tendency in an autosomal dominant manner (Singleton et al., 2003; Ibanez et al., 2004). It would seem therefore that either mutated or excessive α-synuclein production can lead to neurodegeneration of the nigrostriatal pathway and early onset parkinsonism. Studies exploring the frequency of α-synuclein mutations among apparently sporadic PD cases find that this gene is not causally relevant for the vast majority of patients (Johnson et al., 2004; Berg et al., 2005a).

Parkin: A second genetic form of parkinsonism inherited in an autosomal recessive fashion has been identified and mapped to chromosome 6q25 (Matsumine et al., 1997), and the gene at that site which is subjected to either partial deletions or point mutations has been named "parkin." Patients with these mutations have degeneration of the substantia nigra pars compacta (SNc) and the locus coeruleus typical of PD but usually without the formation of Lewy bodies that represent the pathological hallmark of the sporadic form of the disease (Hayashi et al., 2000). It would seem therefore that at least one form of neurodegeneration that leads to dopaminergic cell death and features of parkinsonism can evolve without the formation of Lewy bodies.

Homozygous mutations in the parkin gene have been found to be responsible for 77% of patients with parkinsonism with an age of onset of 20 years or younger, but only 3% of patients with an onset between 30 and 45 years of age (Lucking et al., 2000). There is some evidence however that compound heterozygotes or individuals with even single parkin mutations can develop later onset PD accompanied by the formation of Lewy bodies (Farrer et al., 2001). The gene product "parkin" is a ubiquitin protein ligase thought to be involved in the degradation of abnormal proteins by the proteasome (McNaught et al., 2001).

PINK1: Further families with multiple cases of young onset PD inherited in an autosomal recessive manner, without an abnormality in the parkin gene, have shown a strong linkage (log of the odds ratio (LOD) score of >4.0) to a gene on the short arm of chromosome 1 (1p35–p36) (Valente et al., 2001), and further work identified the gene responsible as PINK1 (Valente et al., 2004). No neuropathological data is available for these families as yet, but it seems that PINK1 localizes to the mitochondria encoding a kinase enzyme and may have a role in protecting against oxidative stress. A further possibility is that PINK1 is involved in cell cycling and apoptotic cell death.

DJ-1: Linkage analysis using another family with early onset autosomal recessive PD (van Duijn et al., 2001) identified a further locus on chromosome 1. This locus has now been mapped and the gene responsible for causing disease among these patients identified as DJ-1 (Bonifati et al., 2003). The precise function of normal DJ-1 and the mechanism by which mutated DJ-1 leads to neurodegeneration remains unknown although the wild type protein is thought to be involved in a process of abnormal protein degradation known as sumoylation and localizes to the mitochondrial matrix (Zhang et al., 2005).

LRRK2: This gene causing autosomal dominant inheritance of PD was identified following linkage to chromosome 12 (Funayama et al., 2002) in a Japanese family, but with low disease penetrance, and then in families from the Basque region (Paisan-Ruiz et al., 2004) and the UK (Nichols et al., 2005). This gene has been found to account for 5%–13% of families with autosomal dominant PD with late onset and typical levodopa (L-dopa) responsive parkinsonian features (Berg et al., 2005b; Khan et al., 2005) but the common mutations in this gene account for perhaps only 1% of the sporadic forms of the disease (Biskup et al., 2005; Kay et al., 2005; Khan et al., 2005). There is evidence that mutant LRRK2 causes neuronal degeneration in cell culture and may interact with parkin in the production of protein aggregates (Smith et al., 2005). Diverse pathology has been seen postmortem, with neuronal loss in the SNc accompanied by Lewy bodies, neurofibrillary tangles, or in the absence of any inclusions (Zimprich et al., 2004). This again questions the precision of both the clinical and pathological definition of PD.

NR4A2: This gene encodes for a nuclear receptor (Nurr-1), necessary for nigral dopaminergic neuronal differentiation, and mutations within it were reported to account for 1 in 10 cases of familial PD but no cases of sporadic PD in one study (Le et al., 2003). In subsequent studies, however, no NR4A2 mutations were found among series of familial PD patients suggesting that the gene has a role among only a small number of such patients perhaps with a common ancestor (Zimprich et al., 2003; Karamohamed et al., 2005).

Mitochondrial dysfunction: A marked deficiency in the activity of complex 1 of the mitochondrial respiratory chain in the nigrostriatal system has been described in a proportion of PD patients (Schapira et al., 1989). Whether this deficiency is due to the presence of neurotoxins or is genetically determined has not been established and many individuals do not have a detectable change in complex 1 activity. Mitochondrial DNA encodes some of the subunits of complex 1, and a high rate of mutations has been observed in the mitochondrial DNA of PD patients compared to that of control patients. Several studies have now suggested that specific mitochondrial haplotypes are protective against PD within large populations with the sporadic form of the disease (van der Walt et al., 2003; Pyle et al., 2005). Mitochondrial dysfunction might lead to elevated production of reactive oxygen species, which leads to the oxidative stress observed in PD tissues (Parker and Swerdlow, 1998). Another hypothesis is that deficiency of ATP production due to mitochondrial dysfunction may lead to failure of the proteasomal proteolytic system (DeMartino and Slaughter, 1999).

Further genetic risks—genome wide linkage screening: (Gwinn-Hardy et al., 2000; Nicholl et al., 2002) Although suggestive linkage is reported in these studies, only one has achieved genome wide significant linkage to a specific region on chromosome 1p32 in late-onset disease (PARK10) (Hicks et al., 2002), although another found some evidence for linkage at five distinct chromosomal regions—PARK2, 17q, 8p, 5q, and 9q (Scott et al., 2001).

Further genetic risks—association studies: The simplicity of association studies has resulted in their frequent use to investigate various candidate genes, such as those coding for enzymes involved in the biotransformation of various chemicals including MPTP (1-methyl-4-phenyl-1,2,3,6-tetrahydropyridine) (Gilham et al., 1997), the most notable of which is Cytochrome P450 2D6 (CYP2D6). The activity of CYP2D6 is genetically determined, with some people having undetectable activity due to two defective alleles, these people being referred to as "poor metabolizers." Three polymorphisms are responsible for 95% of poor metabolizer phenotypes in Caucasians. A meta-analysis of available studies shows an overall risk of borderline significance (OR ~ 1.47) for the poor metabolizer status of the CYP2D6 enzyme and PD (Rostami-Hodjegan et al., 1998). It has been proposed that poor metabolizers are genetically susceptible to PD because of an impaired ability to detoxify neurotoxins that are metabolized by CYP2D6. In addition to CYP2D6, many other genes have been associated with PD in numerous studies. A review of all PD polymorphism association studies excluding CYP2D6 was published in 2000 (Tan et al., 2000). Attempts have been made to perform genome-wide association studies but no potential candidates have withstood attempts at replication (Foltynie et al., 2005; Maraganore et al., 2005).

Although far from being completely elucidated, there are numerous clues emerging from epidemiological and genetic studies that dopaminergic neuronal cell death occurs due to a combination of genetic and environmental factors operating on an interconnected pathway. This pathway seems to be intimately dependent on intact mitochondrial functioning and the ability to deal with oxidative stressors together with efficient metabolism of abnormal proteins by the proteasome or other protein metabolic pathways such as sumoylation. Failure of these systems due either to genetic polymorphisms or mutations, or exposure to certain neurotoxins seems to lead to progressive neuronal degeneration. Dopaminergic neurons in the (SNc) seem to be particularly vulnerable early on, the reason for which is unclear although the disease ultimately progresses to a more widespread cerebral neurodegeneration (see Section 9.5).

9.3   CLINICAL FEATURES OF PD

Two studies have sought to identify the optimum combination of clinical criteria for making the diagnosis of PD, by calculating the sensitivity, specificity, and positive predictive value for 15 possible features among 100 patients with pathologic evidence of diagnosis (Ward and Gibb, 1990; Hughes et al., 1992a). Application of strict clinical criteria improves diagnostic accuracy to 82%, but also leads to exclusion of more than 30% of pathologically genuine PD patients (Hughes et al., 1992a). Nevertheless, as a result of comparisons against the pathological definition of PD, there has been widespread adoption of the UK Brain Bank criteria (Hughes et al., 1992b) (Table 9.1) to make a clinical diagnosis of PD. Recent reevaluation of the clinical diagnoses of PD by the authors of the brain bank criteria suggests that although the positive predictive value of clinical diagnoses is high (98.6%), false negative cases indeed suggest a broader clinical picture of disease than previously thought (Hughes et al., 1992a, 2002).

The cornerstone of these clinical criteria is that the patients exhibit "bradykinesia" in some form or other. Slowness of movement or poverty of movement (hypokinesia) can enable the experienced clinician to diagnose "parkinsonism" with minimal further

TABLE 9.1
UK Parkinson's Disease Society Brain Bank Criteria for PD

At least two of the four cardinal signs
Bradykinesia, Resting tremor, Rigidity, Postural Instability
One sign must be bradykinesia, and there must be at least three supporting criteria—unilateral onset, rest
tremor, progressive nature, asymmetry, levodopa responsive, L-Dopa induced chorea, course >10 years
And no exclusion criteria—repeated strokes or head injuries, encephalitis, oculogyric crises, neuroleptic
exposure, more than one affected relative, sustained remission, unilateral after 3 years, gaze palsy,
cerebellar signs, severe autonomic signs, early dementia, Babinski sign, cerebral tumor, negative
response to L-dopa, MPTP exposure.

examination or investigation. However, a range of neurological disorders affecting movement and muscles as well as musculoskeletal problems can lead to difficulties performing simple motor tasks. The absence of *any* form of bradykinesia however excludes the diagnosis of PD according to the brain bank criteria.

The characteristic tremor of PD is most obvious when the affected limb is at rest, and the term "pill rolling" has been coined to describe the classical appearance of movements of the thumb and index finger in PD although this is not always seen. In addition, it is becoming more apparent that there are a set of patients with isolated asymmetric postural or resting tremors without obvious bradykinesia who may progress to more typical forms of PD responsive to dopaminergic therapy, many years after the onset of their disease (Pal et al., 2002).

Rigidity is most often referred to as "stiffness" by the patient. Similar to some forms of arthritis, stiffness may be worse in the mornings, and can, for example, lead to the mistaken diagnosis of a frozen shoulder. The coexistence of tremor with increased resistance to passive movements can be felt as a ratchet-like sensation known as cogwheeling. Patients frequently comment on other classical motor symptoms such as dragging a foot during walking, loss of manual dexterity, a change in handwriting—usually smaller (micrographia), and writing that becomes spidery or deteriorates over the course of a page. As the illness progresses patients frequently describe a difficulty with turning over in bed and episodes of freezing to the spot especially if confronted with tight spaces or narrow doorways, or a stooped shuffling gait with a sense of poor balance or instability.

As a result of long-term exposure to dopaminergic replacement for PD, patients frequently develop motor complications of the disease making assessments and treatment problematic. Fluctuations in motor symptoms in response to individual doses of medication, sometimes predictable but other times of sudden onset, follow the ongoing loss of dopaminergic nerve terminals which in the earlier stages were able to buffer the release of dopamine into the synapse following L-Dopa treatment. Prolonged drug treatment can also result in the development of dopa-induced dyskinesias characterized by writhing involuntary movements similar to chorea, and frequently necessitating changes in drug dosing or regimes.

In PD, there may also be mild disturbance of the autonomic nerves causing postural hypotension which may be made worse by medication. Other autonomic symptoms such as constipation, poor urinary stream, and erectile dysfunction are not uncommonly seen. It has also been noted that patients with PD frequently have disturbed sleep for some years before the development of other symptoms, such as acting out dreams that occur during REM sleep (in which there is normally complete loss of muscle tone—so-called REM sleep behavior disorder).

Cognitive disturbance in the context of PD has also been rigorously explored. In the early stages of PD, approximately one-third of patients perform poorly in detailed cognitive tasks but the majority of cases are unaware of cognitive problems (Foltynie et al., 2004). With extended follow up of cases, it seems that between 20% (Brown and Marsden, 1984) and 75% (Aarsland et al., 2003) of PD patients will ultimately develop dementia. It is currently impossible to predict whether patients will go on to develop a dementia based on their initial symptoms of PD although there is some suggestion that tremor dominant patients develop less dementia (Marttila and Rinne, 1976; Hughes et al., 1993).

About 40% of patients with Parkinson's disease experience a degree of depression at the time of diagnosis or during the course of the disease. This commonly occurs following diagnosis and in some cases is no more than a natural reaction to this and is both transient and often mild. It has been questioned whether the later development of depression and hallucinations occur as a secondary consequence of the motor symptoms of Parkinson's disease or drug treatments but it has also been suggested that the underlying neurodegenerative disease process is also the cause of cognitive impairment, depression, and hallucinations in PD patients.

9.4   PATHOLOGY OF PD

Large clinico-pathological studies have taken place in several countries in an attempt to reach a consensus pathological definition for Parkinson's disease. PD has been defined as the selective degeneration of pigmented, dopaminergic neurons of the (SNc) and other brainstem nuclei, with the presence of α-synuclein positive staining cytoplasmic inclusions, (known as Lewy bodies) in the surviving neurons (Bernheimer et al., 1973; Gibb and Lees, 1988; Fearnley and Lees, 1991; Forno, 1996; Spillantini et al., 1997). The death of the SNc dopaminergic neurons leads to loss of the nigrostriatal projection and subsequent abnormal patterns of neuronal firing from the striatum to the globus pallidus, subthalamic nucleus, and then to the thalamus. The degeneration of pigmented neurons in PD patients tends to be most marked in the ventrolateral region of the SNc, and this regional selectivity has been proposed as a basis for distinguishing between PD and other neurodegenerative diseases of the basal ganglia (Fearnley and Lees, 1991).

Most importantly, the presence of Lewy bodies has been adopted as a means for discriminating between PD and the Parkinson plus diseases in which these are absent. Lewy bodies were first found to be eosinophilic using hematoxylin and eosin stains on midbrain sections. They were then found to contain ubiquitin and subsequently been found to be largely composed of α-synuclein (Spillantini et al., 1997). This observation coincided closely with the identification of a kindred of patients with a mutation in the α-synuclein gene and the early onset of parkinsonism (Polymeropoulos et al., 1997).

A more recent study by Braak et al. (2002) has suggested that the pathological process passes through stages initially involving the lower brainstem and olfactory bulb, then progressing to the midbrain causing the motor symptoms of PD, before finally involving the cerebral cortex. If this pattern of progression is genuine, it raises the question why some patients present with a cortical dementia due to Lewy body deposition (Dementia with Lewy bodies) years before developing any motor symptoms or signs of PD.

Patients with a mutation in the parkin gene have also been the subject of pathological studies finding that although they develop the neurodegeneration within the SNc and also a clinical phenotype of parkinsonism, they do not develop Lewy bodies. This observation may be very informative regarding the mechanisms of cell death in PD. It seems likely that Lewy bodies are not the cause of neuronal cell degeneration, but may be a cellular attempt to cope with the formation of abnormal proteins such as α-synuclein.

9.5 PATHOGENESIS OF PD

9.5.1 α-SYNUCLEIN AND THE LOSS OF DOPAMINERGIC CELLS IN PD

The synuclein family of proteins was initially identified in 1988 from the electric organ synapse of *Torpedo californicum* (Maroteaux et al., 1988). α-Synuclein is identical to the nonAβ component of Alzheimer's disease amyloid precursor protein (NACP) (Ueda et al., 1993), but the 140 and 134 amino acid α and β isoforms of the protein were only subsequently identified in human cerebral cortex (Jakes et al., 1994). The human protein is 95% homologous to SYN-1 (or synuclein-1) isolated from rodent brain (Maroteaux and Scheller, 1991) (see Figure 9.1).

In the late 1990s, a remarkable convergence of research served to propel this once obscure protein into the limelight. On the one hand, the transmission of PD in a rare kindred with autosomal dominant disease was found to be due to a point mutation in α-synuclein—the A53T mutation in the Contursi kindred (Polymeropoulos et al., 1997). On the other hand, α-synuclein was found to be the major constituent of the Lewy body in sporadic disease (Spillantini et al., 1997). Together these findings suggested that a mutation in α-synuclein was sufficient to cause disease and that this protein was intimately associated with the pathology of the common form of PD. There followed an explosion of interest in α-synuclein, with the generation of transgenic animal models and further work that served to confirm the protein's central role in the pathogenesis of PD, although many controversies still remain.

α-Synuclein is located presynaptically (Maroteaux et al., 1988), and a role in synaptic plasticity has been suggested. It is known to interact with intracellular molecules, functioning as a chaperone to 14-3-3 protein, protein kinases, parkin, Tat binding protein, tau, β-synuclein, and so on (reviewed in Perez and Hastings (2004)). Moreover, it appears to inhibit the enzyme activity of tyrosine hydroxylase, the rate limiting enzyme in the production of dopamine (Perez et al., 2002). In PD, it is not clear whether α-synuclein causes its effects via a toxic gain of function—a proposal that is supported by the autosomal dominant inheritance of rare genetic forms of the disease, and also from the observation that α-synuclein knockout mice

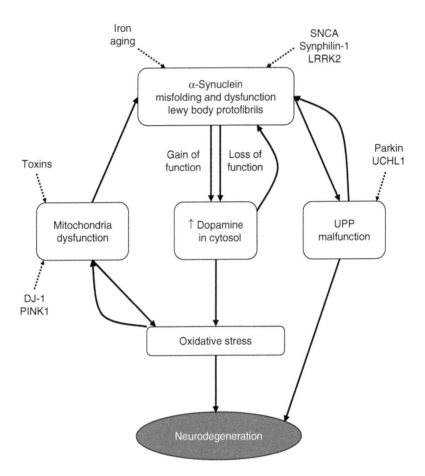

FIGURE 9.1 Pathways to Parkinson's disease. Sporadic PD is caused by dysfunction in, and interaction of, a number of cell pathways. (Gene mutations causing Mendelian diseases are SNCA, synphilin-1, LRRK2, Parkin, UCHL1, DJ-1 and PINK1; UPP, ubiquitin–proteasome pathway. For simplicity not all interactions are shown.)

have only subtle dopaminergic deficits (Abeliovich et al., 2000). The mechanism of the effect is not at present clear, but there are a number of possibilities, including the formation of protofibrils. Alternatively a loss-of-function mechanism of action has been proposed, possibly through its normal ability to suppress the activity of tyrosine hydroxylase (Perez et al., 2002), which if unchecked will increase intracellular dopamine, resulting in cumulative oxidative damage, and perhaps neuronal death.

Of course central to explaining the role of α-synuclein is determining which of its molecular forms is toxic and whether the final production of Lewy bodies is in fact a last ditch attempt at protection or represents a toxic end point. At present the arguments are numerous, and the answer remains unclear (Harrower et al., 2005). Of further interest with respect to α-synuclein and PD is that duplication (Ibanez et al., 2004) or triplication (Singleton et al., 2003) of the α-synuclein gene locus, causes PD to occur earlier and is more severe. In animal models, simple over expression of α-synuclein can cause dopaminergic pathology (Feany and Bender, 2000; Masliah et al., 2000; Kirik et al., 2002) although this is not the case in all disease models (Maries et al., 2003).

9.5.2 THE ROLE OF THE PROTEASOME IN PD

Abnormal α-synuclein is predominantly, although not exclusively, cleared by the proteasome, and mutations in parkin, (an E3 ubiquitin ligase essential for correct function of the ubiquitin–proteasome system) cause autosomal recessive PD (Kitada et al., 2000; Giasson and Lee, 2003). Furthermore, UCHL-1 (ubiquitin carboxy-hydrolase L-1) recycles ubiquitin by cleaving carboxy-terminal ends of ubiquitinated proteins and also causes PD if abnormal (Leroy et al., 1998). It has been shown that α-synuclein inhibits proteasomal activity (Stefanis et al., 2001; Snyder et al., 2003), which is reduced in the nigra of the PD brain (McNaught et al., 2003), and in vitro degradation of mutant α-synuclein by the proteasome is slower than wild type (Bennett et al., 1999).

These reports predict a central role for the ubiquitin–proteasome pathway (UPP) in PD pathogenesis, malfunction causing increased α-synuclein concentration, and damage either by the protofibrils or aggregates as described earlier.

9.5.3 OXIDATIVE STRESS AND THE SELECTIVE VULNERABILITY OF DOPAMINERGIC NEURONS

Nigral neurons have a high load of abnormal protein and low proteasomal activity (in end stage specimens examined at postmortem), but the final straw sealing their selective vulnerability in PD is likely to be the dopamine-rich environment. Dopamine can auto-oxidize at normal pH to form toxic reactive oxygen species: hydroxyl radicals, superoxide anions, and dopamine–quinone species (Graham, 1978). The high iron content in neuromelanin cells of the nigra contributes by catalyzing the breakdown of innocuous hydrogen peroxide to hydroxyl radicals. These reactive oxygen species can oxidize DNA, proteins, and lipids, affecting their function in many ways, for example, by increasing membrane permeability to calcium ions and therefore increasing excitotoxic cell damage. Interestingly, it has been shown that reactive oxygen species can form covalent adducts with α-synuclein, inhibiting the generation of fibrils from protofibrils, thus potentially increasing toxicity (Conway et al., 2001).

This oxidative pressure is normally kept in check by antioxidant systems, and by packaging free dopamine into vesicles for safe storage. However, in PD, although up regulated in the nigra, there remains a relative lack of glutathione, superoxide dismutase, and catalase that normally inactivate reactive oxygen species (Lotharius and Brundin, 2002). Furthermore, as mentioned above, the production of synaptic vesicles is likely to be abnormal, probably as a result of aberrant α-synuclein, resulting in increased cytoplasmic free dopamine. As well as oxidative damage and the stabilization of α-synuclein protofibrils, this cytoplasmic dopamine can affect mitochondrial respiration causing opening of the mitochondrial permeability transition pore which can result in release of Cytochrome c—the first step on the apoptotic cascade to cell death (Berman and Hastings, 1999).

9.6 BIOMARKERS AND PD

9.6.1 DIAGNOSIS

Even in highly specialized centers the sensitivity of the clinical diagnosis of PD in *symptomatic* patients is only just over 90% (Hughes et al., 2002) and is likely to be far less in other settings. Thus any way of distinguishing true idiopathic PD from the numerous causes of a similar clinical syndrome would be beneficial, enabling better clinical management and allowing cheaper, more powerful clinical trials. There is also a need for *presymptomatic* diagnosis, as potential neuroprotective agents should be administered as soon as possible (Stocchi and Olanow, 2003). It is known that approximately 50% of dopaminergic nigral cells are lost before clinical expression of PD becomes apparent (Fearnley and Lees, 1991) and it seems that this presympto-matic phase of PD lasts approximately 5 years (Fearnley and Lees, 1991; Morrish et al., 1996; Marek et al., 2001).

9.6.2 DISEASE PROGRESSION AND THE EFFECT OF TREATMENT

Biomarkers (possibly different to those used for diagnosis) would be useful in longi-tudinal clinical treatment trials as surrogate end points, which if adequately validated, would help provide a degree of objectivity and potentially enable a reduction in the duration of a trial and increase its power. There are a number of problems that any such biomarker would need to overcome. For example symptoms fluctuate over a time scale of minutes to hours whereas the pathology presumably does not. Furthermore, some putative neuroprotective agents (e.g., dopamine agonists) have symptomatic effects unrelated to their potential disease modifying action, so a biomarker should be able to reliably detect retarded pathological progression without relying on symptoms.

9.6.3 DISEASE HETEROGENEITY

Whatever approach to PD is taken, the disease itself appears extremely hetero-geneous: there are many genetic and environmental etiologies, giving rise to a broad spectrum of pathologies involving, to a greater or lesser extent, different neurotransmitters, resulting in a clinical phenotype that varies enormously in age of onset, key symptoms, and speed of progression. If a biomarker is able to identify subgroups within the disease it may also enable selection of people likely to respond well to particular treatments, improving the cost effectiveness of new treatment trials and increasing the clinical success of these new treatments.

As a result of these differing needs of biomarkers, there is some confusion as to what properties any such biomarker should possess, and this will clearly differ depending on what one is looking at in PD and this is perhaps best illustrated with functional imaging.

9.6.4 FUNCTIONAL IMAGING: AN EXAMPLE OF A PD BIOMARKER

PET and SPECT imaging have a number of capabilities: following the decline in neurotransmitter function, examining metabolic activity (using ^{18}F-fluorodeoxy-glucose), or monitoring regional cerebral blood flow. Patients are injected with a small amount of a radioactive compound (^{18}F-dopa or ^{123}I-FP-CIT), which are taken up in specific areas in the brain and can then be visualized by using specialized scanners. The development of new ligands promises to expand this repertoire to allow detection of the microglial response to cell loss, (e.g., ^{11}C-PK11195, Cicchetti et al., 2002; Gerhard et al., 2003) or accumulation of proteins such as α-synuclein or β-amyloid (Zhuang et al., 2001).

To date functional imaging has been used in a number of different ways in PD.

Presymptomatic detection: There is significant variation between the PET or SPECT images from normal subjects, but nonetheless there is hope (and some evidence) that these techniques may enable detection of presymptomatic dopaminergic cell loss. Thus twin studies have revealed a dopaminergic deficit in asymptomatic twins of patients with PD (Burn et al., 1992) and there is some evidence that subjects with preclinical imaging abnormalities actually go on to develop PD (Piccini et al., 1999b). Furthermore, dopaminergic dysfunction has been shown in four people after taking MPTP before the onset of symptoms (Calne et al., 1985; Laihinen et al., 2000).

Distinguishing PD from its imitators in difficult symptomatic patients: Currently most studies on this issue are not particularly representative of real life, where the challenge is to distinguish the different forms of Parkinsonism when the diagnosis is in doubt—in most functional imaging studies the patients are readily clinically distinguishable. Furthermore, the clinical diagnosis becomes the gold standard against which the scan result is compared even though we know that our clinical diagnosis is not totally accurate. Accepting these limitations there are some relatively promising early trials. Thus, ^{18}F-dopa PET diagnosis (based on caudate and putamen dopaminergic function) corresponded to the clinical diagnosis of PD in 64% of patients, increasing to 70% for PSP but was far less for multiple system atrophy (Burn et al., 1994).

Disease progression: PET and SPECT data from dopamine agonist trials have been interpreted (by some) as proof of neuroprotection and demonstration of the power of these techniques as biomarkers—they are worth considering because they illustrate some of the obstacles in interpreting the results of any biomarker.

In the REAL PET study, the ^{18}F-dopa influx constant in the putamen was measured in PD patients randomized to receive either L-dopa or ropinirole over 2 years. At baseline, 11% of patients with the clinical diagnosis of PD were felt to have normal images from the caudate and putamen. In the remainder there was a significantly slowed deterioration of putamen ^{18}F-dopa uptake on ropinirole compared to L-dopa (-13% vs. -20%, $p = 0.02$)—suggesting that ropinirole might have caused about a 30% slowing of the progression of PD over that period (Whone et al., 2003). However, it is possible that L-dopa altered the uptake of ^{18}F-dopa and therefore changed the ^{18}F-dopa influx constant, which would clearly affect the interpretation of these results (Fahn, 2002). A similar result was seen in the CALM-PD study using ^{123}I-β-CIT (2β-carboxy-methoxy-3β(4-iodophenyl)tropane) SPECT (Marek et al., 2002).

9.7   TREATING PARKINSON'S DISEASE

Dopamine replacement has been the major treatment strategy since the dopamine precursor L-Dopa was first used in the late 1960s (Cotzias and Schiffer, 1967), and it has a dramatic benefit especially in the early stages of the disease. L-Dopa is combined with peripherally acting amino acid decarboxylase (AADC) inhibitors to minimize peripheral breakdown of the drug (see Chapter 10). Following its wide-spread introduction, it soon became apparent that L-Dopa was associated with two significant complications, which ultimately limits its effectiveness. Firstly, patients experience a wearing-off phenomenon and secondly, they develop L-Dopa-induced dyskinesias (LID). These involuntary movements occur in 50%–75% of patients on L-Dopa after 5–10 years of treatment (Marsden and Parkes, 1976) and are reported to be less common in patients on other, slightly less effective, antiparkinsonian medi-cations such as dopaminergic agonists and anticholinergics (Hauser and Olanow, 1993; Olanow et al., 1994). These major side effects of prolonged L-Dopa therapy, in conjunction with its failure to alter the natural history of the disease and the nonmotor problems in PD (including psychiatric and cognitive deficits) has led to the search for more effective therapies. This includes better symptomatic medica-tions along with neuroprotective agents and restorative cell therapies, and it is these we shall now concentrate on.

9.7.1 Improved Dopaminergic Drug Delivery

The success of surgical approaches such as deep brain stimulation in treating advanced PD and LID offers much hope, but these procedures are only currently available in limited centers and the long-term efficacy and complications of this procedure is still uncertain (Lyons et al., 2004), so a search for better drug therapies continues. However, to improve medical therapies in PD, a better understanding of the factors that give rise to abnormal neuronal firing patterns within the basal ganglia needs to be obtained (Bezard et al., 2001). In this respect, it is now widely believed that the development of these abnormal neuronal firing patterns in LID is secondary to pulsatile stimulation of striatal dopamine receptors (Chase et al., 1989; Obeso et al., 1994), which relates to the type of medication being used as well as the stage of disease at which it is being used. Therefore, one therapeutic approach to reduce LID has focused on reducing the pulsatility of dopaminergic stimulation including the preferential use of long-acting dopamine agonists as well as using agents that block the breakdown of L-Dopa, (e.g., entacapone, a specific catechol-O-methyl transferase (COMT) inhibitor acting peripherally (Parkinson Study Group, 1997; Olanow and Obeso, 2000)). In terms of disease stage, one obviously has little control. In the early stages, treatment with L-Dopa provides good symptomatic relief and there are no LID, as it is thought that sufficient dopaminergic striatal terminals exist that can buffer the fluxes in local dopamine concentrations. However, as the disease progresses there is a loss of these dopaminergic terminals and L-Dopa is taken up by other decarboxylase-containing cells, most notably serotonergic terminals (Rabey et al., 2000). These terminals have no suitable storage mechanism for dopamine and this molecule simply leaks out into the surrounding environment leading to indiscriminate stimulation of postsynaptic receptors. This unregulated neural system leads to inappropriate nonphysiological stimulation of dopamine receptors, which has several profound effects on the medium spiny output neurons of the striatum. This includes changes in preproenkephalin, preprotachykinin and various opioids, dynorphin, and neurotensin levels (Young et al., 1986; Gerfen et al., 1990; Engber et al., 1991; Herrero et al., 1995; Piccini et al., 1997), which suggests altered gene transcription is occurring in treated PD patients which may underlie the development of the LID. In addition there is possible abnormal dopamine receptor plasticity (Picconi et al., 2003) as well as dopamine–glutamate receptor signaling in LID (Centonze et al., 2003). Using this knowledge of a glutamatergic role in LID, work in animal models has shown that antagonism of NMDA receptors could be beneficial in the treatment of LID (Papa and Chase, 1996). This has been extended into clinical research using the nonselective NMDA receptor antagonist, amantadine (Verhagen et al., 1998b), as well as dextromethorphan (Verhagen et al., 1998a).

It is thus apparent that drugs that can interact with the medium spiny neuron receptors may have a role in modifying the intracellular signaling cascades that presumably act in an unregulated fashion in the striatum of patients with LID. Some of these receptors have already begun to receive attention, generally as a result of the disturbances discovered in various gene products as described above. Two trials of opioid antagonists in PD patients have shown improvements in LID without deterioration in other symptoms (Trabucchi et al., 1982; Sandyk and Snider, 1986) although these findings are not universal (Nutt et al., 1978; Rascol et al., 2000). Other promising results that improve motor function in animal models, presumably by interacting through the striatal neuron receptors, have been reported with adenosine A_{2A} and α-2 adrenoreceptor drugs (Kanda et al., 1998; Grondin et al., 2000; Bara-Jimenez et al., 2003; Hauser et al., 2003; Savola et al., 2003) but large scale clinical trials have yet to be conducted. Recent attention has also focused on the serotonergic system in PD, which is felt to exert an inhibitory influence on dopamine release (Dray et al., 1978). Although clinical trials in patients with LID have had mixed results (Hammerstad et al., 1986; Kleedorfer et al., 1991) renewed interest has come from a recent case of PD in which marked improvements in switching from "off" (marked features of Parkinsonism) to "on" (improved mobility and dexterity) were seen by having taken the illicit drug 3,4-methylenedioxy-N-methylamphetamine (MDMA or *Ecstasy*).

9.7.2 NEUROPROTECTION

Neuroprotection is not a new concept in PD and probably the most well known example is the monoamine oxidase-B (MAO-B) inhibitor, selegiline. Interest in the neuroprotective capabilities of this drug stemmed from experimental studies in which its application in vitro or in vivo prior to MPTP treatment afforded a significant degree of neuroprotection, by inhibiting the formation of the neurotoxic 1-Methyl-4-phenyl-pyridinium ion (MPP^+). As a result a prospective randomized, double-blind, placebo-controlled study evaluating the effect of selegiline on progression in early PD patients was undertaken (Olanow et al., 1995; Shoulson et al., 2002), on the grounds that an $MPTP–MPP^+$ like toxic compound may exist in the environment and cause PD. This DATATOP study demonstrated a significant delay in the deterioration of parkinsonian signs and symptoms, which was initially taken as evidence for a neuroprotective action of selegiline. However, subsequently the trial result was reinterpreted as showing that the drug had a symptomatic effect by inhibiting the breakdown of endogenously produced dopamine. Nevertheless, a recent trial using another MAO-B inhibitor, rasagiline, has produced data suggesting neuroprotection (Parkinson Study Group, 2004; Rascol et al., 2005).

Other prospective neuroprotective agents that have been proposed in PD include controversially the dopamine agonists ropinirole (Schrag et al., 2002; Whone et al., 2003) and pramipexole (see above). An additional group of possible neuroprotective drugs are antiglutamatergic agents such as riluzole (Turski et al., 1991; Boireau et al., 1994; Barneoud et al., 1996), and clinical trials are underway to explore this further, although pilot studies from small groups of patients have not yet reported significant symptomatic effects (Jankovic and Hunter, 2002; Braz et al., 2004). Finally, other less-specific compounds have been used in this capacity in PD including Coenzyme Q10, which at high dosage appears to slow the progressive deterioration of functions without obvious complications (Shults et al., 2002, 2004).

9.7.3 NEUROTROPHIC FACTORS

The use of neurotrophic factors to rescue populations of neurons has been widely explored for a number of neurological conditions and in PD the slow degeneration of dopaminergic neurons within the SNc makes it an attractive target for this type of therapeutic approach. Although a large number of neurotrophic factors have been explored in PD (Bradford et al., 1999), most attention has concentrated on glial cell line-derived neurotrophic factor (GDNF). The discovery of this dopaminotrophic factor in 1993 led to a series of subsequent studies that have demonstrated that GDNF has both a neuroprotective and restorative capacity in animal models of PD (Hurelbrink and Barker, 2001, 2004). Nevertheless, intracerebroventricular GDNF administration in PD proved disappointing in clinics (Nutt et al., 2003), possibly because GDNF did not reach the putamen and SNc where dopaminergic loss is found and indeed the single postmortem case of a patient receiving such an infusion showed no evidence of dopaminergic cell rescue or fiber sprouting (Kordower et al., 1999). Alternative forms of GDNF delivery, such as through viral vectors or directly into the brain parenchyma (Kordower et al., 2000) may prove more effective. Clinical trials have recently commenced to investigate intraparenchymal administration of GDNF in PD patients (Gill et al., 2003; Patel et al., 2005), and the results are controversial with the initial studies of chronic infusion of GDNF directly into the putamen demonstrating significant long-term clinical improvements. Side effects were also reported to be limited, and the clinical improvements were correlated in part to changes in nigrostriatal dopamine levels measured using PET in a single postmortem case. However, recently this result has not been replicated in a double blind placebo controlled trial. This study failed to show a significant benefit for GDNF at 6 months post infusion and this coupled to some evidence of GDNF induced Purkinje cell loss in the cerebellum of nonhuman primates and the development of antiGDNF antibodies in some patients, has led to the trial being abandoned (Nutt et al., 2003; Lang et al., 2006).

9.7.4 CELL REPLACEMENT THERAPIES

As a chronic and relatively focal neurodegenerative disorder, PD presents the most promising neurological disorder to be potentially cured by cell replacement therapy.

The aim of PD cell replacement therapy is to restore nigrostriatal dopaminergic transmission by replacing the loss of dopaminergic cells in the SNc. Fetal ventral mesencephalon (VM) has long been exploited as the major source of tissue in PD cell replacement therapy. Indeed some PD patients receiving intraputamenal transplantation of human fetal VM containing the developing dopamine cells of the SNc have had a major benefit (Brundin et al., 2000; Freed et al., 2001; Bjorklund et al., 2003). Long-term transplant survival is also reported and in one patient this has been documented up to a decade after implantation with the graft still functioning and producing dopamine (Piccini et al., 1999a). In contrast to these encouraging examples, other clinical trials have demonstrated no overall improvements in UPDRS (Unified Parkinson's Disease Rating scale) motor scores, although PET scanning evidence suggests graft-induced dopamine elevations in some patients (Freed et al., 2001; Olanow et al., 2003). Moreover, a significant number of patients have developed off-state dyskinesias after transplantation in some trials similar to those seen in untransplanted, advanced PD patients on L-Dopa. The heterogeneity from these results may stem from the lack of a standardized protocol for carrying out the cell replacement procedure in PD. Nevertheless, these studies have suggested some important criteria that must be fulfilled to achieve effective therapy. For example, the current transplantation protocol is most effective in patients with less severe disease ($\leq$49 UPDRS score at baseline). Also functional benefits are not observed without a high level of surviving dopaminergic cells, which is presently achievable only with a high number of donor fetuses for transplantation (>3 per side). This latter point is very important because findings from experimental animals suggest that the level of dopaminergic cell survival in the transplant is positively correlated to their ability to reduce motor abnormalities (Rioux et al., 1991; Nakao et al., 1994, 1995; Brecknell et al., 1996). At the present time, however, the efficiency of dopaminergic cell survival following transplantation is poor, with only around 5%–10% of the fetal nigral dopaminergic neuron surviving the tissue preparation and grafting procedure (Brundin et al., 1988; Bjorklund, 1993; Nakao et al., 1994; Barker et al., 1996; Kordower et al., 1998; Zawada et al., 1998; Schierle et al., 1999).

In spite of some promising results, cell therapies involving primary VM dopaminergic neural transplants are unlikely to be the optimal cure for PD because of the ethical and practical problems with using this tissue. As a result, the search for alternative sources of cells has been sought. Potential candidates for the next generation of donor tissue in cell replacement therapy includes autologous tissue from patients such as carotid body, sympathetic ganglion neurons, adrenal medulla plus peripheral nerve, as well as the recent breakthrough in mesenchymal stem cell research. Autologous transplantation has the advantage of being free from the problems of graft rejection as well as being associated with less ethical and safety issues. Another possibility is the use of dopaminergic neurons derived from neural precursor cells or embryonic stem cell, as both in theory provide a virtually unlimited, standardized source of donor cells. Immunohistochemical characterization of the isolated embryonic rat VM indicates that they contain undifferentiated precursor cells (Peaire et al., 2003). These precursor cells can be mechanically dissociated, plated on culture dishes, and grown in serum-free medium supplemented with FGF-2 (Fibroblast growth factor-2) where they proliferate and, upon removal of mitogens, differentiate into nestin-positive progenitor cells capable of developing into dopaminergic neurons with extensive fiber outgrowth (Studer et al., 1998; Ling et al., 1998; Potter et al., 1999; Sawamoto et al., 2001; Peaire et al., 2003; Storch et al., 2003). The number of tyrosine hydroxylase (TH)-positive neurons derived from this cell source can be further increased using a combination of cytokines, GDNF, mesencephalic membrane fragments, and striatal culture conditioned media (Ling et al., 1998; Potter et al., 1999; Grothe et al., 2004). Similar protocols can also be applied with dopaminergic neural precursor cells of both human and porcine origin (Storch et al., 2001; Sanchez-Pernaute et al., 2001; Riaz et al., 2002; Armstrong et al., 2003), and at least with human cells they can remain in an undifferentiated state for up to 11 months in vitro using hypoxia and Epidermal Growth Factor (EGF) and FGF-2 as mitogens (Storch et al., 2001) while still retaining the capacity to differentiate into normal midbrain dopaminergic phenotype (Studer et al., 1998; Storch et al., 2001, 2003; Riaz et al., 2002). Transplantation of these expanded precursor cells can also produce functional benefits in vivo and can alleviate behavioral deficits in animal models of PD (Studer et al., 1998; Sawamoto et al., 2001).

Embryonic stem (ES) cells have always been a promising alternative source of tissue for PD cell replacement therapy. Embryonic stem cells are self-renewing, multipotent cells derived from the inner cell mass of blastocyst (Deacon et al., 1998). Naive ES cells, when grafted in low density, develop into normal mesencephalic dopaminergic-like neurons in situ and are able to exert a gradual functional recovery in a rat model of PD (Bjorklund et al., 2002). Nevertheless, the rate of spontaneous differentiation into the dopaminergic lineage from ES cells is low, and 20% of the animals receiving ES cell graft developed teratomas, a fact that currently represents the major concern in the translation of this type of strategy to the clinics. Nevertheless, the risk of tumor formation may be reduced if ES cells are differentiated in vitro before transplantation. The conversion of rodent or primate ES cells to dopaminergic neurons in culture has been reported using a variety of approaches. One approach by McKay and colleagues is a five-stage protocol using defined factors (such as sonic hedgehog, FGF-8, and ascorbic acid) (Lee et al., 2000) and involves the dissociation and maintenance of murine ES cells; formation of embryoid bodies; selection of neural precursors; and the expansion and differentiation to dopaminergic phenotypes. Under optimal culture conditions up to 33.9% of the cells were found to be dopaminergic (Lee et al., 2000). In parallel works, it was discovered that some bone marrow-derived cell lines have a potent stromal cell-derived inducing activity (SDIA) of neuronal differentiation (Kawasaki et al., 2002; Barberi et al., 2003; Takagi et al., 2005). By coculturing dissociated ES cells with these feeder cells, differentiation into dopaminergic phenotype can be induced to levels of up to ~50% of the total cell population. Furthermore, the efficiency of dopaminergic differentiation from both methods can be doubled by overexpressing Nurr1 in donor ES cells (Kim et al., 2002; Chung et al., 2002). These ES cell-derived dopaminergic neurons from rodents and primates have also been investigated in vivo from both the phenotypic and functional perspectives. These cells not only express key dopaminergic neuronal markers and release dopamine, but when transplanted into either rodent or primate models of PD they become integrated into the host striatum with amelioration of motor dysfunction (Kawasaki et al., 2002; Kim et al., 2002; Barberi et al., 2003; Nishimura et al., 2003; Yoshizaki et al., 2004; Takagi et al., 2005). Neurons derived from ES cells are postmitotic, but their derivation in vitro from a heterogeneous population of cell types means that some of the cells may still have the potential for tumor formation (Nishimura et al., 2003). Attempts to eliminate these nonneural or undifferentiated dividing cells have been undertaken using, for example, EGFP-conjugated Tau-protein "knock-in" ES cells followed by fluorescence selection and transplantation (Wernig et al., 2002). In spite of the enrichment of the neuronal population with such an approach, it nevertheless involves genetic modification which may limit its clinical application. An alternative approach to ES cell-derived neuronal purification is to utilize the specificity of a neuron-specific cell adhesion molecule L1, which selects only ES-derived neurons (Jungling et al., 2003). A different approach to deal with the tumorgenic potential of ES cells involves the use of the encapsulation technique, which provides a physical barrier to uncontrolled proliferation from the cell transplant (Li et al., 1998). An additional advantage to this strategy of using encapsulated cells for ES cell-based therapy arises from a recent report of low-level MHC-I (major histocompatibility complex-I) on human ES cells (Drukker et al., 2002)—a situation that would reduce their immunogenicity and thus the ability to elicit an immune reaction in allotransplanted humans. However, this having been said, the clinical application of cell encapsulation is far from straightforward for technological as well as biological reasons.

The search for appropriate tissues for autologous transplantation in PD has made some progress over the past few years. Of these, perhaps the adoption of an approach using precursor or ES-derived cells allows for the in vitro expansion of cell numbers followed by enrichment of dopaminergic neurons, which should reduce the amount of primary human tissue needed for therapeutic grafting, offers the most potential. Moreover, with these cells, the adoption of a defined protocol ensures not only homogeneity of cells within the grafts but also provides an opportunity to manipulate the cultured cells to optimize neuronal viability and neurite outgrowth after transplantation. However, the inherent practical restrictions, such as the limited number of passages that some neural precursor cells can survive in vitro and the concerns of tumor formation with ES cells, coupled to problems of controlling their fate long-term, currently limits their clinical applicability.

9.8 CONCLUSION

Parkinson's disease is a common and complex disorder for which there is no obvious cause in the majority of cases, although there are emerging clues from epidemiological and genetic studies. This has enabled better modeling of pathogenic pathways that may lead to novel disease modifying therapies, assuming we can better diagnose and monitor disease progression using biomarkers. Indeed there are a number of very effective therapies, especially in the early stages of the disease, which target core pathological events, namely the loss of the dopaminergic nigrostriatal pathway. However, with time these therapies start to fail and produce their own unique side effects and we have discussed some of the possible reasons for this. As a result, novel approaches are being sought that attempt to either substitute for or rescue the remaining dopaminergic cells, or increase their number by transplantation. We have discussed each of these strategies to highlight the range of approaches that are being undertaken, and in particular have concentrated on cell therapy approaches in the treatment of PD. We have not discussed at length a range of other approaches including viral vectors and cell lines engineered to deliver trophic factors or dopamine, but nevertheless have presented an account that highlights those areas that have entered, or are about to enter, the clinical arena. As to which of these approaches will ultimately succeed, only time will tell.

ACKNOWLEDGMENTS

The authors' own work has been supported by the Wellcome Trust, Parkinson's Disease Society and MRC.

REFERENCES

Aarsland, D., Andersen, K., Larsen, J.P., Lolk, A. and Kragh-Sorensen, P. (2003) Prevalence and characteristics of dementia in Parkinson disease: an 8-year prospective study. *Archives of Neurology*, **60**, 387–392.

Abbott, R.D., Ross, G.W., White, L.R., Sanderson, W.T., Burchfiel, C.M., Kashon, M. et al. (2003) Environmental, life-style, and physical precursors of clinical Parkinson's disease: recent findings from the Honolulu-Asia Aging Study. *Journal of Neurology*, **250** (Suppl 3), III30–III39.

Abeliovich, A., Schmitz, Y., Farinas, I., Choi-Lundberg, D., Ho, W.H., Castillo, P.E. et al. (2000) Mice lacking alpha-synuclein display functional deficits in the nigrostriatal dopamine system. *Neuron*, **25**, 239–252.

Acosta, J., Calderon, E. and Obeso, J.A. (1989) Prevalence of Parkinson's disease and Essential Tremor in a village of southern Spain. *Neurology*, **39**, 181.

Armstrong, R.J., Tyers, P., Jain, M., Richards, A., Dunnett, S.B., Rosser, A.E. et al. (2003) Transplantation of expanded neural precursor cells from the developing pig ventral mesencephalon in a rat model of Parkinson's disease. *Experimental Brain Research*, **151**, 204–217.

Ascherio, A., Zhang, S.M., Hernan, M.A., Kawachi, I., Colditz, G.A., Speizer, F.E. et al. (2001) Prospective study of caffeine consumption and risk of Parkinson's disease in men and women. *Annals of Neurology*, **50**, 56–63.

Ashok, P.P., Radhakrishnan, K., Sridharan, R. and Mousa, M.E. (1986) Epidemiology of Parkinson's disease in Benghazi, North-East Libya. *Clinical Neurology and Neuro-surgery*, **88**, 109–113.

Attia, R.N., Ben Hamida, M., Mrabet, A., Larnaout, A., Samoud, S., Ben Hamda, A. et al. (1993) Prevalence study of neurologic disorders in Kelibia (Tunisia). *Neuroepidemiology*, **12**, 285–299.

Bara-Jimenez, W., Sherzai, A., Dimitrova, T., Favit, A., Bibbiani, F., Gillespie, M. et al. (2003) Adenosine A(2A) receptor antagonist treatment of Parkinson's disease. *Neurology*, **61**, 293–296.

Barberi, T., Klivenyi, P., Calingasan, N.Y., Lee, H., Kawamata, H., Loonam, K. et al. (2003) Neural subtype specification of fertilization and nuclear transfer embryonic stem cells and application in parkinsonian mice. *Nature Biotechnology*, **21**, 1200–1207.

Barker, R.A., Dunnett, S.B., Faissner, A. and Fawcett, J.W. (1996) The time course of loss of dopaminergic neurons and the gliotic reaction surrounding grafts of embryonic mesencephalon to the striatum. *Experimental Neurology*, **141**, 79–93.

Barneoud, P., Mazadier, M., Miquet, J.M., Parmentier, S., Dudebat, P., Doble, A. et al. (1996) Neuroprotective effects of riluzole on a model of Parkinson's disease in the rat. *Neuroscience*, **74**, 971–983.

Baron, J.A. (1986) Cigarette smoking and Parkinson's disease. *Neurology*, **36**, 1490–1496.

Bennett, M.C., Bishop, J.F., Leng, Y., Chock, P.B., Chase, T.N. and Mouradian, M.M. (1999) Degradation of alpha-synuclein by proteasome. *Journal of Biological Chemistry*, **274**, 33855–33858.

Berg, D., Niwar, M., Maass, S., Zimprich, A., Moller, J.C., Wuellner, U. et al. (2005a) Alpha-synuclein and Parkinson's disease: implications from the screening of more than 1,900 patients. *Movement Disorders*, **20**, 1191–1194.

Berg, D., Schweitzer, K., Leitner, P., Zimprich, A., Lichtner, P., Belcredi, P. et al. (2005b) Type and frequency of mutations in the LRRK2 gene in familial and sporadic Parkinson's disease. *Brain*, **128**, 3000–3011.

Berman, S.B. and Hastings, T.G. (1999) Dopamine oxidation alters mitochondrial respiration and induces permeability transition in brain mitochondria: implications for Parkinson's disease. *Journal of Neurochemistry*, **73**, 1127–1137.

Bernheimer, H., Birkmayer, W., Hornykiewicz, O., Jellinger, K. and Seitelberger, F. (1973) Brain dopamine and the syndromes of Parkinson and Huntington. Clinical, morphological and neurochemical correlations. *Journal of Neurological Sciences*, **20**, 415–455.

Bezard, E., Brotchie, J.M. and Gross, C.E. (2001) Pathophysiology of levodopa-induced dyskinesia, potential for new therapies. *Nature Reviews Neuroscience*, **2**, 577–588.

Biskup, S., Mueller, J.C., Sharma, M., Lichtner, P., Zimprich, A., Berg, D. et al. (2005) Common variants of LRRK2 are not associated with sporadic Parkinson's disease. *Annals of Neurology*, **58**, 905–908.

Bjorklund, A. (1993) Neurobiology. Better cells for brain repair. *Nature*, **362**, 414–415.

Bjorklund, L.M., Sanchez-Pernaute, R., Chung, S., Andersson, T., Chen, I.Y., McNaught, K. S. et al. (2002) Embryonic stem cells develop into functional dopaminergic neurons after transplantation in a Parkinson rat model. *Proceedings of the National Academy of Sciences of the United States of America*, **99**, 2344–2349.

Bjorklund, A., Dunnett, S.B., Brundin, P., Stoessl, A.J., Freed, C.R., Breeze, R.E. et al. (2003) Neural transplantation for the treatment of Parkinson's disease. *Lancet Neurology*, **2**, 437–445.

Boireau, A., Dubedat, P., Bordier, F., Peny, C., Miquet, J.M., Durand, G. et al. (1994) Riluzole and experimental parkinsonism, antagonism of MPTP-induced decrease in central dopamine levels in mice. *Neuroreport*, **5**, 2657–2660.

Bonifati, V., Rizzu, P., van Baren, M.J., Schaap, O., Breedveld, G.J., Krieger, E., et al. (2003) Mutations in the DJ-1 gene associated with autosomal recessive early-onset parkinsonism. *Science*, **299**, 256–259.

Braak, H., Del Tredici, K., Bratzke, H., Hamm-Clement, J., Sandmann-Keil, D. and Rub, U. (2002) Staging of the intracerebral inclusion body pathology associated with idiopathic Parkinson's disease (preclinical and clinical stages). *Journal of Neurology*, **249** (Suppl 3), III/1–III/5.

Bradford, H.F., Zhou, J., Pliego-Rivero, B., Stern, G.M. and Jauniaux, E. (1999) Neurotrophins in the pathogenesis and potential treatment of Parkinson's disease. *Advances in Neurology*, **80**, 19–25.

Braz, C.A., Borges, V. and Ferraz, H.B. (2004) Effect of riluzole on dyskinesia and duration of the on state in Parkinson disease patients, a double-blind, placebo-controlled pilot study. *Clinical Neuropharmacology*, **27**, 25–29.

Brecknell, J.E., Haque, N.S., Du, J.S., Muir, E.M., Fidler, P.S., Hlavin, M.L. et al. (1996) Functional and anatomical reconstruction of the 6-hydroxydopamine lesioned nigrostriatal system of the adult rat. *Neuroscience*, **71**, 913–925.

Brown, R.G. and Marsden C.D. (1984) How common is dementia in Parkinson's disease? *Lancet*, **2**, 1262–1265.

Brundin, P., Barbin, G., Strecker, R.E., Isacson, O., Prochiantz, A. and Bjorklund, A. (1988) Survival and function of dissociated rat dopamine neurones grafted at different developmental stages or after being cultured in vitro. *Brain Research*, **467**, 233–243.

Brundin, P., Pogarell, O., Hagell, P., Piccini, P., Widner, H. and Schrag, A. et al. (2000) Bilateral caudate and putamen grafts of embryonic mesencephalic tissue treated with lazaroids in Parkinson's disease. *Brain*, **123**, 1380–1390.

Burn, D.J., Mark, M.H., Playford, E.D., Maraganore, D.M., Zimmerman, T.R., Duvoisin, R.C. et al. (1992) Parkinson's disease in twins studied with 18F-dopa and positron emission tomography. *Neurology*, **42**, 1894–1900.

Burn, D.J., Sawle, G.V. and Brooks, D.J. (1994) Differential diagnosis of Parkinson's disease, multiple system atrophy, and Steele-Richardson-Olszewski syndrome, discriminant analysis of striatal 18F-dopa PET data. *Journal of Neurology Neurosurgery and Psychiatry*, **57**, 278–284.

Calne, D.B., Langston, J.W., Martin, W.R., Stoessl, A.J., Ruth, T.J., Adam, M.J. et al. (1985) Positron emission tomography after MPTP, observations relating to the cause of Parkinson's disease. *Nature*, **317**, 246–248.

Centonze, D., Gubellini, P., Pisani, A., Bernardi, G. and Calabresi, P. (2003) Dopamine, acetylcholine and nitric oxide systems interact to induce corticostriatal synaptic plasticity. *Reviews in the Neurosciences*, **14**, 207–216.

Chase, T.N., Baronti, F., Fabbrini, G., Heuser, I.J., Juncos, J.L. and Mouradian, M.M. (1989) Rationale for continuous dopaminomimetic therapy of Parkinson's disease. *Neurology*, **39**, 7–10.

Chen, R.C., Chang, S.F., Su, C.L., Chen, T.H., Yen, M.F., Wu, H.M. et al. (2001) Prevalence, incidence, and mortality of PD, a door-to-door survey in Ilan county, Taiwan. *Neurology*, **57**, 1679–1686.

Chen, H., Zhang, S.M., Hernan, M.A., Willett, W.C. and Ascherio, A. (2003) Dietary intakes of fat and risk of Parkinson's disease. *American Journal of Epidemiology*, **157**, 1007–1014.

Chung, S., Sonntag, K.C., Andersson, T., Bjorklund, L.M., Park, J.J., Kim, D.W. et al. (2002) Genetic engineering of mouse embryonic stem cells by Nurr1 enhances differentiation and maturation into dopaminergic neurons. *European Journal of Neuroscience*, **16**, 1829–1838.

Cicchetti, F., Brownell, A.L., Williams, K., Chen, Y.I., Livni, E. and Isacson, O. (2002) Neuroinflammation of the nigrostriatal pathway during progressive 6-OHDA dopamine degeneration in rats monitored by immunohistochemistry and PET imaging. *European Journal of Neuroscience*, **15**, 991–998.

Conway, K.A., Rochet, J.C., Bieganski, R.M. and Lansbury, P.T. Jr. (2001) Kinetic stabilization of the alpha-synuclein protofibril by a dopamine-alpha-synuclein adduct. *Science*, **294**, 1346–1349.

Cotzias, G.C. and Schiffer, L.M. (1967) Aromatic amino acids and modification of parkinsonism. *New England Journal of Medicine*, **276**, 374–379.

de Rijk, M.C., Tzourio, C., Breteler, M.M., Dartigues, J.F., Amaducci, L., Lopez Pousa, S. et al. (1997) Prevalence of parkinsonism and Parkinson's disease in Europe: the EUROPARKINSON Collaborative Study. European community concerted action on the epidemiology of Parkinson's disease. *Journal of Neurology Neurosurgery and Psychiatry*, **62**, 10–15.

Deacon, T., Dinsmore, J., Costantini, L.C., Ratliff, J. and Isacson, O. (1998) Blastula-stage stem cells can differentiate into dopaminergic and serotonergic neurons after transplantation. *Experimental Neurology*, **149**, 28–41.

DeMartino, G.N. and Slaughter, C.A. (1999) The proteasome, a novel protease regulated by multiple mechanisms. *Journal of Biological Chemistry*, **274**, 22123–22126.

Dray, A., Davies, J., Oakley, N.R., Tongroach, P. and Vellucci, S. (1978) The dorsal and medial raphe projections to the substantia nigra in the rat: electrophysiological, biochemical and behavioural observations. *Brain Research*, **151**, 431–442.

Drukker, M., Katz, G., Urbach, A., Schuldiner, M., Markel, G., Itskovitz-Eldor, J. et al. (2002) Characterization of the expression of MHC proteins in human embryonic stem cells. *Proceedings of the National Academy of Sciences of the United States of America*, **99**, 9864–9869.

Duvoisin, R.C., Eldridge, R., Williams, A., Nutt, J. and Calne, D. (1981) Twin study of Parkinson disease. *Neurology*, **31**, 77–80.

Elbaz, A., Grigoletto, F., Baldereschi, M., Breteler, M.M., Manubens-Bertran, J.M., Lopez Pousa, S. et al. (1999) Familial aggregation of Parkinson's disease: a population-based case-control study in Europe. EUROPARKINSON Study Group. *Neurology*, **52**, 1876–1882.

Engber, T.M., Susel, Z., Kuo, S., Gerfen, C.R. and Chase, T.N. (1991) Levodopa replacement therapy alters enzyme activities in striatum and neuropeptide content in striatal output regions of 6-hydroxydopamine lesioned rats. *Brain Research*, **552**, 113–118.

Fahn, S. (2002) Results of the ELLDOPA (Earlier vs Later Levodopa) study. *Movement Disorders*, **17** (Suppl 5), S13–S14.

Farrer, M., Chan, P., Chen, R., Tan, L., Lincoln, S., Hernandez, D. et al. (2001) Lewy bodies and parkinsonism in families with parkin mutations. *Annals of Neurology*, **50**, 293–300.

Feany, M.B. and Bender, W.W. (2000) A Drosophila model of Parkinson's disease. *Nature*, **404**, 394–398.

Fearnley, J.M. and Lees, A.J. (1991) Ageing and Parkinson's disease: substantia nigra regional selectivity. *Brain*, **114**, 2283–2301.

Ferrarese, C., Tremolizzo, L., Rigoldi, M., Sala, G., Begni, B., Brighina, L. et al. (2001) Decreased platelet glutamate uptake and genetic risk factors in patients with Parkinson's disease. *Neurological Sciences*, **22**, 65–66.

Foltynie, T., Brayne, C.E., Robbins, T.W. and Barker, R.A. (2004) The cognitive ability of an incident cohort of Parkinson's patients in the UK. The CamPaIGN study. *Brain*, **127**, 550–560.

Foltynie, T., Hicks, A., Sawcer, S., Jonasdottir, A., Setakis, E., Maranian, M. et al. (2005) A genome wide linkage disequilibrium screen in Parkinson's disease. *Journal of Neurology*, **252**, 597–602.

Forno, L.S. (1996) Neuropathology of Parkinson's disease. *Journal of Neuropathology and Experimental Neurology*, **55**, 259–272.

Freed, C.R., Greene, P.E., Breeze, R.E., Tsai, W.Y., DuMouchel, W., Kao, R. et al. (2001) Transplantation of embryonic dopamine neurons for severe Parkinson's disease. *New England Journal of Medicine*, **344**, 710–719.

Funayama, M., Hasegawa, K., Kowa, H., Saito, M., Tsuji, S. and Obata, F. (2002) A new locus for Parkinson's disease (PARK8) maps to chromosome 12p11.2–q13.1. *Annals of Neurology*, **51**, 296–301.

Gerfen, C.R., Engber, T.M., Mahan, L.C., Susel, Z., Chase, T.N., Monsma, F.J. et al. (1990) D1 and D2 dopamine receptor-regulated gene expression of striatonigral and striatopallidal neurons. *Science*, **250**, 1429–1432.

Gerhard, A., Banati, R.B., Goerres, G.B., Cagnin, A., Myers, R., Gunn, R.N. et al. (2003) [11C](R)-PK11195 PET imaging of microglial activation in multiple system atrophy. *Neurology*, **61**, 686–689.

Giasson, B.I. and Lee, V.M. (2003) Are ubiquitination pathways central to Parkinson's disease? *Cell*, **114**, 1–8.

Gibb, W.R. and Lees, A.J. (1988) The relevance of the Lewy body to the pathogenesis of idiopathic Parkinson's disease. *Journal of Neurology Neurosurgery and Psychiatry*, **51**, 745–752.

Gilham, D.E., Cairns, W., Paine, M.J., Modi, S., Poulsom, R., Roberts, G.C. et al. (1997) Metabolism of MPTP by cytochrome P4502D6 and the demonstration of 2D6 mRNA in human foetal and adult brain by in situ hybridization. *Xenobiotica*, **27**, 111–125.

Gill, S.S., Patel, N.K., Hotton, G.R., O'Sullivan, K., McCarter, R., Bunnage, M. et al. (2003) Direct brain infusion of glial cell line-derived neurotrophic factor in Parkinson disease. *Nature Medicine*, **9**, 589–595.

Graham, D.G. (1978) Oxidative pathways for catecholamines in the genesis of neuromelanin and cytotoxic quinones. *Molecular Pharmacology*, **14**, 633–643.

Grondin, R., Hadj, T.A., Doan, V.D., Ladure, P. and Bedard, P.J. (2000) Noradrenoceptor antagonism with idazoxan improves L-dopa-induced dyskinesias in MPTP monkeys. *Naunyn Schmiedebergs Archives Pharmacology*, **361**, 181–186.

Grothe, C., Timmer, M., Scholz, T., Winkler, C., Nikkah, G., Claus, P. et al. (2004) Fibroblast growth factor-20 promotes the differentiation of Nurr1-overexpressing neural stem cells into tyrosine hydroxylase-positive neurons. *Neurobiology of Disease*, **17**, 163–170.

Gwinn-Hardy, K.A., Crook, R., Lincoln, S., Adler, C.H., Caviness, J.N., Hardy, J. et al. (2000) A kindred with Parkinson's disease not showing genetic linkage to established loci. *Neurology*, **54**, 504–507.

Hammerstad, J.P., Carter, J., Nutt, J.G., Casten, G.C., Shrotriya, R.C., Alms, D.R. et al. (1986) Buspirone in Parkinson's disease. *Clinical Neuropharmacology*, **9**, 556–560.

Harrower, T.P., Michell, A.W. and Barker, R.A. (2005) Lewy bodies in Parkinson's disease: protectors or perpetrators? *Experimental Neurology*, **195**, 1–6.

Hauser, R.A. and Olanow, C.W. (1993) Orobuccal dyskinesia associated with trihexyphenidyl therapy in a patient with Parkinson's disease. *Movement Disorders*, **8**, 512–514.

Hauser, R.A., Hubble, J.P. and Truong, D.D. (2003) Randomized trial of the adenosine A(2A) receptor antagonist istradefylline in advanced PD. *Neurology*, **61**, 297–303.

Hayashi, S., Wakabayashi, K. and Ishikawa, A. (2000) An autopsy case of autosomal-recessive juvenile parkinsonism with a homozygous exon 4 deletion in the parkin gene. *Movement Disorders*, **15**, 884–888.

Hernan, M.A., Zhang, S.M., Rueda-deCastro, A.M., Colditz, G.A., Speizer, F.E. and Ascherio, A. (2001) Cigarette smoking and the incidence of Parkinson's disease in two prospective studies. *Annals of Neurology*, **50**, 780–786.

Herrero, M.T., Augood, S.J., Hirsch, E.C., Javoy Agid, F., Luquin, M.R., Agid, Y. et al. (1995) Effects of L-DOPA on preproenkephalin and preprotachykinin gene expression in the MPTP-treated monkey striatum. *Neuroscience*, **68**, 1189–1198.

Hicks, A.A., Petursson, H., Jonsson, T., Stefansson, H., Jonasdottir, H.S., Sainz, J. et al. (2002) A susceptibility gene for late-onset idiopathic Parkinson's disease. *Annals of Neurology*, **52**, 549–555.

Hughes, A.J., Ben Shlomo, Y., Daniel, S.E. and Lees, A.J. (1992a) What features improve the accuracy of clinical diagnosis in Parkinson's disease: a clinicopathologic study. *Neurology*, **42**, 1142–1146.

Hughes, A.J., Daniel, S.E., Kilford, L. and Lees, A.J. (1992b) Accuracy of clinical diagnosis of idiopathic Parkinson's disease: a clinico-pathological study of 100 cases. *Journal of Neurology Neurosurgery and Psychiatry*, **55**, 181–184.

Hughes, A.J., Daniel, S.E., Blankson, S. and Lees, A.J. (1993) A clinicopathologic study of 100 cases of Parkinson's disease. *Archives of Neurology*, **50**, 140–148.

Hughes, A.J., Daniel, S.E., Ben Shlomo, Y. and Lees, A.J. (2002) The accuracy of diagnosis of parkinsonian syndromes in a specialist movement disorder service. *Brain*, **125**, 861–870.

Hurelbrink, C.B. and Barker, R.A. (2001) Prospects for the treatment of Parkinson's disease using neurotrophic factors. *Expert Opinion on Pharmacotherapy*, **2**, 1531–1543.

Hurelbrink, C.B. and Barker, R.A. (2004) The potential of GDNF as a treatment for Parkinson's disease. *Experimental Neurology*, **185**, 1–6.

Ibanez, P., Bonnet, A.M., Debarges, B., Lohmann, E., Tison, F., Pollak, P. et al. (2004) Causal relation between alpha-synuclein gene duplication and familial Parkinson's disease. *Lancet*, **364**, 1169–1171.

Jakes, R., Spillantini, M.G. and Goedert, M. (1994) Identification of two distinct synucleins from human brain. *FEBS Letters*, **345**, 27–32.

Jankovic, J. and Hunter, C. (2002) A double-blind, placebo-controlled and longitudinal study of riluzole in early Parkinson's disease. *Parkinsonism and Related Disorders*, **8**, 271–276.

Jellinger, K. (1987) The pathology of parkinsonism. In Marsden, C.D. and Fahn, S. (eds). *Movement Disorders*. Butterworth Heineman, London.

Johnson, J., Hague, S.M., Hanson, M., Gibson, A., Wilson, K.E., Evans, E.W. et al. (2004) SNCA multiplication is not a common cause of Parkinson disease or dementia with Lewy bodies. *Neurology*, **63**, 554–556.

Johnson, W.G., Hodge, S.E. and Duvoisin, R. (1990) Twin studies and the genetics of Parkinson's disease—a reappraisal. *Movement Disorders*, **5**, 187–194.

Jungling, K., Nagler, K., Pfrieger, F.W. and Gottmann, K. (2003) Purification of embryonic stem cell-derived neurons by immunoisolation. *FASEB Journal*, **17**, 2100–2102.

Kanda, T., Jackson, M.J., Smith, L.A., Pearce, R.K., Nakamura, J., Kase, H. et al. (1998) Adenosine A2A antagonist: a novel antiparkinsonian agent that does not provoke dyskinesia in parkinsonian monkeys. *Annals of Neurology*, **43**, 507–513.

Karamohamed, S., Golbe, L.I., Mark, M.H., Lazzarini, A.M., Suchowersky, O., Labelle, N. et al. (2005) Absence of previously reported variants in the SCNA (G88C and G209A), NR4A2 (T291D and T245G) and the DJ-1 (T497C) genes in familial Parkinson's disease from the GenePD study. *Movement Disorders*, **20**, 1188–1191.

Kawasaki, H., Suemori, H., Mizuseki, K., Watanabe, K., Urano, F., Ichinose, H. et al. (2002) Generation of dopaminergic neurons and pigmented epithelia from primate ES cells by stromal cell-derived inducing activity. *Proceedings of the National Academy of Sciences of the United States of America*, **99**, 1580–1585.

Kay, D.M., Zabetian, C.P., Factor, S.A., Nutt, J.G., Samii, A., Griffith, A. et al. (2005) Parkinson's disease and LRRK2: frequency of a common mutation in U.S. movement disorder clinics. *Movement Disorders*, **21**,519–23

Khan, N.L., Jain, S., Lynch, J.M., Pavese, N., Abou-Sleiman, P., Holton, J.L. et al. (2005) Mutations in the gene LRRK2 encoding dardarin (PARK8) cause familial Parkinson's disease: clinical, pathological, olfactory and functional imaging and genetic data. *Brain*, **128**, 2786–2796.

Kim, J.H., Auerbach, J.M., Rodriguez-Gomez, J.A., Velasco, I., Gavin, D., Lumelsky, N. et al. (2002) Dopamine neurons derived from embryonic stem cells function in an animal model of Parkinson's disease. *Nature*, **418**, 50–56.

Kirik, D., Rosenblad, C., Burger, C., Lundberg, C., Johansen, T.E., Muzyczka, N. et al. (2002) Parkinson-like neurodegeneration induced by targeted overexpression of alpha-synuclein in the nigrostriatal system. *Journal of Neuroscience*, **22**, 2780–2791.

Kitada, T., Asakawa, S., Matsumine, H., Hattori, N., Shimura, H., Minoshima, S. et al. (2000) Progress in the clinical and molecular genetics of familial parkinsonism. *Neurogenetics*, **2**, 207–218.

Kleedorfer, B., Lees, A.J. and Stern, G.M. (1991) Buspirone in the treatment of levodopa induced dyskinesias. *Journal of Neurology Neurosurgery and Psychiatry*, **54**, 376–377.

Kordower, J.H., Freeman, T.B., Chen, E.Y., Mufson, E.J., Sanberg, P.R., Hauser, R.A. et al. (1998) Fetal nigral grafts survive and mediate clinical benefit in a patient with Parkinson's disease. *Movement Disorders*, **13**, 383–393.

Kordower, J.H., Palfi, S., Chen, E.Y., Ma, S.Y., Sendera, T., Cochran, E.J. et al. (1999) Clinicopathological findings following intraventricular glial-derived neurotrophic factor treatment in a patient with Parkinson's disease. *Annals of Neurology*, **46**, 419–424.

Kordower, J.H., Emborg, M.E., Bloch, J., Ma, S.Y., Chu, Y., Levanthal, L., et al. (2000) Neurodegeneration prevented by lentiviral vector delivery of GDNF in primate models of Parkinson's disease. *Science*, **290**, 767–773.

Kruger, R., Kuhn, W., Muller, T., Woitalla, D., Graeber, M., Kosel, S. et al. (1998) Ala30Pro mutation in the gene encoding alpha-synuclein in Parkinson's disease [letter]. *Nature Genetics*, **18**, 106–108.

Kuopio, A., Marttila, R.J., Helenius, H. and Rinne, U.K. (2001) Familial occurrence of Parkinson's disease in a community-based case-control study. *Parkinsonism and Related Disorders*, **7**, 297–303.

Laihinen, A., Ruottinen, H., Rinne, J.O., Haaparanta, M., Bergman, J., Solin, O. et al. (2000) Risk for Parkinson's disease: twin studies for the detection of asymptomatic subjects using [18F]6-fluorodopa PET. *Journal of Neurology*, **247** (Suppl 2), II110–II113.

Lang, A.E., Gill, S., Patel, N.K., Lozano, A., Nutt, J.G., Penn, R. et al. (2006) Randomized controlled trial of intraputamenal glial cell line-derived neurotrophic factor infusion in Parkinson disease. *Annals of Neurology*, **59**, 459–466.

Le, W.D., Xu, P., Jankovic, J., Jiang, H., Appel, S.H., Smith, R.G. et al. (2003) Mutations in NR4A2 associated with familial Parkinson disease. *Nature Genetics*, **33**, 85–89.

Lee, S.H., Lumelsky, N., Studer, L., Auerbach, J.M. and McKay, R.D. (2000) Efficient generation of midbrain and hindbrain neurons from mouse embryonic stem cells. *Nature Biotechnology*, **18**, 675–679.

Leroy, E., Boyer, R., Auburger, G., Leube, B., Ulm, G., Mezey, E. et al. (1998) The ubiquitin pathway in Parkinson's disease. *Nature*, **395**, 451–452.

Li, R.H., White, M., Williams, S. and Hazlett, T. (1998) Poly(vinyl alcohol) synthetic polymer foams as scaffolds for cell encapsulation. *Journal of Biomaterials Science Polymer Edition*, **9**, 239–258.

Ling, Z.D., Potter, E.D., Lipton, J.W. and Carvey, P.M. (1998) Differentiation of mesencephalic progenitor cells into dopaminergic neurons by cytokines. *Experimental Neurology*, **149**, 411–423.

Lotharius, J. and Brundin, P. (2002) Pathogenesis of Parkinson's disease: dopamine, vesicles and alpha-synuclein. *Nature Reviews Neuroscience*, **3**, 932–942.

Lucking, C.B., Durr, A., Bonifati, V., Vaughan, J., De Michele, G., Gasser, T. et al. (2000) Association between early-onset Parkinson's disease and mutations in the parkin gene. French Parkinson's Disease Genetics Study Group. *New England Journal of Medicine*, **342**, 1560–1567.

Lyons, K.E., Wilkinson, S.B., Overman, J. and Pahwa, R. (2004) Surgical and hardware complications of subthalamic stimulation: a series of 160 procedures. *Neurology*, **63**, 612–616.

Maggio, R., Riva, M., Vaglini, F., Fornai, F., Molteni, R., Armogida, M. et al. (1998) Nicotine prevents experimental parkinsonism in rodents and induces striatal increase of neurotrophic factors. *Journal of Neurochemistry*, **71**, 2439–2446.

Maraganore, D.M., de Andrade, M., Lesnick, T.G., Strain, K.J., Farrer, M.J., Rocca, W.A. et al. (2005) High-resolution whole-genome association study of Parkinson disease. *American Journal of Human Genetics*, **77**, 685–693.

Marder, K., Tang, M.X., Mejia, H., Alfaro, B., Cote, L., Louis, E. et al. (1996) Risk of Parkinson's disease among first-degree relatives: a community-based study. *Neurology*, **47**, 155–160.

Marek, K., Innis, R., van Dyck, C., Fussell, B., Early, M., Eberly, S. et al. (2001) [123I]beta-CIT SPECT imaging assessment of the rate of Parkinson's disease progression. *Neurology*, **57**, 2089–2094.

Marek, K., Seibyl, J., Shoulson, I., Holloway, R., Kieburtz, K., McDermott, M. et al. (2002) Dopamine transporter brain imaging to assess the effects of pramipexole vs levodopa on Parkinson disease progression. *Journal of the American Medical Association*, **287**, 1653–1661.

Maries, E., Dass, B., Collier, T.J., Kordower, J.H. and Steece-Collier, K. (2003) The role of alpha-synuclein in Parkinson's disease: insights from animal models. *Nature Reviews Neuroscience*, **4**, 727–738.

Maroteaux, L. and Scheller, R.H. (1991) The rat brain synucleins; family of proteins transiently associated with neuronal membrane. *Brain Research. Molecular Brain Research*, **11**, 335–343.

Maroteaux, L., Campanelli, J.T. and Scheller, R.H. (1988) Synuclein: a neuron-specific protein localized to the nucleus and presynaptic nerve terminal. *Journal of Neuroscience*, **8**, 2804–2815.

Marsden, C.D. (1987) Parkinson's disease in twins [letter]. *Journal of Neurology Neurosurgery and Psychiatry*, **50**, 105–106.

Marsden, C.D. and Parkes, J.D. (1976) "On-off" effects in patients with Parkinson's disease on chronic levodopa therapy. *Lancet*, **1**, 292–296.

Marttila, R.J. and Rinne, U.K. (1976) Dementia in Parkinson's disease. *Acta Neurologica Scandinavia*, **54**, 431–441.

Marttila, R.J., Kaprio, J., Koskenvuo, M. and Rinne, U.K. (1988) Parkinson's disease in a nationwide twin cohort. *Neurology*, **38**, 1217–1219.

Masliah, E., Rockenstein, E., Veinbergs, I., Mallory, M., Hashimoto, M., Takeda, A. et al. (2000) Dopaminergic loss and inclusion body formation in alpha-synuclein mice: implications for neurodegenerative disorders. *Science*, **287**, 1265–1269.

Matsumine, H., Saito, M., Shimoda, M.S., Tanaka, H., Ishikawa, A., Nakagawa, H.Y. et al. (1997) Localization of a gene for an autosomal recessive form of juvenile Parkinsonism to chromosome 6q25.2-27. *American Journal of Human Genetics*, **60**, 588–596.

McNaught, K.S., Olanow, C.W., Halliwell, B., Isacson, O. and Jenner, P. (2001) Failure of the ubiquitin-proteasome system in Parkinson's disease. *Nature Reviews Neuroscience*, **2**, 589–594.

McNaught, K.S., Belizaire, R., Isacson, O., Jenner, P. and Olanow, C.W. (2003) Altered proteasomal function in sporadic Parkinson's disease. *Experimental Neurology*, **179**, 38–46.

Milanov, I., Kmetski, T.S., Lyons, K.E. and Koller, W.C. (2000) Prevalence of Parkinson's disease in Bulgarian Gypsies. *Neuroepidemiology*, **19**, 206–209.

Morens, D.M., Davis, J.W., Grandinetti, A., Ross, G.W., Popper, J.S. and White, L.R. (1996) Epidemiologic observations on Parkinson's disease: incidence and mortality in a prospective study of middle-aged men. *Neurology*, **46**, 1044–1050.

Morgante, L., Rocca, W.A., Di Rosa, A.E., De Domenico, P., Grigoletto, F., Meneghini, F. et al. (1992) Prevalence of Parkinson's disease and other types of parkinsonism: a door-to-door survey in three Sicilian municipalities. The Sicilian Neuro-Epidemiologic Study (SNES) Group. *Neurology*, **42**, 1901–1907.

Morrish, P.K., Sawle, G.V. and Brooks, D.J. (1996) An [18F]dopa-PET and clinical study of the rate of progression in Parkinson's disease. *Brain*, **119 (Pt 2)**, 585–591.

Nakao, N., Frodl, E.M., Duan, W.M., Widner, H. and Brundin, P. (1994) Lazaroids improve the survival of grafted rat embryonic dopamine neurons. *Proceedings of the National Academy of Sciences of the United States of America*, **91**, 12408–12412.

Nakao, N., Frodl, E.M., Widner, H., Carlson, E., Eggerding, F.A., Epstein, C.J. et al. (1995) Overexpressing Cu/Zn superoxide dismutase enhances survival of transplanted neurons in a rat model of Parkinson's disease. *Nature Medicine*, **1**, 226–231.

Nicholl, D.J., Vaughan, J.R., Khan, N.L., Ho, S.L., Aldous, D.E., Lincoln, S. et al. (2002) Two large British kindreds with familial Parkinson's disease: a clinico-pathological and genetic study. *Brain*, **125**, 44–57.

Nichols, W.C., Pankratz, N., Hernandez, D., Paisan Ruiz, C., Jain, S., Halter, C.A. et al. (2005) Genetic screening for a single common LRRK2 mutation in familial Parkinson's disease. *Lancet*, **365**, 410–412.

Nishimura, F., Yoshikawa, M., Kanda, S., Nonaka, M., Yokota, H., Shiroi, A. et al. (2003) Potential use of embryonic stem cells for the treatment of mouse parkinsonian models: improved behavior by transplantation of in vitro differentiated dopaminergic neurons from embryonic stem cells. *Stem Cells*, **21**, 171–180.

Nutt, J.G., Rosin, A.J., Eisler, T., Calne, D.B. and Chase, T.N. (1978) Effect of an opiate antagonist on movement disorders. *Archives of Neurology*, **35**, 810–811.

Nutt, J.G., Burchiel, K.J., Comella, C.L., Janokovic, J., Lang, A.E., Laws, E.R. et al. (2003) Randomized, double-blind trial of glial cell line-derived neurotrophic factor (GDNF) in PD. *Neurology*, **60**, 69–73.

Obeso, J.A., Grandas, F., Herrero, M.T. and Horowski, R. (1994) The role of pulsatile versus continuous dopamine receptor stimulation for functional recovery in Parkinson's disease. *European Journal of Neuroscience*, **6**, 889–897.

Olanow, C.W. and Obeso, J.A. (2000) Pulsatile stimulation of dopamine receptors and levodopa-induced motor complications in Parkinson's disease: implications for the early use of COMT inhibitors. *Neurology*, **55**, S72–S77.

Olanow, C.W., Fahn, S., Muenter, M., Klawans, H., Hurtig, H., Stern, M. et al. (1994) A multicenter double-blind placebo-controlled trial of pergolide as an adjunct to Sinemet in Parkinson's disease. *Movement Disorders*, **9**, 40–47.

Olanow, C.W., Hauser, R.A., Gauger, L., Malapira, T., Koller, W., Hubble, J. et al. (1995) The effect of deprenyl and levodopa on the progression of Parkinson's disease. *Annals of Neurology*, **38**, 771–777.

Olanow, C.W., Goetz, C.G., Kordower, J.H., Stoessl, A.J., Sossi, V., Brin, M.F. et al. (2003) A double-blind controlled trial of bilateral fetal nigral transplantation in Parkinson's disease. *Annals of Neurology*, **54**, 403–414.

Paganini-Hill, A. (2001) Risk factors for Parkinson's disease: the leisure world cohort study. *Neuroepidemiology*, **20**, 118–124.

Paisan-Ruiz, C., Jain, S., Evans, E.W., Gilks, W.P., Simon, J., van der Brug, M. et al. (2004) Cloning of the gene containing mutations that cause PARK8-linked Parkinson's disease. *Neuron*, **44**, 595–600.

Pal, S., Pens, R., Brooks, D.J., Rao, C.S. and RayChaudhuri, K. (2002) Isolated inherited dominantly postural tremor with or without resting tremor: a variant of long latency tremulous Parkinson's disease? A clinical follow up study. *Association of British Neurologists Spring Meeting*. 3-5-2002.

Papa, S.M. and Chase, T.N. (1996) Levodopa-induced dyskinesias improved by a glutamate antagonist in Parkinsonian monkeys. *Annals of Neurology*, **39**, 574–578.

Parker, W.D. and Swerdlow, R.H. (1998) Mitochondrial dysfunction in idiopathic Parkinson disease. *American Journal of Human Genetics*, **62**, 758–762.

Parkinson, J. (1817) *An Essay on the Shaking Palsy*. London, Whittingham and Rowland.

Parkinson Study Group. (1997) Entacapone improves motor fluctuations in levodopa-treated Parkinson's disease patients. *Annals of Neurology*, **42**, 747–755.

Parkinson Study Group. (2004) A controlled, randomized, delayed-start study of rasagiline in early Parkinson disease. *Archives of Neurology*, **61**, 561–566.

Patel, N.K., Bunnage, M., Plaha, P., Svendsen, C.N., Heywood, P. and Gill, S.S. (2005) Intraputamenal infusion of glial cell line-derived neurotrophic factor in PD: a two-year outcome study. *Annals of Neurology*, **57**, 298–302.

Paulson, GW. (1992) Addiction to nicotine is due to high intrinsic levels of dopamine. *Medical Hypotheses*, **38**, 206–207.

Peaire, A.E., Takeshima, T., Johnston, J.M., Isoe, K., Nakashima, K. and Commissiong, J.W. (2003) Production of dopaminergic neurons for cell therapy in the treatment of Parkinson's disease. *Journal of Neuroscience Methods*, **124**, 61–74.

Perez, R.G. and Hastings, T.G. (2004) Could a loss of alpha-synuclein function put dopaminergic neurons at risk? *Journal of Neurochemistry*, **89**, 1318–1324.

Perez, R.G., Waymire, J.C., Lin, E., Liu, J.J., Guo, F. and Zigmond, M.J. (2002) A role for alpha-synuclein in the regulation of dopamine biosynthesis. *Journal of Neuroscience*, **22**, 3090–3099.

Piccini, P., Weeks, R.A. and Brooks, D.J. (1997) Alterations in opioid receptor binding in Parkinson's disease patients with levodopa-induced dyskinesias. *Annals of Neurology*, **42**, 720–726.

Piccini, P., Brooks, D.J., Bjorklund, A., Gunn, R.N., Grasby, P.M., Rimoldi, O. et al. (1999a) Dopamine release from nigral transplants visualized in vivo in a Parkinson's patient. *Nature Neuroscience*, **2**, 1137–1140.

Piccini, P., Burn, D.J., Ceravolo, R., Maraganore, D. and Brooks, D.J. (1999b) The role of inheritance in sporadic Parkinson's disease: evidence from a longitudinal study of dopaminergic function in twins. *Annals of Neurology*, **45**, 577–582.

Picconi, B., Centonze, D., Hakansson, K., Bernardi, G., Greengard, P., Fisone, G. et al. (2003) Loss of bidirectional striatal synaptic plasticity in L-DOPA-induced dyskinesia. *Nature Neuroscience*, **6**, 501–506.

Polymeropoulos, M.H., Higgins, J.J., Golbe, L.I., Johnson, W.G., Ide, S.E., Di Iorio, G. et al. (1996) Mapping of a gene for Parkinson's disease to chromosome 4q21-q23. *Science*, **274**, 1197–1199.

Polymeropoulos, M.H., Lavedan, C., Leroy, E., Ide, S.E., Dehejia, A., Dutra, A. et al. (1997) Mutation in the alpha-synuclein gene identified in families with Parkinson's disease. *Science*, **276**, 2045–2047.

Potter, E.D., Ling, Z.D. and Carvey, P.M. (1999) Cytokine-induced conversion of mesencephalic-derived progenitor cells into dopamine neurons. *Cell Tissue Research*, **296**, 235–246.

Priyadarshi, A., Khuder, S.A., Schaub, E.A. and Shrivastava, S. (2000) A meta-analysis of Parkinson's disease and exposure to pesticides. *Neurotoxicology*, **21**, 435–440.

Pyle, A., Foltynie, T., Tiangyou, W., Lambert, C., Keers, S.M., Allcock, L.M. et al. (2005) Mitochondrial DNA haplogroup cluster UKJT reduces the risk of PD. *Annals of Neurology*, **57**, 564–567.

Rabey, J.M., Sagi, I., Huberman, M., Melemed, E., Korczyn, A., Giladi, N. et al. (2000) Rasagiline mesylate, a new MAO-B inhibitor for the treatment of Parkinson's disease: a double-blind study as adjunctive therapy to levodopa. *Clinical Neuropharmacology*, **23**, 324–330.

Rascol, O., Brooks, D.J., Korczyn, A.D., De Deyn, P.P., Clarke, C.E., Lang, A.E. (2000) A five-year study of the incidence of dyskinesia in patients with early Parkinson's disease who were treated with ropinirole or levodopa. 056 Study Group. *New England Journal of Medicine*, **342**, 1484–1491.

Rascol, O., Brooks, D.J., Melamed, E., Oertel, W., Poewe, W., Stocchi, F. et al. (2005) Rasagiline as an adjunct to levodopa in patients with Parkinson's disease and motor fluctuations (LARGO, Lasting effect in Adjunct therapy with Rasagiline Given Once daily, study): a randomised, double-blind, parallel-group trial. *Lancet*, **365**, 947–954.

Riaz, S.S., Jauniaux, E., Stern, G.M. and Bradford, H.F. (2002) The controlled conversion of human neural progenitor cells derived from foetal ventral mesencephalon into dopaminergic neurons in vitro. *Brain Research. Developmental Brain Research*, **136**, 27–34.

Rioux, L., Gaudin, D.P., Bui, L.K., Gregoire, L., DiPaolo, T. and Bedard, P.J. (1991) Correlation of functional recovery after a 6-hydroxydopamine lesion with survival of grafted fetal neurons and release of dopamine in the striatum of the rat. *Neuroscience*, **40**, 123–131.

Rostami-Hodjegan, A., Lennard, M.S., Woods, H.F. and Tucker, G.T. (1998) Meta-analysis of studies of the CYP2D6 polymorphism in relation to lung cancer and Parkinson's disease. *Pharmacogenetics*, **8**, 227–238.

Sanchez-Pernaute, R., Studer, L., Bankiewicz, K.S., Major, E.O. and McKay, R.D. (2001) In vitro generation and transplantation of precursor-derived human dopamine neurons. *Journal of Neuroscience Research*, **65**, 284–288.

Sandyk, R. and Snider, S.R. (1986) Naloxone treatment of L-dopa-induced dyskinesias in Parkinson's disease. *American Journal of Psychiatry*, **143**, 118.

Savola, J.M., Hill, M., Engstrom, M., Merivuori, H., Wurster, S., McGuire, S.G. et al. (2003) Fipamezole (JP-1730) is a potent alpha2 adrenergic receptor antagonist that reduces levodopa-induced dyskinesia in the MPTP-lesioned primate model of Parkinson's disease. *Movement Disorders*, **18**, 872–883.

Sawamoto, K., Nakao, N., Kakishita, K., Ogawa, Y., Toyama, Y., Yamamoto, A. et al. (2001) Generation of dopaminergic neurons in the adult brain from mesencephalic precursor cells labeled with a nestin-GFP transgene. *Journal of Neuroscience*, **21**, 3895–3903.

Schapira, A.H., Cooper, J.M., Dexter, D., Jenner, P., Clark, J.B. and Marsden, C.D. (1989) Mitochondrial complex I deficiency in Parkinson's disease. *Lancet*, **1**, 1269.

Schierle, G.S., Hansson, O., Leist, M., Nicotera, P., Widner, H. and Brundin, P. (1999) Caspase inhibition reduces apoptosis and increases survival of nigral transplants. *Nature Medicine*, **5**, 97–100.

Schoenberg, B.S., Anderson, D.W. and Haerer, A.F. (1985) Prevalence of Parkinson's disease in the biracial population of Copiah County, Mississippi. *Neurology*, **35**, 841–845.

Schrag, A., Keens, J. and Warner, J. (2002) Ropinirole for the treatment of tremor in early Parkinson's disease. *European Journal of Neurology*, **9**, 253–257.

Scott, W.K., Nance, M.A., Watts, R.L., Hubble, J.P., Koller, W.C., Lyons, K. et al. (2001) Complete genomic screen in Parkinson disease: evidence for multiple genes. *Journal of the American Medical Association*, **286**, 2239–2244.

Shoulson, I., Oakes, D., Fahn, S., Lang, A., Langston, J.W., LeWitt, P. et al. (2002) Impact of sustained deprenyl (selegiline) in levodopa-treated Parkinson's disease: a randomized placebo-controlled extension of the deprenyl and tocopherol antioxidative therapy of parkinsonism trial. *Annals of Neurology*, **51**, 604–612.

Shults, C.W., Oakes, D., Kieburtz, K., Beal, M.F., Haas, R., Plumb, S. et al. (2002) Effects of coenzyme Q10 in early Parkinson disease: evidence of slowing of the functional decline. *Archives of Neurology*, **59**, 1541–1550.

Shults, C.W., Beal, M.F., Song, D. and Fontaine, D. (2004) Pilot trial of high dosages of coenzyme Q10 in patients with Parkinson's disease. *Experimental Neurology*, **188**, 491–494.

Singleton, A.B., Farrer, M., Johnson, J., Singleton, A., Hague, S., Kachergus, J. et al. (2003) alpha-Synuclein locus triplication causes Parkinson's disease. *Science*, **302**, 841.

Smith, W.W., Pei, Z., Jiang, H., Moore, D.J., Liang, Y., West, A.B. et al. (2005) Leucine-rich repeat kinase 2 (LRRK2) interacts with parkin, and mutant LRRK2 induces neuronal degeneration. *Proceedings of the National Academy of Sciences of the United States of America*, **102**, 18676–18681.

Snyder, H., Mensah, K., Theisler, C., Lee, J., Matouschek, A. and Wolozin, B. (2003) Aggregated and monomeric alpha-synuclein bind to the S6′ proteasomal protein and inhibit proteasomal function. *Journal of Biological Chemistry*, **278**, 11753–11759.

Spillantini, M.G., Schmidt, M.L., Lee, V.M., Trojanowski, J.Q., Jakes, R. and Goedert, M. (1997) Alpha-synuclein in Lewy bodies. *Nature*, **388**, 839–840.

Stefanis, L., Larsen, K.E., Rideout, H.J., Sulzer, D. and Greene, L.A. (2001) Expression of A53T mutant but not wild-type alpha-synuclein in PC12 cells induces alterations of the ubiquitin-dependent degradation system, loss of dopamine release, and autophagic cell death. *Journal of Neuroscience*, **21**, 9549–9560.

Stocchi, F. and Olanow, C.W. (2003) Neuroprotection in Parkinson's disease: clinical trials. *Annals of Neurology*, **53** (Suppl 3), S87–S97.

Storch, A., Paul, G., Csete, M., Boehm, B.O., Carvey, P.M., Kupsch, A. et al. (2001) Long-term proliferation and dopaminergic differentiation of human mesencephalic neural precursor cells. *Experimental Neurology*, **170**, 317–325.

Storch, A., Lester, H.A., Boehm, B.O. and Schwarz, J. (2003) Functional characterization of dopaminergic neurons derived from rodent mesencephalic progenitor cells. *Journal of Chemical Neuroanatomy*, **26**, 133–142.

Studer, L., Tabar, V. and McKay, R.D. (1998) Transplantation of expanded mesencephalic precursors leads to recovery in parkinsonian rats. *Nature Neuroscience*, **1**, 290–295.

Takagi, Y., Takahashi, J., Saiki, H., Morizane, A., Hayashi, T., Kishi, Y. et al. (2005) Dopaminergic neurons generated from monkey embryonic stem cells function in a Parkinson primate model. *Journal of Clinical Investigation*, **115**, 102–109.

Tan, E.K., Khajavi, M., Thornby, J.I., Nagamitsu, S., Jankovic, J. and Ashizawa, T. (2000) Variability and validity of polymorphism association studies in Parkinson's disease. *Neurology*, **55**, 533–538.

Tanner, C.M. and Ben Shlomo, Y. (1999) Epidemiology of Parkinson's disease. *Advances in Neurology*, **80**, 153–159.

Tanner, C.M., Ottman, R., Goldman, S.M., Ellenberg, J., Chan, P., Mayeux, R. et al. (1999) Parkinson disease in twins: an etiologic study. *Journal of the American Medical Association*, **281**, 341–346.

Tekle-Haimanot, R., Abebe, M., Gebre-Mariam, A., Forsgren, L., Hejibel, J., Holmgren, G. et al. (1990) Community-based study of neurological disorders in rural central Ethiopia. *Neuroepidemiology*, **9**, 263–277.

Trabucchi, M., Bassi, S. and Frattola, L. (1982) Effect of naloxone on the 'on-off' syndrome in patients receiving long-term levodopa therapy. *Archives of Neurology*, **39**, 120–121.

Turski, L., Bressler, K., Rettig, K.J., Loschmann, P.A. and Wachtel, H. (1991) Protection of substantia nigra from MPP+ neurotoxicity by *N*-methyl-D-aspartate antagonists. *Nature*, **349**, 414–418.

Ueda, K., Fukushima, H., Masliah, E., Xia, Y., Iwai, A., Yoshimoto, M. et al. (1993) Molecular cloning of cDNA encoding an unrecognized component of amyloid in Alzheimer disease. *Proceedings of the National Academy of Sciences of the United States of America*, **90**, 11282–11286.

Valente, E.M., Bentivoglio, A.R., Dixon, P.H., Ferraris, A., Ialongo, T., Frontali, M. et al. (2001) Localization of a Novel Locus for Autosomal Recessive Early-Onset Parkinsonism, PARK6, on Human Chromosome 1p35-p36. *American Journal of Human Genetics*, **68**, 895–900.

Valente, E.M., Abou-Sleiman, P.M., Caputo, V., Muqit, M.M., Harvey, K., Gispert, S. et al. (2004) Hereditary early-onset Parkinson's disease caused by mutations in PINK1. *Science*, **304**, 1158–1160.

Van Den Eeden, S.K., Tanner, C.M., Bernstein, A.L., Fross, R.D., Leimpeter, A., Bloch, D.A. et al. (2003) Incidence of Parkinson's disease: variation by age, gender, and race/ethnicity. *American Journal of Epidemiology*, **157**, 1015–1022.

van der Walt, J.M., Nicodemus, K.K., Martin, E.R., Scott, W.K., Nance, M.A., Watts, R.L. et al. (2003) Mitochondrial polymorphisms significantly reduce the risk of Parkinson disease. *American Journal of Human Genetics*, **72**, 804–811.

van Duijn, C.M., Dekker, M.C., Bonifati, V., Galjaard, R.J., Houwing-Dusitermaat, J.J., Snijders, P.J. et al. (2001) Park7, a novel locus for autosomal recessive early-onset parkinsonism, on chromosome 1p36. *American Journal of Human Genetics*, **69**, 629–634.

Verhagen, M.L., Del Dotto, P., Natte, R., van den, M.P. and Chase, T.N. (1998a) Dextromethorphan improves levodopa-induced dyskinesias in Parkinson's disease. *Neurology*, **51**, 203–206.

Verhagen, M.L., Del Dotto, P., van den, M.P., Fang, J., Mouradian, M.M. and Chase, T.N. (1998b) Amantadine as treatment for dyskinesias and motor fluctuations in Parkinson's disease. *Neurology*, **50**, 1323–1326.

Ward, C.D., Duvoisin, R.C., Ince, S.E., Nutt, J.D., Eldridge, R. and Calne, D.B. (1983) Parkinson's disease in 65 pairs of twins and in a set of quadruplets. *Neurology*, **33**, 815–824.

Ward, C.D. and Gibb, W.R. (1990) Research diagnostic criteria for Parkinson's disease. *Advances in Neurology*, **53**, 245–249.

Wernig, M., Tucker, K.L., Gornik, V., Schneiders, A., Buschwald, R., Wiestler, O.D. et al. (2002) Tau EGFP embryonic stem cells: an efficient tool for neuronal lineage selection and transplantation. *Journal of Neuroscience Research*, **69**, 918–924.

Whone, A.L., Watts, R.L., Stoessl, A.J., Davis, M., Reske, S., Nahmias, C. et al. (2003) Slower progression of Parkinson's disease with ropinirole versus levodopa: the REAL-PET study. *Annals of Neurology*, **54**, 93–101.

Yoshizaki, T., Inaji, M., Kouike, H., Shimazaki, T., Sawamoto, K., Ando, K. et al. (2004) Isolation and transplantation of dopaminergic neurons generated from mouse embryonic stem cells. *Neuroscience Letters*, **363**, 33–37.

Young, W.S., Bonner, T.I. and Brann, M.R. (1986) Mesencephalic dopamine neurons regulate the expression of neuropeptide mRNAs in the rat forebrain. *Proceedings of the National Academy of Sciences of the United States of America*, **83**, 9827–9831.

Zawada, W.M., Zastrow, D.J., Clarkson, E.D., Adams, F.S., Bell, K.P. and Freed, C.R. (1998) Growth factors improve immediate survival of embryonic dopamine neurons after transplantation into rats. *Brain Research*, **786**, 96–103.

Zhang, L., Shimoji, M., Thomas, B., Moore, D.J., Yu, S.W., Marapudi, N.I. et al. (2005) Mitochondrial localization of the Parkinson's disease related protein DJ-1: implications for pathogenesis. *Human Molecular Genetics*, **14**, 2063–2073.

Zhang, S.M., Hernan, M.A., Chen, H., Spiegelman, D., Willett, W.C. and Ascherio, A. (2002) Intakes of vitamins E and C, carotenoids, vitamin supplements, and PD risk. *Neurology*, **59**, 1161–1169.

Zhuang, Z.P., Kung, M.P., Hou, C., Plossl, K., Skovronsky, D., Gur, T.L. et al. (2001) IBOX (2-(4-dimethylaminophenyl)-6-iodobenzoxazole): a ligand for imaging amyloid plaques in the brain. *Nuclear Medicine and Biology*, **28**, 887–894.

Zimprich, A., Asmus, F., Leitner, P., Castro, M., Bereznai, B., Homann, N. et al. (2003) Point mutations in exon 1 of the NR4A2 gene are not a major cause of familial Parkinson's disease. *Neurogenetics*, 4, 219–220.

Zimprich, A., Biskup, S., Leitner, P., Lichtner, P., Farrer, M., Lincoln, S. et al. (2004) Mutations in LRRK2 cause autosomal-dominant parkinsonism with pleomorphic pathology. *Neuron*, **44**, 601–607.

10 Restoring Dopamine Levels

*P. Nuno Palma, Maria João Bonifácio,
Luís Almeida, and Patrício Soares-da-Silva*

CONTENTS

This chapter reviews the subject of restoring dopamine levels in PD while employing a number of new therapeutic approaches that ease the conversion of levodopa to dopamine at the target region in the brain and facilitate the action of the amine at receptor sites. Main focus will concern the enzymes that affect levodopa catabolism (aromatic L-amino acid decarboxylase and COMT) and dopamine oxidative metabolism (MAO).

10.1 INTRODUCTION

Parkinson's disease (PD) is a chronic degenerative disease with a prevalence of just over 1 per 1000 patients and increasing incidence at older ages. It is characterized by tremor, rigidity, bradykinesia, and postural instability. Other primary symptoms are slowness, stiffness, and the inability to initiate movements. The symptoms of PD become apparent after about 240,000 nigral cells die (60% of the total). The peak onset is 60 years and PD progresses slowly over 10–20 years. In normal, unaffected, people approximately 2400 nigral cells die yearly. Thus if someone lives 100 years they are at risk for PD. In PD the nigral cell loss accelerates—more than 2400 cells die each year. It is unknown why the loss accelerates. Paralleling the nigral cell loss, there is a loss of the neurotransmitter dopamine in another region of the brain: the striatum.

The discoveries by Carlsson et al. (1957) showing that 3,4-dihydroxyphenylalanine (DOPA) can reverse akinesia induced by catecholamine depletion in mice, and by Ehringer and Hornykiewicz (1960) (see Hornykiewicz [2006] for review) demonstrating for the first time striatal dopamine deficiency in PD, provided the background for the first clinical trials with DOPA. Initial trials led to inconsistent results, but some years later Cotzias et al. (1967) reported the impressive antiparkinsonian effects of long-term oral D, L-DOPA, establishing the basis for the modern pharmacotherapy of PD. The improved efficacy of the combination of levodopa (L-DOPA) with a peripheral decarboxylase inhibitor and the major adverse effects of the treatment, including involuntary movements, motor fluctuations, and mental symptoms, were described shortly thereafter (Cotzias et al., 1969). Levodopa therapy radically changed the life history of patients afflicted with PD to such an extent that a large scale, controlled, prospective study comparing levodopa versus placebo has never been conducted.

Since its introduction in 1960s, levodopa has been the mainstay of treatment for PD. Unlike dopamine itself, levodopa crosses the blood–brain barrier (BBB). It is then decarboxylated to dopamine and is released by presynaptic terminals in the striatum, where it replenishes the dopaminergic deficiency that is characteristic of PD. Given orally, levodopa is approximately 95% decarboxylated by the intestinal and liver aromatic amino acid decarboxylase (AADC) to dopamine, which, as mentioned above, cannot cross the BBB. The combination of levodopa with a decarboxylase inhibitor (carbidopa or benserazide), which is unable to penetrate the CNS, diminishes the decarboxylation of levodopa to dopamine in peripheral tissues and permits a greater proportion of levodopa to reach target sites in the striatum. This minimizes the peripheral side effects of levodopa and dopamine (nausea, hypotension, etc.) although the same dosage is used.

Levodopa is the most effective symptomatic agent in the treatment of PD and the "gold standard" against which new agents must be compared. However, there remain two areas of controversy: (1) whether levodopa is toxic, and (2) whether levodopa directly causes motor complications. Levodopa is toxic to cultured dopamine neurons, and this may be a problem in PD where there is evidence of oxidative stress in the *substantia nigra*. However, there is little firm evidence to suggest that levodopa is toxic in vivo or in PD. Levodopa is also associated with motor complications and increasing evidence suggests that they are related, at least in part, to the short half-life of the drug (and its potential to induce pulsatile stimulation of dopamine receptors) rather than to specific properties of the molecule. Treatment strategies that provide more continuous stimulation of dopamine receptors provide reduced motor complications in PD patients (Olanow et al., 2004). These studies raise the possibility that more continuous and physiological delivery of levodopa might reduce the risk of motor complications.

The absorption of levodopa occurs primarily in the duodenum and jejunum by means of an amino acid carrier-mediated transport system. Factors that decrease gastric emptying, such as food intake, gastric acidity, and anticholinergic medication, can delay the delivery of levodopa to the small intestine, allowing more time for peripheral decarboxylation. Protein intake may interfere with levodopa treatment because neutral amino acids will compete with levodopa for transport across the gut and the BBB. The levodopa metabolite 3-*O*-methyl-levodopa (3-OMD) may also compete with levodopa for transport across these membranes.

Catechol-*O*-methyltransferase (COMT) is an enzyme that degrades levodopa, this being particularly evident when given with decarboxylase inhibitors. COMT inhibitors prevent the normal O-methylation of levodopa to its metabolite 3-OMD, a reaction that diverts a portion of the levodopa from conversion to dopamine. By limiting this "metabolic loss" of levodopa to 3-OMD, COMT inhibitors increase the availability of levodopa for dopamine production. COMT inhibitors, taken orally in combination with levodopa and carbidopa or benserazide, reduce peripheral 3-OMD formation from levodopa, enabling more levodopa to enter the brain. Carbidopa and benserazide are also substrates for COMT, and COMT inhibitors may prolong the action of these decarboxylase inhibitors.

Monoamine oxidase (MAO) plays an essential role in the catabolism of monoamines in the brain. With the discovery of subtypes of MAO came the possibility of developing MAO inhibitors that show specificity for each subtype, the MAO A and MAO B, which differ in their chemical structure, distribution and, most importantly, their specific substrates. MAO B inhibition causes a reduction in the enzymatic degradation of dopamine and, thus, prolongation of the availability of dopamine in neurons in the *substantia nigra*.

10.2 AROMATIC AMINO ACID DECARBOXYLASE

10.2.1 Physiological Role

Aromatic L-amino acid decarboxylase (AADC, EC 4.1.1.28), first described by Holtz in 1938, is a pyridoxal 5′-phosphate-dependent enzyme that catalyzes the decarboxylation of several aromatic L-amino acids. Major substrates include levodopa, 5-hydroxytryptophan, L-tyrosine, and L-phenylalanine, the reaction products being the respective amines and CO_2 (Zhu and Juorio, 1995). AADC is a key player in monoaminergic transmission, due to its role in both the catecholamine and 5-hydroxytryptamine synthetic pathways, and also in the synthesis of trace amines which have been recently suggested to function as endogenous neuromodulators (Berry, 2004). The efficacy of the levodopa therapy in PD is due to the brain AADC, since it is the enzyme, in the brain, responsible for all dopamine synthesized from exogenous levodopa (Zhu and Juorio, 1995).

10.2.2 Gene and Tissue Distribution

In humans, the AADC gene is located in chromosome 7, loci p12.1–p12.3, and consists of 15 exons spanning over 85 kbp (Craig et al., 1992; Sumi-Ichinose et al., 1992). Expression of AADC gene is found in both catecholaminergic and serotonergic neurons, in glial cells, chromaffin cells, in liver, kidney, pancreas, gastrointestinal tract, and lung (Berry et al., 1996). Regulation of tissue-specificity appears to be due to the usage of alternative promoters and alternative splicing at the 5' untranslated region of the transcripts. In effect, two transcripts were described, one present in neuronal tissue and another in non-neuronal tissue, with the same coding sequence and differing only in the 5' untranslated region (Ichinose et al., 1992; Jahng et al., 1996). Some exceptions have been reported such as a transcript found in both neuronal and non-neuronal tissues encoding a shorter protein that has yet to be proved that it is active (O'Malley et al., 1995), and also a transcript with another functional 5' untranslated region (Djali et al., 1998).

10.2.3 Enzyme Activity

Physiologically, AADC is considered to be a soluble cellular protein. However, there are reports suggesting an association with membranes (Zhu and Juorio, 1995; Poulikakos et al., 2001). Concerning enzyme activities, in mammals the highest activities are found in liver, kidney, and adrenal medulla (Zhu and Juorio, 1995), although the physiological significance of the presence of AADC in these tissues is not yet completely clear. It might be related to general detoxification in liver and dopamine synthesis in kidney.

AADC activity is subject to ontogenic regulation; it increases after birth and diminishes with age. Furthermore, AADC activity is modulated in response to several stimuli by either up- or down-regulation of AADC mRNA expression, with consequent de novo protein synthesis, or by the activation of second-messenger pathways and posttranslational modifications. The latter involves phosphorylation by cyclic AMP-dependent protein kinase (Duchemin et al., 2000), cGMP-dependent kinase (Hadjiconstantinou et al., 2003), and eventually others. Modulators of AADC activity include dopamine agonists and antagonists, a2-adrenoreceptor antagonists, MAO inhibitors, dexamethasone, prostaglandin E2, interleukin 1b, and amphetamine among others (Berry et al., 1996).

10.2.4 AADC Inhibition

AADC inhibitors were initially developed for the therapy of hypertension; however, it was noticed that one of these molecules improved the effect of levodopa in PD

(Pletscher and DaPrada, 1993). The rationale for the administration of AADC inhibitors to PD patients lies upon the fact that peripheral AADC activity is much higher than in the brain, and therefore the administered levodopa is predominantly decarboxylated at the periphery. The resultant dopamine increase in the circulation leads to severe side effects such as cardiac arrhythmias, hypotension, nausea, vomiting, etc. There are currently two AADC inhibitors used in clinical practice, carbidopa (Merck) and benserazide (Roche), with the latter being about 10-fold more potent than the former (Da Prada et al., 1987). Both molecules are peripheral inhibitors although, at high concentrations, benserazide becomes a central inhibitor.

The two molecules are hydrazine analogs of levodopa (Figure 10.1). The structure of the complex between carbidopa and pig kidney AADC has been solved by x-ray crystallography, to a resolution of 2.25 Å (Burkhard et al., 2001). This structure shows the inhibitor bound to the enzyme by forming a hydrazone linkage with the pyridoxal-5′-phosphate cofactor (Figure 10.2), mimicking the external aldimine enzyme–substrate intermediate. Although no structural information about the complex with benserazide is available, it is suggested that the third hydroxyl group present in the aromatic ring of benserazide could form hydrogen bonds to two structural water molecules identified at the active site, without displacing them, therefore contributing to the higher affinity of this inhibitor (Burkhard et al., 2001).

Although no other AADC inhibitors have yet reached the market to challenge carbidopa or benserazide, several other molecules have been designed and showed to

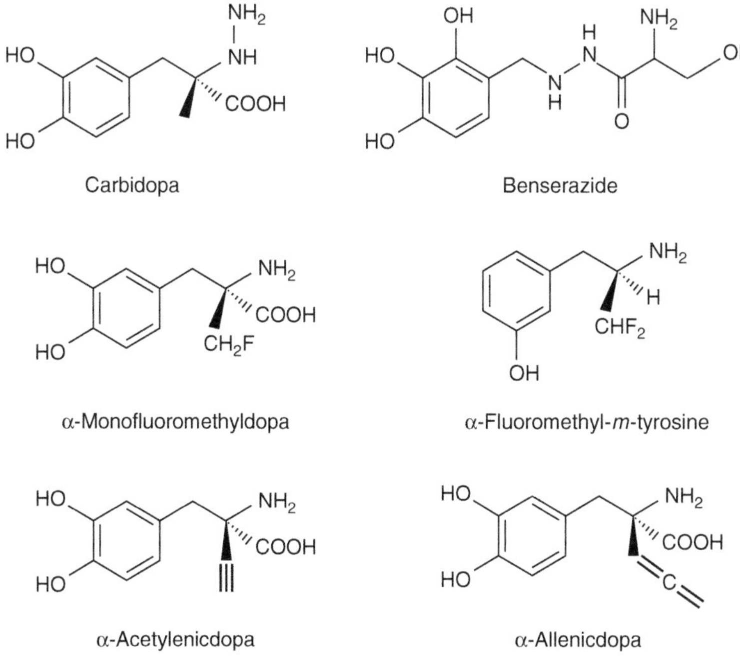

FIGURE 10.1 Examples of AADC inhibitors.

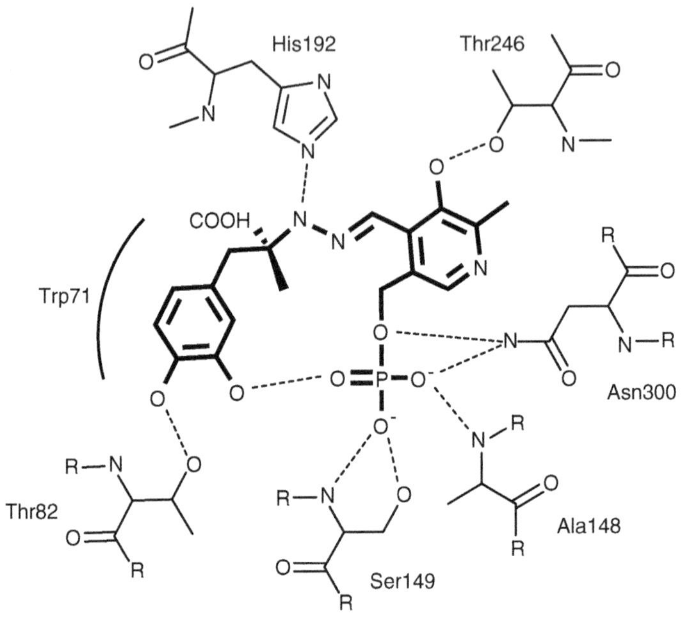

FIGURE 10.2 Schematic representation of the covalent complex between carbidopa and pyridoxal-5′-phosphate (thick structure), at the active site of AADC (thin structures). A hydrogen-bonding network is shown as dashed lines.

be highly selective and pharmacologically effective inhibitors of the AADC enzyme. Among the potent and selective inhibitors are a series of halomethyl derivatives of levodopa, of which, α-monofluoromethyldopa (Figure 10.1) was shown to be a potent, irreversible inhibitor of AADC. Unfortunately, however, it was shown to readily cross the BBB, leading to inhibition of both central and peripheral enzyme (Bey et al., 1980). These molecules act as pseudo-substrates, purportedly generating an enzyme-activated electrophilic intermediate that would eventually form an irreversible adduct with the pyridoxal cofactor and a proteic nucleophile (Maycock et al., 1980). Additional classes of AADC inhibitors include the product-based irreversible inhibitor α-fluoromethyl-*m*-tyrosine and compounds containing functionalities which have been used to inhibit other pyridoxal-dependent enzymes, such as α-acetylenicdopa and α-allenicdopa (Ribereau-Gayon et al., 1979; Castelhano et al., 1984).

10.2.5 CLINICAL EFFICACY OF AADC INHIBITORS

When levodopa is administered to humans, it is absorbed by means of the large neutral amino acid uptake system across the gastrointestinal barrier and it is rapidly metabolized by decarboxylation (70% of the oral dose), followed by O-methylation (about 10% of the oral dose), transamination, and oxidation (Bonifati and Meco, 1999).

Absorption is dependent upon several factors, namely the rate of gastric emptying which is influenced by food composition, the competition with other amino acids for the uptake transporter, and some drug interactions (Deleu et al., 2002). To reach the brain, levodopa must be transported through the BBB, again by the large neutral amino acid uptake system, and as a result, it is subject to competition by other amino acids (Cedarbaum, 1987).

At a 300 mg dose, maximum plasma concentrations (C_{max}) (1.8 μg/L) are reached at 1.4 h (t_{max}), the terminal elimination half-life ($T_{1/2}$) is 1.1 h, and the oral bioavailability is about 33% (Deleu et al., 2002). Nevertheless, only about 1% levodopa reaches the brain. With the coadministration of either carbidopa or benserazide an increase in C_{max}, in the area under the curve (AUC) and in the $T_{1/2}$ is observed as well as a twofold to threefold increase in the levodopa oral bioavailability. As a result the levodopa dose can be reduced by as much as 70%–80% still maintaining the clinical efficacy. Furthermore, the adverse side effects due to the increase in peripheral dopamine are greatly reduced, since circulating dopamine returns to physiological levels (Bonifati and Meco, 1999; Deleu et al., 2002). Nonetheless, under these circumstances, there is an increase in circulating 3-OMD that accumulates in plasma, potentially competing with levodopa for the BBB uptake transporter, and only about 5%–10% levodopa reaches the brain (Männistö and Kaakkola, 1990). Concerning the pharmacokinetics of the AADC inhibitors, both carbidopa and benserazide are not completely absorbed, with an oral bioavailability that ranges 40%–70% with carbidopa and 70% with benserazide. While carbidopa metabolites are pharmacologically inactive, one of the benserazide metabolites, trihydroxybenzylhydrazine, is a potent AADC inhibitor (Cedarbaum, 1987).

The combination of levodopa with a peripheral AADC inhibitor remains the most effective treatment for PD; however, patients under levodopa chronic treatment usually develop motor complications, such as dyskinesias, and "wearing-off" and "ON/OFF" fluctuations. These complications are associated with fluctuations in levodopa plasma levels probably due to not only the pharmacokinetics of levodopa but also to the progression of the nigrostriatal dopaminergic neuron degeneration (Stocchi, 2005). Other therapeutic strategies must therefore be sought to maintain a sustained levodopa concentration in plasma.

10.3   CATECHOL-O-METHYLTRANSFERASE

10.3.1   PHYSIOLOGICAL ROLE

COMT (EC 2.1.1.6), initially described by Axelrod and Tomchick (1958), catalyzes the transference of a methyl group from S-adenosyl-L-methionine (SAM) to a hydroxyl group of a catechol substrate in the presence of magnesium, the reaction products being the O-methylated catechol and S-adenosyl-L-homocysteine (Guldberg and Marsden, 1975). COMT substrates comprise a wide variety of endogenous and exogenous catechol derivatives such as catecholamines, their hydroxylated metabolites, catecholestrogens, ascorbic acid, dietary phytochemicals, and therapeutic compounds (Guldberg and Marsden, 1975; Männistö et al., 1992a; Zhu et al., 2000). The major physiological role of the O-methylation reaction is the inactivation of biologically active or toxic catechols. Of particular relevance is the methylation of levodopa to 3-OMD in patients afflicted with PD (Sharpless and McCann, 1971; Männistö and Kaakkola, 1990). Recent studies also suggest a relevant role of COMT in modulating prefrontal dopamine neurotransmission (Egan et al., 2002; Huotari et al., 2002; Akil et al., 2003; Bilder et al., 2004).

10.3.2 GENE AND TISSUE DISTRIBUTION

COMT is widely distributed in living organisms being present in all taxonomic groups, prokaryotes, and eukaryotes, namely bacteria (Vilbois et al., 1994; Dhar et al., 2000; Kim et al., 2004), yeast (Veser 1987; Veser et al., 1979), plants (Guldberg and Marsden, 1975; Legrand et al., 1976), and animals (Guldberg and Marsden, 1975), invertebrates and vertebrates. Accordingly, queries at the National Center for Biotechnology Information databases (Wheeler et al., 2006) reveal that the COMT gene has been found and characterized in a variety of species, such as human, monkey, dog, rat, mouse, horse, chicken, and *Schizosaccharomyces pombe*, and it is conserved in the fungi/metazoa group. Two molecular forms of the enzyme have been described in mammals, a soluble form (S-COMT) and another associated with membranes (MB-COMT) (Borchardt et al., 1974; Jeffery and Roth, 1984; Grossman et al., 1985), both encoded by a single COMT gene (Salminen et al., 1990; Lundström et al., 1991; Winqvist et al., 1992).

10.3.3 ENZYME ACTIVITY

MB-COMT and S-COMT have in common similar affinities for SAM (Jeffery and Roth, 1987; Lotta et al., 1995), the requirement for magnesium, the inhibition by calcium, and the similar optimal pH for activity (Männistö and Kaakkola, 1999). The affinity for substrates can, however, be significantly different. MB-COMT has a distinctively higher affinity for catecholamines (10- to 100-fold higher) than S-COMT, and this characteristic appears to be horizontal to different species. The reason for this difference is not yet clarified; however, it is probably due to inter-actions between the protein and the membrane through lipids and other membrane proteins (Lotta et al., 1992; Bonifácio et al., 2000).

10.3.4 COMT STRUCTURE AND INHIBITORS

The rat recombinant S-COMT was successfully co-crystallized, for the first time by Vidgren and collaborators (1991, 1994), in complex with the co-substrate SAM, one magnesium ion and the competitive inhibitor 3,5-dinitrocatechol. This structure revealed the detailed atomic arrangement of the active site and encouraged a series of structural and theoretical studies of the catalytic and inhibition mechanisms of COMT. Overall, this protein is composed of one single domain with α/β-folded structure, containing eight α-helices arranged around a central mixed β-sheet (Figure 10.3). This topology represents a variation of the Rossman fold and is characteristic of the catalytic domain of other SAM-dependent methyltransferases. Indeed, the SAM binding region of COMT shows a high degree of structural homology with

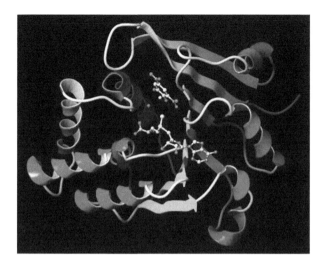

FIGURE 10.3 Schematic representation of the three-dimensional structure of COMT. The *S*-adenosyl-L-methionine (SAM) co-substrate, the inhibitor 3,5-dinitrocatechol (35DNC), the magnesium ion and coordinated water molecules are depicted.

known nucleotide binding proteins, namely DNA- and RNA-methyltransferases (O'Gara et al., 1995; Schluckebier et al., 1995), which suggest a common evolutionary origin.

The active site of COMT consists of the SAM-binding region and the actual catalytic site, which binds one magnesium ion and the catechol substrate (Figure 10.4). While the co-substrate is buried within the structure of the enzyme, the catalytic site is a shallow groove located on the outer surface of COMT. The analysis of the x-ray structure and of kinetics experiments made it possible to establish a sequential ordered mechanism, in which the different compounds associate during the reaction cycle (Lotta et al., 1995): SAM binds first, followed by Mg^{2+}, and finally, the catechol substrate.

The factors affecting the substrate binding affinity and turnover rate in the reaction catalyzed by S-COMT have been studied by Lautala and collaborators (2001). In this QSAR study, the authors utilized a diverse set of 46 catecholic substrates of the enzyme and concluded that the most prevalent factor, that may cause lowering of the rate (or inhibition) of the O-methylation reaction, was the electronic substituent effect on the reacting hydroxyl. Introducing an electron withdrawing substituent in the catechol ring has the effect of lowering the pK_a of the catechol hydroxyls and stabilizing the catecholate anion, through charge delocalization, within the enzyme–substrate complex. In this case, the negative charge of the catecholate and the positive charge of the methylsulfonium are annihilated through the transition state. Stabilization of the Michaelis complex rather than the transition state leads to an increase of the activation energy and hence, a decrease of the rate of the reaction. The electronic effects of substituents on the catechol hydroxyls have

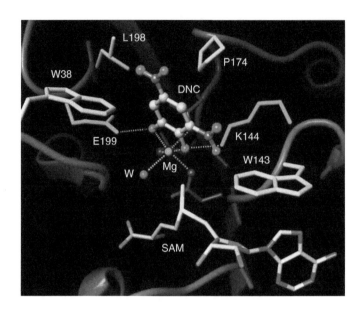

FIGURE 10.4 Close-up view of the catalytic site of COMT.

also been successfully utilized to explain the high affinity and lack of reactivity of the potent second-generation nitrocatechol COMT inhibitors (Taskinen et al., 1989; Lotta et al., 1992; Shinagawa, 1992) (see Section 10.3.4.1). The effectiveness of COMT substrates was also found to depend on the steric fit, flexibility, and hydrophobicity of the side chain substituents (Lautala et al., 2001; Sipila and Taskinen, 2004). To a certain degree, an increase of the steric bulkiness and hydrophobicity of the catechol substituents extending out of the catalytic pocket can decrease K_m and increase the turnover rate, as long as they do not conflict with the gatekeeper residues.

10.3.4.1   First-Generation COMT Inhibitors

Subsequent to the first purification and characterization of COMT, in the late 1950s (Axelrod and Tomchick, 1958), several classes of COMT inhibitors were identified. A comprehensive review of the pharmacological properties of these first-generation COMT inhibitors was published by Guldberg and Marsden (1975). Moreover, several other noncatecholic compounds like ascorbic acid, tropolones and derivatives of 8-hydroxyquinolines, and 3-hydroxylated pyrones and pyridones were identified as COMT inhibitors (Figure 10.5).

10.3.4.2   Second-Generation COMT Inhibitors

The interest in COMT, especially as a therapeutic target, was strongly revived with the discovery of second-generation COMT inhibitors, in the late 1980s. Two independent groups developed simultaneously a new class of disubstituted catechols as a

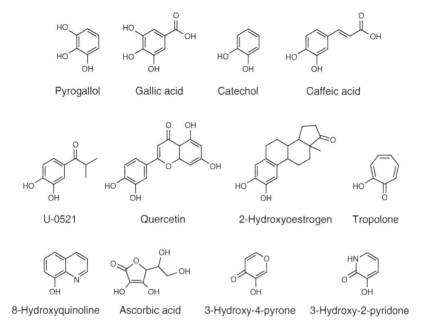

FIGURE 10.5 Representative base structures of some first-generation COMT inhibitors.

new generation of potent and selective COMT inhibitors (Bäckström et al., 1989; Borgulya et al., 1989; Männistö and Kaakkola, 1989).

Of the initial series of nitrocatechol COMT inhibitors, nitecapone, entacapone, and tolcapone (Figure 10.6) were characterized in detail. Some discrepancies may be

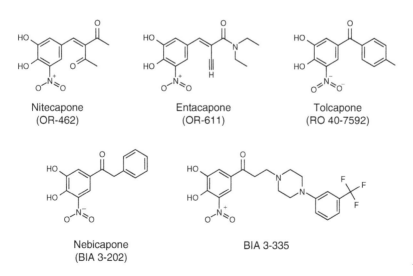

FIGURE 10.6 Nitrocatechol-type second-generation COMT inhibitors.

found among the reported values of inhibition constants K_i for the three compounds (Schultz and Nissinen, 1989; Zurcher et al., 1990b; Nissinen et al., 1992; Lotta et al., 1995). Because of the tight-binding characteristics of nitrocatechol-type inhibitors, typical kinetic data analyses are complicated and, in certain circumstances, the determination of inhibition parameters is highly dependent on the enzyme preparations and assay conditions. A direct comparison of the three inhibitors indicated values of K_i of 1.02, 0.30, and 0.27 nM for nitecapone, entacapone, and tolcapone, respectively (Lotta et al., 1995).

It has been demonstrated that these COMT inhibitors alter the metabolism of levodopa and effectively potentiate the action of orally administered levodopa plus an inhibitor of the AADC, such as carbidopa or benserazide. Indeed, the levels of 3-OMD are markedly reduced and the bioavailability of levodopa in the plasma is increased, when levodopa plus AADC inhibitor are orally administered to rats, together with nitecapone (Nissinen et al., 1988), entacapone (Nissinen et al., 1992), or tolcapone (Borgulya et al., 1989; Zurcher et al., 1990a, 1990b).

More recently, several structure–activity relationship studies undertaken by another research group led to the development of two new nitrocatecholic COMT inhibitors, with superior pharmacological characteristics. Nebicapone (Learmonth et al., 2002) and BIA 3-335 (Learmonth et al., 2004) (Figure 10.6) incorporate the nitrocatechol pharmacophore and show the typical tight-binding characteristics of other nitrocatechols, being potent and reversible competitive inhibitors of COMT (Bonifacio et al., 2003).

10.3.4.3 Late Atypical COMT Inhibitors

Almost simultaneous to the discovery of the first potent nitrocatecholic COMT inhibitors, CGP 28014, a pyridine compound (Figure 10.7), was found to reduce the levels of homovanillic acid (HVA) and to increase those of dihydroxyphenylacetic acid (DOPAC) in the rat striatum, after oral or intraperitoneal administration (Waldmeier et al., 1990a, 1990b). Interestingly, this compound, which, like tropolone, lacks the typical catechol structure, does not significantly inhibit COMT in vitro and, when administered orally to rats, shows a weak inhibition of the formation of 3-OMD in the periphery (Männistö et al., 1992a). However, it mimics the effects of centrally active COMT inhibitors, by decreasing the levels of striatal HVA and 3-methoxytyramine (3-MT) (after clorgyline treatment), when given alone, and reducing the formation of 3-OMD, after administration of levodopa (Männistö et al., 1992a). Therefore, CGP 29014 behaves like a COMT inhibitor-like compound but, as opposed to other

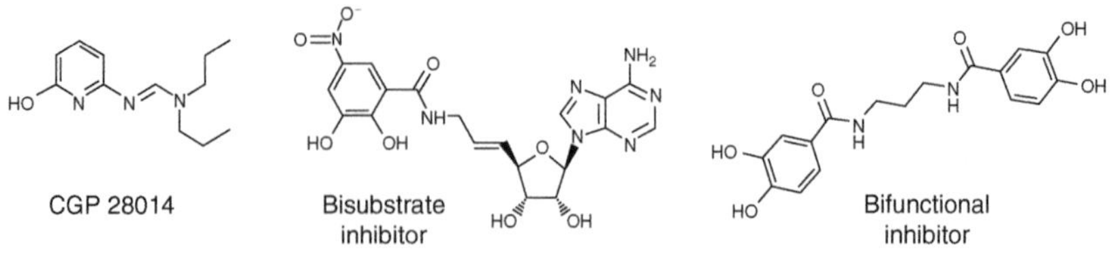

CGP 28014   Bisubstrate inhibitor   Bifunctional inhibitor

FIGURE 10.7 Atypical late COMT inhibitors.

peripherally selective or broadly selective COMT inhibitors, it appears to act preferentially in the brain (Männistö et al., 1992b). The exact mechanism of action of CGP 29014 remains, however, to be clarified.

10.3.5 COMT Inhibition in Parkinson's Disease

Tolcapone and entacapone are currently approved in many countries as adjunctive to levodopa therapy in Parkinson's disease, and nebicapone is currently in clinical development. Nitecapone and CGP-28014 were not developed for PD or any other clinical indication, and therefore are not addressed in the following review of existing data of COMT inhibitors in humans.

10.3.6 Pharmacokinetics of COMT Inhibitors

Following oral administration, tolcapone, entacapone, and nebicapone are rapidly absorbed, but oral bioavailability is not complete due to significant first-pass metabolism, most likely in the liver. The presence of food does not significantly affect the pharmacokinetics of these COMT inhibitors. Oral bioavailability ranges between approximately 35% (Heikkinen et al., 2001) for entacapone and 60% for tolcapone (Jorga, 1998). C_{max} is reached within 0.5–2 h and elimination is rapid, with apparent $T_{1/2}$ between 1.6 and 3.4 h.

All COMT inhibitors are abundantly metabolized, mainly in the liver. The major metabolic pathways undergone by these compounds are conjugation reactions, mostly with glucuronic acid. For tolcapone and nebicapone, the major early metabolite present in plasma is the glucuronide conjugate. Entacapone is not methylated by COMT in humans, whereas tolcapone and nebicapone are converted to 3-O-methyl-tolcapone and 3-O-methyl-nebicapone, respectively (Jorga et al., 1999; Loureiro et al., 2006).

10.3.7 Pharmacodynamics of COMT Inhibitors

All COMT inhibitors have been shown to dose-dependently and reversibly inhibit human erythrocyte S-COMT activity. Time to maximum inhibition is similarly rapid (less than 2 h) between the different COMT inhibitors, but the level of inhibition and the time for recovery of enzyme activity may differ. Tolcapone and nebicapone were shown to have a relatively similar COMT inhibition profile, and both cause a more profound and sustained inhibition than entacapone. Following oral doses of 100 and 200 mg, maximum COMT inhibition is, respectively, 72% and 80% for tolcapone (Dingemanse et al., 1995) and, respectively, 69% and 80% for nebicapone (Silveira et al., 2003). With a 200 mg dose, time to recovery of COMT also does not differ between tolcapone and nebicapone and with both drugs COMT activity returns to baseline approximately 18 h following administration (Dingemanse et al., 1995; Silveira et al., 2003). With an entacapone 200 mg dose, maximum COMT inhibition is 65% and enzyme activity recovery is fully attained within 8 h (Keranen et al., 1994).

10.3.8 Clinical Efficacy of COMT Inhibitors

Similarly to what occurs in healthy volunteers, tolcapone, entacapone, and nebicapone increase levodopa bioavailability in PD patients by significantly increasing levodopa AUC without significantly changing C_{max} and T_{max}. The net result is a more sustained and less fluctuating levodopa level, which has beneficial consequences (Chase, 1998; Olanow and Obeso, 2000; Olanow et al., 2000). In patients with advanced and fluctuating disease, which is characterized by the wearing-OFF phenomenon (end-of-dose deterioration) treated with levodopa/AADC inhibitor, the use of COMT inhibitors was shown to increase the ON time (i.e., to prolong the motor response to levodopa) and to reduce the OFF time, with relevant improvement on the activities of daily living and general quality of life. Patients with nonfluctuating PD may also benefit from the administration of COMT inhibitors as revealed by improvements in daily living activities and motor function.

Head-to-head studies comparing different COMT inhibitors are missing and data from noncomparative studies are difficult to compare, due to the different study designs and dosage regimens used. A comparison of two separate, simultaneous, long-term open label studies, one for tolcapone and the other for entacapone, suggested that tolcapone has a greater and longer efficacy with regard to motor symptoms, "OFF" time, and change in levodopa requirements than entacapone (Factor et al., 2001). The only published head-to-head comparison performed according to a randomized, double-blind, and controlled design is a phase IIa study in PD patients with the wearing-OFF phenomenon (Ferreira et al., 2006). The study adopted a four-way crossover design and nebicapone 75 mg, nebicapone 150 mg, entacapone 200 mg, and placebo were administered concomitantly with levodopa/carbidopa during four sequential treatment periods of 6–9 days each. The sequence of treatments occurred according to randomization and the number of daily doses ranged from 4 to 6. Globally, nebicapone 75 mg was shown to improve motor function similar to entacapone 200 mg, and nebicapone 150 mg was shown to be superior, and the efficacy results are consistent with the results reported for COMT inhibition and levodopa pharmacokinetics (Ferreira et al., 2006).

10.3.9 Safety and Toxicity of COMT Inhibitors

The adverse effects associated with these nitrocatechol-type COMT inhibitors can be grouped into two categories—those related to the dopaminergic potentiation and those that are nondopaminergic related. Within the dopaminergic related effects, dyskinesia is by far the major adverse effect, followed by nausea and dizziness. Other common adverse events include anorexia, vomiting, orthostatic hypotension, sleep disorders, somnolence, and hallucinations. Within the second group, the most common adverse event, whose mechanism is at present unknown, is diarrhea, and it is usually more severe with tolcapone than with the other COMT inhibitors. Abdominal pain, headache, and urine discoloration are also observed with tolcapone and entacapone, the latter due to the presence of the compounds and metabolites.

In phase III clinical trials alterations in liver enzyme levels were reported for 1%–3% patients treated with tolcapone, leading to the necessity to closely follow liver function. After marketing authorization, cases of serious hepatotoxicity were reported, some of them in the form of fatal fulminant hepatitis (Olanow, 2000). Cases of neuroleptic malignant syndrome have also been reported during tolcapone treatment (Keating and Lyseng-Williamson, 2005). Regarding entacapone, no cases of hepatitis or other serious liver failure cases have been reported in phase III trials or after market authorization. Increases in liver enzyme levels are rare, and cannot be definitely attributed to entacapone (Brooks et al., 2005). Although no signs of concern arise from the existing data, clinical experience with nebicapone is still limited, which precludes reliably judging its liver safety profile.

The mechanisms by which tolcapone triggers severe hepatotoxicity are still under discussion. Tolcapone was demonstrated to be an uncoupler of oxidative phosphorylation in vitro (Nissinen et al., 1997) and possibly also in vivo (Haasio et al., 2002a, 2002b). In addition, acute toxicity appears to be due to the compound itself and not due to any reactive or toxic metabolites, thus reforcing its potential in disrupting mitochondrial function (Borroni et al., 2001). There is, however, a report showing formation of glutathione adducts from the amine and acetoamine metabolites in liver microsomes (Smith et al., 2003). Interestingly, tolcapone was shown to impair energy production in neuroblastoma cells with or without a functional respiratory chain, thus suggesting that its toxicity may also involve a mechanism independent of oxidative phosphorylation (Korlipara et al., 2004). Furthermore, the toxicity of tolcapone appears to be observed mainly at concentrations of compound much higher than those observed in the plasma of patients, suggesting that variations in the metabolism of the compound, due to genetic factors, co-medications, and co-morbidities could lead to an increased risk of toxicity (Borroni et al., 2001; Acuna et al., 2002; Borges, 2005).

10.4 MONOAMINE OXIDASE

10.4.1 PHYSIOLOGICAL ROLE

MAO (EC 1.4.3.4) is a flavoprotein that catalyzes the oxidative deamination of biogenic amines to the respective aldehydes and free amines with the production of hydrogen peroxide. MAO substrates include 5-hydroxytryptamine, dopamine, noradrenaline, adrenaline, dietary amines, and xenobiotic amines (Shih et al., 1999). The amines are initially oxidized to an imine product with reduction of the enzyme FAD cofactor (Figure 10.8). The cofactor is then reoxidized by molecular oxygen to generate hydrogen peroxide. The imine is subsequently hydrolyzed to aldehyde and ammonia in a nonenzymatic reaction.

It is an important enzyme for normal brain development, playing a significant role in the metabolism of exogenous amines and in the regulation of neurotransmitter levels and contents of intracellular amine stores (Youdim et al., 2006). Furthermore, the reaction products of MAO-catalyzed oxidations, such as hydrogen peroxide, could be involved in the progress of neurodegenerative disorders such as PD through

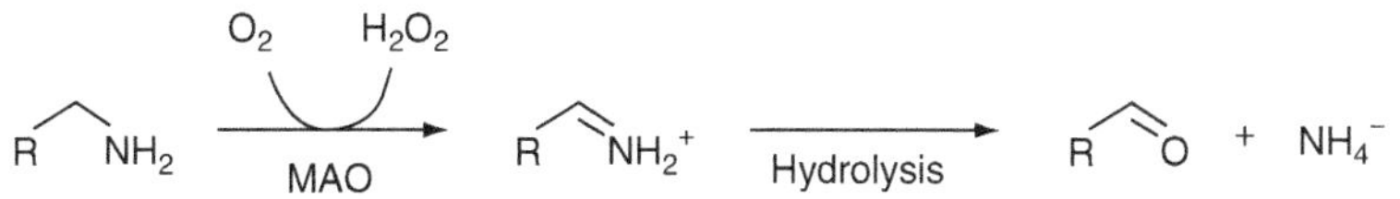

FIGURE 10.8 Scheme of the overall oxidative deamination of a primary amine by MAO.

mitochondrial damage (Hauptmann et al., 1996). It is, therefore, likely that MAO might be involved in the modulation of cellular function/pathology (Billett, 2004).

10.4.2 GENE AND TISSUE DISTRIBUTION

Two forms of the enzyme have been described, MAO A and MAO B, originally distinguished by substrate specificity and inhibitor sensitivity, with distinct functions in neurotransmitter metabolism and behavior (Shih et al., 1999). In fact, MAO A and B are encoded by two distinct genes, each with 15 exons and identical exon–intron organization, localized to chromosome X (Xp11.2–11.4), in adjacent positions, but organized on a tail to tail orientation (Shih et al., 1999). The regulation of MAO A and MAO B gene expression is different and complex, with the core promoters showing a distinct cis-acting element organization (Chen, 2004).

MAO A and B are co-localized in most human tissues, but with differential cellular and tissue-specific expression. High levels of both MAO A and B are found in duodenal villi and crypts, liver, and kidney. However, MAO A is the only form present in fibroblasts and placenta, whereas MAO B is specific to platelets and lymphocytes (Billett, 2004). In the central nervous system, MAO A is predominantly found in catecholaminergic neurons, while MAO B is mainly found in serotonergic and histaminergic neurons and glial cells. Enzyme levels are not, however, directly correlated with enzymatic activity; in fact, MAO A activity predominates in the periphery while MAO B activity predominates in the brain.

MAO A and B activities also vary through ontogenesis; fetal liver has the highest activities of both enzymes, with MAO B activity similar to that of an adult liver, while MAO A activity is lower in adults. On the other hand, in fetal brain MAO A activity appears before that of MAO B, though after birth the activity of the latter increases with age with little or no variation in MAO A activity (Nicotra et al., 2004). The increase in MAO B activity in human brain with aging is probably due to the proliferation of glial cells, although other factors should also be involved (Fowler et al., 1997). It is speculated that in Parkinsonian patients the increased oxidation of dopamine could be associated with the loss of dopaminergic neurons (Shih et al., 1999).

10.4.3 MAO STRUCTURE AND INHIBITORS

At the cellular level, MAOs are membrane-bound proteins, tightly associated to the outer mitochondrial membrane. Human MAO A and MAO B share approximately 70% amino acid sequence identity (Bach et al., 1988), but differ in their substrate specificities and sensitivity to inhibitors. MAO B acts preferentially on small hydrophobic amines, such as 2-phenylethylamine and benzylamine, and is selectively inhibited by selegiline, whereas MAO A carries out the degradation of bulkier endogenous amine neurotransmitters, such as serotonin, and is selectively inhibited by clorgyline (Youdim et al., 2006) (Figure 10.9).

During the last few years, the x-ray structures of human and rat MAO enzymes were obtained in complex with a number of reversible and irreversible inhibitors, some of them also selective for either one or the other isozyme forms. These structures have contributed significantly to a better understanding of the enzyme–inhibitor

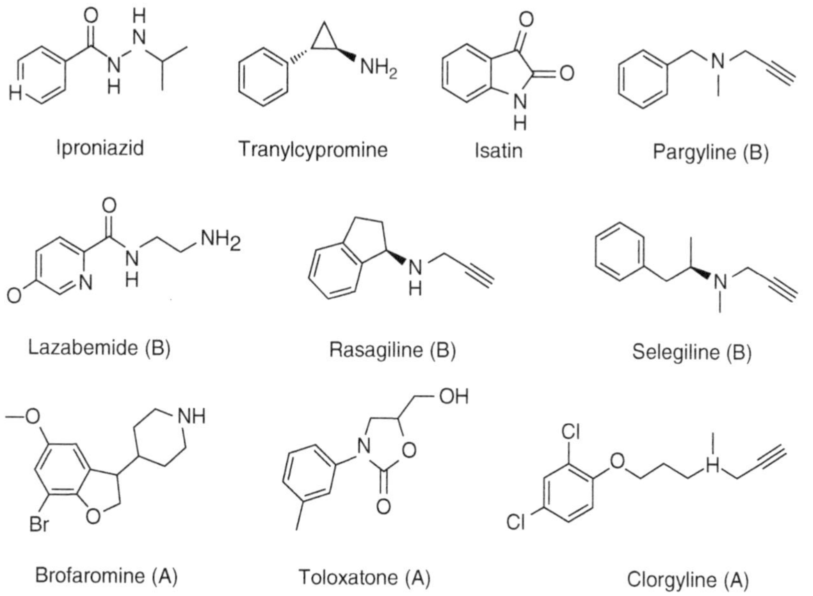

FIGURE 10.9 Chemical structures of nonselective and selective MAO A (A) and MAO B (B) inhibitors.

interactions and are paving the way to a structure-based design of new and improved inhibitors.

MAO (527 amino acids in h-MAO A versus 520 in h-MAO B) is a globular protein exhibiting an overall topology also found in several flavoproteins (Fraaije and Mattevi, 2000). An α-helix located near the C-terminal of the enzyme is probably responsible for anchoring to the mitochondrial outer membrane, although other regions are proposed to interact with the membrane, even if superficially. Interestingly, despite the high degree of homology between the two isoforms, human MAO B crystallizes in a dimeric form, in contrast with human MAO A, which is found as a monomer in the crystal structure. The different oligomerization states are not likely to result from crystal packing artifacts, as they are in agreement with the hydrodynamic behaviors under analytical ultracentrifugation experiments (De Colibus et al., 2005). One single critical amino acid mutation Glu151 (MAO B) to Lys151 (MAO A) at the monomer–monomer interface would allegedly destabilize the dimeric state, resulting in the observed monomeric form (Andres et al., 2004; De Colibus et al., 2005). The implications of the observed oligomerization states on the functional characteristics of the different forms of the enzyme are not, however, well understood.

The structures of the two isozymes differ in the shape and configuration of their respective substrate-binding sites. Human MAO A has a substrate cavity of about 550 Å^3, which may accommodate the relative bulkier substrates. On the other hand, the human MAO B substrate cavity is shaped like an hourglass, with a constriction formed by the side chains of Tyr326, Ile199, Leu171, and Phe168, which separates the actual catalytic pocket (a small cavity of about 490 Å^3) from the entrance chamber,

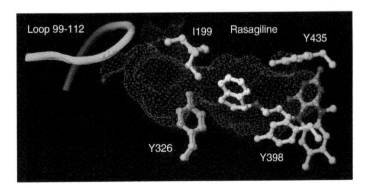

FIGURE 10.10 (See color insert following page 296.) Structure of the active site of human MAO B in complex with rasagiline (pdb code: 1S2Q). The surface of the cavity is shown as a purple dot mesh, containing rasagiline (white) covalently linked to the isoalloxazine ring system of FAD (green). Key amino acid residues are shown in cyan, including the "gating switch" loop 99–112 (in tube representation). The ligand-binding cavity (right) is separated from the smaller entrance cavity (left) by a ring of amino acid side chains, including Ile199 and Tyr326. The conformation of these side chains may vary in different crystal structures depending on the bound inhibitor. This is exemplified with the side chain conformations taken from the crystallographic complex of MAO B with the small inhibitor isatin, (pdb code: 1OJA), as superimposed on the current structure. (From Binda, C., Li, M., Hubalek, F., Restelli, N., Edmondson, D.E., and Mattevi, A., *Proc. Natl. Acad. Sci. U.S.A.*, 100, 9750, 2003. With permission.)

near the protein surface (Figure 10.10). The entry to this cavity is modulated by a flexible loop 99–112, which functions as a "gating switch." The catalytic site, at the bottom of the substrate-binding cavity, is lined by a cage of hydrophobic residues (e.g., Tyr398, Tyr435), just above the FAD isoalloxazine ring system. By analogy with the structures of MAO-inhibitor complexes, one may speculate that, in the case of substrates, the methylene carbon adjacent to the reactive amine group is positioned on the *Re* face of the flavin ring, near the N5 atom, while the amine nitrogen can be accommodated in a vicinal small hydrophilic crevice (Binda et al., 2003).

The surface of the cavity is shown as a purple dot mesh, containing rasagiline (white) covalently linked to the isoalloxazine ring system of FAD (green, in Figure 10.10). Key amino acid residues are shown in cyan, including the "gating switch" loop 99–112. The ligand-binding cavity (right) is separated from the smaller entrance cavity (left) by a ring of amino acid side chains, including Ile199 and Tyr326. The conformation of these side chains may vary in different crystal structures depending on the bound inhibitor. This is exemplified by the side chain conformations taken from the crystallographic complex of MAO B with the small inhibitor isatin, (pdb code: 1OJA) (Binda et al., 2003), as superimposed on the current structure.

The shape of the human MAO B substrate cavity displays an interesting adaptive plasticity, as exemplified by the crystallographic structures of the enzyme complexed with the noncovalent reversible inhibitors isatin and 1,4-diphenyl-2-butene (Binda et al., 2003).

The small isatin molecule is totally enclosed within the small substrate-binding cavity, where the side chains of Ile199 and Tyr326 (orange, in Figure 10.10) adopt a rotamer conformation that result in the separation of the entrance cavity from the substrate-binding site. On the other hand, a large ligand like 1,4-diphenyl-2-butene may induce a rotation of the side chain of Ile199 (cyan, in Figure 10.10), so that the two cavities become fused and form one large single pocket of about 700 Å^3 (Binda et al., 2003; De Colibus et al., 2005), which is spanned by the long inhibitor molecule. This observed plasticity of the MAO B ligand-binding cavity suggests a greater flexibility to accommodate larger ligands and may be exploited in the design of new potent and more selective MAO B inhibitors (Binda et al., 2004a).

Several irreversible MAO inhibitors belonging to the class of propargylamines have shown greatest relevance in clinical applications related to neurodegenerative disorders. The crystal structures of human MAO B in complex with the selective and irreversible MAO B inhibitors pargyline (Binda et al., 2002), rasagiline (Binda et al., 2004b), and selegiline (L-deprenyl) (De Colibus et al., 2005) were determined. Furthermore, in recent years, the selective irreversible MAO A inhibitor clorgyline was also co-crystallized with the human (De Colibus et al., 2005) and rat (Ma et al., 2004) MAO A isozymes. All of these inhibitors share a common mechanism of action, which involves the formation of a covalent bond between the N5 Nitrogen of the FAD isoalloxazine ring and the electrophilic terminal carbon of the propargylamine, resulting in a stable flavocyanine adduct (Figure 10.10).

The analysis of the x-ray structures of human MAO B with rasagiline (Figure 10.10) and related compounds (Binda et al., 2004b) unveils the stereochemical environment of the bound inhibitors and provides some guidelines for improving the potency and specificity of newly designed inhibitors, by introducing appropriate substituents at specific sites on the rasagiline scaffold. Also, the presence of a network of conserved crystallographic water molecules suggests that they should be taken into consideration in future inhibitor design and modeling studies. On the other hand, the availability of the structures of the human MAO A and MAO B in complex with clorgyline and selegiline, respectively, provides an opportunity to compare the detailed three-dimensional structures of the two isozymes and to gain insights on the basis of substrate/inhibitor selectivity (De Colibus et al., 2005). The two enzymes are assumed to share the same catalytic mechanism, which is supported by the highly conserved "aromatic cage" formed by the isoalloxazine rings and the two tyrosine residues at the catalytic site (De Colibus et al., 2005). In contrast, most differences between the two substrate-binding sites are located in the area of the pocket, opposite to the flavin, which governs substrate recognition. The shapes and sizes of the active site cavities differ as a result of a combination of both amino acid replacements and conformational alterations. The co-crystallized selegiline and clorgyline inhibitors could not be interchanged between the two isozymes without major, and possibly energetically penalizing, structural rearrangements.

A comparative study of a series of 3-, 4-, and 7-substituted coumarins, as rat brain MAO inhibitors, was recently conducted, using docking and 3D-QSAR methodologies (Catto et al., 2006). In this study, the structure of the rat MAO A and a model of that of rat MAO B were used to investigate the structural and chemical determinants of the inhibitory potency against each MAO isoforms, as well as those responsible for MAO B versus MAO A inhibition selectivity. Taking advantage of the knowledge of the enzyme structures, the study enables the identification of the physicochemical nature and spatial location of the main binding interactions modulating the enzymatic affinity and selectivity.

10.4.4 CLINICAL STUDIES WITH MAO INHIBITORS

In the middle twentieth century, iproniazid (see Figure 10.9) was found to inhibit MAO and to possess antidepressant effects. Several hydrazine derivatives were then developed for the therapy of depression but, although efficacious, these molecules had severe hepatotoxic effects. Other nonhydrazine irreversible MAO inhibitors were developed, such as tranylcypromine and clorgyline, however, although devoid of liver toxicity, they can lead to severe, sometimes fatal, hypertensive responses (Youdim et al., 2006). This side effect (the Cheese effect) is triggered by the release of noradrenaline as a consequence of the increase in circulating tyramine, or other sympathomimetic amines, derived from the diet that is not metabolized by the irreversibly inhibited MAO A (Youdim and Bakhle, 2006). More recently, the treatment of depression and anxiety disorders gained a new impetus with the development of the reversible MAO A inhibitors (RIMAs) such as moclobemide, toloxatone, and brofaromine. These inhibitors are devoid of the hypertensive side effects since, on the periphery, they can be displaced from the enzyme by the dietary amines (Riederer et al., 2004).

10.4.5 MAO B INHIBITORS IN PARKINSON'S DISEASE

Selective MAO B inhibitors were developed for the therapy of PD based on the rationale that selective MAO B inhibition should increase the bioavailability of central dopamine (Riederer et al., 2004). Several MAO B inhibitors have been developed, including lazabemide (Da Prada et al., 1990) and the propargylamine derivatives selegiline (L-deprenyl) (Knoll and Magyar, 1972) and rasagiline (Finberg et al., 1981). The latter have been approved by the competent authorities for the therapy of PD, both as monotherapy for the control of motor symptoms or in combination with the standard levodopa therapy for the delay of disease progression and treatment of motor complications (Youdim and Bakhle, 2006). Both selegiline and rasagiline are irreversible inhibitors of MAO B with similar potency in vitro and similar selectivity concerning MAO enzymes; on the other hand rasagiline appears to be more potent than selegiline in vivo (Youdim et al., 2001). Rasagiline was shown to possess neuroprotective and anti-apoptotic properties in several in vitro and animal models, that are not entirely due to the inhibition of the enzyme (Youdim et al., 2005), but evidence of neuroprotection in PD patients is still lacking (Rascol et al., 2005).

10.4.6 PHARMACOKINETICS OF MAO B INHIBITORS

Upon oral administration, both selegiline and rasagiline are rapidly absorbed and widely distributed into body tissues and cross the BBB. Following oral administration, selegiline reaches maximum plasma concentrations within 1 h in fasting subjects. Because of extensive metabolization to L-desmethylselegiline, L-metamphetamine and L-amphetamine first pass through the gut wall and liver, but oral bioavailability is only 10%. An elimination half-life of about 10 h for selegiline has been reported at steady-state with a dosage of 10 mg daily. Elimination half-lives of the L-desmethylselegiline, L-metamphetamine, and L-amphetamine metabolites are respectively 2.0, 20.5, and 17.7 h. To overcome the extensive first-pass metabolism, to decrease the formation of amphetamine metabolites (which may account for part of the neuropsychiatric adverse effects, especially in elderly patients), and to reduce the potential for tyramine interactions, alternative modes of selegiline delivery have been developed, such as transdermal and buccal administration (Clarke et al., 2003).

Maximum plasma concentrations of rasagiline are reached at approximately 0.5 h after an oral dose of 2 mg, and elimination half-life is about 0.6–2 h. The absolute bioavailability is 36%, indicating that rasagiline undergoes a significant first-pass metabolism, but inversely to what occurs with selegiline no amphetamine metabolites are formed. Its major metabolite 1-(R)-aminoindan is not pharmacologically active (Chen and Swope, 2005). The kidney is the major elimination pathway for both selegiline and rasagiline metabolites (Mahmood, 1997; Chen and Swope, 2005).

10.4.7 Pharmacodynamics of MAO B Inhibitors

Both selegiline and rasagiline irreversibly and almost completely inhibit MAO B. Because of the irreversibility of the inhibition, the recovery of enzyme activity relies upon de novo enzyme synthesis, which is tissue dependent. Platelet MAO B activity returns to baseline levels after 2 weeks from the last administration of rasagiline (Thebault et al., 2004); however, it is reported that the half-life for the recovery of brain MAO B after selegiline inhibition is 40 days (Fowler et al., 1994).

Selegiline is a relatively selective MAO B inhibitor, but selectivity is not absolute and inhibition of MAO A occurs at relatively high doses. At relatively low doses (i.e., 10 mg), selegiline inhibits cerebral MAO B while having little effect on MAO A in the gastrointestinal tract and liver. At high doses, the selectivity of selegiline for MAO B decreases, causing inhibition of MAO B and MAO A. In a study in rats, the degree of selectivity of rasagiline for inhibition of MAO B as opposed to MAO A was similar to that of selegiline (Youdim et al., 2001).

10.4.8 Clinical Efficacy of MAO B Inhibitors

Several large clinical trials were performed with selegiline (Parkinson-Study-Group, 1989, 1996a, 1996b; Larsen and Boas, 1997; Palhagen et al., 1998; Przuntek et al., 1999; Shoulson et al., 2002) or rasagiline (Parkinson-Study-Group, 2002, 2004, 2005; Rascol et al., 2005) either administered as monotherapy in early untreated PD patients to evaluate the efficacy in motor symptom control, or as adjuncts to the standard levodopa/AADC inhibitor therapy to assess efficacy in treating motor complications, and in preventing clinical progression.

In general, both selegiline and rasagiline, when used as monotherapy, improved disability and Unified Parkinson's Disease Rating Scale (UPDRS) motor scores as compared to placebo groups and are efficacious in delaying the introduction of levodopa treatment in early PD patients. Results from the TEMPO study indicate that an early initiation on rasagiline has more beneficial effects, translated by better UPDRS scores, than a delayed initiation on the compound (Parkinson-Study-Group, 2004). These results imply a possible neuroprotective effect by rasagiline (Chen and Ly, 2006). Selegiline was shown to delay the need for symptomatic therapy in untreated PD patients in the DATATOP study, but interpretation was confounded by its symptomatic effects. A clinical trial utilizing a delayed start design demonstrated that patients initiated on rasagiline at baseline are improved at 1 year in comparison to patients initiated on placebo and switched to rasagiline at 6 months even though both groups were on the same treatment for the last 6 months of the study. These results argue against the benefit being due to a symptomatic effect and are consistent with rasagiline having a protective effect (Olanow, 2006). However, there is insufficient data to draw definite conclusions on the possibility that rasagiline or selegiline may delay the progression rate of PD (Youdim et al., 2006). A neuroprotective therapy that slows or stops disease progression is actually a major unmet medical need in Parkinson's disease. Current evidence indicates that cell death in Parkinson's disease occurs, at least in part, by way of a signal-mediated apoptotic process. This raises the possibility that anti-apoptotic agents might be neuroprotective in Parkinson's disease. Propargylamines have been demonstrated to be potent anti-apoptotic agents in both in vitro and in vivo studies, presumably by maintaining glyceraldehyde-3-phosphate dehydrogenase (GAPDH) as a dimer and thereby preventing its nuclear translocation where it blocks upregulation of anti-apoptotic proteins.

Concerning the use of these molecules as adjuncts to levodopa therapy, the results are less clear. Both selegiline and rasagiline lead to a reduction of the levodopa dose required for the symptomatic relief of PD. Long-term selegiline treatment seems to diminish ON–OFF motor fluctuations and freezing of gait but also to increase the susceptibility for dyskinesias (Shoulson et al., 2002). There is insufficient evidence to prove a significant effect of selegiline in the treatment of motor complications or for the prevention of motor complications (Riederer et al., 2004).

Rasagiline significantly reduces OFF-time in patients with motor fluctuations. Furthermore in the LARGO study it was shown that 1 mg rasagiline per day reduced OFF-time and increased ON-time without concomitant increase of dyskinesias and improved motor symptom control (reduction of gait freezing). These effects were similar to those obtained with entacapone (200 mg with each levodopa dose) in the same study (Rascol et al., 2005). In the PRESTO study, with a longer treatment period, a significant increase in dyskinesias was observed with the 1 mg dose (Parkinson-Study-Group, 2005).

10.4.9 SAFETY AND TOXICITY OF MAO B INHIBITORS

In the therapeutic dosages recommended for Parkinson's disease, selegiline and rasagiline generally are well tolerated. The most frequent side effects observed with selegiline monotherapy were insomnia, dizziness, nausea, euphoria, and headache. When combined with levodopa, many of the adverse effects result from increased dopaminergic activity and include exacerbation of dyskinesias (which usually occurs within 2 weeks after initiating selegiline therapy and usually is mitigated when dosage of levodopa is reduced), confusion, nausea, vomiting, orthostatic hypotension, and hallucinations. Most of the neuropsychiatric adverse events reported with selegiline presumably result from increased dopaminergic activity; however, especially in elderly patients, some of them may be due to the selegiline amphetamine metabolites and their occurrence decreases when transdermal or buccal modes of administration are used (Clarke et al., 2003). Because of the risk of adverse effects associated with nonselective inhibition of MAO, selegiline should not be used in oral doses above 10 mg daily. Occurrence of hypertensive crisis following ingestion of foods containing large amounts of tyramine (e.g., aged cheese or red wine) is rare in patients receiving selegiline at the dosage usually recommended in PD patients, that is, 10 mg daily. At dosages exceeding 10 mg daily, selegiline inhibits both MAO A and MAO B, and the likelihood of hypertensive crises increases.

During the rasagiline monotherapy clinical trials, the most reported adverse effects with the recommended dosage (1 mg daily) were headache, flu syndrome, dyspepsia, arthralgia, vertigo, and depression; the most common adverse effects reported by patients treated with rasagiline 2 mg daily reflected the increased dopaminergic effect: vomiting, abnormal dreams, sleep disorder, somnolence, and ataxia. When rasagiline was used concomitantly with levodopa, the most common adverse effects were dyskinesia, postural hypotension, abdominal pain, abnormal dreams, constipation, and weight loss. Since there were no reports of tyramine/rasagiline interaction in clinical trials conducted without tyramine restriction, it appears that rasagiline can be used safely without dietary tyramine restrictions. Concomitant administration of highly serotoninergic drugs and MAO inhibitors (including selegiline and rasagiline) may result in serotonin syndrome.

REFERENCES

Acuna, G., Foernzler, D., Leong, D., Rabbia, M., Smit, R., Dorflinger, E. et al. (2002) Pharmacogenetic analysis of adverse drug effect reveals genetic variant for susceptibility to liver toxicity. *The Pharmacogenomics Journal*, **2**, 327–334.

Akil, M., Kolachana, B.S., Rothmond, D.A., Hyde, T.M., Weinberger, D.R., and Kleinman, J.E. (2003) Catechol-*O*-methyltransferase genotype and dopamine regulation in the human brain. *Journal of Neuroscience*, **23**, 2008–2013.

Andres, A.M., Soldevila, M., Navarro, A., Kidd, K.K., Oliva, B., and Bertranpetit, J. (2004) Positive selection in MAOA gene is human exclusive: determination of the putative amino acid change selected in the human lineage. *Human Genetics*, **115**, 377–386.

Axelrod, J. and Tomchick, R. (1958) Enzymatic *O*-methylation of epinephrine and other catechols. *Journal of Biological Chemistry*, **233**, 702–705.

Bach, A.W., Lan, N.C., Johnson, D.L., Abell, C.W., Bembenek, M.E., Kwan, S.W. et al. (1988) cDNA cloning of human liver monoamine oxidase A and B: molecular basis of differences in enzymatic properties. *Proceedings of the National Academy of Sciences of the United States of America*, **85**, 4934–4938.

Bäckström, R., Honkanen, E., Pippuri, A., Kairisalo, P., Pystynen, J., Heinola, K. et al. (1989) Synthesis of some novel potent and selective catechol-*O*-methyltransferase inhibitors. *Journal of Medicinal Chemistry*, **32**, 841–846.

Berry, M.D. (2004) Mammalian central nervous system trace amines. Pharmacologic amphetamines, physiologic neuromodulators. *Journal of Neurochemistry*, **90**, 257–271.

Berry, M.D., Juorio, A.V., Li, X.M., and Boulton, A.A. (1996) Aromatic L-amino acid decarboxylase: a neglected and misunderstood enzyme. *Neurochemical Research*, **21**, 1075–1087.

Bey, P., Jung, M.J., Koch-Weser, J., Palfreyman, M.G., Sjoerdsma, A., Wagner, J. et al. (1980) Further studies on the inhibition of monoamine synthesis by monofluoromethyl-dopa. *British Journal of Pharmacology*, **70**, 571–576.

Bilder, R.M., Volavka, J., Lachman, H.M., and Grace, A.A. (2004) The catechol-*O*-methyl-transferase polymorphism: relations to the tonic-phasic dopamine hypothesis and neu-ropsychiatric phenotypes. *Neuropsychopharmacology*, **29**, 1943–1961.

Billett, E.E. (2004) Monoamine oxidase (MAO) in human peripheral tissues. *Neurotoxicology*, **25**, 139–148.

Binda, C., Newton-Vinson, P., Hubalek, F., Edmondson, D.E., and Mattevi, A. (2002) Structure of human monoamine oxidase B, a drug target for the treatment of neuro-logical disorders. *Nature Structural Biology*, **9**, 22–26.

Binda, C., Li, M., Hubalek, F., Restelli, N., Edmondson, D.E., and Mattevi, A. (2003) Insights into the mode of inhibition of human mitochondrial monoamine oxidase B from high-resolution crystal structures. *Proceedings of the National Academy of Sciences of the United States of America*, **100**, 9750–9755.

Binda, C., Hubalek, F., Li, M., Edmondson, D.E., and Mattevi, A. (2004a) Crystal structure of human monoamine oxidase B, a drug target enzyme monotopically inserted into the mitochondrial outer membrane. *FEBS Letters*, **564**, 225–228.

Binda, C., Hubalek, F., Li, M., Herzig, Y., Sterling, J., Edmondson, D.E. et al. (2004b) Crystal structures of monoamine oxidase B in complex with four inhibitors of the *N*-propargylaminoindan class. *Journal of Medicinal Chemistry*, **47**, 1767–1774.

Bonifácio, M.J., Vieira-Coelho, M.A., Borges, N., and Soares-da-Silva, P. (2000) Kinetics of rat brain and liver solubilized membrane-bound-catechol-*O*-methyltransferase. *Archives of Biochemistry and Biophysics*, **384**, 261–367.

Bonifacio, M.J., Vieira-Coelho, M.A., and Soares-da-Silva, P. (2003) Kinetic inhibitory profile of BIA 3-202, a novel fast tight-binding, reversible and competitive catechol-*O*-methyltransferase inhibitor. *European Journal of Pharmacology*, **460**, 163–170.

Bonifati, V. and Meco, G. (1999) New, selective catechol-*O*-methyltransferase inhibitors as therapeutic agents in Parkinson's disease. *Pharmacology & Therapeutics*, **81**, 1–36.

Borchardt, R.T., Cheng, C.F., and Cooke, P.H. (1974) The purification and kinetic properties of liver microsomal-catechol-*O*-methyltransferase. *Life Sciences*, **14**, 1089–1100.

Borges, N. (2005) Tolcapone in Parkinson's disease: liver toxicity and clinical efficacy. *Expert Opinion in Drug Safety*, **4**, 69–73.

Borgulya, J., Bruderer, H., Bernauer, K., Zurcher, G., and Da Prada, M. (1989) Catechol-*O*-methyltransferase—inhibiting pyrocatechol derivatives: synthesis and structure-activity studies. *Helvetica Chimica Acta*, **72**, 952–968.

Borroni, E., Cesura, A.M., Gatti, S., and Gasser, R. (2001) A preclinical re-evaluation of the safety profile of tolcapone. *Functional Neurology*, **16**, 125–134.

Brooks, D.J., Agid, Y., Eggert, K., Widner, H., Ostergaard, K., and Holopainen, A. (2005) Treatment of end-of-dose wearing-off in Parkinson's disease: stalevo (levodopa/carbidopa/entacapone) and levodopa/DDCI given in combination with Comtess/Comtan (entacapone) provide equivalent improvements in symptom control superior to that of traditional levodo-pa/DDCI treatment. *European Neurology*, **53**, 197–202.

Burkhard, P., Dominici, P., Borri-Voltattorni, C., Jansonius, J.N., and Malashkevich, V.N. (2001) Structural insight into Parkinson's disease treatment from drug-inhibited DOPA decarboxylase. *Nature Structural Biology*, **8**, 963–967.

Carlsson, A., Lindqvist, M., and Magnusson, T. (1957) 3,4-Dihydroxyphenylalanine and 5-hydroxytryptophan as reserpine antagonists. *Nature*, **180**, 1200.

Castelhano, A.L., Pliura, D.H., Taylor, G.J., Hsieh, K.C., and Krantz, A. (1984) Allenic suicide substrates. New inhibitors of vitamin B6 linked decarboxylases. *Journal of the American Chemical Society*, **106**, 2734–2735.

Catto, M., Nicolotti, O., Leonetti, F., Carotti, A., Favia, A.D., Soto-Otero, R. et al. (2006) Structural insights into monoamine oxidase inhibitory potency and selectivity of 7-substituted coumarins from ligand- and target-based approaches. *Journal of Medicinal Chemistry*, **49**, 4912–4925.

Cedarbaum, J.M. (1987) Clinical pharmacokinetics of anti-parkinsonian drugs. *Clinical Pharmacokinetics*, **13**, 141–178.

Chase, T.N. (1998) The significance of continuous dopaminergic stimulation in the treatment of Parkinson's disease. *Drugs*, **55**, 1–9.

Chen, K. (2004) Organization of MAO A and MAO B promoters and regulation of gene expression. *Neurotoxicology*, **25**, 31–36.

Chen, J.J. and Ly, A.V. (2006) Rasagiline: a second-generation monoamine oxidase type-B inhibitor for the treatment of Parkinson's disease. *American Journal of Health-System Pharmacy*, **63**, 915–928.

Chen, J.J. and Swope, D.M. (2005) Clinical pharmacology of rasagiline: a novel, second-generation propargylamine for the treatment of Parkinson disease. *Journal of Clinical Pharmacology*, **45**, 878–894.

Clarke, A., Brewer, F., Johnson, E.S., Mallard, N., Hartig, F., Taylor, S. et al. (2003) A new formulation of selegiline: improved bioavailability and selectivity for MAO-B inhibition. *Journal of Neural Transmission*, **110**, 1241–1255.

Cotzias, G.C., Van Woert, M.H., and Schiffer, L.M. (1967) Aromatic amino acids and modification of parkinsonism. *New England Journal of Medicine*, **276**, 374–379.

Cotzias, G.C., Papavasiliou, P.S., and Gellene, R. (1969) Modification of Parkinsonism—chronic treatment with L-dopa. *New England Journal of Medicine*, **280**, 337–345.

Craig, S.P., Thai, A.L., Weber, M., and Craig, I.W. (1992) Localisation of the gene for human aromatic L-amino acid decarboxylase (DDC) to chromosome 7p13→p11 by in situ hybridisation. *Cytogenetics and Cell Genetics*, **61**, 114–116.

Da Prada, M., Kettler, R., Zurcher, G., Schaffner, R., and Haefely, W. (1987) Inhibition of decarboxylase and levels of dopa and 3-*O*-methyldopa: a comparative study of benserazide versus carbidopa in rodents and of Madopar standard versus Madopar HBS in volunteers. *European Neurology Supplement*, **27**, 9–20.

Da Prada, M., Kettler, R., Keller, H.H., Cesura, A.M., Richards, J.G., Saura Marti, J. et al. (1990) From moclobemide to Ro 19–6327 and Ro 41–1049: the development of a new class of reversible, selective MAO-A and MAO-B inhibitors. *Journal of Neural Transmission Supplementum*, **29**, 279–292.

De Colibus, L., Li, M., Binda, C., Lustig, A., Edmondson, D.E., and Mattevi, A. (2005) Three-dimensional structure of human monoamine oxidase A (MAO A): relation to the structures of rat MAO A and human MAO B. *Proceedings of the National Academy of Sciences of the United States of America*, **102**, 12684–12689.

Deleu, D., Northway, M.G., and Hanssens, Y. (2002) Clinical pharmacokinetic and pharmacodynamic properties of drugs used in the treatment of Parkinson's disease. *Clinical Pharmacokinetics*, **41**, 261–309.

Dhar, K. and Rosazza, J.P.N. (2000) Purification and characterization of *Streptomyces griseus* catechol *O*-methyltransferase. *Applied and Environmental Microbiology*, **66**, 4877–4882.

Dingemanse, J., Jorga, K.M., Schmitt, M., Gieschke, R., Fotteler, B., Zurcher, G. et al. (1995) Integrated pharmacokinetics and pharmacodynamics of the novel catechol *O*-methyltransferase inhibitor tolcapone during first administration to humans. *Clinical Pharmacology and Therapeutics*, **57**, 508–517.

Djali, P.K., Dimaline, R., and Dockray, G.J. (1998) Novel form of L-aromatic amino acid decarboxylase mRNA in rat antral mucosa. *Experimental Physiology*, **83**, 617–627.

Duchemin, A.M., Berry, M.D., Neff, N.H., and Hadjiconstantinou, M. (2000) Phosphorylation and activation of brain aromatic L-amino acid decarboxylase by cyclic AMP-dependent protein kinase. *Journal of Neurochemistry*, **75**, 725–731.

Egan, M., Goldman, D., and Weinberger, D. (2002) The human genome: mutations. *American Journal of Psychiatry*, **159**, 12.

Ehringer, H. and Hornykiewicz, O. (1960) Distribution of noradrenaline and dopamine (3-hydroxytyramine) in the human brain and their behavior in diseases of the extrapyramidal system. *Klinische Wochenschrift*, **38**, 1236–1239.

Factor, S.A., Molho, E.S., Feustel, P.J., Brown, D.L., and Evans, S.M. (2001) Long-term comparative experience with tolcapone and entacapone in advanced Parkinson's disease. *Clinical Neuropharmacology*, **24**, 295–299.

Ferreira, J., Rosa, M.M., Coelho, M., Cunha, L., Januário, C., Machado, C. et al. (2006) A double-blind, randomised, placebo- and entacapone-controlled study to investigate the effect of nebicapone (BIA 3-202) on the levodopa pharmacokinetics, COMT activity and motor response in Parkinson disease patients. *Movement Disorders*, **21** (Suppl 15), S644–S645.

Finberg, J.P., Tenne, M., and Youdim, M.B. (1981) Tyramine antagonistic properties of AGN 1135, an irreversible inhibitor of monoamine oxidase type B. *British Journal of Pharmacology*, **73**, 65–74.

Fowler, J.S., Volkow, N.D., Logan, J., Wang, G.J., MacGregor, R.R., Schyler, D. et al. (1994) Slow recovery of human brain MAO B after L-deprenyl (Selegeline) withdrawal. *Synapse*, **18**, 86–93.

Fowler, J.S., Volkow, N.D., Wang, G.J., Logan, J., Pappas, N., Shea, C. et al. (1997) Age-related increases in brain monoamine oxidase B in living healthy human subjects. *Neurobiology of Aging*, **18**, 431–435.

Fraaije, M.W. and Mattevi, A. (2000) Flavoenzymes: diverse catalysts with recurrent features. *Trends in Biochemical Sciences*, **25**, 126–132.

Grossman, M.H., Creveling, C.R., Rybczynski, R., Braverman, M., Isersky, C., and Breakefield, X.O. (1985) Soluble and particulate forms of rat catechol-*O*-methyltransferase distinguished by gel electrophoresis and immune fixation. *Journal of Neurochemistry*, **44**, 421–432.

Guldberg, H.C. and Marsden, C.A. (1975) Catechol-*O*-methyl transferase: pharmacological aspects and physiological role. *Pharmacological Reviews*, **27**, 135–206.

Haasio, K., Koponen, A., Penttila, K.E., and Nissinen, E. (2002a) Effects of entacapone and tolcapone on mitochondrial membrane potential. *European Journal of Pharmacology*, **453**, 21–26.

Haasio, K., Nissinen, E., Sopanen, L., and Heinonen, E.H. (2002b) Different toxicological profile of two COMT inhibitors in vivo: the role of uncoupling effects. *Journal of Neural Transmission*, **109**, 1391–1401.

Hadjiconstantinou, M., Duchemin, A.M., and Neff, N.H. (2003) Cyclic GMP-dependent protein kinase phosphorylates and activates brain aromatic L-amino acid decarboxylase. *Journal of Neurochemistry*, **85**, 40–41.

Hauptmann, N., Grimsby, J., Shih, J.C., and Cadenas, E. (1996) The metabolism of tyramine by monoamine oxidase A/B causes oxidative damage to mitochondrial DNA. *Archives of Biochemistry and Biophysics*, **335**, 295–304.

Heikkinen, H., Saraheimo, M., Antila, S., Ottoila, P., and Pentikainen, P.J. (2001) Pharmacokinetics of entacapone, a peripherally acting catechol-*O*-methyltransferase inhibitor, in man. A study using a stable isotope technique. *European Journal of Clinical Pharmacology*, **56**, 821–826.

Hornykiewicz, O. (2006) The discovery of dopamine deficiency in the parkinsonian brain. *Journal of Neural Transmission Supplementum*, **70**, 9–15.

Huotari, M., Santha, M., Lucas, L.R., Karayiorgou, M., Gogos, J.A., and Mannisto, P.T. (2002) Effect of dopamine uptake inhibition on brain catecholamine levels and locomotion in catechol-*O*-methyltransferase-disrupted mice. *Journal of Pharmacology and Experimental Therapeutics*, **303**, 1309–1316.

Ichinose, H., Sumi-Ichinose, C., Ohye, T., Hagino, Y., Fujita, K., and Nagatsu, T. (1992) Tissue-specific alternative splicing of the first exon generates two types of mRNAs in human aromatic L-amino acid decarboxylase. *Biochemistry*, **31**, 11546–11550.

Jahng, J.W., Wessel, T.C., Houpt, T.A., Son, J.H., and Joh, T.H. (1996) Alternate promoters in the rat aromatic L-amino acid decarboxylase gene for neuronal and nonneuronal expression: an in situ hybridization study. *Journal of Neurochemistry*, **66**, 14–19.

Jeffery, D.R. and Roth, J.A. (1984) Characterization of membrane-bound and soluble catechol-*O*-methyltransferase from human frontal cortex. *Journal of Neurochemistry*, **42**, 826–832.

Jeffery, D.R. and Roth, J.A. (1987) Kinetic reaction mechanism for magnesium binding to membrane-bound and soluble catechol *O*-methyltransferase. *Biochemistry*, **26**, 2955–2958.

Jorga, K.M. (1998) Pharmacokinetics, pharmacodynamics, and tolerability of tolcapone: a review of early studies in volunteers. *Neurology*, **50**, S31–S38.

Jorga, K., Fotteler, B., Heizmann, P., and Gasser, R. (1999) Metabolism and excretion of tolcapone, a novel inhibitor of catechol-*O*-methyltransferase. *British Journal of Clinical Pharmacology*, **48**, 513–520.

Keating, G.M. and Lyseng-Williamson, K.A. (2005) Tolcapone: a review of its use in the management of Parkinson's disease. *CNS Drugs*, **19**, 165–184.

Keranen, T., Gordin, A., Karlsson, M., Korpela, K., Pentikainen, P.J., Rita, H. et al. (1994) Inhibition of soluble catechol-*O*-methyltransferase and single-dose pharmacokinetics after oral and intravenous administration of entacapone. *European Journal of Clinical Pharmacology*, **46**, 151–157.

Kim, Y.H., Moody, J.D., Freeman, J.P., Brezna, B., Engesser, K.H., and Cerniglia, C.E. (2004) Evidence for the existence of PAH-quinone reductase and catechol-*O*-methyltransferase in *Mycobacterium vanbaalenii* PYR-1. *Journal of Industrial Microbiology & Biotechnology*, **31**, 507–516.

Knoll, J. and Magyar, K. (1972) Some puzzling pharmacological effects of monoamine oxidase inhibitors. *Advances Biochemistry and Psychopharmacology*, **5**, 393–408.

Korlipara, L.P., Cooper, J.M., and Schapira, A.H. (2004) Differences in toxicity of the catechol-*O*-methyl transferase inhibitors, tolcapone and entacapone to cultured human neuroblastoma cells. *Neuropharmacology*, **46**, 562–569.

Larsen, J.P. and Boas, J. (1997) The effects of early selegiline therapy on long-term levodopa treatment and parkinsonian disability: an interim analysis of a Norwegian–Danish 5-year study. Norwegian-Danish Study Group. *Movement Disorders*, **12**, 175–182.

Lautala, P., Ulmanen, I., and Taskinen, J. (2001) Molecular mechanisms controlling the rate and specificity of catechol *O*-methylation by human soluble catechol *O*-methyltransferase. *Molecular Pharmacology*, **59**, 393–402.

Learmonth, D.A., Vieira-Coelho, M.A., Benes, J., Alves, P.C., Borges, N., Freitas, A.P. et al. (2002) Synthesis of 1-(3,4-dihydroxy-5-nitrophenyl)-2-phenyl-ethanone and derivatives as potent and long-acting peripheral inhibitors of catechol-*O*-methyltransferase. *Journal of Medicinal Chemistry*, **45**, 685–695.

Learmonth, D.A., Palma, P.N., Vieira-Coelho, M.A., and Soares-da-Silva, P. (2004) Synthesis, biological evaluation, and molecular modeling studies of a novel, peripherally selective inhibitor of catechol-*O*-methyltransferase. *Journal of Medicinal Chemistry*, **47**, 6207–6217.

Legrand, M., Fritig, B., and Hirth, L. (1976) Catechol *O*-methyltransferases of tobacco: evidence for several enzymes with para- and meta-*O*-methylating activities. *FEBS Letters*, **70**, 131–136.

Lotta, T., Taskinen, J., Backstrom, R., and Nissinen, E. (1992) PLS modelling of structure-activity relationships of catechol *O*-methyltransferase inhibitors. *Journal of Computer-Aided Molecular Design*, **6**, 253–272.

Lotta, T., Vidgren, J., Tilgmann, C., Ulmanen, I., Melen, K., Julkunen, I. et al. (1995) Kinetics of human soluble and membrane-bound catechol *O*-methyltransferase: a revised mechanism and description of the thermolabile variant of the enzyme. *Biochemistry*, **34**, 4202–4210.

Loureiro, A.I., Bonifacio, M.J., Fernandes-Lopes, C., Almeida, L., Wright, L.C., and Soares-Da-Silva, P. (2006) Human metabolism of nebicapone (BIA 3–202), a novel catechol-o-methyltransferase inhibitor: characterization of in vitro glucuronidation. *Drug Metabolism and Disposition: The Biological Fate of Chemicals*, **34**, 1856–1862.

Lundström, K., Salminen, M., Jalanko, A., Savolainen, R., and Ulmanen, I. (1991) Cloning and characterization of human placental catechol-*O*-methyltransferase cDNA. *DNA and Cell Biology*, **10**, 181–189.

Ma, J., Yoshimura, M., Yamashita, E., Nakagawa, A., Ito, A., and Tsukihara, T. (2004) Structure of rat monoamine oxidase A and its specific recognitions for substrates and inhibitors. *Journal of Molecular Biology*, **338**, 103–114.

Mahmood, I. (1997) Clinical pharmacokinetics and pharmacodynamics of selegiline. An update. *Clinical Pharmacokinetics*, **33**, 91–102.

Männistö, P.T. and Kaakkola, S. (1989) New selective COMT inhibitors: useful adjuncts for Parkinson's disease? *Trends in Pharmacological Sciences*, **10**, 54–56.

Männistö, P.T. and Kaakkola, S. (1990) Rationale for selective COMT inhibitors as adjuncts in the drug treatment of Parkinson's disease. *Pharmacology & Toxicology*, **66**, 317–323.

Männistö, P.T. and Kaakkola, S. (1999) Catechol-*O*-methyltransferase (COMT): biochemistry, molecular biology, pharmacology, and clinical efficacy of the new selective COMT inhibitors. *Pharmacological Reviews*, **51**, 593–628.

Männistö, P.T., Tuomainen, P., and Tuominen, R.K. (1992a) Different in vivo properties of three new inhibitors of catechol *O*-methyltransferase in the rat. *British Journal of Pharmacology*, **105**, 569–574.

Männistö, P.T., Ulmanen, I., Lundstrom, K., Taskinen, J., Tenhunen, J., Tilgmann, C. et al. (1992b) Characteristics of catechol *O*-methyl-transferase (COMT) and properties of selective COMT inhibitors. *Progress in Drug Research*, **39**, 291–350.

Maycock, A.L., Aster, S.D., and Patchett, A.A. (1980) Inactivation of 3-(3,4-dihydroxyphenyl) alanine decarboxylase by 2-(fluoromethyl)-3-(3,4-dihydroxyphenyl)alanine. *Biochemistry*, **19**, 709–718.

Nicotra, A., Pierucci, F., Parvez, H., and Senatori, O. (2004) Monoamine oxidase expression during development and aging. *Neurotoxicology*, **25**, 155–165.

Nissinen, E., Linden, I.B., Schultz, E., Kaakkola, S., Mannisto, P.T., and Pohto, P. (1988) Inhibition of catechol-*O*-methyltransferase activity by two novel disubstituted catechols in the rat [published erratum appears in *European Journal of Pharmacology* 1988 Nov 22; 157(2–3):244]. *European Journal of Pharmacology*, **153**, 263–269.

Nissinen, E., Linden, I.B., Schultz, E., and Pohto, P. (1992) Biochemical and pharmacological properties of a peripherally acting catechol-*O*-methyltransferase inhibitor entacapone. *Naunyn-Schmiedebergs Archives of Pharmacology*, **346**, 262–266.

Nissinen, E., Kaheinen, P., Penttila, K.E., Kaivola, J., and Linden, I.B. (1997) Entacapone, a novel catechol-*O*-methyltransferase inhibitor for Parkinson's disease, does not impair mitochondrial energy production. *European Journal of Pharmacology*, **340**, 287–294.

O'Gara, M., McCloy, K., Malone, T., and Cheng, X. (1995) Structure-based sequence alignment of three AdoMet-dependent DNA methyltransferases. *Gene*, **157**, 135–138.

O'Malley, K.L., Harmon, S., Moffat, M., Uhland-Smith, A., and Wong, S. (1995) The human aromatic L-amino acid decarboxylase gene can be alternatively spliced to generate unique protein isoforms. *Journal of Neurochemistry*, **65**, 2409–2416.

Olanow, C.W. (2000) Tolcapone and hepatotoxic effects. *Archives of Neurology*, **57**, 263–267.

Olanow, C.W. (2006) Rationale for considering that propargylamines might be neuroprotective in Parkinson's disease. *Neurology*, **66**, S69–S79.

Olanow, C.W. and Obeso, J.A. (2000) Preventing levodopa-induced dyskinesias. *Annals of Neurology*, **47** (4, Suppl 1), S167–S176.

Olanow, W., Schapira, A.H., and Rascol, O. (2000) Continuous dopamine-receptor stimulation in early Parkinson's disease. *Trends in Neurosciences*, **23**, S117–S126.

Olanow, C.W., Agid, Y., Mizuno, Y., Albanese, A., Bonuccelli, U., Damier, P. et al. (2004) Levodopa in the treatment of Parkinson's disease: current controversies. *Movement Disorders*, **19**, 997–1005.

Palhagen, S., Heinonen, E.H., Hagglund, J., Kaugesaar, T., Kontants, H., Maki-Ikola, O. et al. (1998) Selegiline delays the onset of disability in de novo parkinsonian patients. Swedish Parkinson Study Group. *Neurology*, **51**, 520–525.

Parkinson-Study-Group. (1989) Effect of deprenyl on the progression of disability in early Parkinson's disease. *New England Journal of Medicine*, **321**, 1364–1371.

Parkinson-Study-Group. (1996a) Impact of deprenyl and tocopherol treatment on Parkinson's disease in DATATOP patients requiring levodopa. *Annals of Neurology*, **39**, 37–45.

Parkinson-Study-Group. (1996b) Impact of deprenyl and tocopherol treatment on Parkinson's disease in DATATOP subjects not requiring levodopa. *Annals of Neurology*, **39**, 29–36.

Parkinson-Study-Group. (2002) A controlled trial of rasagiline in early Parkinson disease: the TEMPO Study. *Archives of Neurology*, **59**, 1937–1943.

Parkinson-Study-Group. (2004) A controlled, randomized, delayed-start study of rasagiline in early Parkinson disease. *Archives of Neurology*, **61**, 561–566.

Parkinson-Study-Group. (2005) A randomized placebo-controlled trial of rasagiline in levodopa-treated patients with Parkinson disease and motor fluctuations: the PRESTO study. *Archives of Neurology*, **62**, 241–248.

Pletscher, A. and DaPrada, M. (1993) Pharmacotherapy of Parkinson's disease: research from 1960 to 1991. *Acta Neurologica Scandinavica Supplementum*, **146**, 26–31.

Poulikakos, P., Vassilacopoulou, D., and Fragoulis, E.G. (2001) L-DOPA decarboxylase association with membranes in mouse brain. *Neurochemical Research*, **26**, 479–485.

Przuntek, H., Conrad, B., Dichgans, J., Kraus, P.H., Krauseneck, P., Pergande, G. et al. (1999) SELEDO: a 5-year long-term trial on the effect of selegiline in early Parkinsonian patients treated with levodopa. *European Journal of Neurology*, **6**, 141–150.

Rascol, O., Brooks, D.J., Melamed, E., Oertel, W., Poewe, W., Stocchi, F. et al. (2005) Rasagiline as an adjunct to levodopa in patients with Parkinson's disease and motor fluctuations (LARGO, Lasting effect in Adjunct therapy with Rasagiline Given Once daily, study): a randomised, double-blind, parallel-group trial. *Lancet*, **365**, 947–954.

Ribereau-Gayon, G., Danzin, C., Palfreyman, M.G., Aubry, M., Wagner, J., Metcalf, B.W. et al. (1979) In vitro and in vivo effects of alpha-acetylenic DOPA and alpha-vinyl DOPA on aromatic L-amino acid decarboxylase. *Biochemical Pharmacology*, **28**, 1331–1335.

Riederer, P., Lachenmayer, L., and Laux, G. (2004) Clinical applications of MAO-inhibitors. *Current Medicinal Chemistry*, **11**, 2033–2043.

Salminen, M., Lundstrom, K., Tilgmann, C., Savolainen, R., Kalkkinen, N., and Ulmanen, I. (1990) Molecular cloning and characterization of rat liver catechol-*O*-methyltransferase. *Gene*, **93**, 241–247.

Schluckebier, G., O'Gara, M., Saenger, W., and Cheng, X. (1995) Universal catalytic domain structure of AdoMet-dependent methyltransferases. *Journal of Molecular Biology*, **247**, 16–20.

Schultz, E. and Nissinen, E. (1989) Inhibition of rat liver and duodenum soluble catechol-*O*-methyltransferase by a tight-binding inhibitor OR-462. *Biochemical Pharmacology*, **38**, 3953–3956.

Sharpless, N.S. and McCann, D.S. (1971) Dopa and 3-*O*-methyldopa in cerebrospinal fluid of Parkinsonism patients during treatment with oral L-dopa. *Clinica Chimica Acta*, **31**, 155–169.

Shih, J.C., Chen, K., and Ridd, M.J. (1999) Monoamine oxidase: from genes to behavior. *Annual Review of Neuroscience*, **22**, 197–217.

Shinagawa, Y. (1992) Molecular orbital studies on the structure-activity relationships of catechol *O*-methyltransferase inhibitors. *Japanese Journal of Pharmacology*, **58**, 95–106.

Shoulson, I., Oakes, D., Fahn, S., Lang, A., Langston, J.W., LeWitt, P. et al. (2002) Impact of sustained deprenyl (selegiline) in levodopa-treated Parkinson's disease: a randomized placebo–controlled extension of the deprenyl and tocopherol antioxidative therapy of parkinsonism trial. *Annals of Neurology*, **51**, 604–612.

Silveira, P., Vaz-Da-Silva, M., Almeida, L., Maia, J., Falcao, A., Loureiro, A. et al. (2003) Pharmacokinetic–pharmacodynamic interaction between BIA 3–202, a novel COMT inhibitor, and levodopa/benserazide. *European Journal of Clinical Pharmacology*, **59**, 603–609.

Sipila, J. and Taskinen, J. (2004) CoMFA modeling of human catechol *O*-methyltransferase enzyme kinetics. *Journal of Chemical Information and Computer Sciences*, **44**, 97–104.

Smith, K.S., Smith, P.L., Heady, T.N., Trugman, J.M., Harman, W.D., and Macdonald, T.L. (2003) In vitro metabolism of tolcapone to reactive intermediates: relevance to tolcapone liver toxicity. *Chemical Research in Toxicology*, **16**, 123–128.

Stocchi, F. (2005) Optimising levodopa therapy for the management of Parkinson's disease. *Journal of Neurology*, **252** (Suppl 4), iv43–iv48.

Sumi-Ichinose, C., Ichinose, H., Takahashi, E., Hori, T., and Nagatsu, T. (1992) Molecular cloning of genomic DNA and chromosomal assignment of the gene for human aromatic L-amino acid decarboxylase, the enzyme for catecholamine and serotonin biosynthesis. *Biochemistry*, **31**, 2229–2238.

Taskinen, J., Vidgren, J., Ovaska, M., Bäckström, R., Pippuri, A., and Nissinen, E. (1989) QSAR and binding model for inhibition of rat liver catechol-*O*-methyltransferase by 1,5-substituted-3,4-dihydroxybenzenes. *Quantitative Structure-Activity Relationships*, **8**, 210–213.

Thebault, J.J., Guillaume, M., and Levy, R. (2004) Tolerability, safety, pharmacodynamics, and pharmacokinetics of rasagiline: a potent, selective, and irreversible monoamine oxidase type B inhibitor. *Pharmacotherapy*, **24**, 1295–1305.

Veser, J. (1987) Kinetics and inhibition studies of catechol *O*-methyltransferase from the yeast *Candida tropicalis*. *Journal of Bacteriology*, **169**, 3696–3700.

Veser, J., Geywitz, P., and Thomas, H. (1979) Purification and properties of a catechol methyltransferase of the yeast *Candida tropicalis*. *Zeitschrift für Naturforschung C*, **34**, 709–714.

Vidgren, J., Tilgmann, C., Lundstrom, K., and Liljas, A. (1991) Crystallization and preliminary X-ray investigation of a recombinant form of rat catechol O-methyltransferase. *Proteins*, **11**, 233–236.

Vidgren, J., Svensson, L.A., and Liljas, A. (1994) Crystal structure of catechol O-methyltransferase. *Nature*, **368**, 354–358.

Vilbois, F., Caspers, P., da Prada, M., Lang, G., Karrer, C., Lahm, H.W. et al. (1994) Mass spectrometric analysis of human soluble catechol O-methyltransferase expressed in *Escherichia coli*. Identification of a product of ribosomal frameshifting and of reactive cysteines involved in S-adenosyl-L-methionine binding. *European Journal of Biochemistry*, **222**, 377–386.

Waldmeier, P.C., Baumann, P.A., Feldtrauer, J.J., Hauser, K., Bittiger, H., Bischoff, S. et al. (1990a) CGP 28014, a new inhibitor of cerebral catechol-O-methylation with a non-catechol structure. *Naunyn-Schmiedebergs Archives of Pharmacology*, **342**, 305–311.

Waldmeier, P.C., De Herdt, P., and Maitre, L. (1990b) Effects of the COMT inhibitor, CGP 28014, on plasma homovanillic acid and O-methylation of exogenous L-dopa in the rat. *Journal of Neural Transmission Supplementum*, **32**, 381–386.

Wheeler, D.L., Barrett, T., Benson, D.A., Bryant, S.H., Canese, K., Chetvernin, V. et al. (2006) Database resources of the National Center for Biotechnology Information. *Nucleic Acids Research*, **34**, D173–D180.

Winqvist, R., Lundstrom, K., Salminen, M., Laatikainen, M., and Ulmanen, I. (1992) The human catechol-O-methyltransferase (COMT) gene maps to band q11.2 of chromosome 22 and shows a frequent RFLP with BglI. *Cytogenetics and Cell Genetics*, **59**, 253–257.

Youdim, M.B. and Bakhle, Y.S. (2006) Monoamine oxidase: isoforms and inhibitors in Parkinson's disease and depressive illness. *British Journal of Pharmacology*, **147** (Suppl 1), S287–S296.

Youdim, M.B., Gross, A., and Finberg, J.P. (2001) Rasagiline [N-propargyl-1R(+)-aminoindan], a selective and potent inhibitor of mitochondrial monoamine oxidase B. *British Journal of Pharmacology*, **132**, 500–506.

Youdim, M.B., Bar Am, O., Yogev-Falach, M., Weinreb, O., Maruyama, W., Naoi, M. et al. (2005) Rasagiline: neurodegeneration, neuroprotection, and mitochondrial permeability transition. *Journal of Neuroscience Research*, **79**, 172–179.

Youdim, M.B., Edmondson, D., and Tipton, K.F. (2006) The therapeutic potential of monoamine oxidase inhibitors. *Nature Reviews Neuroscience*, **7**, 295–309.

Zhu, M.-Y. and Juorio, A.V. (1995) Aromatic L-amino acid decarboxylase: biological characterization and functional role. *General Pharmacology*, **26**, 681–696.

Zhu, B.T., Patel, U.K., Cai, M.X.X., and Conney, A.H. (2000) O-methylation of tea polyphenols catalyzed by human placental cytosolic catechol-O-methyltransferase. *Drug Metabolism & Disposition*, **28**, 1024–1030.

Zurcher, G., Colzi, A., and Da Prada, M. (1990a) Ro 40–7592: inhibition of COMT in rat brain and extracerebral tissues. *Journal of Neural Transmission Supplementum*, **32**, 375–380.

Zurcher, G., Keller, H.H., Kettler, R., Borgulya, J., Bonetti, E.P., Eigenmann, R. et al. (1990b) Ro 40–7592, a novel, very potent, and orally active inhibitor of catechol-O-methyltransferase: a pharmacological study in rats. *Advances in Neurology*, **53**, 497–503.

Part III

Huntington's Disease

11 Huntington's Disease

Claire-Anne Gutekunst and Fran Norflus

CONTENTS

11.1 INTRODUCTION

Huntington's Disease (HD) is a fatal progressive neurodegenerative disorder that was named after Dr. George Huntington, a U.S. physician who first described the symptoms and hereditary characteristics of HD in 1872. HD symptoms usually occur at middle age, however, there is a juvenile form of the disease with onset as early as 2 years of age (Nance and Myers, 2001). Early symptoms of HD include involuntary twitching and clumsiness, and emotional and cognitive disturbances. As the disease advances, concentration and short-term memory diminish and involuntary movements of the head, trunk, and limbs increase. Walking, speaking, and swallowing abilities worsen. Affected individuals are rapidly disabled by early functional decline, require care and supervision, and die within 15–25 years from onset because of complications such as choking or heart failure. There is presently no therapy to delay the onset or slow progression of the disease, and the current medical treatments primarily focus on alleviating symptoms and optimizing function. In this chapter, we will review the neuropathology and genetics of HD, and then briefly describe how protein misfolding has been thought to play a role in its neuropathogenesis and has guided the development of therapeutic approaches.

HD has been diagnosed on all continents, and in various geographic and ethnic populations worldwide from industrialized countries to more remote untouched areas of the world (Novelletto et al., 1994; Almqvist et al., 1995; McCusker et al., 2000; Murgod et al., 2001; Saleem et al., 2003; Okun and Thommi, 2004; Tumas et al., 2004). The frequency of HD varies among different populations, ranging from an estimated 4 to 10 individuals per 100,000, with the highest prevalence in Europe and North America. The availability of genetic testing for HD has made it possible to identify a previously unrecognized symptomatic population of HD, including those with an atypical presentation or patients without a family history of HD (Almqvist et al., 2001).

11.2 HD NEUROPATHOLOGY

The earliest and most striking neuropathological changes in human HD are found in the neostriatum (Lange et al., 1976; Graveland et al., 1985; Ferrante et al., 1991; Forno, 1992; Sapp et al., 1995) and the cerebral cortex (Hedreen et al., 1991; Jackson et al., 1995), however, neuronal loss has been identified in many other regions of the brain including the hippocampus (Spargo et al., 1993), cerebellum (Lange et al., 1976; Jeste et al., 1984), and the hypothalamus (Kremer et al., 1990, 1991). In the striatum, there is gross atrophy of the caudate nucleus and putamen accompanied by marked neuronal loss and astrogliosis (Figure 11.1).

There is a dorsal–ventral progression of neuronal death, with the dorsal striatum affected earliest and relative sparing of the ventromedial striatum and nucleus accumbens. The progressive macroscopic and microscopic neuropathological changes of the caudate or putamen region was used to establish the Vonsattel grading system, which closely correlates with the extent of clinical dysfunction (Vonsattel

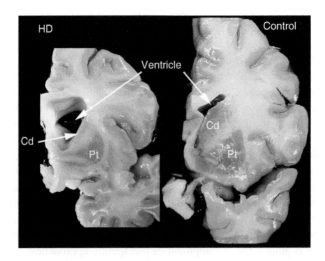

FIGURE 11.1 Photographs of coronal sections showing the caudate and putamen regions in postmortem brains from a grade 3 HD patient and age-matched control. Note the extensive enlargement of the ventricle in HD and atrophy of the caudate nucleus.

et al., 1985). Extensive neurochemical studies of the striatal cell types in HD have shown that medium spiny projection neurons are particularly vulnerable to degeneration, whereas other cell types tend to be spared (Ferrante et al., 1985, 1987a, 1987b; Morton et al., 1993; Richfield and Herkenham, 1994). Cortical atrophy is progressive and most prominent in the somatosensory cortex (Rosas et al., 2002) with loss of neurons in layers V and VI (Hedreen et al., 1991). Striatal and cortical degeneration is largely responsible for the motor symptoms of HD. Cerebral cortex degeneration is also suspected to participate in the nonmotor symptoms of HD, such as dementia, irritability, apathy, and depression (Hedreen et al., 1991). Hypothalamic degeneration is seen in the lateral tuberal nucleus of the hypothalamus of end-stage patients (Kremer et al., 1990, 1991), accompanied by a progressive loss of orexin-positive neurons (Petersen et al., 2005) and may underlie the abnormal sleep–wake cycle seen in HD (Petersen et al., 2005).

11.3 HD MUTATION

HD is inherited in an autosomal dominant fashion. In 1983, HD was linked to a mutation located on chromosome 4p16.3 (Gusella et al., 1983). Ten years later, the mutation was identified as the abnormal repetition of the trinucleotide repeat CAG located in the coding region of the first exon of the *IT15* or *HD* gene (HDCRG, 1993). In the normal population, the number of CAG repeats ranges from 6 to 35, whereas individuals affected by HD carry at least one allele with 35–141 CAG repeats (Duyao et al., 1993; HDCRG, 1993).

The age of onset for HD has been correlated to the number of CAG repeats, with longer repeat length correlating with early onset (HDCRG, 1993). Although this correlation is strong, it is widely acknowledged that the repeat size is a poor predictor of age of onset. The age at which symptoms first appear varies largely between people carrying the same number of CAG repeats suggesting the influence of other genetic, epigenetic, and environmental factors (Wexler et al., 2004). Genetic modifiers of age of onset include polymorphisms in the genes encoding for methylenetetrahydrofolate reductase (Brune et al., 2004), variations in single nucleotide polymorphisms (SNPs) in the genes coding for the transcription factor p53, and human caspase-activated DNase (Chattopadhyay et al., 2005) as well as modifier genes at 6p23–24, 6q24–26, 18q22, and 4p16 loci (Li et al., 2003, 2006; Djousse et al., 2004). One study shows that the age of onset is influenced by the size of the normal repeat, which lessens the expression of the disease among HD-affected persons with large expanded CAG repeats (Djousse et al., 2003).

Although HD affects both sexes equally, most juvenile onset HD is inherited through the male germ line. This "anticipation effect" is due to meiotic instability of the trinucleotide repeat in males, which causes expansion and contraction of the repeat stretch (Duyao et al., 1993; Ranen et al., 1995). During maternal transmission, similar numbers of expansions and contractions are seen, but the shifts are small, ranging from 1 to 3 repeats. For paternal transmission, there are more expansion events than contractions, and changes in repeat numbers tend to result in large increases in the size of the CAG expansion (Duyao et al., 1993).

11.4   HUNTINGTIN LOCALIZATION AND NORMAL FUNCTION

11.4.1   LOCALIZATION

The gene product of the HD gene is a 340 kDa soluble protein (3141 amino acids) that has been named huntingtin (HDCRG, 1993) that contains no reported homology with other proteins and only a few known sequence motifs. Huntingtin is widely expressed in cells throughout the body with highest expression levels detected in the brain and testis (Trottier et al., 1995). In the brain its expression is highest in neurons (Figure 11.2), where it is detected mostly in the perikarya, dendrites, axons, and dendritic spines (Gutekunst et al., 1995). At the subcellular level, huntingtin is found associated with the cytoplasmic surface of a variety of organelles, including transport vesicles, synaptic vesicles, and mitochondria (DiFiglia et al., 1995; Trottier et al.,

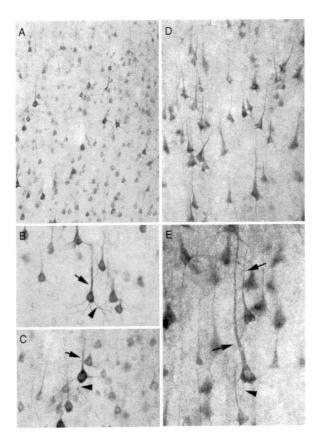

FIGURE 11.2 Huntingtin immunocytochemistry in normal human and nonhuman primate. Micrographs showing huntingtin labeling of cerebral cortex from a rhesus monkey (A–C) and a normal human (D–E). Huntingtin is present mostly in the perikarya, dendrites (*arrows*), and axons (*arrowheads*) of neurons.

1995; Bhide et al., 1996; Sapp et al., 1997; Velier et al., 1998) as well as along microtubules (Gutekunst et al., 1995).

11.4.2 FUNCTIONS

A variety of functions have been postulated for huntingtin based on its subcellular localization and the identification of several of its binding partners and the results of studies in which huntingtin expression levels were altered (see Cattaneo et al. [2005] for review). A direct role for huntingtin in anchoring or transporting other molecules or organelles along microtubules was suggested following electron microscopy localization of huntingtin alongside microtubules (Gutekunst et al., 1995). This role was further confirmed by the finding that Huntingtin Associated Protein 1 (HAP1) associates with dynactin p150 glued, an accessory protein for the microtubule motor protein dynein (Engelender et al., 1997; Li et al., 1998), and proposed to participate in endosomal trafficking (Li et al., 2002b). Several other huntingtin interactors are known to participate in cellular trafficking. HIP1 (Huntingtin Interacting Protein 1) is a component of clathrin-coated vesicles that plays a role in clathrin-mediated internalization of glutamate α-amino-5-hydroxy-3-methyl-4-isoxazole propionic acid (AMPA) receptors (Metzler et al., 2003). HIP14 (Huntingtin Interacting Protein 14), a neuronal palmitoyl transferase, enhances palmitoylation-dependent vesicular trafficking of several proteins involved in the dynamic assembly of components that control vesicle trafficking and synaptic vesicle function (Singaraja et al., 2002; Huang et al., 2004; Yanai et al., 2006). HAP40 (Huntingtin Associated Protein 40), an interactor to Rab5, plays a role in the trafficking of early endosomes along microtu-bules (Peters and Ross, 2001; Pal et al., 2006). Finally, another huntingtin-interacting phosphoprotein, PACSIN1 (protein kinase C and casein kinase substrate in neurons 1), is involved in synaptic vesicle recycling (Modregger et al., 2002). A direct role in fast axonal transport was shown in Drosophila in which decreased levels of hunting-tin resulted in accumulation of organelles within fly larval nerves (Gunawardena et al., 2003; Lee et al., 2004). These findings were further confirmed in mouse striatal neurons in which reducing wild-type huntingtin levels 50% below normal decreased mitochondrial movements (Trushina et al., 2004).

Normal huntingtin is protective in vitro against a variety of insults including serum deprivation, the mitochondrial toxin 3-nitropropionic acid (3-NP), an inhibitor of mitochondrial complex I activity which has been used in modeling HD (Brouillet et al., 1995), the over expression of Bcl-X_L-Bcl-2-associated death promoter (BAD) (Rigamonti et al., 2000, 2001), and the toxicity of mutant huntingtin (Ho et al., 2001). The protective effect of huntingtin was also demonstrated in vivo following ischemic injury (Zhang et al., 2003) and excitotoxicity (Leavitt et al., 2006). Normal huntingtin works via inhibition of pro-caspase 9, an enzyme required for apoptosis (Rigamonti et al., 2001) and its interaction with pro-apoptotic proteins such as HIP1 and the HIP1-interacting protein, HIPPI (Hackam et al., 2000; Gervais et al., 2002). This protective function of huntingtin has led to the hypothesis that HD pathogenesis results from the loss of the beneficial effects of wild-type huntingtin (Zuccato et al., 2001). Another way huntingtin helps neurons survive is by stimulating the produc-tion and release of the brain-derived neurotrophic factor (BDNF) by both regulating

BDNF transport along the cortico-striatal axons and acting at the level of Bdnf gene transcription (Zuccato et al., 2001, 2003; Gauthier et al., 2004). Wild-type huntingtin controls Bdnf gene transcription by binding and sequestering in the cytoplasm the repressor element-1 transcription factor/neuron restrictive silencer factor (REST/NRSF), a transcription factor that binds to a neuron restrictive silencer element (NRSE) present in the promoter region of the Bdnf gene (Zuccato et al., 2003). Since NRSE is present in other genes, it is possible that wild-type huntingtin also regulates the levels of these other NRSE-containing genes via its binding to REST/NRSE. Huntingtin can also regulate transcription of GC-rich elements containing genes by interacting with Sp1, a transcription factor that activates transcription by binding to the GC-rich regions of promoters (Li et al., 2002a).

In addition to a potential role in transcription regulation, huntingtin has also been proposed to act as a transcriptional factor itself. This is largely based on its polyglutamine (polyQ) content and its binding to many transcription factors including the nuclear receptor corepressor N-CoR (Boutell et al., 1999), the transcriptional corepressor C-terminal binding protein (CtBP) (Kegel et al., 2002), the Gln-Ala repeat transcriptional activator CA150 (Holbert et al., 2001), the transcription factor HYPB (Passani et al., 2000), and the basic helix–loop–helix transcription factor, NeuroD (Marcora et al., 2003). Huntingtin has also been implicated in iron homeostasis (Hilditch-Maguire et al., 2000) and calcium-signaling pathways via interaction with HAP1 and inositol 1,4,5-trisphosphate receptor (InsP3R1) (Tang et al., 2003, 2004).

Huntingtin plays a crucial role in early embryogenesis, as evidenced by the fact that mice that are homozygous for a targeted mutation in the HD gene die at approximately embryonic day (E) 7.5 (Duyao et al., 1995; Nasir et al., 1995; Zeitlin et al., 1995). Wild-type blastocysts injected with embryonic stem (ES) cells obtained from homozygous knockout huntingtin (HD$^{-/-}$) embryos produced viable chimeric embryos, whereas HD$^{-/-}$ blastocysts injected with wild-type ES cells did not, showing that the primary defect of huntingtin depletion is in extra-embryonic tissues (Dragatsis et al., 1998). It was proposed that huntingtin might participate in intracellular trafficking of nutrients in early embryos (Dragatsis et al., 1998). The rescued embryos developed to adulthood and were found to have huntingtin HD$^{-/-}$ neurons throughout their brain showing that many brain regions do not require huntingtin for their normal functioning (Reiner et al., 2001). This is further suggested by the finding that HD$^{-/-}$ ES cells are able to differentiate into functional postmitotic neurons in vitro (Metzler et al., 1999). In another study, however, decreased levels of huntingtin to about half-normal levels caused brain developmental defects (White et al., 1997) as well as neurodegeneration and behavioral deficits (Dragatsis et al., 1998). Using a conditional Cre/lox promoter-specific strategy, it was shown that huntingtin depletion in the forebrain and testis of adult mice led to apoptosis, neurodegeneration, and sterility (Dragatsis et al., 2000).

11.5 HUNTINGTIN MISFOLDING AND AGGREGATION IN HD

The CAG repeat in exon 1 of the HD gene encodes for a polyQ stretch in the aminoterminal region of huntingtin (Figure 11.3). Abnormal expansion of this polyQ stretch results in aberrant folding and processing of the mutant protein. Both wild-type and

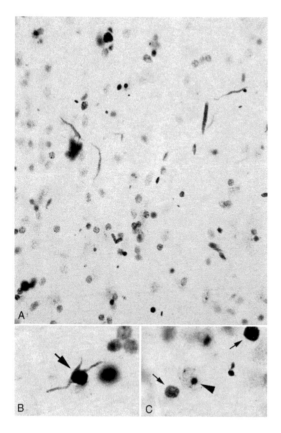

FIGURE 11.3 Huntingtin aggregates in an HD brain. Sections through cerebral cortex from a grade 1 HD case were immunostained with an antibody raised against the N-terminal portion of huntingtin that specifically labels huntingtin aggregates (light grey). Sections have been counterstained with Cresyl violet to show cellular nuclei (black). Huntingtin aggregates are visible as small to large puncta in the neuropil (A), processes (A and B), and neuronal nuclei (C). In (C), a nuclear aggregate is present in the nucleus of a neuron (*arrowhead*) but not of the adjacent glial cells (*arrows*).

mutant huntingtin are cleaved by the pro-apoptotic enzyme caspase-3, releasing an N-terminal fragment containing the polyQ stretch (Goldberg et al., 1996; Martindale et al., 1998; Wellington et al., 1998). The proteolytic N-terminal fragments contain expanded polyQ form insoluble aggregates, which are also referred to as inclusions or huntingtin bodies (Figure 11.4). These aggregates become ubiquitinated and are visible in the processes, cytoplasm, and nuclei of neurons in the brains of HD patients (DiFiglia et al., 1997; Gutekunst et al., 1999; van Roon-Mom et al., 2006) as well as in brains and nonneuronal tissues of HD transgenic animal models (Satyal et al., 2000; Tanaka et al., 2004).

It was proposed that elongated polyQ peptides beyond a critical length of 40 glutamines induce a phase change from random coils to hydrogen-bonded hairpins

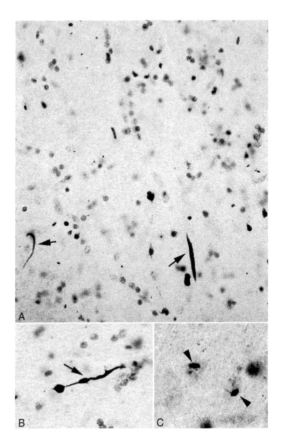

FIGURE 11.4 Micrographs showing ubiquitin staining of aggregates in the cerebral cortex from a grade 1 HD case (light grey). Sections have been counterstained with Cresyl violet to show cellular nuclei (black). Ubiquitin-positive aggregates are visible in the processes (A and B) and nuclei of neurons (C).

that assemble into β-sheet structures that are also referred to as polar zippers (Perutz et al., 1994). Aggregation is dependent on the length of the polyQ (Li and Li, 1998). The protein region flanking the polyQ stretch can affect the type of aggregate being formed, ultimately affecting toxicity (Duennwald et al., 2006), which might explain the difference in cell specific toxicity seen in the context of each disease.

Mutant huntingtin has been shown to be more prone to cleavage, probably due to its misfolding causing increased accessibility of protease cleavage sites to proteases (Toneff et al., 2002). Huntingtin is also cleaved by caspase-2 at amino acid 552 and binding of caspase-2 to huntingtin is polyQ repeat length dependent (Hermel et al., 2004). Another protease capable of releasing the polyQ-containing N-terminal fragment of huntingtin is the calcium-dependent nonlysosomal cysteine protease calpain (Kim et al., 2001; Gafni and Ellerby, 2002; Gafni et al., 2004). Here again, the rate of cleavage of huntingtin by calpain is polyQ length dependent (Gafni and Ellerby, 2002).

Autophagy has been shown to play a role in normal huntingtin processing and may be important in the removal of mutant huntingtin, especially in early stages of HD. Autophagy contributes to the catabolism and cleavage of huntingtin by activating cathepsin D and caspase-3. However, mutant huntingtin is more resistant than wild-type huntingtin to degradation by cathepsin D, thereby contributing to HD pathogenesis (Kegel et al., 2000; Kim et al., 2006).

11.6 AGGREGATE INTERACTING PROTEINS

Aggregates containing misfolded huntingtin also contain a number of critical cellular components. These include normal huntingtin interactors, which also bind to mutant huntingtin, proteins that only bind to mutant huntingtin, and yet another class of proteins that are simply being entrapped as the aggregates are forming. However, huntingtin interactors do not automatically qualify as aggregate interactors. For example, HAP1 is not present in the abnormal intranuclear and neuritic aggregates containing the N-terminal fragment of mutant huntingtin (Gutekunst et al., 1998).

A number of investigators have argued that by sequestering particular proteins, aggregates interfere with their normal localization and functioning thereafter leading to cell death. For example, mutant N-terminal huntingtin can recruit proteins containing polyQ tracts into aggregates, such as the N-terminal proteolytic fragments of normal huntingtin as well as the transcription factors, TATA-binding protein (TBP) (Huang et al., 1998; Matsumoto et al., 2006) and CREB-binding protein (CBP) (Kazantsev et al., 1999; Steffan et al., 2001; Matsumoto et al., 2006). Recruitment of TBP and CBP into the aggregates might interfere with the transcription of genes they normally control. An additional example is the binding of the nuclear receptor corepressor N-CoR to mutant huntingtin (Boutell et al., 1999). N-CoR, which is normally present both in the nucleus and the cytoplasm of neurons, is predominantly found in the cytoplasm of neurons in the cortex and caudate of HD patients, suggesting that binding to mutant huntingtin prevents normal nuclear localization and function of the repressor (Boutell et al., 1999). In addition to transcription factors, aggregates also accumulate a number of molecular chaperones such as Hsp40, Hdj-2, Hsc70, Hsp70, ubiquitin, and proteasome components implicating protein folding and degradation in the pathogenesis of the disease (Jana et al., 2000, 2001).

To systematically identify proteins that might be recruited into huntingtin-containing aggregates, several groups selected the use of proteomics. In one study, aggregates were induced by transient expression of the N-terminal region of mutant huntingtin fused to an enhanced version of the green fluorescent protein (eGFP) in HEK293 kidney cells (Suhr et al., 2001). The proteins that they isolated were proteins that had previously been associated with polyQ aggregates such as ubiquitin, huntingtin N-terminal polyQ-containing fragments, native normal huntingtin, Hsp70, the transcription factor TBP, MEF-2a, the tumor suppressor protein p53, and the p53-regulating protein mdm-2. In addition, they identified the 68 kD light neurofilament protein (NFL), actin, the RNA-dependent ATPase DEAD box protein Dbp-5, which is involved in a variety of mRNA-processing and import or export functions and RanBP3-a, a member of a family of Ran-GTPase-binding proteins, NPCP-Nup62 and NPCP-NupHMW (Suhr et al., 2001). A second study purified aggregates from stably transfected lines of Neuro 2A cells expressing exon 1 of huntingtin-containing 150Q fused to eGFP either with or without a nuclear localization signal to force the formation of either cytoplasmic or nuclear aggregates (Doi et al., 2004; Mitsui et al., 2006). In addition to the proteins identified in the previous study, they also found that nuclear aggregates contain the ubiquitin-interacting proteins ubiquilin 1 and 2 and Tollip (Doi et al., 2004). The identification of molecular targets or interactors offers opportunities for the development of potential therapeutic strategies aimed at preventing or ameliorating HD and perhaps other polyglutamine repeat diseases.

11.7 HD MODELS

To try to understand the mechanisms underlying HD, various animal models have been generated in the hope of replicating motor and cognitive impairments, and neuropathological features seen in the human disease. Transgenic mice and rats have been generated using a variety of promoters and constructs ranging from expression of the first 63 amino acids of huntingtin (Mangiarini et al., 1996; von Horsten et al., 2003) to expression of longer portions (171 amino acids) (Schilling et al., 1999) or even full-length huntingtin (Reddy et al., 1998; Hodgson et al., 1999) containing polyglutamine repeats of various magnitude. These later models show extensive and progressive neuronal injury and death found in the striatum that more closely models human neuropathology. Knock-in models have also been created with different numbers of CAG repeats introduced into the mouse huntingtin gene (Menalled et al., 2003).

Knock-in strategies have the advantage of creating mouse lines most analogous to human HD at the molecular level, and having normal and expanded mouse genes controlled by the endogenous mouse promoter. So far, three knock-in models have been developed (Levine et al., 1999; Shelbourne et al., 1999; Wheeler et al., 1999, 2000). In addition, the localized in vivo expression of mutant huntingtin in the brain has been achieved in the rat using a viral vector approach (Senut et al., 2000). Models have also been generated in lower species including *Drosophila melanogaster* (Jackson et al., 1998; Gunawardena et al., 2003; Lee et al., 2004; Kretzschmar et al., 2005), *Caenorhabditis elegans* (Faber et al., 1999; Parker et al., 2001; Brignull et al., 2006), and zebrafish *Danio renio* (Miller et al., 2005). In flies, mutant huntingtin was expressed in the photoreceptors of the eye resulting in the formation of nuclear aggregates and neuronal degeneration. In nematodes, mutant huntingtin expressed in sensory neurons led to functional impairment that preceded neurodegeneration, cell death, and protein aggregates (Faber et al., 1999). In another nematode model, mutant huntingtin expression in the posterior lateral microtubule (PLM) touch neurons produced touch deficit but no cell death (Parker et al., 2001). Finally, in zebrafish, mutant huntingtin expression led to early abnormal morphology, developmental delay and gross abnormalities in body plan, and differentiation as well as the formation of aggregates and cell death (Miller et al., 2005).

Having these different models is important for both practical and scientific reasons. No one model thus far completely recapitulates all the symptoms and neuropathological features of human HD. Transgenic models expressing small fragments of huntingtin have the most dramatic and rapid phenotype but may not show selective vulnerability. Models using larger fragments of huntingtin are more similar pathologically to human HD but also have more slowly developing phenotypes, which can be a disadvantage in experiments that test whether drugs can reverse the phenotype and pathological features of the model. The current animal models have greatly assisted in dissecting and understanding the molecular pathology of HD and are being used in testing therapeutic strategies.

11.8 THERAPEUTIC APPROACHES

Neuroprotective therapeutic approaches that specifically target aggregate formation in HD have been tested both in vitro and in vivo (see Table 11.1). The different cellular models of HD that have been used are COS1 cells, PC12 cells, Neuro2a cells, human embryonic kidney 293 cells, HdhQ111/Q111 striatal cells, and hippocampal slice cultures from R6/2 mice. A variety of compounds were identified that inhibited polyQ aggregation in these models. These included mAb 1C2, congo red, thioflavine S, chrysamine G, direct fast yellow (Heiser et al., 2000; Smith et al., 2001), geldenamycin (Sittler et al., 2001), the chaperonin TRiC (CCT) (Kitamura et al., 2006; Tam et al., 2006), glucocorticoid receptor (Diamond et al., 2000), creatine (Ferrante et al., 2000; Andreassen et al., 2001; Smith et al., 2001; Dedeoglu et al., 2003), various benzothiazoles and derivatives including PGL-135 and riluzole (Heiser et al., 2002; Hockly et al., 2006), the transglutaminase inhibitor cystamine (Zainelli et al., 2005), and intrabodies (Colby et al., 2004; Wolfgang et al., 2005). In another study, 10 compounds were identified that inhibit aggregation in HdhQ111/Q111 striatal cells including gossypol, gambogic acid, juglone, celastrol, sanguinarine, and anthralin. Of these, both juglone and celastrol were effective in reversing the abnormal cellular localization of full-length mutant huntingtin observed in mutant HdhQ111/Q111 striatal cells (Wang et al., 2005).

Other compounds producing effective decreases in both aggregation and survival in vitro include the chaperone proteins GroEl, GroES, Hsp104, and Hsp70 (Carmichael et al., 2000, 2002; Miller et al., 2005; Novoselova et al., 2005), clioquinol, a metal-binding compound (Nguyen et al., 2005), modified single-stranded oligonucleotides (Parekh-Olmedo et al., 2004), the disaccharide trehalose (Tanaka et al., 2004, 2005), as well as the biologically active small compound C2–8 (Zhang et al., 2005). In another study, biologically active small molecules were screened in an intracellular aggregation assay based on fluorescence resonance energy transfer (FRET). Further analysis showed that five of these compounds were found to inhibit aggregation in an HD model of PC12 cells. These compounds were Ac-YAD-cmk (inhibits caspase 1, limited caspase 5 inhibition), levonordefrin (α-adrenergic agonist), EGF inhibitor 3 (tyrosine kinase inhibitors), nadolol (β-adrenergic receptor antagonist), and fosfosal (nonacetylated salicylic acid derivative) (Desai et al., 2006).

Some of these same compounds were also found to inhibit aggregation and increase survival in yeast and Drosophila models of HD (see Table 11.1). These potentially therapeutic agents include CCT (Behrends et al., 2006; Tam et al., 2006), C2–8 (Zhang et al., 2005), Ac-YAD-cmk, levonordefrin, EGF inhibitor 3, nadolol (β-adrenergic

TABLE 11.1

Effects of Various Compounds on Inhibition of Aggregate Formation in Different in vitro and in vivo Systems

Compound or Modifier	HD Cell Model	Yeast and Drosophila	Mice (R6/2)	Reference
Congo Red, Chrysamine G, thioflavine S, direct fast yellow	↓af	NR	NR	Heiser et al. (2000), Smith et al. (2001)
Geldanamycin	↓af	NR	NR	Sittler et al. (2001)
Chaperonin TRiC	↓af	↓af, ↑ls	NR	Behrends et al. (2006), Kitamura et al. (2006), Tam et al. (2006)
Glucocorticoid receptor	↓af	NR	NR	Diamond et al. (2000)
Creatine	↓af	NR	↓af, ↓np, ↑b, ↑ls	Ferrante et al. (2000), Andreassen et al. (2001), Smith et al. (2001), Dedeoglu et al. (2003)
Benzothiazoles	↓af	NR		Heiser et al. (2002), Hockly et al. (2006)
Intrabodies	↓af, ↓cd	↓af, ↑ls, ↓nd	NR	Colby et al. (2004), Wolfgang et al. (2005)
Gossypol, gambogic acid, juglone, celastrol, sanguinarine and anthralin	↓af	NR	NR	Wang et al. (2005)
Chaperones	↓af, ↓cd	↓af	NR	Carmichael et al. (2000, 2002), Krobitsch and Lindquist (2000), Miller et al. (2005), Novoselova et al. (2005)
Clioquinol	↓af, ↓cd	NR	↓af, ↓np, ↑b, ↑ls	Nguyen et al. (2005)

Single-stranded oligonucleotides	↓af, ↓cd	NR	NR	Parekh-Olmedo et al. (2004)
Trehalose	↓af, ↓cd	NR	↓af, ↓np, ↑b, ↑ls	Tanaka et al. (2004, 2005)
C2-8	↓af, ↓cd	↓af, ↑ls, ↓nd	NR	Zhang et al. (2005)
Ac-YVAD-cmk; levonordefrin; EGF inhibitor 3; nadolol; fosfosal	↓af	↓af, ↑ls, ↓nd	NR	Desai et al. (2006)
dHDJ1, dTPR2	↓cd	↓af, ↑ls	NR	Kazemi-Esfarjani and Benzer (2000)
CoQ	NR	NR	↓af, ↓np, ↑b, ↑ls	Ferrante et al. (2002), Stack et al. (2006), Smith et al. (2006)
Minocycline	NR	NR	↓af, ↓np, ↑b, ↑ls	Chen et al. (2000), Hersch et al. (2003), Stack et al. (2006)
Remacimide	NR	NR	↓af, ↓np, ↑b, ↑ls	Schilling et al. (2001), Ferrante et al. (2002)
Anti-oxidant BN82451	NR	NR	↓af, ↓np, ↑b, ↑ls	Klivenyi et al. (2003)
Cystamine	↓af, ↓cd	NR	↓af, ↓np, ↑b, ↑ls	Dedeoglu et al. (2002), Karpuj et al. (2002), Zainelli et al. (2005)
SUMOylation	↓af	Worsened pathology	NR	Bence et al. (2001), Steffan et al. (2004)
Ubiquitin or proteosome	↓af	Improved pathology	NR	Bence et al. (2001), Steffan et al. (2004)
Palmitoylation	↓af	NR	NR	Yanai et al. (2006)
Rapamycin and analogue CCI-779	↓af	↓nd	NR	Ravikumar et al. (2004)
Apaf-1	↓af	↓af	NR	Sang et al. (2005)
HSF	↓af	NR	↓af (in muscles), ↑ls	Fujimoto et al. (2005)

Note: ↓af: decreased aggregate formation; ↓cd: decreased cell death; NR: not reported; ↑ls: increased lifespan; ↓nd: decreased neurodegeneration; ↓np: decreased neuropathology; ↑b: increased performance on behavioral tests.

receptor antagonist), fosfosal (nonacetylated salicylic acid derivative) (Desai et al., 2006), and intrabodies (Colby et al., 2004; Wolfgang et al., 2005). Additional compounds were also tested in these models. In Drosophila, polyglutamine-dependent aggregation and toxicity in the eye was suppressed by dHDJ1, a homologue of the human heat shock protein 40/HDJ1 and by dTPR2, a homologue of the human tetratricopeptide repeat protein 2 (Kazemi-Esfarjani and Benzer, 2000). It was suggested that the suppression effect might be exerted through their chaperone-related J domains-regulating protein–protein interaction. In yeast, the expression of chaperone proteins was also shown to regulate aggregation as a deletion of the Hsp104 gene-reduced aggregate formation (Krobitsch and Lindquist, 2000).

Although these mechanistic approaches are encouraging, their applicability and involvement in the neuropathogenesis of HD and other polyglutamine diseases remains to be seen. To further understand their mode of action and potential applicability, a variety of compounds have been tested in the R6/2 mouse model of HD (see Table 11.1) and shown to decrease aggregate formation, decrease striatal atrophy, decrease weight loss, improve motor functions, and extend lifespan. These compounds include the uridine pro-drug 2′,3′,5′-tri-O-acetyluridine (PN401) (Saydoff et al., 2006), the disaccharide trehalose (Tanaka et al., 2004, 2005), clioquinol (Nguyen et al., 2005), the anti-oxidant BN82451 (Klivenyi et al., 2003), creatine (Ferrante et al., 2000; Andreassen et al., 2001; Smith et al., 2001; Dedeoglu et al., 2003), the transglutaminase inhibitor cystamine (Figure 11.5) (Dedeoglu et al., 2002; Karpuj et al., 2002), and the mitochondrial cofactor CoQ10 (Ferrante et al., 2002; Smith et al., 2006; Stack et al., 2006). Additionally, when CoQ10 was administered with the tetracycline derivative minocycline (Chen et al., 2000; Hersch et al., 2003; Stack et al., 2006) or with remacimide (Ferrante et al., 2002), the treatments were more effective than with either treatment alone. Remacimide and CoQ also improved rotarod performance in the N171-82Q mouse model of HD but did not extend survival (Schilling et al., 2001). Overexpression of yeast hsp104 reduced polyglutamine aggregation and prolonged survival of the N171-82Q transgenic mouse model of HD (Vacher et al., 2005).

Not only do a variety of drugs alter aggregate formation but posttranslational modifications of huntingtin and the activity of the proteosome also play a role. SUMOylation is the covalent attachment of SUMO (small ubiquitin-like modifier) to lysine residues. SUMOylation of huntingtin in cell culture reduced its ability to form aggregates and also increased transcriptional repression (Steffan et al., 2004). The ubiquitin-conjugating enzyme E2-25K (Hip2) is also involved in aggregation as studies show that mutants of this enzyme resulted in decreased in aggregate formation (de Pril et al., 2006). Addition of the lipid palmitate is another posttranslational modification that is important for the proper function and targeting of proteins. Studies show that an expansion of the polyQ region in huntingtin decreases palmitoylation and increases both inclusion formation and neuronal toxicity (Yanai et al., 2006). A variety of cell culture experiments have shown that inhibition of the ubiquitin or proteosome degradation pathway resulted in aggregation of huntingtin (Bence et al., 2001; Jana et al., 2001; Waelter et al., 2001). However, in a Drosophila HD model, SUMOylation made the pathology worse, whereas ubiquitination improved the pathology (Bence et al., 2001; Steffan et al., 2004). In the R6/2

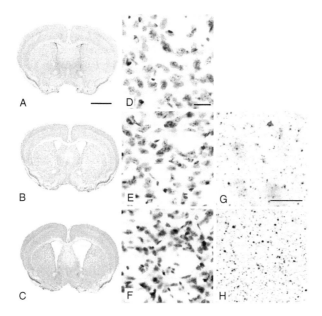

FIGURE 11.5 Neuroprotection with cystamine treatment. Photomicrographs of coronal sections through the striatum in a wild-type littermate mouse (A), cystamine-treated (B), and untreated (C) R6/2 HD transgenic mice at 90 days and corresponding Nissl-stained sections (D–F). There is gross atrophy of the brain in the untreated R6/2 mouse with enlargement of the lateral ventricles (C). In contrast, the cystamine-treated mouse (B) shows significantly less atrophy and ventricular enlargement. The neuronal size in the unsupplemented R6/2 mouse (F) is significantly smaller than that of the neurons in the cystamine-treated R6/2 mouse (E) and control (A). Sections immunostained with N-terminal huntingtin antibodies show a decrease in the load of aggregates in cystamine-treated mice (G) compared with untreated mice (H). Scale bars: (A–C) 2 mm; (D–F) 50 μm; (G and H) 100 μm. (From Dedeoglu, A., Kubilus, J.K., Jeitner, T.M., Matson, S.A., Bogdanov, M., Kowall, N.W. et al., *J. Neurosci.*, 22, 8942, 2002. With permission.)

mouse, studies did not show any impairment in proteosomal activity (Bett et al., 2006).

Alteration of other genes and proteins also affects aggregate formation. The mammalian target of rapamycin (mTOR) is sequestered in huntingtin aggregates. Rapamycin decreased aggregate formation and cell death in cellular models of HD. It also protected against neurodegeneration in a Drosophila model. Furthermore, the rapamycin analogue CCI-779 improved behavior and decreased aggregate formation in the N171-82Q mouse model (Ravikumar et al., 2004). In another study, mutant huntingtin aggregation was suppressed in Drosophila that lacked the dark gene, the human homologue of the apoptotic gene Apaf-1 (Sang et al., 2005). Huntingtin also colocalized with Apaf-1 in a mouse model of HD and in human brain tissue supporting the role for this gene in altering aggregate formation and increasing pathogenesis of the disease (Sang et al., 2005). Another factor, heat shock transcription factor 1 (HSF1), suppressed polyglutamine inclusion formation possibly by regulating expression of unknown genes, as well as major Hsps. R6/2 mice crossed with transgenic mice expressing an active HSF1 (HSF1Tg) had decreased inclusion formation in muscles and lessened weight loss. Lifespan of R6/2/HSF1Tg mice was significantly improved, although active HSF1 is not expressed in the brain (Fujimoto et al., 2005).

11.9 MECHANISMS OF CELL DEATH—THE ROLE OF AGGREGATES IN PATHOGENESIS

Connections have been hypothesized between the genetic mutation, cellular metabolic dysfunction, secondary oxidative damage, excitotoxicity, and eventual apoptosis in HD. Although there is extensive evidence for each of these mechanisms occurring from many studies in humans and experimental models (Browne et al., 1999; del Hoyo et al., 2006), it is not at all clear whether these processes contribute to the initiation of neurodegeneration or are more involved in later stages of cell death. It is possible that the toxic properties of mutant huntingtin initiate a cascade of molecular events with metabolic, oxidative, and apoptotic mechanisms coming into play only very late. Earlier steps in cellular toxicity might involve alterations in the normal function of huntingtin, alterations in its normal interactions with other molecules, or new interactions with molecules it normally does not associate with.

Insoluble aggregate formation was initially proposed to participate in the neuropathogenesis of the disease. However, a growing number of in vitro and in vivo studies suggest that aggregates might actually be part of a coping mechanism put forth by the cells to isolate the abnormal polyglutamine-containing proteins and prevent their toxic effects. For example, in striatal neuronal cultures, blocking the nuclear localization of mutant huntingtin suppresses its ability to form intranuclear inclusions and reduces cell toxicity (Saudou et al., 1998). In addition, as we have described in Section 11.8, several treatments that reduced the size and number of aggregates in vivo and in vitro did not have any quantifiable effect on survival of the cell or organism. Histone deacetylase (HDAC) inhibitors that resulted in behavioral improvement had no effect on aggregate size and number (Ferrante et al., 2003), and transglutaminase inhibitors which increased aggregate burden also decreased weight loss and mortality rate (Mastroberardino et al., 2002).

Another indication that insoluble aggregates might not be the primary cause of neuropathology comes from studies of postmortem HD brain tissue sections. In these studies, the distribution of huntingtin aggregates does not match well with neuropathology. For example, although huntingtin aggregates are present within the cerebral cortex in great numbers, they are actually quite rare in the striatum (Gutekunst et al., 1999). Furthermore, within the striatum, most of the nuclear aggregates were observed in spared NADPH-diaphorase interneurons, with few or no aggregates found within vulnerable spiny striatal neurons (Kuemmerle et al., 1999). One possibility is that aggregates are so toxic to vulnerable cell populations that degeneration occurs before aggregates are large enough to be identified at the resolution of light microscopic immunocytochemistry.

Moreover, in some of the animal models where full length mutant huntingtin is expressed, the behavioral defects precede the appearance of large aggregates

(Menalled et al., 2003; Van Raamsdonk et al., 2005), and it has been proposed that soluble intermediates of mutant huntingtin might be more toxic than insoluble aggregates. Furthermore, mutant huntingtin may undergo other modifications in addition to aggregation that could be important for its toxic effect.

11.10 CONCLUSION

This chapter has emphasized the importance of protein aggregates. The results from these studies have significant implications not only for HD but also for the understanding of other diseases that form protein aggregates such as Alzheimer's and Parkinson's disease.

The role of aggregates in HD is currently controversial. It has been shown that aggregates sequester critical proteins and prevent their normal function and thus contribute to the pathogenesis of the disease. The question of whether aggregates initiate the neurodegeneration seen in the disease or whether they may initiate a cascade of events but are not the primary cause of the disease remains unanswered. The results of drug studies on aggregate formation have created more questions. In some cases, decreasing aggregate formation has increased survival, whereas in other studies it has not had this effect. In some cases, aggregate formation may even be protective by isolating misfolded and abnormal proteins and preventing their toxic effects. In any event, the relationship between aggregate formation and pathogenesis deserves more study. Whether or not they increase or decrease survival, studies show that they do play a role. By more precisely defining their exact mechanism of action, scientists will be better able to devise therapies not only for HD but for other neurodegenerative diseases, as well.

REFERENCES

Almqvist E., Spence N., Nichol K., Andrew S.E., Vesa J., Peltonen L. et al. (1995) Ancestral differences in the distribution of the delta 2642 glutamic acid polymorphism is associated with varying CAG repeat lengths on normal chromosomes: Insights into the genetic evolution of Huntington disease. *Human Molecular Genetics*, 4, 207–214.

Almqvist E.W., Elterman D.S., MacLeod P.M. and Hayden M.R. (2001) High incidence rate and absent family histories in one quarter of patients newly diagnosed with Huntington disease in British Columbia. *Clinical Genetics*, 60, 198–205.

Andreassen O.A., Dedeoglu A., Ferrante R.J., Jenkins B.G., Ferrante K.L., Thomas M. et al. (2001) Creatine increase survival and delays motor symptoms in a transgenic animal model of Huntington's disease. *Neurobiology of Disease*, 8, 479–491.

Behrends C., Langer C.A., Boteva R., Bottcher U.M., Stemp M.J., Schaffar G. et al. (2006) Chaperonin TRiC promotes the assembly of polyQ expansion proteins into nontoxic oligomers. *Molecular Cell*, 23, 887–897.

Bence N.F., Sampat R.M. and Kopito R.R. (2001) Impairment of the ubiquitin-proteasome system by protein aggregation. *Science*, 292, 1552–1555.

Bett J.S., Goellner G.M., Woodman B., Pratt G., Rechsteiner M. and Bates G.P. (2006) Proteasome impairment does not contribute to pathogenesis in R6/2 Huntington's disease mice: Exclusion of proteasome activator REGgamma as a therapeutic target. *Human Molecular Genetics*, 15, 33–44.

Bhide P.G., Day M., Sapp E., Schwarz C., Sheth A., Kim J. et al. (1996) Expression of normal and mutant huntingtin in the developing brain. *Journal of Neuroscience*, 16, 5523–5535.

Boutell J.M., Thomas P., Neal J.W., Weston V.J., Duce J., Harper P.S. et al. (1999) Aberrant interactions of transcriptional repressor proteins with the Huntington's disease gene product, huntingtin. *Human Molecular Genetics*, 8, 1647–1655.

Brignull H.R., Morley J.F., Garcia S.M. and Morimoto R.I. (2006) Modeling polyglutamine pathogenesis in C. elegans. *Methods in Enzymology*, 412, 256–282.

Brouillet E., Hantraye P., Ferrante R.J., Dolan R., Leroy-Willig A., Kowall N.W. et al. (1995) Chronic mitochondrial energy impairment produces selective striatal degeneration and abnormal choreiform movements in primates. *Proceedings of the National Academy of Sciences of the United States of America*, 92, 7105–7109.

Browne S.E., Ferrante R.J. and Beal M.F. (1999) Oxidative stress in Huntington's disease. *Brain Pathology*, 9, 147–163.

Brune N., Andrich J., Gencik M., Saft C., Muller T., Valentin S. et al. (2004) Methyltetrahydrofolate reductase polymorphism influences onset of Huntington's disease. *Journal of Neural Transmission* (Suppl), 105–110.

Carmichael J., Chatellier J., Woolfson A., Milstein C., Fersht A.R. and Rubinsztein D.C. (2000) Bacterial and yeast chaperones reduce both aggregate formation and cell death in mammalian cell models of Huntington's disease. *Proceedings of the National Academy of Sciences of the United States of America*, 97, 9701–9705.

Carmichael J., Vacher C. and Rubinsztein D.C. (2002) The bacterial chaperonin GroEL requires GroES to reduce aggregation and cell death in a COS-7 cell model of Huntington's disease. *Neuroscience Letters*, 330, 270–274.

Cattaneo E., Zuccato C. and Tartari M. (2005) Normal huntingtin function: An alternative approach to Huntington's disease. *Nature Reviews Neuroscience*, 6, 919–930.

Chattopadhyay B., Baksi K., Mukhopadhyay S. and Bhattacharyya N.P. (2005) Modulation of age at onset of Huntington disease patients by variations in TP53 and human caspase activated DNase (hCAD) genes. *Neuroscience Letters*, 374, 81–86.

Chen M., Ona V.O., Li M., Ferrante R.J., Fink K.B., Zhu S. et al. (2000) Minocycline inhibits caspase-1 and caspase-3 expression and delays mortality in a transgenic mouse model of Huntington disease. *Nature Medicine*, 6, 797–801.

Colby D.W., Chu Y., Cassady J.P., Duennwald M., Zazulak H., Webster J.M. et al. (2004) Potent inhibition of huntingtin aggregation and cytotoxicity by a disulfide bond-free single-domain intracellular antibody. *Proceedings of the National Academy of Sciences of the United States of America*, 101, 17616–17621.

de Pril R., Fischer D.F., Roos R.A. and van Leeuwen F.W. (2006) Ubiquitin-conjugating enzyme E2-25K increases aggregate formation and cell death in polyglutamine diseases. *Molecular and Cell Neuroscience*, 34, 10–19.

Dedeoglu A., Kubilus J.K., Jeitner T.M., Matson S.A., Bogdanov M., Kowall N.W. et al. (2002) Therapeutic effects of cystamine in a murine model of Huntington's disease. *Journal of Neuroscience*, 22, 8942–8950.

Dedeoglu A., Kubilus J.K., Yang L., Ferrante K.L., Hersch S.M., Beal M.F. et al. (2003) Creatine therapy provides neuroprotection after onset of clinical symptoms in Huntington's disease transgenic mice. *Journal of Neurochemistry*, 85, 1359–1367.

del Hoyo P., Garcia-Redondo A., de Bustos F., Molina J.A., Sayed Y., Alonso-Navarro H. et al. (2006) Oxidative stress in skin fibroblasts cultures of patients with Huntington's disease. *Neurochemistry Research*, 31, 1103–1109.

Desai U.A., Pallos J., Ma A.A., Stockwell B.R., Thompson L.M., Marsh J.L. et al. (2006) Biologically active molecules that reduce polyglutamine aggregation and toxicity. *Human and Molecular Genetics*, 15, 2114–2124.

Diamond M.I., Robinson M.R. and Yamamoto K.R. (2000) Regulation of expanded poly-glutamine protein aggregation and nuclear localization by the glucocorticoid receptor. *Proceedings of the National Academy of Sciences of the United States of America*, 97, 657–661.

DiFiglia M., Sapp E., Chase K., Schwarz C., Meloni A., Young C. et al. (1995) Huntingtin is a cytoplasmic protein associated with vesicles in human and rat brain neurons. *Neuron*, 14, 1075–1081.

DiFiglia M., Sapp E., Chase K.O., Davies S.W., Bates G.P., Vonsattel J.P. et al. (1997) Aggregation of huntingtin in neuronal intranuclear inclusions and dystrophic neurites in brain. *Science*, 277, 1990–1993.

Djousse L., Knowlton B., Hayden M., Almqvist E.W., Brinkman R., Ross C. et al. (2003) Interaction of normal and expanded CAG repeat sizes influences age at onset of Huntington disease. *American Journal of Medical Genetics A*, 119, 279–282.

Djousse L., Knowlton B., Hayden M.R., Almqvist E.W., Brinkman R.R., Ross C.A. et al. (2004) Evidence for a modifier of onset age in Huntington disease linked to the HD gene in 4p16. *Neurogenetics*, 5, 109–114.

Doi H., Mitsui K., Kurosawa M., Machida Y., Kuroiwa Y. and Nukina N. (2004) Identification of ubiquitin-interacting proteins in purified polyglutamine aggregates. *FEBS Letters*, 571, 171–176.

Dragatsis I., Efstratiadis A. and Zeitlin S. (1998) Mouse mutant embryos lacking huntingtin are rescued from lethality by wild-type extraembryonic tissues. *Development*, 125, 1529–1539.

Dragatsis I., Levine M.S. and Zeitlin S. (2000) Inactivation of Hdh in the brain and testis results in progressive neurodegeneration and sterility in mice. *Nature Genetics*, 26, 300–306.

Duennwald M.L., Jagadish S., Muchowski P.J. and Lindquist S. (2006) Flanking sequences profoundly alter polyglutamine toxicity in yeast. *Proceedings of the National Academy of Sciences of the United States of America*, 103, 11045–11050.

Duyao M., Ambrose C., Myers R., Novelletto A., Persichetti F., Frontali M. et al. (1993) Trinucleotide repeat length instability and age of onset in Huntington's disease. *Nature Genetics*, 4, 387–392.

Duyao M.P., Auerbach A.B., Ryan A., Persichetti F., Barnes G.T., McNeil S.M. et al. (1995) Inactivation of the mouse Huntington's disease gene homolog Hdh. *Science*, 269, 407–410.

Engelender S., Sharp A.H., Colomer V., Tokito M.K., Lanahan A., Worley P. et al. (1997) Huntingtin-associated protein 1 (HAP1) interacts with the p150Glued subunit of dynactin. *Human Molecular Genetics*, 6, 2205–2212.

Faber P.W., Alter J.R., MacDonald M.E. and Hart A.C. (1999) Polyglutamine-mediated dysfunction and apoptotic death of a Caenorhabditis elegans sensory neuron. *Proceedings of the National Academy of Sciences of the United States of America*, 96, 179–184.

Ferrante R.J., Kowall N.W., Beal M.F., Richardson E.P., Jr., Bird E.D. and Martin J.B. (1985) Selective sparing of a class of striatal neurons in Huntington's disease. *Science*, 230, 561–563.

Ferrante R.J., Beal M.F., Kowall N.W., Richardson E.P., Jr. and Martin J.B. (1987a) Sparing of acetylcholinesterase-containing striatal neurons in Huntington's disease. *Brain Research*, 411, 162–166.

Ferrante R.J., Kowall N.W., Beal M.F., Martin J.B., Bird E.D. and Richardson E.P., Jr. (1987b) Morphologic and histochemical characteristics of a spared subset of striatal neurons in Huntington's disease. *Journal of Neuropathology and Experimental Neurology*, 46, 12–27.

Ferrante R.J., Kowall N.W. and Richardson E.P., Jr. (1991) Proliferative and degenerative changes in striatal spiny neurons in Huntington's disease: A combined study using the section-Golgi method and calbindin D28k immunocytochemistry. *Journal of Neuroscience*, 11, 3877–3887.

Ferrante R.J., Andreassen O.A., Jenkins B.G., Dedeoglu A., Kuemmerle S., Kubilus J.K. et al. (2000) Neuroprotective effects of creatine in a transgenic mouse model of Huntington's disease. *Journal of Neuroscience*, 20, 4389–4397.

Ferrante R.J., Andreassen O.A., Dedeoglu A., Ferrante K.L., Jenkins B.G., Hersch S.M. et al. (2002) Therapeutic effects of coenzyme Q10 and remacemide in transgenic mouse models of Huntington's disease. *Journal of Neuroscience*, 22, 1592–1599.

Ferrante R.J., Kubilus J.K., Lee J., Ryu H., Beesen A., Zucker B. et al. (2003) Histone deacetylase inhibition by sodium butyrate chemotherapy ameliorates the neurodegenerative phenotype in Huntington's disease mice. *Journal of Neuroscience*, 23, 9418–9427.

Forno L.S. (1992) Neuropathologic features of Parkinson's, Huntington's, and Alzheimer's diseases. *Annals of the New York Academy of Science*, 648, 6–16.

Fujimoto M., Takaki E., Hayashi T., Kitaura Y., Tanaka Y., Inouye S. et al. (2005) Active HSF1 significantly suppresses polyglutamine aggregate formation in cellular and mouse models. *Journal of Biological Chemistry*, 280, 34908–34916.

Gafni J. and Ellerby L.M. (2002) Calpain activation in Huntington's disease. *Journal of Neuroscience*, 22, 4842–4849.

Gafni J., Hermel E., Young J.E., Wellington C.L., Hayden M.R. and Ellerby L.M. (2004) Inhibition of calpain cleavage of huntingtin reduces toxicity: Accumulation of calpain/caspase fragments in the nucleus. *Journal of Biological Chemistry*, 279, 20211–20220.

Gauthier L.R., Charrin B.C., Borrell-Pages M., Dompierre J.P., Rangone H., Cordelieres F.P. et al. (2004) Huntingtin controls neurotrophic support and survival of neurons by enhancing BDNF vesicular transport along microtubules. *Cell*, 118, 127–138.

Gervais F.G., Singaraja R., Xanthoudakis S., Gutekunst C.A., Leavitt B.R., Metzler M. et al. (2002) Recruitment and activation of caspase-8 by the Huntingtin-interacting protein Hip-1 and a novel partner Hippi. *Nature Cell Biology*, 4, 95–105.

Goldberg Y.P., Nicholson D.W., Rasper D.M., Kalchman M.A., Koide H.B., Graham R.K. et al. (1996) Cleavage of huntingtin by apopain, a proapoptotic cysteine protease, is modulated by the polyglutamine tract. *Nature Genetics*, 13, 442–449.

Graveland G.A., Williams R.S. and DiFiglia M. (1985) Evidence for degenerative and regenerative changes in neostriatal spiny neurons in Huntington's disease. *Science*, 227, 770–773.

Gunawardena S., Her L.S., Brusch R.G., Laymon R.A., Niesman I.R., Gordesky-Gold B. et al. (2003) Disruption of axonal transport by loss of huntingtin or expression of pathogenic polyQ proteins in Drosophila. *Neuron*, 40, 25–40.

Gusella J.F., Wexler N.S., Conneally P.M., Naylor S.L., Anderson M.A., Tanzi R.E. et al. (1983) A polymorphic DNA marker genetically linked to Huntington's disease. *Nature*, 306, 234–238.

Gutekunst C.A., Levey A.I., Heilman C.J., Whaley W.L., Yi H., Nash N.R. et al. (1995) Identification and localization of huntingtin in brain and human lymphoblastoid cell lines with anti-fusion protein antibodies. *Proceedings of the National Academy of Sciences of the United States of America*, 92, 8710–8714.

Gutekunst C.A., Li S.H., Yi H., Ferrante R.J., Li X.J. and Hersch S.M. (1998) The cellular and subcellular localization of huntingtin-associated protein 1 (HAP1): Comparison with huntingtin in rat and human. *Journal of Neuroscience*, 18, 7674–7686.

Gutekunst C.A., Li S.H., Yi H., Mulroy J.S., Kuemmerle S., Jones R. et al. (1999) Nuclear and neuropil aggregates in Huntington's disease: Relationship to neuropathology. *Journal of Neuroscience*, 19, 2522–2534.

Hackam A.S., Yassa A.S., Singaraja R., Metzler M., Gutekunst C.A., Gan L. et al. (2000) Huntingtin interacting protein 1 induces apoptosis via a novel caspase-dependent death effector domain. *Journal of Biological Chemistry*, 275, 41299–41308.

HDCRG. (1993) A novel gene containing a trinucleotide repeat that is expanded and unstable on Huntington's disease chromosomes. The Huntington's Disease Collaborative Research Group. *Cell*, 72, 971–983.

Hedreen J.C., Peyser C.E., Folstein S.E. and Ross C.A. (1991) Neuronal loss in layers V and VI of cerebral cortex in Huntington's disease. *Neuroscience Letters*, 133, 257–261.

Heiser V., Scherzinger E., Boeddrich A., Nordhoff E., Lurz R., Schugardt N. et al. (2000) Inhibition of huntingtin fibrillogenesis by specific antibodies and small molecules: Implications for Huntington's disease therapy. *Proceedings of the National Academy of Sciences of the United States of America*, 97, 6739–6744.

Heiser V., Engemann S., Brocker W., Dunkel I., Boeddrich A., Waelter S. et al. (2002) Identification of benzothiazoles as potential polyglutamine aggregation inhibitors of Huntington's disease by using an automated filter retardation assay. *Proceedings of the National Academy of Sciences of the United States of America*, 99 (Suppl 4), 16400–16406.

Hermel E., Gafni J., Propp S.S., Leavitt B.R., Wellington C.L., Young J.E. et al. (2004) Specific caspase interactions and amplification are involved in selective neuronal vulnerability in Huntington's disease. *Cell Death and Differentiation*, 11, 424–438.

Hersch S., Fink K., Vonsattel J.P. and Friedlander R.M. (2003) Minocycline is protective in a mouse model of Huntington's disease. *Annals of Neurology*, 54, 841; author reply 842–843.

Hilditch-Maguire P., Trettel F., Passani L.A., Auerbach A., Persichetti F. and MacDonald M.E. (2000) Huntingtin: An iron-regulated protein essential for normal nuclear and perinuclear organelles. *Human Molecular Genetics*, 9, 2789–2797.

Ho L.W., Brown R., Maxwell M., Wyttenbach A. and Rubinsztein D.C. (2001) Wild type Huntingtin reduces the cellular toxicity of mutant Huntingtin in mammalian cell models of Huntington's disease. *Journal of Medical Genetics*, 38, 450–452.

Hockly E., Tse J., Barker A.L., Moolman D.L., Beunard J.L., Revington A.P. et al. (2006) Evaluation of the benzothiazole aggregation inhibitors riluzole and PGL-135 as therapeutics for Huntington's disease. *Neurobiology of Disease*, 21, 228–236.

Hodgson J.G., Agopyan N., Gutekunst C.A., Leavitt B.R., LePiane F., Singaraja R. et al. (1999) A YAC mouse model for Huntington's disease with full-length mutant huntingtin, cytoplasmic toxicity, and selective striatal neurodegeneration. *Neuron*, 23, 181–192.

Holbert S., Denghien I., Kiechle T., Rosenblatt A., Wellington C., Hayden M.R. et al. (2001) The Gln-Ala repeat transcriptional activator CA150 interacts with huntingtin: Neuropathologic and genetic evidence for a role in Huntington's disease pathogenesis. *Proceedings of the National Academy of Sciences of the United States of America*, 98, 1811–1816.

Huang C.C., Faber P.W., Persichetti F., Mittal V., Vonsattel J.P., MacDonald M.E. et al. (1998) Amyloid formation by mutant huntingtin: Threshold, progressivity and recruitment of normal polyglutamine proteins. *Somatic Cell and Molecular Genetics*, 24, 217–233.

Huang K., Yanai A., Kang R., Arstikaitis P., Singaraja R.R., Metzler M. et al. (2004) Huntingtin-interacting protein HIP14 is a palmitoyl transferase involved in palmitoylation and trafficking of multiple neuronal proteins. *Neuron*, 44, 977–986.

Jackson M., Gentleman S., Lennox G., Ward L., Gray T., Randall K. et al. (1995) The cortical neuritic pathology of Huntington's disease. *Neuropathology and Applied Neurobiology*, 21, 18–26.

Jackson G.R., Salecker I., Dong X., Yao X., Arnheim N., Faber P.W. et al. (1998) Polyglutamine-expanded human huntingtin transgenes induce degeneration of Drosophila photoreceptor neurons. *Neuron*, 21, 633–642.

Jana N.R., Tanaka M., Wang G. and Nukina N. (2000) Polyglutamine length-dependent interaction of Hsp40 and Hsp70 family chaperones with truncated N-terminal huntingtin: Their role in suppression of aggregation and cellular toxicity. *Human Molecular Genetics*, 9, 2009–2018.

Jana N.R., Zemskov E.A., Wang G. and Nukina N. (2001) Altered proteasomal function due to the expression of polyglutamine-expanded truncated N-terminal huntingtin induces apoptosis by caspase activation through mitochondrial cytochrome c release. *Human Molecular Genetics*, 10, 1049–1059.

Jeste D.V., Barban L. and Parisi J. (1984) Reduced Purkinje cell density in Huntington's disease. *Experimental Neurology*, 85, 78–86.

Karpuj M.V., Becher M.W. and Steinman L. (2002) Evidence for a role for transglutaminase in Huntington's disease and the potential therapeutic implications. *Neurochemistry International*, 40, 31–36.

Kazantsev A., Preisinger E., Dranovsky A., Goldgaber D. and Housman D. (1999) Insoluble detergent-resistant aggregates form between pathological and nonpathological lengths of polyglutamine in mammalian cells. *Proceedings of the National Academy of Sciences of the United States of America*, 96, 11404–11409.

Kazemi-Esfarjani P. and Benzer S. (2000) Genetic suppression of polyglutamine toxicity in Drosophila. *Science*, 287, 1837–1840.

Kegel K.B., Kim M., Sapp E., McIntyre C., Castano J.G., Aronin N. et al. (2000) Huntingtin expression stimulates endosomal-lysosomal activity, endosome tubulation, and autophagy. *Journal of Neuroscience*, 20, 7268–7278.

Kegel K.B., Meloni A.R., Yi Y., Kim Y.J., Doyle E., Cuiffo B.G. et al. (2002) Huntingtin is present in the nucleus, interacts with the transcriptional corepressor C-terminal binding protein, and represses transcription. *Journal of Biological Chemistry*, 277, 7466–7476.

Kim Y.J., Yi Y., Sapp E., Wang Y., Cuiffo B., Kegel K.B. et al. (2001) Caspase 3-cleaved N-terminal fragments of wild-type and mutant huntingtin are present in normal and Huntington's disease brains, associate with membranes, and undergo calpain-dependent proteolysis. *Proceedings of the National Academy of Sciences of the United States of America*, 98, 12784–12789.

Kim Y.J., Sapp E., Cuiffo B.G., Sobin L., Yoder J., Kegel K.B. et al. (2006) Lysosomal proteases are involved in generation of N-terminal huntingtin fragments. *Neurobiology of Disease*, 22, 346–356.

Kitamura A., Kubota H., Pack C.G., Matsumoto G., Hirayama S., Takahashi Y. et al. (2006) Cytosolic chaperonin prevents polyglutamine toxicity with altering the aggregation state. *Nature Cell Biology*, 8, 1163–1170.

Klivenyi P., Ferrante R.J., Gardian G., Browne S., Chabrier P.E. and Beal M.F. (2003) Increased survival and neuroprotective effects of BN82451 in a transgenic mouse model of Huntington's disease. *Journal of Neurochemistry*, 86, 267–272.

Kremer H.P., Roos R.A., Dingjan G., Marani E. and Bots G.T. (1990) Atrophy of the hypothalamic lateral tuberal nucleus in Huntington's disease. *Journal of Neuropathology and Experimental Neurology*, 49, 371–382.

Kremer H.P., Roos R.A., Dingjan G.M., Bots G.T., Bruyn G.W. and Hofman M.A. (1991) The hypothalamic lateral tuberal nucleus and the characteristics of neuronal loss in Huntington's disease. *Neuroscience Letters*, 132, 101–104.

Kretzschmar D., Tschape J., Bettencourt Da Cruz A., Asan E., Poeck B., Strauss R. et al. (2005) Glial and neuronal expression of polyglutamine proteins induce behavioral changes and aggregate formation in Drosophila. *Glia*, 49, 59–72.

Krobitsch S. and Lindquist S. (2000) Aggregation of huntingtin in yeast varies with the length of the polyglutamine expansion and the expression of chaperone proteins. *Proceedings of the National Academy of Sciences of the United States of America*, 97, 1589–1594.

Kuemmerle S., Gutekunst C.A., Klein A.M., Li X.J., Li S.H., Beal M.F. et al. (1999) Huntington aggregates may not predict neuronal death in Huntington's disease. *Annals of Neurology*, 46, 842–849.

Lange H., Thorner G., Hopf A. and Schroder K.F. (1976) Morphometric studies of the neuro-pathological changes in choreatic diseases. *Journal of Neurological Science*, 28, 401–425.

Leavitt B.R., van Raamsdonk J.M., Shehadeh J., Fernandes H., Murphy Z., Graham R.K. et al. (2006) Wild-type huntingtin protects neurons from excitotoxicity. *Journal of Neuro-chemistry*, 96, 1121–1129.

Lee W.C., Yoshihara M. and Littleton J.T. (2004) Cytoplasmic aggregates trap polyglutamine-containing proteins and block axonal transport in a Drosophila model of Huntington's disease. *Proceedings of the National Academy of Sciences of the United States of America*, 101, 3224–3229.

Levine M.S., Klapstein G.J., Koppel A., Gruen E., Cepeda C., Vargas M.E. et al. (1999) Enhanced sensitivity to N-methyl-D-aspartate receptor activation in transgenic and knockin mouse models of Huntington's disease. *Journal of Neuroscience Research*, 58, 515–532.

Li S.H. and Li X.J. (1998) Aggregation of N-terminal huntingtin is dependent on the length of its glutamine repeats. *Human and Molecular Genetics*, 7, 777–782.

Li S.H., Gutekunst C.A., Hersch S.M. and Li X.J. (1998) Interaction of huntingtin-associated protein with dynactin P150Glued. *Journal of Neuroscience*, 18, 1261–1269.

Li S.H., Cheng A.L., Zhou H., Lam S., Rao M., Li H. et al. (2002a) Interaction of Huntington disease protein with transcriptional activator Sp1. *Molecular and Cell Biology*, 22, 1277–1287.

Li Y., Chin L.S., Levey A.I. and Li L. (2002b) Huntingtin-associated protein 1 interacts with hepatocyte growth factor-regulated tyrosine kinase substrate and functions in endosomal trafficking. *Journal of Biological Chemistry*, 277, 28212–28221.

Li J.L., Hayden M.R., Almqvist E.W., Brinkman R.R., Durr A., Dode C. et al. (2003) A genome scan for modifiers of age at onset in Huntington disease: The HD MAPS study. *American Journal of Human Genetics*, 73, 682–687.

Li J.L., Hayden M.R., Warby S.C., Durr A., Morrison P.J., Nance M. et al. (2006) Genome-wide significance for a modifier of age at neurological onset in Huntington's disease at 6q23–24: The HD MAPS study. *BMC Medical Genetics*, 7, 71.

Mangiarini L., Sathasivam K., Seller M., Cozens B., Harper A., Hetherington C. et al. (1996) Exon 1 of the HD gene with an expanded CAG repeat is sufficient to cause a progressive neurological phenotype in transgenic mice. *Cell*, 87, 493–506.

Marcora E., Gowan K. and Lee J.E. (2003) Stimulation of NeuroD activity by huntingtin and huntingtin-associated proteins HAP1 and MLK2. *Proceedings of the National Academy of Sciences of the United States of America*, 100, 9578–9583.

Martindale D., Hackam A., Wieczorek A., Ellerby L., Wellington C., McCutcheon K. et al. (1998) Length of huntingtin and its polyglutamine tract influences localization and frequency of intracellular aggregates. *Nature Genetics*, 18, 150–154.

Mastroberardino P.G., Iannicola C., Nardacci R., Bernassola F., De Laurenzi V., Melino G. et al. (2002) 'Tissue' transglutaminase ablation reduces neuronal death and prolongs survival in a mouse model of Huntington's disease. *Cell Death and Differentiation*, 9, 873–880.

Matsumoto G., Kim S. and Morimoto R.I. (2006) Huntingtin and mutant SOD1 form aggregate structures with distinct molecular properties in human cells. *Journal of Biological Chemistry*, 281, 4477–4485.

McCusker E.A., Casse R.F., Graham S.J., Williams D.B. and Lazarus R. (2000) Prevalence of Huntington disease in New South Wales in 1996. *Medical Journal of Australia*, 173, 187–190.

Menalled L.B., Sison J.D., Dragatsis I., Zeitlin S. and Chesselet M.F. (2003) Time course of early motor and neuropathological anomalies in a knock-in mouse model of Huntington's disease with 140 CAG repeats. *Journal of Comparative Neurology*, 465, 11–26.

Metzler M., Chen N., Helgason C.D., Graham R.K., Nichol K., McCutcheon K. et al. (1999) Life without huntingtin: Normal differentiation into functional neurons. *Journal of Neurochemistry*, 72, 1009–1018.

Metzler M., Li B., Gan L., Georgiou J., Gutekunst C.A., Wang Y. et al. (2003) Disruption of the endocytic protein HIP1 results in neurological deficits and decreased AMPA receptor trafficking. *EMBO Journal*, 22, 3254–3266.

Miller V.M., Nelson R.F., Gouvion C.M., Williams A., Rodriguez-Lebron E., Harper S.Q. et al. (2005) CHIP suppresses polyglutamine aggregation and toxicity in vitro and in vivo. *Journal of Neuroscience*, 25, 9152–9161.

Mitsui K., Doi H. and Nukina N. (2006) Proteomics of polyglutamine aggregates. *Methods in Enzymology*, 412, 63–76.

Modregger J., DiProspero N.A., Charles V., Tagle D.A. and Plomann M. (2002) PACSIN 1 interacts with huntingtin and is absent from synaptic varicosities in presymptomatic Huntington's disease brains. *Human Molecular Genetics*, 11, 2547–2558.

Morton A.J., Nicholson L.F. and Faull R.L. (1993) Compartmental loss of NADPH diaphorase in the neuropil of the human striatum in Huntington's disease. *Neuroscience*, 53, 159–168.

Murgod U.A., Saleem Q., Anand A., Brahmachari S.K., Jain S. and Muthane U.B. (2001) A clinical study of patients with genetically confirmed Huntington's disease from India. *Journal of Neurological Science*, 190, 73–78.

Nance M.A. and Myers R.H. (2001) Juvenile onset Huntington's disease—clinical and research perspectives. *Mental Retardation and Developmental Disabilities Research Reviews*, 7, 153–157.

Nasir J., Floresco S.B., O'Kusky J.R., Diewert V.M., Richman J.M., Zeisler J. et al. (1995) Targeted disruption of the Huntington's disease gene results in embryonic lethality and behavioral and morphological changes in heterozygotes. *Cell*, 81, 811–823.

Nguyen T., Hamby A. and Massa S.M. (2005) Clioquinol down-regulates mutant huntingtin expression in vitro and mitigates pathology in a Huntington's disease mouse model. *Proceedings of the National Academy of Sciences of the United States of America*, 102, 11840–11845.

Novelletto A., Persichetti F., Sabbadini G., Mandich P., Bellone E., Ajmar F. et al. (1994) Analysis of the trinucleotide repeat expansion in Italian families affected with Huntington disease. *Human Molecular Genetics*, 3, 93–98.

Novoselova T.V., Margulis B.A., Novoselov S.S., Sapozhnikov A.M., van der Spuy J., Cheetham M.E. et al. (2005) Treatment with extracellular HSP70/HSC70 protein can reduce polyglutamine toxicity and aggregation. *Journal of Neurochemistry*, 94, 597–606.

Okun M.S. and Thommi N. (2004) Americo Negrette (1924 to 2003): Diagnosing Huntington disease in Venezuela. *Neurology*, 63, 340–343.

Pal A., Severin F., Lommer B., Shevchenko A. and Zerial M. (2006) Huntingtin-HAP40 complex is a novel Rab5 effector that regulates early endosome motility and is up-regulated in Huntington's disease. *Journal of Cell Biology*, 172, 605–618.

Parekh-Olmedo H., Wang J., Gusella J.F. and Kmiec E.B. (2004) Modified single-stranded oligonucleotides inhibit aggregate formation and toxicity induced by expanded poly-glutamine. *Journal of Molecular Neuroscience*, 24, 257–267.

Parker J.A., Connolly J.B., Wellington C., Hayden M., Dausset J. and Neri C. (2001) Expanded polyglutamines in Caenorhabditis elegans cause axonal abnormalities and severe dysfunction of PLM mechanosensory neurons without cell death. *Proceedings of the National Academy of Sciences of the United States of America*, 98, 13318–13323.

Passani L.A., Bedford M.T., Faber P.W., McGinnis K.M., Sharp A.H., Gusella J.F. et al. (2000) Huntingtin's WW domain partners in Huntington's disease post-mortem brain fulfill genetic criteria for direct involvement in Huntington's disease pathogenesis. *Human and Molecular Genetics*, 9, 2175–2182.

Perutz M.F., Johnson T., Suzuki M. and Finch J.T. (1994) Glutamine repeats as polar zippers: Their possible role in inherited neurodegenerative diseases. *Proceedings of the National Academy of Sciences of the United States of America*, 91, 5355–5358.

Peters M.F. and Ross C.A. (2001) Isolation of a 40-kDa Huntingtin-associated protein. *Journal of Biological Chemistry*, 276, 3188–3194.

Petersen A., Gil J., Maat-Schieman M.L., Bjorkqvist M., Tanila H., Araujo I.M. et al. (2005) Orexin loss in Huntington's disease. *Human and Molecular Genetics*, 14, 39–47.

Ranen N.G., Stine O.C., Abbott M.H., Sherr M., Codori A.M., Franz M.L. et al. (1995) Anticipation and instability of IT-15 (CAG)n repeats in parent-offspring pairs with Huntington disease. *American Journal of Human Genetics*, 57, 593–602.

Ravikumar B., Vacher C., Berger Z., Davies J.E., Luo S., Oroz L.G. et al. (2004) Inhibition of mTOR induces autophagy and reduces toxicity of polyglutamine expansions in fly and mouse models of Huntington disease. *Nature Genetics*, 36, 585–595.

Reddy P.H., Williams M., Charles V., Garrett L., Pike-Buchanan L., Whetsell W.O., Jr. et al. (1998) Behavioural abnormalities and selective neuronal loss in HD transgenic mice expressing mutated full-length HD cDNA. *Nature Genetics*, 20, 198–202.

Reiner A., Del Mar N., Meade C.A., Yang H., Dragatsis I., Zeitlin S. et al. (2001) Neurons lacking huntingtin differentially colonize brain and survive in chimeric mice. *Journal of Neuroscience*, 21, 7608–7619.

Richfield E.K. and Herkenham M. (1994) Selective vulnerability in Huntington's disease: Preferential loss of cannabinoid receptors in lateral globus pallidus. *Annals of Neurology*, 36, 577–584.

Rigamonti D., Bauer J.H., De-Fraja C., Conti L., Sipione S., Sciorati C. et al. (2000) Wild-type huntingtin protects from apoptosis upstream of caspase-3. *Journal of Neuroscience*, 20, 3705–3713.

Rigamonti D., Sipione S., Goffredo D., Zuccato C., Fossale E. and Cattaneo E. (2001) Huntingtin's neuroprotective activity occurs via inhibition of procaspase-9 processing. *Journal of Biological Chemistry*, 276, 14545–14548.

Rosas H.D., Liu A.K., Hersch S., Glessner M., Ferrante R.J., Salat D.H. et al. (2002) Regional and progressive thinning of the cortical ribbon in Huntington's disease. *Neurology*, 58, 695–701.

Saleem Q., Roy S., Murgood U., Saxena R., Verma I.C., Anand A. et al. (2003) Molecular analysis of Huntington's disease and linked polymorphisms in the Indian population. *Acta Neurologica Scandinavica*, 108, 281–286.

Sang T.K., Li C., Liu W., Rodriguez A., Abrams J.M., Zipursky S.L. et al. (2005) Inactivation of Drosophila Apaf-1 related killer suppresses formation of polyglutamine aggregates and blocks polyglutamine pathogenesis. *Human Molecular Genetics*, 14, 357–372.

Sapp E., Ge P., Aizawa H., Bird E., Penney J., Young A.B. et al. (1995) Evidence for a preferential loss of enkephalin immunoreactivity in the external globus pallidus in low grade Huntington's disease using high resolution image analysis. *Neuroscience*, 64, 397–404.

Sapp E., Schwarz C., Chase K., Bhide P.G., Young A.B., Penney J. et al. (1997) Huntingtin localization in brains of normal and Huntington's disease patients. *Annals of Neurology*, 42, 604–612.

Satyal S.H., Schmidt E., Kitagawa K., Sondheimer N., Lindquist S., Kramer J.M. et al. (2000) Polyglutamine aggregates alter protein folding homeostasis in Caenorhabditis elegans. *Proceedings of the National Academy of Sciences of the United States of America*, 97, 5750–5755.

Saudou F., Finkbeiner S., Devys D. and Greenberg M.E. (1998) Huntingtin acts in the nucleus to induce apoptosis but death does not correlate with the formation of intranuclear inclusions. *Cell*, 95, 55–66.

Saydoff J.A., Garcia R.A., Browne S.E., Liu L., Sheng J., Brenneman D. et al. (2006) Oral uridine pro-drug PN401 is neuroprotective in the R6/2 and N171-82Q mouse models of Huntington's disease. *Neurobiology of Disease*, 24, 455–465.

Schilling G., Becher M.W., Sharp A.H., Jinnah H.A., Duan K., Kotzuk J.A. et al. (1999) Intranuclear inclusions and neuritic aggregates in transgenic mice expressing a mutant N-terminal fragment of huntingtin. *Human Molecular Genetics*, 8, 397–407.

Schilling G., Coonfield M.L., Ross C.A. and Borchelt D.R. (2001) Coenzyme Q10 and remacemide hydrochloride ameliorate motor deficits in a Huntington's disease transgenic mouse model. *Neuroscience Letters*, 315, 149–153.

Senut M.C., Suhr S.T., Kaspar B. and Gage F.H. (2000) Intraneuronal aggregate formation and cell death after viral expression of expanded polyglutamine tracts in the adult rat brain. *Journal of Neuroscience*, 20, 219–229.

Shelbourne P.F., Killeen N., Hevner R.F., Johnston H.M., Tecott L., Lewandoski M. et al. (1999) A Huntington's disease CAG expansion at the murine Hdh locus is unstable and associated with behavioural abnormalities in mice. *Human Molecular Genetics*, 8, 763–774.

Singaraja R.R., Hadano S., Metzler M., Givan S., Wellington C.L., Warby S. et al. (2002) HIP14, a novel ankyrin domain-containing protein, links huntingtin to intracellular trafficking and endocytosis. *Human Molecular Genetics*, 11, 2815–2828.

Sittler A., Lurz R., Lueder G., Priller J., Lehrach H., Hayer-Hartl M.K. et al. (2001) Geldanamycin activates a heat shock response and inhibits huntingtin aggregation in a cell culture model of Huntington's disease. *Human Molecular Genetics*, 10, 1307–1315.

Smith D.L., Portier R., Woodman B., Hockly E., Mahal A., Klunk W.E. et al. (2001) Inhibition of polyglutamine aggregation in R6/2 HD brain slices-complex dose-response profiles. *Neurobiology of Disease*, 8, 1017–1026.

Smith K.M., Matson S., Matson W.R., Cormier K., Del Signore S.J., Hagerty S.W. et al. (2006) Dose ranging and efficacy study of high-dose coenzyme Q10 formulations in Huntington's disease mice. *Biochimica et Biophysica Acta*, 1762, 616–626.

Spargo E., Everall I.P. and Lantos P.L. (1993) Neuronal loss in the hippocampus in Huntington's disease: A comparison with HIV infection. *Journal of Neurology, Neurosurgery and Psychiatry*, 56, 487–491.

Stack E.C., Smith K.M., Ryu H., Cormier K., Chen M., Hagerty S.W. et al. (2006) Combination therapy using minocycline and coenzyme Q10 in R6/2 transgenic Huntington's disease mice. *Biochimica et Biophysica Acta*, 1762, 373–380.

Steffan J.S., Bodai L., Pallos J., Poelman M., McCampbell A., Apostol B.L. et al. (2001) Histone deacetylase inhibitors arrest polyglutamine-dependent neurodegeneration in Drosophila. *Nature*, 413, 739–743.

Steffan J.S., Agrawal N., Pallos J., Rockabrand E., Trotman L.C., Slepko N. et al. (2004) SUMO modification of Huntingtin and Huntington's disease pathology. *Science*, 304, 100–104.

Suhr S.T., Senut M.C., Whitelegge J.P., Faull K.F., Cuizon D.B. and Gage F.H. (2001) Identities of sequestered proteins in aggregates from cells with induced polyglutamine expression. *Journal of Cell Biology*, 153, 283–294.

Tam S., Geller R., Spiess C. and Frydman J. (2006) The chaperonin TRiC controls polyglutamine aggregation and toxicity through subunit-specific interactions. *Nature Cell Biology*, 8, 1155–1162.

Tanaka M., Machida Y., Niu S., Ikeda T., Jana N.R., Doi H. et al. (2004) Trehalose alleviates polyglutamine-mediated pathology in a mouse model of Huntington disease. *Nature Medicine*, 10, 148–154.

Tanaka M., Machida Y. and Nukina N. (2005) A novel therapeutic strategy for polyglutamine diseases by stabilizing aggregation-prone proteins with small molecules. *Journal of Molecular Medicine*, 83, 343–352.

Tang T.S., Tu H., Chan E.Y., Maximov A., Wang Z., Wellington C.L. et al. (2003) Huntingtin and huntingtin-associated protein 1 influence neuronal calcium signaling mediated by inositol-(1,4,5) triphosphate receptor type 1. *Neuron*, 39, 227–239.

Tang T.S., Tu H., Orban P.C., Chan E.Y., Hayden M.R. and Bezprozvanny I. (2004) HAP1 facilitates effects of mutant huntingtin on inositol 1,4,5-trisphosphate-induced Ca release in primary culture of striatal medium spiny neurons. *European Journal of Neuroscience*, 20, 1779–1787.

Toneff T., Mende-Mueller L., Wu Y., Hwang S.R., Bundey R., Thompson L.M. et al. (2002) Comparison of huntingtin proteolytic fragments in human lymphoblast cell lines and human brain. *Journal of Neurochemistry*, 82, 84–92.

Trottier Y., Devys D., Imbert G., Saudou F., An I., Lutz Y. et al. (1995) Cellular localization of the Huntington's disease protein and discrimination of the normal and mutated form. *Nature Genetics*, 10, 104–110.

Trushina E., Dyer R.B., Badger J.D., 2nd, Ure D., Eide L., Tran D.D. et al. (2004) Mutant huntingtin impairs axonal trafficking in mammalian neurons in vivo and in vitro. *Molecular Cell Biology*, 24, 8195–8209.

Tumas V., Camargos S.T., Jalali P.S., Galesso Ade P. and Marques W., Jr. (2004) Internal consistency of a Brazilian version of the unified Huntington's disease rating scale. *Arquivos de Neuro-psiquiatria*, 62, 977–982.

Vacher C., Garcia-Oroz L. and Rubinsztein D.C. (2005) Overexpression of yeast hsp104 reduces polyglutamine aggregation and prolongs survival of a transgenic mouse model of Huntington's disease. *Human Molecular Genetics*, 14, 3425–3433.

Van Raamsdonk J.M., Pearson J., Slow E.J., Hossain S.M., Leavitt B.R. and Hayden M.R. (2005) Cognitive dysfunction precedes neuropathology and motor abnormalities in the YAC128 mouse model of Huntington's disease. *Journal of Neuroscience*, 25, 4169–4180.

van Roon-Mom W.M., Hogg V.M., Tippett L.J. and Faull R.L. (2006) Aggregate distribution in frontal and motor cortex in Huntington's disease brain. *Neuroreport*, 17, 667–670.

Velier J., Kim M., Schwarz C., Kim T.W., Sapp E., Chase K. et al. (1998) Wild-type and mutant huntingtins function in vesicle trafficking in the secretory and endocytic pathways. *Experimental Neurology*, 152, 34–40.

von Horsten S., Schmitt I., Nguyen H.P., Holzmann C., Schmidt T., Walther T. et al. (2003) Transgenic rat model of Huntington's disease. *Human and Molecular Genetics*, 12, 617–624.

Vonsattel J.P., Myers R.H., Stevens T.J., Ferrante R.J., Bird E.D. and Richardson E.P., Jr. (1985) Neuropathological classification of Huntington's disease. *Journal of Neuropathology and Experimental Neurology*, 44, 559–577.

Waelter S., Boeddrich A., Lurz R., Scherzinger E., Lueder G., Lehrach H. et al. (2001) Accumulation of mutant huntingtin fragments in aggresome-like inclusion bodies as a result of insufficient protein degradation. *Molecular Biology of the Cell*, 12, 1393–1407.

Wang J., Gines S., MacDonald M.E. and Gusella J.F. (2005) Reversal of a full-length mutant huntingtin neuronal cell phenotype by chemical inhibitors of polyglutamine-mediated aggregation. *BMC Neuroscience*, 6, 1.

Wellington C.L., Ellerby L.M., Hackam A.S., Margolis R.L., Trifiro M.A., Singaraja R. et al. (1998) Caspase cleavage of gene products associated with triplet expansion disorders generates truncated fragments containing the polyglutamine tract. *Journal of Biological Chemistry*, 273, 9158–9167.

Wexler N.S., Lorimer J., Porter J., Gomez F., Moskowitz C., Shackell E. et al. (2004) Venezuelan kindreds reveal that genetic and environmental factors modulate Huntington's disease age of onset. *Proceedings of the National Academy of Sciences of the United States of America*, 101, 3498–3503.

Wheeler V.C., Auerbach W., White J.K., Srinidhi J., Auerbach A., Ryan A. et al. (1999) Length-dependent gametic CAG repeat instability in the Huntington's disease knock-in mouse. *Human and Molecular Genetics*, 8, 115–122.

Wheeler V.C., White J.K., Gutekunst C.A., Vrbanac V., Weaver M., Li X.J. et al. (2000) Long glutamine tracts cause nuclear localization of a novel form of huntingtin in medium spiny striatal neurons in HdhQ92 and HdhQ111 knock-in mice. *Human and Molecular Genetics*, 9, 503–513.

White J.K., Auerbach W., Duyao M.P., Vonsattel J.P., Gusella J.F., Joyner A.L. et al. (1997) Huntingtin is required for neurogenesis and is not impaired by the Huntington's disease CAG expansion. *Nature Genetics*, 17, 404–410.

Wolfgang W.J., Miller T.W., Webster J.M., Huston J.S., Thompson L.M., Marsh J.L. et al. (2005) Suppression of Huntington's disease pathology in Drosophila by human single-chain Fv antibodies. *Proceedings of the National Academy of Sciences of the United States of America*, 102, 11563–11568.

Yanai A., Huang K., Kang R., Singaraja R.R., Arstikaitis P., Gan L. et al. (2006) Palmitoylation of huntingtin by HIP14 is essential for its trafficking and function. *Nature Neuroscience*, 9, 824–831.

Zainelli G.M., Dudek N.L., Ross C.A., Kim S.Y. and Muma N.A. (2005) Mutant huntingtin protein: A substrate for transglutaminase 1, 2, and 3. *Journal of Neuropathology and Experimental Neurology*, 64, 58–65.

Zeitlin S., Liu J.P., Chapman D.L., Papaioannou V.E. and Efstratiadis A. (1995) Increased apoptosis and early embryonic lethality in mice nullizygous for the Huntington's disease gene homologue. *Nature Genetics*, 11, 155–163.

Zhang Y., Li M., Drozda M., Chen M., Ren S., Mejia Sanchez R.O. et al. (2003) Depletion of wild-type huntingtin in mouse models of neurologic diseases. *Journal of Neurochemistry*, 87, 101–106.

Zhang X., Smith D.L., Meriin A.B., Engemann S., Russel D.E., Roark M. et al. (2005) A potent small molecule inhibits polyglutamine aggregation in Huntington's disease neurons and suppresses neurodegeneration in vivo. *Proceedings of the National Academy of Sciences of the United States of America*, 102, 892–897.

Zuccato C., Ciammola A., Rigamonti D., Leavitt B.R., Goffredo D., Conti L. et al. (2001) Loss of huntingtin-mediated BDNF gene transcription in Huntington's disease. *Science*, 293, 493–498.

Zuccato C., Tartari M., Crotti A., Goffredo D., Valenza M., Conti L. et al. (2003) Huntingtin interacts with REST/NRSF to modulate the transcription of NRSE-controlled neuronal genes. *Nature Genetics*, 35, 76–83.

Part IV

Amyotrophic Lateral Sclerosis

12 Amyotrophic Lateral Sclerosis (Motor Neuron Disease)

Teresa Sanelli, Janice Robertson,
Avi Chakabartty, and Michael J. Strong

CONTENTS

12.1 INTRODUCTION

Amyotrophic lateral sclerosis (ALS) is a progressive adult-onset neurodegenerative disorder affecting primarily motor neurons of the brain and spinal cord, which accounts for approximately 10% of adult-onset motor neuron disorders (Strong, 2003). For the majority of patients, ALS is relentlessly progressive, with the median survival from symptom onset ranging on average from 3 to 5 years (Strong, 2004). In approximately 20% of patients, longer term survival may occur.

While the majority of ALS cases are sporadic (sALS), 5%–10% are familial (fALS). Although ALS is a disease of aging in which peak incidence rates occur in the sixth and seventh decades of life, disease onset in young adults (third and fourth decades) occurs not infrequently and may be associated with a better survival. In the pre-menopausal years, males exceed females in a 2:1 ratio, becoming 1:1 thereafter. Current prevalence rates for ALS are relatively uniform worldwide and in the order of 8–10/100,000.

ALS commonly manifests first with limb involvement, although a third will have a bulbar symptom onset in which speech and swallowing difficulties are prominent. The clinical features of ALS involve progressive signs and symptoms of lower motor neuron dysfunction that include muscle atrophy, weakness, cramps, and fasciculations (muscle twitching). These occur in association with signs of upper motor neuron (typically the corticospinal tracts) dysfunction, including spasticity and pathologically hyperactive deep tendon reflexes. For the most part, these features occur in the absence of clinically significant sensory deficits (Cudkowicz et al., 2004). As the disease progresses, patients become paralyzed because of severe, diffuse muscle atrophy. In many, the involvement of muscle groups innervated by the lower brainstem gives rise to severe impairments in swallowing, resulting in the need for enteral nutrition. Death most commonly results from respiratory failure.

Although ALS is traditionally considered to be a selective motor system disorder, it is increasingly evident that the disease process can include nonmotor neuronal degeneration (Ince et al., 1998). Included in this is a spectrum of cognitive and behavioral deficits reflective of frontotemporal lobar degeneration (FLTD). These syndromes can include a cognitive disorder, behavioral disorder, dysexecutive syndrome or a true frontotemporal dementia (FTD) as defined by the Neary criteria (Strong et al., 1996; Neary et al., 1998; Wilson et al., 2001; Strong, 2003).

12.2 CLINICAL PHENOTYPES

12.2.1 Familial Amyotrophic Lateral Sclerosis

12.2.1.1 fALS Associated with Copper/Zinc Superoxide Dismutase (SOD1) Mutations

Mutations of the copper/zinc superoxide dismutase (SOD1) gene on chromosome 21 at position q22 are associated with approximately 15% of all fALS (Table 12.1) cases

TABLE 12.1

Genetics of Amyotrophic Lateral Sclerosis

ALS Variants	Inheritance Pattern	Linkage	Gene Defect	Age at Onset	Clinical Features
ALS 1	AD	21q22.1	SOD1	Predominantly adult	Classic ALS phenotype
ALS 2	AR	2q33-q35	ALSin	Juvenile	First or second decade of life; spastic pseudobulbar syndrome with spastic paraplegia; distal amyotrophy; slow progression
ALS 3	AD	18q21	Unknown	Adult	Classic ALS phenotype; non-SOD1-linked
ALS 4	AD	9q34	Senataxin	Juvenile	Onset typically in second decade, slow progression with normal life span; distal weakness, and amyotrophy; severe loss of motor neurons in brainstem and spinal cord
ALS 5	AR	15q15.1-q21.1	Unknown	Juvenile	Absence of pseudobulbar features; distal amyotrophy; minor spasticity; long-term survival
ALS 6	AD	16q12.1-q12.2	Unknown	Adult	Majority with short duration; rarely FTD
ALS 7	AD	20ptel	Unknown	Adult	Linkage in only two family members; similar phenotype to chromosome 16 linked
ALS 8	AD	20q13.33	VAPB	Adult	Slowly progressive disorder associated with fasciculations, cramps, and postural tremor
FTD-MND	AD	9q21-a22	Unknown	Adult	ALS with dementia with features of frontotemporal degeneration
FTDP-17	AD	17q21	Tau	Adult	Disinhibition-dementia-parkinsonism-amyotrophy syndrome
ALS-FTD "San Francisco family A"	AD	17	Not linked to Tau	Adult	Prominent tau positive and alpha-synuclein positive inclusions
ALS with bulbar onset		Unknown		Juvenile	Prominent early-onset bulbar dysfunction; slow progression; dementia

Source: Adapted from Strong, M.J., McLean, J. and Julien, J.-P., *Amyotrophic Lateral Sclerosis*, H. Mitsumoto, S. Przedborski, P. Gordon and M. De Bene, eds, Marcel Dekker Inc., London, 2005b.

FTD (frontotemporal dementia); SOD1 (copper/zinc SOD1); VAPB (vesicle-associated membrane protein B).

(Rosen et al., 1993; Kato et al., 2000; Cudkowicz et al., 2004). Although the most common involves a missense mutation of alanine for valine in codon 4 (A4V) (Ince et al., 1998), over 110 mutations are now recognized affecting 60% of the amino acids distributed throughout the protein (for an updated listing, see http://www.alsod.org) (Andersen, 2006). For the majority, inheritance is in an autosomal dominant pattern although autosomal recessive variants have been identified (i.e., the D90A mutation) (Andersen et al., 1997). There are differences in phenotypes that result from various SOD1 mutations, predominantly with regard to survival times. For example, the A4V mutation results in shorter survival, whereas G37R, G41D, and G93C mutations result in longer survival (Cudkowicz et al., 1997; Juneja et al., 1997).

The SOD enzymes catalyze the disproportionation reaction where two molecules of superoxide anion (O_2^-) are converted into O_2 and H_2O_2. Three isoforms of SOD have been identified: Copper/Zinc-binding homodimeric cytosolic SOD1 (32 kDa), Mn-dependent homotetrameric mitochondrial SOD2 (88 kDa), and Copper/Zinc-binding homotetrameric extracellular SOD3 (135 kDa). Each is encoded by an individual gene on different chromosomes. SOD2 is located on 6q21 and SOD3 is located on 4pter-q21 (Shibata et al., 2000a). SOD1 is encoded by a gene located at chromosome 21q22.1. In homodimeric SOD1, each subunit consists of an eight-stranded β-barrel and two large loops that are denoted the electrostatic loop and the metal-binding loop (Weisiger and Fridovich, 1973; Fridovich, 1986).

The mechanism(s) by which SOD1 mutations give rise to motor neuron degeneration is controversial (reviewed in detail below). Although some SOD1 mutations are associated with decreased SOD1 activity, this is not the case for the majority (Deng et al., 1993; Borchelt et al., 1994). One proposed mechanism by which mutant SOD1 (mtSOD1) is believed to acquire a toxic gain of function is through aberrant binding of substrates leading to the formation of toxic species such as peroxynitrite that produce oxidative neuronal injury, by either nitrating or oxidizing critical proteins (Beckman et al., 1993, 2001). Another postulates that a very high propensity for aggregation is the root cause of the toxic gain-of-function exhibited by the SOD1 mutants (Bruijn et al., 1998). The conundrum here is that over 110 different mutations spread throughout the molecule, all result in the same disease. As will be reviewed in Section 12.2, the commonality of all of the SOD1 mutants is their propensity to aggregate.

12.2.1.2 fALS Not Associated with SOD1 Mutations

In addition to SOD1 mutations, ALS-associated gene linkages have been found on chromosome 2q33 (alsin; ALS type 2) (Hadano et al., 2001; Yang et al., 2001), chromosome 15q15–q21 (ALS 5 variant) (Hentati et al., 1998), chromosome 9q34 (senataxin: ALS 4) (Chance et al., 1998), and chromosome 20q13.3 (vesicle-associated membrane protein B (VAPB); ALS8) (Nishimura et al., 2004). For the majority of inherited variants of ALS, the affected genes have yet to be identified.

Although the function of alsin is not fully understood, mutations in the *Alsin* gene are associated with a juvenile autosomal recessive disorder with slow progression. Several functions have been proposed based on its structure and the domains found within the area spanning chromosome 2q33–q35 (Hadano et al., 2001; Yang et al., 2001). The long form of alsin (alsinLF) contains three guanine nucleotide exchange factor (GEF) domains consisting of a regulator of chromatin condensation 1 (RCC1), a Rho guanine nucleotide exchange factor (RhoGEF), and vacuolar protein sorting 9 (VPS9) domains. These findings suggest that alsin is possibly involved in signal transduction, membrane transport, and cytoskeletal organization. It is noteworthy that alsinLF has been shown to be neuroprotective against mtSOD1-mediated cell death in a motor neuron hybridoma cell line (Kanekura et al., 2005). Alternative splicing of alsin produces a short transcript. When a deletion occurs in both transcripts, the result is ALS2. However, when homozygous deletions occur, which affect only the shorter transcript, the result is juvenile onset slowly progressive spastic paraparesis (juvenile primary lateral sclerosis) (Majoor-Krakauer et al., 2003).

12.2.1.3 Genetic Association with sALS

Although the majority of mutations in SOD1 are associated with *f*ALS, they have also been observed in apparently *s*ALS (Jackson et al., 1997; Shaw et al., 1998), with several authors suggesting that upward of 12% of patients with sALS may harbor such mutations (Orrell et al., 1997; Majoor-Krakauer et al., 2003).

A number of susceptibility genes have been identified that may increase the risk of *s*ALS. These include coding deletions or insertions of the KSP repeat region of the NFH gene (Figlewicz et al., 1994; Tomkins et al., 1998; Al-Chalabi et al., 1999), microdeletions of mitochondrial DNA encoding cytochrome *c* oxidase (Comi et al., 1998), RNA processing errors in the glutamate transporter EAAT2 (Aoki et al., 1998; Jackson et al., 1999), an abnormal copy number of the motor neuron survival gene SMN or decreased protein levels of SMN (Corcia et al., 2002; Veldink et al., 2005), gene deletions of the chromosome 5q13-linked neuronal apoptosis inhibitory protein (NAIP) gene (Jackson et al., 1996), and frameshift mutations in peripherin (Gros-Louis et al., 2004; Leung et al., 2004). There is also an increased frequency of the cytochrome P450 debrisoquine hydroxylase CYP2D6(B) allele associated with a "poor drug metabolizer" phenotype in *s*ALS (Siddons et al., 1996).

The *ApoE* gene may also be a determinant of disease phenotype in that the presence of an ε3/ε4 genotype is associated with bulbar symptom onset of *s*ALS (Al-Chalabi et al., 1996), whereas the ε2/ε3 genotype is associated with a limb-onset variant (Lacomblez et al., 2002). The expression of the *ε4* allele may be associated with an earlier onset of disease, although there are conflicting reports (Mui et al., 1995; Drory et al., 2001).

Similarly, vascular endothelial growth factor (VEGF) haplotype may be associated with specific ALS phenotypes, although this observation remains controversial (Lambrechts et al., 2003; Van Den Bosch et al., 2004; Van Vught et al., 2005). When administered intracerebroventricularly in the SOD1 mutation rat model of ALS, a significant improvement in survival was observed (Storkebaum et al., 2004). Although the mechanism by which VEGF-mediates neuroprotection is unknown, it may have a role in reduction and prevention of ischemic motor neuron death (Lambrechts et al., 2003).

12.2.2 WESTERN PACIFIC VARIANT OF ALS

A hyperendemic focus of ALS has been described in the western Pacific, including the Chamorro population of Guam and Rota, and the Kii Peninsula. Here, prevalence rates have been as high as 100/100,000, but have declined in the last 20 years to 25–50/100,000 (Wiederholt, 1999). This variant of ALS can occur either in isolation or in combination with Parkinsonism or dementia, or as a Parkinsonism–dementia complex (PDC) (Elizan et al., 1986). Unique to this variant of ALS, intraneuronal-paired helical filaments (PHFs) containing hyperphosphorylated microtubule-associated tau protein, are observed (Esclaire et al., 1999). Although the etiology of the Western Pacific variant of ALS remains elusive, contemporary theories have focused on the role of environmental neurotoxins in a genetically susceptible population (Esclaire et al., 1999; Armon, 2003).

12.2.3 ALS WITH FRONTOTEMPORAL LOBAR DEGENERATION

Although traditionally ALS is considered to be a selective motor system disorder, increasingly it is clear that the disease process of ALS can include nonmotor neuronal degeneration (Ince et al., 1998; Ince and Morris, 2006). Among this is a FTLD marked by both frontal and temporal lobar atrophy, vacuolization of cortical layers II and III (superficial linear spongiosus) and neuronal loss (Wilson et al., 2001). The pathognomic lesion is believed to be the presence of ubiquitin-immunoreactive, tau- and α-synuclein negative intraneuronal inclusions in the dentate gyrus, entorhinal cortex, and layers two and three of the frontal cortex. In addition, a number of authors have observed tau-immunoreactive intraneuronal and glial aggregates in cognitively impaired or demented ALS patients (Yang et al., 2003).

The observation of tau protein aggregation in ALS suggests that, for at least a population of ALS patients, there is overlap with disorders of tau metabolism (tauopathy). The majority of the degenerative tauopathies present either as a FTLD or as Alzheimer's disease. Tau deposition in these disorders ranges from discrete intraneuronal aggregates to neurofibrillary tangles (NFTs), glial (both astrocytic and oligodendroglial) and extraneuronal (neuropil threads, argentophilic granules).

The potential for an FTLD to be associated with amyotrophy has been highlighted by the occurrence of an FTLD with Parkinsonism (FTDP) (Hutton et al., 1998). A significant proportion of families with FTDP harbor mutations in the microtubule-associated tau protein gene on chromosome 17 (and are thus termed FTDP-17). Interestingly, non-tau linked FTDP-17 has also been recently described. An additional FTLD/ALS locus has been linked to chromosome 9, although tau protein aggregation is not a major component of this variant (Morita et al., 2006; Vance et al., 2006).

12.3 PATHOBIOLOGY OF ALS

ALS is a syndromic, multisystems disorder (Strong, 2001) and thus it is not surprising that at death multiple metabolic derangements are evident. Whether all of these processes are key to the initial pathogenesis of ALS is less clear. Among the many neurochemical abnormalities identified to date in ALS, key pathogenic mechanisms include reactive oxygen species (ROS)/free radical-induced toxicity, intermediate filament (IF) disorganization (including neurofilaments and peripherin), excitotoxicity, protein misfolding and protein aggregation.

The evidence of ROS involvement in ALS has included the observation that protein carbonyl levels (protein adducts of reactive aldehydes as a marker of oxidative damage to proteins) in ALS spinal cords are twice that of normal and neurological disease controls (Ince et al., 1998). Both peroxynitrite and nitrotyrosine, the end product of reactive nitrating species formation, are observed within spinal motor neurons in ALS (Chou et al., 1996a).

As will be discussed, the stoichiometry of NF protein expression is disturbed in degenerating motor neurons in ALS. This is of importance in that transgenic mice in which NF protein expression is altered through either the knocking out or over-expression of one or more members of the IF proteins, including NF, disrupts motor neuron function (Julien, 1995; Julien et al., 1995; Strong et al., 2003). This will be reviewed in detail subsequently.

Excitotoxicity refers to toxicity caused by an excitatory amino acid neurotransmitter, such as glutamate or N-methyl-D-aspartate (NMDA), and results when excessive concentrations of excitatory neurotransmitters are present in the synaptic cleft (Li et al., 1998; Tymianski et al., 1998; Kovacs et al., 2001). The neurodegeneration in ALS has been attributed to this mechanism in a number of studies (Delfs et al., 1997; Hayes et al., 2000; Saroff et al., 2000; Sinor et al., 2000; Laslo et al., 2001). Of note, mutations in the glutamate transporter have been observed in a significant proportion of ALS patients, suggesting that the motor neurons in ALS may be at risk for excitotoxic injury due to a failure to readily clear glutamate from the synaptic cleft (Lin et al., 1998; Trotti et al., 1999, 2001). Although these findings would be consistent with the observation of elevated cerebrospinal fluid glutamate levels in ALS patients, not all authors have been able to replicate these results (Rothstein et al., 1990; Meyer et al., 1998; Honig et al., 2000).

12.3.1 Protein Aggregation in ALS

The pathology of sALS and fALS are similar and include aggregation and inclusions of various proteins in both neuronal and nonneuronal cells, neuronal degeneration throughout the motor cortex and the spinal cord, as well as evidence of inflammation and immune system involvement (Strong et al., 2005a; Moisse and Strong, 2006). The neuropathology of ALS includes numerous protein inclusions and aggregates that are defined on the basis of their morphological appearance, location, and composition (Figure 12.1). The major types include Bunina bodies, ubiquitin positive inclusions, hyaline conglomerate inclusions (HCIs)/neurofilament (NF) inclusions and other filamentous inclusions (peripherin, Tau), and mtSOD1 aggregates (Ince et al., 1998; Sasaki et al., 2006).

12.3.1.1 Bunina Bodies

Although Bunina bodies were first described in fALS, they are found in 80%–100% of sALS cases (Bunina, 1962; Ince et al., 1998; Wada et al., 1999). Bunina bodies are neuronal eosinophilic 2–5 μm in diameter inclusions that are surrounded by a clear

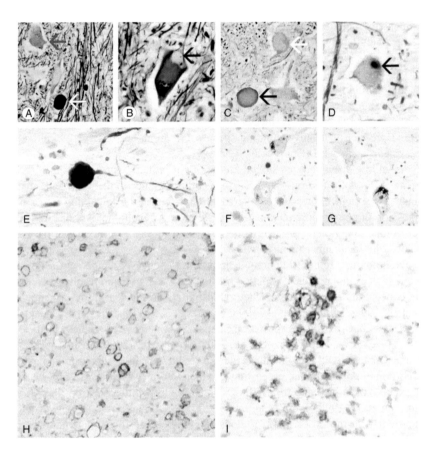

FIGURE 12.1 Cytoskeletal protein aggregation in ALS. (A) Using silver staining (Biel-chowsky staining), proteinaceous aggregates in ALS can appear as intensely argentophilic, well-circumscribed neuroaxonal spheroids (white arrow), as are less intensely staining, more amorphous aggregates (grey arrow) (lumbar spinal cord, 20× magnification before reproduc-tion). (B) Intensely staining, punctate aggregates, corresponding to Bunina bodies observed on routine stains, are also readily apparent (B, arrow, 40× magnification). Neuroaxonal spheroids are intensely immunoreactive with a monoclonal antibody recognizing highly phosphorylated NF-H (C, 40× magnification; black arrow), whereas perikaryal NF aggregates are more amorphous and inhomogeneous in immunostaining (white arrow). Peripherin immunoreactiv-ity colocalizes to dense perikaryal aggregates (D, arrow) and prominently to neuroaxonal spheroids (E). Ubiquitin immunoreactive threads and aggregates (F and G) are invariably observed as early markers of aberrant protein aggregate in otherwise healthy appearing motor neurons. Microglial activation is also a prominent feature of the neuropathology of ALS, including the appearance of phagocytic microglia within degenerating corticospinal tracts (H) and in the region of the ventral horn (I) where both phagocytic and activated, ramified microglia are evident (HLA-Dr immunoreactivity, 40× magnification before reproduction). All immunostains developed with 3,4-diaminobenzidine (DAB), yielding a brown reaction product. (Modified from Strong, M.J., Kesavapany, S. and Pant, H.C., *J. Neuropathol. Exp. Neurol.*, 64, 649, 2005a. With permission.)

halo and are intensely immunostained by antibodies to the cysteine protease inhibitor cystatin C. These inclusions are observed in both spinal motor neurons and in neurons of Clarke's column (Okamoto et al., 1993). They are composed of electron-dense amorphous or granular material, with vesicular structures and neurofilaments (NFs) (Bunina, 1962; Wada et al., 1999). The expression of the lysosomal cysteine protease cathepsin B is low in Bunina body-containing motor neurons in ALS when compared with motor neurons without Bunina bodies, suggesting that the presence of Bunina bodies may attenuate the pro-apoptotic effect of cathepsin B (Kikuchi et al., 2003).

12.3.1.2 Ubiquitin Immunoreactive Inclusions

Ubiquitin immunoreactive inclusions are present in all variants of ALS (Piao et al., 2000; Wood et al., 2003). Morphologically, ubiquitin immunoreactive inclusions in both spinal motor neurons and motor neurons of the brain stem in ALS can appear as filamentous or skein-like structures, as rounded and compact aggregates (Lewy-body like), or as a combination of both (Ince et al., 1998). Using immunohistochemical techniques, ubiquitin inclusions have been shown to colocalize both SOD1 and NF proteins. These inclusions are also immunoreactive to the Ring finger type E3 Ub ligase, Dorfin, suggesting that the ubiquitinization may be involved in the targeting of intracellular proteins for proteolytic degradation (Leigh et al., 1989; Ciechanover, 1993; Niwa et al., 2002). The presence of increased numbers of lysosomes in the immediate vicinity of the ubiquitin inclusions also supports this concept (Kato et al., 2000).

12.3.1.3 Hyaline Conglomerate Inclusions

HCIs are a pathological marker of both fALS and sALS (Chou et al., 1998b). These inclusions are glassy or hyaline inclusions observed with hemotoxylin and eosin staining (Wood et al., 2003). Ultrastructurally, HCIs contain both phosphorylated and nonphosphorylated NF (Kondo et al., 1986; Leigh et al., 1989). Among the enzymes and their metabolic products that have been found to colocalize with HCIs, both neuronal nitric oxide synthase (nNOS) and SOD1 have been observed (Chou et al., 1996a, 1996b, 1998b). Chou et al. found that both serpin-serine proteases, and advanced glycation end products (AGEs) were also localized to HCIs, as were other markers of oxidative protein modification (Chou et al., 1998a, 1998b), suggesting that the posttranslational modification of NF or other proteins are associated with HCI formation.

12.4 MECHANISMS OF PROTEIN AGGREGATION IN ALS

12.4.1 INTERMEDIATE FILAMENT PROTEINS

12.4.1.1 Neurofilament

It is not surprising, given the nature of inclusions in ALS, that cytoskeletal disorganization has been postulated as a key component of the pathogenesis of motor neuron degeneration (Lariviere and Julien, 2004). Given the relative size of motor neurons compared with other neuronal populations and the need for lengthy axonal

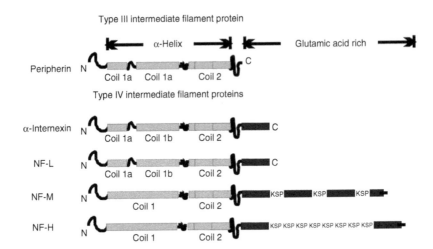

FIGURE 12.2 Intermediate filaments. The intermediate filaments share a common highly conserved α-helical core region that forms double-stranded coiled-coils flanked by head (amino)- and tail (carboxy)-domains. (From Strong, M.J., McLean, J. and Julien, J.-P., *Amyotrophic Lateral Sclerosis*, H. Mitsumoto, S. Przedborski, P. Gordon and M. De Bene, eds, Marcel Dekker, London, 2005b. With permission.)

projections, coupled with the abundance of NF proteins, motor neurons are at particular risk of NF aggregate formation. Neurofilaments are members of the IF family of proteins (Figure 12.2), and consist of an alpha helical core of 310 amino acids capable of forming double-stranded coiled-coils (Strong, 1999). Based on molecular weight determined by SDS–PAGE, three NF proteins can be identified: the high molecular weight NF (NFH; 200 kDa), the medium molecular weight NF (NFM; 160 kDa), and the low molecular weight NF (NFL; 68 kDa). Neurofilaments assemble as obligate heteropolymers. Critical to this process is the initial formation of NFL homo- or heteropolymers upon which NFM and NFH associate (Ching and Liem, 1993; Lee et al., 1993; Leung and Liem, 1996).

N-terminus phosphorylation of NFL regulates NF triplet extension, promoting NFL disassembly, and preventing NFL polymerization (Hisanaga et al., 1990, 1994; Nakamura et al., 1990; Heins et al., 1993). This reversibility of the NFL assembly is key to the formation of the NF triplet (NFL, NFM, and NFH) (Nixon and Shea, 1992). Alterations in the head/amino terminus region of NFL, which includes N-terminus deletion of the NFL protein and a missense mutation in the N-terminus of NFL, result in NF protein assembly impairment, as well as aggregation (Gill et al., 1990; Schmalbruch et al., 1991; Lee et al., 1993). NFH phosphorylation has also been shown to be crucial in maintaining neuronal cytoskeleton, as it was shown that removing blocking of phosphate groups on NFH, in vitro, lead to neuronal cytoskeletal collapse (Durham, 1992). NFH also regulates the axonal caliber through its ability, via phosphorylation, to cross-link the NF triplet proteins.

Perturbations in the NF triplet by either under- or over-expression of one of the subunits leads to the formation of NF aggregates in vivo and in vitro

(Table 12.2). Over-expression of either the wild-type (*wt*) NFH or NFL subunit in mice leads to progressive motor neuropathy and amyotrophy associated with NF aggregation in spinal motor neurons (Cote et al., 1993; Xu et al., 1993). Although somewhat surprisingly this did not result in neuronal loss, arguing against aggregation per se being detrimental, we have shown that the appearance of NF aggregates in these models is a transient phenomenon (McLean et al., 2005). NFM over-expression in mice also leads to aggregation and impairment of axonal growth in lower motor neurons (Wong et al., 1995). NFL knock-out mice are observed to develop a motor degeneration phenotype associated with NF aggregates (Julien and Beaulieu, 2000).

We have examined a potential role for aggregates in further sensitizing motor neurons to excitotoxicity. Using dissociated motor neurons derived from fetal transgenic mice that express the human NFL (*h*NFL +/+), we observed that NMDA-mediated calcium entry was dysregulated in NF aggregate bearing neurons when compared with intact motor neurons derived from control, nontransgenic mice. The presence of NF aggregates was associated with a failure of neuronal nNOS to localize to the NMDA receptor, thus preventing nNOS from inhibiting NMDA-mediated calcium influx (Arundine et al., 2003; Sanelli et al., 2004). Instead, nNOS translocated to the aggregate, suggesting that aggregates can disrupt neuronal homeostasis by sequestering proteins.

12.4.1.2 Peripherin

Peripherin is a 58 kDa type III intermediate filament protein that is neuron-specific and whose expression is upregulated in regenerating motor neurons following spinal cord transsection (Portier et al., 1983, 1993; Troy et al., 1990; Wong and Oblinger, 1990; Migheli et al., 1993). Peripherin is normally highly abundant in autonomic and peripheral sensory neurons, with only very low amounts in spinal motor neurons (Parysek et al., 1988; Brody et al., 1989; Escurat et al., 1990). In ALS, however, peripherin immunoreactivity is observed to colocalize to axonal spheroids in degenerating motor neurons (Corbo and Hays, 1992; Migheli et al., 1993). It is also present in motor neuron perikaryal and axonal aggregates of transgenic mutant G37R SOD1 mice (Tu et al., 1996). Beaulieu et al. used the SW13 cell line that is void of endogenous intermediate filament proteins, to show that peripherin is able to self-assemble to establish an intermediate filament network or dimerize with any of the NF subunits (Beaulieu et al., 1999b). In the absence of NFL, peripherin and NFH or NFM interaction results in aggregation and a disorganized network (Beaulieu et al., 1999b).

The over-expression of peripherin in mice is associated with a late-onset selective motor neuron death in which the depletion of NFL further accelerates disease and neuronal death (Beaulieu et al., 1999b, 2000; Robertson et al., 2001). Before symptom onset and neuronal death, both perikaryal and axonal aggregates containing peripherin, NFH, NFM, and mitochondria are present. Of note, these peripherin aggregates bear a greater morphological similarity to those observed in ALS than those that are present in mice that over-express NFH or NFL alone (Beaulieu et al., 1999b). Although the mechanism of neuronal death in peripherin over-expressing transgenics is not known, it has been proposed that it may be due to a blockade of

TABLE 12.2
Murine Transgenic Models of Targeted Disruption of Neuronal Intermediate Filament Gene Expression

Transgenic Mouse	Age (months)	Mouse Strain	Neuropathologic Effect	Axonal Loss	Effects on IF Content	Phenotype
NF-L						
Murine NF-L (−/−; +/−)	2–3	C57BL/6	≥50% Reduction in ventral root axonal caliber	~20% loss	Reduction in both NF-M and NF-H protein levels in sciatic nerves and brain homogenates; significant loss of myelinated axons	Normal development, no overt phenotype; reduced rate of regeneration postaxotomy
Murine NF-L (+/+)	6	B6AF$_1$J	Massive perikaryal aggregates of NF; axonal NF aggregates	None	Increased NF density, no alteration in NF-L or NF-H expression levels	Variable motor phenotype
Human NF-L (+/+)	P70 (adult)	C57BL/6	None	None	None	No overt phenotype
Human NF-L		C57BL/6	Prominent perikaryal NF-L immunostaining	None stated	None stated	None stated
NF-M						
Murine NF-M (−/−)	4	129 Sv/J bred to C57BL/6	≥50% Reduction in ventral root axonal caliber; loss of large diameter myelinated axons	~10% loss	Significant reduction in NF-L protein levels; concomitant increase in NF-H protein levels in brain but not spinal cord	Normal development; No overt phenotype
Human NF-M (+/+)	3–12	C57BL/6	Age-dependent prominent NF aggregate formation		Increased levels of NF-L expression; reduced phosphorylation of NF-H; increased axonal NF density	Age-dependent deficits on reference and memory tasks

NF-H						
Murine NF-H (−/−)	4	129 Sv/J bred to C57BL/6	Significant reduction in the development of large-diameter myelinated axons with concomitant increase in smaller diameter fibers	No	Approximately 10%–25% reduction in NF-L protein level	Normal development; no overt phenotype
Murine NF-H (+/+)	4 weeks, 9 weeks, and 1 year	C5BL6/C3	NF-H dose-dependent induction of severe axonal and perikaryal NF aggregate formation in spinal cord, reduction in axon caliber	Reduced large diameter axons if perikaryal aggregates formed	Reduced NF-L and NF-M molar ratios	Normal phenotype; no evidence of neurogenic atrophy (up to 2 years)
Human NF-H +/+	4	C57BL6/C3H	Prominent perikaryal NF aggregate formation; dying back axonopathy	Severe axonopathy	Elevated level of NF-H expression	Progressive motor neuropathy with death by 1 year
Peripherin						
Knockout murine			Normal	No	Normal	
Murine peripherin (+/+)	6–10	C57BL/C3H	diffuse, increased peripherin immunoreactivity in perikarya and neuritis; peripherin-immunoreactive aggregates in aged mice	~35% motor axon loss, age-dependent	Normal	Late-onset (>2 years) motor dysfunction
α-Internexin						
Murine (−/−)	3	C57BL6/129 Sv B6CBA F,J	Normal	No	Normal	No overt phenotype
Murine (+/+)	12–18		Enhanced α-internexin immunostaining in cerebellum, neocortex, and thalamus	No	Normal	Reduced motor coordination and balance on rotorod

Source: Adapted from Strong, M.J., McLean, J. and Julien, J.-P., *Amyotrophic Lateral Sclerosis*, H. Mitsumoto, S. Przedborski, P. Gordon, and M. De Bene, eds, Marcel Dekker Inc., London, 2005b. With permission.

intracellular transport caused by the aggregation or sequestering of key cellular constituents within these aggregates, including mitochondria (Beaulieu et al., 1999b). Robertson and colleagues demonstrated in vitro, using motor neurons derived from peripherin over-expressing mice in which intraneuronal protein aggregates form, that the extent of cell death was enhanced when motor neurons were cocultured with dorsal root ganglion neurons (Robertson et al., 2003).

Of potentially greater importance to the pathogenesis of ALS has been the observation that peripherin can exist in three isoforms due to alternative splicing (Per 58, Per 56, and Per 61, based on protein size on SDS–PAGE) (Landon et al., 1989, 2000). The Per 61 isoform is assembly incompetent and neurotoxic. It is of interest therefore that Per 61 expression is upregulated in spinal motor neurons of G37R mtSOD1 mice, but not in peripherin transgenic mice or controls. In addition, Per 61 antibody labeled spheroids and motor neurons in fALS cases, but not in controls (Robertson et al., 2003).

12.4.1.3 α-Internexin

α-Internexin is a type IV IF protein expressed in developing neurons more abundantly and earlier than NFs. α-Internexin is expressed throughout the CNS (Fliegner and Liem, 1991). In the adult CNS, α-internexin colocalizes with NFs in axons although it is expressed at much lower levels (Kaplan et al., 1990). While NFs are obligate heteropolymers, α-internexin is able to self-assemble and co-assemble with each NF (Ching and Liem, 1993), properties that imply α-internexin has a role in the stability of small axons, acting as a scaffold on which NFs may co-assemble during development (Ching et al., 1999). The over-expression of rat α-internexin in transgenic mice is associated with a progressive motor degeneration and neuronal loss with death within 3 months of age (Ching et al., 1999). Affected mice also develop axonal swellings in Purkinje cells as well as IF and organelle accumulations in large pyramidal neurons in the neocortex and in the ventral nuclei of the thalamus (Ching et al., 1999). Conversely, α-internexin knock-out mice showed no motor or pathological phenotype, suggesting that type IV IFs do not necessarily have a role in axonal growth during development (Levavasseur et al., 1999).

12.4.1.4 Transgenic Mice with Altered Neuronal IF Expression

Although the phenotype of mice expressing alterations in neuronal IF protein expression develop a limited phenotype, or none at all, motor neuron pathology replicating aspects of the pathology of ALS are produced in many (Table 12.2).

The key issue in the neuronal IF models of motor neuron degeneration relates to the observation that altering the stoichiometry of expression of the individual nIFs relative to one another results in aggregate formation. For example, in the NFL knockout mice (NFL $-/-$), motor neuron aggregates NFM and NFH are formed, in spite of a reduced level of expression of both NFH and NFM (Robertson et al., 2002). The over-expression of peripherin leads to motor neuron degeneration associated with NF accumulation (Beaulieu et al., 1999a). It is of note that the over-expression of NFL in transgenic mice that over-express NFH results in decreased perikaryal aggregate formation and decreased motor neuron death (Meier et al.,

1999), whereas knocking out NFL in transgenic mice that over-express peripherin results in a marked increase in the extent of neuronal death (Beaulieu et al., 1999a). Both the sets of studies highlight the critical nature of maintaining IF stoichiometry to the integrity of the motor neuron.

12.4.2 SOD1 AGGREGATES IN ALS

12.4.2.1 Mechanism of SOD1 Neurotoxicity

Since the discovery of the potential role of mutant copper/zinc superoxide dismutase (mtSOD1) in *f*ALS (Rosen et al., 1993), and the creation of the transgenic mtSOD1 mouse (Gurney et al., 1994), the topic of the mechanism of mtSOD1 neurotoxicity has been one of much theorization. Motor neurons are unique with significantly higher levels of SOD1 expression than observed in other neuronal populations, thus placing them at risk for mtSOD1-mediated damage (Bergeron et al., 1996).

SOD1 contains copper/zinc-binding residues that are essential for catalytic activity of the enzyme, as well as for imparting stability to the dimer interface (Tainer et al., 1982; Deng et al., 1993; Siddique et al., 1997). The β-barrel domain of SOD1 contains the motif known as the "Greek Key" (due to its structure resembling a Greek key pattern). This motif has been implicated in maintaining stability of the SOD1 dimer (Getzoff et al., 1989; Deng et al., 1993). It has been observed, however, that engineered SOD1 molecules with permuted structures that contained interruptions in the Greek Key loops were only slightly destabilized (Boissinot et al., 1997). Dimerization of the enzyme is essential for its function. Monomeric SOD1 is deficient in allowing access to superoxide at the active site, as well as having only 10% the activity of the dimeric enzyme (Banci et al., 1998). Furthermore, it was shown by Rakhit et al. that oxidative damage to SOD1 leads to SOD1 monomerization, followed by its aggregation (Rakhit et al., 2004).

The predominant theories with regard to the mechanism of mtSOD1 toxicity can be arbitrarily divided into either "loss of function" or "gain of function" theories. With regard to the loss of function theory, it has been hypothesized that mtSDO1 toxicity results from the inability of mtSOD1 to scavenge superoxide, leading to excess superoxide, and to a depletion of the glutathione peroxidase and catalase enzymes, which would normally eliminate the excessive peroxide and hydroxyl radicals. In doing so, the neuron would be at risk of severe oxidative injury. However, the failure of SOD1 knock-out mice to develop either motor impairment or neuronal loss argues against this (Reaume et al., 1996).

The "gain of function" theories include an enhanced rate of reactive nitrating species formation, the formation of toxic aggregates, and the formation of toxic mtSOD1 monomers. With regard to aggregate formation, it is theorized that SOD1 mutations result in a destabilization of the active site, allowing access to unidentified targets that are not normally accessible (Brown, 1998; Bruijn et al., 1998). Aberrant binding of these targets could result in aggregation that is neurotoxic (Shibata et al., 1999, 2000a, 2000b; Kato et al., 1999, 2000). In vitro, not only mtSOD1, but also zinc-deficient wild type SOD1, is highly prone to aggregate upon oxidation, with the latter occurring predominantly on histidine residues that bind active site metals (Rakhit et al., 2002). This capacity to form both neuronal and extraneuronal aggregates has been confirmed in a number of in vitro and in vivo models (Bruijn et al., 1997; Bruening et al., 1999; Migheli et al., 1999; Takehisa et al., 2001; Sanelli et al., 2004).

Of potential therapeutic importance is the observation, in both mtSOD1 mice and cultured cells expressing mtSOD1, that a decrease in chaperoning activity and the up-regulation of heat shock protein 70 (HSP 70), a protein chaperone, is associated with decreased aggregation (Bruening et al., 1999). This indicates a potential for chaperones to stabilize mtSOD1, preventing misfolding and aggregation.

There is also evidence of significant oxidative injury associated with the expression of mtSOD1, including posttranslational modifications of SOD1 itself that can be associated with enhanced SOD1 aggregation and oxidative damage. This includes the formation of AGEs of mtSOD1 fibrils that are associated with increased aggregation and greater toxicity (Kato et al., 2000). N^{ε}-carboxymethyl-lysine (CML) immunoreactivity, as a marker of lipid peroxidation, colocalizes with HCIs in spinal motor neurons in fALS cases harboring the A4V SOD1 mutation (Shibata et al., 2000b). SOD1 itself may be a target of oxidative modification due to its high intracellular concentration (Kato et al., 1997), a concept supported by the observation that G93A SOD1 transgenic mice have heavily oxidized mtSOD1 (Andrus et al., 1998). Valentine and Hart (2003) propose that mtSOD1 is inactivated by hydrogen peroxide, as it oxidizes a specific residue in the metal-binding active site of mtSOD1, thereby causing compromised metal-binding and aggregation of SOD1.

Distinct from aggregate formation, fALS-associated SOD1 mutations will impact on a diverse range of SOD1 functions. This includes destabilizing the metal-binding and electrostatic loops or the dimer interface of SOD1 (Deng et al., 1993), and an increased accessibility of the Copper/Zinc active site, leading to enhanced production of reactive nitrating species (Beckman et al., 1993; Brown, 1998; Cookson and Shaw, 1999).

Another proposed gain of function of mtSOD1 is to alter IF protein expression through interacting with mRNA. Ge et al. found that mtSOD1 but not wild type SOD1, was able to decrease NFL mRNA stability in a dose-dependant manner by binding directly to the 3' untranslated region (UTR) of NFL mRNA (Ge et al., 2005). In addition, p190RhoGEF, an RNA-binding protein involved in the aggregation of murine NFL, appears to colocalize with NFL aggregates as an early feature of the neurodegeneration of the SOD1^{G93A} mouse (Lin et al., 2005).

12.4.2.2   Transgenic Models of Mutant SOD1 Toxicity

12.4.2.2.1   Murine Models

The most frequently studied transgenic mtSOD1 mouse models include the G93A (Gurney et al., 1994; Dal Canto and Gurney, 1995), the G37R (Wong et al., 1995), the G85R (Bruijn et al., 1997), and the D90A (Bergemalm et al., 2006) transgenic mice (Table 12.3). By far the most studied is the G93A mtSOD1 mouse. A critical feature of the interpretation of these mice lies in recognizing that disease onset, duration, and severity of disease will vary based on not only the mtSOD1 expressed and copy number, but even upon the background mouse strain upon which the transgenic mouse has been developed (Wong et al., 1995; Dal Canto and Gurney, 1997; Heiman-Patterson et al., 2005).

TABLE 12.3

Transgenic Models of Mutant SOD1 Expression

Mutation	Transgene Origin	Strain	SOD1 Activity	Onset[a]	Duration[b]	Reference
Murine						
Gly^{37}-Arg (G37R)	Human	B6/C	Increased	105	ND	Wong et al. (1995)
Gly^{85}-Arg (G85R)	Human	B6/C	Same	240–420	7–14	Bruijn et al. (1997)
Gly^{86}-Arg (G86R)	Mouse	FVB	Same	90–120	3–30	Ripps et al. (1995)
Asp^{90}-Ala (D90A)	Human	ND	Increased	100	Slow	Bergemalm et al. (2006)
Gly^{93}-Ala (G93A)	Human	B6/SJL	Increased	90	30	Gurney et al. (1994)
Gly^{93}-Arg (G93R)	Human	ND	ND	238	11	Friedlander et al. (1997)
Rat						
Gly^{93}-Ala (G93A)	Human	SD	Increased	123	8	Nagai et al. (2001) and Aoki et al. (2005)
His^{46}-Arg (H46R)	Human	SD	Decreased	145–172	24–37	Nagai et al. (2001) and Aoki et al. (2005)

[a] Onset (days of age).
[b] Duration: survival from first clinical sign.

Generally, the motor neuron degeneration in these mice manifests first as tremulous movement in the hind limbs followed by amyotrophy, paralysis, and death (Shibata et al., 2000b). It is noteworthy that the development of motor neuron degeneration in these mice is also dependant upon the expression of the mtSOD1 in nonneuronal cells, including microglia and astrocytes (Pramatarova et al., 2001). The latter is associated with a reduced expression of the glutamate transporter, EAAT2, similar to that seen in ALS and thus of relevance to the excitotoxicity hypothesis of ALS (Schiffer et al., 1996; Hall et al., 1998; Cleveland and Rothstein, 2001).

SOD1 aggregates, observed in both motor neurons and astrocytes in the G93A transgenic mouse, are immunoreactive for ubiquitin and GFAP (Bruijn et al., 1997; Dal Canto and Gurney, 1997; Shibata et al., 1999). In these mice, motor neuron dysfunction is associated with an impairment in axonal transport (Zhang et al., 1997).

12.4.2.2.2   Rat Models

Two rat models of mtSOD1 (G93A) expression have been developed in which motor neuron degeneration is observed (Nagai et al., 2001; Aoki et al., 2005) (Table 12.3). Both develop a selective loss of motor neurons in the spinal cord in association with HCIs and astrocytic hyaline inclusions. Typically, these animals present with weakness in one hindlimb, which then progresses to the other hindlimb and forelimbs, followed by feeding impairment and death (Matsumoto et al., 2006). Different, however, from the mtSOD1 expressing mice, mtSOD1 rats have a relatively rapid course of decline with survival ranging between 13 and 24 days following the first signs of motor dysfunction. The advantage of the rat models lies in their larger size, thus allowing testing of therapeutic strategies that are more invasive, including the delivery of glial-derived neurotrophic factor (GDNF) via human neural progenitor cells (Klein et al., 2005), and the use of intracerebroventricular VEGF (Storkebaum et al., 2005).

12.4.3   Tau Protein Aggregates in ALS with Cognitive Impairment (ALSci)

Although tau aggregation is most commonly associated with Alzheimer's disease and a subpopulation of patients with a FTLD, there has been an increasing recognition that tau aggregates can be observed in ALS (Mitsuyama, 2000; Yoshida, 2004; Mott et al., 2005) (Figure 12.3). Although the full extent to which alterations in tau metabolism can be observed in ALS, this is particularly true of ALS patients in whom cognitive impairment is evident (Yang et al., 2003).

The primary function of tau protein is to bind to microtubules and promote their stabilization (Weingarten et al., 1975). It is also a cytoskeletal scaffolding protein that is involved in signaling, organelle transport, and cell growth (Dinoto et al., 2005). Six tau isoforms exist due to alternative splicing of chromosome 17 mRNA (Figure 12.4). Tau aggregation in Alzheimer's disease may occur due to abnormal phosphorylation (Perez et al., 2002; Lee et al., 2005), glutaminase-mediated cross-linking (Miller and Anderton, 1986; Singer et al., 2002), glycation (Smith et al., 1995), ubiquitination (Manetto et al., 1988), and mutation (Dinoto et al., 2005) (see Chapter 6 for a detailed discussion on tau protein).

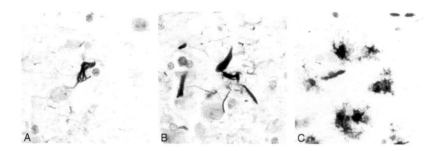

FIGURE 12.3 Illustrative tau pathology observed in ALS with cognitive impairment (ALSci). Archival, paraffin-embedded tissue stained with the Gallyas–Braak silver stain (sensitive to abnormal phosphoprotein aggregation). Protein aggregation is observed as fine filamentous structures (black staining) within neurons (A), as extraneuronal neuropil threads (B) and within astrocytes (C) (magnification before reproduction 40×).

Both astrocytic and neuronal inclusions that are immunoreactive to phosphorylated tau protein have been observed in a proportion of both ALS and ALSci cases, although more numerous in ALSci (Yang et al., 2003). Although total tau mRNA in both ALS and ALSci was significantly elevated, the normal ratio of 3R:4R isoform expression is maintained. More recently, tau isolated from ALS cases have been observed to be hyperphosphorylated (Strong et al., 2006).

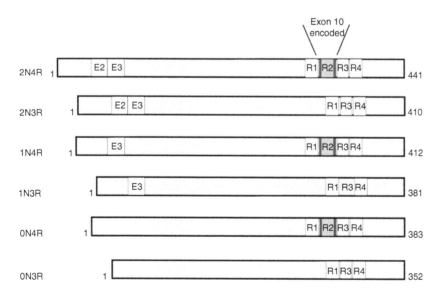

FIGURE 12.4 Tau isoforms. Six human tau isoforms are generated by alternative splicing of the tau gene, which includes 16 exons. Alternative splicing of the E2, E3, and E10 exons produce the six isoforms, with E10 encoding an 18 amino acid microtubule-binding repeat. The isoforms range from 352 to 441 amino acids in length.

12.4.3.1 Transgenic Models of Altered Tau Expression

Both in vitro and in vivo models in which tau protein is either over-expressed or selectively expressed within neurons have been developed. The over-expression of tau in neuronal cell lines and cortical neurons lead to NF aggregates with accumulation of organelles (mitochondria) (Stamer et al., 2002).

The critical interrelationship between tau proteins and other cytoskeletal proteins (e.g., microtubules, NF) in determining the phenotype of tau aggregation has been elegantly delineated in a series of transgenic mice studies. Mice expressing the human 4R/2N isoform develop a prominent motor phenotype with an axonopathy marker by neuroaxonal spheroids, but no evidence of tau protein aggregates (Spittaels et al., 1999; Terwel et al., 2005). However, when the FTLD-17 associated tau-P301L mutation is introduced into these mice, the formation of the axonal dilatation and the severity of the motor deficit are reduced. A potential explanation for this reversal of the axonal stasis is the reduced affinity of the tau-P301L mutation for microtubules (Barghorn et al., 2000; Tanemura et al., 2002). In contrast, transgenic mice expressing only tau-P301L, while developing minor motor deficits, are striking in the development of aggregates of hyperphosphorylated tau in cortical and hippocampal neurons within somatodendritic compartments (Lewis et al., 2000; Gotz et al., 2001). The authors postulate that the differences in pathology between the two transgenic models relate to the intrinsic differences between the tau isoforms being expressed. Thus, the axonopathy of the 4R/2N mice is postulated to relate to the excessive binding of tau to microtubules (simply as a reflection of the level of expression of tau and its natural affinity for microtubules), leading to an inhibition of axonal transport. The observation that tau phosphorylation by GSK-3β in the 4R/2N transgenic mice reverses this process, and thus corrects the axonopathy, supports this hypothesis (Spittaels et al., 2000). Conversely, the reduced binding of tau-P301L to microtubules, mediated through reduced tau phosphorylation, leads to an increased pool of tau to aggregate spontaneously (Allen et al., 2002).

Of particular interest to a discussion of the cooccurrence of aggregates of tau and NF, observed in some cases of ALSci, is the observation that the cross-breeding of transgenic mice that over-express the shortest tau isoform with knock-out mice devoid of either NFL or NFH, attenuates the tau transgenic phenotype. Thus, when the shortest human tau isoform alone is expressed, mice develop an age-dependant CNS pathology with motor weakness associated with intraneuronal inclusions in cortical, brainstem, and spinal motor neurons (Ishihara et al., 1999). In the latter, spinal motor axonal degeneration is observed in association with reduced axonal transport and diminished microtubule numbers. Although crossbreeding these mice with knock-out mice devoid of NFL or NFH did not completely inhibit this process, a dramatic attenuation in the number of spinal motor neuron tau aggregates was observed (Ishihara et al., 2001). This was associated with an improvement in survival, reduced weight loss, and a less severe motor phenotype.

Thus, there is considerable evidence that altering tau metabolism gives rise to phenotypes in transgenic mice in which at least a component is centered upon the motor neurons. Altering this phenotype by altering NF expression, a process reminiscent of that described within spinal motor neurons in ALS, attenuates this process.

Perhaps the relationship between tau aggregation in ALSci and motor neuron degeneration without motor neuron tau aggregation is more than a coincidence.

12.5 PHARMACOTHERAPY

Currently, most, if not all studies of potential drug therapies for ALS are conducted using mtSOD1 mice, typically the G93A SOD1 transgenics. An example of such a process and its relative limitations was the development of the antiglutamatergic agent Riluzole as a treatment for ALS. Although to date, this is the only FDA approved pharmaceutical approved for the treatment of ALS, having increased survival in the G93ASOD1 mouse by an additional 10–15 days, it has not proven to have the same magnitude of effect in humans (Meininger et al., 2000). Of interest, however, Riluzole inhibits the aggregation of Huntingtin protein in an in vivo model of Huntington disease (Heiser et al., 2002; Schiefer et al., 2002).

The tetracycline antibiotic derivative minocycline has been postulated to be of potential benefit in ALS due to its ability to inhibit both cytochrome c release and microglial activation, as well as potential neuroprotective effects (Tikka and Koistinaho, 2001; Tikka et al., 2001b). It has been shown to be neuroprotective in an in vitro model of excitotoxicity (Tikka et al., 2001a) and in that induced in vitro by ALS-derived cerebrospinal fluid (Tikka et al., 2002). In the G37R mSOD1 mice, minocycline increased life span by 21–25 days (Kriz et al., 2002; Van Den Bosch et al., 2002). It is currently in clinical trials.

The transgenic mouse models of altered IF expression have been used predominantly to determine whether altering the relative expression ratios of the individual IFs impacts on disease phenotype or pathology. The observation that in SOD1^{G37R} mice in which gene replacement therapy was used to remove the tail domains of either NFM or NFH, including the majority of C-terminus phosphorylation sites, significantly prolonged survival suggests that altering the neuronal IF expression represents a viable research direction (Lobsiger et al., 2005). One potential view of these results is that the formation of IF aggregates may not necessarily be detrimental to the cell.

The use of small inhibitory RNA technologies to suppress the expression of mtSOD1 in transgenic mice is rapidly becoming a viable therapeutic target. In this, the use of sequence specific RNA interference (RNAi) can result in the posttranscriptional inactivation of gene expression by promoting specific endonucleolytic cleavage of mRNA targets. This is of specific relevance in that such techniques can lead to silencing of *mtSOD1* gene expression, while leaving expression of the wild-type SOD1 intact (Ding et al., 2003; Maxwell et al., 2004; Xia et al., 2005). RNAi can be introduced in mammalian cells using either small interfering RNAs (siRNA) or short hairpin RNAs (shRNA), the latter being processed into siRNA in cells. Using a lentiviral vector that produces RNAi-mediated silencing of mtSOD1 (G93A) expression, the rate of disease progression can be significantly retarded (Ralph et al., 2005; Raoul et al., 2005). Similar results were obtained in crossing anti-SOD1 siRNA transgenic mice expressing high levels of siRNA with mtSOD1 (G93A) transgenic mice (Saito et al., 2005).

Pharmacotherapies targeted at the disruption of aggregate formation in mtSOD1 expressing transgenic mice represent a potential valuable investigative pathway for the treatment of ALS. For instance, the observation that the oxidation of SOD1 leads to SOD1 monomerization and aggregation (Rakhit et al., 2004) suggests a potential avenue to be pursued with regard to mtSOD1 aggregate modifying therapies. Ghadge et al., using mtSOD1 transfected PC12 cells examined the effect of D-penicillamine and trientine, both of which are calcium chelators (Ghadge et al., 1997). These chelators were originally shown to be beneficial for G93A mice (Hottinger et al., 1997; Nagano et al., 1999). Both D-penicillamine- and trientine-inhibited cell death in mtSOD1-expressing PC12 cells (Ghadge et al., 1997). Similarly, copper chaperone of SOD1 (CCS) inhibitors may also be beneficial (Corson et al., 1998). Although CCS inhibitors may prevent excessive binding of Cu to mtSOD1 and thus minimize the formation of toxic species, knocking out the *CCS* gene in mtSOD1 expressing mice did not alter disease onset or progression (Subramaniam et al., 2002).

One of the most promising treatments is the administration of chaperone proteins, such as HSP 70. Chaperone protein over-expression increased survival and function of motor neuron cultures (Takeuchi et al., 2002; Shaw, 2005), and a drug that enhances expression of HSPs, Arimocramol, when given to G93A mice, increased their life span by 22% (Kieran et al., 2004). HSPs have also been found to decrease mtSOD1 aggregation, as well as aggregation in both Huntington's disease and spinobulbar muscular atrophy (Bruening et al., 1999; Cleveland and Rothstein, 2001; Takeuchi et al., 2002).

Other potential candidates include therapies that have been tested for other neurodegenerative disorders. Khlistunova et al. have shown that in a cell model of taupathies, small molecules of 100–150 MW in size, are able to both inhibit and disrupt tau aggregation, which resulted in a 40%–50% decrease in tau filamentous aggregates (Khlistunova et al., 2006).

12.6 CONCLUSION

Although the presence of protein aggregates and inclusions in ALS is indisputable, there continues to be controversy as to the role, if any, of these aggregates in mediating the disease process. Are such aggregates merely bystanders that are generated as the consequence of the disease process? It is hard, however, to imagine that the aggregation of proteins within both neuronal and nonneuronal cells in ALS is anything but fundamental to the disease process of ALS, and that a significant component of the biology of ALS reflects a proteinopathy. The mechanisms underlying this proteinopathy are as varied as the nature of the proteins aggregating. These include such seemingly divergent processes as alterations in stoichiometry of IF expression giving rise to NF aggregates (either on the basis of alterations in NFL mRNA stability or in the presence of altered peripherin isoform expression), mutations giving rise to the preferential formation of SOD1 monomers with both toxic and aggregating properties, posttranslational modifications as a reflection of oxidative stress that induce significant structural modifications, or alterations in phosphorylation state yielding highly insoluble protein aggregates as observed in tau aggregation.

The key issue remains therefore not whether such aggregates are critical neuropathological hallmarks of ALS, but rather whether their formation is of direct toxicity to the neuron. In the case of mtSOD1 protein expression in *f*ALS, the answer is clearly that the formation of aggregates is at minimum, a key surrogate marker of a biologically relevant alteration in protein expression. The use of siRNA methodologies to suppress mutant protein expression, resulting in significant improvement in survival of murine models of *f*ALS harboring SOD1 mutations supports such a concept. Using this as a pivotal example, therapies directed toward modifying protein aggregation in ALS should be predicted to enhance disease survival.

ACKNOWLEDGMENTS

MJS research supported by the ALS Association (USA), the ALS Society of Canada, the Muscular Dystrophy Association (Tuscon), and the Scottish Rite Charitable Foundation. TS is supported by a research fellowship from the Canadian Institutes of Health Research (CIHR).

REFERENCES

Al-Chalabi, A., Enayat, Z.E., Bakker, M.C., Sham, P.C., Ball, D.M., Shaw, C.E. et al. (1996) Association of apolipoprotein E epsilon 4 allele with bulbar-onset motor neuron disease. *Lancet*, 347, 159–160.

Al-Chalabi, A., Andersen, P.M., Nilsson, P., Chioz, B., Andersson, J.L., Russ, C. et al. (1999) Deletions of the heavy neurofilament subunit tail in amyotrophic lateral sclerosis. *Human Molecular Genetics*, 8, 157–164.

Allen, B., Ingram, E., Takao, M., Smith, M.J., Jakes, R, Virdee, K. et al. (2002) Abundant tau filaments and nonapoptotic neurodegeneration in transgenic mice expressing human P301S tau protein. *Journal of Neuroscience*, 22, 9340–9351.

Andersen, P.M. (2006) Amyotrophic lateral sclerosis associated with mutations in the CuZn superoxide dismutase gene. *Current Neurology and Neuroscience Reports*, 6, 37–46.

Andersen, P.M., Nilsson, P., Keranen, M.L., Forsgren, L., Hagglund, J., Karlsborg, M. et al. (1997) Phenotypic heterogeneity in motor neuron disease patients with CuZn-superoxide dismutase mutations in Scandinavia. *Brain*, 120, 1723–1737.

Andrus, P.K., Fleck, T.J., Gurney, M.E. and Hall, E.D. (1998) Protein oxidative damage in a transgenic mouse model of familial amyotrophic lateral sclerosis. *Journal of Neurochemistry*, 71, 2041–2048.

Aoki, M., Lin, C.L., Rothstein, J.D., Geller, B.A., Hosler, B.A., Munsat, T.L. et al. (1998) Mutations in the glutamate transporter EAAT2 gene do not cause abnormal EAAT2 transcripts in amyotrophic lateral sclerosis. *Annals of Neurology*, 43, 645–653.

Aoki, M., Kato, S., Nagai, M. and Itoyama, Y. (2005) Development of a rat model of amyotrophic lateral sclerosis expressing a human SOD1 transgene. *Neuropathology*, 25, 365–370.

Armon, C. (2003) An evidence-based medicine approach to the evaluation of the role of exogenous risk factors in sporadic amyotrophic lateral sclerosis. *Neuroepidemiology*, 22, 217–228.

Arundine, M., Sanelli, T., He, B.P. and Strong, M.J. (2003) NMDA induces NOS 1 translocation to the cell membrane in NGF-differentiated PC 12 cells. *Brain Research*, 976, 149–158.

Banci, L., Benedetto, M., Bertini, I., Del Conte, R., Piccioli, M. and Viezzoli, M.S. (1998) Solution structure of reduced monomeric Q133M2 copper, zinc superoxide dismutase (SOD). Why is SOD a dimeric enzyme? *Biochemistry*, 37, 11780–11791.

Barghorn, S., Zheng-Fischhofer, Q., Ackmann, M., Biernat, J., von Bergen, M., Mandelkow, E.M. et al. (2000) Structure, microtubule interactions, and paired helical filament aggregation by tau mutants of frontotemporal dementias. *Biochemistry*, 39, 11714–11721.

Beaulieu, J.M., Nguyen, M.D. and Julien, J.P. (1999a) Late onset of motor neurons in mice overexpressing wild-type peripherin. *Journal of Cell Biology*, 147, 531–544.

Beaulieu, J.M., Robertson, J. and Julien, J.P. (1999b) Interactions between peripherin and neurofilaments in cultured cells: disruption of peripherin assembly by the NF-M and NF-H subunits. *Biochemistry and Cell Biology*, 77, 41–45.

Beaulieu, J.M., Jacomy, H. and Julien, J.P. (2000) Formation of intermediate filament protein aggregates with disparate effects in two transgenic mouse models lacking the neurofilament light subunit. *Journal of Neuroscience*, 20, 5321–5328.

Beckman, J.S., Carson, M., Smith, C.D. and Koppenol, W.H. (1993) ALS, SOD and peroxynitrite. *Nature*, 364, 584.

Beckman, J.S., Estevez, A.G., Crow, J.P. and Barbeito, L. (2001) Superoxide dismutase and the death of motoneurons in ALS. *Trends in Neuroscience*, 24, S15–S20.

Bergemalm, D., Jonsson, P.A., Graffmo, K.S., Andersen, P.M., Brannstrom, T., Rehnmark, A. and Marklund, S.L. (2006) Overloading of stable and exclusion of unstable human superoxide dismutase-1 variants in mitochondria of murine amyotrophic lateral sclerosis models. *Journal of Neuroscience*, 26, 4147–4154.

Bergeron, C., Petrunka, C. and Weyer, L. (1996) Copper/zinc superoxide dismutase expression in the human central nervous system. Correlation with selective neuronal vulnerability. *American Journal of Pathology*, 148, 273–279.

Boissinot, M., Karnas, S., Lepock, J.R., Cabelli, D.E., Tainer, J.A., Getzoff, E.D. et al. (1997) Function of the Greek key connection analysed using circular permutants of superoxide dismutase. *EMBO Journal*, 16, 2171–2178.

Borchelt, D.R., Lee, M.K., Slunt, H.S., Guarnieri, M., Xu, Z.S., Wong, P.C. et al. (1994) Superoxide dismutase 1 with mutations linked to familial amyotrophic lateral sclerosis possesses significant activity. *Proceedings of the National Academy of Sciences of the United States of America*, 91, 8292–8296.

Brody, B.A., Ley, C.A. and Parysek, L.M. (1989) Selective distribution of the 57 kDa neural intermediate filament protein in the rat CNS. *Journal of Neuroscience*, 9, 2391–2401.

Brown Jr., R.H. (1998) SOD1 aggregates in ALS: cause, correlate or consequence? *Nature Medicine*, 4, 1362–1364.

Bruening, W., Roy, J., Giasson, B., Figlewicz, D.A., Mushynski, W.E. and Durham, H.D. (1999) Up-regulation of protein chaperones preserves viability of cells expressing toxic Cu/Zn-superoxide dismutase mutants associated with amyotrophic lateral sclerosis. *Journal of Neurochemistry*, 72, 693–699.

Bruijn, L.I., Becher, M.W., Lee, M.K., Anderson, K.L., Jenkins, N.A., Copeland, N.G. et al. (1997) ALS-linked SOD1 mutant G85R mediates damage to astrocytes and promotes rapidly progressive disease with SOD1-containing inclusions. *Neuron*, 18, 327–338.

Bruijn, L.I., Houseweart, M.K., Kato, S., Anderson, K.L., Anderson, S.D., Ohama, E. et al. (1998) Aggregation and motor neuron toxicity of an ALS-linked SOD1 mutant independent from wild-type SOD1. *Science*, 281, 1851–1854.

Bunina, T.L. (1962) On intracellular inclusions in familial amyotrophic lateral sclerosis. *Zh Nevropatol Psikhiatr Im S S Korsakova*, 62, 1293–1299.

Chance, P.F., Rabin, B.A., Ryan, S.G., Ding, Y., Scavina, M., Crain, B. et al. (1998) Linkage of the gene for an autosomal dominant form of juvenile amyotrophic lateral sclerosis to chromosome 9q34. *American Journal of Human Genetics*, 62, 633–640.

Ching, G.Y. and Liem, R.K. (1993) Assembly of type IV neuronal intermediate filaments in nonneuronal cells in the absence of preexisting cytoplasmic intermediate filaments. *Journal of Cell Biology*, 122, 1323–1335.

Ching, G.Y., Chien, C.L., Flores, R. and Liem, R.K. (1999) Overexpression of alpha-internexin causes abnormal neurofilamentous accumulations and motor coordination deficits in transgenic mice. *Journal of Neuroscience* 19, 2974–2986.

Chou, S.M., Wang, H.S. and Komai, K. (1996a) Colocalization of NOS and SOD1 in neurofilament accumulation within motor neurons of amyotrophic lateral sclerosis: an immunohistochemical study. *Journal of Chemical Neuroanatomy*, 10, 249–258.

Chou, S.M., Wang, H.S. and Taniguchi, A. (1996b) Role of SOD-1 and nitric oxide/cyclic GMP cascade on neurofilament aggregation in ALS/MND. *Journal of Neurological Science*, 139 Suppl., 16–26.

Chou, S.M., Taniguchi, A., Wang, H.S. and Festoff, B.W. (1998a) Serpin-serine protease-like complexes within neurofilament conglomerates of motoneurons in amyotrophic lateral sclerosis. *Journal of Neurological Science*, 160 Suppl. 1, S73–S79.

Chou, S.M., Wang, H.S., Taniguchi, A. and Bucala, R. (1998b) Advanced glycation end-products in neurofilament conglomeration of motoneurons in familial and sporadic amyotrophic lateral sclerosis. *Molecular Medicine*, 4, 324–332.

Ciechanover, A. (1993) The ubiquitin-mediated proteolytic pathway. *Brain Pathology*, 3, 67–75.

Cleveland, D.W. and Rothstein, J.D. (2001) From Charcot to Lou Gehrig: deciphering selective motor neuron death in ALS. *Nature Reviews Neuroscience*, 2, 806–819.

Comi, G.P., Bordoni, A., Salani, S., Franceschina, L., Sciacco, M., Prelle, A. et al. (1998) Cytochrome c oxidase subunit I microdeletion in a patient with motor neuron disease. *Annals of Neurology*, 43, 110–116.

Cookson, M.R. and Shaw, P.J. (1999) Oxidative stress and motor neurone disease. *Brain Pathology*, 9, 165–186.

Corbo, M. and Hays, A.P. (1992) Peripherin and neurofilament protein coexist in spinal spheroids of motor neuron disease. *Journal of Neuropathology and Experimental Neurology*, 51, 531–537.

Corcia, P., Mayeux-Portas, V., Khoris, J., de Toffel, B., Autret, A., Muh, J.P. et al. (2002) Abnormal SMN1 gene copy number is a susceptibility factor for amyotrophic lateral sclerosis. *Annals of Neurology*, 51, 243–246.

Corson, L.B., Strain, J.J., Culotta, V.C. and Cleveland, D.W. (1998) Chaperone-facilitated copper binding is a property common to several classes of familial amyotrophic lateral sclerosis-linked superoxide dismutase mutants. *Proceedings of the National Academy of Sciences of the United States of America*, 95, 6361–6366.

Cote, F., Collard, J.F. and Julien, J.P. (1993) Progressive neuronopathy in transgenic mice expressing the human neurofilament heavy gene: a mouse model of amyotrophic lateral sclerosis. *Cell*, 73, 35–46.

Cudkowicz, M.E., Kenna-Yasek, D., Sapp, P.E., Chin, W., Geller, B., Hayden, D.L. et al. (1997) Epidemiology of mutations in superoxide dismutase in amyotrophic lateral sclerosis. *Annals of Neurology*, 41, 210–221.

Cudkowicz, M., Qureshi, M. and Shefner, J. (2004) Measures and markers in amyotrophic lateral sclerosis. *NeuroRx*, 1, 273–283.

Dal Canto, M.C. and Gurney, M.E. (1995) Neuropathological changes in two lines of mice carrying a transgene for mutant human Cu,Zn SOD, and in mice overexpressing wild type human SOD: a model of familial amyotrophic lateral sclerosis (FALS). *Brain Research*, 676, 25–40.

Dal Canto, M.C. and Gurney, M.E. (1997) A low expressor line of transgenic mice carrying a mutant human Cu, Zn superoxide dismutase (SOD1) gene develops pathological changes that most closely resemble those in human amyotrophic lateral sclerosis. *Acta Neuropathology (Berl)*, 93, 537–550.

Delfs, J.R., Saroff, D.M., Nishida, Y., Friend, J. and Geula, C. (1997) Effects of NMDA and its antagonists on ventral horn cholinergic neurons in organotypic roller tube spinal cord cultures. *Journal of Neural Transmission*, 104, 31–51.

Deng, H.X., Hentati, A., Tainer, J.A., Iqbal, Z., Cayabyab, A., Hung, W.Y. et al. (1993) Amyotrophic lateral sclerosis and structural defects in Cu,Zn superoxide dismutase. *Science*, 261, 1047–1051.

Ding, H., Schwarz, D.S., Keene, A., Affar, E.B., Fenton, L., Xia, X. et al. (2003) Selective silencing by RNAi of a dominant allele that causes amyotrophic lateral sclerosis. *Aging Cell*, 2, 209–217.

Dinoto, L., Deture, M.A. and Purich, D.L. (2005) Structural insights into Alzheimer filament assembly pathways based on site-directed mutagenesis and S-glutathionylation of three-repeat neuronal Tau protein. *Microscopy Research and Technique*, 67, 156–163.

Drory, V.E., Birnbaum, M., Korczyn, A.D. and Chapman, J. (2001) Association of APOE epsilon4 allele with survival in amyotrophic lateral sclerosis. *Journal of Neurological Science*, 190, 17–20.

Durham, H.D. (1992) An antibody against hyperphosphorylated neurofilament proteins collapses the neurofilament network in motor neurons but not in dorsal root ganglion cells. *Journal of Neuropathology and Experimental Neurology*, 51, 287–297.

Elizan, T.S., Sroka, H., Maker, H., Smith, H. and Yahr, M.D. (1986) Dementia in idiopathic Parkinson's disease. Variables associated with its occurrence in 203 patients. *Journal of Neural Transmission*, 65, 285–302.

Esclaire, F., Kisby, G., Spencer, P., Milne, J., Lesort, M. and Hugon, J. (1999) The Guam cycad toxin methylazoxymethanol damages neuronal DNA and modulates tau mRNA expression and excitotoxicity. *Experimental Neurology*, 155, 11–21.

Escurat, M., Djabali, K., Gumpel, M., Gros, F. and Portier, M.M. (1990) Differential expression of two neuronal intermediate-filament proteins, peripherin and the low-molecular-mass neurofilament protein (NF-L), during the development of the rat. *Journal of Neuroscience*, 10, 764–784.

Figlewicz, D.A., Krizus, A., Martinoli, M.G., Meininger, V., Dib, M., Rouleau, G.A. et al. (1994) Variants of the heavy neurofilament subunit are associated with the development of amyotrophic lateral sclerosis. *Human Molecular Genetics*, 3, 1757–1761.

Fliegner, K.H. and Liem, R.K. (1991) Cellular and molecular biology of neuronal intermediate filaments. *International Review of Cytology*, 131, 109–167.

Fridovich, I. (1986) Superoxide dismutases. *Advances in Enzymology Related Areas of Molecular Biology*, 58, 61–97.

Friedlander, R.M., Brown, R.H., Gagliardini, V., Wang, J. and Yuan, J. (1997) Inhibition of ICE slows ALS in mice. *Nature*, 388, 31.

Ge, W.W., Wen, W., Strong, W., Leystra-Lantz, C. and Strong, M.J. (2005) Mutant copperzinc superoxide dismutase binds to and destabilizes human low molecular weight neurofilament mRNA. *Journal of Biological Chemistry*, 280, 118–124.

Getzoff, E.D., Tainer, J.A., Stempien, M.M., Bell, G.I. and Hallewell, R.A. (1989) Evolution of CuZn superoxide dismutase and the Greek key beta-barrel structural motif. *Proteins*, 5, 322–336.

Ghadge, G.D., Lee, J.P., Bindokas, V.P., Jordan, J., Ma, L., Miller, R.J. and Roos, R.P. (1997) Mutant superoxide dismutase-1-linked familial amyotrophic lateral sclerosis: molecular mechanisms of neuronal death and protection. *Journal of Neuroscience*, 17, 8756–8766.

Gill, S.R., Wong, P.C., Monteiro, M.J. and Cleveland, D.W. (1990) Assembly properties of dominant and recessive mutations in the small mouse neurofilament (NF-L) subunit. *Journal of Cell Biology*, 111, 2005–2019.

Gotz, J., Chen, F., Barmettler, R. and Nitsch, R.M. (2001) Tau filament formation in transgenic mice expressing P301L tau. *Journal of Biological Chemistry*, 276, 529–534.

Gros-Louis, F., Lariviere, R., Gowing, G., Laurent, S., Camu, W., Bouchard, J.P. et al. (2004) A frameshift deletion in peripherin gene associated with amyotrophic lateral sclerosis. *Journal of Biological Chemistry*, 279, 45951–45956.

Gurney, M.E., Pu, H., Chiu, A.Y., Dal Canto, M.C., Polchow, C.Y., Alexander, D.D. et al. (1994) Motor neuron degeneration in mice that express a human Cu,Zn superoxide dismutase mutation. *Science*, 264, 1772–1775.

Hadano, S., Hand, C.K., Osuga, H., Yanagisawa, Y., Otomo, A., Devon, R.S. et al. (2001) A gene encoding a putative GTPase regulator is mutated in familial amyotrophic lateral sclerosis 2. *Nature Genetics*, 29, 166–173.

Hall, E.D., Oostveen, J.A. and Gurney, M.E. (1998) Relationship of microglial and astrocytic activation to disease onset and progression in a transgenic model of familial ALS. *Glia*, 23, 249–256.

Hayes, G.M., Fox, R.M., Cuzner, M.L. and Griffin, G.E. (2000) Human rotation-mediated fetal mixed brain cell aggregate culture: characterization and *N*-methyl-d-aspartate toxicity. *Neuroscience Letters*, 287, 146–150.

Heiman-Patterson, T.D., Deitch, J.S., Blankenhorn, E.P., Erwin, K.L., Perreault, M.J., Alexander, B.K. et al. (2005) Background and gender effects on survival in the TgN (SOD1-G93A)1Gur mouse model of ALS. *Journal of Neurological Science*, 236, 1–7.

Heins, S., Wong, P.C., Muller, S., Goldie, K., Cleveland, D.W. and Aebi, U. (1993) The rod domain of NF-L determines neurofilament architecture, whereas the end domains specify filament assembly and network formation. *Journal of Cell Biology*, 123, 1517–1533.

Heiser, V., Engemann, S., Brocker, W., Dunkel, I., Boeddrich, A., Waelter, S. et al. (2002) Identification of benzothiazoles as potential polyglutamine aggregation inhibitors of Huntington's disease by using an automated filter retardation assay. *Proceedings of the National Academy of Sciences of the United States of America*, 99 Suppl. 4, 16400–16406.

Hentati, A., Ouahchi, K., Pericak-Vance, M.A., Nijhawan, D., Ahmad, A., Yang, Y. et al. (1998) Linkage of a commoner form of recessive amyotrophic lateral sclerosis to chromosome 15q15–q22 markers. *Neurogenetics*, 2, 55–60.

Hisanaga, S., Gonda, Y., Inagaki, M., Ikai, A. and Hirokawa, N. (1990) Effects of phosphorylation of the neurofilament 1 protein on filamentous structures. *Cell Regulation*, 1, 237–248.

Hisanaga, S., Matsuoka, Y., Nishizawa, K., Saito, T., Inagaki, M. and Hirokawa, N. (1994) Phosphorylation of native and reassembled neurofilaments composed of NF-L, NF-M, and NF-H by the catalytic subunit of cAMP-dependent protein kinase. *Molecular Biology of the Cell*, 5, 161–172.

Honig, L.S., Chambliss, D.D., Bigio, E.H., Carroll, S.L. and Elliott, J.L. (2000) Glutamate transporter EAAT2 splice variants occur not only in ALS, but also in AD and controls. *Neurology*, 55, 1082–1088.

Hottinger, A.F., Fine, E.G., Gurney, M.E., Zurn, A.D. and Aebischer, P. (1997) The copper chelator d-penicillamine delays onset of disease and extends survival in a transgenic mouse model of familial amyotrophic lateral sclerosis. *European Journal of Neuroscience*, 9, 1548–1551.

Hutton, M., Lendon, C.L., Rizzu, P., Baker, M., Froelich, S., Houlden, H. et al. (1998) Association of missense and 5'-splice-site mutations in tau with the inherited dementia FTDP-17. *Nature*, 393, 702–705.

Ince, P.G. and Morris, J.C. (2006) Demystifying lobar degenerations: tauopathies vs Gehrigopathies. *Neurology*, 66, 8–9.

Ince, P.G., Lowe, J. and Shaw, P.J. (1998) Amyotrophic lateral sclerosis: current issues in classification, pathogenesis and molecular pathology. *Neuropathology and Applied Neurobiology*, 24, 104–117.

Ishihara, T., Hong, M., Zhang, B., Nakagawa, Y., Lee, M.K., Trojanowski, J.Q. et al. (1999) Age-dependent emergence and progression of a tauopathy in transgenic mice overexpressing the shortest human tau isoform. *Neuron*, 24, 751–762.

Ishihara, T., Higuchi, M., Zhang, B., Yoshiyama, Y., Hong, M., Trojanowski, J.Q. et al. (2001) Attenuated neurodegenerative disease phenotype in tau transgenic mouse lacking neurofilaments. *Journal of Neuroscience*, 21, 6026–6035.

Jackson, M., Morrison, K.E., Al-Chalabi, A., Bakker, M. and Leigh, P.N. (1996) Analysis of chromosome 5q13 genes in amyotrophic lateral sclerosis: homozygous NAIP deletion in a sporadic case. *Annals of Neurology*, 39, 796–800.

Jackson, M., Al-Chalabi, A., Enayat, Z.E., Chioza, B., Leigh, P.N. and Morrison, K.E. (1997) Copper/zinc superoxide dismutase 1 and sporadic amyotrophic lateral sclerosis: analysis of 155 cases and identification of a novel insertion mutation. *Annals of Neurology*, 42, 803–807.

Jackson, M., Steers, G., Leigh, P.N. and Morrison, K.E. (1999) Polymorphisms in the glutamate transporter gene EAAT2 in European ALS patients. *Journal of Neurology*, 246, 1140–1144.

Julien, J.P. (1995) A role for neurofilaments in the pathogenesis of amyotrophic lateral sclerosis. *Biochemistry and Cell Biology*, 73, 593–597.

Julien, J.P. and Beaulieu, J.M. (2000) Cytoskeletal abnormalities in amyotrophic lateral sclerosis: beneficial or detrimental effects? *Journal of Neurological Sciences*, 180, 7–14.

Julien, J.P., Cote, F. and Collard, J.F. (1995) Mice overexpressing the human neurofilament heavy gene as a model of ALS. *Neurobiology of Aging*, 16, 487–490.

Juneja, T., Pericak-Vance, M.A., Laing, N.G., Dave, S. and Siddique, T. (1997) Prognosis in familial amyotrophic lateral sclerosis: progression and survival in patients with glu100gly and ala4val mutations in Cu,Zn superoxide dismutase. *Neurology*, 48, 55–57.

Kanekura, K., Hashimoto, Y., Kita, Y., Sasabe, J., Aiso, S., Nishimoto, I. et al. (2005) A Rac1/phosphatidylinositol 3-kinase/Akt3 anti-apoptotic pathway, triggered by AlsinLF, the product of the ALS2 gene, antagonizes Cu/Zn-superoxide dismutase (SOD1) mutant-induced motoneuronal cell death. *Journal of Biological Chemistry*, 280, 4532–4543.

Kaplan, M.P., Chin, S.S., Fliegner, K.H. and Liem, R.K. (1990) Alpha-internexin, a novel neuronal intermediate filament protein, precedes the low molecular weight neurofilament protein (NF-L) in the developing rat brain. *Journal of Neuroscience*, 10, 2735–2748.

Kato, S., Hayashi, H., Nakashima, K., Nanba, E., Kato, M., Hirano, A. et al. (1997) Pathological characterization of astrocytic hyaline inclusions in familial amyotrophic lateral sclerosis. *American Journal of Pathology*, 151, 611–620.

Kato, S., Saito, M., Hirano, A. and Ohama, E. (1999) Recent advances in research on neuropathological aspects of familial amyotrophic lateral sclerosis with superoxide dismutase 1 gene mutations: neuronal Lewy body-like hyaline inclusions and astrocytic hyaline inclusions. *Histology and Histopathology*, 14, 973–989.

Kato, S., Takikawa, M., Nakashima, K., Hirano, A., Cleveland, D.W., Kusaka, H. et al. (2000) New consensus research on neuropathological aspects of familial amyotrophic lateral sclerosis with superoxide dismutase 1 (SOD1) gene mutations: inclusions containing SOD1 in neurons and astrocytes. *Amyotrophic Lateral Sclerosis and Other Motor Neuron Disorders*, 1, 163–184.

Khlistunova, I., Biernat, J., Wang, Y., Pickhardt, M., von Bergen, M., Gazova, Z. et al. (2006) Inducible expression of Tau repeat domain in cell models of tauopathy: aggregation is toxic to cells but can be reversed by inhibitor drugs. *Journal of Biological Chemistry*, 281, 1205–1214.

Kieran, D., Kalmar, B., Dick, J.R., Riddoch-Contreras, J., Burnstock, G. and Greensmith, L. (2004) Treatment with arimoclomol, a coinducer of heat shock proteins, delays disease progression in ALS mice. *Nature Medicine*, 10, 402–405.

Kikuchi, H., Yamada, T., Furuya, H., Doh-ura, K., Ohyagi, Y., Iwaki, T. et al. (2003) Involvement of cathepsin B in the motor neuron degeneration of amyotrophic lateral sclerosis. *Acta Neuropathology (Berl)*, 105, 462–468.

Klein, S.M., Behrstock, S., McHugh, J., Hoffmann, K., Wallace, K., Suzuki, M. et al. (2005) GDNF delivery using human neural progenitor cells in a rat model of ALS. *Human Gene Therapy*, 16, 509–521.

Kondo, A., Iwaki, T., Tateishi, J., Kirimoto, K., Morimoto, T. and Oomura, I. (1986) Accumulation of neurofilaments in a sporadic case of amyotrophic lateral sclerosis. *Japanese Journal of Psychiatry and Neurology*, 40, 677–684.

Kovacs, A.D., Cebers, G., Cebere, A., Moreira, T. and Liljequist, S. (2001) Cortical and striatal neuronal cultures of the same embryonic origin show intrinsic differences in glutamate receptor expression and vulnerability to excitotoxicity. *Experimental Neurology*, 168, 47–62.

Kriz, J., Nguyen, M.D. and Julien, J.P. (2002) Minocycline slows disease progression in a mouse model of amyotrophic lateral sclerosis. *Neurobiology of Disease*, 10, 268–278.

Lacomblez, L., Doppler, V., Beucler, I., Costes, G., Salachas, F., Raisonnier, A. et al. (2002) APOE: a potential marker of disease progression in ALS. *Neurology*, 58, 1112–1114.

Lambrechts, D., Storkebaum, E., Morimoto, M., Del-Favero, J., Desmet, F., Marklund, S.L. et al. (2003) VEGF is a modifier of amyotrophic lateral sclerosis in mice and humans and protects motoneurons against ischemic death. *Nature Genetics*, 34, 383–394.

Landon, F., Lemonnier, M., Benarous, R., Huc, C., Fiszman, M., Gros, F. and Portier, M.M. (1989) Multiple mRNAs encode peripherin, a neuronal intermediate filament protein. *EMBO Journal*, 8, 1719–1726.

Landon, F., Wolff, A. and de Nechaud, B. (2000) Mouse peripherin isoforms. *Biology of the Cell*, 92, 397–407.

Lariviere, R.C. and Julien, J.P. (2004) Functions of intermediate filaments in neuronal development and disease. *Journal of Neurobiology*, 58, 131–148.

Laslo, P., Lipski, J. and Funk, G.D. (2001) Differential expression of Group I metabotropic glutamate receptors in motoneurons at low and high risk for degeneration in ALS. *Neuroreport*, 12, 1903–1908.

Lee, M.K., Xu, Z., Wong, P.C. and Cleveland, D.W. (1993) Neurofilaments are obligate heteropolymers in vivo. *Journal of Cell Biology*, 122, 1337–1350.

Lee, H.G., Perry, G., Moreira, P.I., Garrett, M.R., Liu, Q., Zhu, X. et al. (2005) Tau phosphorylation in Alzheimer's disease: pathogen or protector? *Trends in Molecular Medicine*, 11, 164–169.

Leigh, P.N., Probst, A., Dale, G.E., Power, D.P., Brion, J.P., Dodson, A. et al. (1989) New aspects of the pathology of neurodegenerative disorders as revealed by ubiquitin antibodies. *Acta Neuropathology (Berl)*, 79, 61–72.

Leung, C.L. and Liem, R.K. (1996) Characterization of interactions between the neurofilament triplet proteins by the yeast two-hybrid system. *Journal of Biological Chemistry*, 271, 14041–14044.

Leung, C.L., He, C.Z., Kaufmann, P., Chin, S.S., Naini, A., Liem, R.K. et al. (2004) A pathogenic peripherin gene mutation in a patient with amyotrophic lateral sclerosis. *Brain Pathology*, 14, 290–296.

Levavasseur, F., Zhu, Q. and Julien, J.P. (1999) No requirement of alpha-internexin for nervous system development and for radial growth of axons. *Brain Research Molecular Brain Research*, 69, 104–112.

Lewis, J., McGowan, E., Rockwood, J., Melrose, H., Nacharaju, P., Van Slegtenhorst, M. et al. (2000) Neurofibrillary tangles, amyotrophy and progressive motor disturbance in mice expressing mutant (P301L) tau protein. *Nature Genetics*, 25, 402–405.

Li, Y., Copin, J.C., Reola, L.F., Calagui, B., Gobbel, G.T., Chen, S.F. et al. (1998) Reduced mitochondrial manganese-superoxide dismutase activity exacerbates glutamate toxicity in cultured cortical neurons. *Brain Research*, 814, 164–170.

Lin, C.L., Bristol, L.A., Jin, L., Dykes-Hoberg, M., Crawford, T., Clawson, L. et al. (1998) Aberrant RNA processing in a neurodegenerative disease: the cause for absent EAAT2, a glutamate transporter, in amyotrophic lateral sclerosis. *Neuron*, 20, 589–602.

Lin, H., Zhai, J. and Schlaepfer, W.W. (2005) RNA-binding protein is involved in aggregation of light neurofilament protein and is implicated in the pathogenesis of motor neuron degeneration. *Human Molecular Genetics*, 14, 3643–3659.

Lobsiger, C.S., Garcia, M.L., Ward, C.M. and Cleveland, D.W. (2005) Altered axonal architecture by removal of the heavily phosphorylated neurofilament tail domains strongly slows superoxide dismutase 1 mutant-mediated ALS. *Proceedings of the National Academy of Sciences of the United States of America*, 102, 10351–10356.

Majoor-Krakauer, D., Willems, P.J. and Hofman, A. (2003) Genetic epidemiology of amyotrophic lateral sclerosis. *Clinical Genetics*, 63, 83–101.

Manetto, V., Perry, G., Tabaton, M., Mulvihill, P., Fried, V.A., Smith, H.T. et al. (1988) Ubiquitin is associated with abnormal cytoplasmic filaments characteristic of neurodegenerative diseases. *Proceedings of the National Academy of Sciences of the United States of America*, 85, 4501–4505.

Matsumoto, A., Okada, Y., Nakamichi, M., Nakamura, M., Toyama, Y., Sobue, G. et al. (2006) Disease progression of human SOD1 (G93A) transgenic ALS model rats. *Journal of Neuroscience Research*, 83, 119–133.

Maxwell, M.M., Pasinelli, P., Kazantsev, A.G. and Brown Jr., R.H. (2004) RNA interference-mediated silencing of mutant superoxide dismutase rescues cyclosporin A-induced death in cultured neuroblastoma cells. *Proceedings of the National Academy of Sciences of the United States of America*, 101, 3178–3183.

McLean, J.R., Sanelli, T.R., Leystra-Lantz, C., He, B.P. and Strong, M.J. (2005) Temporal profiles of neuronal degeneration, glial proliferation, and cell death in hNFL(+/+) and NFL(−/−)mice. *Glia*, 52, 59–69.

Meier, J., Couillard-Despres, S., Jacomy, H., Gravel, C. and Julien, J.P. (1999) Extra neurofilament NF-L subunits rescue motor neuron disease caused by overexpression of the human NF-H gene in mice. *Journal of Neuropathology and Experimental Neurology*, 58, 1099–1110.

Meininger, V., Lacomblez, L. and Salachas, F. (2000) What has changed with riluzole? *Journal of Neurology*, 247, 19–22.

Meyer, T., Munch, C., Volkel, H., Booms, P. and Ludolph, A.C. (1998) The EAAT2 (GLT-1) gene in motor neuron disease: absence of mutations in amyotrophic lateral sclerosis and a point mutation in patients with hereditary spastic paraplegia. *Journal of Neurology Neurosurgery and Psychiatry*, 65, 594–596.

Migheli, A., Pezzulo, T., Attanasio, A. and Schiffer, D. (1993) Peripherin immunoreactive structures in amyotrophic lateral sclerosis. *Laboratory Investigation*, 68, 185–191.

Migheli, A., Cordera, S., Bendotti, C., Atzori, C., Piva, R. and Schiffer, D. (1999) S-100beta protein is upregulated in astrocytes and motor neurons in the spinal cord of patients with amyotrophic lateral sclerosis. *Neuroscience Letters*, 261, 25–28.

Miller, C.C. and Anderton, B.H. (1986) Transglutaminase and the neuronal cytoskeleton in Alzheimer's disease. *Journal of Neurochemistry*, 46, 1912–1922.

Mitsuyama, Y. (2000) Dementia with motor neuron disease. *Neuropathology*, 20, S79–S81.

Moisse, K. and Strong, M.J. (2006) Innate immunity in amyotrophic lateral sclerosis. *Biochimica et Biophysica Acta*, 1762, 1083–1093.

Morita, M., Al-Chalabi, A., Andersen, P.M., Hosler, B., Sapp, P., Englund, E. et al. (2006) A locus on chromosome 9p confers susceptibility to ALS and frontotemporal dementia. *Neurology*, 66, 839–844.

Mott, R.T., Dickson, D.W., Trojanowski, J.Q., Zhukareva, V., Lee, V.M., Forman, M. et al. (2005) Neuropathologic, biochemical, and molecular characterization of the frontotemporal dementias. *Journal of Neuropathology and Experimental Neurology*, 64, 420–428.

Mui, S., Rebeck, G.W., Kenna-Yasek, D., Hyman, B.T. and Brown Jr., R.H. (1995) Apolipoprotein E epsilon 4 allele is not associated with earlier age at onset in amyotrophic lateral sclerosis. *Annals of Neurology*, 38, 460–463.

Nagai, M., Aoki, M., Miyoshi, I., Kato, M., Pasinelli, P., Kasai, N. et al. (2001) Rats expressing human cytosolic copper-zinc superoxide dismutase transgenes with amyotrophic lateral sclerosis: associated mutations develop motor neuron disease. *Journal of Neuroscience*, 21, 9246–9254.

Nagano, S., Ogawa, Y., Yanagihara, T. and Sakoda, S. (1999) Benefit of a combined treatment with trientine and ascorbate in familial amyotrophic lateral sclerosis model mice. *Neuroscience Letters*, 265, 159–162.

Nakamura, Y., Takeda, M., Angelides, K.J., Tanaka, T., Tada, K. and Nishimura, T. (1990) Effect of phosphorylation on 68 KDa neurofilament subunit protein assembly by the cyclic AMP dependent protein kinase in vitro. *Biochemical and Biophysical Research Communications*, 169, 744–750.

Neary, D., Snowden, J.S., Gustafson, L., Passant, U., Stuss, D., Black, S. et al. (1998) Frontotemporal lobar degeneration: a consensus on clinical diagnostic criteria. *Neurology*, 51, 1546–1554.

Nishimura, A.L., Mitne-Neto, M., Silva, H.C., Richieri-Costa, A., Middleton, S., Cascio, D. et al. (2004) A mutation in the vesicle-trafficking protein VAPB causes late-onset spinal muscular atrophy and amyotrophic lateral sclerosis. *American Journal of Human Genetics*, 75, 822–831.

Niwa, J., Ishigaki, S., Hishikawa, N., Yamamoto, M., Doyu, M., Murata, S. et al. (2002) Dorfin ubiquitylates mutant SOD1 and prevents mutant SOD1-mediated neurotoxicity. *Journal of Biological Chemistry*, 277, 36793–36798.

Nixon, R.A. and Shea, T.B. (1992) Dynamics of neuronal intermediate filaments: a developmental perspective. *Cell Motility and the Cytoskeleton*, 22, 81–91.

Okamoto, K., Hirai, S., Amari, M., Watanabe, M. and Sakurai, A. (1993) Bunina bodies in amyotrophic lateral sclerosis immunostained with rabbit anti-cystatin C serum. *Neuroscience Letters*, 162, 125–128.

Orrell, R.W., Habgood, J.J., de Belleroche, J.S. and Lane, R.J. (1997) The relationship of spinal muscular atrophy to motor neuron disease: investigation of SMN and NAIP gene deletions in sporadic and familial ALS. *Journal of Neurological Science*, 145, 55–61.

Parysek, L.M., Chisholm, R.L., Ley, C.A. and Goldman, R.D. (1988) A type III intermediate filament gene is expressed in mature neurons. *Neuron*, 1, 395–401.

Perez, M., Hernandez, F., Gomez-Ramos, A., Smith, M., Perry, G. and Avila, J. (2002) Formation of aberrant phosphotau fibrillar polymers in neural cultured cells. *European Journal of Biochemistry*, 269, 1484–1489.

Piao, Y.S., Wakabayashi, K., Hayashi, S., Yoshimoto, M. and Takahashi, H. (2000) Aggregation of alpha-synuclein/NACP in the neuronal and glial cells in diffuse Lewy body disease: a survey of six patients. *Clinical Neuropathology*, 19, 163–169.

Portier, M.M., de Nechaud, B. and Gros, F. (1983) Peripherin, a new member of the intermediate filament protein family. *Developmental Neuroscience*, 6, 335–344.

Portier, M.M., Escurat, M., Landon, F., Djabali, K. and Bousquet, O. (1993) Peripherin and neurofilaments: expression and role during neural development. *Comptes Rendus de l'Academie des Sciences Serie III*, 316, 1124–1140.

Pramatarova, A., Laganiere, J., Roussel, J., Brisebois, K. and Rouleau, G.A. (2001) Neuron-specific expression of mutant superoxide dismutase 1 in transgenic mice does not lead to motor impairment. *Journal of Neuroscience*, 21, 3369–3374.

Rakhit, R., Cunningham, P., Furtos-Matei, A., Dahan, S., Qi, X.F., Crow, J.P. et al. (2002) Oxidation-induced misfolding and aggregation of superoxide dismutase and its implications for amyotrophic lateral sclerosis. *Journal of Biological Chemistry*, 277, 47551–47556.

Rakhit, R., Crow, J.P., Lepock, J.R., Kondejewski, L.H., Cashman, N.R., Chakrabartty, A. (2004) Monomeric Cu,Zn-superoxide dismutase is a common misfolding intermediate in the oxidation models of sporadic and familial amyotrophic lateral sclerosis. *Journal of Biological Chemistry*, 279, 15499–15504.

Ralph, G.S., Radcliffe, P.A., Day, D.M., Carthy, J.M., Leroux, M.A., Lee, D.C.P. et al. (2005) Silencing mutant SOD1 using RNAi protects against neurodegeneration and extends survival in an ALS model. *Nature Medicine*, 11, 429–433.

Raoul, C., Abbas-Terki, T., Bensadoun, J.-C., Guillot, S., Haase, G., Szulc, J. et al. (2005) Lentiviral-mediated silencing of SOD1 through RNA interference retards disease onset and progression in a mouse model of ALS. *Nature Medicine*, 11, 423–428.

Reaume, A.G., Elliott, J.L., Hoffman, E.K., Kowall, N.W., Ferrante, R.J., Siwek, D.F. et al (1996) Motor neurons in Cu/Zn superoxide dismutase-deficient mice develop normally but exhibit enhanced cell death after axonal injury. *Nature Genetics*, 13, 43–47.

Ripps, M.E., Huntley, G.W., Hof, P.R., Morrison, J.H. and Gordon, J.W. (1995) Transgenic mice expressing an altered murine superoxide dismutase gene provide an animal model of amyotrophic lateral sclerosis. *Proceedings of the National Academy of Sciences of the United States of America*, 92, 689–693.

Robertson, J., Beaulieu, J.M., Doroudchi, M.M., Durham, H.D., Julien, J.P. and Mushynski, W.E. (2001) Apoptotic death of neurons exhibiting peripherin aggregates is mediated by the proinflammatory cytokine tumor necrosis factor-alpha. *Journal of Cell Biology*, 155, 217–226.

Robertson, J., Kriz, J., Nguyen, M.D. and Julien, J.P. (2002) Pathways to motor neuron degeneration in transgenic mouse models. *Biochimie*, 84, 1151–1160.

Robertson, J., Doroudchi, M.M., Nguyen, M.D., Durham, H.D., Strong, M.J., Shaw, G. et al. (2003) A neurotoxic peripherin splice variant in a mouse model of ALS. *Journal of Cell Biology*, 160, 939–949.

Rosen, D.R., Siddique, T., Patterson, D., Figlewicz, D.A., Sapp, P., Hentati, A. et al. (1993) Mutations in Cu/Zn superoxide dismutase gene are associated with familial amyotrophic lateral sclerosis. *Nature*, 362, 59–62.

Rothstein, J.D., Tsai, G., Kuncl, R.W., Clawson, L., Cornblath, D.R., Drachman, D.B. et al. (1990) Abnormal excitatory amino acid metabolism in amyotrophic lateral sclerosis. *Annals of Neurology*, 28, 18–25.

Saito, Y., Yokota, T., Mitani, T., Ito, K., Anzai, M., Miyagishi, M. et al. (2005) Transgenic small interfering RNA halts amyotrophic lateral sclerosis in a mouse model. *Journal of Biological Chemistry*, 280, 42826–42830.

Sanelli, T.R., Sopper, M.M. and Strong, M.J. (2004) Sequestration of nNOS in neurofilamentous aggregate bearing neurons in vitro leads to enhanced NMDA-mediated calcium influx. *Brain Research*, 1004, 8–17.

Saroff, D., Delfs, J., Kuznetsov, D. and Geula, C. (2000) Selective vulnerability of spinal cord motor neurons to non-NMDA toxicity. *Neuroreport*, 11, 1117–1121.

Sasaki, S., Komori, T. and Iwata, M. (2006) Neuronal inclusions in sporadic motor neuron disease are negative for alpha-synuclein. *Neuroscience Letters*, 397, 15–19.

Schiefer, J., Landwehrmeyer, G.B., Luesse, H.G., Sprunken, A., Puls, C., Milkereit, A. et al. (2002) Riluzole prolongs survival time and alters nuclear inclusion formation in a transgenic mouse model of Huntington's disease. *Movement Disorders*, 17, 748–757.

Schiffer, D., Cordera, S., Cavalla, P. and Migheli, A. (1996) Reactive astrogliosis of the spinal cord in amyotrophic lateral sclerosis. *Journal of Neurological Science*, 139 Suppl., 27–33.

Schmalbruch, H., Jensen, H.J., Bjaerg, M., Kamieniecka, Z. and Kurland, L. (1991) A new mouse mutant with progressive motor neuronopathy. *Journal of Neuropathology and Experimental Neurology*, 50, 192–204.

Shaw, P.J. (2005) Molecular and cellular pathways of neurodegeneration in motor neurone disease. *Journal of Neurology Neurosurgery and Psychiatry*, 76, 1046–1057.

Shaw, C.E., Enayat, Z.E., Chioza, B.A., Al-Chalabi, A., Radunovic, A., Powell, J.F. et al. (1998) Mutations in all five exons of SOD-1 may cause ALS. *Annals of Neurology*, 43, 390–394.

Shibata, N., Hirano, A., Kato, S., Nagai, R., Horiuchi, S., Komori, T. et al. (1999) Advanced glycation endproducts are deposited in neuronal hyaline inclusions: a study on familial amyotrophic lateral sclerosis with superoxide dismutase-1 mutation. *Acta Neuropathology (Berl)*, 97, 240–246.

Shibata, N., Hirano, A., Yamamoto, T., Kato, Y. and Kobayashi, M. (2000a) Superoxide dismutase-1 mutation-related neurotoxicity in familial amyotrophic lateral sclerosis. *Amyotrophic Lateral Sclerosis and Other Motor Neuron Disorders*, 1, 143–161.

Shibata, N., Nagai, R., Miyata, S., Jono, T., Horiuchi, S., Hirano, A. et al. (2000b) Nonoxidative protein glycation is implicated in familial amyotrophic lateral sclerosis with superoxide dismutase-1 mutation. *Acta Neuropathology (Berl)*, 100, 275–284.

Siddique, T., Nijhawan, D. and Hentati, A. (1997) Familial amyotrophic lateral sclerosis. *Journal of Neural Transmissions*, 49 Suppl., 219–233.

Siddons, M.A., Pickering-Brown, S.M., Mann, D.M., Owen, F. and Cooper, P.N. (1996) Debrisoquine hydroxylase gene polymorphism frequencies in patients with amyotrophic lateral sclerosis. *Neuroscience Letters*, 208, 65–68.

Singer, S.M., Zainelli, G.M., Norlund, M.A., Lee, J.M. and Muma, N.A. (2002) Transglutaminase bonds in neurofibrillary tangles and paired helical filament tau early in Alzheimer's disease. *Neurochemistry International*, 40, 17–30.

Sinor, J.D., Du, S., Venneti, S., Blitzblau, R.C., Leszkiewicz, D.N., Rosenberg, P.A. and Aizenman, E. (2000) NMDA and glutamate evoke excitotoxicity at distinct cellular locations in rat cortical neurons in vitro. *Journal of Neuroscience*, 20, 8831–8837.

Smith, M.A., Monnier, V.M., Sayre, L.M. and Perry, G. (1995) Amyloidosis, advanced glycation end products and Alzheimer disease. *Neuroreport*, 6, 1595–1596.

Spittaels, K., Van den Haute, C., Van Dorpe, J., Bruynseels, K., Vandezande, K., Laenen, I. et al. (1999) Prominent axonopathy in the brain and spinal cord of transgenic mice overexpressing four-repeat human tau protein. *American Journal of Pathology*, 155, 2153–2165.

Spittaels, K., Van den Haute, C., Van Dorpe, J., Geerts, H., Mercken, M., Bruynseels, K. et al. (2000) Glycogen synthase kinase-3beta phosphorylates protein tau and rescues the axonopathy in the central nervous system of human four-repeat tau transgenic mice. *Journal of Biological Chemistry*, 275, 41340–41349.

Stamer, K., Vogel, R., Thies, E., Mandelkow, E. and Mandelkow, E.M. (2002) Tau blocks traffic of organelles, neurofilaments, and APP vesicles in neurons and enhances oxidative stress. *Journal of Cell Biology*, 156, 1051–1063.

Storkebaum, E., Lambrechts, D. and Carmeliet, P. (2004) VEGF: once regarded as a specific angiogenic factor, now implicated in neuroprotection. *Bioessays*, 26, 943–954.

Storkebaum, E., Lambrechts, D., Dewerchin, M., Moreno-Murciano, M.P., Appelmans, S., Oh, H. et al. (2005) Treatment of motoneuron degeneration by intracerebroventricular delivery of VEGF in a rat model of ALS. *Nature Neuroscience*, 8, 85–92.

Strong, M.J. (1999) Neurofilament metabolism in sporadic amyotrophic lateral sclerosis. *Journal of Neurological Science*, 169, 170–177.

Strong, M.J. (2001) Progress in clinical neurosciences: the evidence for ALS as a multisystems disorder of limited phenotype expression. *Canadian Journal of Neurological Science*, 28, 283–298.

Strong, M.J. (2003) The basic aspects of therapeutics in amyotrophic lateral sclerosis. *Pharmacology & Therapeutics*, 98, 379–414.

Strong, M.J. (2004) Amyotrophic lateral sclerosis: contemporary concepts in etiopathogenesis and pharmacotherapy. *Expert Opinion in Investigational Drugs*, 13, 1593–1614.

Strong, M.J., Grace, G.M., Orange, J.B. and Leeper, H.A. (1996) Cognition, language, and speech in amyotrophic lateral sclerosis: a review. *Journal of Clinical and Experimental Neuropsychology*, 18, 291–303.

Strong, M.J., Sopper, M. and He, B.P. (2003) In vitro reactive nitrating species toxicity in dissociated spinal motor neurons from NFL $(-/-)$ and hNFL $(+/+)$ transgenic mice. *Amyotrophic Lateral Sclerosis and Other Motor Neuron Disorders*, 4, 81–89.

Strong, M.J., Kesavapany, S. and Pant, H.C. (2005a) The pathobiology of amyotrophic lateral sclerosis: a proteinopathy? *Journal of Neuropathology and Experimental Neurology*, 64, 649–664.

Strong, M.J., McLean, J. and Julien, J.-P. (2005b) Cytoskeleton in amyotrophic lateral sclerosis, In *Amyotrophic Lateral Sclerosis*, Mitsumoto, H., Przedborski, S., Gordon, P., and De Bene, M. (eds). Marcel Dekker Inc., London.

Strong, M.J., Yang, W., Strong, W.L., Leystra-Lantz, C., Jaffe, H. and Pant, H.C. (2006) Tau protein hyperphosphorylation in sporadic ALS with cognitive impairment. *Neurology*, 66, 1770–1771.

Subramaniam, J.R., Lyons, W.E., Liu, J., Bartnikas, T.B., Rothstein, J. et al. (2002) Mutant SOD1 causes motor neuron disease independent of copper chaperone-mediated copper loading. *Nature Neuroscience*, 5, 301–307.

Tainer, J.A., Getzoff, E.D., Beem, K.M., Richardson, J.S. and Richardson, D.C. (1982) Determination and analysis of the 2 A-structure of copper, zinc superoxide dismutase. *Journal of Molecular Biology*, 160, 181–217.

Takehisa, Y., Ujike, H., Ishizu, H., Terada, S., Haraguchi, T., Tanaka, Y. et al. (2001) Familial amyotrophic lateral sclerosis with a novel Leu126Ser mutation in the copper/zinc superoxide dismutase gene showing mild clinical features and lewy body-like hyaline inclusions. *Archives of Neurology*, 58, 736–740.

Takeuchi, H., Kobayashi, Y., Yoshihara, T., Niwa, J., Doyu, M., Ohtsuka, K. and Sobue, G. (2002) Hsp70 and Hsp40 improve neurite outgrowth and suppress intracytoplasmic aggregate formation in cultured neuronal cells expressing mutant SOD1. *Brain Research*, 949, 11–22.

Tanemura, K., Murayama, M., Akagi, T., Hashikawa, T., Tominaga, T., Ichikawa, M. et al. (2002) Neurodegeneration with tau accumulation in a transgenic mouse expressing V337M human tau. *Journal of Neuroscience*, 22, 133–141.

Terwel, D., Lasrado, R., Snauwaert, J., Vandeweert, E., Van Haesendonck, C., Borghgraef, P. and Van Leuven, F. (2005) Changed conformation of mutant Tau-P301L underlies the moribund tauopathy, absent in progressive, nonlethal axonopathy of Tau-4R/2N transgenic mice. *Journal of Biological Chemistry*, 280, 3963–3973.

Tikka, T.M. and Koistinaho, J.E. (2001) Minocycline provides neuroprotection against N-methyl-d-aspartate neurotoxicity by inhibiting microglia. *Journal of Immunology*, 166, 7527–7533.

Tikka, T., Fiebich, B.L., Goldsteins, G., Keinanen, R. and Koistinaho, J. (2001a) Minocycline, a tetracycline derivative, is neuroprotective against excitotoxicity by inhibiting activation and proliferation of microglia. *Journal of Neuroscience*, 21, 2580–2588.

Tikka, T., Usenius, T., Tenhunen, M., Keinanen, R. and Koistinaho, J. (2001b) Tetracycline derivatives and ceftriaxone, a cephalosporin antibiotic, protect neurons against apoptosis induced by ionizing radiation. *Journal of Neurochemistry*, 78, 1409–1414.

Tikka, T.M., Vartiainen, N.E., Goldsteins, G., Oja, S.S., Andersen, P.M. et al. (2002) Minocycline prevents neurotoxicity induced by cerebrospinal fluid from patients with motor neurone disease. *Brain*, 125, 722–731.

Tomkins, J., Usher, P., Slade, J.Y., Ince, P.G., Curtis, A., Bushby, K. and Shaw, P.J. (1998) Novel insertion in the KSP region of the neurofilament heavy gene in amyotrophic lateral sclerosis (ALS). *Neuroreport*, 9, 3967–3970.

Trotti, D., Rolfs, A., Danbolt, N.C., Brown Jr., R.H., and Hediger, M.A. (1999) SOD1 mutants linked to amyotrophic lateral sclerosis selectively inactivate a glial glutamate transporter. *Nature Neuroscience*, 2, 848.

Trotti, D., Aoki, M., Pasinelli, P., Berger, U.V., Danbolt, N.C., Brown Jr., R.H., et al. (2001) Amyotrophic lateral sclerosis-linked glutamate transporter mutant has impaired glutamate clearance capacity. *Journal of Biological Chemistry*, 276, 576–582.

Troy, C.M., Muma, N.A., Greene, L.A., Price, D.L. and Shelanski, M.L. (1990) Regulation of peripherin and neurofilament expression in regenerating rat motor neurons. *Brain Research*, 529, 232–238.

Tu, P.H., Raju, P., Robinson, K.A., Gurney, M.E., Trojanowski, J.Q. and Lee, V.M. (1996) Transgenic mice carrying a human mutant superoxide dismutase transgene develop neuronal cytoskeletal pathology resembling human amyotrophic lateral sclerosis lesions. *Proceedings of the National Academy of Sciences of the United States of America*, 93, 3155–3160.

Tymianski, M., Sattler, R., Zabramski, J.M. and Spetzler, R.F. (1998) Characterization of neuroprotection from excitotoxicity by moderate and profound hypothermia in cultured cortical neurons unmasks a temperature-insensitive component of glutamate neurotoxicity. *Journal of Cerebral Blood Flow and Metabolism*, 18, 848–867.

Valentine, J.S. and Hart, P.J. (2003) Misfolded CuZnSOD and amyotrophic lateral sclerosis. *Proceedings of the National Academy of Sciences of the United States of America*, 100, 3617–3622.

Van Den Bosch, L., Tilkin, P., Lemmens, G. and Robberecht, W. (2002) Minocycline delays disease onset and mortality in a transgenic model of ALS. *Neuroreport*, 13, 1067–1070.

Van Den Bosch, L., Storkebaum, E., Vleminckx, V., Moons, L., Vanopdenbosch, L., Scheveneels, W. et al. (2004) Effects of vascular endothelial growth factor (VEGF) on motor neuron degeneration. *Neurobiology Disorders*, 17, 21–28.

Van Vught, P.W., Sutedja, N.A., Veldink, J.H., Koeleman, B.P., Groeneveld, G.J. et al. (2005) Lack of association between VEGF polymorphisms and ALS in a Dutch population. *Neurology*, 65, 1643–1645.

Vance, C., Al-Chalabi, A., Ruddy, D., Smith, B.N., Hu, X., Sreedharan, J. et al. (2006) Familial amyotrophic lateral sclerosis with frontotemporal dementia is linked to a locus on chromosome 9p13.2–21.3. *Brain,* 129, 868–876.

Veldink, J.H., Kalmijn, S., Van der Hout, A.H., Lemmink, H.H., Groeneveld, G.J., Lummen, C. et al. (2005) SMN genotypes producing less SMN protein increase susceptibility to and severity of sporadic ALS. *Neurology,* 65, 820–825.

Wada, M., Uchihara, T., Nakamura, A. and Oyanagi, K. (1999) Bunina bodies in amyotrophic lateral sclerosis on Guam: a histochemical, immunohistochemical and ultrastructural investigation. *Acta Neuropathology (Berl),* 98, 150–156.

Weingarten, M.D., Lockwood, A.H., Hwo, S.Y. and Kirschner, M.W. (1975) A protein factor essential for microtubule assembly. *Proceedings of the National Academy of Sciences of the United States of America,* 72, 1858–1862.

Weisiger, R.A. and Fridovich, I. (1973) Superoxide dismutase. Organelle specificity. *Journal of Biological Chemistry,* 248, 3582–3592.

Wiederholt, W.C. (1999) Neuroepidemiologic research initiatives on Guam: past and present. *Neuroepidemiology,* 18, 279–291.

Wilson, C.M., Grace, G.M., Munoz, D.G., He, B.P. and Strong, M.J. (2001) Cognitive impairment in sporadic ALS: a pathologic continuum underlying a multisystem disorder. *Neurology,* 57, 651–657.

Wong, J. and Oblinger, M.M. (1990) Differential regulation of peripherin and neurofilament gene expression in regenerating rat DRG neurons. *Journal of Neuroscience Research,* 27, 332–341.

Wong, P.C., Marszalek, J., Crawford, T.O., Xu, Z., Hsieh, S.T., Griffin, J.W. et al. (1995) Increasing neurofilament subunit NF-M expression reduces axonal NF-H, inhibits radial growth, and results in neurofilamentous accumulation in motor neurons. *Journal of Cell Biology,* 130, 1413–1422.

Wood, J.D., Beaujeux, T.P. and Shaw, P.J. (2003) Protein aggregation in motor neurone disorders. *Neuropathology and Applied Neurobiology,* 29, 529–545.

Xia, X.G., Zhou, H., Zhou, S., Yu, Y., Wu, R. and Xu, Z. (2005) An RNAi strategy for treatment of amyotrophic lateral sclerosis caused by mutant Cu,Zn superoxide dismutase. *Journal of Neurochemistry,* 92, 362–367.

Xu, Z., Cork, L.C., Griffin, J.W. and Cleveland, D.W. (1993) Increased expression of neurofilament subunit NF-L produces morphological alterations that resemble the pathology of human motor neuron disease. *Cell,* 73, 23–33.

Yang, Y., Hentati, A., Deng, H.X., Dabbagh, O., Sasaki, T., Hirano, M. et al. (2001) The gene encoding alsin, a protein with three guanine-nucleotide exchange factor domains, is mutated in a form of recessive amyotrophic lateral sclerosis. *Nature Genetics,* 29, 160–165.

Yang, W., Sopper, M.M., Leystra-Lantz, C. and Strong, M.J. (2003) Microtubule-associated tau protein positive neuronal and glial inclusions in ALS. *Neurology,* 61, 1766–1773.

Yoshida, M. (2004) Amyotrophic lateral sclerosis with dementia: the clinicopathological spectrum. *Neuropathology,* 24, 87–102.

Zhang, B., Tu, P., Abtahian, F., Trojanowski, J.Q. and Lee, V.M. (1997) Neurofilaments and orthograde transport are reduced in ventral root axons of transgenic mice that express human SOD1 with a G93A mutation. *Journal of Cell Biology,* 139, 1307–1315.

Part V

Transmissible Spongiform
Encephalopathies

13 Transmissible Spongiform Encephalopathies

Michael D. Geschwind and Giuseppe Legname

CONTENTS

Prion (pree-ahn) diseases, also referred to as transmissible spongiform encephalopathies (TSEs), are a group of fatal neurodegenerative diseases that are unique in that they occur in humans in three modes: sporadic (spontaneous or of unknown etiology), genetic, and acquired or infectious. No other disease is known to occur in these three ways. The discovery by Stanley B. Prusiner that the accumulation of a misshapen protein, the prion, can cause disease has led to a new way of thinking about these and other neurodegenerative diseases. The field of prion disease has made remarkable advances over the past few years, possibly more so than for most other neurodegenerative diseases, including Alzheimer's disease. Despite many advances in our understanding of the biology of prions, prion diseases remain mysterious and still incurable, with new forms in humans (Will et al., 1996) and in domestic animals (Casalone et al., 2004) and ongoing epidemics in wild animals (Williams, 2005).

13.1   WHAT ARE PRIONS?

Prion diseases occur in humans and animals. Prion diseases in animals include scrapie in sheep and goats, bovine spongiform encephalopathy (BSE), transmissible mink encephalopathy, chronic wasting disease (CWD) of deer, elk, and moose, and exotic ungulate encephalopathy (Williams, 2005).

For many years, prion diseases were mistakenly thought to be due to "slow viruses," in part due to the transmissibility of the diseases and the long incubation period between exposure and symptom onset (Gajdusek, 1977; Brown et al., 1986). Further research, however, found that the infectious agent did not contain nucleic acid, a component of viruses. Treating prion contaminated material with methods that would typically inactivate viruses and other microorganisms did not prevent these diseases from being experimentally transmitted; yet methods that denatured or destroyed proteins prevented transmission, thus providing strong evidence that the causative agent was a protein (Gajdusek et al., 1977; Prusiner, 1982). In 1997, Stanley B. Prusiner received the Nobel Prize in Medicine and Physiology for his work on identifying the prion (Prusiner, 1998a).

Prion diseases are now known to be caused by changes in the conformation of the endogenous prion protein, PrP^C (the normal cellular form of the prion protein, PrP), into an abnormally shaped, disease-causing form of PrP called the prion, or PrP^{Sc} (in which Sc stands for scrapie, the prion disease of sheep and goats). PrP^C contains three α-helixes (spirals) and little β sheet (flat) structure, whereas PrP^{Sc} has less α-helical content and mostly β sheet structure (Prusiner, 1998b, 2001). PrP^C and PrP^{Sc} have the same primary polypeptide sequence, but different secondary and tertiary structures. PrP^{Sc} is produced by conversion (rearranging the folding) of existing PrP^C into PrP^{Sc}. The process by which PrP^{Sc} is made from PrP^C is not completely understood. It is believed that this occurs through PrP^C coming into contact with PrP^{Sc} and thereby PrP^C is induced to change into the shape of PrP^{Sc} (Prusiner, 1998a). Although it is clear that PrP^C is necessary for prion disease, there is still debate regarding whether other proteins or molecules are involved in the conformational change in vivo (Telling et al., 1995; Wong et al., 2001b; Deleault et al., 2003). The recent discovery that recombinant mouse (Mo) PrP, expressed in *Escherichia coli*, may be infectious to mice when polymerized into amyloid fibrils has opened new avenues of research in the prion field (Legname et al., 2004). In fact, characterizations of these newly generated synthetic prions showed novel features associated with neuropathological changes in mouse models of prion disease (Legname et al., 2005). These conformational changes appeared to confer increasing stability to PrP^{Sc}, as measured by the amount of guanidine hydrochloride necessary to completely unfold PrP^{Sc}. Moreover, a direct correlation is found when the measure of stability of any particular isolates is expressed as a function of mouse survival times to the disease (Legname et al., 2006). The latter work has indicated that PrP^{Sc} may exist as a large number of different conformers and may be persuasive evidence for the molecular basis for the existence of prion strains (Legname et al., 2006).

13.2 FUNCTION OF PrP^C

13.2.1 Protein Structure

The precise function of PrP^C has not yet been elucidated, however because it is evolutionarily conserved, it probably plays an important role in neuronal develop-ment and physiology (Kanaani et al., 2005). In humans, it is encoded by the *PRNP* gene located on the short arm of chromosome 20 (Oesch et al., 1985; Basler et al., 1986). The mature PrP^C protein is attached to the outer cell membrane by a glycosylphosphatidylinositol (GPI) anchor (Borchelt et al., 1992, 1993; Taraboulos et al., 1992). Transmembrane forms of PrP^C have been identified although their function is also debated (Hay et al., 1987; Hegde et al., 1998, 1999). In mice the *PRNP* gene is located in chromosome 2 and a single exon, exon 3, contains the open reading frame (ORF) encoding for the prion protein (Basler et al., 1986; Westaway et al., 1987). The primary structure of the protein consists of 254 amino acids. Two signal sequences are present in the pre-pro-protein: an endoplasmic reticulum trans-location signal peptide is at the amino terminal, and a GPI peptide signal sequence is at the carboxy terminus, for correct anchoring to the plasma membrane. The mature 209 residues protein possesses a single disulphide bridge between Cys179 and Cys213 (mouse PrP numbering), and two sites for Asn-linked glycosylation within the carboxy-terminal region (Figure 13.1).

13.2.2 Function of PrP^C Derived from Knockout Mice and Cell Models

Several lines of knockout mice have been developed in which the prion gene has been deleted. In these models, typically either the entire ORF of exon 3 of *PRNP* or

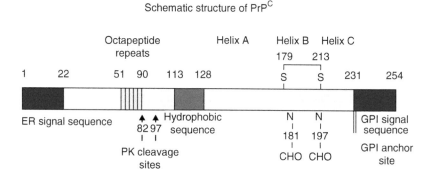

FIGURE 13.1 Structure of the prion protein, PrP^c.

the ORF as well as flanking sequences are deleted (Weissmann and Flechsig, 2003). Mice devoid of PrP^C cannot be infected with, nor can they replicate prions, providing strong evidence that PrP^C is necessary for prion disease (Bueler et al., 1993; Prusiner et al., 1993; Katamine et al., 1998). Knockout mice due to deletion of the ORF (abbreviation: *Prnp*$^{0/0}$ or *Prnp*$^{-/-}$) appear to develop and reproduce normally (Bueler et al., 1992; Manson et al., 1994), but further evaluation of these mice found several abnormalities. Although clinically asymptomatic, they develop peripheral nerve demyelination (Nishida et al., 1999), have increased susceptibility to ischemic brain injury (Spudich et al., 2005), altered sleep and circadian rhythm (Tobler et al., 1996, 1997), and altered hippocampal neuropathology and physiology, including deficits in hippocampal-dependent spatial learning and hippocampal synaptic plasticity (Colling et al., 1997; Criado et al., 2005). Mice or cell lines devoid of prion PrP^C, *Prnp*$^{0/0}$, are also more susceptible to oxidative stress and PrP^C appears to play a neuroprotective role in cellular response to hypoxic-ischemic injury (Kuwahara et al., 1999; Klamt et al., 2001; Wong et al., 2001a; Brown et al., 2002; Miele et al., 2002; McLennan et al., 2004; Spudich et al., 2005; Weise et al., 2006). Knockout mice in which a larger region of PrP^C, extending beyond the ORF, is deleted also develop normally, but develop ataxia and Purkinje cell loss later in life (Sakaguchi et al., 1996; Katamine et al., 1998; Rossi et al., 2001). For an excellent review on PrP^C knock-out mice, see Weissmann et al. (2003).

13.2.3 FUNCTION OF PrP^C DERIVED FROM BIOCHEMICAL INTERACTION WITH OTHER PROTEINS

Many attempts to define a physiological role of PrP^C have focused on identifying interacting molecules. PrP^C has been found to interact with a number of proteins, including bacterial HSP-60 (Edenhofer et al., 1996), the 37 kDa/67 kDa laminin receptor (Rieger et al., 1997), laminin (Graner et al., 2000), neural cell adhesion molecule (NCAM) (Schmitt-Ulms et al., 2001), and the Grb2 protein, which is central in many signal transduction pathways (Spielhaupter and Schätzl, 2001). In search of interacting ligands, one of us (G.L.) showed the staining of several regions of the brain using novel PrP-Fc fusion proteins. In particular, the granule cell layer of the cerebellum showed intense staining, suggesting the presence of a PrP^C-interacting ligand(s) in this region (Legname et al., 2002). In one study, a PrP-Fc fusion protein was shown to moderately increase neurite extension and neuronal survival in primary cultures of mouse cerebellar granule cells (Chen et al., 2003; Santuccione et al., 2005). In another report, recombinant (rec) PrP was incubated with primary cultures of embryonic rat hippocampal neurons and it was shown to promote neuronal polarization and synaptic formation. Only full-length PrP led to the dramatic effect of inducing rapid neuritogenesis while a truncated form of PrP, lacking residues 23–88, failed to promote differentiation (Kanaani et al., 2005). The latter study suggests that experimental efforts should be focused on the N-terminal domain of PrP^C, which is known to contain motifs capable of coordinating metal ions. PrP^C is known to bind copper and may have a copper-binding function in vivo (Brown et al., 1997). As mentioned earlier with PrP knockouts, PrP^C may be neuroprotective either as consequence of its primary function or as concomitant mechanism with other proteins. PrP^C appears to be upregulated after cerebral ischemia, and overexpression of PrP^C in an ischemia mouse model seems to be protective (Shyu et al., 2005). Overexpression of PrP^C in transgenic mice leads to a dose-dependent disease characterized by truncal ataxia, hindlimb paralysis, and tremors; these mice had a necrotizing myopathy of skeletal muscle, a demyelinating polyneuropathy, and focal vacuolation of the central nervous system (Westaway et al., 1994; DeArmond and Prusiner, 2003). There are conflicting data on whether PrP^C may be pro- or anti-apoptotic (Kuwahara et al., 1999; Chabry et al., 2003; Spudich et al., 2005; Weise et al., 2006). PrP^C may also play a role in neuronal excitability (Mallucci et al., 2002), neuritigenesis (Kuwahara et al., 1999; Kanaani et al., 2005; Santuccione et al., 2005), and signal transduction (Spudich et al., 2005). Nevertheless, the precise roles of PrP^C are still the subject of ongoing controversy.

13.3   HUMAN PRION DISEASE

In 1921 and 1923, Alfons Jakob published four papers describing five unusual cases of rapidly progressive dementia. In one paper, he stated that his cases were nearly identical to a case described by Hans Creutzfeldt in 1920 (Creutzfeldt, 1989; Jakob, 1989; Katscher, 1998). This disease was referred to for the next few decades as Jakob–Creutzfeldt disease (Manuelidis and Manuelidis, 1989; Katscher, 1998). For years, the disease was called Jakob–Creutzfeldt disease until Clarence J. Gibbs, a prominent researcher in the field, started using the term Creutzfeldt–Jakob disease because the acronym was closer to his own initials (Gibbs, 1992). In retrospect, the cases that Jakob described were very different than Creutzfeldt's case and only two of Jakob's five cases actually had the disease that we now call Creutzfeldt–Jakob disease. Creutzfeldt's case also did not have what we now call CJD (Masters, 1989). Thus, the name of the prion disease should be Jakob's disease or possibly Jakob–Creutzfeldt disease. Unfortunately, the term JC disease or JCD currently may be confused with the neurological condition caused by the JC virus, progressive multifocal leukoencephalopathy. In this chapter, we use the terms Jakob–Creutzfeldt disease and CJD for Jakob's disease.

The most common form of human prion disease, sporadic CJD (sCJD), to the best of our knowledge occurs spontaneously—we do not know the cause. sCJD accounts for about 85% of human prion disease. Genetic forms account for 10%–15% and acquired cases account for less than 1% of human prion diseases. In most Western countries, prion diseases occur at a rate of 1–2 per million/year. In a recent study of prion disease, mortality rates from 1999 to 2002 in nine European countries, in addition to Canada and Australia, were 1.67 per million for all forms of prion disease and 1.39 for sCJD (Ladogana et al., 2005).

13.3.1   Sporadic Prion Disease

13.3.1.1   sCJD

Median survival of sCJD is reported to be about 4 months and mean of about 5–8 months, with about 85%–90% of cases surviving 1 year or less from disease onset

(Brown et al., 1994; Pocchiari et al., 2004; Collins et al., 2006). In our own cohort of more than 200 referred cases, mean survival is about 12 months (unpublished data), much longer than that reported in the prior literature. This may be because we typically based the first symptom (disease onset) on extensive patient and family interviews occurring over several days, in addition to medical record reviews. Through this labor-intensive approach, we typically found the first symptom to be often weeks or months earlier than stated in medical records, thus increasing disease duration (Rabinovici et al., 2006). In a cohort of 114 sCJD patients, we found the most common first symptom was cognitive (49% of patients), followed equally by behavioral (24%), constitutional (24%), and cerebellar (21%). Motor (noncerebellar; 10%), sensory (9%), and visual (7%) signs or symptoms were less common. The distribution of symptoms within each of the five most common major categories is shown in Figure 13.2 (Rabinovici et al., 2006).

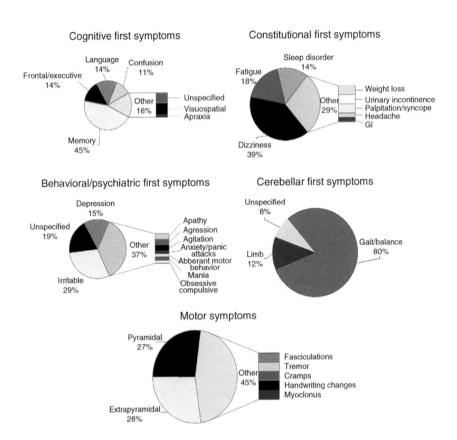

FIGURE 13.2 First symptoms in sCJD. This figure shows the relative distribution of symptoms within each of five major symptom categories: cognitive, behavioral, motor (noncerebellar), constitutional, and cerebellar. See Rabinovici et al. (2006) for more background.

13.3.1.1.1 Diagnosis of sCJD

Clinical Criteria for sCJD. The two most commonly used diagnostic criteria for sCJD are those from WHO (1998) and Masters et al. (1979) (see Table 13.1) . These criteria are divided into three categories of diagnostic certainty: definite, probable, or possible. Definite always requires pathological evidence, including the presence of PrP^{Sc} (Kretzschmar et al., 1996). Master's criteria are very sensitive, but unfortunately not very specific; they are so broad that most neurodegenerative dementias and many other nonprion rapidly progressive dementing conditions (RPDs) will fulfill them. Many patients with neurodegenerative conditions other than prion disease, such as corticobasal degeneration (CBD), progressive supranuclear palsy (PSP), dementia with Lewy bodies (DLB), and multiple system atrophy (MSA), would fulfill these criteria. WHO criteria are more specific and have been widely accepted, but they too have significant short-comings. Neither a typical EEG nor the 14-3-3 protein is required for WHO possible sCJD diagnosis. By requiring at least two symptoms, WHO criteria are more specific than Masters', but they also have other deficiencies. WHO criteria include akinetic mutism, which occurs at the very end-stage of prion disease, too late for any potential treatment to be effective. Furthermore, cerebellar

TABLE 13.1
Several Diagnostic Criteria for Probable sCJD

Masters (Masters et al., 1979) Require Progressive Dementia with Any ≥1 of the Following:	WHO (WHO, 1998) Require Progressive Dementia with ≥2 of the Following:	UCSF Require Progressive Dementia with ≥2 of the Following:
Myoclonus	Myoclonus	Myoclonus
Pyramidal signs	Visual or cerebellar disturbance	Pyramidal/extrapyramidal dysfunction
A characteristic EEG	Pyramidal/extrapyramidal dysfunction	Visual disturbance
Cerebellar signs	Akinetic mutism	Cerebellar signs
Extrapyramidal signs	*AND*	Akinetic mutism
	A typical EEG during an illness of any duration *and/or*	Higher focal cortical sign[a]
	A positive 14-3-3 CSF assay and a clinical duration to death <2 years;	*AND*
	AND routine investigations should not suggest an alternative diagnosis	A typical EEG or MRI
		AND Routine investigations should not suggest an alternative diagnosis

[a] For example, aphasia, neglect, apraxia or acalculia.

and visual symptoms are combined as "visual/cerebellar," yet the neuroanatomy and circuitry of cerebellar and visual symptomatology are distinct and there is no evidence that these two symptoms co-occur. Similarly, as we describe later, the 14-3-3 protein and the EEG both lack sensitivity and specificity, which can lead to not diagnosing treatable disorders, while missing many cases of CJD as well.

We have been using criteria at UCSF that are modifications of 1998 WHO Revised criteria. UCSF criteria separate out visual and cerebellar symptoms and add a category of other focal cortical symptoms (such as aphasia, neglect, or apraxia) and replace the 14-3-3 protein test with a brain MRI consistent with sCJD (Young et al., 2005) (see Table 13.1). These UCSF criteria are still undergoing development to allow earlier diagnosis while maintaining high specificity.

Putative Biomarkers for sCJD.   Certain ancillary tests have been reported as having utility in the diagnosis of CJD. Some of these tests include EEG, MRI, and CSF levels of 14-3-3 protein, total tau (T-tau), neuron specific enolase (NSE), and S-100 proteins. The earliest EEG finding in CJD is focal slowing and later periodic sharp wave complexes (PSWCs) (1–2 Hz periodic sharp waves, epileptiform discharges or triphasic waves) may appear. These complexes may be focal or diffuse. Serial EEGs, weeks apart, are often required to show the evolution of these PSWCs and they may not appear until late stages of the disease. The sensitivity and specificity of these changes for CJD have varied greatly in the literature with sensitivity varying between 50% and 66% and specificity from 74% to 91% among pathology-proven subjects (Pals et al., 1999; Zerr et al., 2000; Steinhoff et al., 2004; Collins et al., 2006). Other conditions with similar EEG findings that may mimic CJD include hepatic encephalopathy, Hashimoto's encephalopathy, and late stages of other neurodegenerative diseases such as Alzheimer's and DLB (Poser et al., 1999). In the proper clinical context and when other conditions with overlapping EEG findings have been ruled out, EEG should have high sensitivity for CJD.

As mentioned earlier, surrogate CSF protein markers are also being used to try to distinguish CJD from other conditions (Hsich et al., 1996; Otto and Wiltfang, 2003; Van Everbroeck et al., 2003; Sanchez-Juan et al., 2006). The reported sensitivity and specificity of 14-3-3 protein in CJD have ranged considerably in the literature, and with each successive large study this appears to show declining sensitivity and specificity (Zerr et al., 1998; Poser et al., 1999; Zerr et al., 2000; Sanchez-Juan et al., 2006), seriously calling into question the utility of this protein as a biomarker for sCJD. Among pathology-proven cases, the reported sensitivity has varied from 53% (Geschwind et al., 2003) to 100% (Hsich et al., 1996; Lemstra et al., 2000), with most larger studies reporting in the mid to high 80th percentile (Beaudry et al., 1999; Zerr et al., 2000; Collins et al., 2006; Sanchez-Juan et al., 2006). The specificity for sCJD has been reported to be in the range of 84% (Zerr et al., 2000; Sanchez-Juan et al., 2006). The 14-3-3 protein comes in seven isoforms, five of which (beta, gamma, epsilon, eta, and zeta) are found in the brain. Most 14-3-3 protein studies in CJD have examined the beta isoform; however, one report suggested that the gamma isoform has higher specificity for CJD (Van Everbroeck et al., 2005).

Proteins other than 14-3-3 that have also been considered surrogate markers for CJD include T-tau, NSE, S-100, and Aβ42 (low levels). Numerous studies have shown quite variable results with all of these CSF proteins in the diagnosis of CJD. Cut-offs for tau levels as a marker for CJD have been 1000, 1200, and 1300 pg/mL, with >1200 or >1300 pg/mL being the most common among studies (Otto et al., 2002; Van Everbroeck et al., 2002; Sanchez-Juan et al., 2006). Some investigators feel that elevated total tau has high sensitivity for CJD and specificity is high with a low ratio of phosphorylated tau to total tau. Unfortunately, these data were derived from mostly non-pathological proven cases (Satoh et al., 2006). One large study suggests that these CSF proteins have higher sensitivity in more rapidly progressive cases (Sanchez-Juan et al., 2006), consistent with the idea that these markers are probably just markers of rapid neuronal injury and may lack true specificity for CJD (Geschwind et al., 2003). It is essential for the clinician to know that several conditions that may clinically mimic CJD can also have a positive 14-3-3 protein, making it necessary to rule out other conditions (Chapman et al., 2000; Geschwind et al., 2003).

MRI of High Diagnostic Value for CJD. FLAIR and particularly diffusion-weighted MRI have high sensitivity (~92%) and specificity (~95%) for CJD. The most common changes seen in decreasing frequency are hyperintensity in the cerebral cortex, basal ganglia (caudate and putamen), and thalamus (Shiga et al., 2004; Young et al., 2005). Abnormalities on MRI may be symmetric or asymmetric (Bavis et al., 2003; Cambier et al., 2003). A brain MRI of a patient with sCJD is shown in Figure 13.3. Note the high signal (brightness) in the striatum (caudate and putamen) and along the cortical gyri.

Brain Biopsy in CJD. Currently, brain biopsy is the only way to make a definitive antemortem diagnosis of CJD. Brain biopsy, however, can be problematic from an infection control standpoint as prion proteins are not removed by standard surgical sterilization methods (Peretz et al., 2006) and there is the risk of transmission to

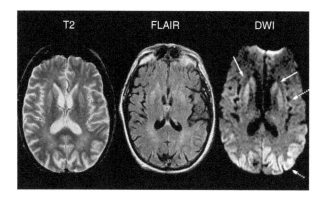

FIGURE 13.3 Brain MRI of sCJD patient. T2, FLAIR, and DWI sequences show hyperintensity in bilateral striatum (caudate and putamen; *solid arrow*) and cortical hyperintensity in bilateral occipital and posterior temporal lobes (*dotted arrows*), with more subtle abnormal hyperintensity in bilateral insula.

operating room personnel. Our own unpublished data suggest that brain biopsy has an 86% sensitivity for diagnosis (MDG unpublished data).

Direct Detection of PrP^{Sc}. To date, there is no test available for antemortem diagnosis of prion diseases; postmortem analysis of CNS (or lymphatic tissues) is required for assessment of the disease in affected humans and animals. Several techniques are used for the identification of PrP^{Sc}. Immunohistochemistry (IHC) is the leading method through which researchers can establish the neuropathological changes in the nervous system. IHC utilizes specific antibodies for detection and localization of neuronal markers in tissue preparation. The most common pathological changes that characterize prion disease are astrocytic gliosis and vacuolation, paralleled by abnormal PrP^{Sc} accumulation in certain regions of the brain (Kretzschmar et al., 1996). A number of tests are commercially available for the immunodetection of PrP^{Sc}; most share a common platform of an enzyme-linked immunosorbent assay in which antibodies against PrP can capture and detect PrP^{Sc}. Many tests exploit the resistance to proteolytic digestion of PrP^{Sc} and utilize specific proteolytic enzymes for the digestion of brain homogenates of suspected cases (Serban et al., 1990; Muller et al., 2000). Among these tests, the conformation-dependent immunoassay (CDI) offers the major advantage of detecting immunochemical differences between PrP^{C} and PrP^{Sc} without the need for proteases to remove PrP^{C}. The CDI is being developed to detect PrP^{Sc} in blood and other bodily fluids (Safar et al., 1998, 2005, 2006). Soto et al. have used methodology, called protein misfolding cyclic amplification (PMCA), to markedly increase the amount of PrP^{Sc} in a sample (Saborio et al., 2001; Soto et al., 2005).

13.3.2 MOLECULAR CLASSIFICATION OF sCJD

13.3.2.1 Codon 129 Polymorphism

Codon 129 in *PRNP* is polymorphic, encoding methionine (M) or valine (V), such that an individual can be MM, VV, or MV. Homozygosity at codon 129 is associated with increased risk for developing prion disease, with greatest risk in persons with codon 129 MM, followed by VV and then MV (Palmer et al., 1991; Laplanche et al., 1994; Parchi et al., 1999b). In addition to affecting an individual's susceptibility to developing prion disease, the codon 129 polymorphism can play a role in how a prion disease presents clinically and pathologically (Parchi et al., 1996, 1999b). Persons with MM and VV not only are more likely to get prion disease than those with codon 129 MV, but when CJD occurs in homozygous patients it tends to be more rapid (Pocchiari et al., 2004). Codon 129 can also affect the phenotype of genetic prion diseases. For example, patients with the D178N mutation (aspartate → asparagine in *PRNP* mutation at codon 178) who have methionine at codon 129 on the same chromosome ("cis") present as fatal familial insomnia (FFI), whereas those with a *cis*-valine present similar to CJD (Kong et al., 2004). Polymorphisms at codon 129 have been associated with increased neuronal vulnerability in other conditions; codon 129 VV in *PRNP* has been linked to increased risk for early-onset AD (Dermaut et al., 2003), earlier cognitive decline (Croes et al., 2003), worsened cognitive function in the elderly (Berr et al., 1998), long-term memory problems in normal subjects, and more severe neurological symptoms in Wilson's disease (Grubenbecher et al., 2006). Homozygous glutamate (E/E) at Codon 219 in *PRNP* is a risk factor for CJD in several Asian populations (Seno et al., 2000; Jeong et al., 2004, 2005). Other polymorphisms, including those at codons 171 and 232, can also modify risk of disease as well (Hoque et al., 1996; Saito et al., 2000; Koide et al., 2002).

13.3.2.2 Prion Typing

The prion type has also been found to play a role in the clinicopathological phenotype of prion disease. Prions (PrP^{Sc}) have been subdivided based on their size after protease treatment and glycosylation pattern. PrP has two glycosylation sites, so it can have none, one, or two sugars moieties attached. Different prion strains have different glycosylation patterns as determined by the percent of PrP^{Sc} without glycosylation, or with mono or diglycosylation. The size of the PrP^{Sc} protein after proteinase K (PK) digestion as determined by sodium dodecyl sulfate polyacrylamide gel electrophoresis (SDS-PAGE) is 19–21 KDa. PrP^{C} is completely digested by PK treatment, whereas most of PrP^{Sc} is resistant to, and only partially digested by PK, with cleavage occurring at either of two positions, at about amino acids 82 or 97. Cleavage at around amino acid 82 leaves a larger protein, called Type 1 that migrates at 21 KDa in its unglycosylated form. Cleavage of PrP^{Sc} near amino acid 97 results in a shorter protein, called Type 2, that migrates at about 19 KDa in its unglycosylated form. It is probably the conformation of the PrP^{Sc} that determines where PrP^{Sc} is cleaved (exposed to PK) and hence its type (Parchi et al., 1996, 1999b).

sCJD has recently been classified into at least six subtypes. In a study of 300 pathology-proven sCJD subjects, investigators subdivided sCJD molecularly into approximately six different forms (or variants) based on an individuals *PRNP* codon 129 polymorphism (MM, MV, or VV) and on their prion type (Type 1 or 2) (Parchi et al., 1996, 1999b). Seventy percent of their patients had Type 1 PrP^{Sc} and had at least one methionine allele (MM or MV; most were MM). These patients presented as classic sCJD, with a rapidly progressive dementia, early and prominent myoclonus, and a classic EEG (Steinhoff et al., 2004). Twenty-five percent of patients had a specific, distinct neuropathology called kuru plaques with significant ataxia and were Type 2 with at least one valine allele (MV2 or VV2). The MM2 variant was associated with either a thalamic form (MM2-T) or a dementing cortical form (MM2-C) of sCJD. The least common form was VV1, comprising less than 1% of their cases; these patients had a progressive dementia without classic EEG findings, and had severe cortical and basal ganglia pathology, that spared the cerebellum and brain stem (Parchi et al., 1999b).

Unfortunately, for several reasons subdividing patients using these categories, particularly based on prion type, is somewhat crude and may be misleading. Many patients do not match clinicopathologically their sCJD variant based on these criteria (unpublished data) (Parchi et al., 1999b). Secondly, even within the same individual, different prion types can be found in separate brain regions (Polymenidou et al.,

2005). Lastly, by using more sensitive methods than standard Western blots, several more prion types have been identified (Safar et al., 1998, 2005). To make matters even more confusing, other prominent prion investigators have developed another method of prion typing; Types 1–3 for sCJD or iCJD and Type 4 for vCJD (Collinge et al., 1996; Collinge, 2005).

13.3.3 SPORADIC FATAL INSOMNIA

Sporadic fatal insomnia (sFI) is a rare form of sporadic prion disease that presents clinically and pathologically similar to FFI. Patients with sFI typically have thalamic symptoms, particularly insomnia and dysautonomia, and cerebellar ataxia. The pathology is mostly thalamic and olivary. As of December 2006, only eight cases have been published in the literature (Mastrianni et al., 1999; Parchi et al., 1999a; Scaravilli et al., 2000; Piao et al., 2005), and we have recently seen a ninth case (Geschwind, unpublished data).

13.4 GENETIC PRION DISEASES

The identification of mutations in the prion gene in several familial diseases, familial CJD (fCJD or gCJD), Gerstmann–Sträussler–Scheinker (GSS), and FFI, showed that prion diseases were unique in the annals of medicine in that they could occur in spontaneous, infectious, and genetic forms. Genetic prion diseases have been divided into three forms based on their clinical and pathological presentation: familial CJD (fCJD) or genetic CJD (gCJD), GSS, and FFI. Some authors separate out from gCJD a fourth type of genetic prion disease—those patients with mutations consisting of base pair insertions (Kovacs et al., 2002; Pocchiari et al., 2004). More than 30 mutations and at least three polymorphisms in the prion gene have been identified (Kong et al., 2004) and new mutations are still being identified. In most cases, *PRNP* mutations are autosomal dominant with complete penetrance; however, in some mutations there may be incomplete penetrance (Kong et al., 2004). Clinically, gCJD typically presents either identical to sCJD, as a rapidly progressive dementia with motor features, or it can present as a dementia progressing over years.

GSS typically presents as either a Parkinsonian or ataxic disease, progressing over years. A rare spastic form of GSS can also occur (Ghetti et al., 1995). Five mutations, at codons 102, 105, 117, 145, 198, and 217, in the ORF of the prion gene, *PRNP*, have been associated with GSS (Ghetti et al., 1995). In GSS, mean age of onset is about 47 years, with average survival in GSS of about 57 months (Kovacs et al., 2002), and more than 75% of cases survive more than 24 months (Pocchiari et al., 2004).

In a large European survey, 60% of genetic prion patients did not report a known family history of prion disease, although closer review often uncovered a family history of Alzheimer's or Parkinson's disease that was likely to be misdiagnosed (Will et al., 1998). Because many cases of gCJD do not have a positive family history, some prefer the term genetic, rather than familial CJD (Pocchiari et al., 2004).

13.5 VARIANT CJD

In 1995, a new form of human prion disease called variant CJD was identified in the United Kingdom (Will et al., 1996). Patients with vCJD tend to be younger than sCJD, with a mean age of 28 and a range of 12–72. Two hundred cases of vCJD have been identified, most in the UK, followed by France. vCJD has been linked to consumption of BSE-contaminated meat (Bruce et al., 1997; Hill et al., 1997a; Scott et al., 1999). The clinical features of vCJD may overlap with sCJD, but there are some general differences. Early in the illness, patients usually experience psychiatric symptoms, which most commonly take the form of depression or, less often, a schizophrenia-like psychosis. This psychiatric prodrome usually precedes by at least 6 months the onset of other neurological features. Patients often have persistent painful paresthesias, although they may also be transient. Motor features typically include ataxia, chorea, myoclonus, and/or dystonia. Dementia eventually occurs as well. As in sCJD, by the time of death, patients become completely immobile and mute and they usually succumb to aspiration pneumonia. Brain MRI in vCJD typically has the pulvinar sign, in which the pulvinar is more hyperintense than the anterior putamen on T2-weighted MRI sequences, a feature that often distinguishes it from sCJD and other forms of prion disease (Zeidler et al., 2000; Collie et al., 2001, 2003). The characteristic neuropathologic profile of vCJD includes, in both the cerebellum and cerebrum, numerous kuru-type amyloid florid plaques of high concentrations of PrP^{Sc} surrounded by vacuoles to give a flower appearance. The PrP^{Sc} type in vCJD is called type 2b; it is a 19 kDa band on Western blot, but it has a different ratio of glycosylated forms than found in type 2a sCJD PrP^{Sc} (Will et al., 1996; Parchi et al., 1999b; Will, 2004). In vCJD, the prion appears to be present at very high levels in the lymphoreticular system. Premortem pathological diagnosis of vCJD can be made by tonsillar or brain biopsy. Postmortem reveals that the prion is found throughout the lymphoreticular and central nervous systems (Will et al., 1996; Hill et al., 1997b, 1999; Hilton et al., 2004). Why vCJD tends to occur in younger patients has not been fully explained, but it may be that BSE prions are more readily acquired across the gastrointestinal track when inflammation is present. As children have a higher incidence of gastritis than adults, they may therefore be more susceptible to consumed BSE prions (Heikenwalder et al., 2005; Seeger et al., 2005).

As of February 2007, four patients have acquired vCJD via blood transfusion from vCJD patients who had donated blood prior to the onset of symptoms (Llewelyn et al., 2004; Peden et al., 2004; UK Health Protection Agency, 2007). Although all patients thus far with clinical vCJD have codon 129 MM, data from transgenic mice models suggest that persons with MV and VV may be susceptible to vCJD, particularly via human to human transmission, such as by blood donation (Bishop et al., 2006).

13.6 DIFFERENTIAL DIAGNOSIS OF PRION DISEASES

The differential for prion diseases is large, as depending upon where prions accumulate in the brain, the symptoms can be quite varied. One category of diseases that is commonly mistaken for prion disease are atypical, rapid forms of other more common neurodegenerative diseases, particularly Alzheimer's disease and atypical Parkinsonian dementias, such as CBD and DLB. Paraneoplastic (e.g., Anti-Hu, CV2, and MaTa) and nonparaneoplastic (e.g., Anti-VGKC, Anti-GAD, and Hashimoto's encephalopathy) autoimmune conditions may also mimic CJD (Ghika-Schmid et al., 1996; Poser et al., 1999; Saiz et al., 1999; Seipelt et al., 1999; Geschwind and Jay, 2003; Chang et al., 2007). Vasculitis may clinically resemble sCJD. Cancers, such as intravascular lymphoma, primary central nervous system lymphoma, or gliomatosis cerebri can all present like sCJD; however, brain MRI should easily differentiate these conditions from prion disease (Carlson, 1996; Bakshi et al., 1999; Heinrich et al., 2005; Slee et al., 2006; Josephson et al., 2007). Subacute infections, such as subacute sclerosing panencephalitis (SSPE) from German measles or rubella can look somewhat like CJD, particularly vCJD in a young person. Lyme and HIV should probably always be ruled out while considering CJD. Toxins, such as bismuth (found in some antidiarrheal agents), when taken in large quantities, can cause a CJD-like clinical picture. Thiamine deficiency causing Wernicke's encephalopathy is readily treatable and should always be considered (Geschwind and Jay, 2003).

13.7   TREATMENT OF PRION DISEASE

Many compounds have been used successfully to remove prions or inhibit their formation in vitro. Such compounds include quinoline derivatives (Korth et al., 2001; Barret et al., 2003; Murakami-Kubo et al., 2004), antibiotics such as doxycycline and tetracycline (Forloni et al., 2002), Congo red and its analogues, (Sellarajah et al., 2004), and various "chemical chaperones" (Tatzelt et al., 1996; Georgieva et al., 2006). Several of these compounds and others have been effective in preventing or delaying disease onset in vivo when mixed with the prions prior to inoculation or given before or at the time of prion inoculation (Ehlers and Diringer, 1984; Kimberlin and Walker, 1986; Tagliavini et al., 1997; Priola et al., 2000; Forloni et al., 2002). Unfortunately, with the animal models that are relatively easy to use and which have short incubation times amenable to research, such as mice, hamsters, voles, there is a very narrow time window between first symptoms and death. In most mice models, there are usually only a few days between the first clear signs of neurologic disease and death or incapacitation. By the time an animal develops symptoms, the disease is so fulminant that no treatment will likely have efficacy. Therefore, investigators often start potential treatment at the midpoint of incubation. For example, if an animal develops symptoms at 110–120 days postinoculation (dpi) with prions and succumbs within a few days of onset, investigators may begin to test a compound at around 60 dpi—the midpoint of incubation. Although several compounds have prevented disease before or at the time of inoculation (Korth and Peters, 2006), no treatment has cured animals when given later in the incubation and only a few compounds have delayed disease onset (Doh-ura et al., 2004).

   To our knowledge, as of February 2007, the only medication in formal treatment trials for prion disease in the world is oral quinacrine, a quinoline derivative used orally for many decades to treat malaria. In in vitro models of prion disease, quinacrine eliminates prions (Doh-ura et al., 2000; Korth et al., 2001) and may even allow recovery of some cellular functions (Sandberg et al., 2004). Two prion studies in mice with oral quinacrine showed no benefit in survival (Collins et al., 2002; Barret et al., 2003). In one study, intracerebrally inoculated mice given low dose oral quinacrine (10 mg/kg/day) at 5 dpi or at 65 dpi (midpoint of incubation) did not show improved survival compared with controls (Collins et al., 2002). In another study, mice inoculated intraperitoneally (IP) with prions and given low dose IP quinacrine (10 mg/kg/day) beginning 1 dpi for 30 days and then sacrificed did not show a larger decrease of PrP^{Sc} accumulation in the spleen than untreated PrP^{Sc}-inoculated control mice. Several other compounds also showed no effect in this in vivo model (Barret et al., 2003).

Because quinacrine has been used in medicine orally for decades and it crosses the blood–brain barrier, compassionate treatment was begun in humans with prion disease. Our data on compassionate use of quinacrine suggested prolonged survival (unpublished), although published observational case reports have shown mixed results with some patients having possible transient improvement (Haik et al., 2004; Nakajima et al., 2004). To determine whether quinacrine is efficacious in human prion disease, two formal studies were initiated in humans, one in the United Kingdom (PRION1 by the Medical Research Council) and another in the United States of America, at our center (UCSF; CJD Quinacrine Treatment Study) (Korth and Peters, 2006). The United Kingdom trial enrolled patients with all forms of human prion disease and was essentially an unblinded, observational study. Subjects were given the option of which of three treatment groups they wished to be entered: (1) Open label quinacrine, (2) No treatment, or (3) Randomized to either quinacrine or no treatment. This study completed enrollment in July 2006. To our knowledge, unfortunately only one patient chose the randomization arm of the study, so the investigators will not be able to determine if quinacrine improves outcomes (survival, cognition, quality of life, etc.), although they will have significant data on the course of CJD and quinacrine toxicity. Our study at UCSF is an NIH-funded, randomized, double-blinded placebo-controlled (delayed treatment start) study of quinacrine in sCJD with the primary outcome being survival from start of treatment (http://www.clinicaltrials.gov and http://memory.ucsf.edu).

The two main mechanisms for clearance of prions are either blocking of conversion of PrP^{C} to PrP^{Sc} or clearance of PrP^{Sc}. Quinacrine appears to work via the former mechanism (Barret et al., 2003). The mechanism of how quinacrine may work in prion disease has not been completely elucidated. It appears to have tropism for the lysosomal compartments, which are involved in removal of proteins, such as PrP^{Sc}, from the cell (Lullmann-Rauch et al., 1996). Quinacrine binds readily to PrP^{C} and appears to prevent conversion of PrP^{C} to PrP^{Sc} (Barret et al., 2003), possibly by binding to the c-terminal helix of PrP^{C} (Vogtherr et al., 2003). Quinacrine appears to have an antioxidant effect (Turnbull et al., 2003). It may also protect PrP^{Sc} infected cells from attack by microglia (Barret et al., 2003). Based on quinacrine and other compounds, numerous other potential treatments are currently being developed (Follette, 2003; May et al., 2003; Murakami-Kubo et al., 2004; Klingenstein et al., 2006; Korth and Peters, 2006).

Several investigators are developing compounds to interfere with the conversion of PrP^{C} into PrP^{Sc}. Tetracyclines may have antiprion activity, possibly through interaction with PrP^{Sc} or PrP^{Sc} fibrils (Forloni et al., 2002; Barret et al., 2003). Human treatment trials using oral doxycycline are planned in Europe.

Many papers have shown that antibodies or single chain fragments of antibodies can eliminate prions in cell culture (Heppner et al., 2001; Peretz et al., 2002; Donofrio et al., 2005; Pankiewicz et al., 2006), but getting antibodies into the brain in sufficient amounts may be a difficult hurdle to overcome (Love, 2001). Vaccination against PrP^C has also been studied and may be helpful in preventing infection, particularly after known exposure (Sigurdsson et al., 2003; White et al., 2003; Goni et al., 2005; Magri et al., 2005; Bade et al., 2006). With all of these immunological methods, however, it is not yet clear what deleterious effects may occur in humans from blocking the normal function of PrP^C.

13.8 CONCLUSIONS

The notion that a misfolded protein can cause disease is no longer considered heretical and is widely accepted in the scientific community. Although we have learned a great deal about the mechanisms of prion diseases over the past few decades, there is much that we do not know. Through greater understanding of disorders of protein misfolding, we hopefully will soon be entering an era of treating these currently uniformly fatal diseases.

REFERENCES

Bade, S., Baier, M., Boetel, T. and Frey, A. (2006) Intranasal immunization of Balb/c mice against prion protein attenuates orally acquired transmissible spongiform encephalopathy. *Vaccine*, **24**, 1242–53.

Bakshi, R., Mazziotta, J.C., Mischel, P.S., Jahan, R., Seligson, D.B. and Vinters, H.V. (1999) Lymphomatosis cerebri presenting as a rapidly progressive dementia: Clinical, neuroimaging and pathologic findings. *Dementia and Geriatric Cognitive Disorders*, **10**, 152–7.

Barret, A., Tagliavini, F., Forloni, G., Bate, C., Salmona, M., Colombo, L. et al. (2003) Evaluation of quinacrine treatment for prion diseases. *Journal of Virology*, **77**, 8462–9.

Basler, K., Oesch, B., Scott, M., Westaway, D., Walchli, M., Groth, D.F. et al. (1986) Scrapie and cellular PrP isoforms are encoded by the same chromosomal gene. *Cell*, **46**, 417–28.

Bavis, J., Reynolds, P., Tegeler, C. and Clark, P. (2003) Asymmetric neuroimaging in Creutzfeldt–Jakob disease: A ruse. *Journal of Neuroimaging*, **13**, 376–9.

Beaudry, P., Cohen, P., Brandel, J.P., Delasnerie-Laupretre, N., Richard, S., Launay, J.M., Laplanche, J.L. (1999) 14-3-3 Protein, neuron-specific enolase, and S-100 protein in cerebrospinal fluid of patients with Creutzfeldt–Jakob disease. *Dementia and Geriatric Cognitive Disorders*, **10**, 40–6.

Berr, C., Richard, F., Dufouil, C., Amant, C., Alperovitch, A. and Amouyel, P. (1998) Polymorphism of the prion protein is associated with cognitive impairment in the elderly: The EVA study. *Neurology*, **51**, 734–7.

Bishop, M.T., Hart, P., Aitchison, L., Baybutt, H.N., Plinston, C., Thomson, V. et al. (2006) Predicting susceptibility and incubation time of human-to-human transmission of vCJD. *Lancet Neurology*, **5**, 393–8.

Borchelt, D.R., Taraboulos, A. and Prusiner, S.B. (1992) Evidence for synthesis of scrapie prion proteins in the endocytic pathway. *Journal of Biological Chemistry*, **267**, 16188–99.

Borchelt, D.R., Rogers, M., Stahl, N., Telling, G. and Prusiner, S.B. (1993) Release of the cellular prion protein from cultured cells after loss of its glycoinositol phospholipid anchor. *Glycobiology*, **3**, 319–29.

Brown, P., Rohwer, R.G. and Gajdusek, D.C. (1986) Newer data on the inactivation of scrapie virus or Creutzfeldt–Jakob disease virus in brain tissue. *Journal of Infectious Disease*, **153**, 1145–8.

Brown, P., Gibbs, C.J., Jr., Rodgers-Johnson, P., Asher, D.M., Sulima, M.P., Bacote, A. et al. (1994) Human spongiform encephalopathy: The National Institutes of Health series of 300 cases of experimentally transmitted disease. *Annals of Neurology*, **35**, 513–29.

Brown, D.R., Qin, K., Herms, J.W., Madlung, A., Manson, J., Strome, R. et al. (1997) The cellular prion protein binds copper in vivo. *Nature*, **390**, 684–7.

Brown, D.R., Nicholas, R.S. and Canevari, L. (2002) Lack of prion protein expression results in a neuronal phenotype sensitive to stress. *Journal of Neuroscience Research*, **67**, 211–24.

Bruce, M.E., Will, R.G., Ironside, J.W., McConnell, I., Drummond, D., Suttie, A. et al. (1997) Transmissions to mice indicate that 'new variant' CJD is caused by the BSE agent. *Nature*, **389**, 498–501.

Bueler, H., Fischer, M., Lang, Y., Bluethmann, H., Lipp, H.P., DeArmond, S.J. et al. (1992) Normal development and behaviour of mice lacking the neuronal cell-surface PrP protein. *Nature*, **356**, 577–82.

Bueler, H., Aguzzi, A., Sailer, A., Greiner, R.A., Autenried, P., Aguet, M. et al. (1993) Mice devoid of PrP are resistant to scrapie. *Cell*, **73**, 1339–47.

Cambier, D.M., Kantarci, K., Worrell, G.A., Westmoreland, B.F. and Aksamit, A.J. (2003) Lateralized and focal clinical, EEG, and FLAIR MRI abnormalities in Creutzfeldt–Jakob disease. *Clinical Neurophysiology*, **114**, 1724–8.

Carlson, B.A. (1996) Rapidly progressive dementia caused by nonenhancing primary lymphoma of the central nervous system. *American Journal of Neuroradiology*, **17**, 1695–7.

Casalone, C., Zanusso, G., Acutis, P., Ferrari, S., Capucci, L., Tagliavini, F. et al. (2004) Identification of a second bovine amyloidotic spongiform encephalopathy: Molecular similarities with sporadic Creutzfeldt–Jakob disease. *Proceedings of the National Academy of Sciences of the United States of America*, **101**, 3065–70.

Chabry, J., Ratsimanohatra, C., Sponne, I., Elena, P.P., Vincent, J.P. and Pillot, T. (2003) In vivo and in vitro neurotoxicity of the human prion protein (PrP) fragment P118–135 independently of PrP expression. *Journal of Neuroscience*, **23**, 462–9.

Chang, C.C., Eggers, S.D., Johnson, J.K., Haman, A., Miller, B.L. and Geschwind, M.D. (2007) Anti-GAD antibody cerebellar ataxia mimicking Creutzfeldt–Jakob disease. *Clinical Neurology and Neurosurgery*, **109**, 54–7.

Chapman, T., McKeel, D.W., Jr. and Morris, J.C. (2000) Misleading results with the 14-3-3 assay for the diagnosis of Creutzfeldt–Jakob disease. *Neurology*, **55**, 1396–7.

Chen, S., Mange, A., Dong, L., Lehmann, S. and Schachner, M. (2003) Prion protein as trans-interacting partner for neurons is involved in neurite outgrowth and neuronal survival. *Molecular and Cellular Neuroscience*, **22**, 227–33.

Collie, D.A., Sellar, R.J., Zeidler, M., Colchester, C.F., Knight, R. and Will, R.G. (2001) MRI of Creuztfeldt-Jakob disease: Imaging features and recommended MRI protocol. *Clinical Radiology*, **56**, 726–39.

Collie, D.A., Summers, D.M., Sellar, R.J., Ironside, J.W., Cooper, S., Zeidler, M. et al. (2003) Diagnosing variant Creutzfeldt–Jakob disease with the pulvinar sign: MR imaging findings in 86 neuropathologically confirmed cases. *American Journal of Neuroradiology*, **24**, 1560–9.

Colling, S.B., Khana, M., Collinge, J. and Jefferys, J.G. (1997) Mossy fibre reorganization in the hippocampus of prion protein null mice. *Brain Research*, **755**, 28–35.

Collinge, J. (2005) Molecular neurology of prion disease. *Journal of Neurology and Neurosurgery and Psychiatry*, **76**, 906–19.

Collinge, J., Sidle, K.C., Meads, J., Ironside, J. and Hill, A.F. (1996) Molecular analysis of prion strain variation and the aetiology of 'new variant' CJD. *Nature*, **383**, 685–90.

Collins, S.J., Lewis, V., Brazier, M., Hill, A.F., Fletcher, A. and Masters, C.L. (2002) Quinacrine does not prolong survival in a murine Creutzfeldt–Jakob disease model. *Annals of Neurology*, **52**, 503–6.

Collins, S.J., Sanchez-Juan, P., Masters, C.L., Klug, G.M., van Duijn, C., Poleggi, A. et al. (2006) Determinants of diagnostic investigation sensitivities across the clinical spectrum of sporadic Creutzfeldt–Jakob disease. *Brain*, **129**, 2278–87.

Creutzfeldt, H.G. (1989) On a particular focal disease of the central nervous system (preliminary communication), 1920. *Alzheimers Disease and Associated Disorders*, **3**, 3–25.

Criado, J.R., Sanchez-Alavez, M., Conti, B., Giacchino, J.L., Wills, D.N., Henriksen, S.J. et al. (2005) Mice devoid of prion protein have cognitive deficits that are rescued by reconstitution of PrP in neurons. *Neurobiology of Disease*, **19**, 255–65.

Croes, E.A., Dermaut, B., Houwing-Duistermaat, J.J., Van den Broeck, M., Cruts, M., Breteler, M.M. et al. (2003) Early cognitive decline is associated with prion protein codon 129 polymorphism. *Annals of Neurology*, **54**, 275–6.

DeArmond, S.J. and Prusiner, S.B. (2003) Perspectives on prion biology, prion disease pathogenesis, and pharmacologic approaches to treatment. *Clinical Laboratory Medicine*, **23**, 1–41.

Deleault, N.R., Lucassen, R.W. and Supattapone, S. (2003) RNA molecules stimulate prion protein conversion. *Nature*, **425**, 717–20.

Dermaut, B., Croes, E., Rademakers, R., Van den Broeck, M., Cruts, M., and Hofman, A. (2003) PRNP Val129 homozygosity increases risk for early onset Alzheimer's disease. *Annals of Neurology*, **53**, 409–12.

Doh-ura, K., Iwaki, T. and Caughey, B. (2000) Lysosomotropic agents and cysteine protease inhibitors inhibit scrapie-associated prion protein accumulation. *Journal of Virology*, **74**, 4894–7.

Doh-ura, K., Ishikawa, K., Murakami-Kubo, I., Sasaki, K., Mohri, S., Race, R., Iwaki, T. (2004) Treatment of transmissible spongiform encephalopathy by intraventricular drug infusion in animal models. *Journal of Virology*, **78**, 4999–5006.

Donofrio, G., Heppner, F.L., Polymenidou, M., Musahl, C. and Aguzzi, A. (2005) Paracrine inhibition of prion propagation by anti-PrP single-chain Fv mini antibodies. *Journal of Virology*, **79**, 8330–8.

Edenhofer, F., Rieger, R., Famulok, M., Wendler, W., Weiss, S. and Winnacker, E.-L. (1996) Prion protein PrP^C interacts with molecular chaperones of the Hsp60 family. *Journal of Virology*, **70**, 4724–8.

Ehlers, B. and Diringer, H. (1984) Dextran sulphate 500 delays and prevents mouse scrapie by impairment of agent replication in spleen. *Journal of General Virology*, **65**, 1325–30.

Follette, P. (2003) New perspectives for prion therapeutics meeting. Prion disease treatment's early promise unravels. *Science*, **299**, 191–2.

Forloni, G., Iussich, S., Awan, T., Colombo, L., Angeretti, N., Girola, L. et al. (2002) Tetracyclines affect prion infectivity. *Proceedings of the National Academy of Sciences of the United States of America*, **99**, 10849–54.

Gajdusek, D.C. (1977) Unconventional viruses and the origin and disappearance of kuru. *Science*, **197**, 943–60.

Gajdusek, D.C., Gibbs, C.J., Jr., Asher, D.M., Brown, P., Diwan, A., Hoffman, P. et al. (1977) Precautions in medical care of, and in handling materials from, patients with transmissible virus dementia (Creutzfeldt–Jakob disease). *New England Journal of Medicine*, **297**, 1253–8.

Georgieva, D., Schwark, D., von Bergen, M., Redecke, L., Genov, N. and Betzel, C. (2006) Interactions of recombinant prions with compounds of therapeutical significance. *Biochemistry and Biophysics Research Communications*, **344**, 463–70.

Geschwind, M.D. and Jay, C. (2003) Assessment of rapidly progressive dementias. Concise review related to Chapter 362: Alzheimer's disease and other primary dementias in Harrison's Textbook of Internal Medicine. McGraw Hill. Accessed: 6/9/2003, http:// harrisons.accessmedicine.com/

Geschwind, M.D., Martindale, J., Miller, D., DeArmond, S.J., Uyehara-Lock, J., Gaskin, D. et al. (2003) Challenging the clinical utility of the 14-3-3 protein for the diagnosis of sporadic Creutzfeldt–Jakob disease. *Archives of Neurology*, **60**, 813–6.

Ghetti, B., Dlouhy, S.R., Giaccone, G., Bugiani, O., Frangione, B., Farlow, M.R. et al. (1995) Gerstmann-Straussler-Scheinker disease and the Indiana kindred. *Brain Pathology*, **5**, 61–75.

Ghika-Schmid, F., Ghika, J., Regli, F., Dworak, N., Bogousslavsky, J., Stadler, C. et al. (1996) Hashimoto's myoclonic encephalopathy: An underdiagnosed treatable condition? *Movement Disorders*, **11**, 555–62.

Gibbs, C.J., Jr. (1992) Spongiform encephalopathies—slow, latent, and temperate virus infections—in retrospect, in *Prion Diseases of Humans and Animals*, Prusiner, S.B., Collinge, J., Powell, J. and Anderton, B (Eds.). Ellis Horwood, London, pp. 53–62.

Goni, F., Knudsen, E., Schreiber, F., Scholtzova, H., Pankiewicz, J., Carp, R. et al. (2005) Mucosal vaccination delays or prevents prion infection via an oral route. *Neuroscience*, **133**, 413–21.

Graner, E., Mercadante, A.F., Zanata, S.M., Martins, V.R., Jay, D.G. and Brentani, R. (2000) Laminin-induced PC-12 cell differentiation is inhibited following laser inactivation of cellular prion protein. *FEBS Letters*, **482**, 257–60.

Grubenbecher, S., Stuve, O., Hefter, H. and Korth, C. (2006) Prion protein gene codon 129 modulates clinical course of neurological Wilson disease. *Neuroreport*, **17**, 549–52.

Haik, S., Brandel, J.P., Salomon, D., Sazdovitch, V., Delasnerie-Laupretre, N., Laplanche, J.L. et al. (2004) Compassionate use of quinacrine in Creutzfeldt–Jakob disease fails to show significant effects. *Neurology*, **63**, 2413–15.

Hay, B., Prusiner, S.B. and Lingappa, V.R. (1987) Evidence for a secretory form of the cellular prion protein. *Biochemistry*, **26**, 8110–15.

Hegde, R.S., Mastrianni, J.A., Scott, M.R., DeFea, K.A., Tremblay, P., Torchia, M. et al. (1998) A transmembrane form of the prion protein in neurodegenerative disease. *Science*, **279**, 827–34.

Hegde, R.S., Tremblay, P., Groth, D., DeArmond, S.J., Prusiner, S.B. and Lingappa, V.R. (1999) Transmissible and genetic prion diseases share a common pathway of neurodegeneration. *Nature*, **402**, 822–6.

Heikenwalder, M., Zeller, N., Seeger, H., Prinz, M., Klohn, P.C., Schwarz, P. et al. (2005) Chronic lymphocytic inflammation specifies the organ tropism of prions. *Science*, **307**, 1107–10.

Heinrich, A., Vogelgesang, S., Kirsch, M. and Khaw, A.V. (2005) Intravascular lymphomatosis presenting as rapidly progressive dementia. *European Neurology*, **54** (1), 55–8.

Heppner, F.L., Musahl, C., Arrighi, I., Klein, M.A., Rulicke, T., Oesch, B. et al. (2001) Prevention of scrapie pathogenesis by transgenic expression of anti-prion protein antibodies. *Science*, **294**, 178–82.

Hill, A.F., Desbruslais, M., Joiner, S., Sidle, K.C., Gowland, I., Collinge, J. et al. (1997a) The same prion strain causes vCJD and BSE. *Nature*, **389**, 448–50, 526.

Hill, A.F., Zeidler, M., Ironside, J. and Collinge, J. (1997b) Diagnosis of new variant Creutzfeldt–Jakob disease by tonsil biopsy. *Lancet*, **349**, 99–100.

Hill, A.F., Butterworth, R.J., Joiner, S., Jackson, G., Rossor, M.N., Thomas, D.J. et al. (1999) Investigation of variant Creutzfeldt–Jakob disease and other human prion diseases with tonsil biopsy samples. *Lancet*, **353**, 183–9.

Hilton, D.A., Ghani, A.C., Conyers, L., Edwards, P., McCardle, L., Ritchie, D., Penney, M. et al. (2004) Prevalence of lymphoreticular prion protein accumulation in UK tissue samples. *Journal of Pathology*, **203**, 733–9.

Hoque, M.Z., Kitamoto, T., Furukawa, H., Muramoto, T. and Tateishi, J. (1996) Mutation in the prion protein gene at codon 232 in Japanese patients with Creutzfeldt–Jakob disease: A clinicopathological, immunohistochemical and transmission study. *Acta Neuropathology (Berlin)*, **92**, 441–6.

Hsich, G., Kenney, K., Gibbs, C.J., Lee, K.H. and Harrington, M.G. (1996) The 14-3-3 brain protein in cerebrospinal fluid as a marker for transmissible spongiform encephalopathies. *New England Journal of Medicine*, **335**, 924–30.

Jakob, A. (1989) Concerning a disorder of the central nervous system clinically resembling multiple sclerosis with remarkable anatomic findings (spastic pseudosclerosis). Report of a fourth case. *Alzheimers Disease and Associated Disorders*, **3**, 26–45.

Jeong, B.H., Nam, J.H., Lee, Y.J., Lee, K.H., Jang, M.K., Carp, R.I. et al. (2004) Polymorphisms of the prion protein gene (PRNP) in a Korean population. *Journal of Human Genetics*, **49**, 319–24.

Jeong, B.H., Lee, K.H., Kim, N.H., Jin, J.K., Kim, J.I., Carp, R.I. et al. (2005) Association of sporadic Creutzfeldt–Jakob disease with homozygous genotypes at PRNP codons 129 and 219 in the Korean population. *Neurogenetics*, **6**, 229–32.

Josephson, S.A., Papanastassiou, A.M., Berger, M.S., Barbaro, N.M., McDermott, M.W., Hilton, J.F. et al. (2007) The diagnostic utility of brain biopsy procedures in patients with rapidly deteriorating neurological conditions or dementia. *Journal of Neurosurgery*, **106**, 72–5.

Kanaani, J., Prusiner, S.B., Diacovo, J., Baekkeskov, S. and Legname, G. (2005) Recombinant prion protein induces rapid polarization and development of synapses in embryonic rat hippocampal neurons in vitro. *Journal of Neurochemistry*, **95**, 1373–86.

Katamine, S., Nishida, N., Sugimoto, T., Noda, T., Sakaguchi, S., Shigematsu, K. et al. (1998) Impaired motor coordination in mice lacking prion protein *Cellular and Molecular Neurobiology*, **18**, 731–42.

Katscher, F. (1998) It's Jakob's disease, not Creutzfeldt's. *Nature*, **393**, 11.

Kimberlin, R.H. and Walker, C.A. (1986) Suppression of scrapie infection in mice by heteropolyanion 23, dextran sulfate, and some other polyanions. *Antimicrobial Agents and Chemotherapy*, **30**, 409–13.

Klamt, F., Dal-Pizzol, F., Conte da Frota, M.J., Walz, R., Andrades, M.E., da Silva, E.G. et al. (2001) Imbalance of antioxidant defense in mice lacking cellular prion protein. *Free Radical Biology Medicine*, **30**, 1137–44.

Klingenstein, R., Lober, S., Kujala, P., Godsave, S., Leliveld, S.R., Gmeiner, P. et al. (2006) Tricyclic antidepressants, quinacrine and a novel, synthetic chimera thereof clear prions by destabilizing detergent-resistant membrane compartments. *Journal of Neurochemistry*, **98**, 748–59.

Koide, T., Ohtake, H., Nakajima, T., Furukawa, H., Sakai, K., Kamei, H. et al. (2002) A patient with dementia with Lewy bodies and codon 232 mutation of PRNP. *Neurology*, **59**, 1619–21.

Kong, Q.K., Surewicz, W.K., Petersen, R.B., Zhou, W., Chen, S.G., Gambetti, P. et al. (2004) Inherited prion diseases, in *Prion Biology and Disease*, 2nd ed., Prusiner, S.B. (Ed.). Cold Spring Harbor Laboratory Press, Cold Spring Harbor, pp. 673–776.

Korth, C. and Peters, P.J. (2006) Emerging pharmacotherapies for Creutzfeldt–Jakob disease. *Archives of Neurology*, **63**, 497–501.

Korth, C., May, B.C.H., Cohen, F.E. and Prusiner, S.B. (2001) Acridine and phenothiazine derivatives as pharmacotherapeutics for prion disease. *Proceedings of the National Academy of Sciences of the United States of America*, **98**, 9836–41.

Kovacs, G.G., Trabattoni, G., Hainfellner, J.A., Ironside, J.W., Knight, R.S. and Budka, H. (2002) Mutations of the prion protein gene phenotypic spectrum. *Journal of Neurology*, **249**, 1567–82.

Kretzschmar, H.A., Ironside, J.W., DeArmond, S.J. and Tateishi, J. (1996) Diagnostic criteria for sporadic Creutzfeldt–Jakob disease. *Archives of Neurology*, **53**, 913–20.

Kuwahara, C., Takeuchi, A.M., Nishimura, T., Haraguchi, K., Kubosaki, A., Matsumoto, Y. et al. (1999) Prions prevent neuronal cell-line death. *Nature*, **400**, 225–6.

Ladogana, A., Puopolo, M., Croes, E.A., Budka, H., Jarius, C., Collins, S. et al. (2005) Mortality from Creutzfeldt–Jakob disease and related disorders in Europe, Australia, and Canada. *Neurology*, **64**, 1586–91.

Laplanche, J.L., Delasnerie-Laupretre, N., Brandel, J.P., Chatelain, J., Beaudry, P., Alperovitch, A. et al. (1994) Molecular genetics of prion diseases in France. French Research Group on Epidemiology of Human Spongiform Encephalopathies. *Neurology*, **44**, 2347–51.

Legname, G., Nelken, P., Guan, Z., Kanyo, Z.F., DeArmond, S.J. and Prusiner, S.B. (2002) Prion and doppel proteins bind to granule cells of the cerebellum. *Proceedings of the National Academy of Sciences of the United States of America*, **99**, 16285–90.

Legname, G., Baskakov, I.V., Nguyen, H.O., Riesner, D., Cohen, F.E., DeArmond, S.J. et al. (2004) Synthetic mammalian prions. *Science*, **305**, 673–6.

Legname, G., Nguyen, H.O., Baskakov, I.V., Cohen, F.E., Dearmond, S.J. and Prusiner, S.B. (2005) Strain-specified characteristics of mouse synthetic prions. *Proceedings of the National Academy of Sciences of the United States of America*, **102**, 2168–73.

Legname, G., Nguyen, H.O., Peretz, D., Cohen, F.E., DeArmond, S.J. and Prusiner, S.B. (2006) Continuum of prion protein structures enciphers a multitude of prion isolate-specified phenotypes. *Proceedings of the National Academy of Sciences of the United States of America*, **103**, 19105–10.

Lemstra, A.W., van Meegan, M.T., Vreyling, J.P., Meijerink, P.H.S., Jansen, G.H., Bulk, S. et al. (2000) 14-3-3 Testing in diagnosing Creutzfeldt–Jakob disease. *Neurology*, **55**, 514–16.

Llewelyn, C.A., Hewitt, P.E., Knight, R.S., Amar, K., Cousens, S., Mackenzie, J. et al. (2004) Possible transmission of variant Creutzfeldt–Jakob disease by blood transfusion. *Lancet*, **363**, 417–21.

Love, R. (2001) Antibodies effective against scrapie infection, report European researchers. *Lancet*, **358**, 816.

Lullmann-Rauch, R., Pods, R. and von Witzendorff, B. (1996) The antimalarials quinacrine and chloroquine induce weak lysosomal storage of sulphated glycosaminoglycans in cell culture and in vivo. *Toxicology*, **110**, 27–37.

Magri, G., Clerici, M., Dall'Ara, P., Biasin, M., Caramelli, M., Casalone, C. et al. (2005) Decrease in pathology and progression of scrapie after immunisation with synthetic prion protein peptides in hamsters. *Vaccine*, **23**, 2862–8.

Mallucci, G.R., Ratte, S., Asante, E.A., Linehan, J., Gowland, I., Jefferys, J.G. et al. (2002) Post-natal knockout of prion protein alters hippocampal CA1 properties, but does not result in neurodegeneration. *EMBO Journal*, **21**, 202–10.

Manson, J.C., Clarke, A.R., Hooper, M.L., Aitchison, L., McConnell, I. and Hope, J. (1994) 129/Ola mice carrying a null mutation in PrP that abolishes mRNA production are developmentally normal. *Molecular Neurobiology*, **8**, 121–7.

Manuelidis, E.E. and Manuelidis, L. (1989) Suggested links between different types of dementias: Creutzfeldt–Jakob disease, Alzheimer disease, and retroviral CNS infections. *Alzheimers Disease and Associated Disorders*, **3**, 100–9.

Masters, C.L. (1989) Creutzfeldt–Jakob disease: Its origins. *Alzheimers Disease and Associated Disorders*, **3**, 46–51.

Masters, C.L., Harris, J.O., Gajdusek, D.C., Gibbs, C.J., Jr., Bernoulli, C. and Asher, D.M. (1979) Creutzfeldt–Jakob disease: Patterns of worldwide occurrence and the significance of familial and sporadic clustering. *Annals of Neurology*, **5**, 177–88.

Mastrianni, J.A., Nixon, R., Layzer, R., Telling, G.C., Han, D., DeArmond, S.J. et al. (1999) Prion protein conformation in a patient with sporadic fatal insomnia. *New England Journal of Medicine*, **340**, 1630–8.

May, B.C., Fafarman, A.T., Hong, S.B., Rogers, M., Deady, L.W., Prusiner, S.B. et al. (2003) Potent inhibition of scrapie prion replication in cultured cells by bis-acridines. *Proceedings of the National Academy of Sciences of the United States of America*, **100**, 3416–21.

McLennan, N.F., Brennan, P.M., McNeill, A., Davies, I., Fotheringham, A., Rennison, K.A. et al. (2004) Prion protein accumulation and neuroprotection in hypoxic brain damage. *American Journal of Pathology*, **165**, 227–35.

Miele, G., Jeffrey, M., Turnbull, D., Manson, J. and Clinton, M. (2002) Ablation of cellular prion protein expression affects mitochondrial numbers and morphology. *Biochemistry and Biophysics Research Communications*, **291**, 372–7.

Muller, W.E., Laplanche, J.L., Ushijima, H. and Schroder, H.C. (2000) Novel approaches in diagnosis and therapy of Creutzfeldt–Jakob disease. *Mechanisms of Ageing and Development*, **116**, 193–218.

Murakami-Kubo, I., Doh-ura, K., Ishikawa, K., Kawatake, S., Sasaki, K., Kira, J. et al. (2004) Quinoline derivatives are therapeutic candidates for transmissible spongiform encephalopathies. *Journal of Virology*, **78**, 1281–8.

Nakajima, M., Yamada, T., Kusuhara, T., Furukawa, H., Takahashi, M., Yamauchi, A. et al. (2004) Results of quinacrine administration to patients with Creutzfeldt–Jakob disease. *Dementia and Geriatric Cognitive Disorders*, **17**, 158–63.

Nishida, N., Tremblay, P., Sugimoto, T., Shigematsu, K., Shirabe, S., Petromilli, C. et al. (1999) A mouse prion protein transgene rescues mice deficient for the prion protein gene from Purkinje cell degeneration and demyelination. *Laboratory Investigations*, **79**, 689–97.

Oesch, B., Westaway, D., Walchli, M., McKinley, M.P., Kent, S.B., Aebersold, R. et al. (1985) A cellular gene encodes scrapie PrP 27–30 protein. *Cell*, **40**, 735–46.

Otto, M. and Wiltfang, J. (2003) Differential diagnosis of neurodegenerative diseases with special emphasis on Creutzfeldt–Jakob disease. *Restorative Neurology Neuroscience*, **21**, 191–209.

Otto, M., Wiltfang, J., Cepek, L., Neumann, M., Mollenhauer, B., Steinacker, P. et al. (2002) Tau protein and 14-3-3 protein in the differential diagnosis of Creutzfeldt–Jakob disease. *Neurology*, **58**, 192–7.

Palmer, M.S., Dryden, A.J., Hughes, J.T. and Collinge, J. (1991) Homozygous prion protein genotype predisposes to sporadic Creutzfeldt–Jakob disease. *Nature*, **352** (6333), 340–2.

Pals, P., Van Everbroeck, B., Sciot, R., Godfraind, C., Robberecht, W., Dom, R. et al. (1999) A retrospective study of Creutzfeldt–Jakob disease in Belgium. *European Journal of Epidemiology*, **15**, 517–9.

Pankiewicz, J., Prelli, F., Sy, M.S., Kascsak, R.J., Kascsak, R.B., Spinner, D.S. et al. (2006) Clearance and prevention of prion infection in cell culture by anti-PrP antibodies. *European Journal of Neuroscience*, **23**, 2635–47.

Parchi, P., Castellani, R., Capellari, S., Ghetti, B., Young, K., Chen, S.G. et al. (1996) Molecular basis of phenotypic variability in sporadic Creutzfeldt–Jakob disease. *Annals of Neurology*, **39**, 767–78.

Parchi, P., Capellari, S., Chin, S., Schwarz, H.B., Schecter, N.P., Butts, J.D. et al. (1999a) A subtype of sporadic prion disease mimicking fatal familial insomnia. *Neurology*, **52**, 1757–63.

Parchi, P., Giese, A., Capellari, S., Brown, P., Schulz-Schaeffer, W., Windl, O. et al. (1999b) Classification of sporadic Creutzfeldt–Jakob disease based on molecular and phenotypic analysis of 300 subjects. *Annals of Neurology*, **46**, 224–33.

Peden, A.H., Head, M.W., Ritchie, D.L., Bell, J.E. and Ironside, J.W. (2004) Preclinical vCJD after blood transfusion in a PRNP codon 129 heterozygous patient. *Lancet*, **364**, 527–9.

Peretz, D., Williamson, R.A., Legname, G., Matsunaga, Y., Vergara, J., Burton, D.R. et al. (2002) A change in the conformation of prions accompanies the emergence of a new prion strain. *Neuron*, **34**, 921–32.

Peretz, D., Supattapone, S., Giles, K., Vergara, J., Freyman, Y., Lessard, P. et al. (2006) Inactivation of prions by acidic sodium dodecyl sulfate. *Journal of Virology*, **80**, 322–31.

Piao, Y.-S., Kakita, A., Watanabe, H., Kitamoto, T. and Takahashi, H. (2005) Sporadic fatal insomnia with spongiform degeneration in the thalamus and widespread PrPSc deposits in the brain. *Neuropathology*, **25**, 144–9.

Pocchiari, M., Puopolo, M., Croes, E.A., Budka, H., Gelpi, E., Collins, S. et al. (2004) Predictors of survival in sporadic Creutzfeldt–Jakob disease and other human transmissible spongiform encephalopathies. *Brain*, **127**, 2348–59.

Polymenidou, M., Stoeck, K., Glatzel, M., Vey, M., Bellon, A. and Aguzzi, A. (2005) Coexistence of multiple PrPSc types in individuals with Creutzfeldt–Jakob disease. *Lancet Neurology*, **4**, 805–14.

Poser, S., Mollenhauer, B., Kraubeta, A., Zerr, I., Steinhoff, B., Schroeter, A. et al. (1999) How to improve the clinical diagnosis of Creutzfeldt–Jakob disease. *Brain*, **122**, 2345–51.

Priola, S.A., Raines, A. and Caughey, W.S. (2000) Porphyrin and phthalocyanine antiscrapie compounds. *Science*, **287**, 1503–6.

Prusiner, S.B. (1982) Novel proteinaceous infectious particles cause scrapie. *Science*, **216**, 136–44.

Prusiner, S.B. (1993) Prion encephalopathies of animals and human. *Developments in Biological Standardization*, **80**, 31–44.

Prusiner, S.B. (1998a) Prions. *Proceedings of the National Academy of Sciences of the United States of America*, **95**, 13363–83.

Prusiner, S.B. (1998b) The prion diseases. *Brain Pathology*, **8**, 499–513.

Prusiner, S.B. (2001) Shattuck lecture—Neurodegenerative diseases and prions. *New England Journal of Medicine*, **344**, 1516–26.

Prusiner, S.B., Groth, D., Serban, A., Koehler, R., Foster, D., Torchia, M. et al. (1993) Ablation of the prion protein (PrP) gene in mice prevents scrapie and facilitates production of anti-PrP antibodies. *Proceedings of the National Academy of Sciences of the United States of America*, **90**, 10608–12.

Rabinovici, G.D., Wang, P.N., Levin, J., Cook, L., Pravdin, M., Davis, J. et al. (2006) First symptom in sporadic Creutzfeldt–Jakob disease. *Neurology*, **66**, 286–7.

Rieger, R., Edenhofer, F., Lasmézas, C.I. and Weiss, S. (1997) The human 37-kDa laminin receptor precursor interacts with the prion protein in eukaryotic cells. *Nature Medicine*, **3**, 1383–8.

Rossi, D., Cozzio, A., Flechsig, E., Klein, M.A., Rulicke, T., Aguzzi, A. et al. (2001) Onset of ataxia and Purkinje cell loss in PrP null mice inversely correlated with Dpl level in brain. *EMBO Journal*, **20**, 694–702.

Saborio, G.P., Permanne, B. and Soto, C. (2001) Sensitive detection of pathological prion protein by cyclic amplification of protein misfolding. *Nature*, **411**, 810–13.

Safar, J., Wille, H., Itri, V., Groth, D., Serban, H., Torchia, M. et al. (1998) Eight prion strains have PrP(Sc) molecules with different conformations. *Nature Medicine*, **4**, 1157–65.

Safar, J.G., Geschwind, M.D., Deering, C., Didorenko, S., Sattavat, M., Sanchez, H. et al. (2005) Diagnosis of human prion disease. *Proceedings of the National Academy of Sciences of the United States of America*, **102**, 3501–6.

Safar, J.G., Wille, H., Geschwind, M.D., Deering, C., Latawiec, D., Serban, A. et al. (2006) Human prions and plasma lipoproteins. *Proceedings of the National Academy of Sciences of the United States of America*, **103**, 11312–17.

Saito, T., Isozumi, K., Komatsumoto, S., Nara, M., Suzuki, K. and Dohura, K. (2000) A case of codon 232 mutation-induced Creutzfeldt–Jakob disease visualized by the MRI-FLAIR images with atypical clinical symptoms. *Rinsho Shinkeigaku*, **40**, 51–4.

Saiz, A., Graus, F., Dalmau, J., Pifarre, A., Marin, C. and Tolosa, E. (1999) Detection of 14-3-3 brain protein in the cerebrospinal fluid of PATIENTS with paraneoplastic neurological disorders. *Annals of Neurology*, **46**, 774–7.

Sakaguchi, S., Katamine, S., Nishida, N., Moriuchi, R., Shigematsu, K., Sugimoto, T. et al. (1996) Loss of cerebellar Purkinje cells in aged mice homozygous for a disrupted PrP gene. *Nature*, **380**, 528–31.

Sanchez-Juan, P., Green, A., Ladogana, A., Cuadrado-Corrales, N., Saanchez-Valle, R., Mitrovaa, E. et al. (2006) CSF tests in the differential diagnosis of Creutzfeldt–Jakob disease. *Neurology*, **67**, 637–43.

Sandberg, M.K., Wallen, P., Wikstrom, M.A. and Kristensson, K. (2004) Scrapie-infected GT1–1 cells show impaired function of voltage-gated N-type calcium channels (Ca(v) 2.2) which is ameliorated by quinacrine treatment. *Neurobiology of Disease*, **15**, 143–51.

Santuccione, A., Sytnyk, V., Leshchyns'ka, I. and Schachner, M. (2005) Prion protein recruits its neuronal receptor NCAM to lipid rafts to activate p59fyn and to enhance neurite outgrowth. *Journal of Cell Biology*, **169**, 341–54.

Satoh, K., Shirabe, S., Eguchi, H., Tsujino, A., Eguchi, K., Satoh, A. et al. (2006) 14-3-3 Protein, total tau and phosphorylated tau in cerebrospinal fluid of patients with Creutzfeldt–Jakob disease and neurodegenerative disease in Japan. *Cellular and Molecular Neurobiology*, **26**, 45–52.

Scaravilli, F., Cordery, R.J., Kretzschmar, H., Gambetti, P., Brink, B., Fritz, V. et al. (2000) Sporadic fatal insomnia: A case study. *Annals of Neurology*, **48**, 665–8.

Schmitt-Ulms, G., Legname, G., Baldwin, M.A., Ball, H.L., Bradon, N., Bosque, P.J. et al. (2001) Binding of neural cell adhesion molecules (N-CAMs) to the cellular prion protein. *Journal of Molecular Biology*, **314**, 1209–25.

Scott, M.R., Will, R., Ironside, J., Nguyen, H.O., Tremblay, P., DeArmond, S.J. et al. (1999) Compelling transgenetic evidence for transmission of bovine spongiform encephalopathy prions to humans. *Proceedings of the National Academy of Sciences of the United States of America*, **96**, 15137–42.

Seeger, H., Heikenwalder, M., Zeller, N., Kranich, J., Schwarz, P., Gaspert, A. et al. (2005) Coincident scrapie infection and nephritis lead to urinary prion excretion. *Science*, **310**, 324–6.

Seipelt, M., Zerr, I., Nau, R., Mollenhauer, B., Kropp, S., Steinhoff, B.J. et al. (1999) Hashimoto's encephalitis as a differential diagnosis of Creutzfeldt–Jakob disease. *Journal of Neurology, Neurosurgery, and Psychiatry*, **66**, 172–6.

Sellarajah, S., Lekishvili, T., Bowring, C., Thompsett, A.R., Rudyk, H., Birkett, C.R. et al. (2004) Synthesis of analogues of Congo red and evaluation of their anti-prion activity. *Journal of Medicinal Chemistry*, **47**, 5515–34.

Seno, H., Tashiro, H., Ishino, H., Inagaki, T., Nagasaki, M. and Morikawa, S. (2000) New haplotype of familial Creutzfeldt–Jakob disease with a codon 200 mutation and a codon 219 polymorphism of the prion protein gene in a Japanese family. *Acta Neuropathology (Berlin)*, **99**, 125–30.

Serban, D., Taraboulos, A., DeArmond, S.J. and Prusiner, S.B. (1990) Rapid detection of Creutzfeldt–Jakob disease and scrapie prion proteins. *Neurology*, **40**, 110–7.

Shiga, Y., Miyazawa, K., Sato, S., Fukushima, R.S., Shibuya, S., Sato, Y. et al. (2004) Diffusion-weighted MRI abnormalities as an early diagnostic marker for Creutzfeldt–Jakob disease. *Neurology*, **163**, 443–9.

Shyu, W.C., Lin, S.Z., Chiang, M.F., Ding, D.C., Li, K.W., Chen, S.F. et al. (2005) Over-expression of PrPC by adenovirus-mediated gene targeting reduces ischemic injury in a stroke rat model. *Journal of Neuroscience*, **25**, 8967–77.

Sigurdsson, E.M., Sy, M.S., Li, R., Scholtzova, H., Kascsak, R.J., Kascsak, R. et al. (2003) Anti-prion antibodies for prophylaxis following prion exposure in mice. *Neuroscience Letters*, **336**, 185–7.

Slee, M., Pretorius, P., Ansorge, O., Stacey, R. and Butterworth, R. (2006) Parkinsonism and dementia due to gliomatosis cerebri mimicking sporadic Creutzfeldt–Jakob disease (CJD). *Journal of Neurology, Neurosurgery, and Psychiatry*, **77**, 283–4.

Soto, C., Anderes, L., Suardi, S., Cardone, F., Castilla, J., Frossard, M.J. et al. (2005) Pre-symptomatic detection of prions by cyclic amplification of protein misfolding. *FEBS Letters*, **579**, 638–42.

Spielhaupter, C. and Schätzl, H.M. (2001) PrP^C directly interacts with proteins involved in signaling pathways. *Journal of Biological Chemistry*, **276**, 44604–12.

Spudich, A., Frigg, R., Kilic, E., Kilic, U., Oesch, B., Raeber, A. et al. (2005) Aggravation of ischemic brain injury by prion protein deficiency: Role of ERK-1/-2 and STAT-1. *Neurobiology of Disease*, **20**, 442–9.

Steinhoff, B.J., Zerr, I., Glatting, M., Schulz-Schaeffer, W., Poser, S. and Kretzschmar, H.A. (2004) Diagnostic value of periodic complexes in Creutzfeldt–Jakob disease. *Annals of Neurology*, **56**, 702–8.

Tagliavini, F., McArthur, R.A., Canciani, B., Giaccone, G., Porro, M., Bugiani, M. et al. (1997) Effectiveness of anthracycline against experimental prion disease in Syrian hamsters. *Science*, **276**, 1119–22.

Taraboulos, A., Jendroska, K., Serban, D., Yang, S.L., DeArmond, S.J. and Prusiner, S.B. (1992) Regional mapping of prion proteins in brain. *Proceedings of the National Academy of Sciences of the United States of America*, **89**, 7620–4.

Tatzelt, J., Prusiner, S.B. and Welch, W.J. (1996) Chemical chaperones interfere with the formation of scrapie prion protein. *EMBO Journal*, **15**, 6363–73.

Telling, G.C., Scott, M., Mastrianni, J., Gabizon, R., Torchia, M., Cohen, F.E. et al. (1995) Prion propagation in mice expressing human and chimeric PrP transgenes implicates the interaction of cellular PrP with another protein. *Cell*, **83**, 79–90.

Tobler, I., Gaus, S.E., Deboer, T., Achermann, P., Fischer, M., Rulicke, T. et al. (1996) Altered circadian activity rhythms and sleep in mice devoid of prion protein. *Nature*, **380**, 639–42.

Tobler, I., Deboer, T. and Fischer, M. (1997) Sleep and sleep regulation in normal and prion protein-deficient mice. *Journal of Neuroscience*, **17**, 1869–79.

Turnbull, S., Tabner, B.J., Brown, D.R. and Allsop, D. (2003) Quinacrine acts as an antioxidant and reduces the toxicity of the prion peptide PrP106–126. *Neuroreport*, **14**, 1743–5.

UK Health Protection Agency. (2007) Variant CJD and blood products. London, Health Protection Agency. Accessed: January 23, 2007. Last Updated: January 18, 2007. http://www.hpa.org.uk/infections/topics_az/cjd/blood_products.htm

Van Everbroeck, B., Green, A., Vanmechelen, E., Vanderstichele, H., Pals, P., Sanchez-Valle, R. et al. (2002) Phosphorylated tau in cerebrospinal fluid as a marker for Creutzfeldt–Jakob disease. *Journal of Neurology, Neurosurgery, and Psychiatry*, **73**, 79–81.

Van Everbroeck, B., Quoilin, S., Boons, J., Martin, J.J. and Cras, P. (2003) A prospective study of CSF markers in 250 patients with possible Creutzfeldt–Jakob disease. *Journal of Neurology, Neurosurgery, and Psychiatry*, **74**, 1210–4.

Van Everbroeck, B.R.J., Boons, J. and Cras, P. (2005) 14-3-3 {gamma}-isoform detection distinguishes sporadic Creutzfeldt–Jakob disease from other dementias. *Journal of Neurology, Neurosurgery, and Psychiatry*, **76**, 100–2.

Vogtherr, M., Grimme, S., Elshorst, B., Jacobs, D.M., Fiebig, K., Griesinger, C. et al. (2003) Antimalarial drug quinacrine binds to C-terminal helix of cellular prion protein. *Journal of Medicinal Chemistry*, **46**, 3563–4.

Weise, J., Sandau, R., Schwarting, S., Crome, O., Wrede, A., Schulz-Schaeffer, W. et al. (2006) Deletion of cellular prion protein results in reduced Akt activation, enhanced postischemic caspase-3 activation, and exacerbation of ischemic brain injury. *Stroke*, **37**, 1296–300.

Weissmann, C. and Flechsig, E. (2003) PrP knock-out and PrP transgenic mice in prion research. *British Medical Bulletin*, **66**, 43–60.

Westaway, D., Goodman, P.A., Mirenda, C.A., McKinley, M.P., Carlson, G.A. and Prusiner, S.B. (1987) Distinct prion proteins in short and long scrapie incubation period mice. *Cell*, **51**, 651–62.

Westaway, D., DeArmond, S.J., Cayetano-Canlas, J., Groth, D., Foster, D., Yang, S.L. et al. (1994) Degeneration of skeletal muscle, peripheral nerves, and the central nervous system in transgenic mice overexpressing wild-type prion proteins. *Cell*, **76**, 117–29.

White, A.R., Enever, P., Tayebi, M., Mushens, R., Linehan, J., Brandner, S. et al. (2003) Monoclonal antibodies inhibit prion replication and delay the development of prion disease. *Nature*, **422**, 80–3.

WHO (1998) Global surveillance, diagnosis and therapy of human transmissible spongiform encephalopathies: Report of a WHO consultation, in *World Health Organization: Emerging and Other Communicable Diseases, Surveillance and Control*, Geneva.

Will, R. (2004) Variant Creutzfeldt–Jakob disease. *Folia Neuropathology*, **42** (Suppl A), 77–83.

Will, R.G., Ironside, J.W., Zeidler, M., Cousens, S.N., Estibeiro, K., Alperovitch, A. et al. (1996) A new variant of Creutzfeldt–Jakob disease in the UK. *Lancet*, **347**, 921–5.

Will, R.G., Alperovitch, A., Poser, S., Pocchiari, M., Hofman, A., Mitrova, E. et al. (1998) Descriptive epidemiology of Creutzfeldt–Jakob disease in six European countries, 1993–1995. EU Collaborative Study Group for CJD. *Annals of Neurology*, **43**, 763–7.

Williams, E.S. (2005) Chronic wasting disease. *Vetinary Pathology*, **42**, 530–49.

Wong, B.S., Liu, T., Li, R., Pan, T., Petersen, R.B., Smith, M.A. et al. (2001a) Increased levels of oxidative stress markers detected in the brains of mice devoid of prion protein. *Journal of Neurochemistry*, **76**, 565–72.

Wong, C., Xiong, L.W., Horiuchi, M., Raymond, L., Wehrly, K., Chesebro, B. et al. (2001b) Sulfated glycans and elevated temperature stimulate PrP(Sc)-dependent cell-free formation of protease-resistant prion protein. *EMBO Journal*, **20**, 377–86.

Young, G.S., Geschwind, M.D., Fischbein, N.J., Martindale, J.L., Henry, R.G., Liu, S. et al. (2005) Diffusion-weighted and fluid-attenuated inversion recovery imaging in Creutzfeldt–Jakob disease: High sensitivity and specificity for diagnosis. *American Journal of Neuroradiology*, **26**, 1551–62.

Zeidler, M., Sellar, R.J., Collie, D.A., Knight, R., Stewart, G., Macleod, M.A. et al. (2000) The pulvinar sign on magnetic resonance imaging in variant Creutzfeldt–Jakob disease. *Lancet*, **355**, 1412–8.

Zerr, I., Bodemer, M., Gefeller, O., Otto, M., Poser, S., Wiltfang, J. et al. (1998) Detection of 14-3-3 protein in the cerebrospinal fluid supports the diagnosis of Creutzfeldt–Jakob disease. *Annals of Neurology*, **43**, 32–40.

Zerr, I., Pocchiari, M., Collins, S., Brandel, J.P., de Pedro Cuesta, J., Knight, R.S. et al. (2000) Analysis of EEG and CSF 14-3-3 proteins as aids to the diagnosis of Creutzfeldt–Jakob disease. *Neurology*, **55**, 811–15.

Part VI

Overview

14 Overview

H. John Smith, Robert D.E. Sewell, and Claire Simons

CONTENTS

The chapters in this book focus on the possible causes of a number of distinctive, commonly occurring neurodegenerative diseases. The main aim of this overview is to highlight any consensus of opinion concerning the initiating factors responsible for the different diseases and the current, as well as, promising future approaches to their treatment.

Unraveling the potential underlying causes of a disease at the molecular level may well pinpoint specific modulatory targets. As a consequence, this could facilitate the rational design of lead compounds which might then be modified to improve their potency, selectivity, pharmacokinetics, and blood–brain barrier penetration thereby yielding therapeutic agents suitable for the clinic.

14.1   INTRODUCTION

Proteins have an important function within the cell not only as enzymes but also as receptors. In addition, they have equally fundamental roles which include regulatory and defense mechanisms, cellular transcription processes, and the maintenance of membrane and cytoskeletal structure. In order to accomplish these functions, they need to be in an appropriate three-dimensional structural conformation. Misfolding of native protein, however, leads to an unusual exposure of the constituent amino acid residues in the monomer which, if uncorrected by the cell's controlling mechanisms, can induce subsequent oligomerization, aggregation, and deposition of polymeric associations mainly through β-sheet formation, that is, fibrils.

Misfolding of normal proteins in the brain, possibly triggered by aging or cellular stress, is currently being considered as an etiological component in a number of sporadic neurodegenerative diseases. The most common of these are Alzheimer's disease (AD) and Parkinson's disease (PD), both of which are characterized by extensive deposits of specific proteins in the brain observed by postmortem examination.

Miscoding of a protein through genetic or environmental factors leads either to single or multipoint mutants. These can resist the natural folding restraints of the wild-type protein with a decrease (or loss) of normal function or a functional gain, not shown by the wild-type protein. A view currently held by many researchers is that an elementary cause of the disease is not misfolding with aggregate formation and associated extracellular or intracellular deposits per se, but rather, misfolding which generates toxic oligomeric species.

Protein misfolding has therefore been viewed as a common causative feature of these diseases but this is not to the exclusion of other, as yet, unclear factors. However, it seems unlikely if the homeostasis of the cell is affected, that there would not be a number of consequences, which would tend to mask the primary origin of the disease. A stark reality that ensues is that if the basic cause is not identified for a known disease, then it will either not be curable or the symptoms may not be readily ameliorated by rational intervention.

The neurodegenerative diseases considered here are AD, PD, amyotrophic lateral sclerosis (ALS, motor neuron disease), Huntington's disease (HD), and Creutzfeld–Jacob disease (CJD), all of which have been linked to misfolding proteins and aggregate deposits.

14.2   ALZHEIMER'S DISEASE

AD is the most common form of dementia and is associated with age and family history. It is distinguished from other related forms of dementia by postmortem examination when amyloid plaques can be observed in the brain.

Neuronal transmembrane amyloid precursor protein (APP) is cleaved by the amyloid cascade involving α-, β-, and γ-secretase to yield peptide fragments. These include $A\beta_{(1-40)}$ (present in the brain at appreciable concentrations), $A\beta_{(1-42)}$ (which is more toxic and amyloidogenic; forming fibrils and extracellular aggregates), as well as nontoxic, soluble $A\beta_{(17-42 \text{ or } 43)}$ peptides. These fragments account for up to 90% of APP metabolism. Soluble oligomeric forms of $A\beta_{(4-12)}$ are more toxic than those of $A\beta_{(1-40 \text{ or } 42)}$ and overall, accumulation of misfolded $A\beta$ species appears to be related to decreased oligomer clearance rather than increased monomer production.

Aggregation of peptides $A\beta_{(1-40 \text{ or } 42)}$ occur by way of β-sheet formation to generate fibrils (which are a component of plaques in AD patients) although they also exist in a significant percentage of cognitively unimpaired aged individuals.

Diminishing $A\beta$ formation via brain γ-secretase inhibition has been a strategy to prevent plaque formation, believed at that time to be the cause of AD, at source. Specific and potent in vitro inhibitors of γ-secretase are known and the feasibility of reducing brain $A\beta$ in animal models has been established. However, side effects arising from lack of selectivity toward γ-secretase and interference with Notch signaling causing gastrointestinal and other serious toxicities have been reported.

An alternative approach to avoid Notch deficiency was based on the observation that selective nonsteroidal anti-inflammatory drugs (NSAIDs) affected the cleavage specificity of γ-secretase. This action entailed a decline in levels of the more amyloidogenic $A\beta_{(1-42)}$, an elevation of $A\beta_{(1-38)}$ species, and no modification of $A\beta_{(1-40)}$ output. Such an NSAID effect seems to be unrelated to inhibition of cyclooxygenase 1/2 (COX1/COX2). This aspect of their action can be designed out of candidates intended as γ-secretase inhibitors to reduce otherwise potential side effects already noted for some COX inhibitors. However, NSAIDs in current clinical use, exhibit poor brain penetration and low potency in decreasing $A\beta_{(1-42)}$ production. In addition to their action on Aβ formation by γ-secretase inhibition, they also have other possibly useful actions to suppress AD pathogenicity.

NSAIDs inhibit in vitro activation of nuclear factor-kappa B (NF-κB) possibly by affecting its binding to DNA. NF-κB, besides having anti-inflammatory and other roles, is activated by Aβ formation and indeed, Aβ toxicity is mediated by the NF-κB pathway. Thus in mice, increased activation of NF-κB in brains overexpressing mutant human APP leads to Aβ deposition.

NSAIDs activate the peroxisome proliferator-activated gamma (PPARγ) system which inhibits expression of inflammatory cytokines by monocytes and this action could also contribute to the beneficial effects of NSAIDs in AD.

An antiAβ aggregation strategy, originally aimed at reducing oligomer or fibril formation, might be potentiated by encouragement of their dissolution into soluble monomeric species. Such a maneuver would promote their subsequent degradation by proteases and their clearance from the brain.

The ability of Aβ to aggregate into a β-sheet conformation and form fibrils is determined by its hydrophobic N-terminal. Alternatively, amyloid formation could be competitively blocked by short peptide aggregation inhibitors with a similar sequence to the Aβ N-terminal. Extending this concept to the sequence in Aβ responsible for misfolding (but with an inserted disrupting proline residue) also showed promise in decreasing Aβ aggregation in animal models. However, the use of peptides as therapeutic agents in general, is hampered by their poor oral absorption (~1%), their rapid degradation in vivo, and potential immunogenicity as foreign proteins. Nonpeptide inhibitors have manifested some success in preventing aggregation in vitro but they possess weak selectivity for the aggregate versus monomer. However, in all the approaches to reduce aggregate levels, there is a dependence on the cell's inherent degradation and efflux mechanisms for clearance of toxic monomers or oligomers.

More recently, thinking has changed from considering plaque formation as the cause of AD to the view that the lesions are not the disease initiators but could be a primary line of defense to clear toxic misfolded monomers or oligomers. Thus, Aβ and tau (see below) accumulation could reflect a physiological response to chronic stress, a possibly universal concept of protection in other neurodegenerative diseases presenting with pathological lesions (Parkinson's disease—Lewy bodies, Pick's diseases—Pick bodies, and Amyotrophic Lateral Sclerosis—motor axon spheroids).

Cholesterol is a major lipid constituent of neuronal membranes and is synthesized in the brain. Statins, 3-hydroxy-3-methylglutaryl coenzyme A reductase (HMG-CoA reductase) inhibitors are well established as clinically beneficial cholesterol-lowering agents. Because AD patients have higher cholesterol levels than normal subjects and reduced cholesterol is associated with a decreased prevalence of AD, this factor has been targeted. Statins may have other beneficial effects in AD apart from lipid-lowering activity. In this context, mevalonic acid is a precursor not only of cholesterol, but also isoprenoids implicated in cell membrane attachment of proteins requiring prenylation prior to membrane association. This is essential for their wide-ranging functions which include cell signaling, differentiation, proliferation, inflammatory mediator production, and numerous other biochemical activities. Hence, inhibition of mevalonate production may indicate a potential use for statins in neurodegenerative and other diseases. In this respect, some, but not all, clinical studies have confirmed a favorable statin effect in AD (in addition to the lipid-lowering action). Statins have also been shown to reduce serum Aβ peptide but only at higher than usual clinical doses.

Recently, however, attention has been focused on the safety profile of certain statins due to a low incidence of liver and muscle toxicity. Hence, the benefits to risk ratio of statin use in AD patients needs to be considered.

Another aspect of postmortem AD brain is the presence of neurofibrillary tangles (NFTs) caused by the aggregation of tau, a microtubule-associated protein that accounts for cytoskeletal integrity. Microtubules transport material between the cell body and distal neuronal components and disassembly leads to a block of these carrier processes followed by cell death. In NFTs, tau (six isoforms) differs from the normal type in its degree of aggregation and modification by phosphorylation. Hyperphosphorylation reduces the binding of tau to the microtubules (with a loss in their stability), and cytosolic modified tau aggregates to form filamentous inclusions having an amyloid β-sheet structure. The concept of microtubule stabilization in AD with known stabilizing agents such as vinca alkaloids, colchicines, and paclitaxel, has been examined as another potential target for AD prevention. Abnormally phosphorylated and aggregated tau is toxic to the neuronal cell by ubiquitin–proteosome pathway inhibition. This leads to stress, the sequestering of proteins (as hyperphosphorylated tau) vital for cell viability, and activation of tyrosine phosphorylation-mediated signal transduction pathways (a "gain of function"). Decreasing tau concentration by increasing its rate of degradation or reducing expression could lower neurofilament formation.

A potential objective for the reduction of tau levels is mTOR, a protein kinase in the pathway concerned with the regulation of protein translation and turnover. This pathway also inhibits protein degradation and several inhibitors of mTOR are known and under investigation.

Hyperphosphorylation of tau is due to excessive cellular protein kinase activity in relation to the reverse step ascribed to phosphoprotein phosphatase. Several kinases have been identified, and although inhibitors are known, selectivity has been a problem. A more promising approach has involved reduction in the reaction of the protein phosphatase, PP2A, specific to tau dephosphorylation through inhibition of an essential supporting enzyme.

The ubiquitin–proteosome system is involved in the degradation of modified tau, and the balance between degradation and proper folding is controlled by the chaperones Hsp70 and Hsp90. Strategies to increase Hsp70 levels by inhibiting Hsp90, which leads to induction of Hsp70, have been described. Such inhibitors reduce phosphorylated and misfolded tau species in cellular models.

Carbonic anhydrase (CA), a zinc metalloenzyme, exists in many isozymic forms distributed throughout the body. Under physiological conditions, it catalyzes the reversible hydration of CO_2 to bicarbonate ions and a proton. It has a number of physiological roles and recent attention has been directed toward reports that CA activators have a conceivable application in the restoration of memory and possibly also as anti-AD agents. There is a significant decline in brain CA accompanying learning and memory loss in AD as well as during aging. Conversely, the synaptic switch of the γ-aminobutyric acid (GABA)-ergic response from inhibitory to excitatory depends on increased HCO_3^- conductance through the $GABA_A$ receptor–channel complex induced by certain CA activators administered in rats. The consequent outcome is a clear improvement in learning and memory capability. This potential field for AD and age-related memory deficits is currently unexplored, although many exogenous (and endogenous) activators are known.

There is evidence linking the involvement of vitamin A and its retinoid derivatives (retinoic acid, RA) as a causal agent in late onset AD (LOAD) because the disease is influenced by the availability of RA in the brain. RA carries out the neurological function of retinoids by combining with its receptor proteins and activating transcription of genes that modulate Aβ plaque production and brain cell death (apoptosis). Disruption of the retinoid signaling pathway has been suggested to cause deposition of Aβ in rat brain. Moreover, upregulation of the retinoid receptor decreases cognitive decline in aged rats supporting the role of RA on cognitive effects.

In LOAD, it is proposed that genetic and environmental factors, modified in the course of natural aging, can cause a reduction in RA signaling in the brain. Potential clinical strategies to increase the RA response include the use of RA agonists (retinoids, rexinoids), with improved selectivity and toxicity profiles, as well as specific CYP2D6 inhibitors (retinoic acid metabolism blocking agents—RAMBAs) to reduce metabolism of RA in the cell, that is, RA build up.

There has been a long-held commonly accepted view that decline in cognitive aspects of AD learning and memory abilities is due to a deficiency in cholinergic transmission linked to a loss of cholinergic neurons in the Nucleus Basalis of Maynert and other regional nuclei (i.e., the "Cholinergic Hypothesis"). Various strategies to restore cholinergic transmission in the brain have been followed: (1) enhancing release of acetylcholine (Ach) at the presynaptic level using nicotinic agonists, M2 muscarinic antagonists, and modulators of voltage-gated channels; (2) activation of postsynaptic M1 mucarinic receptors by synthetic agents; and (3) suppressing Ach breakdown by inhibition of acetylcholine esterase (AChE).

The most studied and clinically advanced strategy in a search for symptomatic treatment of AD is that employing AChE inhibitors. Additionally, there is evidence that cholinergic activity might be involved in the processing of amyloid precursor protein. Because AChE may accelerate Aβ plaque deposition, AChE inhibition might not only reduce cognitive decline but also protect neurons and moderate the course of the disease.

The three AChE inhibitors with FDA approval for the treatment of AD are donepezil, rivastigimine, and galanthamine. Clinical trial results with these compounds have been variable, either minimal benefit being seen or effectiveness and good tolerability reported. A recent review has concluded that these inhibitors can delay cognitive impairment for at least 6 months and that therapeutic benefits could last for up to 2 years in patients with moderate AD.

14.3 PARKINSON'S DISEASE

PD is a common neurodegenerative illness. The symptoms include movement disorders namely, bradykinesia (slowness of movement), rigidity and resting tremor, owing to the loss of dopaminergic function in the nigrostriatal pathway, and the formation of intracellular aggregates of α-synuclein-positive Lewy bodies in surviving neurons. The cause of the disease is not known but several environmental as well as genetic factors have been identified.

α-Synuclein is located presynaptically and interacts with several intracellular molecules in addition to inhibiting tyrosine hydroxylase, the rate-limiting enzyme in the synthesis of dopamine. It may cause its effects through a "toxic gain of function" instigating the formation of protofibrils. Alternatively, a possibility exists that by "loss of function" through inhibition of dopamine synthesis, intracellular dopamine levels would increase. An outcome would be associated oxidative damage and neuronal death owing to generation of reactive oxygen species (ROS) such as hydroxyl radicals, superoxide anions, and dopamine–quinone species. Dopamine auto-oxidizes at physiological pH to form toxic ROS which may in turn oxidize DNA, proteins, and lipids, so affecting their function. This oxidative process is normally controlled by the body's antioxidant systems, but in PD, there is a relative deficiency of the required components, that is, glutathione, superoxide dismutase, and catalase.

Over expression of α-synuclein in some disease models has been reported to cause dopaminergic pathology. Malfunctioning of the ubiquitin–proteosome pathway which removes α-synuclein, may induce increased protein levels and damage by subsequent aggregation.

The symptoms of PD are alleviated, but the cause of this disease is not targeted by dopamine replacement therapy, a strategy to increase agonist concentrations at remaining intact neurons. Levodopa (L-Dopa), which crosses the blood–brain barrier, is orally administered in combination with a peripherally acting aromatic amino acid decarboxylase (AADC) inhibitor to minimize peripheral conversion of the drug to dopamine and its subsequent breakdown by monoamine oxidases (MAOs). However, the effectiveness of L-Dopa is reduced on prolonged administration ("wearing off") and L-Dopa induced dyskinesia (involuntary movement at rest) develops. To circumvent this element of L-Dopa therapy, long-acting dopamine agonists, which do not have a pulsatile action at the dopamine receptor, have been developed.

In clinical studies, early rather than delayed PD treatment initiation on the MAO-B inhibitor rasagiline reduces the breakdown of endogenously produced dopamine and also has some beneficial effects thought to signify a neuroprotective propensity. Selegiline does delay the need for symptomatic therapy in untreated PD sufferers, but the outcome interpretation is obscured by its symptomatic effects. Therefore, an effective neuroprotective treatment capable of slowing or stopping disease progress remains an unmet therapeutic need in PD.

Several new approaches to tackling PD include the use of neurotrophic factors to protect and restore neuronal populations, a strategy possible in this disease where slow degeneration of dopaminergic neurons occurs. Hence, in PD animal models, the glial cell line-derived neurotrophic factor (GDNF) exhibits not only a neuroprotective action but also has a restorative capacity.

Cell transplant therapies using developing dopaminergic neurons to replace dopamine in the brain have given variable results. This may originate from the low efficiency (5%–10%) of neuronal survival during preparation and grafting. The clinical application of embryonic stem cells could be limited because of their tumorigenic inclination and this is under study. Potential candidates have been described for the next generation of donor tissue to be used in these potential replacement therapies.

14.4 HUNTINGTON'S DISEASE

HD, where neuronal loss occurs in several parts of the brain, is a fatal progressive neurodegenerative disease, usually occurring in middle age. Early symptoms of involuntary twitching and clumsiness progress to memory impairment, general disablement, and death as a result of complications, for example, choking or heart failure. Current medical treatment can only alleviate symptoms and does not delay the progression of the disease. HD is linked to a gene mutation identified as an abnormal repetition of the trinucleotide repeat CAG, where 35–141 CAG repeats occur instead of the 6–36 repeats in the normal population.

The gene product is a protein, huntingtin, which is highly expressed in the brain and is associated with the cytoplasmic surface of a variety of organelles. Huntingtin functions by anchoring and transporting cell constituents along microtubules as well as through several other interactions with entities involved in cellular trafficking. The protective effect of huntingtin in vivo following ischemic injury and excitotoxicity has led to the view that the pathogenesis stems from loss of beneficial effects of the natural protein. Huntingtin facilitates neuronal survival by stimulating the production and release of brain-derived neurotrophic factor (BDNF). It has also been proposed as a transcription factor in view of its polyglutamine (polyQ) content and its binding characteristics to a variety of transcription factors.

The CAG repeat in the HD gene encodes for a polyQ stretch in the N-terminal of huntingtin, and abnormal expression of this stretch in the mutant protein results in aberrant folding and processing. Normal and mutant forms of huntingtin are cleared by the proapoptotic enzyme caspase-3 among others, which releases the N-terminal fragments containing the polyQ stretch. These fragments form insoluble aggregates which become ubiquitinated and are visible in brain neurons as inclusions of huntingtin bodies. Elongated polyQ peptides beyond 40 glutamines are thought to be assembled in β-sheet structures. These aggregates contain misfolded huntingtin and also include a number of critical cellular components likely to interfere with their normal localization or functioning, thereby leading to cell death.

Identification of potential targets associated with the disease process has led to strategies aimed at elaborating therapeutics for the prevention or reduction of symptoms of the disease. Neuroprotective approaches with agents that specifically target huntingtin aggregate formation have been screened both in vitro and in vivo, a variety of compounds having been identified that inhibit polyQ aggregation in these models. However, their clinical assessment in HD remains to be investigated. Although relationships have been hypothesized between genetic mutation, metabolic dysfunction, secondary oxidative damage, and eventual apoptosis in HD (and there is supporting evidence for these mechanisms), it is unclear as to whether these processes contribute to the initiation or later stages of cell death. Initiation may be through alterations in normal functions of huntingtin in the mutant triggering a cascade of events, which are followed by metabolic, oxidative, and apoptotic mechanisms.

Recent in vitro and in vivo studies suggest that aggregate formation might actually be part of the housekeeping mechanism by the cells to isolate abnormal polyQ-containing proteins and prevent their toxic effects. This view accords with studies on postmortem HD brain tissue sections where the distribution of huntingtin aggregates does not correlate with neuropathology.

14.5 AMYOTROPHIC LATERAL SCLEROSIS

ALS, commonly known as motor neuron disease, is a progressively neurodegenerative disorder in adults, affecting motor neurons of the brain and spinal cord. The symptoms are comprised of muscle atrophy, weakness, cramps, and twitching of the lower limbs, with progression of the disease to paralysis and death.

Familial ALS (fALS) is associated with mutations of the copper/zinc superoxide dismutase (SOD1) gene, although mutations (mtSOD1) also occur in sporadic ALS (sALS). However, up to 110 different mutations of SOD1 result in the same disease. The function of normal SOD1 is to catalyze the disproportionation of two molecules of superoxide anion (O_2^-) to O_2 and H_2O_2. Motor neurons have higher levels of SOD1 expression than other neuronal entities, placing them at risk of mtSOD1 damage through "loss of function" where toxicity results from inability to scavenge superoxide. This leads to an excess of superoxide and is accompanied by depletion of glutathione peroxidase and catalase, which would normally eliminate excessive peroxide and hydroxyl radicals along with their neurotoxic effects.

Alternatively, it has been considered that mtSOD1 has a toxic "gain of function" by abnormal binding of substrates, leading to the formation of a toxic species, peroxynitrite, producing neuronal injury by nitration or oxidation of critical proteins. ALS is a multisystem disorder and it is not clear which processes are mainly responsible for the initial pathogenesis. Candidates include ROS (free radicals)—induced toxicity, intermediate filament disorganization, excitotoxicity, or protein misfolding and aggregation (a familiar culprit for neurodegenerative diseases in general). ROS involvement in oxidative protein damage is increased in the spinal cords of ALS patients and peroxynitrite and nitrotyrosine appear within spinal motor neurons. Excitotoxicity and neurodegeneration occur when there are excessive intrasynaptic levels of the excitatory amino acid neurotransmitter, glutamate.

The pathology of sALS and fALS are similar and involve the aggregation and inclusion of various proteins in neuronal (and other) cells, for example, Bunina bodies, ubiquitin positive inclusions, (neuro)filamentous inclusions (peripherins, tau), and mtSOD1.

It is not clear whether protein aggregates, as neuronal inclusions in ALS, could be generated as mediators or a consequence of the disease process, rather than as part of a mechanism to remove their direct toxic effects on the neuron. mtSOD1 expression in fALS suggests that the formation of aggregates is a relevant surrogate marker of modified cell protein expression. This is supported by studies using siRNA which disclose a significant improvement in the survival time of mtSOD1 mice in models of fALS.

Potential therapies for ALS have been screened using mtSOD1 transgenic mice. The antiglutamatergic agent riluzole (approved by the FDA for ALS) in humans, has not reflected the promising results seen in mice. Furthermore, the antibiotics minocycline and tetracycline are neuroprotective in in vitro models and the latter is currently undergoing clinical trials. Moreover, the use of RNA binding inhibitors (RNAi) to suppress expression of mtSOD1 in transgenic mice revealed that disease progression may be significantly retarded.

14.6 CREUTZFELDT–JACOB DISEASE

Prion diseases, also known as transmissible spongiform encephalopathies (TSEs), are a group of fatal neurodegenerative diseases which present in humans not only as sporadic and genetic forms but also as an acquired form as the result of infection. The disease occurs in animals as scrapie (sheep, goat), bovine spongiform encephalopathy (BSE, cattle), and other forms. The human type of prion disease is known as CJD, with sporadic CJD accounting for 85% of cases. The symptoms are cognitive, behavioral, constitutional, and cerebellar.

Prion diseases are caused by a conformation change in the structure of the endogenous brain prion protein PrP^c (prion protein, cellular) into an abnormally shaped disease-causing prion, PrP^{sc} (scrapie), where a β-sheet structure dominates. The misfolding process is not clear and even the precise function of normal PrP^c is unknown, though it is assumed to play an important role in neuronal development. In addition, distinguishing CJD from related diseases can only be confirmed by postmortem immunohistochemical analysis of CNS or lymphatic tissue to characterize PrP^c and PrP^{sc}.

A new form of human prion disease, variant CJD (vCJD) was identified a decade ago and occurs mainly in the UK and France. Patients with vCJD tend to be younger and the disease is linked to consumption of BSE-contaminated food with clinical features generally different to sporadic CJD. The cerebellum and cerebrum possess plaques containing high concentrations of PrP^{sc} surrounded by vacuoles as well as high levels of PrP^{sc} in the lymphoreticular system. Postmortem pathological diagnosis of vCJD can be made by tonsillar or brain biopsy.

There are no treatments available to alleviate or cure the disease but some agents (quinoline derivatives, the antibiotics doxycycline and tetracycline) have been used successfully to remove PrP^{sc} or inhibit its formation in vitro. Some are effective in preventing or delaying disease onset in vivo in animals but this depends on a favorable inoculation regimen for the agent and PrP^{sc}.

Oral quinacrine is the only medication currently in clinical trials. Although it eliminates prions in vitro and may allow recovery of some cellular functions, it does not improve survival in mice. However, a human study has suggested that quinacrine prolongs survival although other reports have given inconsistent results with possible transient improvements in some patients.

The two main mechanisms for eliminating PrP^{sc} are either by blocking conversion of PrP^{c} to PrP^{sc} or via clearance of PrP^{sc}. Quinacrine appears to operate by the former mechanism but its action at the molecular level is not understood. Tetracyclines possibly interact with PrP^{sc} or PrP^{sc} fibrils and human trials are planned with oral doxycycline. Another strategy is based on the knowledge that antibodies or single-chain fragments of antibodies can eliminate PrP^{sc} in cell culture.

Index

A

Aβ aggregates
  approaches, 226
  cytotoxicity properties, 159–161
  inhibitors, 156–157, 161–163
  as key elements, 46
  targeting, 224–225
Aβ assembly
  clearance, 226–227
  individual reactions, 217–218
  misfolding, 211–212
  polymerization, 229
  polymerization reduction, 209
  properties of, 213
  targeting, 220
Abeta peptides
  CSF levels, 85
  extracellular aggregates of, 70
  formation and aggregation, 75–78
  toxicity, 82
Aβ peptides
  aggregation, 11, 32–35
  in brain, 212, 214–215
  folding pathways, 216–217
  forms of, 215–216
  synthesis, 159
  various species, 152–153
Acetycholine pharmacology, 96
Acetylcholine esterase (AChE), 551
  clinical assessment, 114
  inhibition, 96–97, 103–104
  irreversible inhibitors, 110–111
  reactivation and aging of, 112–114
  release enhancers, 97–103
  reversible inhibitors, 108–110
  structure of, 104–108
α-Acetylenicdopa, 419
AChE inhibitors, 114–115
  natural, 115–119
  synthetic, 120–123
  in use, 124–127

Acridines, 122
Acting aromatic amino acid carboxylase
  (AADC), 552
Advanced glycation end products
  (AGEs), 487
Aggregate toxicity, 2, 4
Aggregation nuclei, 9
α-Allenicdopa, 419
Alzheimer's disease (AD), 518, 530,
  548–552
  and amyloid plaques, 147–149
  and animal models, 79–82, 228–229
  biomarkers, 82–84
  brain inflammation, 175
  and cholestrol, 70–71
  clinical trials, 229–231
  and CSF levels, 84
  diagnosis, 73, 75
  early onset, 11
  epidemiology, 70–71
  genetic risk factors, 71–72
  molecular chaperones levels, 39
  and NSAIDs, 174–177
  pathogenesis, 31–32
  pathology, 75–76
  and physical activity, 70
  prodromol form, 72–73
  and retinoids, 338–339
  risk factors, 70
  statistics, 71
  and tau lesions, 295–297
  therapeutic approaches, 153–156
  therapies, 85–86
Amino acid sequences
  and amyloid toxicity, 31
  of catalytic portions, 193
  features, 27
  modified, 3
  protein folding, 4–6, 24
  thermodynamics, 210–211
γ-Aminobutyric acid (GABA), 551
Amphetamine, 418

T - #0305 - 071024 - C4 - 234/156/26 - PB - 9780367388126 - Gloss Lamination